Atomic and Electronic Structure of Solids

This text is a modern treatment of the theory of solids. The core of the book deals with the physics of electron and phonon states in crystals and how they determine the structure and properties of the solid.

The discussion uses the single-electron picture as a starting point and covers electronic and optical phenomena, magnetism and superconductivity. There is also an extensive treatment of defects in solids, including point defects, dislocations, surfaces and interfaces. A number of modern topics where the theory of solids applies are also explored, including quasicrystals, amorphous solids, polymers, metal and semiconductor clusters, carbon nanotubes and biological macromolecules. Numerous examples are presented in detail and each chapter is accompanied by problems and suggested further readings. An extensive set of appendices provides the necessary background for deriving all the results discussed in the main body of the text.

The level of theoretical treatment is appropriate for first-year graduate students of physics, chemistry and materials science and engineering, but the book will also serve as a reference for scientists and researchers in these fields.

Efthimios Kaxiras received his PhD in theoretical physics at the Massachusetts Institute of Technology, and worked as a Postdoctoral Fellow at the IBM T. J. Watson Research Laboratory in Yorktown Heights. He joined Harvard University in 1991, where he is currently a Professor of Physics and the Gordon McKay Professor of Applied Physics. He has worked on theoretical modeling of the properties of solids, including their surfaces and defects; he has published extensively in refereed journals, as well as several invited review articles and book chapters. He has co-organized a number of scientific meetings and co-edited three volumes of conference proceedings. He is a member of the American Physical Society, the American Chemical Society, the Materials Research Society, Sigma Xi-Scientific Research Society, and a Chartered Member of the Institute of Physics (London).

Atomic and Electronic Structure of Solids

Efthimios Kaxiras

PUBLISHED BY THE PRESS SYNDICATE OF THE UNIVERSITY OF CAMBRIDGE
The Pitt Building, Trumpington Street, Cambridge, United Kingdom

CAMBRIDGE UNIVERSITY PRESS
The Edinburgh Building, Cambridge CB2 2RU, UK
40 West 20th Street, New York, NY 10011-4211, USA
477 Williamstown Road, Port Melbourne, VIC 3207, Australia
Ruiz de Alarcón 13, 28014 Madrid, Spain
Dock House, The Waterfront, Cape Town 8001, South Africa

http://www.cambridge.org

First published 2003

Printed in the United Kingdom at the University Press, Cambridge

Typeface Times 11/14 pt   *System* LaTeX 2_ε [TB]

A catalogue record for this book is available from the British Library

ISBN 0 521 81010 8 hardback
ISBN 0 521 52339 7 paperback

I dedicate this book to three great physics teachers:
Evangelos Anastassakis, who inspired me to become a physicist,
John Joannopoulos, who taught me how to think like one, and
Lefteris Economou, who vastly expanded my physicist's horizon.

Contents

Preface

This book is addressed to first-year graduate students in physics, chemistry, materials science and engineering. It discusses the atomic and electronic structure of solids. Traditional textbooks on solid state physics contain a large amount of useful information about the properties of solids, as well as extensive discussions of the relevant physics, but tend to be overwhelming as introductory texts. This book is an attempt to introduce the single-particle picture of solids in an accessible and self-contained manner. The theoretical derivations start at a basic level and go through the necessary steps for obtaining key results, while some details of the derivations are relegated to problems, with proper guiding hints. The exposition of the theory is accompanied by worked-out examples and additional problems at the end of chapters.

The book addresses mostly *theoretical* concepts and tools relevant to the physics of solids; there is no attempt to provide a thorough account of related experimental facts. This choice was made in order to keep the book within a limit that allows its contents to be covered in a reasonably short period (one or two semesters; see more detailed instructions below). There are many sources covering the experimental side of the field, which the student is strongly encouraged to explore if not already familiar with it. The suggestions for further reading at the end of chapters can serve as a starting point for exploring the experimental literature. There are also selected references to original research articles that laid the foundations of the topics discussed, as well as to more recent work, in the hope of exciting the student's interest for further exploration. Instead of providing a comprehensive list of references, the reader is typically directed toward review articles and monographs which contain more advanced treatments and a more extended bibliography.

As already mentioned, the treatment is mostly restricted to the single-particle picture. The meaning of this is clarified and its advantages and limitations are described in great detail in the second chapter. Briefly, the electrons responsible for the cohesion of a solid interact through long-range Coulomb forces both with the nuclei of the solid and with all the other electrons. This leads to a very complex many-electron state which is difficult to describe quantitatively. In certain limits, and for certain classes of phenomena, it is feasible to describe the solid in terms of an approximate picture involving "single electrons", which interact with the other electrons through an average field. In fact, these "single-electron" states do not correspond to physical electron states (hence the quotes). This picture, although based on approximations that cannot be systematically improved, turns out to be extremely useful and remarkably realistic for many, but not all, situations. There are several phenomena – superconductivity and certain aspects of magnetic phenomena being prime examples – where the collective behavior of electrons in a solid is essential in understanding the nature of the beast (or beauty). In these cases the "single-electron" picture is not adequate, and a full many-body approach is necessary. The phenomena involved in the many-body picture require an approach and a theoretical formalism beyond what is covered here; typically, these topics constitute the subject of a second course on the theory of solids.

The book is divided into two parts. The first part, called Crystalline solids, consists of eight chapters and includes material that I consider essential in understanding the physics of solids. The discussion is based on crystals, which offer a convenient model for studying macroscopic numbers of atoms assembled to form a solid. In this part, the first five chapters develop the theoretical basis for the single-electron picture and give several applications of this picture, for solids in which atoms are frozen in space. Chapter 6 develops the tools for understanding the motion of atoms in crystals through the language of phonons. Chapters 7 and 8 are devoted to magnetic phenomena and superconductivity, respectively. The purpose of these last two chapters is to give a glimpse of interesting phenomena in solids which go beyond the single-electron picture. Although more advanced, these topics have become an essential part of the physics of solids and must be included in a general introduction to the field. I have tried to keep the discussion in these two chapters at a relatively simple level, avoiding, for example, the introduction of tools like second quantization, Green's functions and Feynman diagrams. The logic of this approach is to make the material accessible to a wide audience, at the cost of not employing a more elegant language familiar to physicists.

The second part of the book consists of five chapters, which contain discussions of defects in crystals (chapters 9, 10 and 11), of non-crystalline solids (chapter 12) and of finite structures (chapter 13). The material in these chapters is more specific than that in the first part of the book, and thus less important from a fundamental point of view. This material, however, is relevant to real solids, as opposed to idealized theoretical concepts such as a perfect crystal. I must make here a clarification on why the very last chapter is devoted to finite structures, a topic not traditionally discussed in the context of solids. Such structures are becoming increasingly important, especially in the field of nanotechnology, where the functional components may be measured in nanometers. Prime examples of such objects are clusters or tubes of carbon (the fullerenes and the carbon nanotubes) and biological structures (the nucleic acids and proteins), which are studied by ever increasing numbers of traditional physicists, chemists and materials scientists, and which are expected to find their way into solid state applications in the not too distant future. Another reason for including a discussion of these systems in a book on solids, is that they *do* have certain common characteristics with traditional crystals, such as a high degree of order. After all, what could be a more relevant example of a regular one-dimensional structure than the human DNA chain which extends for three billion base-pairs with essentially perfect stacking, even though it is not rigid in the traditional sense?

This second part of the book contains material closer to actual research topics in the modern theory of solids. In deciding what to include in this part, I have drawn mostly from my own research experience. This is the reason for omitting some important topics, such as the physics of metal alloys. My excuse for such omissions is that the intent was to write a modern textbook on the physics of solids, with representative examples of current applications, rather than an encyclopedic compilation of research topics. Despite such omissions, I hope that the scope of what *is* covered is broad enough to offer a satisfactory representation of the field.

Finally, a few comments about the details of the contents. I have strived to make the discussion of topics in the book as self-contained as possible. For this reason, I have included unusually extensive appendices in what constitutes a third part of the book. Four of these appendices, on classical electrodynamics, quantum mechanics, thermodynamics and statistical mechanics, contain all the information necessary to derive from very basic principles the results of the first part of the book. The appendix on elasticity theory contains the background information relevant to the discussion of line defects and the mechanical properties of solids. The appendix on the Madelung energy provides a detailed account of an important term in the total energy of solids, which was deemed overly technical to include in the first part. Finally, the appendix on mathematical tools reviews a number of formulae, techniques and tricks which are used extensively throughout the text. The material in the second part of the book could not be made equally self-contained by the addition of appendices, because of its more specialized nature. I have made an effort to provide enough references for the interested reader to pursue in more detail any topic covered in the second part. An appendix at the end includes Nobel prize citations relevant to work mentioned in the text, as an indication of how vibrant the field has been and continues to be. The appendices may seem excessively long by usual standards, but I hope that a good fraction of the readers will find them useful.

Some final comments on notation and figures: I have made a conscious effort to provide a consistent notation for all the equations throughout the text. Given the breadth of topics covered, this was not a trivial task and I was occasionally forced to make unconventional choices in order to avoid using the same symbol for two different physical quantities. Some of these are: the choice of Ω for the volume so that the more traditional symbol V could be reserved for the potential energy; the choice of Θ for the enthalpy so that the more traditional symbol H could be reserved for the magnetic field; the choice of Y for Young's modulus so that the more traditional symbol E could be reserved for the energy; the introduction of a subscript in the symbol for the divergence, $\nabla_{\mathbf{r}}$ or $\nabla_{\mathbf{k}}$, so that the variable of differentiation would be unambiguous even if, on certain occasions, this is redundant information. I have also made extensive use of superscripts, which are often in parentheses to differentiate them from exponents, in order to make the meaning of symbols more transparent. Lastly, I decided to draw all the figures "by hand" (using software tools), rather than to reproduce figures from the literature, even when discussing classic experimental or theoretical results. The purpose of this choice is to maintain, to the extent possible, the feeling of immediacy in the figures as I would have drawn them on the blackboard, pointing out important features rather than being faithful to details. I hope that the result is not disagreeable, given my admittedly limited drawing abilities. Exceptions are the set of figures on electronic structure of metals and semiconductors in chapter 4 (Figs. 4.6–4.12), which were produced by Yannis Remediakis, and the figure of the KcsA protein in chapter 13 (Fig. 13.30), which was provided by Pavlos Maragakis.

The book has been constructed to serve two purposes. (a) For students with adequate background in the basic fields of physics (electromagnetism, quantum mechanics, thermodynamics and statistical mechanics), the first part represents a comprehensive introduction to the single-particle theory of solids and can be covered in a one-semester course. As an indication of the degree of familiarity with basic physics expected of the reader, I have included sample problems in the corresponding appendices; the readers who can tackle these problems easily can proceed directly to the main text covered in the first part. My own teaching experience indicates that approximately 40 hours of lectures (roughly five per chapter) are adequate for a brisk, but not unreasonable, covering of this part. Material from the second part can be used selectively as illustrative examples of how the basic concepts are applied to realistic situations. This can be done in the form of special assignments, or as projects at the end of the one-semester course.

(b) For students without graduate level training in the basic fields of physics mentioned above, the entire book can serve as the basis for a full-year course. The material in the first part can be covered at a more leisurely pace, with short introductions of the important physics background where needed, using the appendices as a guide. The material of the second part of the book can then be covered, selectively or in its entirety as time permits, in the remainder of the full-year course.

Acknowledgments

The discussion of many topics in this book, especially the chapters that deal with symmetries of the crystalline state and band structure methods, was inspired to a great extent by the lectures of John Joannopoulos who first introduced me to this subject. I hope the presentation of these topics here does justice to his meticulous and inspired teaching.

In my two-decade-long journey through the physics of solids, I had the good fortune to interact with a great number of colleagues, from all of whom I have learned a tremendous amount. In roughly chronological order in which I came to know them, they are: John Joannopoulos, Karin Rabe, Alex Antonelli, Dung-Hai Lee, Yaneer Bar-Yam, Eugen Tarnow, David Vanderbilt, Oscar Alerhand, Bob Meade, George Turner, Andy Rappe, Michael Payne, Jim Chelikowsky, Marvin Cohen, Jim Chadi, Steven Louie, Stratos Manousakis, Kosal Pandey, Norton Lang, Jerry Tersoff, Phaedon Avouris, In-When Lyo, Ruud Tromp, Matt Copel, Bob Hamers, Randy Feenstra, Ken Shih, Franz Himpsel, Joe Demuth, Sokrates Pantelides, Pantelis Kelires, Peter Blöchl, Dimitri Papaconstantopoulos, Barry Klein, Jeremy Broughton, Warren Pickett, David Singh, Michael Mehl, Koblar Jackson, Mark Pederson, Steve Erwin, Larry Boyer, Joe Feldman, Daryl Hess, Joe Serene, Russ Hemley, John Weeks, Ellen Williams, Bert Halperin, Henry Ehrenreich, Daniel Fisher, David Nelson, Paul Martin, Jene Golovchenko, Bill Paul, Eric Heller, Cynthia Friend, Roy Gordon, Howard Stone, Charlie Lieber, Eric Mazur, Mike Aziz, Jim Rice, Frans Spaepen, John Hutchinson, Michael Tinkham, Ike Silvera, Peter Pershan, Bob Westervelt, Venky Narayanamurti, George Whitesides, Charlie Marcus, Leo Kouwenhoven, Martin Karplus, Dan Branton, Dave Weitz, Eugene Demler, Uzi Landman, Andy Zangwill, Peter Feibelman, Priya Vashishta, Rajiv Kalia, Mark Gyure, Russ Cafflisch, Dimitri Vvedensky, Jenna Zink, Bill Carter, Lloyd Whitman, Stan Williams, Dimitri Maroudas, Nick Kioussis, Michael Duesbery, Sidney Yip, Farid Abraham, Shi-Yu Wu, John Wilkins, Ladislas Kubin, Rob Phillips, Bill Curtin, Alan Needleman, Michael Ortiz, Emily Carter,

John Smith, Klaus Kern, Oliver Leifeld, Lefteris Economou, Nikos Flytzanis, Stavros Farantos, George Tsironis, Grigoris Athanasiou, Panos Tzanetakis, Kostas Fotakis, George Theodorou, José Soler, Thomas Frauenheim, Riad Manaa, Doros Theodorou, Vassilis Pontikis and Sauro Succi. Certain of these individuals played not only the role of a colleague or collaborator, but also the role of a mentor at various stages of my career: they are, John Joannopoulos, Kosal Pandey, Dimitri Papaconstantopoulos, Henry Ehrenreich, Bert Halperin and Sidney Yip; I am particularly indebted to them for guidance and advice, as well as for sharing with me their deep knowledge of physics.

I was also very fortunate to work with many talented graduate and undergraduate students, including Yumin Juan, Linda Zeger, Normand Modine, Martin Bazant, Noam Bernstein, Greg Smith, Nick Choly, Ryan Barnett, Sohrab Ismail-Beigi, Jonah Erlebacher, Melvin Chen, Tim Mueller, Yuemin Sun, Joao Justo, Maurice de Koning, Yannis Remediakis, Helen Eisenberg, Trevor Bass, and with a very select group of Postdoctoral Fellows and Visiting Scholars, including Daniel Kandel, Laszlo Barabàsi, Gil Zumbach, Umesh Waghmare, Ellad Tadmmor, Vasily Bulatov, Kyeongjae Cho, Marcus Elstner, Ickjin Park, Hanchul Kim, Olivier Politano, Paul Maragakis, Dionisios Margetis, Daniel Orlikowski, Qiang Cui and Gang Lu. I hope that they have learned from me a small fraction of what I have learned from them over the last dozen years.

Last but not least, I owe a huge debt of gratitude to my wife, Eleni, who encouraged me to turn my original class notes into the present book and supported me with patience and humor throughout this endeavor.

The merits of the book, to a great extent, must be attributed to the generous input of friends and colleagues, while its shortcomings are the exclusive responsibility of the author. Pointing out these shortcomings to me would be greatly appreciated.

Cambridge, Massachusetts, October 2001

Part I

Crystalline solids

If, in some cataclysm, all of scientific knowledge were to be destroyed, and only one sentence passed on to the next generation of creatures, what statement would contain the most information in the fewest words? I believe it is the atomic hypothesis that all things are made of atoms – little particles that move around in perpetual motion, attracting each other when they are a little distance apart, but repelling upon being squeezed into one another. In that one sentence, there is an enormous amount of information about the world, if just a little imagination and thinking are applied.

(R. P. Feynman, *The Feynman Lectures on Physics*)

Solids are the physical objects with which we come into contact continuously in our everyday life. Solids are composed of atoms. This was first postulated by the ancient Greek philosopher Demokritos, but was established scientifically in the 20th century. The atoms ($\alpha\tau o\mu\alpha$ = indivisible units) that Demokritos conceived bear no resemblance to what we know today to be the basic units from which all solids and molecules are built. Nevertheless, this postulate is one of the greatest feats of the human intellect, especially since it was not motivated by direct experimental evidence but was the result of pure logical deduction.

There is an amazing degree of regularity in the structure of solids. Many solids are crystalline in nature, that is, the atoms are arranged in a regular three-dimensional periodic pattern. There is a wide variety of crystal structures formed by different elements and by different combinations of elements. However, the mere fact that a number of atoms of order 10^{24} (Avogadro's number) in a solid of size 1 cm^3 are arranged in essentially a perfect periodic array is quite extraordinary. In some cases it has taken geological times and pressures to form certain crystalline solids, such as diamonds. Consisting of carbon and found in mines, diamonds represent the hardest substance known, but, surprisingly, they do not represent the ground state equilibrium structure of this element. In many other cases, near perfect macroscopic crystals can be formed by simply melting and then slowly cooling a substance in the laboratory. There are also many ordinary solids we encounter in everyday life in which there exists a surprising degree of crystallinity. For example, a bar of soap, a chocolate bar, candles, sugar or salt grains, even bones in the human body, are composed of crystallites of sizes between 0.5 and 50 μm. In these examples, what determines the properties of the material is not so much the structure of individual crystallites but their relative orientation and the structure of boundaries between them. Even in this case, however, the nature of a boundary between two crystallites is ultimately dictated by the structure of the crystal grains on either side of it, as we discuss in chapter 11.

The existence of crystals has provided a tremendous boost to the study of solids, since a crystalline solid can be analyzed by considering what happens in a single unit of the crystal (referred to as the unit cell), which is then repeated periodically in all three dimensions to form the idealized perfect and infinite solid. The unit cell contains typically one or a few atoms, which are called the basis. The points in space that are equivalent by translations form the so called Bravais lattice . The Bravais lattice and the basis associated with each unit cell determine the crystal. This regularity has made it possible to develop powerful analytical tools and to use clever experimental techniques to study the properties of solids.

Real solids obviously do not extend to infinity in all three dimensions – they terminate on surfaces, which in a sense represent two-dimensional defects in the perfect crystalline structure. For all practical purposes the surfaces constitute a very small perturbation in typical solids, since the ratio of atoms on the surface to atoms in the bulk is typically $1 : 10^8$. The idealized picture of atoms in the bulk behaving as if they belonged to an infinite periodic solid, is therefore a reasonable one. In fact, even very high quality crystals contain plenty of one-dimensional or zero-dimensional defects in their bulk as well. It is actually the presence of such defects that renders solids useful, because the manipulation of defects makes it possible to alter the properties of the ideal crystal, which in perfect form would have a much more limited range of properties. Nevertheless, these defects exist in relatively small concentrations in the host crystal, and as such can be studied in a perturbative manner, with the ideal crystal being the base or, in a terminology that physicists often use, the "vacuum" state . If solids lacked any degree of order in their structure, study of them would be much more complicated. There are many solids that are not crystalline, with some famous examples being glasses, or amorphous semiconductors. Even in these solids, there exists a high degree of *local* order in their structure, often very reminiscent of the local arrangement of atoms in their crystalline counterparts. As a consequence, many of the notions advanced to describe disordered solids are extensions of, or use as a point of reference, ideas developed for crystalline solids. All this justifies the prominent role that the study of crystals plays in the study of solids.

It is a widely held belief that the crystalline state represents the ground state structure of solids, even though there is no theoretical proof of this statement. A collection of 10^{24} atoms has an immense number of almost equivalent ordered or disordered metastable states in which it can exist, but only one lowest energy crystalline state; and the atoms can find this state in relatively short time scales and with relatively very few mistakes! If one considers the fact that atomic motion is quite difficult and rare in the dense atomic packing characteristic of crystals, so that the atoms have little chance to correct an error in their placement, the existence of crystals becomes even more impressive.

The above discussion emphasizes how convenient it has proven for scientists that atoms like to form crystalline solids. Accordingly, we will use crystals as the basis for studying general concepts of bonding in solids, and we will devote the first part of the book to the study of the structure and properties of crystals.

1

Atomic structure of crystals

Solids exhibit an extremely wide range of properties, which is what makes them so useful and indispensable to mankind. While our familiarity with many different types of solids makes this fact seem unimpressive, it is indeed extraordinary when we consider its origin. The origin of all the properties of solids is nothing more than the interaction between electrons in the outer shells of the atoms, the so called *valence* electrons. These electrons interact among themselves and with the nuclei of the constituent atoms. In this first chapter we will give a general description of these interactions and their relation to the structure and the properties of solids.

The extremely wide range of the properties of solids is surprising because most of them are made up from a relatively small subset of the elements in the Periodic Table: about 20 or 30 elements, out of more than 100 total, are encountered in most common solids. Moreover, most solids contain only very few of these elements, from one to half a dozen or so. Despite this relative simplicity in composition, solids exhibit a huge variety of properties over ranges that differ by many orders of magnitude. It is quite extraordinary that even among solids which are composed of single elements, physical properties can differ by many orders of magnitude.

One example is the ability of solids to conduct electricity, which is measured by their electrical resistivity. Some typical single-element metallic solids (such as Ag, Cu, Al), have room-temperature resistivities of 1–5$\mu\Omega$·cm, while some metallic alloys (like nichrome) have resistivities of $10^2\mu\Omega$·cm. All these solids are considered good conductors of electrical current. Certain single-element solids (like C, Si, Ge) have room-temperature resistivities ranging from $3.5 \times 10^3\mu\Omega$·cm (for graphitic C) to $2.3 \times 10^{11}\mu\Omega$·cm (for Si), and they are considered semimetals or semiconductors. Finally, certain common solids like wood (with a rather complex structure and chemical composition) or quartz (with a rather simple structure and composed of two elements, Si and O), have room-temperature resistivities of 10^{16}–$10^{19}\mu\Omega$·cm (for wood) to $10^{25}\mu\Omega$·cm (for quartz). These solids are considered insulators. The range of electrical resistivities covers an astonishing 25 orders of magnitude!

Another example has to do with the mechanical properties of solids. Solids are classified as ductile when they yield plastically when stressed, or brittle when they do not yield easily, but instead break when stressed. A useful measure of this behavior is the yield stress σ_Y, which is the stress up to which the solid behaves as a linear elastic medium when stressed, that is, it returns to its original state when the external stress is removed. Yield stresses in solids, measured in units of MPa, range from 40 in Al, a rather soft and ductile metal, to 5×10^4 in diamond, the hardest material, a brittle insulator. The yield stresses of common steels range from 200–2000 MPa. Again we see an impressive range of more than three orders of magnitude in how a solid responds to an external agent, in this case a mechanical stress.

Naively, one might expect that the origin of the widely different properties of solids is related to great differences in the concentration of atoms, and correspondingly that of electrons. This is far from the truth. Concentrations of atoms in a solid range from 10^{22} cm^{-3} in Cs, a representative alkali metal, to 17×10^{22} cm^{-3} in C, a representative covalently bonded solid. Anywhere from one to a dozen valence electrons per atom participate actively in determining the properties of solids. These considerations give a range of atomic concentrations of roughly 20, and of electron concentrations[1] of roughly 100. These ranges are nowhere close to the ranges of yield stresses and electrical resistivities mentioned above. Rather, the variation of the properties of solids has to do with the specific ways in which the valence electrons of the constituent atoms interact when these atoms are brought together at distances of a few angstroms (1 Å$= 10^{-10}$ m $= 10^{-1}$ nm). Typical distances between nearest neighbor atoms in solids range from 1.5 to 3 Å. The way in which the valence electrons interact determines the atomic structure, and this in turn determines all the other properties of the solid, including mechanical, electrical, optical, thermal and magnetic properties.

1.1 Building crystals from atoms

The structure of crystals can be understood to some extent by taking a close look at the properties of the atoms from which they are composed. We can identify several broad categories of atoms, depending on the nature of electrons that participate actively in the formation of the solid. The electrons in the outermost shells of the isolated atom are the ones that interact strongly with similar electrons in neighboring atoms; as already mentioned these are called valence electrons. The remaining electrons of the atom are tightly bound to the nucleus, their wavefunctions (orbitals)

[1] The highest concentration of atoms does not correspond to the highest number of valence electrons per atom.

do not extend far from the position of the nucleus, and they are very little affected when the atom is surrounded by its neighbors in the solid. These are called the core electrons. For most purposes it is quite reasonable to neglect the presence of the core electrons as far as the solid is concerned, and consider how the valence electrons behave. We will discuss below the crystal structure of various solids based on the properties of electronic states of the constituent atoms. We are only concerned here with the basic features of the crystal structures that the various atoms form, such as number of nearest neighbors, without paying close attention to details; these will come later. Finally, we will only concern ourselves with the low-temperature structures, which correspond to the lowest energy static configuration; dynamical effects, which can produce a different structure at higher temperatures, will not be considered [1]. We begin our discussion with those solids formed by atoms of one element only, called elemental solids, and then proceed to more complicated structures involving several types of atoms. Some basic properties of the elemental solids are collected in the Periodic Table (pp. 8, 9), where we list:

- The crystal structure of the most common phase. The acronyms for the crystal structures that appear in the Table stand for: BCC = body-centered cubic, FCC = face-centered cubic, HCP = hexagonal-close-packed, GRA = graphite, TET = tetragonal, DIA = diamond, CUB = cubic, MCL = monoclinic, ORC = orthorhombic, RHL = rhombohedral. Selected shapes of the corresponding crystal unit cells are shown in Fig. 1.1.

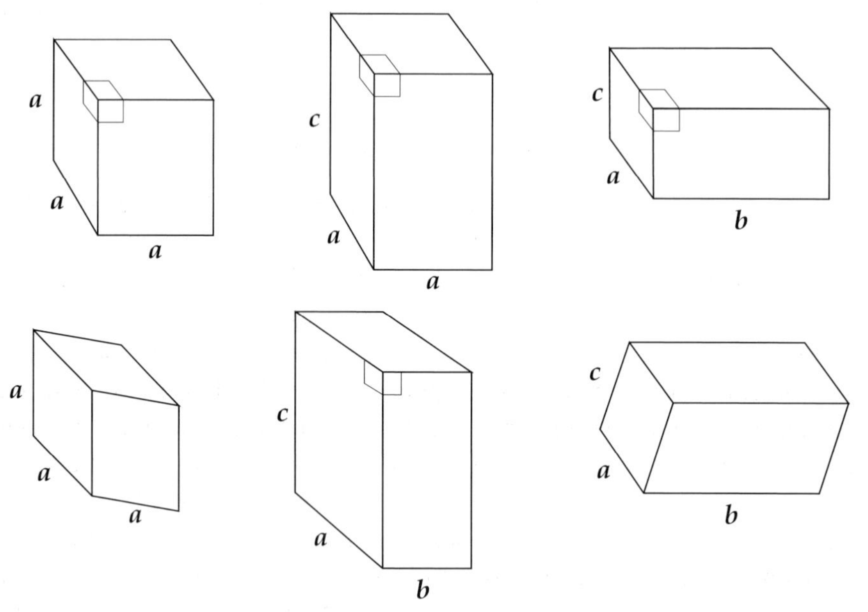

Figure 1.1. Shapes of the unit cells in some lattices that appear in Periodic Table. **Top row:** cubic, tetragonal, orthorhombic. **Bottom row:** rhombohedral, monoclinic, triclinic. The corners in thin lines indicate right angles between edges.

- The covalent radius in units of angstroms, Å, which is a measure of the typical distance of an atom to its neighbors; specifically, the sum of covalent radii of two nearest neighbor atoms give their preferred distance in the solid.
- The melting temperature in millielectronvolts (1 meV $= 10^{-3}$ eV $= 11.604$ K). The melting temperature provides a measure of how much kinetic energy is required to break the rigid structure of the solid. This unconventional choice of units for the melting temperature is meant to facilitate the discussion of cohesion and stability of solids. Typical values of the cohesive energy of solids are in the range of a few electronvolts (see Tables 5.4 and 5.5), which means that the melting temperature is only a small fraction of the cohesive energy, typically a few percent.
- The atomic concentration of the most common crystal phase in 10^{22} cm^{-3}.
- The electrical resistivity in units of micro-ohm-centimeters, μΩ·cm; for most elemental solids the resistivity is of order 1–100 in these units, except for some good insulators which have resistivities 10^3(k), 10^6(M) or 10^9(G) times higher.

The natural units for various physical quantities in the context of the structure of solids and the names of unit multiples are collected in two tables at the end of the book (see Appendix I).

The columns of the Periodic Table correspond to different valence electron configurations, which follow a smooth progression as the s, p, d and f shells are being filled. There are a few exceptions in this progression, which are indicated by asterisks denoting that the higher angular momentum level is filled in preference to the lower one (for example, the valence electronic configuration of Cu, marked by one asterisk, is s^1d^{10} instead of s^2d^9; that of Pd, marked by two asterisks, is s^0d^{10} instead of s^2d^8, etc.).

1.1.1 Atoms with no valence electrons

The first category consists of those elements which have no valence electrons. These are the atoms with all their electronic shells completely filled, which in gaseous form are very inert chemically, i.e. the noble elements He, Ne, Ar, Kr and Xe. When these atoms are brought together to form solids they interact very weakly. Their outer electrons are not disturbed much since they are essentially core electrons, and the weak interaction is the result of slight polarization of the electronic wavefunction in one atom due to the presence of other atoms around it. Fortunately, the interaction is attractive. This interaction is referred to as "fluctuating dipole" or van der Waals interaction. Since the interaction is weak, the solids are not very stable and they have very low melting temperatures, well below room temperature. The main concern of the atoms in forming such solids is to have as many neighbors as possible, in order to maximize the cohesion since all interactions are attractive. The crystal structure that corresponds to this atomic arrangement is one of the close-packing geometries, that is, arrangements which allow the closest packing of hard spheres.

symbol	Li
atomic number	3
name	Lithium
crystal structure / covalent radius (Å)	BCC 1.23
melting point (meV)	39.08
atomic concentration (10^{22}cm^{-3})	4.70
resistivity (μΩ cm)	9.4

I-A s^1	II-A s^2	III-B s^2d^1	IV-B s^2d^2	V-B s^2d^3	VI-B s^2d^4	VII-B s^2d^5	VIII s^2d^6	VIII s^2d^7
Li 3 Lithium BCC 1.23 39.08 4.70 9.4	Be 4 Beryllium HCP 0.90 134.4 12.1 3.3							
Na 11 Sodium BCC 1.54 8.42 2.65 4.75	Mg 12 Magnesium HCP 1.36 79.54 4.30 4.46							
K 19 Potassium BCC 2.03 28.98 1.40 21.6	Ca 20 Calcium FCC 1.74 96.09 2.30 3.7	Sc 21 Scandium HCP 1.44 156.3 4.27 51	Ti 22 Titanium HCP 1.32 167.3 5.66 47.8	V 23 Vanadium BCC 1.22 188.1 7.22 24.8	Cr* 24 Chromium BCC 1.18 187.9 8.33 12.9	Mn 25 Manganese CUB 1.17 130.9 8.18 139	Fe 26 Iron BCC 1.17 156.1 8.50 9.71	Co 27 Cobalt HCP 1.16 152.4 8.97 6.34
Rb 37 Rubidium BCC 2.16 26.89 1.15 12.1	Sr 38 Strontium FCC 1.91 96.49 1.78 22.8	Y 39 Yttrium HCP 1.62 154.7 3.02 60	Zr 40 Zirconium HCP 1.45 183.4 4.29 41.4	Nb* 41 Niobium BCC 1.34 237.0 5.56 15.2	Mo* 42 Molybdenum BCC 1.30 249.6 6.42 5.17	Tc 43 Technetium HCP 1.28 234.2 7.04 14	Ru* 44 Ruthenium HCP 1.25 224.7 7.36 7.2	Rh* 45 Rhodium FCC 1.25 192.9 7.26 4.5
Cs 55 Cesium BCC 2.35 25.97 0.91 20	Ba 56 Barium BCC 1.98 86.18 1.60 50	La 57 Lanthanum HCP 1.69 102.8 2.70 80	Ha 72 Hafnium HCP 1.44 215.4 4.52 35.1	Ta 73 Tantalum BCC 1.34 281.7 5.55 13.5	W 74 Wolframium BCC 1.30 317.4 6.30 5.6	Re 75 Rhenium HCP 1.28 297.6 6.80 19.3	Os 76 Osmium HCP 1.26 285.9 7.14 8.1	Ir 77 Iridium FCC 1.27 234.7 7.06 5.1

$f^2d^0s^2$	$f^3d^0s^2$	$f^4d^0s^2$	$f^5d^0s^2$	$f^6d^0s^2$	$f^7d^0s^2$
Ce 58 Cerium FCC 1.65 92.3 2.91 85.4	Pr 59 Praseodymium HCP 1.65 103.8 2.92 68.0	Nd 60 Neodymium HCP 1.64 110.6 2.93 64.3	Pm 61 Promethium	Sm 62 Samarium RHL 1.62 115.9 3.03 105.0	Eu 63 Europium BCC 1.85 94.4 3.04 91.0

The particular crystal structure that noble-element atoms assume in solid form is called face-centered cubic (FCC). Each atom has 12 equidistant nearest neighbors in this structure, which is shown in Fig. 1.2.

Thus, in the simplest case, atoms that have no valence electrons at all behave like hard spheres which attract each other with weak forces, but are not deformed. They

VIII	I-B	II-B	III–A	IV–A	V–A	VI–A	VII–A	Noble
s^2d^8	s^2d^9	s^2d^{10}	s^2p^1	s^2p^2	s^2p^3	s^2p^4	s^2p^5	s^2p^6
			B [5] Boron TET 0.82 202.3 13.0 4 M	C [6] Carbon GRA 0.77 338.1 17.6 1.4 G	N [7] Nitrogen HCP 0.70 28.98	O [8] Oxygen CUB 0.66 28.24	F [9] Fluorine MCL 0.64 28.14	Ne [10] Neon FCC 25.64 4.36
			Al [13] Aluminum FCC 1.18 80.44 6.02 2.67	Si [14] Silicon DIA 1.17 145.4 5.00 230 G	P [15] Phosphorus CUB 1.10 59.38	S [16] Sulfur ORC 1.04 33.45	Cl [17] Chlorine ORC 0.99 38.37	Ar [18] Argon FCC 30.76 2.66
Ni [28] Nickel FCC 1.15 148.9 9.14 6.84	Cu* [29] Copper FCC 1.17 116.9 8.45 1.67	Zn [30] Zinc HCP 1.25 59.68 6.55 5.92	Ga [31] Gallium ORC 1.26 26.09 5.10	Ge [32] Germanium DIA 1.22 104.3 4.42 47 M	As [33] Arsenic RHL 1.21 93.93 4.65 12 M	Se [34] Selenium HCP 1.17 42.23 3.67	Br [35] Bromine ORC 1.14 46.45 2.36	Kr [36] Krypton FCC 33.57 2.17
Pd** [46] Palladium FCC 1.28 157.5 6.80 9.93	Ag* [47] Silver FCC 1.34 106.4 5.85 1.63	Cd [48] Cadmium HCP 1.48 51.19 4.64 6.83	In [49] Indium TET 1.44 37.02 3.83 8.37	Sn [50] Tin TET 1.40 43.53 2.91 11	Sb [51] Antimony RHL 1.41 77.88 3.31 39	Te [52] Tellurium HCP 1.37 62.26 2.94 160 k	I [53] Iodine ORC 1.33 56.88 2.36	Xe [54] Xenon FCC 37.43 1.64
Pt** [78] Platinum FCC 1.30 175.9 6.62 9.85	Au* [79] Gold FCC 1.34 115.2 5.90 2.12	Hg [80] Mercury RHL 1.49 20.18 4.26 96	Tl [81] Thallium HCP 1.48 49.68 3.50 18	Pb [82] Lead FCC 1.47 51.75 3.30 20.6	Bi [83] Bismuth RHL 1.34 46.91 2.82 107	Po [84] Pollonium	At [85] Astatine	Rn [86] Radon

$f^7d^1s^2$	$f^8d^1s^2$	$f^{10}d^0s^2$	$f^{11}d^0s^2$	$f^{12}d^0s^2$	$f^{13}d^0s^2$	$f^{14}d^0s^2$	$f^{14}d^1s^2$
Gd [64] Gadolinium HCP 1.61 136.6 3.02 131.0	Tb [65] Terbium HCP 1.59 140.7 3.22 114.5	Dy [66] Dysprosium HCP 1.59 144.8 3.17 92.6	Ho [67] Holmium HCP 1.58 150.2 3.22 81.4	Er [68] Erbium HCP 1.57 154.7 3.26 86.0	Tm [69] Thulium HCP 1.56 156.7 3.32 67.6	Yb [70] Ytterbium FCC 94.5 3.02 25.1	Lu [71] Lutetium HCP 1.56 166.2 3.39 58.2

form weakly bonded solids in the FCC structure, in which the attractive interactions are optimized by maximizing the number of nearest neighbors in a close packing arrangement. The only exception to this rule is He, in which the attractive interaction between atoms is so weak that it is overwhelmed by the zero-point motion of the atoms. Unless we apply external pressure to enhance this attractive interaction,

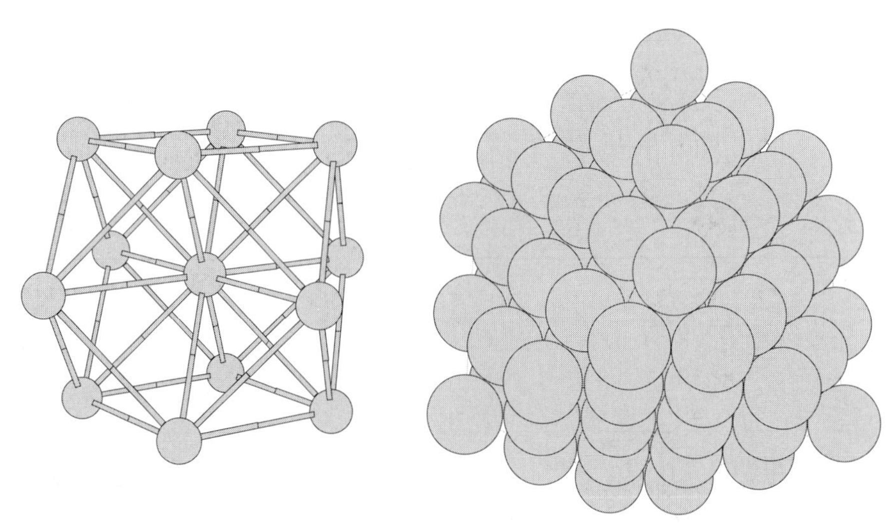

Figure 1.2. **Left:** one atom and its 12 neighbors in the face-centered cubic (FCC) lattice; the size of the spheres representing atoms is chosen so as to make the neighbors and their distances apparent. **Right:** a portion of the three-dimensional FCC lattice; the size of the spheres is chosen so as to indicate the close-packing nature of this lattice.

He remains a liquid. This is also an indication that in some cases it will prove unavoidable to treat the nuclei as quantum particles (see also the discussion below about hydrogen).

The other close-packing arrangement of hard spheres is the hexagonal structure (HCP for hexagonal-close-packed), with 12 neighbors which are separated into two groups of six atoms each: the first group forms a planar six-member ring surrounding an atom at the center, while the second group consists of two equilateral triangles, one above and one below the six-member ring, with the central atom situated above or below the geometrical center of each equilateral triangle, as shown in Fig. 1.3. The HCP structure bears a certain relation to FCC: we can view both structures as planes of spheres closely packed in two dimensions, which gives a hexagonal lattice; for close packing in three dimensions the successive planes must be situated so that a sphere in one plane sits at the center of a triangle formed by three spheres in the previous plane. There are two ways to form such a stacking of hexagonal close-packed planes: $...ABCABC...$, and $...ABABAB...$, where A, B, C represent the three possible relative positions of spheres in successive planes according to the rules of close packing, as illustrated in Fig. 1.4. The first sequence corresponds to the FCC lattice, the second to the HCP lattice.

An interesting variation of the close-packing theme of the FCC and HCP lattices is the following: consider two interpenetrating such lattices, that is, two FCC or two HCP lattices, arranged so that in the resulting crystal the atoms in each sublattice

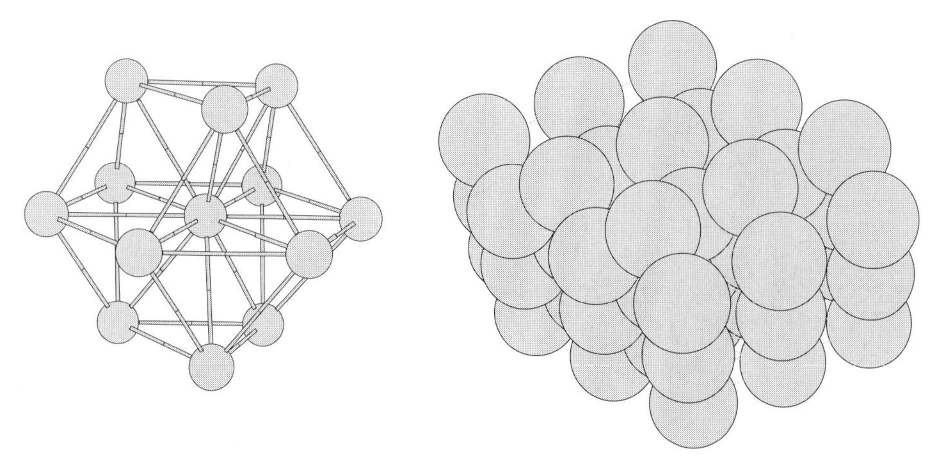

Figure 1.3. **Left:** one atom and its 12 neighbors in the hexagonal-close-packed (HCP) lattice; the size of the spheres representing atoms is chosen so as to make the neighbors and their distances apparent. **Right:** a portion of the three-dimensional HCP lattice; the size of the spheres is chosen so as to indicate the close-packing nature of this lattice.

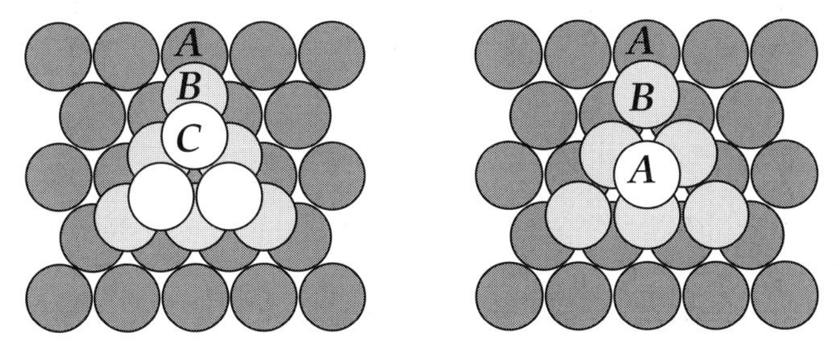

Figure 1.4. The two possible close packings of spheres: **Left:** the ...$ABCABC$... stacking corresponding to the FCC crystal. **Right:** the ...$ABABAB$... stacking corresponding to the HCP crystal. The lattices are viewed along the direction of stacking of the hexagonal-close-packed planes.

have as nearest equidistant neighbors atoms belonging to the other sublattice. These arrangements give rise to the diamond lattice or the zincblende lattice (when the two original lattices are FCC) and to the wurtzite lattice (when the two original lattices are HCP). This is illustrated in Fig. 1.5. Interestingly, in both cases each atom finds itself at the center of a tetrahedron with exactly four nearest neighbors. Since the nearest neighbors are exactly the same, these two types of lattices differ only in the relative positions of second (or farther) neighbors. It should be evident that the combination of two close-packed lattices cannot produce another close-packed lattice. Consequently, the diamond, zincblende and wurtzite lattices are encountered in covalent or ionic structures in which four-fold coordination is preferred. For example: tetravalent group IV elements such as C, Si, Ge form the diamond lattice; combinations of two different group IV elements or complementary elements

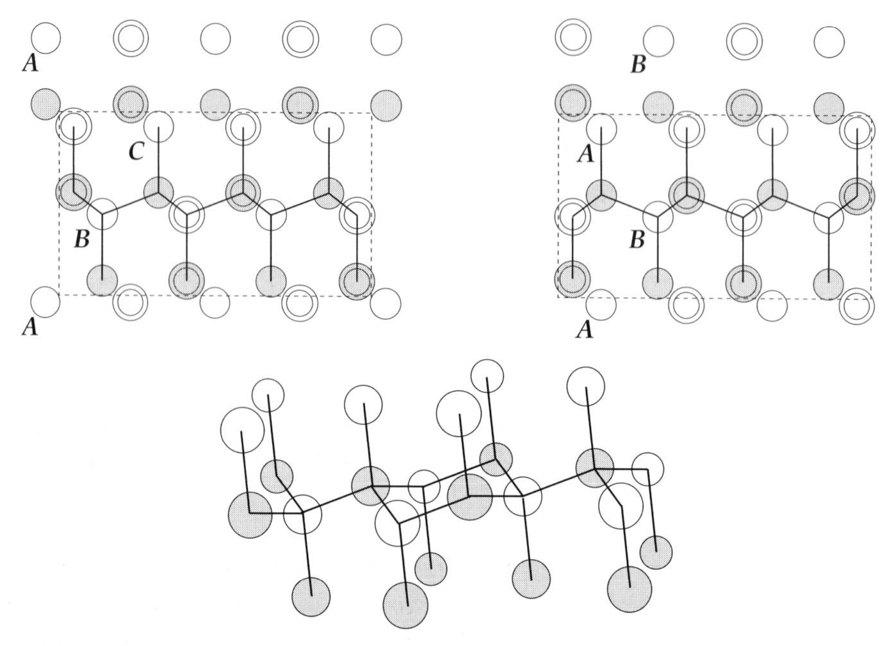

Figure 1.5. **Top:** illustration of two interpenetrating FCC (left) or HCP (right) lattices; these correspond to the diamond (or zincblende) and the wurtzite lattices, respectively. The lattices are viewed from the side, with the vertical direction corresponding to the direction along which close-packed planes of the FCC or HCP lattices would be stacked (see Fig. 1.4). The two original lattices are denoted by sets of white and shaded circles. All the circles of medium size would lie on the plane of the paper, while the circles of slightly smaller and slightly larger size (which are superimposed in this view) lie on planes behind and in front of the plane of the paper. Lines joining the circles indicate covalent bonds between nearest neighbor atoms. **Bottom:** a perspective view of a portion of the diamond (or zincblende) lattice, showing the tetrahedral coordination of all the atoms; this is the area enclosed by the dashed rectangle in the top panel, left side (a corresponding area can also be identified in the wurtzite lattice, upon reflection).

(such as group III–group V, group II–group VI, group I–group VII) form the zincblende lattice; certain combinations of group III–group V elements form the wurtzite lattice. These structures are discussed in more detail below. A variation of the wurtzite lattice is also encountered in ice and is due to hydrogen bonding.

Yet another version of the close-packing arrangement is the icosahedral structure. In this case an atom again has 12 equidistant neighbors, which are at the apexes of an icosahedron. The icosahedron is one of the Platonic solids in which all the faces are perfect planar shapes; in the case of the icosahedron, the faces are 20 equilateral triangles. The icosahedron has 12 apexes arranged in five-fold symmetric rings,[2] as shown in Fig. 1.6. In fact, it turns out that the icosahedral arrangement is optimal for close packing of a small number of atoms, but it is not possible to fill

[2] An n-fold symmetry means that rotation by $2\pi/n$ around an axis leaves the structure invariant.

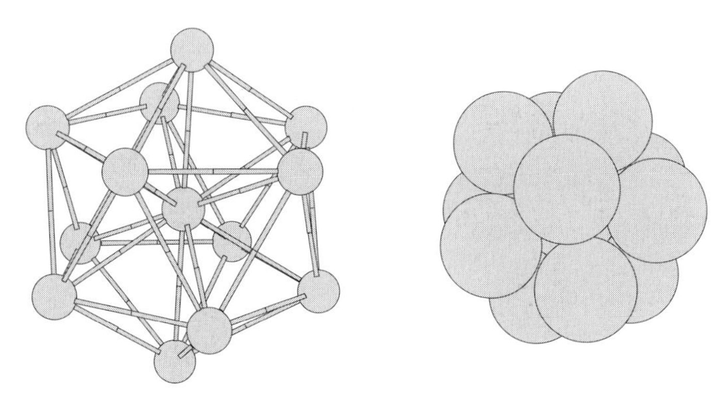

Figure 1.6. **Left:** one atom and its 12 neighbors in the icosahedral structure; the size of the spheres representing atoms is chosen so as to make the neighbors and their distances apparent. **Right:** a rendition of the icosahedron that illustrates its close-packing nature; this structure cannot be extended to form a periodic solid in three-dimensional space.

three-dimensional space in a periodic fashion with icosahedral symmetry. This fact is a simple geometrical consequence (see also chapter 3 on crystal symmetries). Based on this observation it was thought that crystals with perfect five-fold (or ten-fold) symmetry could not exist, unless defects were introduced to allow for deviations from the perfect symmetry [2–4]. The discovery of solids that exhibited five-fold or ten-fold symmetry in their diffraction patterns, in the mid 1980s [5], caused quite a sensation. These solids were named "quasicrystals", and their study created a new exciting subfield in condensed matter physics. They are discussed in more detail in chapter 12.

1.1.2 Atoms with s valence electrons

The second category consists of atoms that have only *s* valence electrons. These are Li, Na, K, Rb and Cs (the alkalis) with one valence electron, and Be, Mg, Ca, Sr and Ba with two valence electrons. The wavefunctions of valence electrons of all these elements extend far from the nucleus. In solids, the valence electron wavefunctions at one site have significant overlap with those at the nearest neighbor sites. Since the *s* states are spherically symmetric, the wavefunctions of valence electrons do not exhibit any particular preference for orientation of the nearest neighbors in space. For the atoms with one and two *s* valence electrons a simplified picture consists of all the valence electrons overlapping strongly, and thus being shared by all the atoms in the solid forming a "sea" of negative charge. The nuclei with their core electrons form ions, which are immersed in this sea of valence electrons. The ions have charge +1 for the alkalis and +2 for the atoms with two *s* valence electrons. The resulting crystal structure is the one which optimizes the electrostatic repulsion

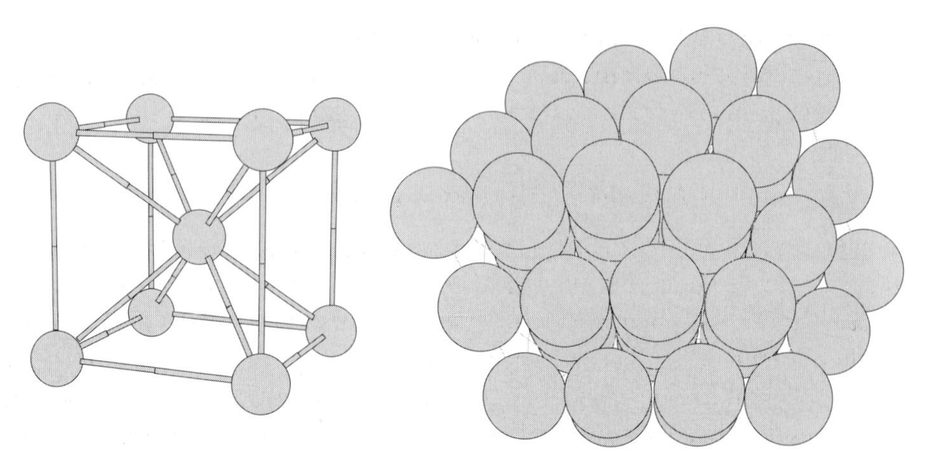

Figure 1.7. **Left:** one atom and its eight neighbors in the body-centered cubic (BCC) lattice; the size of the spheres representing atoms is chosen so as to make the neighbors and their distances apparent. **Right:** a portion of the three-dimensional BCC lattice; the size of the spheres is chosen so as to indicate the almost close-packing nature of this lattice.

of the positively charged ions with their attraction by the sea of electrons. The actual structures are body-centered cubic (BCC) for all the alkalis, and FCC or HCP for the two-s-valence-electron atoms, except Ba, which prefers the BCC structure. In the BCC structure each atom has eight equidistant nearest neighbors as shown in Fig. 1.7, which is the second highest number of nearest neighbors in a simple crystalline structure, after FCC and HCP.

One point deserves further clarification: we mentioned that the valence electrons have significant overlap with the electrons in neighboring atoms, and thus they are shared by all atoms in the solid, forming a sea of electrons. It may seem somewhat puzzling that we can jump from one statement – the overlap of electron orbitals in nearby atoms – to the other – the sharing of valence electrons by all atoms in the solid. The physical symmetry which allows us to make this jump is the periodicity of the crystalline lattice. This symmetry is the main feature of the external potential that the valence electrons feel in the bulk of a crystal: they are subjected to a periodic potential in space, in all three dimensions, which for all practical purposes extends to infinity – an idealized situation we discussed earlier. Just like in any quantum mechanical system, the electronic wavefunctions must obey the symmetry of the external potential, which means that the wavefunctions themselves must be periodic up to a phase. The mathematical formulation of this statement is called Bloch's theorem and will be considered in detail later. A periodic wavefunction implies that if two atoms in the crystal share an electronic state due to overlap between atomic orbitals, then all equivalent atoms of the crystal share the same state equally, that is, the electronic state is delocalized over the entire solid. This behavior is central to the physics of solids, and represents a feature that is qualitatively different from what happens in atoms and molecules, where electronic states are localized (except in certain large molecules that possess symmetries akin to lattice periodicity).

1.1.3 Atoms with s and p valence electrons

The next level of complexity in crystal structure arises from atoms that have both s and p valence electrons. The individual p states are not spherically symmetric so they can form linear combinations with the s states that have directional character: a single p state has two lobes of opposite sign pointing in diametrically opposite directions. The s and p states, illustrated in Fig. 1.8, can then serve as the new basis for representing electron wavefunctions, and their overlap with neighboring wavefunctions of the same type can lead to interesting ways of arranging the atoms into a stable crystalline lattice (see Appendix B on the character of atomic orbitals).

In the following we will use the symbols $s(\mathbf{r})$, $p_l(\mathbf{r})$, $d_m(\mathbf{r})$, to denote atomic orbitals as they would exist in an isolated atom, which are functions of $\mathbf{r}$. When they are related to an atom A at position $\mathbf{R}_A$, these become functions of $\mathbf{r} - \mathbf{R}_A$ and are denoted by $s^A(\mathbf{r})$, $p_l^A(\mathbf{r})$, $d_m^A(\mathbf{r})$. We use $\phi_i^A(\mathbf{r})(i = 1, 2, \ldots)$ to denote linear combinations of the atomic orbitals at site A, and $\psi^n(\mathbf{r})(n = a, b)$ for combinations of $\phi_i^X(\mathbf{r})$'s $(X = A, B, \ldots; i = 1, 2, \ldots)$ which are appropriate for the description of electronic states in the crystal.

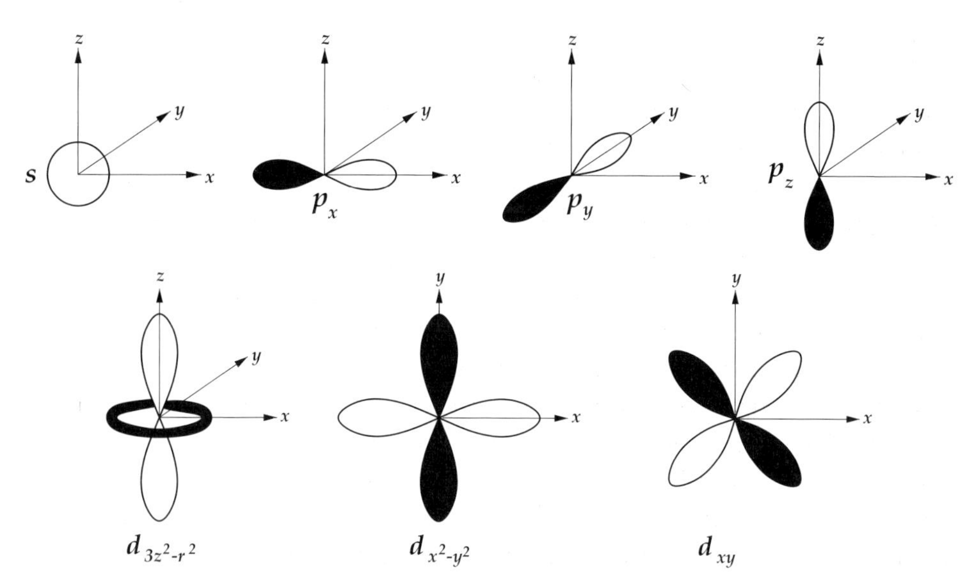

Figure 1.8. Representation of the character of s, p, d atomic orbitals. The lobes of opposite sign in the p_x, p_y, p_z and $d_{x^2-y^2}$, d_{xy} orbitals are shown shaded black and white. The d_{yz}, d_{zx} orbitals are similar to the d_{xy} orbital, but lie on the yz and zx planes.

The possibility of combining these atomic orbitals to form covalent bonds in a crystal is illustrated by the following two-dimensional example. For an atom, labeled A, with states s^A, p_x^A, p_y^A, p_z^A which are orthonormal, we consider first the linear combinations which constitute a new orthonormal basis of atomic orbitals:

$$\begin{aligned}
\phi_1^A &= \frac{1}{\sqrt{3}}s^A + \frac{\sqrt{2}}{\sqrt{3}}p_x^A \\
\phi_2^A &= \frac{1}{\sqrt{3}}s^A - \frac{1}{\sqrt{6}}p_x^A + \frac{1}{\sqrt{2}}p_y^A \\
\phi_3^A &= \frac{1}{\sqrt{3}}s^A - \frac{1}{\sqrt{6}}p_x^A - \frac{1}{\sqrt{2}}p_y^A \\
\phi_4^A &= p_z^A
\end{aligned} \tag{1.1}$$

The first three orbitals, $\phi_1^A, \phi_2^A, \phi_3^A$ point along three directions on the xy plane separated by 120°, while the last one, ϕ_4^A, points in a direction perpendicular to the xy plane, as shown in Fig. 1.9. It is easy to show that, if the atomic orbitals are orthonormal, and the states s^A, $p_i^A(i = x, y, z)$ have energies ϵ_s and ϵ_p, then the states $\phi_k(k = 1, 2, 3)$ have energy $(\epsilon_s + 2\epsilon_p)/3$; these states, since they are composed of one s and two p atomic orbitals, are called sp^2 orbitals. Imagine now a second identical atom, which we label B, with the following linear combinations:

$$\begin{aligned}
\phi_1^B &= \frac{1}{\sqrt{3}}s^B - \frac{\sqrt{2}}{\sqrt{3}}p_x^B \\
\phi_2^B &= \frac{1}{\sqrt{3}}s^B + \frac{1}{\sqrt{6}}p_x^B - \frac{1}{\sqrt{2}}p_y^B \\
\phi_3^B &= \frac{1}{\sqrt{3}}s^B + \frac{1}{\sqrt{6}}p_x^B + \frac{1}{\sqrt{2}}p_y^B \\
\phi_4^B &= p_z^B
\end{aligned} \tag{1.2}$$

The orbitals $\phi_1^B, \phi_2^B, \phi_3^B$ also point along three directions on the xy plane separated by 120°, but in the opposite sense (rotated by 180°) from those of atom A. For example, ϕ_1^A points along the $+\hat{\mathbf{x}}$ direction, while ϕ_1^B points along the $-\hat{\mathbf{x}}$ direction. Now imagine that we place atoms A and B next to each other along the x axis, first atom A and to its right atom B, at a distance a. We arrange the distance so that there is significant overlap between orbitals ϕ_1^A and ϕ_1^B, which are pointing toward each other, thereby maximizing the interaction between these two orbitals. Let us assume that in the neutral isolated state of the atom we can occupy each of these orbitals by one electron; note that this is *not* the ground state of the atom. We can form two linear combinations, $\psi_1^b = \frac{1}{2}(\phi_1^A + \phi_1^B)$ and $\psi_1^a = \frac{1}{2}(\phi_1^A - \phi_1^B)$ of which the first

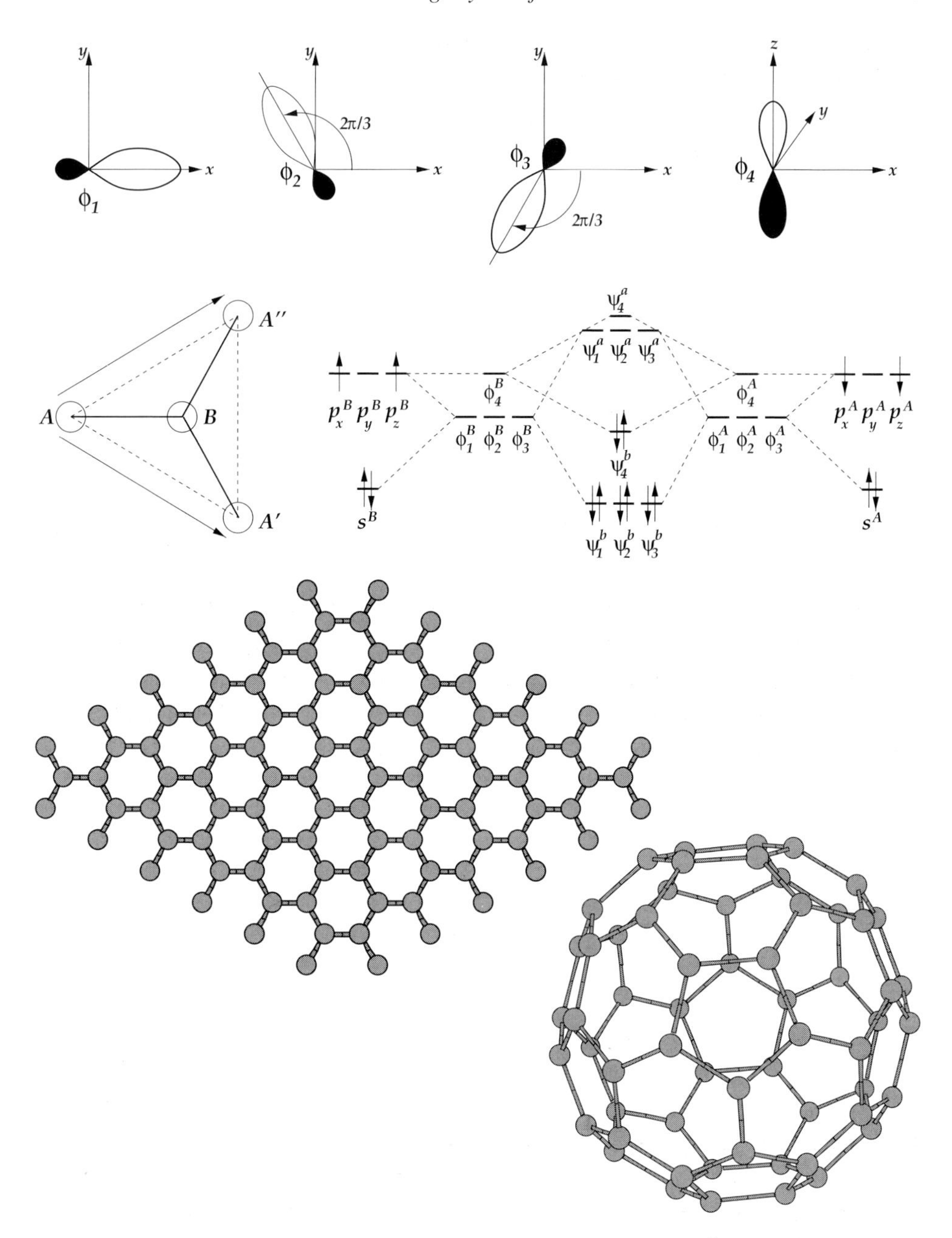

Figure 1.9. Illustration of covalent bonding in graphite. **Top:** the sp^2 linear combinations of s and p atomic orbitals (defined in Eq. (1.1)). **Middle:** the arrangement of atoms on a plane with B at the center of an equilateral triangle formed by A, A', A'' (the arrows connect equivalent atoms); the energy level diagram for the s, p atomic states, their sp^2 linear combinations (ϕ_i^A and ϕ_i^B) and the bonding (ψ_i^b) and antibonding (ψ_i^a) states (up–down arrows indicate electrons spins). **Bottom:** the graphitic plane (honeycomb lattice) and the C_{60} molecule.

maximizes the overlap and the second has a node at the midpoint between atoms A and B. As usual, we expect the symmetric linear combination of single-particle orbitals (called the bonding state) to have lower energy than the antisymmetric one (called the antibonding state) in the system of the two atoms; this is a general feature of how combinations of single-particle orbitals behave (see Problem 2). The exact energy of the bonding and antibonding states will depend on the overlap of the orbitals ϕ_1^A, ϕ_1^B. We can place two electrons, one from each atomic orbital, in the symmetric linear combination because of their spin degree of freedom; this is based on the assumption that the spin wavefunction of the two electrons is antisymmetric (a spin singlet), so that the total wavefunction, the product of the spatial and spin parts, is antisymmetric upon exchange of their coordinates, as it should be due to their fermionic nature. Through this exercise we have managed to lower the energy of the system, since the energy of ψ^b is lower than the energy of ϕ_1^A or ϕ_1^B. This is the essence of the chemical bond between two atoms, which in this case is called a covalent σ bond.

Imagine next that we repeat this exercise: we take another atom with the same linear combinations of orbitals as A, which we will call A', and place it in the direction of the vector $\frac{1}{2}\hat{\mathbf{x}} - \frac{\sqrt{3}}{2}\hat{\mathbf{y}}$ relative to the position of atom B, and at the same distance a as atom A from B. Due to our choice of orbitals, ϕ_2^B and $\phi_2^{A'}$ will be pointing toward each other. We can form symmetric and antisymmetric combinations from them, occupy the symmetric (lower energy) one with two electrons as before and create a second σ bond between atoms B and A'. Finally we repeat this procedure with a third atom A'' placed along the direction of the vector $\frac{1}{2}\hat{\mathbf{x}} + \frac{\sqrt{3}}{2}\hat{\mathbf{y}}$ relative to the position of atom B, and at the same distance a as the previous two neighbors. Through the same procedure we can form a third σ bond between atoms B and A'', by forming the symmetric and antisymmetric linear combinations of the orbitals ϕ_3^B and $\phi_3^{A''}$. Now, as far as atom B is concerned, its three neighbors are exactly equivalent, so we consider the vectors that connect them as the repeat vectors at which equivalent atoms in the crystal should exist. If we place atoms of type A at all the possible integer multiples of these vectors, we form a lattice. To complete the lattice we have to place atoms of type B also at all the possible integer multiples of the same vectors, relative to the position of the original atom B. The resulting lattice is called the honeycomb lattice. Each atom of type A is surrounded by three atoms of type B and vice versa, as illustrated in Fig. 1.9. Though this example may seem oversimplified, it actually corresponds to the structure of graphite, one of the most stable crystalline solids. In graphite, planes of C atoms in the honeycomb lattice are placed on top of each other to form a three-dimensional solid, but the interaction between planes is rather weak (similar to the van der Waals interaction). An indication of this weak bonding between planes compared to the in-plane bonds is that the distance between nearest neighbor atoms on a plane is 1.42 Å, whereas the distance between successive planes is 3.35 Å, a factor of 2.36 larger.

What about the orbitals p_z (or ϕ_4), which so far have not been used? If each atom had only three valence electrons, then these orbitals would be left empty since they have higher energy than the orbitals ϕ_1, ϕ_2, ϕ_3, which are linear combinations of s and p orbitals (the original s atomic orbitals have lower energy than p). In the case of C, each atom has four valence electrons so there is one electron left per atom when all the σ bonds have been formed. These electrons remain in the p_z orbitals, which are perpendicular to the xy plane and thus parallel to each other. Symmetric and antisymmetric combinations of neighboring p_z^A and p_z^B orbitals can also be formed (the states ψ_4^b, ψ_4^a, respectively), and the energy can be lowered by occupying the symmetric combination. In this case the overlap between neighboring p_z orbitals is significantly smaller and the corresponding gain in energy significantly less than in σ bonds. This is referred to as a π bond, which is generally weaker than a σ bond. Carbon is a special case, in which the π bonds are almost as strong as the σ bonds.

An intriguing variation of this theme is a structure that contains pentagonal rings as well as the regular hexagons of the honeycomb lattice, while maintaining the three-fold coordination and bonding of the graphitic plane. The presence of pentagons introduces curvature in the structure, and the right combination of pentagonal and hexagonal rings produces the almost perfect sphere, shown in Fig. 1.9. This structure actually exists in nature! It was discovered in 1985 and it has revolutionized carbon chemistry and physics – its discoverers, R. F. Curl, H. W. Kroto and R. E. Smalley, received the 1996 Nobel prize for Chemistry. Many more interesting variations of this structure have also been produced, including "onions" – spheres within spheres – and "tubes" – cylindrical arrangements of three-fold coordinated carbon atoms. The tubes in particular seem promising for applications in technologically and biologically relevant systems. These structures have been nicknamed after Buckminster Fuller, an American scientist and practical inventor of the early 20th century, who designed architectural domes based on pentagons and hexagons; the nicknames are buckminsterfullerene or bucky-ball for C_{60}, bucky-onions, and bucky-tubes. The physics of these structures will be discussed in detail in chapter 13.

There is a different way of forming bonds between C atoms: consider the following linear combinations of the s and p atomic orbitals for atom A:

$$\begin{aligned}
\phi_1^A &= \frac{1}{2}[s^A - p_x^A - p_y^A - p_z^A] \\
\phi_2^A &= \frac{1}{2}[s^A + p_x^A - p_y^A + p_z^A] \\
\phi_3^A &= \frac{1}{2}[s^A + p_x^A + p_y^A - p_z^A] \\
\phi_4^A &= \frac{1}{2}[s^A - p_x^A + p_y^A + p_z^A]
\end{aligned} \tag{1.3}$$

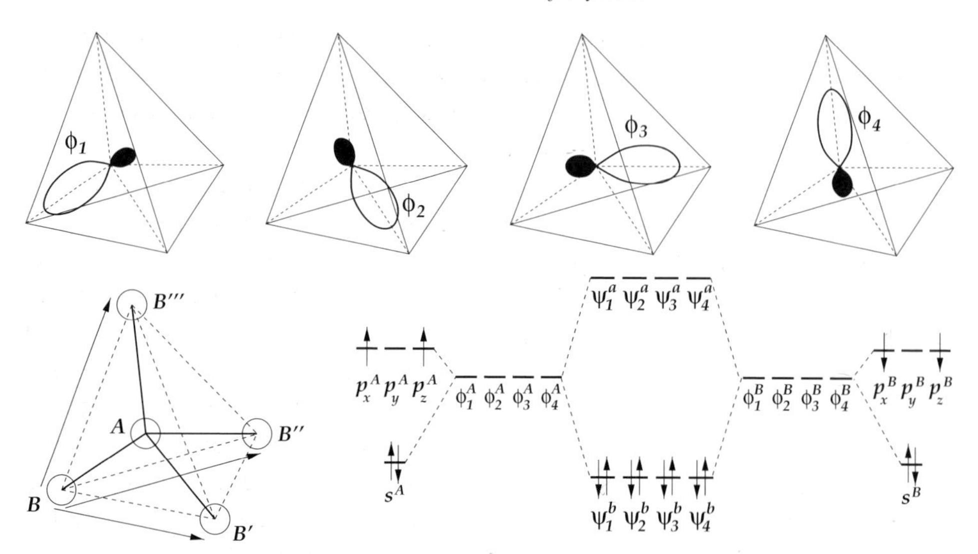

Figure 1.10. Illustration of covalent bonding in diamond. **Top panel:** representation of the sp^3 linear combinations of s and p atomic orbitals appropriate for the diamond structure, as defined in Eq. (1.3), using the same convention as in Fig. 1.8. **Bottom panel:** on the left side, the arrangement of atoms in the three-dimensional diamond lattice; an atom A is at the center of a regular tetrahedron (dashed lines) formed by equivalent B, B', B'', B''' atoms; the three arrows are the vectors that connect equivalent atoms. On the right side, the energy level diagram for the s, p atomic states, their sp^3 linear combinations (ϕ_i^A and ϕ_i^B) and the bonding (ψ_i^b) and antibonding (ψ_i^a) states. The up–down arrows indicate occupation by electrons in the two possible spin states. For a perspective view of the diamond lattice, see Fig. 1.5.

It is easy to show that the energy of these states, which are degenerate, is equal to $(\epsilon_s + 3\epsilon_p)/4$, where ϵ_s, ϵ_p are the energies of the original s and p atomic orbitals; the new states, which are composed of one s and three p orbitals, are called sp^3 orbitals. These orbitals point along the directions from the center to the corners of a regular tetrahedron, as illustrated in Fig. 1.10. We can now imagine placing atoms B, B', B'', B''' at the corners of the tetrahedron, with which we associate linear combinations of s and p orbitals just like those for atom A, but having all the signs of the p orbitals reversed:

$$\begin{aligned}
\phi_1^B &= \frac{1}{2}[s^B + p_x^B + p_y^B + p_z^B] \\
\phi_2^B &= \frac{1}{2}[s^B - p_x^B + p_y^B - p_z^B] \\
\phi_3^B &= \frac{1}{2}[s^B - p_x^B - p_y^B + p_z^B] \\
\phi_4^B &= \frac{1}{2}[s^B + p_x^B - p_y^B - p_z^B]
\end{aligned} \tag{1.4}$$

Then we will have a situation where the ϕ orbitals on neighboring A and B atoms will be pointing toward each other, and we can form symmetric and antisymmetric combinations of those, ψ^b, ψ^a, respectively, to create four σ bonds around atom A. The exact energy of the ψ orbitals will depend on the overlap between the ϕ^A and ϕ^B orbitals; for sufficiently strong overlap, we can expect the energy of the ψ^b states to be lower than the original s atomic orbitals and those of the ψ^a states to be higher than the original p atomic orbitals, as shown schematically in Fig. 1.10. The vectors connecting the equivalent B, B', B'', B''' atoms define the repeat vectors at which atoms must be placed to form an infinite crystal. By placing both A-type and B-type atoms at all the possible integer multiples of these vectors we create the diamond lattice, shown in Fig. 1.10. This is the other stable form of bulk C. Since C has four valence electrons and each atom at the center of a tetrahedron forms four σ bonds with its neighbors, all electrons are taken up by the bonding states. This results in a very stable and strong three-dimensional crystal. Surprisingly, graphite has a somewhat lower internal energy than diamond, that is, the thermodynamically stable solid form of carbon is the soft, black, cheap graphite rather than the very strong, brilliant and very expensive diamond crystal!

The diamond lattice, with four neighbors per atom, is relatively open compared to the close-packed lattices. Its stability comes from the very strong covalent bonds formed between the atoms. Two other elements with four valence s and p electrons, namely Si and Ge, also crystallize in the diamond, but not the graphite, lattice. There are two more elements with four valence s and p electrons in the Periodic Table, Sn and Pb. Sn forms crystal structures that are distorted variants of the diamond lattice, since its σ bonds are not as strong as those of the other group-IV-A elements, and it can gain some energy by increasing the number of neighbors (from four to six) at the expense of perfect tetrahedral σ bonds. Pb, on the other hand, behaves more like a metal, preferring to optimize the number of neighbors, and forms the FCC crystal (see also below). Interestingly, elements with only three valence s and p electrons, like B, Al, Ga, In and Tl, do not form the graphite structure, as alluded above. They instead form more complex structures in which they try to optimize bonding given their relatively small number of valence electrons per atom. Some examples: the common structural unit for B is the icosahedron, shown in Fig. 1.6, and such units are close packed to form the solid; Al forms the FCC crystal and is the representative metal with s and p electrons and a close-packed structure; Ga forms quite complicated crystal structures with six or seven near neighbors (not all of them at the same distance); In forms a distorted version of the cubic close packing in which the 12 neighbors are split into a group of four and another group of eight equidistant atoms. None of these structures can be easily described in terms of the notions introduced above to handle s and p valence electrons, demonstrating the limitations of this simple approach.

Of the other elements in the Periodic Table with s and p valence electrons, those with five electrons, N, P, As, Sb and Bi, tend to form complex structures where atoms have three σ bonds to their neighbors but not in a planar configuration. A characteristic structure is one in which the three p valence electrons participate in covalent bonding while the two s electrons form a filled state which does not contribute much to the cohesion of the solid; this filled state is called the "lone pair" state. If the covalent bonds were composed of purely p orbitals the bond angles between nearest neighbors would be 90°; instead, the covalent bonds in these structures are a combination of s and p orbitals with predominant p character, and the bond angles are somewhere between 120° (sp^2 bonding) and 90° (pure p bonding), as illustrated in Fig. 1.11. The structure of solid P is represented by this kind of atomic arrangement. In this structure, the covalent bonds are arranged in puckered hexagons which form planes, and the planes are stacked on top of each other to form the solid. The interaction between planes is much weaker than that between atoms on a single plane: an indication of this difference in bonding is the fact that the distance between nearest neighbors in a plane is 2.17 Å while the closest distance between atoms on successive planes is 3.87 Å, almost a factor of 2 larger. The structures of As, Sb and Bi follow the same general pattern with three-fold bonded atoms, but in those solids there exist additional covalent bonds between the planes of puckered atoms so that the structure is not clearly planar as is the case for P. An exception to this general tendency is nitrogen, the lightest element with five valence electrons which forms a crystal composed of nitrogen molecules; the N_2 unit is particularly stable.

The elements with six s and p valence electrons, O, S, Se, Te and Po, tend to form molecular-like ring or chain structures with two nearest neighbors per atom, which are then packed to form three-dimensional crystals. These rings or chains are puckered and form bonds at angles that try to satisfy bonding requirements analogous to those described for the solids with four s and p valence electrons. Examples of such units are shown in Fig. 1.12. Since these elements have a valence of 6, they tend to keep four of their electrons in one filled s and one filled p orbital and form covalent bonds to two neighbors with their other two p orbitals. This picture is somewhat oversimplified, since significant hybridization takes place between s and p orbitals that participate in bonding, so that the preferred angle between the bonding orbitals is not 90°, as pure p bonding would imply, but ranges between 102° and 108°. Typical distances between nearest neighbor atoms in the rings or the chains are 2.06 Å for S, 2.32 Å for Se and 2.86 Å for Te, while typical distances between atoms in successive units are 3.50 Å for S, 3.46 Å for Se and 3.74 Å for Te; that is, the ratio of distances between atoms within a bonding unit and across bonding units is 1.7 for S, 1.5 for Se and 1.3 for Te. An exception

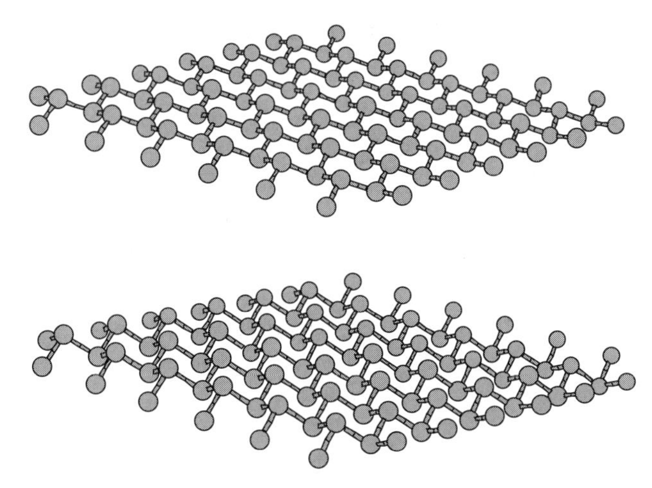

Figure 1.11. The layers of buckled atoms that correspond to the structure of group-V elements: all atoms are three-fold coordinated as in a graphitic plane, but the bond angles between nearest neighbors are not 120° and hence the atoms do not lie on the plane. For illustration two levels of buckling are shown: in the first structure the bond angles are 108°, in the second 95°. The planes are stacked on top of each other as in graphite to form the 3D solids.

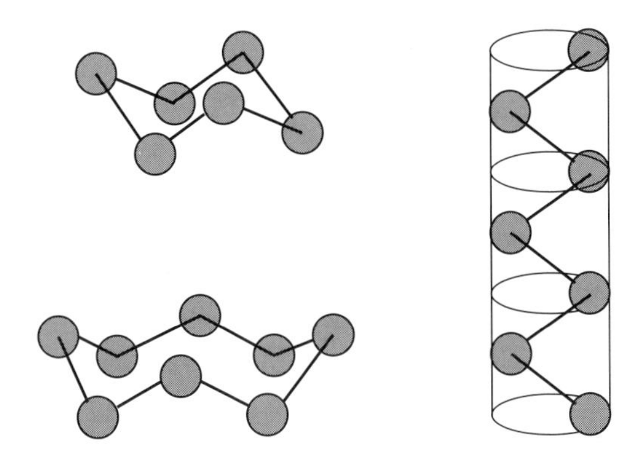

Figure 1.12. Characteristic units that appear in the solid forms of S, Se and Te: six-fold rings (S), eight-fold rings (Se) and one-dimensional chains (Se and Te). The solids are formed by close packing of these units.

to this general tendency is oxygen, the lightest element with six valence electrons which forms a crystal composed of oxygen molecules; the O_2 unit is particularly stable. The theme of diatomic molecules as the basic unit of the crystal, already mentioned for nitrogen and oxygen, is common in elements with seven s and p valence electrons also: chlorine, bromine and iodine form solids by close packing of diatomic molecules.

1.1.4 Atoms with *s* and *d* valence electrons

This category includes all the atoms in the middle columns of the Periodic Table, that is, columns I-B–VII-B and VIII. The *d* orbitals in atoms have directional nature like the *p* orbitals. However, since there are five *d* orbitals it is difficult to construct linear combinations with *s* orbitals that would neatly point toward nearest neighbors in three-dimensional space and produce a crystal with simple σ bonds. Moreover, the *d* valence orbitals typically lie lower in energy than the *s* valence orbitals and therefore do not participate as much in bonding (see for example the discussion about Ag, in chapter 4). Note that *d* orbitals *can* form strong covalent bonds by combining with *p* orbitals of other elements, as we discuss in a subsequent section. Thus, elements with *s* and *d* valence electrons tend to form solids where the *s* electrons are shared among all atoms in the lattice, just like elements with one or two *s* valence electrons. These elements form space-filling close-packed crystals, of the FCC, HCP and BCC type. There are very few exceptions to this general tendency, namely Mn, which forms a very complex structure with a cubic lattice and a very large number of atoms (58) in the unit cell, and Hg, which forms a low-symmetry rhombohedral structure. Even those structures, however, are slight variations of the basic close-packing structures already mentioned. For instance, the Mn structure, in which atoms have from 12 to 16 neighbors, is a slight variation of the BCC structure. The crystals formed by most of these elements typically have metallic character.

1.1.5 Atoms with *s*, *d* and *f* valence electrons

The same general trends are found in the rare earth elements, which are grouped in the lanthanides (atomic numbers 58–71) and the actinides (atomic numbers 90 and beyond) . Of those we discuss briefly the lanthanides as the more common of the rare earths that are found in solids. In these elements the *f* shell is gradually filled as the atomic number increases, starting with an occupation of two electrons in Ce and completing the shell with 14 electrons in Lu. The *f* orbitals have directional character which is even more complicated than that of *p* or *d* orbitals. The solids formed by these elements are typically close-packed structures such as FCC and HCP, with a couple of exceptions (Sm which has rhombohedral structure and Eu which has BCC structure). They are metallic solids with high atomic densities. However, more interesting are structures formed by these elements and other elements of the Periodic Table, in which the complex character of the *f* orbitals can be exploited in combination with orbitals from neighboring atoms to form strong bonds. Alternatively, these elements are used as dopants in complicated crystals, where they donate some of their electrons to states formed by other atoms. One such example is discussed in the following sections.

1.1.6 Solids with two types of atoms

Some of the most interesting and useful solids involve two types of atoms, which in some ways are complementary. One example that comes immediately to mind are solids composed of atoms in the first (group I-A) and seventh (group VII-A) columns of the Periodic Table, which have one and seven valence electrons, respectively. Solids composed of such elements are referred to as "alkali halides". It is natural to expect that the atom with one valence electron will lose it to the more electronegative atom with the seven valence electrons, which then acquires a closed electronic shell, completing the s and p levels. This of course leads to one positively and one negatively charged ion, which are repeated periodically in space to form a lattice. The easiest way to arrange such atoms is at the corners of a cube, with alternating corners occupied by atoms of the opposite type. This arrangement results in the sodium chloride (NaCl) or rock-salt structure, one of the most common crystals. Many combinations of group I-A and group VII-A atoms form this kind of crystal. In this case each ion has six nearest neighbors of the opposite type. A different way to arrange the ions is to have one ion at the center of a cube formed by ions of the opposite type. This arrangement forms two interpenetrating cubic lattices and is known as the cesium chloride (CsCl) structure. In this case each ion has eight nearest neighbors of the opposite type. Several combinations of group I-A and group VII-A atoms crystallize in this structure. Since in all these structures the group I-A atoms lose their s valence electron to the group VII-A atoms, this type of crystal is representative of ionic bonding. Both of these ionic structures are shown in Fig. 1.13.

Another way of achieving a stable lattice composed of two kinds of ions with opposite sign, is to place them in the two interpenetrating FCC sublattices of the diamond lattice, described earlier. In this case each ion has four nearest neighbors of the opposite type, as shown in Fig. 1.14. Many combinations of atoms in the I-B column of the Periodic Table and group VII-B atoms crystallize in this structure, which is called the zincblende structure from the German term for ZnS, the representative solid with this lattice.

The elements in the I-B column have a filled d-shell (ten electrons) and one extra s valence electron, so it is natural to expect them to behave in some ways similar to the alkali metals. However, the small number of neighbors in this structure, as opposed to the rock-salt and cesium chloride structures, suggest that the cohesion of these solids cannot be attributed to simple ionic bonding alone. In fact, this becomes more pronounced when atoms from the second (group II-B) and sixth (group VI-A)

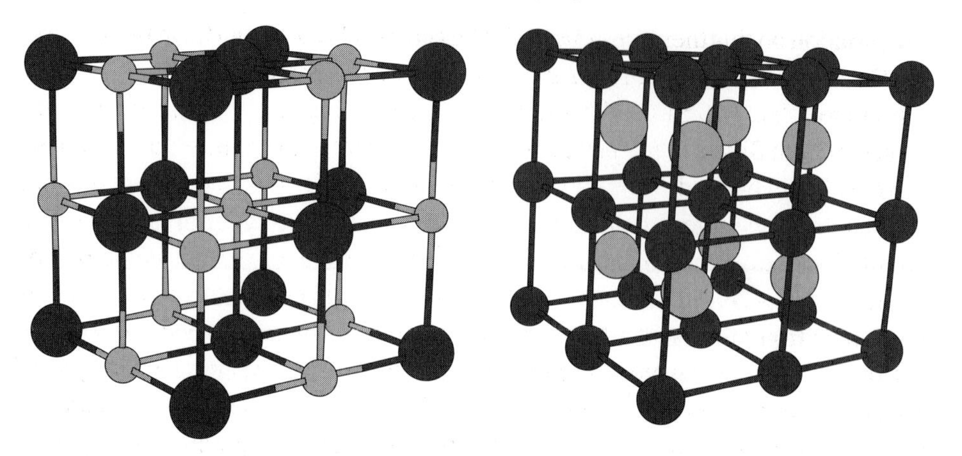

Figure 1.13. **Left:** the rock-salt, NaCl, structure, in which the ions form a simple cubic lattice with each ion surrounded by six neighbors of the opposite type. **Right:** the CsCl structure, in which the ions form a body-centered cubic lattice with each ion surrounded by eight neighbors of the opposite type.

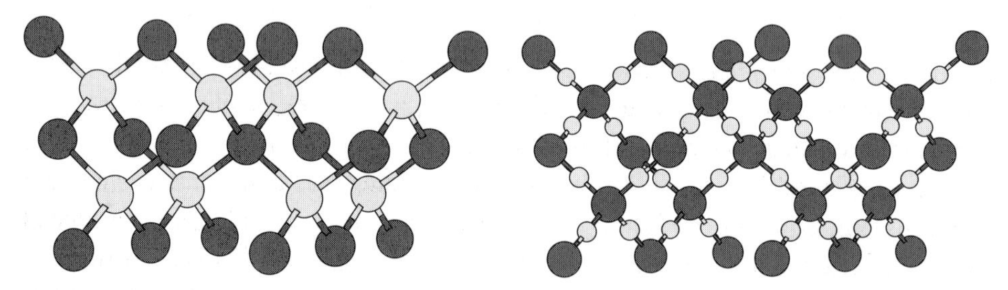

Figure 1.14. **Left:** the zincblende lattice in which every atom is surrounded by four neighbors of the opposite type, in mixed ionic and covalent bonding; several III-V, II-VI and IV-IV solids exist in this lattice. **Right:** a representative SiO_2 structure, in which each Si atom has four O neighbors, and each O atom has two Si neighbors.

columns of the Periodic Table form the zincblende structure (ZnS itself is the prime example). In this case we would have to assume that the group II atoms lose their two electrons to the group VI atoms, but since the electronegativity difference is not as great between these two types of elements as between group I-A and group VII-A elements, something more than ionic bonding must be involved. Indeed the crystals of group II and group VI atoms in the zincblende structure are good examples of mixed ionic and covalent bonding. This trend extends to one more class of solids: group III-A and group V-A atoms also form zincblende crystals, for example AlP, GaAs, InSb, etc. In this case, the bonding is even more tilted toward covalent character, similar to the case of group IV atoms which form the diamond lattice. Finally, there are combinations of two group IV atoms that form the zincblende structure; some interesting examples are SiC and GeSi alloys.

A variation on this theme is a class of solids composed of Si and O. In these solids, each Si atom has four O neighbors and is situated at the center of a tetrahedron, while each O atom has two Si neighbors, as illustrated in Fig. 1.14. In this manner the valence of both Si and O are perfectly satisfied, so that the structure can be thought of as covalently bonded. Due to the large electronegativity of O, the covalent bonds are polarized to a large extent, so that the two types of atoms can be considered as partially ionized. This results again in a mixture of covalent and ionic bonding. The tetrahedra of Si–O atoms are very stable units and the relative positions of atoms in a tetrahedron are essentially fixed. The position of these tetrahedra relative to each other, however, can be changed with little cost in energy, because this type of structural distortion involves only a slight bending of bond angles, without changing bond lengths. This freedom in relative tetrahedron orientation produces a very wide variety of solids based on this structural unit, including amorphous structures, such as common glass, and structures with many open spaces in them, such as the zeolites.

1.1.7 Hydrogen: a special one-s-valence-electron atom

So far we have left H out of the discussion. This is because H is a special case: it has no core electrons. Its interaction with the other elements, as well as between H atoms, is unusual, because when H tries to share its one valence s electron with other atoms, what is left is a bare proton rather than a nucleus shielded partially by core electrons. The proton is an ion much smaller in size than the other ions produced by stripping the valence electrons from atoms: its size is 10^{-15} m, five orders of magnitude smaller than typical ions, which have a size of order 1 Å. It also has the smallest mass, which gives it a special character: in all other cases (except for He) we can consider the ions as classical particles, due to their large mass, while in the case of hydrogen, its light mass implies a large zero-point motion which makes it necessary to take into account the quantum nature of the proton's motion. Yet another difference between hydrogen and all other elements is the fact that its s valence electron is very strongly bound to the nucleus: the ionization energy is 13.6 eV, whereas typical ionization energies of valence electrons in other elements are in the range 1–2 eV. Due to its special character, H forms a special type of bond called "hydrogen bond". This is encountered in many structures composed of molecules that contain H atoms, such as organic molecules and water.

The solid in which hydrogen bonding plays the most crucial role is ice. Ice forms many complex phases [6]; in its ordinary phase called Ih, the H_2O molecules are placed so that the O atoms occupy the sites of a wurtzite lattice (see Fig. 1.5), while the H atoms are along lines that join O atoms [7]. There are two H atoms attached to each O atom by short covalent bonds (of length 1.00 Å), while the distance between O atoms is 2.75 Å. There is one H atom along each line joining two O atoms. The

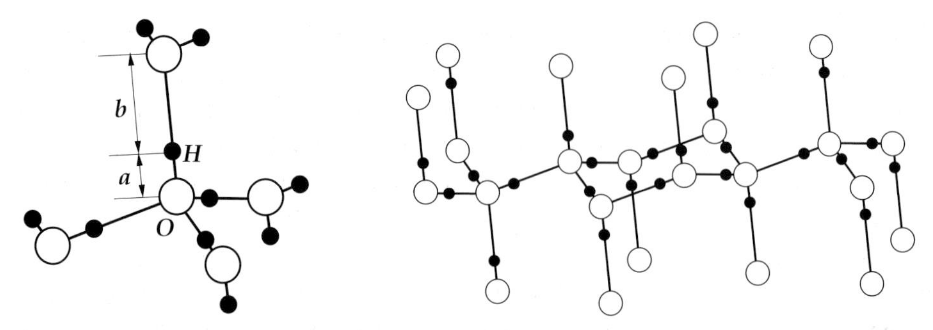

Figure 1.15. **Left:** illustration of hydrogen bonding between water molecules in ice: the O atom is at the center of a tetrahedron fromed by other O atoms, and the H atoms are along the directions joining the center to the corners of the tetrahedron. The O–H covalent bond distance is $a = 1.00$ Å, while the H–O hydrogen bond distance is $b = 1.75$ Å. The relative position of atoms is not given to scale, in order to make it easier to visualize which H atoms are attached by covalent bonds to the O atoms. **Right:** illustration of the structure of Ih ice: the O atoms sit at the sites of a wurtzite lattice (compare with Fig. 1.5) and the H atoms are along the lines joining O atoms; there is one H atom along each such line, and two H atoms are bonded by short covalent bonds to each O atom.

bond between a H atom and an O atom to which it is *not* covalently bonded is called a hydrogen bond, and, in this system, has length 1.75 Å; it is these hydrogen bonds that give stability to the crystal. This is illustrated in Fig. 1.15. The hydrogen bond is much weaker than the covalent bond between H and O in the water molecule: the energy of the hydrogen bond is 0.3 eV, while that of the covalent H–O bond is 4.8 eV. There are many ways of arranging the H atoms within these constraints for a fixed lattice of O atoms, giving rise to a large configurational entropy. Other forms of ice have different lattices, but this motif of local bonding is common.

Within the atomic orbital picture discussed earlier for solids with s and p electrons, we can construct a simple argument to rationalize hydrogen bonding in the case of ice. The O atom has six valence electrons in its s and p shells and therefore needs two more electrons to complete its electronic structure. The two H atoms that are attached to it to form the water molecule provide these two extra electrons, at the cost of an anisotropic bonding arrangement (a completed electronic shell should be isotropic, as in the case of Ne which has two more electrons than O). The cores of the H atoms (the protons), having lost their electrons to O, experience a Coulomb repulsion. The most favorable structure for the molecule which optimizes this repulsion would be to place the two H atoms in diametrically opposite positions relative to the O atom, but this would involve only one p orbital of the O atom to which both H atoms would bond. This is an unfavorable situation as far as formation of covalent bonds is concerned, because it is not possible to form two covalent bonds with only one p orbital and two electrons from the O atom. A compromise between the desire to form strong covalent bonds and the repulsion between the H cores is the formation of four sp^3 hybrids from the orbitals of the O atom, two of which form covalent bonds with the H atoms, while the other two are filled by two electrons each. This produces a tetrahedral structure with two lobes which have more positive charge (the two sp^3 orbitals to which the H atoms are bonded) than the other two lobes (the two sp^3 orbitals which are occupied by two electrons each). It is natural to expect that bringing similar molecular units together would produce some attraction between the lobes of opposite charge in neighboring units. This is precisely the arrangement of molecules in the structure of ice discussed above and shown in Fig. 1.15. This rationalization, however, is somewhat misleading as it suggests that the hydrogen bond, corresponding to the attraction between oppositely charged lobes of the H_2O tetrahedra, is essentially ionic. In fact, the hydrogen bond has significant covalent character as well: the two types of orbitals pointing toward each other form bonding (symmetric) and antibonding (antisymmetric) combinations leading to covalent bonds between them. This point of view was originally suggested by Pauling [8] and has remained controversial until recently, when sophisticated scattering experiments and quantum mechanical calculations provided convincing evidence in its support [9].

The solid phases of pure hydrogen are also unusual. At low pressure and temperature, H is expected to form a crystal composed of H_2 molecules in which every molecule behaves almost like an inert unit, with very weak interactions to the other molecules. At higher pressure, H is supposed to form an atomic solid when the molecules have approached each other enough so that their electronic distributions are forced to overlap strongly [10]. However, the conditions of pressure and temperature at which this transition occurs, and the structure of the ensuing atomic solid, are still a subject of active research [11–13]. The latest estimates are that it takes more than 3 Mbar of pressure to form the atomic H solid, which can only be reached under very special conditions in the laboratory, and which has been achieved only in the 1990s. There is considerable debate about what the crystal structure at this pressure should be, and although the BCC structure seems to be the most likely phase, by analogy to all other alkalis, this has not been unambiguously proven to date.

1.1.8 Solids with many types of atoms

If we allow several types of atoms to participate in the formation of a crystal, many more possibilities open up. There are indeed many solids with complex composition, but the types of bonding that occur in these situations are variants of

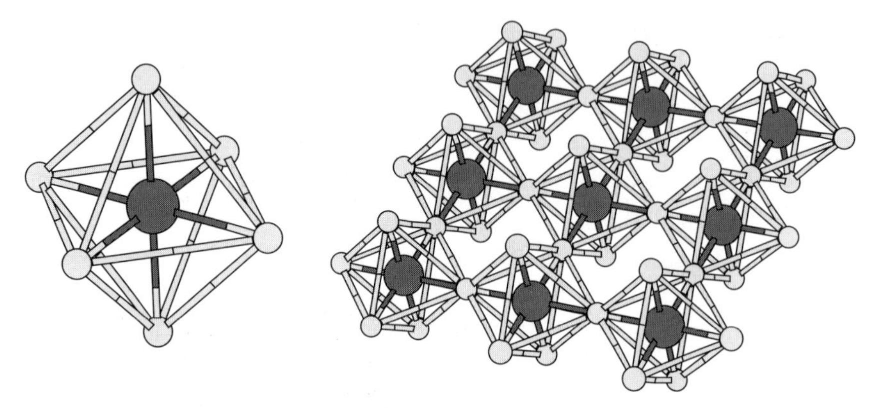

Figure 1.16. **Left:** a Cu atom surrounded by six O atoms, which form an octahedron; the Cu–O atoms are bonded by strong covalent bonds. **Right:** a set of corner-sharing O octahedra, forming a two-dimensional square lattice. The octahedra can also be joined at the remaining apexes to form a fully three-dimensional lattice. The empty spaces between the octahedra can accommodate atoms which are easily ionized, to produce a mixed covalent–ionic structure.

the types we have already discussed: metallic, covalent, ionic, van der Waals and hydrogen bonding. In many situations, several of these types of bonding are present simultaneously.

One interesting example of such complex structures is the class of ceramic materials in which high-temperature superconductivity (HTSC) was observed in the mid-1980s (this discovery, by J. G. Bednorz and K. A. Müller, was recongnized by the 1987 Nobel prize for Physics). In these materials strong covalent bonding between Cu and O forms one-dimensional or two-dimensional structures where the basic building block is oxygen octahedra; rare earth atoms are then placed at hollow positions of these backbond structures, and become partially ionized giving rise to mixed ionic and covalent bonding (see, for example, Fig. 1.16).

The motif of oxygen octahedra with a metal atom at the center to which the O atoms are covalently bonded, supplemented by atoms which are easily ionized, is also the basis for a class of structures called "perovskites". The chemical formula of perovskites is ABO_3, where A is the easily ionized element and B the element which is bonded to the oxygens. The basic unit is shown in Fig. 1.17: bonding in the xy plane is accomplished through the overlap between the p_x and p_y orbitals of the first (O_1) and second (O_2) oxygen atoms, respectively, and the $d_{x^2-y^2}$ orbital of B; bonding along the z axis is accomplished through the overlap between the p_z orbital of the third (O_3) oxygen atom and the $d_{3z^2-r^2}$ orbital of B (see Fig. 1.8 for the nature of these p and d orbitals). The A atoms provide the necessary number of electrons to satisfy all the covalent bonds. Thus, the overall bonding involves

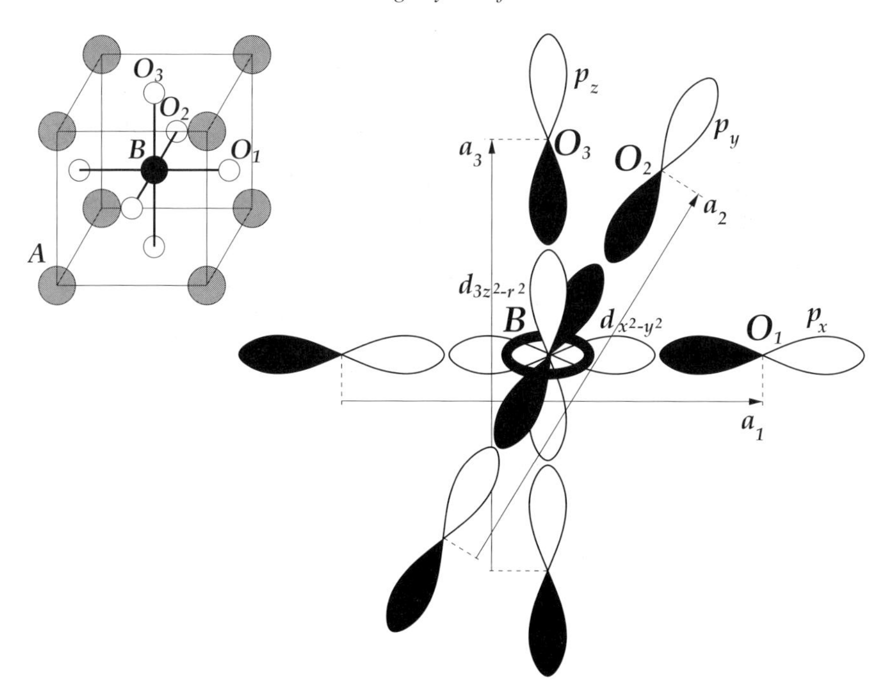

Figure 1.17. The basic structural unit of perovskites ABO_3 (upper left) and the atomic orbitals that contribute to covalent bonding. The three distinct oxygen atoms in the unit cell are labeled O_1, O_2, O_3 (shown as the small open circles in the structural unit); the remaining oxygen atoms are related to those by the repeat vectors of the crystal, indicated as $\mathbf{a}_1$, $\mathbf{a}_2$, $\mathbf{a}_3$. The six oxygen atoms form an octahedron at the center of which sits the B atom. The thin lines outline the cubic unit cell, while the thicker lines between the oxygen atoms and B represent the covalent bonds in the structural unit. The p_x, p_y, p_z orbitals of the three O atoms and the $d_{x^2-y^2}$, $d_{3z^2-r^2}$ orbitals of the B atoms that participate in the formation of covalent bonds in the octahedron are shown schematically.

both strong covalent character between B and O, as well as ionic character between the B–O units and the A atoms. The complexity of the structure gives rise to several interesting properties, such as ferroelectricity, that is, the ability of the solid to acquire and maintain an internal dipole moment. The dipole moment is associated with a displacement of the B atom away from the center of the octahedron, which breaks the symmetry of the cubic lattice. These solids have very intriguing behavior: when external pressure is applied on them it tends to change the shape of the unit cell of the crystal and therefore produces an electrical response since it affects the internal dipole moment; conversely, an external electric field can also affect the internal dipole moment and the solid changes its shape to accommodate it. This coupling of mechanical and electrical responses is very useful for practical applications, such as sensors and actuators and non-volatile memories. The solids that exhibit this behavior are called piezoelectrics; some examples are $CaTiO_3$

(calcium titanate), $PbTiO_3$ (lead titanate), $BaTiO_3$ (barium titanate), $PbZrO_3$ (lead zirconate).

Another example of complex solids is the class of crystals formed by fullerene clusters and alkali metals: there is strong covalent bonding between C atoms in each fullerene cluster, weak van der Waals bonding between the fullerenes, and ionic bonding between the alkali atoms and the fullerene units. The clusters act just like the group VII atoms in ionic solids, taking up the electrons of the alkali atoms and becoming ionized. It is intriguing that these solids also exhibit superconductivity at relatively high temperatures!

1.2 Bonding in solids

In our discussion on the formation of solids from atoms we encountered five general types of bonding in solids:

(1) *Van der Waals bonding*, which is formed by atoms that do not have valence electrons available for sharing (the noble elements), and is rather weak; the solids produced in this way are not particularly stable.
(2) *Metallic bonding*, which is formed when electrons are shared by all the atoms in the solid, producing a uniform "sea" of negative charge; the solids produced in this way are the usual metals.
(3) *Covalent bonding*, which is formed when electrons in well defined directional orbitals, which can be thought of as linear combinations of the original atomic orbitals, have strong overlap with similar orbitals in neighboring atoms; the solids produced in this way are semiconductors or insulators.
(4) *Ionic bonding*, which is formed when two different types of atoms are combined, one that prefers to lose some of its valence electrons and become a positive ion, and one that prefers to grab electrons from other atoms and become a negative ion. Combinations of such elements are I–VII, II–VI, and III–V. In the first case bonding is purely ionic, in the other two there is a degree of covalent bonding present.
(5) *Hydrogen bonding*, which is formed when H is present, due to its lack of core electrons, its light mass and high ionization energy.

For some of these cases, it is possible to estimate the strength of bonding without involving a detailed description of the electronic behavior. Specifically, for van der Waals bonding and for purely ionic bonding it is sufficient to assume simple classical models. For van der Waals bonding, one assumes that there is an attractive potential between the atoms which behaves like $\sim r^{-6}$ with distance r between atoms (this behavior can actually be derived from perturbation theory, see Problem 4). The potential must become repulsive at very short range, as the electronic densities of the two atoms start overlapping, but electrons have no incentive to form bonding states (as was the case in covalent bonding) since all electronic shells are already

Table 1.1. *Parameters for the Lennard–Jones potential for noble gases.*
For the calculation of $\hbar\omega$ using the Lennard–Jones parameters see the following discussion and Table 1.2.

	Ne	Ar	Kr	Xe
ϵ (meV)	3.1	10.4	14.0	20.0
a (Å)	2.74	3.40	3.65	3.98
$\hbar\omega$ (meV)	2.213	2.310	1.722	1.510

Original sources: see Ashcroft and Mermin [14].

filled. For convenience the attractive part is taken to be proportional to r^{-12}, which gives the famous Lennard–Jones 6–12 potential:

$$V_{LJ}(r) = 4\epsilon\left[\left(\frac{a}{r}\right)^{12} - \left(\frac{a}{r}\right)^{6}\right] \tag{1.5}$$

with ϵ and a constants that determine the energy and length scales. These have been determined for the different elements by referring to the thermodynamic properties of the noble gases; the parameters for the usual noble gas elements are shown in Table 1.1.

Use of this potential can then provide a quantitative measure of cohesion in these solids. One measure of the strength of these potentials is the vibrational frequency that would correspond to a harmonic oscillator potential with the same curvature at the minimum; this is indicative of the stiffness of the bond between atoms. In Table 1.1 we present the frequencies corresponding to the Lennard–Jones potentials of the common noble gas elements (see following discussion and Table 1.2 for the relation between this frequency and the Lennard–Jones potential parameters). For comparison, the vibrational frequency of the H_2 molecule, the simplest type of covalent bond between two atoms, is about 500 meV, more than two orders of magnitude larger; the Lennard–Jones potentials for the noble gases correspond to very soft bonds indeed!

A potential of similar nature, also used to describe effective interactions between atoms, is the Morse potential:

$$V_M(r) = \epsilon\left[e^{-2(r-r_0)/b} - 2e^{-(r-r_0)/b}\right] \tag{1.6}$$

where again ϵ and b are the constants that determine the energy and length scales and r_0 is the position of the minimum energy. It is instructive to compare these two potentials with the harmonic oscillator potential, which has the same minimum and

Table 1.2. *Comparison of the three effective potentials, Lennard–Jones $V_{LJ}(r)$, Morse $V_M(r)$, and harmonic oscillator $V_{HO}(r)$.*

The relations between the parameters that ensure the three potentials have the same minimum and curvature at the minimum are also given (the parameters of the Morse and harmonic oscillator potentials are expressed in terms of the Lennard–Jones parameters).

	$V_{LJ}(r)$	$V_M(r)$	$V_{HO}(r)$
Potential	$4\epsilon\left[\left(\frac{a}{r}\right)^{12}-\left(\frac{a}{r}\right)^{6}\right]$	$\epsilon\left[e^{-2(r-r_0)/b}-2e^{-(r-r_0)/b}\right]$	$-\epsilon+\frac{1}{2}m\omega^2(r-r_0)^2$
V_{min}	$-\epsilon$	$-\epsilon$	$-\epsilon$
r_{min}	$(2^{\frac{1}{6}})a$	r_0	r_0
$V''(r_{min})$	$(72/2^{\frac{1}{3}})(\epsilon/a^2)$	$2(\epsilon/b^2)$	$m\omega^2$
Relations		$r_0=(2^{\frac{1}{6}})a$	$r_0=(2^{\frac{1}{6}})a$
		$b=(2^{\frac{1}{6}}/6)a$	$\omega=(432^{\frac{1}{3}})\sqrt{\epsilon/ma^2}$

curvature, given by:

$$V_{HO}(r)=-\epsilon+\frac{1}{2}m\omega^2(r-r_0)^2 \tag{1.7}$$

with ω the frequency, m the mass of the particle in the potential and r_0 the position of the minimum.

The definitions of the three potentials are such that they all have the same value of the energy at their minimum, namely $-\epsilon$. The relations between the values of the other parameters which ensure that the minimum in the energy occurs at the same value of r and that the curvature at the minimum is the same are given in Table 1.2; a plot of the three potentials with these parameters is given in Fig. 1.18. The harmonic oscillator potential is what we would expect near the equilibrium of any normal interaction potential. The other two potentials extend the range far from the minimum; both potentials have a much sharper increase of the energy for distances shorter than the equilibrium value, and a much weaker increase of the energy for distances larger than the equilibrium value, relative to the harmonic oscillator. The overall behavior of the two potentials is quite similar. One advantage of the Morse potential is that it can be solved exactly, by analogy to the harmonic oscillator potential (see Appendix B). This allows a comparison between the energy levels associated with this potential and the corresponding energy levels of the harmonic oscillator; the latter are given by:

$$E_n^{HO}=\left(n+\frac{1}{2}\right)\hbar\omega \tag{1.8}$$

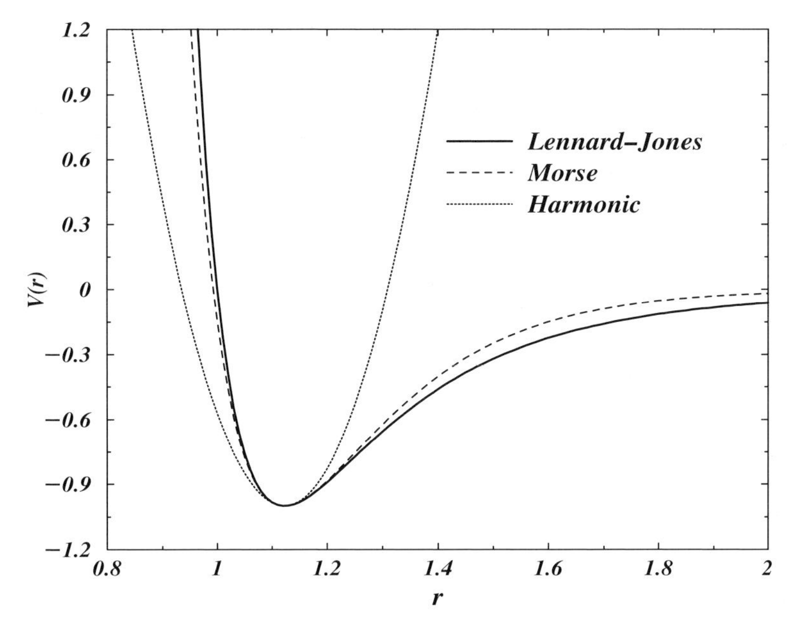

Figure 1.18. The three effective potentials discussed in the text, Lennard–Jones Eq. (1.5), Morse Eq. (1.6) and harmonic oscillator Eq. (1.7), with same minimum and curvature at the minimum. The energy is given in units of ϵ and the distance in units of a, the two parameters of the Lennard–Jones potential.

with n the integer index of the levels, whereas those of the Morse potential are given by:

$$E_n^M = \left(n + \frac{1}{2}\right)\hbar\omega\left[1 - \frac{\hbar\omega}{4\epsilon}\left(n + \frac{1}{2}\right)\right] \tag{1.9}$$

for the parameters defined in Table 1.2. We thus see that the spacing of levels in the Morse potential is smaller than in the corresponding harmonic oscillator, and that it becomes progressively smaller as the index of the levels increases. This is expected from the behavior of the potential mentioned above, and in particular from its asymptotic approach to zero for large distances. Since the Lennard–Jones potential has an overall shape similar to the Morse potential, we expect its energy levels to behave in the same manner.

For purely ionic bonding, one assumes that what keeps the crystal together is the attractive interaction between the positively and negatively charged ions, again in a purely classical picture. For the ionic solids with rock-salt, cesium chloride and zincblende lattices we have discussed already, it is possible to calculate the cohesive energy, which only depends on the ionic charges, the crystal structure and the distance between ions. This is called the Madelung energy. The only difficulty is that the summation converges very slowly, because the interaction potential

(Coulomb) is long range. In fact, formally this sum does not converge, and any simple way of summing successive terms gives results that depend on the choice of terms. The formal way for treating periodic structures, which we will develop in chapter 3, makes the calculation of the Madelung energy through the Ewald summation trick much more efficient (see Appendix F).

The other types of bonding, metallic, covalent and mixed bonding, are much more difficult to describe quantitatively. For metallic bonding, even if we think of the electrons as a uniform sea, we need to know the energy of this uniform "liquid" of fermions, which is not a trivial matter. This will be the subject of the next chapter. In addition to the electronic contributions, we have to consider the energy of the positive ions that exist in the uniform negative background of the electron sea. This is another Madelung sum, which converges very slowly. As far as covalent bonding is concerned, although the approach we used by combining atomic orbitals is conceptually simple, much more information is required to render it a realistic tool for calculations, and the electron interactions again come into play in an important way. This will also be discussed in detail in subsequent chapters.

The descriptions that we mentioned for the metallic and covalent solids are also referred to by more technical terms. The metallic sea of electrons paradigm is referred to as the "jellium" model in the extreme case when the ions (atoms stripped of their valence electrons) are considered to form a uniform positive background; in this limit the crystal itself does not play an important role, other than it provides the background for forming the electronic sea. The description of the covalent bonding paradigm is referred to as the Linear Combination of Atomic Orbitals (LCAO) approach, since it relies on the use of a basis of atomic orbitals in linear combinations that make the bonding arrangement transparent, as was explained above for the graphite and diamond lattices. We will revisit these notions in more detail.

Further reading

We collect here a number of general books on the physics of solids. Material in these books goes well beyond the topics covered in the present chapter and is relevant to many other topics covered in subsequent chapters.

1. *Solid State Physics*, N.W. Ashcroft and N.D. Mermin (Saunders College Publishing, Philadelphia, 1976).
   This is a comprehensive and indispensable source on the physics of solids; it provides an inspired coverage of most topics that had been the focus of research up to its publication.
2. *Introduction to Solid State Theory*, O.Madelung (Springer-Verlag, Berlin, Heidelberg, 1981).

This book represents a balanced introduction to the theoretical formalism needed for the study of solids at an advanced level; it covers both the single-particle and the many-body pictures.

3. *Basic Notions of Condensed Matter Physics*, P.W. Anderson (Benjamin-Cummings Publishing, Menlo Park, 1984).
4. *The Solid State*, A. Guinier and R. Jullien (Oxford University Press, Oxford, 1989).
5. *Electronic Structure of Materials*, A.P. Sutton (Oxford University Press, Oxford, 1993).
   This book is a modern account of the physics of solids, with an emphasis on topics relevant to materials science and technological applications.
6. *Bonding and Structure of Molecules and Solids*, D. Pettifor (Oxford University Press, Oxford, 1995).
7. *Introduction to Solid State Physics*, C. Kittel (7th edn, J. Wiley, New York, 1996).
   This is one of the standard introductory texts in the physics of solids with a wealth of useful information, covering a very broad range of topics at an introductory level.
8. *Quantum Theory of Matter: A Novel Introduction*, A. Modinos (J. Wiley, New York, 1996).
   This is a fresh look at the physics of condensed matter, emphasizing both the physical and chemical aspects of bonding in solids and molecules.
9. *Quantum Theory of Solids*, C. Kittel (J. Wiley, New York, 1963).
   This is an older text with an advanced treatment of the physics of solids.
10. *Solid State Theory*, W.A. Harrison (McGraw-Hill, New York, 1970).
    This is an older but extensive treatment of the physics of solids.
11. *Principles of the Theory of Solids*, J.M. Ziman (Cambridge University Press, Cambridge, 1972).
    This is an older treatment with many useful physical insights.
12. *Theoretical Solid State Physics*, W. Jones and N.H. March (J. Wiley, London, 1973).
    This is an older but very comprehensive two-volume work, presenting the physics of solids, covering many interesting topics.
13. *The Nature of the Chemical Bond and the Structure of Molecules and Solids*, L. Pauling (Cornell University Press, Ithaca, 1960).
    This is a classic treatment of the nature of bonding between atoms. It discusses extensively bonding in molecules but there is also a rich variety of topics relevant to the bonding in solids.
14. *Crystal Structures*, R.W.G. Wyckoff (J. Wiley, New York, 1963).
    This is a very useful compilation of all the structures of elemental solids and a wide variety of common compounds.
15. *The Structure of the Elements*, J. Donohue (J. Wiley, New York, 1974).
    This is a useful compilation of crystal structures for elemental solids.

Problems

1. The three ionic lattices, rock-salt, cesium chloride and zincblende, which we have discussed in this chapter, are called bipartite lattices, because they include two equivalent sites per unit cell which can be occupied by the different ions, so that each ion type is completely surrounded by the other. Describe the corresponding bipartite

lattices in two dimensions. Are they all different from each other? Try to obtain the Madelung energy for one of them, and show how the calculation is sensitive to the way in which the infinite sum is truncated.

2. We wish to demonstrate in a simple one-dimensional example that symmetric and antisymmetric combinations of single-particle orbitals give rise to bonding and antibonding states.[3] We begin with an atom consisting of an ion of charge $+e$ and a single valence electron: the electron–ion interaction potential is $-e^2/|x|$, arising from the ion which is situated at $x = 0$; we will take the normalized wavefunction for the ground state of the electron to be

$$\phi_0(x) = \frac{1}{\sqrt{a}} \mathrm{e}^{-|x|/a}$$

where a is a constant. We next consider two such atoms, the first ion at $x = -b/2$ and the second at $x = +b/2$, with b the distance between them (also referred to as the "bond length"). From the two single-particle states associated with the electrons in each atom,

$$\phi_1(x) = \frac{1}{\sqrt{a}} \mathrm{e}^{-|x-b/2|/a}, \qquad \phi_2(x) = \frac{1}{\sqrt{a}} \mathrm{e}^{-|x+b/2|/a}$$

we construct the symmetric (+) and antisymmetric (−) combinations:

$$\phi^{(\pm)}(x) = \frac{1}{\sqrt{\lambda^{(\pm)}}} \left[\mathrm{e}^{-|x-b/2|/a} \pm \mathrm{e}^{-|x+b/2|/a} \right]$$

with $\lambda^{(\pm)}$ the normalization factors

$$\lambda^{(\pm)} = 2a \left[1 \pm \mathrm{e}^{-b/a} \left(1 + \frac{b}{a} \right) \right].$$

Show that the difference between the probability of the electron being in the symmetric or the antisymmetric state rather than in the average of the states $\phi_1(x)$ and $\phi_2(x)$, is given by:

$$\delta n^{(\pm)}(x) = \pm \frac{1}{\lambda^{(\pm)}} \left[2\phi_1(x)\phi_2(x) - \mathrm{e}^{-b/a} \left(1 + \frac{b}{a} \right) \left(|\phi_1(x)|^2 + |\phi_2(x)|^2 \right) \right]$$

A plot of the probabilities $|\phi^{(\pm)}(x)|^2$ and the differences $\delta n^{(\pm)}(x)$ is given in Fig. 1.19 for $b = 2.5a$. Using this plot, interpret the bonding character of state $\phi^{(+)}(x)$ and the antibonding character of state $\phi^{(-)}(x)$, taking into account the enhanced Coulomb attraction between the electron and the two ions in the region $-b/2 < x < +b/2$. Make sure to take into account possible changes in the kinetic energy and show that they do not affect the argument about the character of these two states.

A common approximation is to take the symmetric and antisymmetric combinations to be defined as:

$$\phi^{(\pm)}(x) = \frac{1}{\sqrt{2a}} \left[\phi_1(x) \pm \phi_2(x) \right],$$

[3] Though seemingly oversimplified, this example is relevant to the hydrogen molecule, which is discussed at length in the next chapter.

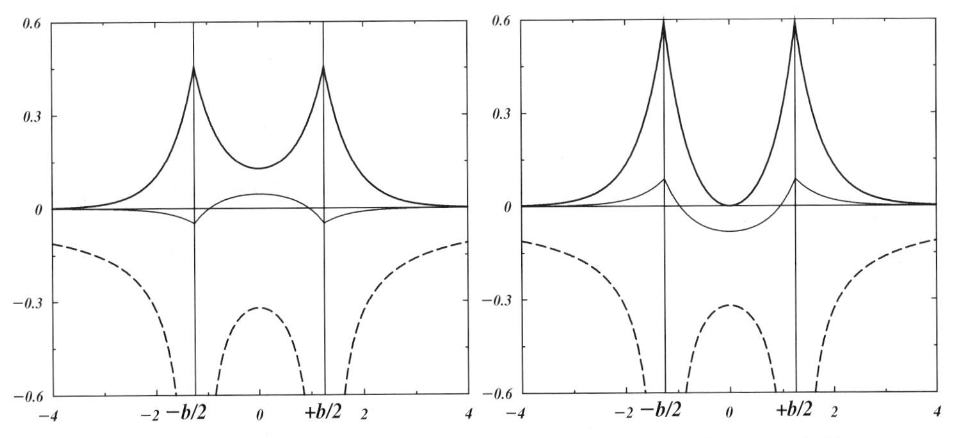

Figure 1.19. Symmetric ((+), left panel) and antisymmetric ((−), right panel) linear combinations of single-particle orbitals: the probability densities $|\phi^{(\pm)}(x)|^2$ are shown by thick solid lines in each case, and the differences $\delta n^{(\pm)}(x)$ between them and the average occupation of states $\phi_1(x)$, $\phi_2(x)$ by thinner solid lines; the dashed lines show the attractive potential of the two ions located at $\pm b/2$ (positions indicated by thin vertical lines). In this example $b = 2.5a$ and x is given in units of a.

that is, $\lambda^{(\pm)} = 2a$, which is reasonable in the limit $b \gg a$. In this limit, show by numerical integration that the gain in potential energy

$$\Delta V \equiv \langle \phi^{(\pm)} | \left[V_1(x) + V_2(x) \right] | \phi^{(\pm)} \rangle - \frac{1}{2} \left[\langle \phi_1 | V_1(x) | \phi_1 \rangle + \langle \phi_2 | V_2(x) | \phi_2 \rangle \right]$$

$$\text{where} \quad V_1(x) = \frac{-e^2}{|x - b/2|}, \qquad V_2(x) = \frac{-e^2}{|x + b/2|}$$

is always negative for $\phi^{(+)}$ and always positive for $\phi^{(-)}$; this again justifies our assertion that $\phi^{(+)}$ corresponds to a bonding state and $\phi^{(-)}$ to an antibonding state.

3. Produce an energy level diagram for the orbitals involved in the formation of the covalent bonds in the water molecule, as described in section 1.1.7. Provide an argument of how different combinations of orbitals than the ones discussed in the text would not produce as favorable a covalent bond between H and O. Describe how the different orbitals combine to form hydrogen bonds in the solid structure of ice.
4. In order to derive the attractive part of the Lennard–Jones potential, we consider two atoms with Z electrons each and filled electronic shells. In the ground state, the atoms will have spherical electronic charge distributions and, when sufficiently far from each other, they will not interact. When they are brought closer together, the two electronic charge distributions will be polarized because each will feel the effect of the ions and electrons of the other. We are assuming that the two atoms are still far enough from each other so that their electronic charge distributions do not overlap, and therefore we can neglect exchange of electrons between them. Thus, it is the

polarization that gives rise to an attractive potential; for this reason this interaction is sometimes also referred to as the "fluctuating dipole interaction".
To model the polarization effect, we consider the interaction potential between the two neutral atoms:

$$V_{int} = \frac{Z^2e^2}{|\mathbf{R}_1 - \mathbf{R}_2|} - \sum_i \frac{Ze^2}{|\mathbf{r}_i^{(1)} - \mathbf{R}_2|} - \sum_j \frac{Ze^2}{|\mathbf{r}_j^{(2)} - \mathbf{R}_1|} + \sum_{ij} \frac{e^2}{|\mathbf{r}_i^{(1)} - \mathbf{r}_j^{(2)}|}$$

where $\mathbf{R}_1, \mathbf{R}_2$ are the positions of the two nuclei and $\mathbf{r}_i^{(1)}, \mathbf{r}_j^{(2)}$ are the sets of electronic coordinates associated with each nucleus. In the above equation, the first term is the repulsion between the two nuclei, the second term is the attraction of the electrons of the first atom to the nucleus of the second, the third term is the attraction of the electrons of the second atom to the nucleus of the first, and the last term is the repulsion between the two sets of electrons in the two different atoms.
From second order perturbation theory, the energy change due to this interaction is given by:

$$\begin{aligned}\Delta E &= \langle \Psi_0^{(1)} \Psi_0^{(2)} | V_{int} | \Psi_0^{(1)} \Psi_0^{(2)} \rangle \\ &+ \sum_{nm} \frac{1}{E_0 - E_{nm}} \left| \langle \Psi_0^{(1)} \Psi_0^{(2)} | V_{int} | \Psi_n^{(1)} \Psi_m^{(2)} \rangle \right|^2\end{aligned} \tag{1.10}$$

where $\Psi_0^{(1)}, \Psi_0^{(2)}$ are the ground-state many-body wavefunctions of the two atoms, $\Psi_n^{(1)}, \Psi_m^{(2)}$ are their excited states, and E_0, E_{nm} are the corresponding energies of the two-atom system in their unperturbed states.
We define the electronic charge density associated with the ground state of each atom through:

$$\begin{aligned}n_0^{(I)}(\mathbf{r}) &= Z \int \left| \Psi_0^{(I)}(\mathbf{r}, \mathbf{r}_2, \mathbf{r}_3, \ldots, \mathbf{r}_Z) \right|^2 d\mathbf{r}_2 d\mathbf{r}_3 \cdots d\mathbf{r}_Z \\ &= \sum_{i=1}^{Z} \int \delta(\mathbf{r} - \mathbf{r}_i) \left| \Psi_0^{(I)}(\mathbf{r}_1, \mathbf{r}_2, \ldots, \mathbf{r}_Z) \right|^2 d\mathbf{r}_1 d\mathbf{r}_2 \cdots d\mathbf{r}_Z\end{aligned} \tag{1.11}$$

with $I = 1, 2$ (the expression for the density $n(\mathbf{r})$ in terms of the many-body wavefunction $|\Psi\rangle$ is discussed in detail in Appendix B). Show that the first order term in ΔE corresponds to the electrostatic interaction energy between the charge density distributions $n_0^{(1)}(\mathbf{r}), n_0^{(2)}(\mathbf{r})$. Assuming that there is no overlap between these two charge densities, show that this term vanishes (the two charge densities in the unperturbed ground state are spherically symmetric).
The wavefunctions involved in the second order term in ΔE will be negligible, unless the electronic coordinates associated with each atom are within the range of non-vanishing charge density. This implies that the distances $|\mathbf{r}_i^{(1)} - \mathbf{R}_1|$ and $|\mathbf{r}_j^{(2)} - \mathbf{R}_2|$ should be small compared with the distance between the atoms $|\mathbf{R}_2 - \mathbf{R}_1|$, which defines the distance at which interactions between the two charge densities become negligible. Show that expanding the interaction potential in the small quantities

$|\mathbf{r}_i^{(1)} - \mathbf{R}_1|/|\mathbf{R}_2 - \mathbf{R}_1|$ and $|\mathbf{r}_j^{(2)} - \mathbf{R}_2|/|\mathbf{R}_2 - \mathbf{R}_1|$, gives, to lowest order:

$$-\frac{e^2}{|\mathbf{R}_2 - \mathbf{R}_1|}\sum_{ij} 3\frac{(\mathbf{r}_i^{(1)} - \mathbf{R}_1)\cdot(\mathbf{R}_2 - \mathbf{R}_1)}{(\mathbf{R}_2 - \mathbf{R}_1)^2}\cdot\frac{(\mathbf{r}_j^{(2)} - \mathbf{R}_2)\cdot(\mathbf{R}_2 - \mathbf{R}_1)}{(\mathbf{R}_2 - \mathbf{R}_1)^2}$$

$$+\frac{e^2}{|\mathbf{R}_2 - \mathbf{R}_1|}\sum_{ij}\frac{(\mathbf{r}_i^{(1)} - \mathbf{R}_1)\cdot(\mathbf{r}_j^{(2)} - \mathbf{R}_2)}{(\mathbf{R}_2 - \mathbf{R}_1)^2} \tag{1.12}$$

Using this expression, show that the leading order term in the energy difference ΔE behaves like $|\mathbf{R}_2 - \mathbf{R}_1|^{-6}$ and is negative. This establishes the origin of the attractive term in the Lennard–Jones potential.

2
The single-particle approximation

In the previous chapter we saw that except for the simplest solids, like those formed by noble elements or by purely ionic combinations which can be described essentially in classical terms, in all other cases we need to consider the behavior of the valence electrons. The following chapters deal with these valence electrons (we will also refer to them as simply "the electrons" in the solid); we will study how their behavior is influenced by, and in turn influences, the ions.

Our goal in this chapter is to establish the basis for the single-particle description of the valence electrons. We will do this by starting with the exact hamiltonian for the solid and introducing approximations in its solution, which lead to sets of single-particle equations for the electronic degrees of freedom in the external potential created by the presence of the ions. Each electron also experiences the presence of other electrons through an effective potential in the single-particle equations; this effective potential encapsulates the many-body nature of the true system in an approximate way. In the last section of this chapter we will provide a formal way for eliminating the core electrons from the picture, while keeping the important effect they have on valence electrons.

2.1 The hamiltonian of the solid

An exact theory for a system of ions and interacting electrons is inherently quantum mechanical, and is based on solving a many-body Schrödinger equation of the form

$$\mathcal{H}\Psi(\{\mathbf{R}_I;\mathbf{r}_i\}) = E\Psi(\{\mathbf{R}_I;\mathbf{r}_i\}) \tag{2.1}$$

where $\mathcal{H}$ is the hamiltonian of the system, containing the kinetic energy operators

$$-\sum_I \frac{\hbar^2}{2M_I}\nabla^2_{\mathbf{R}_I} - \sum_i \frac{\hbar^2}{2m_e}\nabla^2_{\mathbf{r}_i} \tag{2.2}$$

and the potential energy due to interactions between the ions and the electrons. In the above equations: $\hbar$ is Planck's constant divided by 2π; M_I is the mass of ion I;

m_e is the mass of the electron; E is the energy of the system; $\Psi(\{\mathbf{R}_I; \mathbf{r}_i\})$ is the many-body wavefunction that describes the state of the system; $\{\mathbf{R}_I\}$ are the positions of the ions; and $\{\mathbf{r}_i\}$ are the variables that describe the electrons. Two electrons at $\mathbf{r}_i, \mathbf{r}_j$ repel one another, which produces a potential energy term

$$\frac{e^2}{\mid \mathbf{r}_i - \mathbf{r}_j \mid} \tag{2.3}$$

where e is the electronic charge. An electron at $\mathbf{r}$ is attracted to each positively charged ion at $\mathbf{R}_I$, producing a potential energy term

$$-\frac{Z_I e^2}{\mid \mathbf{R}_I - \mathbf{r} \mid} \tag{2.4}$$

where Z_I is the valence charge of this ion (nucleus plus core electrons). The total external potential experienced by an electron due to the presence of the ions is

$$V_{ion}(\mathbf{r}) = -\sum_I \frac{Z_I e^2}{\mid \mathbf{R}_I - \mathbf{r} \mid} \tag{2.5}$$

Two ions at positions $\mathbf{R}_I, \mathbf{R}_J$ also repel one another giving rise to a potential energy term

$$\frac{Z_I Z_J e^2}{\mid \mathbf{R}_I - \mathbf{R}_J \mid} \tag{2.6}$$

Typically, we can think of the ions as moving slowly in space and the electrons responding instantaneously to any ionic motion, so that Ψ has an explicit dependence on the electronic degrees of freedom alone: this is known as the Born–Oppenheimer approximation. Its validity is based on the huge difference of mass between ions and electrons (three to five orders of magnitude), making the former behave like classical particles. The only exception to this, noted in the previous chapter, are the lightest elements (especially H), where the ions have to be treated as quantum mechanical particles. We can then omit the quantum mechanical term for the kinetic energy of the ions, and take their kinetic energy into account as a classical contribution. If the ions are at rest, the hamiltonian of the system becomes

$$\mathcal{H} = -\sum_i \frac{\hbar^2}{2m_e}\nabla^2_{\mathbf{r}_i} - \sum_{iI} \frac{Z_I e^2}{\mid \mathbf{R}_I - \mathbf{r}_i \mid} + \frac{1}{2}\sum_{ij(j\neq i)} \frac{e^2}{\mid \mathbf{r}_i - \mathbf{r}_j \mid}$$
$$+ \frac{1}{2}\sum_{IJ(J\neq I)} \frac{Z_I Z_J e^2}{\mid \mathbf{R}_I - \mathbf{R}_J \mid} \tag{2.7}$$

In the following we will neglect for the moment the last term, which as far as the electron degrees of freedom are concerned is simply a constant. We discuss how this constant can be calculated for crystals in Appendix F (you will recognize in this term the Madelung energy of the ions, mentioned in chapter 1). The hamiltonian then takes the form

$$\mathcal{H} = -\sum_i \frac{\hbar^2}{2m_e}\nabla^2_{\mathbf{r}_i} + \sum_i V_{ion}(\mathbf{r}_i) + \frac{e^2}{2}\sum_{ij(j\neq i)} \frac{1}{\mid \mathbf{r}_i - \mathbf{r}_j \mid} \tag{2.8}$$

with the ionic potential that every electron experiences $V_{ion}(\mathbf{r})$ defined in Eq. (2.5).

Even with this simplification, however, solving for $\Psi(\{\mathbf{r}_i\})$ is an extremely difficult task, because of the nature of the electrons. If two electrons of the same spin interchange positions, Ψ must change sign; this is known as the "exchange" property, and is a manifestation of the Pauli exclusion principle. Moreover, each electron is affected by the motion of every other electron in the system; this is known as the "correlation" property. It is possible to produce a simpler, approximate picture, in which we describe the system as a collection of classical ions and essentially single quantum mechanical particles that reproduce the behavior of the electrons: this is the single-particle picture. It is an appropriate description when the effects of exchange and correlation are not crucial for describing the phenomena we are interested in. Such phenomena include, for example, optical excitations in solids, the conduction of electricity in the usual ohmic manner, and all properties of solids that have to do with cohesion (such as mechanical properties). Phenomena which are outside the scope of the single-particle picture include all the situations where electron exchange and correlation effects are crucial, such as superconductivity, transport in high magnetic fields (the quantum Hall effects), etc.

In developing the one-electron picture of solids, we will not neglect the exchange and correlation effects between electrons, we will simply take them into account in an average way; this is often referred to as a mean-field approximation for the electron–electron interactions. To do this, we have to pass from the many-body picture to an equivalent one-electron picture. We will first derive equations that look like single-particle equations, and then try to explore their meaning.

2.2 The Hartree and Hartree–Fock approximations

2.2.1 The Hartree approximation

The simplest approach is to assume a specific form for the many-body wavefunction which would be appropriate if the electrons were non-interacting particles, namely

$$\Psi^H(\{\mathbf{r}_i\}) = \phi_1(\mathbf{r}_1)\phi_2(\mathbf{r}_2)\cdots\phi_N(\mathbf{r}_N) \tag{2.9}$$

with the index i running over all electrons. The wavefunctions $\phi_i(\mathbf{r}_i)$ are states in which the individual electrons would be if this were a realistic approximation. These are single-particle states, normalized to unity. This is known as the Hartree approximation (hence the superscript H). With this approximation, the total energy of the system becomes

$$
\begin{aligned}
E^H &= \langle \Psi^H \mid \mathcal{H} \mid \Psi^H \rangle \\
&= \sum_i \langle \phi_i \mid \frac{-\hbar^2 \nabla_{\mathbf{r}}^2}{2m_e} + V_{ion}(\mathbf{r}) \mid \phi_i \rangle + \frac{e^2}{2} \sum_{ij(j \neq i)} \langle \phi_i \phi_j \mid \frac{1}{\mid \mathbf{r} - \mathbf{r}' \mid} \mid \phi_i \phi_j \rangle
\end{aligned}
\tag{2.10}
$$

Using a variational argument, we obtain from this the single-particle Hartree equations:

$$
\left[\frac{-\hbar^2 \nabla_{\mathbf{r}}^2}{2m_e} + V_{ion}(\mathbf{r}) + e^2 \sum_{j \neq i} \langle \phi_j \mid \frac{1}{\mid \mathbf{r} - \mathbf{r}' \mid} \mid \phi_j \rangle \right] \phi_i(\mathbf{r}) = \epsilon_i \phi_i(\mathbf{r}) \tag{2.11}
$$

where the constants ϵ_i are Lagrange multipliers introduced to take into account the normalization of the single-particle states ϕ_i (the bra $\langle \phi_i |$ and ket $| \phi_i \rangle$ notation for single-particle states and its extension to many-particle states constructed as products of single-particle states is discussed in Appendix B). Each orbital $\phi_i(\mathbf{r}_i)$ can then be determined by solving the corresponding single-particle Schrödinger equation, if all the other orbitals $\phi_j(\mathbf{r}_j)$, $j \neq i$ were known. In principle, this problem of self-consistency, i.e. the fact that the equation for one ϕ_i depends on all the other ϕ_j's, can be solved iteratively. We assume a set of ϕ_i's, use these to construct the single-particle hamiltonian, which allows us to solve the equations for each new ϕ_i; we then compare the resulting ϕ_i's with the original ones, and modify the original ϕ_i's so that they resemble more the new ϕ_i's. This cycle is continued until input and output ϕ_i's are the same up to a tolerance δ_{tol}, as illustrated in Fig. 2.1 (in this example, the comparison of input and output wavefunctions is made through the densities, as would be natural in Density Functional Theory, discussed below).

The more important problem is to determine how realistic the solution is. We can make the original trial ϕ's orthogonal, and maintain the orthogonality at each cycle of the self-consistency iteration to make sure the final ϕ's are also orthogonal. Then we would have a set of orbitals that would look like single particles, each $\phi_i(\mathbf{r})$ experiencing the ionic potential $V_{ion}(\mathbf{r})$ as well as a potential due to the presence of all other electrons, $V_i^H(\mathbf{r})$ given by

$$
V_i^H(\mathbf{r}) = +e^2 \sum_{j \neq i} \langle \phi_j \mid \frac{1}{\mid \mathbf{r} - \mathbf{r}' \mid} \mid \phi_j \rangle \tag{2.12}
$$

1. $\boxed{\text{Choose } \phi_i^{(in)}(\mathbf{r})}$

$\downarrow$

2. $\boxed{\text{Construct } \rho^{(in)}(\mathbf{r}) = \sum_i |\phi_i^{(in)}(\mathbf{r})|^2, \ V^{sp}(\mathbf{r}, \rho^{(in)}(\mathbf{r}))}$

$\downarrow$

3. $\boxed{\text{Solve } \left[-\frac{\hbar^2}{2m_e}\nabla_{\mathbf{r}}^2 + V^{sp}(\mathbf{r}, \rho^{(in)}(\mathbf{r}))\right] \phi_i^{(out)}(\mathbf{r}) = \epsilon_i^{(out)} \phi_i^{(out)}(\mathbf{r})}$

$\downarrow$

4. $\boxed{\text{Construct } \rho^{(out)}(\mathbf{r}) = \sum_i |\phi_i^{(out)}(\mathbf{r})|^2}$

$\downarrow$

5. $\boxed{\text{Compare } \rho^{(out)}(\mathbf{r}) \text{ to } \rho^{(in)}(\mathbf{r})}$

$\boxed{\text{If } |\rho^{(in)}(\mathbf{r}) - \rho^{(out)}(\mathbf{r})| < \delta_{tol} \rightarrow \texttt{STOP}; \text{ else } \phi_i^{(in)}(\mathbf{r}) = \phi_i^{(out)}(\mathbf{r}), \quad \texttt{GOTO 2.}}$

Figure 2.1. Schematic representation of iterative solution of coupled single-particle equations. This kind of operation is easily implemented on the computer.

This is known as the Hartree potential and includes only the Coulomb repulsion between electrons. The potential is different for each particle. It is a mean-field approximation to the electron–electron interaction, taking into account the electronic charge only, which is a severe simplification.

2.2.2 Example of a variational calculation

We will demonstrate the variational derivation of single-particle states in the case of the Hartree approximation, where the energy is given by Eq. (2.10), starting with the many-body wavefunction of Eq. (2.9). We assume that this state is a stationary state of the system, so that any variation in the wavefunction will give a zero variation in the energy (this is equivalent to the statement that the derivative of a function at an extremum is zero). We can take the variation in the wavefunction to be of the form $\langle \delta\phi_i|$, subject to the constraint that $\langle \phi_i|\phi_i\rangle = 1$, which can be taken into account by introducing a Lagrange multiplier ϵ_i:

$$\delta \left[E^H - \sum_i \epsilon_i \left(\langle \phi_i|\phi_i\rangle - 1 \right) \right] = 0 \qquad (2.13)$$

Notice that the variations of the bra and the ket of ϕ_i are considered to be independent of each other; this is allowed because the wavefunctions are complex quantities, so varying the bra and the ket independently is equivalent to varying the real and imaginary parts of a complex variable independently, which is legitimate since they represent independent components (for a more detailed justification of this see, for example, Ref. [15]). The above variation then produces

$$\langle\delta\phi_i| - \frac{\hbar^2\nabla_{\mathbf{r}}^2}{2m_e} + V_{ion}(\mathbf{r})|\phi_i\rangle + e^2\sum_{j\neq i}\langle\delta\phi_i\phi_j|\frac{1}{|\mathbf{r}-\mathbf{r}'|}|\phi_i\phi_j\rangle - \epsilon_i\langle\delta\phi_i|\phi_i\rangle$$

$$= \langle\delta\phi_i|\left[-\frac{\hbar^2\nabla_{\mathbf{r}}^2}{2m_e} + V_{ion}(\mathbf{r}) + e^2\sum_{j\neq i}\langle\phi_j|\frac{1}{|\mathbf{r}-\mathbf{r}'|}|\phi_j\rangle - \epsilon_i\right]|\phi_i\rangle = 0$$

Since this has to be true for any variation $\langle\delta\phi_i|$, we conclude that

$$\left[-\frac{\hbar^2\nabla_{\mathbf{r}}^2}{2m_e} + V_{ion}(\mathbf{r}) + e^2\sum_{j\neq i}\langle\phi_j|\frac{1}{|\mathbf{r}-\mathbf{r}'|}|\phi_j\rangle\right]\phi_i(\mathbf{r}) = \epsilon_i\phi_i(\mathbf{r})$$

which is the Hartree single-particle equation, Eq. (2.11).

2.2.3 The Hartree–Fock approximation

The next level of sophistication is to try to incorporate the fermionic nature of electrons in the many-body wavefunction $\Psi(\{\mathbf{r}_i\})$. To this end, we can choose a wavefunction which is a properly antisymmetrized version of the Hartree wavefunction, that is, it changes sign when the coordinates of two electrons are interchanged. This is known as the Hartree–Fock approximation. For simplicity we will neglect the spin of electrons and keep only the spatial degrees of freedom. This does not imply any serious restriction; in fact, at the Hartree–Fock level it is a simple matter to include explicitly the spin degrees of freedom, by considering electrons with up and down spins at position $\mathbf{r}$. Combining then Hartree-type wavefunctions to form a properly antisymmetrized wavefunction for the system, we obtain the determinant (first introduced by Slater [16]):

$$\Psi^{HF}(\{\mathbf{r}_i\}) = \frac{1}{\sqrt{N!}}\begin{vmatrix} \phi_1(\mathbf{r}_1) & \phi_1(\mathbf{r}_2) & \cdots & \phi_1(\mathbf{r}_N) \\ \phi_2(\mathbf{r}_1) & \phi_2(\mathbf{r}_2) & \cdots & \phi_2(\mathbf{r}_N) \\ \cdot & \cdot & & \cdot \\ \cdot & \cdot & & \cdot \\ \cdot & \cdot & & \cdot \\ \phi_N(\mathbf{r}_1) & \phi_N(\mathbf{r}_2) & \cdots & \phi_N(\mathbf{r}_N) \end{vmatrix} \tag{2.14}$$

where N is the total number of electrons. This has the desired property, since interchanging the position of two electrons is equivalent to interchanging the corresponding columns in the determinant, which changes its sign.

The total energy with the Hartree–Fock wavefunction is

$$
\begin{aligned}
E^{HF} &= \langle \Psi^{HF} \mid \mathcal{H} \mid \Psi^{HF} \rangle \\
&= \sum_i \langle \phi_i \mid \frac{-\hbar^2 \nabla_{\mathbf{r}}^2}{2m_e} + V_{ion}(\mathbf{r}) \mid \phi_i \rangle \\
&+ \frac{e^2}{2} \sum_{ij(j \neq i)} \langle \phi_i \phi_j \mid \frac{1}{\mid \mathbf{r} - \mathbf{r}' \mid} \mid \phi_i \phi_j \rangle \\
&- \frac{e^2}{2} \sum_{ij(j \neq i)} \langle \phi_i \phi_j \mid \frac{1}{\mid \mathbf{r} - \mathbf{r}' \mid} \mid \phi_j \phi_i \rangle
\end{aligned} \tag{2.15}
$$

and the single-particle Hartree–Fock equations , obtained by a variational calculation, are

$$
\left[\frac{-\hbar^2 \nabla_{\mathbf{r}}^2}{2m_e} + V_{ion}(\mathbf{r}) + V_i^H(\mathbf{r}) \right] \phi_i(\mathbf{r}) - e^2 \sum_{j \neq i} \langle \phi_j \mid \frac{1}{\mid \mathbf{r} - \mathbf{r}' \mid} \mid \phi_i \rangle \phi_j(\mathbf{r}) = \epsilon_i \phi_i(\mathbf{r}) \tag{2.16}
$$

This equation has one extra term compared with the Hartree equation, the last one, which is called the "exchange" term. The exchange term describes the effects of exchange between electrons, which we put in the Hartree–Fock many-particle wavefunction by construction. This term has the peculiar character that it cannot be written simply as $V_i^X(\mathbf{r}_i)\phi_i(\mathbf{r}_i)$ (in the following we use the superscript X to denote "exchange"). It is instructive to try to put this term in such a form, by multiplying and dividing by the proper factors. First we express the Hartree term in a different way: define the single-particle and the total densities as

$$
\rho_i(\mathbf{r}) = \mid \phi_i(\mathbf{r}) \mid^2 \tag{2.17}
$$

$$
\rho(\mathbf{r}) = \sum_i \rho_i(\mathbf{r}) \tag{2.18}
$$

so that the Hartree potential takes the form

$$
V_i^H(\mathbf{r}) = e^2 \sum_{j \neq i} \int \frac{\rho_j(\mathbf{r}')}{\mid \mathbf{r} - \mathbf{r}' \mid} \mathrm{d}\mathbf{r}' = e^2 \int \frac{\rho(\mathbf{r}') - \rho_i(\mathbf{r}')}{\mid \mathbf{r} - \mathbf{r}' \mid} \mathrm{d}\mathbf{r}' \tag{2.19}
$$

Now construct the single-particle exchange density to be

$$
\rho_i^X(\mathbf{r}, \mathbf{r}') = \sum_{j \neq i} \frac{\phi_i(\mathbf{r}')\phi_i^*(\mathbf{r})\phi_j(\mathbf{r})\phi_j^*(\mathbf{r}')}{\phi_i(\mathbf{r})\phi_i^*(\mathbf{r})} \tag{2.20}
$$

Then the single-particle Hartree–Fock equations take the form

$$
\left[\frac{-\hbar^2 \nabla_{\mathbf{r}}^2}{2m_e} + V_{ion}(\mathbf{r}) + V_i^H(\mathbf{r}) + V_i^X(\mathbf{r}) \right] \phi_i(\mathbf{r}) = \epsilon_i \phi_i(\mathbf{r}) \tag{2.21}
$$

with the exchange potential, in analogy with the Hartree potential, given by

$$V_i^X(\mathbf{r}) = -e^2 \int \frac{\rho_i^X(\mathbf{r}, \mathbf{r}')}{| \mathbf{r} - \mathbf{r}' |} \mathrm{d}\mathbf{r}' \tag{2.22}$$

The Hartree and exchange potentials give the following potential for electron–electron interaction in the Hartree–Fock approximation:

$$V_i^{HF}(\mathbf{r}) = e^2 \int \frac{\rho(\mathbf{r}')}{| \mathbf{r} - \mathbf{r}' |} \mathrm{d}\mathbf{r}' - e^2 \int \frac{\rho_i(\mathbf{r}') + \rho_i^X(\mathbf{r}, \mathbf{r}')}{| \mathbf{r} - \mathbf{r}' |} \mathrm{d}\mathbf{r}' \tag{2.23}$$

which can be written, with the help of the Hartree–Fock density

$$\rho_i^{HF}(\mathbf{r}, \mathbf{r}') = \sum_j \frac{\phi_i(\mathbf{r}')\phi_i^*(\mathbf{r})\phi_j(\mathbf{r})\phi_j^*(\mathbf{r}')}{\phi_i(\mathbf{r})\phi_i^*(\mathbf{r})} \tag{2.24}$$

as the following expression for the total electron–electron interaction potential:

$$V_i^{HF}(\mathbf{r}) = e^2 \int \frac{\rho(\mathbf{r}') - \rho_i^{HF}(\mathbf{r}, \mathbf{r}')}{| \mathbf{r} - \mathbf{r}' |} \mathrm{d}\mathbf{r}' \tag{2.25}$$

The first term is the total Coulomb repulsion potential of electrons common for all states $\phi_i(\mathbf{r})$, while the second term is the effect of fermionic exchange, and is different for each state $\phi_i(\mathbf{r})$.

2.3 Hartree–Fock theory of free electrons

To elucidate the physical meaning of the approximations introduced above we will consider the simplest possible case, that is one in which the ionic potential is a uniformly distributed positive background . This is referred to as the jellium model. In this case, the electronic states must also reflect this symmetry of the potential, which is uniform, so they must be plane waves:

$$\phi_i(\mathbf{r}) = \frac{1}{\sqrt{\Omega}} \mathrm{e}^{\mathrm{i}\mathbf{k}_i \cdot \mathbf{r}} \tag{2.26}$$

where Ω is the volume of the solid and $\mathbf{k}_i$ is the wave-vector which characterizes state ϕ_i. Since the wave-vectors suffice to characterize the single-particle states, we will use those as the only index, i.e. $\phi_i \rightarrow \phi_{\mathbf{k}}$. Plane waves are actually a very convenient and useful basis for expressing various physical quantities. In particular, they allow the use of Fourier transform techniques, which simplify the calculations. In the following we will be using relations implied by the Fourier transform method, which are proven in Appendix G.

We also define certain useful quantities related to the density of the uniform electron gas: the wave-vectors have a range of values from zero up to some maximum magnitude k_F, the Fermi momentum, which is related to the density $n = N/\Omega$ through

$$n = \frac{k_F^3}{3\pi^2} \tag{2.27}$$

(see Appendix D, Eq. (D.10)). The Fermi energy is given in terms of the Fermi momentum

$$\epsilon_F = \frac{\hbar^2 k_F^2}{2m_e} \tag{2.28}$$

It is often useful to express equations in terms of another quantity, r_s, which is defined as the radius of the sphere whose volume corresponds to the average volume per electron:

$$\frac{4\pi}{3} r_s^3 = \frac{\Omega}{N} = n^{-1} = \frac{3\pi^2}{k_F^3} \tag{2.29}$$

and r_s is typically measured in atomic units (the Bohr radius, $a_0 = 0.529177$ Å). This gives the following expression for k_F:

$$k_F = \frac{(9\pi/4)^{1/3}}{r_s} \Longrightarrow k_F a_0 = \frac{(9\pi/4)^{1/3}}{(r_s/a_0)} \tag{2.30}$$

where the last expression contains the dimensionless combinations of variables $k_F a_0$ and r_s/a_0. If the electrons had only kinetic energy, the total energy of the system would be given by

$$E^{kin} = \frac{\Omega}{\pi^2} \frac{\hbar^2 k_F^5}{10 m_e} \Longrightarrow \frac{E^{kin}}{N} = \frac{3}{5}\epsilon_F \tag{2.31}$$

(see Appendix D, Eq. (D.12)). Finally, we introduce the unit of energy rydberg (Ry), which is the natural unit for energies in solids,

$$\frac{\hbar^2}{2m_e a_0^2} = \frac{e^2}{2a_0} = 1 \text{ Ry} \tag{2.32}$$

With the electrons represented by plane waves, the electronic density must be uniform and equal to the ionic density. These two terms, the uniform positive ionic charge and the uniform negative electronic charge of equal density, cancel each other. The only terms remaining in the single-particle equation are the kinetic energy and the part of $V_i^{HF}(\mathbf{r})$ corresponding to exchange, which arises from $\rho_i^{HF}(\mathbf{r}, \mathbf{r}')$:

$$\left[\frac{-\hbar^2 \nabla_{\mathbf{r}}^2}{2m_e} - e^2 \int \frac{\rho_{\mathbf{k}}^{HF}(\mathbf{r}, \mathbf{r}')}{|\mathbf{r} - \mathbf{r}'|} d\mathbf{r}'\right] \phi_{\mathbf{k}}(\mathbf{r}) = \epsilon_{\mathbf{k}} \phi_{\mathbf{k}}(\mathbf{r}) \tag{2.33}$$

We have asserted above that the behavior of electrons in this system is described by plane waves; we prove this statement next. Plane waves are of course eigenfunctions of the kinetic energy operator:

$$-\frac{\hbar^2\nabla^2}{2m_e}\frac{1}{\sqrt{\Omega}}\mathrm{e}^{\mathrm{i}\mathbf{k}\cdot\mathbf{r}} = \frac{\hbar^2\mathbf{k}^2}{2m_e}\frac{1}{\sqrt{\Omega}}\mathrm{e}^{\mathrm{i}\mathbf{k}\cdot\mathbf{r}} \tag{2.34}$$

so that all we need to show is that they are also eigenfunctions of the second term in the hamiltonian of Eq. (2.33). Using Eq. (2.24) for $\rho_{\mathbf{k}}^{HF}(\mathbf{r},\mathbf{r}')$ we obtain

$$\left[\int \frac{-e^2\rho_{\mathbf{k}}^{HF}(\mathbf{r},\mathbf{r}')}{|\mathbf{r}-\mathbf{r}'|}\mathrm{d}\mathbf{r}'\right]\phi_{\mathbf{k}}(\mathbf{r}) = \frac{-e^2}{\sqrt{\Omega}}\int \frac{\rho_{\mathbf{k}}^{HF}(\mathbf{r},\mathbf{r}')}{|\mathbf{r}-\mathbf{r}'|}\mathrm{d}\mathbf{r}'\mathrm{e}^{\mathrm{i}\mathbf{k}\cdot\mathbf{r}}$$
$$= \frac{-e^2}{\sqrt{\Omega}}\sum_{\mathbf{k}'}\int \frac{\phi_{\mathbf{k}}(\mathbf{r}')\phi_{\mathbf{k}}^*(\mathbf{r})\phi_{\mathbf{k}'}(\mathbf{r})\phi_{\mathbf{k}'}^*(\mathbf{r}')}{\phi_{\mathbf{k}}(\mathbf{r})\phi_{\mathbf{k}}^*(\mathbf{r})}\frac{1}{|\mathbf{r}-\mathbf{r}'|}\mathrm{d}\mathbf{r}'\mathrm{e}^{\mathrm{i}\mathbf{k}\cdot\mathbf{r}}$$
$$= \frac{-e^2}{\sqrt{\Omega}}\sum_{\mathbf{k}'}\int \frac{\mathrm{e}^{-\mathrm{i}(\mathbf{k}-\mathbf{k}')\cdot(\mathbf{r}-\mathbf{r}')}}{\Omega}\frac{1}{|\mathbf{r}-\mathbf{r}'|}\mathrm{d}\mathbf{r}'\mathrm{e}^{\mathrm{i}\mathbf{k}\cdot\mathbf{r}}$$

Expressing $1/\mid\mathbf{r}-\mathbf{r}'\mid$ in terms of its Fourier transform provides a convenient way for evaluating the last sum once it has been turned into an integral using Eq. (D.8). The inverse Fourier transform of $1/|\mathbf{r}-\mathbf{r}'|$ turns out to be

$$\frac{1}{|\mathbf{r}-\mathbf{r}'|} = \int \frac{\mathrm{d}\mathbf{q}}{(2\pi)^3}\frac{4\pi}{q^2}\mathrm{e}^{\mathrm{i}\mathbf{q}\cdot(\mathbf{r}-\mathbf{r}')} \tag{2.35}$$

as proven in Appendix G. Substituting this expression into the previous equation we obtain

$$\frac{-e^2}{\sqrt{\Omega}}\sum_{\mathbf{k}'}\int \frac{\mathrm{e}^{-\mathrm{i}(\mathbf{k}-\mathbf{k}')\cdot(\mathbf{r}-\mathbf{r}')}}{\Omega}\int \frac{\mathrm{d}\mathbf{q}}{(2\pi)^3}\frac{4\pi}{q^2}\mathrm{e}^{\mathrm{i}\mathbf{q}\cdot(\mathbf{r}-\mathbf{r}')}\mathrm{d}\mathbf{r}'\mathrm{e}^{\mathrm{i}\mathbf{k}\cdot\mathbf{r}}$$
$$= \frac{-4\pi e^2}{\sqrt{\Omega}}\int_{k'<k_{\mathrm{F}}}\frac{\mathrm{d}\mathbf{k}'}{(2\pi)^3}\int \mathrm{d}\mathbf{q}\frac{1}{q^2}\left[\frac{1}{(2\pi)^3}\int \mathrm{e}^{-\mathrm{i}(\mathbf{k}-\mathbf{k}'-\mathbf{q})\cdot(\mathbf{r}-\mathbf{r}')}\mathrm{d}\mathbf{r}'\right]\mathrm{e}^{\mathrm{i}\mathbf{k}\cdot\mathbf{r}} \tag{2.36}$$

At this point it will be necessary to employ the Fourier transform representation of the δ-function, which is derived in Appendix G; this representation allows us to identify the quantity in square brackets in the last expression with a δ-function in momentum space,

$$\frac{1}{(2\pi)^3}\int \mathrm{e}^{-\mathrm{i}(\mathbf{k}-\mathbf{k}'-\mathbf{q})\cdot(\mathbf{r}-\mathbf{r}')}\mathrm{d}\mathbf{r}' = \delta(\mathbf{q}-(\mathbf{k}-\mathbf{k}')) \tag{2.37}$$

which upon integration over $\mathbf{q}$ gives

$$\frac{-4\pi e^2}{\sqrt{\Omega}}\left[\int_{k'<k_{\mathrm{F}}}\frac{\mathrm{d}\mathbf{k}'}{(2\pi)^3}\frac{1}{|\mathbf{k}-\mathbf{k}'|^2}\right]\mathrm{e}^{\mathrm{i}\mathbf{k}\cdot\mathbf{r}} = -\frac{e^2}{\pi}k_{\mathrm{F}}F(k/k_{\mathrm{F}})\frac{1}{\sqrt{\Omega}}\mathrm{e}^{\mathrm{i}\mathbf{k}\cdot\mathbf{r}} \tag{2.38}$$

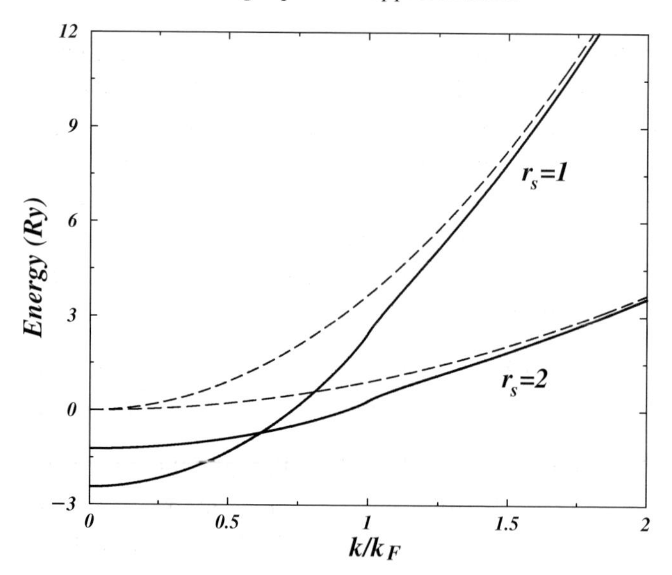

Figure 2.2. Energy (in rydbergs) of individual single-particle states as a function of momentum $k/k_{\rm F}$ (with $k_{\rm F}$ = Fermi momentum), as given by Eq. (2.41), for two different values of $r_s = 1, 2$ (in a_0). The dashed curves give the kinetic energy contribution (first term on the right-hand side of Eq. (2.41)).

with $k = |\mathbf{k}|$, where the function $F(x)$ is defined as

$$F(x) = 1 + \frac{1 - x^2}{2x} \ln \left| \frac{1 + x}{1 - x} \right| \tag{2.39}$$

This completes the proof that plane waves are eigenfunctions of the single-particle hamiltonian in Eq. (2.33).

With this result, the energy of single-particle state $\phi_{\mathbf{k}}(\mathbf{r})$ is given by

$$\epsilon_{\mathbf{k}} = \frac{\hbar^2 k^2}{2m_e} - \frac{e^2}{\pi} k_{\rm F} F(k/k_{\rm F}) \tag{2.40}$$

which, using the variable r_s introduced in Eq. (2.29) and the definition of the Ry for the energy unit given in Eq. (2.32), can be rewritten in the following form:

$$\epsilon_{\mathbf{k}} = \left[\left(\frac{(9\pi/4)^{1/3}}{r_s/a_0} \right)^2 (k/k_{\rm F})^2 - \frac{2}{\pi} \left(\frac{(9\pi/4)^{1/3}}{r_s/a_0} \right) F(k/k_{\rm F}) \right] \text{Ry} \tag{2.41}$$

The behavior of the energy $\epsilon_{\mathbf{k}}$ as function of the momentum (in units of $k_{\rm F}$) is illustrated in Fig. 2.2 for two different values of r_s.

This is an intriguing result: it shows that, even though plane waves are eigenstates of this hypothetical system, due to the exchange interaction the energy of state $\phi_{\mathbf{k}}$

is not simply $\hbar^2 k^2/2m_e$, as might be expected for non-interacting particles; it also contains the term proportional to $F(k/k_{\rm F})$ in Eq. (2.40). This term has interesting behavior at $|\mathbf{k}| = k_{\rm F}$, as is evident in Fig. 2.2. It also gives a lower energy than the non-interacting electron case for all values of $\mathbf{k}$, an effect which is more pronounced for small values of r_s (see Fig. 2.2). Thus, the electron–electron interaction included at the Hartree–Fock level lowers the energy of the system significantly. We can calculate the total energy of this system by summing the single-particle energies over $\mathbf{k}$ up to momentum $k_{\rm F}$:

$$E^{HF} = 2\sum_{k<k_{\rm F}} \frac{\hbar^2 \mathbf{k}^2}{2m_e} - \frac{e^2 k_{\rm F}}{\pi} \sum_{k<k_{\rm F}} \left[1 + \frac{k_{\rm F}^2 - k^2}{2kk_{\rm F}} \ln\left|\frac{k_{\rm F}+k}{k_{\rm F}-k}\right|\right] \tag{2.42}$$

Notice that we must include a factor of 2 for the spin of the electrons in both summations. This is indeed explicitly done for the kinetic energy part (see Appendix D where the expression for E^{kin} was derived). But for the second term, which represents the effective electron-electron interaction due to exchange, this factor of 2 is canceled by a factor of $1/2$ needed to compensate double counting of the effective interaction in the sum of $\epsilon_{\mathbf{k}}$'s: remember that this effective interaction is contained in the HF single-particle equations Eq. (2.16) as the sum over all states other than state i, so if we simply sum all these contributions contained in the $\epsilon_{\mathbf{k}}$'s we will be counting each contribution twice. Turning the second term in the above equation into an integral through the usual procedure, we can evaluate the sum to find

$$\frac{E^{HF}}{N} = \frac{3}{5}\epsilon_{\rm F} - \frac{3}{4}\frac{e^2 k_{\rm F}}{\pi} \tag{2.43}$$

which quantifies by how much the effective electron–electron interaction due to exchange lowers the energy of the system relative to the kinetic energy alone. Using the expression of $k_{\rm F}$ in terms of r_s, Eq. (2.30), and expressing everything in rydbergs with the help of Eq. (2.32), we obtain

$$\frac{E^{HF}}{N} = \left[\frac{2.21}{(r_s/a_0)^2} - \frac{0.916}{(r_s/a_0)}\right] \text{Ry} \tag{2.44}$$

This result should be compared with the expansion for the exact energy of the electron gas in the high-density limit (low r_s/a_0 values), first obtained by Gell-Mann and Brueckner (for details and original references see Ref. [17]),

$$\frac{E}{N} = \left[\frac{2.21}{(r_s/a_0)^2} - \frac{0.916}{(r_s/a_0)} + 0.0622\ln(r_s/a_0) - 0.096 + \mathcal{O}(r_s/a_0)\right] \text{Ry} \tag{2.45}$$

It is quite remarkable that the Hartree–Fock approximation, based on an *ad hoc* expression for the many-body wavefunction, captures the first two terms in the exact expansion of the total energy. Of course in real situations this may not be very helpful, since in typical metals (r_s/a_0) varies between 2 and 6.

Another interesting point is that we can express the potential due to exchange in a way that involves the density. This potential will give rise to the second term on the right-hand side of Eq. (2.44), namely

$$\frac{E^X}{N} = -\frac{0.916}{(r_s/a_0)}\,\mathrm{Ry} \tag{2.46}$$

which, using the expressions for r_s discussed earlier, can be written as

$$\frac{E^X}{N} = -\frac{3e^2}{4}\left(\frac{3}{\pi}\right)^{1/3} n^{1/3} = -1.477[a_0^3 n]^{1/3}\,\mathrm{Ry} \tag{2.47}$$

One of the most insightful proposals in the early calculations of the properties of solids, due to Slater [18], was to generalize this term for situations where the density is not constant, that is, a system with non-homogeneous distribution of electrons. In this case the exchange energy would arise from a potential energy term in the single-particle hamiltonian which will have the form

$$V^X(\mathbf{r}) = -\frac{3e^2}{2}\left(\frac{3}{\pi}\right)^{1/3}[n(\mathbf{r})]^{1/3} = -\frac{3e^2}{2\pi}k_F(\mathbf{r}) = -2.954[a_0^3 n(\mathbf{r})]^{1/3}\,\mathrm{Ry} \tag{2.48}$$

where an extra factor of 2 is introduced to account for the fact that a variational derivation gives rise to a potential term in the single-particle equations which is twice as large as the corresponding energy term; conversely, when one calculates the total energy by summing terms in the single-particle equations, a factor of 1/2 must be introduced to account for double-counting of interactions. In the last equation, the density, and hence the Fermi momentum, have become functions of $\mathbf{r}$, i.e. they can be non-homogeneous. There is actually good justification to use such a term in single-particle equations in order to describe the exchange contribution, although the values of the constants involved are different from Slater's (see also section 2.5 on Density Functional Theory).

2.4 The hydrogen molecule

In order to demonstrate the difficulty of including explicitly all the interaction effects in a system with more than one electron, we discuss briefly a model of the hydrogen molecule. This molecule consists of two protons and two electrons, so it is the simplest possible system for studying electron–electron interactions in a realistic manner.[1] We begin by defining the hamiltonian for a single hydrogen atom:

$$h_1(\mathbf{r}_1) = -\frac{\hbar^2 \nabla^2_{\mathbf{r}_1}}{2m_e} - \frac{e^2}{| \mathbf{r}_1 - \mathbf{R}_1 |} \tag{2.49}$$

where $\mathbf{R}_1$ is the position of the first proton. The wavefunction for this hamiltonian is $s(\mathbf{r}_1 - \mathbf{R}_1) = s_1(\mathbf{r}_1)$, the ground state of the hydrogen atom with energy ϵ_0. Similarly, an atom at position $\mathbf{R}_2$ (far from $\mathbf{R}_1$) will have the hamiltonian

$$h_2(\mathbf{r}_2) = -\frac{\hbar^2 \nabla^2_{\mathbf{r}_2}}{2m_e} - \frac{e^2}{| \mathbf{r}_2 - \mathbf{R}_2 |} \tag{2.50}$$

and the wavefunction $s(\mathbf{r}_2 - \mathbf{R}_2) = s_2(\mathbf{r}_2)$. When the two protons are very far away, the two electrons do not interact and the two electronic wavefunctions are the same, only centered at different points in space. When the atoms are brought together, the new hamiltonian becomes

$$\mathcal{H}(\mathbf{r}_1, \mathbf{r}_2) = -\frac{\hbar^2 \nabla^2_{\mathbf{r}_1}}{2m_e} - \frac{e^2}{| \mathbf{r}_1 - \mathbf{R}_1 |} - \frac{\hbar^2 \nabla^2_{\mathbf{r}_2}}{2m_e} - \frac{e^2}{| \mathbf{r}_2 - \mathbf{R}_2 |}$$
$$+ \left[-\frac{e^2}{| \mathbf{R}_1 - \mathbf{r}_2 |} - \frac{e^2}{| \mathbf{r}_1 - \mathbf{R}_2 |} + \frac{e^2}{| \mathbf{r}_1 - \mathbf{r}_2 |} + \frac{e^2}{| \mathbf{R}_1 - \mathbf{R}_2 |} \right] \tag{2.51}$$

where the last four terms represent electron–proton attraction between the electron in one atom and the proton in the other (the cross terms), and electron–electron, and proton–proton repulsion. As we have done so far, we will ignore the proton–proton repulsion, (last term in Eq. (2.51)), since it is only a constant term as far as the electrons are concerned, and it does not change the character of the electronic wavefunction. This is equivalent to applying the Born–Oppenheimer approximation to the problem and neglecting the quantum nature of the protons, even though we mentioned in chapter 1 that this may not be appropriate for hydrogen. The justification for using this approximation here is that we are concentrating our attention on the electron–electron interactions in the simplest possible model rather than attempting to give a realistic picture of the system as a whole. Solving for the wavefunction $\Psi(\mathbf{r}_1, \mathbf{r}_2)$ of this new hamiltonian analytically is already an impossible task. We will attempt to do this approximately, using the orbitals $s_1(\mathbf{r})$ and $s_2(\mathbf{r})$ as a convenient basis.

If we were dealing with a single electron, then this electron would see the following hamiltonian, in the presence of the two protons:

$$\mathcal{H}^{sp}(\mathbf{r}) = -\frac{\hbar^2 \nabla^2_{\mathbf{r}}}{2m_e} - \frac{e^2}{| \mathbf{r} - \mathbf{R}_1 |} - \frac{e^2}{| \mathbf{r} - \mathbf{R}_2 |} \tag{2.52}$$

[1] This is a conveniently simple model for illustrating electron exchange and correlation effects. It is discussed in several of the textbooks mentioned in chapter 1.

We define the expectation value of this hamiltonian in the state $s_1(\mathbf{r})$ or $s_2(\mathbf{r})$ to be

$$\epsilon \equiv \langle s_1|\mathcal{H}^{sp}|s_1\rangle = \langle s_2|\mathcal{H}^{sp}|s_2\rangle \tag{2.53}$$

Notice that $s_1(\mathbf{r})$ or $s_2(\mathbf{r})$ are not eigenstates of $\mathcal{H}^{sp}(\mathbf{r})$, and $\epsilon \neq \epsilon_0$. Also notice that we can write the total hamiltonian as

$$\mathcal{H}(\mathbf{r}_1, \mathbf{r}_2) = \mathcal{H}^{sp}(\mathbf{r}_1) + \mathcal{H}^{sp}(\mathbf{r}_2) + \frac{e^2}{|\mathbf{r}_1 - \mathbf{r}_2|} \tag{2.54}$$

We call the very last term in this expression, the electron–electron repulsion, the "interaction" term. It will prove convenient within the $s_1(\mathbf{r}_1)$, $s_2(\mathbf{r}_2)$ basis to define the so called "hopping" matrix elements

$$t \equiv -\langle s_1|\mathcal{H}^{sp}|s_2\rangle = -\langle s_2|\mathcal{H}^{sp}|s_1\rangle \tag{2.55}$$

where we can choose the phases in the wavefunctions $s_1(\mathbf{r})$, $s_2(\mathbf{r})$ to make sure that t is a real positive number. These matrix elements describe the probability of one electron "hopping" from state $s_1(\mathbf{r})$ to $s_2(\mathbf{r})$ (or vice versa), within the single-particle hamiltonian $\mathcal{H}^{sp}(\mathbf{r})$. A different term we can define is the "on-site" repulsive interaction between two electrons, which arises from the interaction term when the two electrons are placed at the same orbital:

$$U \equiv \langle s_1 s_1|\frac{e^2}{|\mathbf{r}_1 - \mathbf{r}_2|}|s_1 s_1\rangle = \langle s_2 s_2|\frac{e^2}{|\mathbf{r}_1 - \mathbf{r}_2|}|s_2 s_2\rangle \tag{2.56}$$

where U is also taken to be a real positive quantity. A model based on these physical quantities, the hopping matrix element and the on-site Coulomb repulsion energy, was introduced originally by Hubbard [18–20]. The model contains the bare essentials for describing electron–electron interactions in solids, and has found many applications, especially in highly correlated electron systems. Despite its apparent simplicity, the Hubbard model has not been solved analytically, and research continues today to try to understand its physics.

Now we want to construct single-particle orbitals for $\mathcal{H}^{sp}(\mathbf{r})$, using as a basis $s_1(\mathbf{r})$ and $s_2(\mathbf{r})$, which reflect the basic symmetry of the hamiltonian, that is, inversion relative to the midpoint of the distance between the two protons (the center of the molecule). There are two such possibilities:

$$\phi_0(\mathbf{r}) = \frac{1}{\sqrt{2}}[s_1(\mathbf{r}) + s_2(\mathbf{r})]$$
$$\phi_1(\mathbf{r}) = \frac{1}{\sqrt{2}}[s_1(\mathbf{r}) - s_2(\mathbf{r})] \tag{2.57}$$

the first being a symmetric and the second an antisymmetric wavefunction, upon inversion with respect to the center of the molecule; these are illustrated in Fig. 2.3.

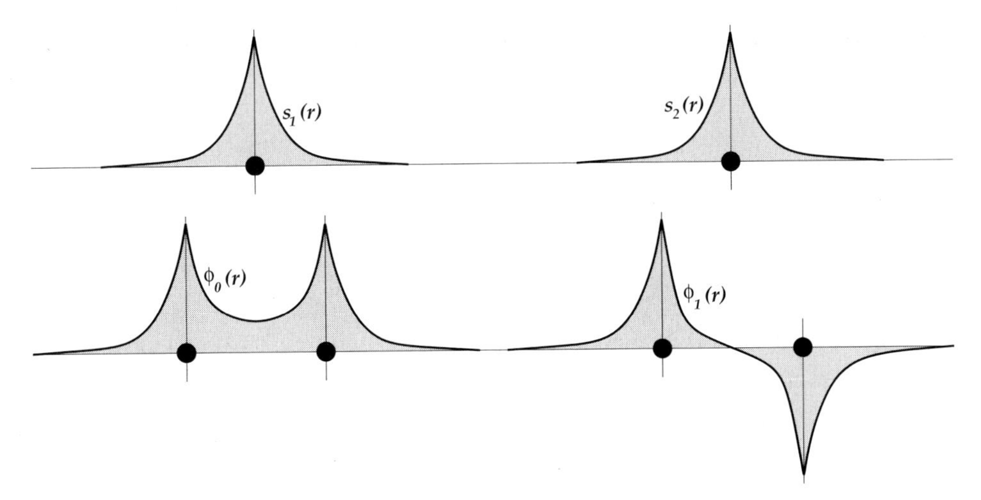

Figure 2.3. Schematic representation of the hydrogen wavefunctions for isolated atoms (top panel), and the linear combinations that preserve the inversion symmetry with respect to the center of the molecule: a symmetric combination (bottom panel, left) and an antisymmetric combination (bottom panel, right). The latter two are the states defined in Eq. (2.57).

The expectation values of $\mathcal{H}^{sp}$ in terms of the $\phi_i (i = 0, 1)$ are

$$\begin{aligned} \langle\phi_0|\mathcal{H}^{sp}(\mathbf{r})|\phi_0\rangle &= \epsilon - t \\ \langle\phi_1|\mathcal{H}^{sp}(\mathbf{r})|\phi_1\rangle &= \epsilon + t \end{aligned} \tag{2.58}$$

Using the single-particle orbitals of Eq. (2.57), we can write three possible Hartree wavefunctions:

$$\Psi_0^H(\mathbf{r}_1, \mathbf{r}_2) = \phi_0(\mathbf{r}_1)\phi_0(\mathbf{r}_2) \tag{2.59}$$

$$\Psi_1^H(\mathbf{r}_1, \mathbf{r}_2) = \phi_1(\mathbf{r}_1)\phi_1(\mathbf{r}_2) \tag{2.60}$$

$$\Psi_2^H(\mathbf{r}_1, \mathbf{r}_2) = \phi_0(\mathbf{r}_1)\phi_1(\mathbf{r}_2) \tag{2.61}$$

Notice that both $\Psi_0^H(\mathbf{r}_1, \mathbf{r}_2)$ and $\Psi_1^H(\mathbf{r}_1, \mathbf{r}_2)$ place the two electrons in the same state; this is allowed because of the electron spin. The expectation values of the two-particle hamiltonian $\mathcal{H}(\mathbf{r}_1, \mathbf{r}_2)$ which contains the interaction term, in terms of the $\Psi_i^H (i = 0, 1)$, are

$$\langle\Psi_0^H|\mathcal{H}(\mathbf{r}_1, \mathbf{r}_2)|\Psi_0^H\rangle = 2(\epsilon - t) + \frac{1}{2}U \tag{2.62}$$

$$\langle\Psi_1^H|\mathcal{H}(\mathbf{r}_1, \mathbf{r}_2)|\Psi_1^H\rangle = 2(\epsilon + t) + \frac{1}{2}U \tag{2.63}$$

Next we try to construct the Hartree–Fock approximation for this problem. We will assume that the total wavefunction has a spin-singlet part, that is, the spin degrees of freedom of the electrons form a totally antisymmetric combination, which multiplies the spatial part of the wavefunction. This tells us that the spatial part of the wavefunction should be totally symmetric. One possible choice is

$$\Psi_0^{HF}(\mathbf{r}_1, \mathbf{r}_2) = \frac{1}{\sqrt{2}} [s_1(\mathbf{r}_1)s_2(\mathbf{r}_2) + s_1(\mathbf{r}_2)s_2(\mathbf{r}_1)] \tag{2.64}$$

The wavefunction $\Psi_0^{HF}(\mathbf{r}_1, \mathbf{r}_2)$ is known as the Heitler–London approximation. Two other possible choices for totally symmetric spatial wavefunctions are

$$\Psi_1^{HF}(\mathbf{r}_1, \mathbf{r}_2) = s_1(\mathbf{r}_1)s_1(\mathbf{r}_2) \tag{2.65}$$

$$\Psi_2^{HF}(\mathbf{r}_1, \mathbf{r}_2) = s_2(\mathbf{r}_1)s_2(\mathbf{r}_2) \tag{2.66}$$

Using the three functions $\Psi_i^{HF} (i = 0, 1, 2)$, we can construct matrix elements of the hamiltonian $\mathcal{H}_{ij} = \langle \Psi_i^{HF} | \mathcal{H} | \Psi_j^{HF} \rangle$ and we can diagonalize this 3×3 matrix to find its eigenvalues and eigenstates. This exercise shows that the ground state energy is

$$E_{GS} = 2\epsilon + \frac{1}{2}U - \sqrt{4t^2 + \frac{1}{4}U^2} \tag{2.67}$$

and that the corresponding wavefunction is

$$\Psi_{GS}(\mathbf{r}_1, \mathbf{r}_2) = \frac{1}{\sqrt{2}\mathcal{N}} \Psi_0^{HF}(\mathbf{r}_1, \mathbf{r}_2) + \frac{1}{2\mathcal{N}} \left[\sqrt{1 + \left(\frac{U}{4t}\right)^2} - \frac{U}{4t} \right] \left(\Psi_1^{HF}(\mathbf{r}_1, \mathbf{r}_2) + \Psi_2^{HF}(\mathbf{r}_1, \mathbf{r}_2) \right) \tag{2.68}$$

where $\mathcal{N}$ is a normalization constant. To the extent that $\Psi_{GS}(\mathbf{r}_1, \mathbf{r}_2)$ involves several Hartree–Fock type wavefunctions, and the corresponding energy is lower than all other approximations we tried, this represents the optimal solution to the problem within our choice of basis, including correlation effects. A study of the ground state as a function of the parameter (U/t) elucidates the effects of correlation between the electrons in this simple model (see Problem 5). A more accurate description should include the excited states of electrons in each atom, which increases significantly the size of the matrices involved. Extending this picture to more complex systems produces an almost exponential increase of the computational difficulty.

2.5 Density Functional Theory

In a series of seminal papers, Hohenberg, Kohn and Sham developed a different way of looking at the problem, which has been called Density Functional

Theory (DFT). The basic ideas of Density Functional Theory are contained in the two original papers of Hohenberg, Kohn and Sham, [22, 23] and are referred to as the Hohenberg–Kohn–Sham theorem. This theory has had a tremendous impact on realistic calculations of the properties of molecules and solids, and its applications to different problems continue to expand. A measure of its importance and success is that its main developer, W. Kohn (a theoretical physicist) shared the 1998 Nobel prize for Chemistry with J.A. Pople (a computational chemist). We will review here the essential ideas behind Density Functional Theory.

The basic concept is that instead of dealing with the many-body Schrödinger equation, Eq. (2.1), which involves the many-body wavefunction $\Psi(\{\mathbf{r}_i\})$, one deals with a formulation of the problem that involves the total density of electrons $n(\mathbf{r})$. This is a huge simplification, since the many-body wavefunction need never be explicitly specified, as was done in the Hartree and Hartree–Fock approximations. Thus, instead of starting with a drastic approximation for the behavior of the system (which is what the Hartree and Hartree–Fock wavefunctions represent), one can develop the appropriate single-particle equations in an exact manner, and then introduce approximations as needed.

In the following discussion we will make use of the density $n(\mathbf{r})$ and the one-particle and two-particle density matrices, denoted by $\gamma(\mathbf{r}, \mathbf{r}')$, $\Gamma(\mathbf{r}, \mathbf{r}'|\mathbf{r}, \mathbf{r}')$, respectively, as expressed through the many-body wavefunction:

$$n(\mathbf{r}) = N \int \Psi^*(\mathbf{r}, \ldots, \mathbf{r}_N)\Psi(\mathbf{r}, \ldots, \mathbf{r}_N) d\mathbf{r}_2 \cdots d\mathbf{r}_N$$

$$\gamma(\mathbf{r}, \mathbf{r}') = N \int \Psi^*(\mathbf{r}, \mathbf{r}_2, \ldots, \mathbf{r}_N)\Psi(\mathbf{r}', \mathbf{r}_2, \ldots, \mathbf{r}_N) d\mathbf{r}_2 \cdots d\mathbf{r}_N$$

$$\Gamma(\mathbf{r}, \mathbf{r}'|\mathbf{r}, \mathbf{r}') = \frac{N(N-1)}{2} \int \Psi^*(\mathbf{r}, \mathbf{r}', \mathbf{r}_3, \ldots, \mathbf{r}_N)\Psi(\mathbf{r}, \mathbf{r}', \mathbf{r}_3, \ldots, \mathbf{r}_N) d\mathbf{r}_3 \cdots d\mathbf{r}_N$$

These quantities are defined in detail in Appendix B, Eqs. (B.13)–(B.15), where their physical meaning is also discussed.

First, we will show that the density $n(\mathbf{r})$ is uniquely defined given an external potential $V(\mathbf{r})$ for the electrons (this of course is identified with the ionic potential). To prove this, suppose that two different external potentials, $V(\mathbf{r})$ and $V'(\mathbf{r})$, give rise to the same density $n(\mathbf{r})$. We will show that this is impossible. We assume that $V(\mathbf{r})$ and $V'(\mathbf{r})$ are different in a non-trivial way, that is, they do not differ merely by a constant. Let E and Ψ be the total energy and wavefunction and E' and Ψ' be the total energy and wavefunction for the systems with hamiltonians $\mathcal{H}$ and $\mathcal{H}'$, respectively, where the first hamiltonian contains $V(\mathbf{r})$ and the second $V'(\mathbf{r})$ as external potential:

$$E = \langle\Psi|\mathcal{H}|\Psi\rangle \tag{2.69}$$

$$E' = \langle\Psi'|\mathcal{H}|\Psi'\rangle \tag{2.70}$$

Then we will have, by the variational principle,

$$E < \langle \Psi' | \mathcal{H} | \Psi' \rangle = \langle \Psi' | \mathcal{H} + V' - V' | \Psi' \rangle = \langle \Psi' | \mathcal{H}' + V - V' | \Psi' \rangle$$
$$= \langle \Psi' | \mathcal{H}' | \Psi' \rangle + \langle \Psi' | (V - V') | \Psi' \rangle$$
$$= E' + \langle \Psi' | (V - V') | \Psi' \rangle \qquad (2.71)$$

where the strict inequality is a consequence of the fact that the two potentials are different in a non-trivial way. Similarly we can prove

$$E' < E - \langle \Psi | (V - V') | \Psi \rangle \qquad (2.72)$$

Adding Eqs. (2.71) and (2.72), we obtain

$$(E + E') < (E + E') + \langle \Psi' | (V - V') | \Psi' \rangle - \langle \Psi | (V - V') | \Psi \rangle \qquad (2.73)$$

But the last two terms on the right-hand side of Eq. (2.73) give

$$\int n'(\mathbf{r})[V(\mathbf{r}) - V'(\mathbf{r})]\mathrm{d}\mathbf{r} - \int n(\mathbf{r})[V(\mathbf{r}) - V'(\mathbf{r})]\mathrm{d}\mathbf{r} = 0 \qquad (2.74)$$

because by assumption the densities $n(\mathbf{r})$ and $n'(\mathbf{r})$ corresponding to the two potentials are the same. This leads to the relation $E + E' < E + E'$, which is obviously wrong; therefore we conclude that our assumption about the densities being the same cannot be correct. This proves that there is a one-to-one correspondence between an external potential $V(\mathbf{r})$ and the density $n(\mathbf{r})$. But the external potential determines the wavefunction, so that the wavefunction must be a unique functional of the density. If we denote as $T + W$ the terms in the hamiltonian other than V, with T representing the kinetic energy and W the electron–electron interaction, we conclude that the expression

$$F[n(\mathbf{r})] = \langle \Psi | (T + W) | \Psi \rangle \qquad (2.75)$$

must be a universal functional of the density, since the terms T and W, the kinetic energy and electron–electron interactions, are common to all solids, and therefore this functional does not depend on anything else other than the electron density (which is determined uniquely by the external potential V that differs from system to system).

From these considerations we conclude that the total energy of the system is a functional of the density, and is given by

$$E[n(\mathbf{r})] = \langle \Psi | \mathcal{H} | \Psi \rangle = F[n(\mathbf{r})] + \int V(\mathbf{r}) n(\mathbf{r}) \mathrm{d}\mathbf{r} \qquad (2.76)$$

From the variational principle we can deduce that this functional attains its minimum for the correct density $n(\mathbf{r})$ corresponding to $V(\mathbf{r})$, since for a given $V(\mathbf{r})$ and any other density $n'(\mathbf{r})$ we would have

$$E[n'(\mathbf{r})] = \langle \Psi' | \mathcal{H} | \Psi' \rangle = F[n'(\mathbf{r})] + \int V(\mathbf{r}) n'(\mathbf{r}) d\mathbf{r} > \langle \Psi | \mathcal{H} | \Psi \rangle = E[n(\mathbf{r})] \tag{2.77}$$

Using our earlier expressions for the one-particle and the two-particle density matrices, we can obtain explicit expressions for $E[n]$ and $F[n]$:

$$E[n(\mathbf{r})] = \langle \Psi | \mathcal{H} | \Psi \rangle = -\frac{\hbar^2}{2m_e} \int \nabla_{\mathbf{r}'}^2 \gamma(\mathbf{r}, \mathbf{r}')|_{\mathbf{r}'=\mathbf{r}} d\mathbf{r}$$
$$+ \int \int \frac{e^2}{| \mathbf{r} - \mathbf{r}' |} \Gamma(\mathbf{r}, \mathbf{r}' | \mathbf{r}, \mathbf{r}') d\mathbf{r} d\mathbf{r}' + \int V(\mathbf{r}) \gamma(\mathbf{r}, \mathbf{r}) d\mathbf{r} \tag{2.78}$$

Now we can attempt to reduce these expressions to a set of single-particle equations, as before. The important difference in the present case is that we do not have to interpret these single-particle states as corresponding to electrons. They represent fictitious fermionic particles with the only requirement that their density is identical to the density of the real electrons. These particles can be considered to be *non-interacting*: this is a very important aspect of the nature of the fictitious particles, which will allow us to simplify things considerably, since their behavior will not be complicated by interactions. The assumption that we are dealing with non-interacting particles can be exploited to express the many-body wavefunction $\Psi(\{\mathbf{r}_i\})$ in the form of a Slater determinant, as in Eq. (2.14). We can then express the various physical quantities in terms of the single-particle orbitals $\phi_i(\mathbf{r})$ that appear in the Slater determinant. We obtain

$$n(\mathbf{r}) = \sum_i |\phi_i(\mathbf{r})|^2 \tag{2.79}$$

$$\gamma(\mathbf{r}, \mathbf{r}') = \sum_i \phi_i^*(\mathbf{r}) \phi_i(\mathbf{r}') \tag{2.80}$$

$$\Gamma(\mathbf{r}, \mathbf{r}' | \mathbf{r}, \mathbf{r}') = \frac{1}{2} \left[n(\mathbf{r}) n(\mathbf{r}') - | \gamma(\mathbf{r}, \mathbf{r}') |^2 \right] \tag{2.81}$$

With the help of Eqs. (2.79)–(2.81), we can express the various terms in the energy functional, which take the form

$$F[n(\mathbf{r})] = T^S[n(\mathbf{r})] + \frac{e^2}{2} \int \int \frac{n(\mathbf{r}) n(\mathbf{r}')}{| \mathbf{r} - \mathbf{r}' |} d\mathbf{r} d\mathbf{r}' + E^{XC}[n(\mathbf{r})] \tag{2.82}$$

In this expression, the first term represents the kinetic energy of the states in the Slater determinant (hence the superscript S). Since the fictitious particles are non-interacting, we can take the kinetic energy to be given by

$$T^S[n(\mathbf{r})] = \sum_i \langle \phi_i | -\frac{\hbar^2}{2m_e} \nabla_{\mathbf{r}}^2 | \phi_i \rangle \tag{2.83}$$

Notice that this expression could not be written down in a simple form had we been dealing with interacting particles. The second term in Eq. (2.82) is the Coulomb interaction, which we separate out from the electron-electron interaction term in the functional $F[n(\mathbf{r})]$. What remains is, by definition, the "exchange-correlation" term, $E^{XC}[n(\mathbf{r})]$. This term includes all the effects of the many-body character of the true electron system; we will deal with it separately below. We can now consider a variation in the density, which we choose to be

$$\delta n(\mathbf{r}) = \delta\phi_i^*(\mathbf{r})\phi_i(\mathbf{r}) \tag{2.84}$$

with the restriction that

$$\int \delta n(\mathbf{r})\mathrm{d}\mathbf{r} = \int \delta\phi_i^*(\mathbf{r})\phi_i(\mathbf{r})\mathrm{d}\mathbf{r} = 0 \tag{2.85}$$

so that the total number of particles does not change; note that $\phi_i(\mathbf{r})$ and $\phi_i^*(\mathbf{r})$ are treated as independent, as far as their variation is concerned. With this choice, and taking the restriction into account through a Lagrange multiplier ϵ_i, we arrive at the following single-particle equations, through a variational argument:

$$\left[-\frac{\hbar^2}{2m_e}\nabla_{\mathbf{r}}^2 + V^{eff}(\mathbf{r}, n(\mathbf{r}))\right]\phi_i(\mathbf{r}) = \epsilon_i\phi_i(\mathbf{r}) \tag{2.86}$$

where the effective potential is given by

$$V^{eff}(\mathbf{r}, n(\mathbf{r})) = V(\mathbf{r}) + e^2\int \frac{n(\mathbf{r}')}{\mid \mathbf{r} - \mathbf{r}' \mid}\mathrm{d}\mathbf{r}' + \frac{\delta E^{XC}[n(\mathbf{r})]}{\delta n(\mathbf{r})} \tag{2.87}$$

with $V(\mathbf{r})$ the external potential due to the ions; the last term is the variational functional derivative[2] of the as yet unspecified functional $E^{XC}[n(\mathbf{r})]$. The single-particle equations Eq. (2.86) are referred to as Kohn–Sham equations and the single-particle orbitals $\phi_i(\mathbf{r})$ that are their solutions are called Kohn–Sham orbitals.

Since the effective potential is a function of the density, which is obtained from Eq. (2.79) and hence depends on all the single-particle states, we will need to solve these equations by iteration until we reach self-consistency. As mentioned earlier, this is not a significant problem. A more pressing issue is the exact form of $E^{XC}[n(\mathbf{r})]$ which is unknown. We can consider the simplest situation, in which the true electronic system is endowed with only one aspect of electron interactions (beyond Coulomb repulsion), that is, the exchange property. As we saw in the case of the Hartree–Fock approximation, which takes into account exchange explicitly, in a uniform system the contribution of exchange to the total energy is

$$E^X = -\frac{3}{4}\frac{e^2}{\pi}k_{\mathrm{F}}N \tag{2.88}$$

[2] For the definition of this term in the context of the present theory see Appendix G.

Since the total number of electrons in the system can be written as $N = \int n d\mathbf{r}$, we can write

$$E^X[n] = -\frac{3}{4}\frac{e^2}{\pi}\int k_F n d\mathbf{r} = -\frac{3}{4}e^2\left(\frac{3}{\pi}\right)^{1/3}\int [n]^{1/3} n d\mathbf{r} \tag{2.89}$$

We can now generalize this to situations where the density is not uniform, and assume that the same expression holds, obtaining:

$$E^X[n(\mathbf{r})] = \int \epsilon^X[n(\mathbf{r})]n(\mathbf{r})d\mathbf{r} \tag{2.90}$$

$$\epsilon^X[n(\mathbf{r})] = -\frac{3}{4}e^2\left(\frac{3}{\pi}\right)^{1/3}[n(\mathbf{r})]^{1/3} \tag{2.91}$$

This allows us to calculate the expression for $\delta E^{XC}[n]/\delta n$ in the case where we are considering only the exchange aspect of the many-body character. We obtain

$$\frac{\delta E^X[n(\mathbf{r})]}{\delta n(\mathbf{r})} = \frac{\partial}{\partial n(\mathbf{r})}\left[\epsilon^X[n(\mathbf{r})]n(\mathbf{r})\right] = \frac{4}{3}\epsilon^X[n(\mathbf{r})] = -e^2\left(\frac{3}{\pi}\right)^{1/3}[n(\mathbf{r})]^{1/3} \tag{2.92}$$

This is remarkably similar to Slater's exchange potential, Eq. (2.48), which was based on an *ad hoc* assumption. In fact it differs from the Slater exchange potential by only a factor of 2/3.

This analysis shows that, if electrons interacted only through the Pauli exclusion principle (hence only the exchange interaction would be needed), we could adopt the exchange potential as derived above for use in the single-particle equations; this would provide an exact solution in terms of the single-particle wavefunctions $\phi_i(\mathbf{r})$. Recall that these are not true electron wavefunctions, but they give rise to exactly the same density as the true electrons. Since the density is the same, we could calculate from the energy functional the total energy of the system. However, we know that electrons are not interacting merely through the exclusion principle, but that they also experience the long-range Coulomb repulsion from each other, as well as the Coulomb attraction to the ions, represented by the external potential $V(\mathbf{r})$; the latter does not complicate the many-body character of the wavefunction Ψ. One part of the Coulomb interaction is included in the effective potential, the second term in Eq. (2.87). The electrons themselves, being interacting particles, experience the effects of this long range interaction since the motion of one affects the motion of each and every other one; this is known as the correlation aspect of the many-body wavefunction. Thus, we cannot assume that the full potential which the fictitious particles experience due to many-body effects, embodied in the last term of the effective potential in Eq. (2.87), can be described by the exchange part we have discussed so far. In an early attempt to account for this, Slater introduced

Table 2.1. *Correlation energy functionals $\epsilon^{cor}[n(\mathbf{r})]$ and exchange-correlation potentials $V^{XC}[n(\mathbf{r})]$ in various models.*

H–L = Hedin–Lundqvist [24], P–Z = Perdew–Zunger [25]. ϵ^X is the pure exchange energy from Eq. (2.91). r_s is measured in units of a_0 and the energy is in rydbergs. The numerical constants have units which depend on the factor of r_s involved with each.

Model	$\epsilon^{cor}[n(\mathbf{r})]$	$V^{XC}[n(\mathbf{r})]$
Exchange	0	$\frac{4}{3}\epsilon^X$
Slater	$(\frac{3}{2}\alpha - 1)\epsilon^X$	$2\alpha\epsilon^X$
Wigner	$A(B + r_s)^{-1}$	
	$A = 0.884, \quad B = 7.8$	
H–L		$\frac{4}{3}\epsilon^X\left[1 + Br_s \ln\left(1 + Ar_s^{-1}\right)\right]$
		$A = 21, \quad B = 0.0368$
P–Z: $r_s < 1$	$A_1 + A_2 r_s + [A_3 + A_4 r_s]\ln(r_s)$	
	$A_1 = -0.096, \quad A_2 = -0.0232$	
	$A_3 = 0.0622, \quad A_4 = 0.004$	
$r_s \geq 1$	$B_1\left[1 + B_2\sqrt{r_s} + B_3 r_s\right]^{-1}$	
	$B_1 = -0.2846, \quad B_2 = 1.0529$	
	$B_3 = 0.3334$	

a "fudge factor" in his expression for the exchange potential, denoted by α (hence the expression $X - \alpha$ potential). This factor is usually taken to be close to, but somewhat smaller than, 1 ($\alpha = 0.7$ is a typical choice). If $\alpha \neq 2/3$, the value required for the potential arising from pure exchange, it is thought that the $X - \alpha$ expression includes in some crude way the effects of both exchange and correlation. It is easy to extract the part that corresponds to correlation in the $X - \alpha$ expression, by comparing Slater's exchange potential, multiplied by the fudge factor α, to the potential involved in the single-particle equations derived from Density Functional Theory (see Table 2.1).

What should $E^{XC}[n(\mathbf{r})]$ be to capture all the many-body effects? This question is the holy grail in electronic structure calculations. So far no completely satisfactory answer has emerged. There are many interesting models, of which we mention a few so that the reader can get a feeling of what is typically involved, but the problem remains an area of active research. In fact it is not likely that any expression which depends on $n(\mathbf{r})$ in a *local* fashion will suffice, since the exchange and correlation effects are inherently non-local in an interacting electron system (see also the discussion below). A collection of proposed expressions for the correlation energies and exchange-correlation potentials is given in Table 2.1. In these expressions the exchange-correlation functional is written as

$$E^{XC}[n(\mathbf{r})] = \int \left(\epsilon^{X}[n(\mathbf{r})] + \epsilon^{cor}[n(\mathbf{r})]\right) n(\mathbf{r})d\mathbf{r} \tag{2.93}$$

and the exchange-correlation potential that appears in the single-particle equations is defined as

$$V^{XC}[n(\mathbf{r})] = \frac{\delta E^{XC}[n(\mathbf{r})]}{\delta n(\mathbf{r})} \tag{2.94}$$

while the pure exchange energy $\epsilon^{X}[n]$ is the expression given in Eq. (2.91). These expressions are usually given in terms of r_s, which is related to the density through Eq. (2.29). The expression proposed by Wigner extrapolates between known limits in r_s, obtained by series expansions (see Problem 7). The parameters that appear in the expression proposed by Hedin and Lundqvist [24] are determined by fitting to the energy of the uniform electron gas, obtained by numerical methods at different densities. A similar type of expression was proposed by Perdew and Zunger [25], which captures the more sophisticated numerical calculations for the uniform electron gas at different densities performed by Ceperley and Alder [26].

The common feature in all these approaches is that E^{XC} depends on $n(\mathbf{r})$ in a *local* fashion, that is, n needs to be evaluated at one point in space at a time. For this reason they are referred to as the Local Density Approximation to Density Functional Theory. This is actually a severe restriction, because even at the exchange level, the functional should be *non-local*, that is, it should depend on $\mathbf{r}$ and $\mathbf{r}'$ simultaneously (recall, for example, the expressions for the exchange density, $\rho_i^X(\mathbf{r}, \mathbf{r}')$, Eq. (2.20)). It is a much more difficult task to develop non-local exchange-correlation functionals. More recently, a concentrated effort has been directed toward producing expressions for E^{XC} that depend not only on the density $n(\mathbf{r})$, but also on its gradients [27]. These expansions tend to work better for finite systems (molecules, surfaces, etc.), but still represent a local approximation to the exchange-correlation functional. Including correlation effects in a realistic manner is exceedingly difficult, as we have already demonstrated for the hydrogen molecule.

2.6 Electrons as quasiparticles

So far we have examined how one can justify the reduction of the many-body equation Eq. (2.1) to a set of single-particle equations. This can be done by introducing certain approximations. In the case of the Hartree and Hartree–Fock approximations, one starts with a guess for the many-body wavefunction, expressed in terms of single-particle states. The resulting single-particle equations are supposed to describe the behavior of electrons as independent particles in an external potential defined by the ions, as well as an external field produced by the presence of all other electrons. In the case of DFT, one can derive exactly the single-particle equations for *non-interacting, fictitious* particles, whose density is the same as the density of real electrons. However, these equations cannot be solved exactly, because the exchange-correlation functional which appears in them is not known explicitly. One can construct approximations to this functional, by comparison with the results of numerical calculations for the electron gas. Once this functional has been determined (in an approximate way), the equations can be solved, and the wavefunctions of the fictitious particles can be determined.

What do the Kohn–Sham single-particle states actually represent? We know that they are not necessarily associated with electrons. A better interpretation is to consider them as *quasiparticles*. This general notion was first introduced by L.D. Landau. Landau's basic idea was that in a complicated system of *strongly* interacting particles, it may still be possible to describe the properties of the system in terms of *weakly* interacting particles, which represent some type of collective excitations of the original particles. Dealing with these weakly interacting particles is much simpler, so describing the system in this language can be very advantageous. In particular, some type of perturbation theory treatment can be formulated when one deals with a weakly interacting system of particles, beginning with the non-interacting particles as the unperturbed state. The fictitious non-interacting particles involved in the Kohn–Sham equations Eq. (2.86) possess some flavor of the quasiparticle notion: in the non-interacting Kohn-Sham picture all the important effects, including kinetic energy, Coulomb repulsion, exchange energy, and of course the ionic potential, are properly taken into account; correlation effects are then added approximately. Although this is not expressed as a systematic perturbative approach, it embodies certain features of such an approach because correlation effects are typically the weakest contribution to the total energy. It is quite extraordinary that many of the measurable properties of real systems can be identified with the behavior of the Kohn-Sham quasiparticles in a direct way. We will see some examples of this in later chapters.

Why is the single-particle approximation successful? We can cite the following three general arguments:

- **The variational principle**: Even when the wavefunctions are not accurate, as in the case of the Hartree–Fock approximation, the total energy is not all that bad. Energy differences between states of the system (corresponding to different atomic configurations, for which the single-particle equations are solved self-consistently each time), turn out to be remarkably good. This is because the optimal set of single-particle states contains most of the physics related to the motion of ions. The case of coherent many-body states for the electrons, where the single-particle approximation fails, concerns much more delicate phenomena in solids. For example, the phenomenon of superconductivity involves energy scales of a few degrees kelvin (or at most $\sim$ 100 K), whereas the motion of atoms in solids (involving changes in atomic positions) involves energies of the order of 1 eV = 11 604 K.

- **The exclusion principle**: The single-particle states are filled up to the Fermi level in order to satisfy the Pauli exclusion principle; these are the optimal states as far as the energy is concerned in the single-particle picture. In order to build a more accurate description of the system, we need to consider virtual excitations of this ground state and include their contribution to the total energy. Since it is the interaction between electrons that we want to capture more accurately, we should try to include in the correction to the total energy the effect of excitations through the Coulomb interaction. By its nature this interaction involves a pair of particles, therefore the lowest order non-vanishing terms in the correction must involve the excitation of two electrons (with energies $\epsilon_i, \epsilon_j > \epsilon_{\mathrm{F}}$) and two holes (with energies $\epsilon_n, \epsilon_m \leq \epsilon_{\mathrm{F}}$); the contribution of such excitations to the total energy will be

$$\Delta E \sim \frac{\left| \langle \phi_i \phi_j | \frac{e^2}{|\mathbf{r}_1 - \mathbf{r}_2|} | \phi_m \phi_n \rangle \right|^2}{\epsilon_i + \epsilon_j - \epsilon_m - \epsilon_n} \tag{2.95}$$

  However, the single-particle states are orthogonal, so that there is no overlap between such states: $\langle \phi_i | \phi_m \rangle = \langle \phi_j | \phi_n \rangle = 0$ (except when $i = m, j = n$ or $i = n, j = m$, which do not correspond to an excitation at all). This indicates that the matrix elements in the numerator of Eq. (2.95) are very small, and the corresponding corrections to the energy are very small. That is, in the single-particle picture, the energy of the system is reasonably well described and corrections to it tend to be small.
- **Screening**: The Coulomb interaction between real electrons is "screened" by the correlated motion of all other electrons: each electron is surrounded by a region where the density of electronic charge is depleted; this is referred to as the "exchange-correlation hole". This forms an effective positive charge cloud which screens an electron from all its neighbors. There is no real hole in the system; the term is just a figurative way of describing the many-body effects in the single-particle picture. The net effect is a weakened interaction between electrons; the weaker the interaction, the closer the system is to the picture of non-interacting single particles. These particles, however, can no longer be identified with individual electrons, since they carry with them the effects of interaction with all other electrons (the exchange-correlation hole). Thus, not only are the energies in the single-particle picture quite good, but the description of electronic behavior is also reasonable.

To illustrate the concept of screening and the exchange-correlation hole, we refer again to the simple example of free electrons in the Hartree–Fock approximation. In this picture, the potential experienced by each electron due to electron–electron interactions is given by Eq. (2.25), which, with plane waves as the single-particle states Eq. (2.26), becomes

$$V_{\mathbf{k}}^{HF}(\mathbf{r}) = e^2 \int \frac{(N/\Omega) - \rho_{\mathbf{k}}^{HF}(\mathbf{r}, \mathbf{r}')}{|\mathbf{r} - \mathbf{r}'|} d\mathbf{r}' \tag{2.96}$$

In this potential, the effect of exchange between electrons is expressed by the second term in the numerator of the integrand, which arises from the Hartree–Fock density defined in Eq. (2.24). For the present simple example, the Hartree–Fock density takes the form:

$$\rho_{\mathbf{k}}^{HF}(\mathbf{r}, \mathbf{r}') = \frac{1}{\Omega} \sum_{\mathbf{k}'} \mathrm{e}^{-\mathrm{i}(\mathbf{k}-\mathbf{k}')\cdot(\mathbf{r}-\mathbf{r}')} \tag{2.97}$$

With the help of Eqs. (D.8) and (G.64), we obtain

$$\rho_{\mathbf{k}}^{HF}(\mathbf{r}, \mathbf{r}') = \mathrm{e}^{-\mathrm{i}\mathbf{k}\cdot(\mathbf{r}-\mathbf{r}')} \delta(\mathbf{r} - \mathbf{r}') \tag{2.98}$$

that is, the electrons in this case experience an exchange hole which is a δ-function centered at the electron position. This is an extreme case of the exchange-correlation hole, in which correlation effects have been neglected and the hole has infinite strength and is located right on the electron itself with zero extent around it. In more realistic treatments, where the electron wavefunctions are not simply plane waves and correlation effects are taken into account, the exchange-correlation hole has finite strength and finite extent around the electron position.

2.6.1 Quasiparticles and collective excitations

There is another notion which is also very important for the description of the properties of solids, that of "collective excitations". In contrast to quasiparticles, these are bosons, they bear no resemblance to constituent particles of a real system, and they involve collective (that is, coherent) motion of many physical particles. We summarize here the most common quasiparticles and collective excitations encountered in solids:

(a) **Electron**: As we discussed above already, this is a quasiparticle consisting of a real electron and the exchange-correlation hole, a cloud of effective charge of opposite sign due to exchange and correlation effects arising from interaction with all other electrons in the system. The electron is a fermion with spin 1/2. The Fermi energy (highest occupied state) is of order 5 eV, and the Fermi velocity ($v_F = \hbar k_{\mathrm{F}}/m_e$) is $\sim 10^8$ cm/sec, that is, it can be treated as a non-relativistic particle. Notice that the mass of this quasiparticle can be different than that of the free electron.

(b) **Hole**: This is a quasiparticle, like the electron, but of opposite charge; it corresponds to the absence of an electron from a single-particle state which lies below the Fermi level. The notion of a hole is particularly convenient when the reference state consists of quasiparticle states that are fully occupied and are separated by an energy gap from the unoccupied states. Perturbations with respect to this reference state, such as missing electrons, can be conveniently discussed in terms of holes. This is, for example, the situation in p-doped semiconductor crystals.

(c) **Polaron**: This is a quasiparticle, like the electron, tied to a distortion of the lattice of ions. Polarons are invoked to describe polar crystals, where the motion of a negatively charged electron distorts the lattice of positive and negative ions around it. Because its motion is coupled to motion of ions, the polaron has a different mass than the electron.

(d) **Exciton**: This is a collective excitation, corresponding to a bound state of an electron and a hole. The binding energy is of order $e^2/(\varepsilon a)$, where ε is the dielectric constant of the material (typically of order 10) and a is the distance between the two quasiparticles (typically a few lattice constants, of order 10 Å), which give for the binding energy ~ 0.1 eV.

(e) **Phonon**: This is a collective excitation, corresponding to coherent motion of all the atoms in the solid. It is a quantized lattice vibration, with a typical energy scale of $\hbar\omega \sim 0.1$ eV.

(f) **Plasmon**: This is a collective excitation of the entire electron gas relative to the lattice of ions; its existence is a manifestation of the long-range nature of the Coulomb interaction. The energy scale of plasmons is $\hbar\omega \sim \hbar\sqrt{4\pi n e^2/m_e}$, where n is the density; for typical densities this gives an energy of order 5–20 eV.

(g) **Magnon**: This is a collective excitation of the spin degrees of freedom on the crystalline lattice. It corresponds to a spin wave, with an energy scale of $\hbar\omega \sim 0.001$–0.1 eV.

In the following chapters we will explore in detail the properties of the some of these quasiparticles and collective excitations.

2.6.2 Thomas–Fermi screening

To put a more quantitative expression to the idea of electrons as screened quasiparticles, consider the following simple model. First, suppose that the electrons are maximally efficient in screening each other's charge. This means that the Coulomb potential of an electron $(-e)/r$ would be quickly suppressed as we move away from the electron's position (here taken to be the origin of the coordinate system). The most efficient way to do this is to multiply the Coulomb potential of the bare electron by a decaying exponential:

$$\Phi^{scr}(\mathbf{r}) = \frac{(-e)}{r}\mathrm{e}^{-k_s r} \tag{2.99}$$

where k_s is the screening inverse length: the potential is negligible at distances larger than $\sim k_s^{-1}$. The response of a system of charged particles to an external field is typically described by the dielectric function, defined as

$$\varepsilon = \frac{\Phi^{ext}}{\Phi} = \frac{\Phi - \Phi^{ind}}{\Phi} = 1 - \frac{\Phi^{ind}}{\Phi} \tag{2.100}$$

where Φ^{ext} is the external field, Φ^{ind} is the field induced by the response of the charged particles, and $\Phi = \Phi^{ext} + \Phi^{ind}$ is the total field. This general definition holds both in real space (with the appropriate expressions in terms of integral equations) and in Fourier space. Now, if the bare Coulomb potential $\Phi^{ext}(\mathbf{r}) = (-e)/r$, whose Fourier transform is $(-e)4\pi/k^2$, is screened by an exponential factor $\mathrm{e}^{-k_s r}$ to become the total potential $\Phi(\mathbf{r}) = (-e)\mathrm{e}^{-k_s r}/r$, its Fourier transform becomes $\Phi(\mathbf{k}) = (-e)4\pi/(k^2 + k_s^2)$ (see Appendix G), and therefore the dielectric function takes the form

$$\varepsilon(\mathbf{k}) = \frac{\Phi^{ext}(\mathbf{k})}{\Phi(\mathbf{k})} = \frac{(-e)4\pi/k^2}{(-e)4\pi/(k^2 + k_s^2)} = 1 + \left(\frac{k_s}{k}\right)^2 \tag{2.101}$$

We can also define the induced charge $\rho^{ind}(\mathbf{r})$ through the Poisson equation:

$$-\nabla^2 \Phi^{ind}(\mathbf{r}) = 4\pi\rho^{ind}(\mathbf{r}) \Longrightarrow \Phi^{ind}(\mathbf{k}) = \frac{4\pi}{|\mathbf{k}|^2}\rho^{ind}(\mathbf{k}) \tag{2.102}$$

which gives for the dielectric function in Fourier space

$$\varepsilon(\mathbf{k}) = 1 - \frac{4\pi}{k^2}\frac{\rho^{ind}(\mathbf{k})}{\Phi(\mathbf{k})} \tag{2.103}$$

For sufficiently weak fields, we take the response of the system to be linear in the total field:

$$\rho^{ind}(\mathbf{k}) = \chi(\mathbf{k})\Phi(\mathbf{k}) \tag{2.104}$$

where the function $\chi(\mathbf{k})$ is the susceptibility or response function. This gives for the dielectric function

$$\varepsilon(\mathbf{k}) = 1 - \frac{4\pi}{k^2}\chi(\mathbf{k}) \tag{2.105}$$

As can be shown straightforwardly using perturbation theory (see Problem 8), for a system of single particles with energy $\epsilon_{\mathbf{k}} = \hbar^2 k^2/2m_e$ and Fermi occupation numbers

$$n(\mathbf{k}) = \frac{1}{\mathrm{e}^{(\epsilon_{\mathbf{k}} - \epsilon_{\mathrm{F}})/k_{\mathrm{B}}T} + 1} \tag{2.106}$$

where ϵ_{F} is the Fermi energy and T the temperature, the response function $\chi(\mathbf{k})$ takes the form

$$\chi(\mathbf{k}) = -e^2 \int \frac{\mathrm{d}\mathbf{k}'}{(2\pi)^3} \frac{n(\mathbf{k}' - \mathbf{k}/2) - n(\mathbf{k}' + \mathbf{k}/2)}{\hbar^2(\mathbf{k} \cdot \mathbf{k}')/2m_e} \tag{2.107}$$

This is called the Lindhard dielectric response function . Notice that, at $T = 0$, in order to have a non-vanishing integrand, one of the occupation numbers must correspond to a state below the Fermi level, and the other to a state above the

Fermi level; that is, we must have $|\mathbf{k}' - \mathbf{k}/2| < k_{\mathrm{F}}$ and $|\mathbf{k}' + \mathbf{k}/2| > k_{\mathrm{F}}$ (or vice versa), since otherwise the occupation numbers are either both 0 or both 1. These considerations indicate that the contributions to the integral will come from electron–hole excitations with total momentum $\mathbf{k}$. At $T = 0$, the integral over $\mathbf{k}'$ in the Lindhard dielectric function can be evaluated to yield

$$\chi_0(\mathbf{k}) = -e^2 \frac{m_e k_{\mathrm{F}}}{2\hbar^2 \pi^2} F(k/2k_{\mathrm{F}}) \tag{2.108}$$

where $F(x)$ is the same function as the one encountered earlier in the discussion of Hartree–Fock single-particle energies, defined in Eq. (2.39). The study of this function provides insight into the behavior of the system of single particles (see Problem 8).

A special limit of the Lindhard response function is known as the Thomas-Fermi screening. This is based on the limit of the Lindhard function for $|\mathbf{k}| \ll k_{\mathrm{F}}$. Consistent with our earlier remarks, this corresponds to electron–hole excitations with small total momentum relative to the Fermi momentum. In this case, expanding the occupation numbers $n(\mathbf{k}' \pm \mathbf{k}/2)$ through their definition in Eq. (2.106) about $\mathbf{k} = 0$, we obtain to first order in $\mathbf{k}$

$$n(\mathbf{k}' \pm \mathbf{k}/2) = n(\mathbf{k}') \mp \frac{\hbar^2}{2m_e}(\mathbf{k}' \cdot \mathbf{k}) \frac{\partial n(\mathbf{k}')}{\partial \epsilon_{\mathrm{F}}} \tag{2.109}$$

This expression, when substituted in the equation for the dielectric constant, gives

$$\varepsilon(\mathbf{k}) = 1 + \left(\frac{l_s^{TF}}{k} \right)^2, \qquad l_s^{TF}(T) = \sqrt{4\pi e^2 \frac{\partial n_T}{\partial \epsilon_{\mathrm{F}}}} \tag{2.110}$$

where l_s^{TF} is the so called Thomas–Fermi screening inverse length, which depends on the temperature T through the occupation n_T, defined as

$$n_T = 2 \int \frac{\mathrm{d}\mathbf{k}}{(2\pi)^3} \frac{1}{\mathrm{e}^{(\epsilon_{\mathbf{k}} - \epsilon_{\mathrm{F}})/k_{\mathrm{B}}T} + 1} \tag{2.111}$$

This quantity represents the full occupation of states in the ground state of the single-particle system at temperature T, and is evidently a function of ϵ_{F}.

By comparing the expression for the dielectric function in the Thomas–Fermi case, Eq. (2.110), with the general result of the exponentially screened potential, Eq. (2.101), we conclude that this approximation corresponds to an exponentially screened potential of the form given in Eq. (2.99) with the Thomas–Fermi screening inverse length l_s^{TF}.

2.7 The ionic potential

So far we have justified why it is possible to turn the many-body Schrödinger equation for electrons in a solid into a set of single-particle equations, which with the proper approximations can be solved to determine the single-particle wavefunctions. When actually solving the single-particle equations we need to specify the ionic potential. As we have emphasized before, we are only dealing with the valence electrons of atoms, while the core electrons are largely unaffected when the atoms are placed in the solid. The separation of the atomic electron density into its core and valence parts is shown in Fig. 2.4 for four elements of column IV of the Periodic Table, namely C, Si, Ge and Pb. It is clear from these figures that the contribution of the valence states to the total electron density is negligible within the core region and dominant beyond it. Because of this difference between valence and core electrons, a highly effective approach has been developed to separate the two sets of states. This approach, known as the *pseudopotential method*, allows us to take the core electrons out of the picture, and at the same time to create a smoother potential for the valence electrons; the work of Phillips and Kleinman first established the theoretical basis of the pseudopotential method [28]. Since this approach is one of the pillars of modern electronic structure theory, we will describe its main ideas here.

In order to develop the pseudopotential for a specific atom we consider it as isolated, and denote by $|\psi^{(n)}\rangle$ the single-particle states which are the solutions of the single-particle equations discussed earlier, as they apply to the case of an isolated atom. In principle, we need to calculate these states for all the electrons of the atom, using as an external potential that of its nucleus. Let us separate explicitly the single-particle states into valence and core sets, identified as $|\psi^{(v)}\rangle$ and $|\psi^{(c)}\rangle$ respectively. These satisfy the Schrödinger type equations

$$\mathcal{H}^{sp}|\psi^{(v)}\rangle = \epsilon^{(v)}|\psi^{(v)}\rangle \tag{2.112}$$

$$\mathcal{H}^{sp}|\psi^{(c)}\rangle = \epsilon^{(c)}|\psi^{(c)}\rangle \tag{2.113}$$

where $\mathcal{H}^{sp}$ is the appropriate single-particle hamiltonian for the atom: it contains a potential V^{sp} which includes the external potential due to the nucleus, as well as all the other terms arising from electron-electron interactions. Now let us define a new set of single-particle valence states $|\tilde{\phi}^{(v)}\rangle$ through the following relation:

$$|\psi^{(v)}\rangle = |\tilde{\phi}^{(v)}\rangle - \sum_c \langle\psi^{(c)}|\tilde{\phi}^{(v)}\rangle|\psi^{(c)}\rangle \tag{2.114}$$

Applying the single-particle hamiltonian $\mathcal{H}^{sp}$ to this equation, we obtain

$$\mathcal{H}^{sp}|\tilde{\phi}^{(v)}\rangle - \sum_c \langle\psi^{(c)}|\tilde{\phi}^{(v)}\rangle\mathcal{H}^{sp}|\psi^{(c)}\rangle = \epsilon^{(v)}\left[|\tilde{\phi}^{(v)}\rangle - \sum_c \langle\psi^{(c)}|\tilde{\phi}^{(v)}\rangle|\psi^{(c)}\rangle\right] \tag{2.115}$$

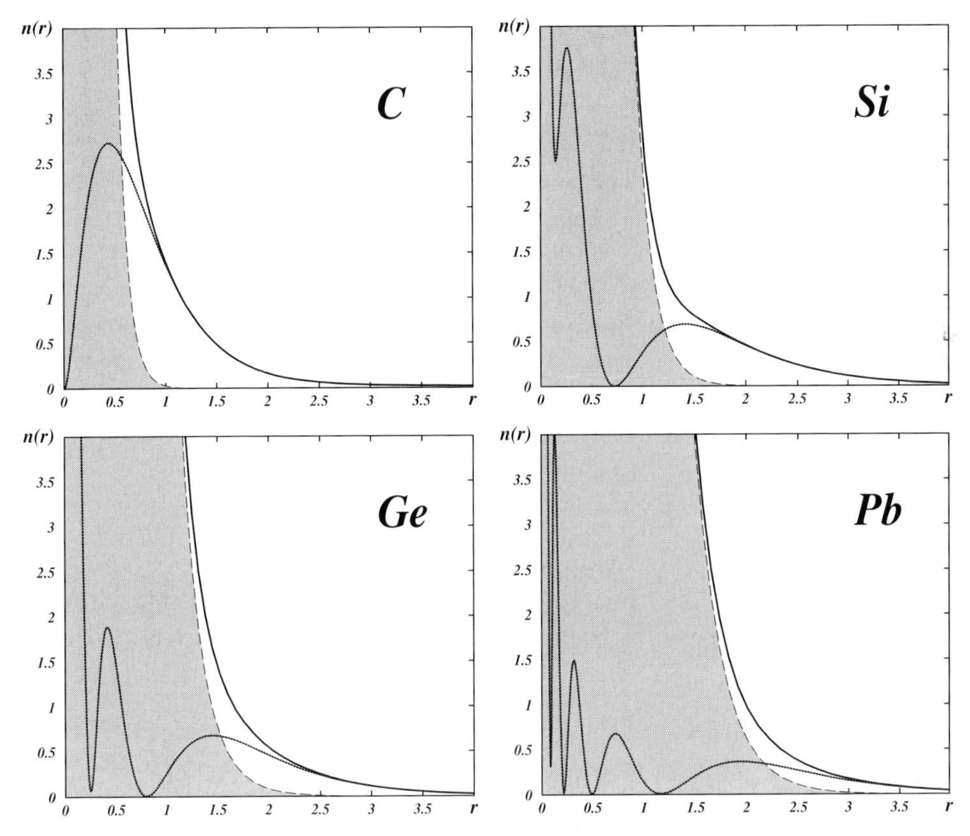

Figure 2.4. Electron densities $n(r)$ as a function of the radial distance from the nucleus r in angstroms, for four elements of column IV of the Periodic Table: C ($Z = 6$, $[1s^2]2s^22p^2$), Si ($Z = 14$, $[1s^22s^22p^6]3s^23p^2$), Ge ($Z = 32$, $[1s^22s^22p^63s^23p^63d^{10}]4s^24p^2$) and Pb ($Z = 82$, $[1s^22s^22p^63s^23p^63d^{10}4s^24p^64d^{10}4f^{14}5s^25p^65d^{10}]6s^26p^2$); the core states are given inside square brackets. In each case, the dashed line with the shaded area underneath it represents the density of core electrons, while the solid line represents the density of valence electrons and the total density of electrons (core plus valence). The core electron density for C is confined approximately below 1.0 Å, for Si below 1.5 Å, for Ge below 2.0 Å, and for Pb below 2.5 Å. In all cases the valence electron density extends well beyond the range of the core electron density and is relatively small within the core. The wiggles that develop in the valence electron densities for Si, Ge and Pb are due to the nodes of the corresponding wavefunctions, which acquire oscillations in order to become orthogonal to core states.

which, taking into account that $\mathcal{H}^{sp}|\psi^{(c)}\rangle = \epsilon^{(c)}|\psi^{(c)}\rangle$, gives

$$\left[\mathcal{H}^{sp} - \sum_c \epsilon^{(c)}|\psi^{(c)}\rangle\langle\psi^{(c)}|\right]|\tilde{\phi}^{(v)}\rangle = \epsilon^{(v)}\left[1 - \sum_c |\psi^{(c)}\rangle\langle\psi^{(c)}|\right]|\tilde{\phi}^{(v)}\rangle$$

$$\Rightarrow \left[\mathcal{H}^{sp} + \sum_c (\epsilon^{(v)} - \epsilon^{(c)})|\psi^{(c)}\rangle\langle\psi^{(c)}|\right]|\tilde{\phi}^{(v)}\rangle = \epsilon^{(v)}|\tilde{\phi}^{(v)}\rangle \tag{2.116}$$

Therefore, the new states $|\tilde{\phi}^{(v)}\rangle$ obey a single-particle equation with a modified potential, but have the same eigenvalues $\epsilon^{(v)}$ as the original valence states $|\psi^{(v)}\rangle$. The modified potential for these states is called the "pseudopotential", given by

$$V^{ps} = V^{sp} + \sum_c (\epsilon^{(v)} - \epsilon^{(c)})|\psi^{(c)}\rangle\langle\psi^{(c)}| \tag{2.117}$$

and, correspondingly, the $|\tilde{\phi}^{(v)}\rangle$'s are called "pseudo-wavefunctions".

Why is this a useful approach? First, consider the definition of the pseudo-wavefunctions through Eq. (2.114): what this definition amounts to is projecting out of the valence wavefunctions any overlap they have with the core wavefunctions. In fact, the quantity

$$\sum_c |\psi^{(c)}\rangle\langle\psi^{(c)}| \tag{2.118}$$

is a projection operator that achieves exactly this result. So the new valence states defined through Eq. (2.114) have zero overlap with core states, but they have the same eigenvalues as the original valence states. Moreover, the potential that these states experience includes, in addition to the regular potential V^{sp}, the term

$$\sum_c (\epsilon^{(v)} - \epsilon^{(c)})|\psi^{(c)}\rangle\langle\psi^{(c)}| \tag{2.119}$$

which is strictly positive, because $\epsilon^{(v)} > \epsilon^{(c)}$ (valence states have by definition higher energy than core states). Thus, this term is repulsive and tends to push the corresponding states $|\tilde{\phi}^{(v)}\rangle$ outside the core. In this sense, the pseudopotential represents the effective potential that valence electrons feel, if the only effect of core electrons were to repel them from the core region. Therefore the pseudo-wavefunctions experience an attractive Coulomb potential which is shielded near the position of the nucleus by the core electrons, so it should be a much smoother potential without the $1/r$ singularity due to the nucleus at the origin. Farther away from the core region, where the core states die exponentially, the potential that the pseudo-wavefunctions experience is the same as the Coulomb potential of an ion, consisting of the nucleus plus the core electrons. In other words, through the pseudopotential formulation we have created a new set of valence states, which experience a weaker potential near the atomic nucleus, but the proper ionic potential away from the core region. Since it is this region in which the valence electrons interact to form bonds that hold the solid together, the pseudo-wavefunctions preserve all the important physics relevant to the behavior of the solid. The fact that they also have exactly the same eigenvalues as the original valence states, also indicates that they faithfully reproduce the behavior of true valence states.

There are some aspects of the pseudopotential, at least in the way that was formulated above, that make it somewhat suspicious. First, it is a non-local potential:

applying it to the state $|\tilde{\phi}^{(v)}\rangle$ gives

$$\sum_c (\epsilon^{(v)} - \epsilon^{(c)})|\psi^{(c)}\rangle\langle\psi^{(c)}|\tilde{\phi}^{(v)}\rangle = \int V^{ps}(\mathbf{r}, \mathbf{r}')\tilde{\phi}^{(v)}(\mathbf{r}')\mathrm{d}\mathbf{r}'$$

$$\Longrightarrow V^{ps}(\mathbf{r}, \mathbf{r}') = \sum_c (\epsilon^{(v)} - \epsilon^{(c)})\psi^{(c)*}(\mathbf{r}')\psi^{(c)}(\mathbf{r}) \qquad (2.120)$$

This certainly complicates things. The pseudopotential also depends on the energy $\epsilon^{(v)}$, as the above relationship demonstrates, which is an unknown quantity if we view Eq. (2.116) as the Schrödinger equation that determines the pseudo-wavefunctions $|\tilde{\phi}^{(v)}\rangle$ and their eigenvalues. Finally, the pseudopotential is not unique. This can be demonstrated by adding any linear combination of $|\psi^{(c)}\rangle$ states to $|\tilde{\phi}^{(v)}\rangle$ to obtain a new state $|\hat{\phi}^{(v)}\rangle$:

$$|\hat{\phi}^{(v)}\rangle = |\tilde{\phi}^{(v)}\rangle + \sum_{c'} \alpha_{c'}|\psi^{(c')}\rangle \qquad (2.121)$$

where α_c are numerical constants. Using $|\tilde{\phi}^{(v)}\rangle = |\hat{\phi}^v\rangle - \sum_{c'} \alpha_{c'}|\psi^{(c')}\rangle$ in Eq. (2.116), we obtain

$$\left[\mathcal{H}^{sp} + \sum_c (\epsilon^{(v)} - \epsilon^{(c)})|\psi^{(c)}\rangle\langle\psi^{(c)}|\right]\left[|\hat{\phi}^{(v)}\rangle - \sum_{c'} \alpha_{c'}|\psi^{(c')}\rangle\right]$$
$$= \epsilon^{(v)}\left[|\hat{\phi}^{(v)}\rangle - \sum_{c'} \alpha_{c'}|\psi^{(c')}\rangle\right]$$

We can now use $\langle\psi^{(c)}|\psi^{(c')}\rangle = \delta_{cc'}$ to reduce the double sum $\sum_c \sum_{c'}$ on the left hand side of this equation to a single sum, and eliminate common terms from both sides to arrive at

$$\left[\mathcal{H}^{sp} + \sum_c (\epsilon^{(v)} - \epsilon^{(c)})|\psi^{(c)}\rangle\langle\psi^{(c)}|\right]|\hat{\phi}^{(v)}\rangle = \epsilon^{(v)}|\hat{\phi}^{(v)}\rangle \qquad (2.122)$$

This shows that the state $|\hat{\phi}^{(v)}\rangle$ obeys exactly the same single-particle equation as the state $|\tilde{\phi}^{(v)}\rangle$, which means it is not uniquely defined, and therefore the pseudopotential is not uniquely defined. All these features may cast a long shadow of doubt on the validity of the pseudopotential construction in the mind of the skeptic (a trait not uncommon among physicists). Practice of this art, however, has shown that these features can actually be exploited to define pseudopotentials that work very well in reproducing the behavior of the valence wavefunctions in the regions outside the core, which are precisely the regions of interest for the physics of solids.

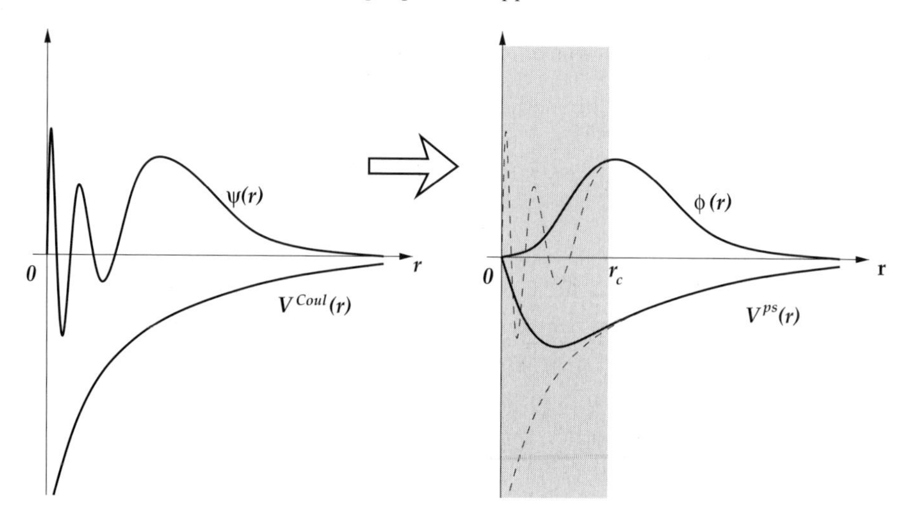

Figure 2.5. Schematic representation of the construction of the pseudo-wavefunction $\phi(r)$ and pseudopotential $V^{ps}(r)$, beginning with the real valence wavefunction $\psi(r)$ and Coulomb potential $V^{Coul}(r)$; r_c is the cutoff radius beyond which the wavefunction and potential are not affected.

As an example, we discuss next how typical pseudopotentials are constructed for modern calculations of the properties of solids [29]. The entire procedure is illustrated schematically in Fig. 2.6. We begin with a self-consistent solution of the single-particle equations for all the electrons in an atom (core and valence). For each valence state of interest, we take the calculated radial wavefunction and keep the tail starting at some point slightly before the last extremum. When atoms are placed at usual interatomic distances in a solid, these valence tails overlap significantly, and the resulting interaction between the corresponding electrons produces binding between the atoms. We want therefore to keep this part of the valence wavefunction as realistic as possible, and we identify it with the tail of the calculated atomic wavefunction. We call the radial distance beyond which this tail extends the "cutoff radius" r_c, so that the region $r < r_c$ corresponds to the core. Inside the core, the behavior of the wavefunction is not as important for the properties of the solid. Therefore, we can construct the pseudo-wavefunction to be a smooth function which has no nodes and goes to zero at the origin, as shown in Fig. 2.5. We can achieve this by taking some combination of smooth functions which we can fit to match the true wavefunction and its first and second derivative at r_c, and approach smoothly to zero at the origin. This hypothetical wavefunction must be normalized properly. Having defined the pseudo-wavefunction, we can invert the Schrödinger equation to obtain the potential which would produce such a wavefunction. This is by definition the desired pseudopotential: it is guaranteed by construction to produce a wavefunction which matches exactly the real atomic

Solve $H^{sp}\psi^{(v)}(r) = [\hat{F} + V^{Coul}(r)]\psi^{(v)}(r) = \epsilon^{(v)}\psi^{(v)}(r)$

↓

Fix pseudo-wavefunction $\phi^{(v)}(r) = \psi^{(v)}(r)$ for $r \geq r_c$

↓

Construct $\phi^{(v)}(r)$ for $0 \leq r < r_c$, under the following conditions:

$\phi^{(v)}(r)$ smooth, nodeless; $\mathrm{d}\phi^{(v)}/\mathrm{d}r$, $\mathrm{d}^2\phi^{(v)}/\mathrm{d}r^2$ continuous at r_c

↓

Normalize pseudo-wavefunction $\phi^{(v)}(r)$ for $0 \leq r < \infty$

↓

Invert $[\hat{F} + V^{ps}(r)]\phi^{(v)}(r) = \epsilon^{(v)}\phi^{(v)}(r)$

↓

$V^{ps}(r) = \epsilon^{(v)} - [\hat{F}\phi^{(v)}(r)]/\phi^{(v)}(r)]$

Figure 2.6. The basic steps in constructing a pseudopotential. $\hat{F}$ is the operator in the single-particle hamiltonian $\mathcal{H}^{sp}$ that contains all other terms except the ionic (external) potential, that is, $\hat{F}$ consists of the kinetic energy operator, the Hartree potential term and the exchange-correlation term. V^{Coul} and V^{ps} are the Coulomb potential and the pseudopotential of the ion.

wavefunction beyond the core region ($r > r_c$), and is smooth and nodeless inside the core region, giving rise to a smooth potential. We can then use this pseudopotential as the appropriate potential for the valence electrons in the solid.

We note here two important points:

(i) The pseudo-wavefunctions can be chosen to be nodeless inside the core, due to the non-uniqueness in the definition of the pseudopotential and the fact that their behavior inside the core is not relevant to the physics of the solid. The true valence wavefunctions have many nodes in order to be orthogonal to core states.
(ii) The nodeless and smooth character of the pseudo-wavefunctions guarantees that the pseudopotentials produced by inversion of the Schrödinger equation are finite and smooth near the origin, instead of having a $1/r$ singularity like the Coulomb potential.

Of course, each valence state will give rise to a different pseudopotential, but this is not a serious complication as far as actual calculations are concerned. All the pseudopotentials corresponding to an atom will have tails that behave like Z^v/r, where Z^v is the valence charge of the atom, that is, the ionic charge for an ion consisting of the nucleus and the core electrons. The huge advantage of the pseudopotential is that now we have to deal with the valence electrons only in the solid (the core electrons are essentially frozen in their atomic wavefunctions), and the pseudopotentials are smooth so that standard numerical methods can be applied (such as Fourier expansions) to solve the single-particle equations. There are several details of the construction of the pseudopotential that require special attention in order to obtain potentials that work and actually simplify calculations of the properties of solids, but we will not go into these details here. Suffice to say that pseudopotential construction is one of the arts of performing reliable and accurate calculations for solids, but through the careful work of many physicists in this field over the last couple of decades there exist now very good pseudopotentials for essentially all elements of interest in the Periodic Table [30]. The modern practice of pseudopotentials has strived to produce in a systematic way potentials that are simultaneously smoother, more accurate and more transferable, for a wide range of elements [30–32].

Further reading

1. *Density Functional Theory*, E.K.U. Gross and R.M. Dreizler, eds., (Plenum Press, New York, 1995).
   This book provides a detailed account of technical issues related to DFT.
2. *Electron Correlations in Molecules and Solids*, P. Fulde (Springer-Verlag, Berlin, 1991).
   This is a comprehensive discussion of the problem of electron correlations in condensed matter, with many useful insights and detailed accounts of theoretical tools.
3. *Pseudopotential Methods in Condensed Matter Applications*, W.E. Pickett, Computer Physics Reports, vol. 9, pp. 115–198 (North Holland, Amsterdam, 1989).
   This is an excellent review of the theory and applications of pseudopotentials.

Problems

1. Use a variational calculation to obtain the Hartree–Fock single-particle equations Eq. (2.16) from the Hartree–Fock many-body wavefunction defined in Eq. (2.14).
2. Show that the quantities ϵ_i appearing in the Hartree–Fock equations, which were introduced as the Lagrange multipliers to preserve the normalization of state $\phi_i(\mathbf{r}_i)$, have the physical meaning of the energy required to remove this state from the system. To do this, find the energy difference between two systems, one with and one without the state $\phi_i(\mathbf{r}_i)$ which have different numbers of electrons, N and $N-1$, respectively; you may assume that N is very large, so that removing the electron in state $\phi_i(\mathbf{r}_i)$ does not affect the other states $\phi_j(\mathbf{r}_j)$.
3. Consider a simple excitation of the ground state of the free-electron system, consisting of taking an electron from a state with momentum $\mathbf{k}_1$ and putting it in a state with

momentum $\mathbf{k}_2$; since the ground state of the system consists of filled single-particle states with momentum up to the Fermi momentum k_F, we must have $|\mathbf{k}_1| \leq k_F$ and $|\mathbf{k}_2| > k_F$. Removing the electron from state $\mathbf{k}_1$ leaves a "hole" in the Fermi sphere, so this excitation is described as an "electron–hole pair". Discuss the relationship between the total excitation energy and the total momentum of the electron–hole pair; show a graph of this relationship in terms of reduced variables, that is, the excitation energy and momentum in units of the Fermi energy ϵ_F and the Fermi momentum k_F. (At this point we are not concerned with the nature of the physical process that can create such an excitation and with how momentum is conserved in this process.)

4. The bulk modulus B of a solid is defined as

$$B = -\Omega \frac{\partial P}{\partial \Omega} = \Omega \frac{\partial^2 E}{\partial \Omega^2} \tag{2.123}$$

where Ω is the volume, P is the pressure, and E is the total energy; this quantity describes how the solid responds to external pressure by changes in its volume. Show that for the uniform electron gas with the kinetic energy and exchange energy terms only, Eq. (2.31) and Eq. (2.46), respectively, the bulk modulus is given by

$$B = \left[\frac{5}{6\pi} \frac{2.21}{(r_s/a_0)^5} - \frac{2}{6\pi} \frac{0.916}{(r_s/a_0)^4} \right] (\text{Ry}/a_0^3) \tag{2.124}$$

or equivalently, in terms of the kinetic and exchange energies,

$$B = \frac{1}{6\pi (r_s/a_0)^3} \left[5\frac{E^{kin}}{N} + 2\frac{E^X}{N} \right] (1/a_0^3) \tag{2.125}$$

Discuss the physical implications of this result for a hypothetical solid that might be reasonably described in terms of the uniform electron gas, and in which the value of (r_s/a_0) is relatively small ($(r_s/a_0) < 1$).

5. We will investigate the model of the hydrogen molecule discussed in the text.

   (a) Consider first the single-particle hamiltonian given in Eq. (2.52); show that its expectation values in terms of the single-particle wavefunctions $\phi_i (i = 0, 1)$ defined in Eq. (2.57) are those given in Eq. (2.58).

   (b) Consider next the two-particle hamiltonian $\mathcal{H}(\mathbf{r}_1, \mathbf{r}_2)$, given in Eq. (2.54), which contains the interaction term; show that its expectation values in terms of the Hartree wavefunctions $\Psi_i^H (i = 0, 1)$ defined in Eqs. (2.59) and (2.60), are those given in Eqs. (2.62) and (2.63), respectively. To derive these results certain matrix elements of the interaction term need to be neglected; under what assumptions is this a reasonable approximation? What would be the expression for the energy if we were to use the wavefunction $\Psi_2^H(\mathbf{r}_1, \mathbf{r}_2)$ defined in Eq. (2.61)?

   (c) Using the Hartree–Fock wavefunctions for this model defined in Eqs. (2.64)–(2.66), construct the matrix elements of the hamiltonian

$$\mathcal{H}_{ij} = \langle \Psi_i^{HF} | \mathcal{H} | \Psi_j^{HF} \rangle$$

   and diagonalize this 3×3 matrix to find the eigenvalues and eigenstates; verify that the ground state energy and wavefunction are those given in Eq. (2.67) and Eq. (2.68), respectively. Here we will assume that the same approximations as those involved in part (b) are applicable.

(d) Find the probability that the two electrons in the ground state, defined by Eq. (2.68), are on the same proton. Give a plot of this result as a function of (U/t) and explain the physical meaning of the answer for the behavior at the small and large limits of this parameter.

6. We want to determine the physical meaning of the quantities ϵ_i in the Density Functional Theory single-particle equations Eq. (2.86). To do this, we express the density as

$$n(\mathbf{r}) = \sum_i n_i |\phi_i(\mathbf{r})|^2 \tag{2.126}$$

where the n_i are real numbers between 0 and 1, called the "filling factors". We take a partial derivative of the total energy with respect to n_i and relate it to ϵ_i. Then we integrate this relation with respect to n_i. What is the physical meaning of the resulting equation?

7. In the extremely low density limit, a system of electrons will form a regular lattice, with each electron occupying a unit cell; this is known as the Wigner crystal. The energy of this crystal has been calculated to be

$$E^{Wigner} = \left[-\frac{3}{(r_s/a_0)} + \frac{3}{(r_s/a_0)^{3/2}} \right] \text{Ry} \tag{2.127}$$

This can be compared with the energy of the electron gas in the Hartree–Fock approximation Eq. (2.44), to which we must add the electrostatic energy (this term is canceled by the uniform positive background of the ions, but here we are considering the electron gas by itself). The electrostatic energy turns out to be

$$E^{es} = -\frac{6}{5}\frac{1}{(r_s/a_0)}\text{Ry} \tag{2.128}$$

Taking the difference between the two energies, E^{Wigner} and $E^{HF} + E^{es}$, we obtain the correlation energy, which is by definition the interaction energy after we have taken into account all the other contributions, kinetic, electrostatic and exchange. Show that the result is compatible with the Wigner correlation energy given in Table 2.1, in the low density (high (r_s/a_0)) limit.

8. We wish to derive the Lindhard dielectric response function for the free-electron gas, using perturbation theory. The charge density is defined in terms of the single-particle wavefunctions as

$$\rho(\mathbf{r}) = (-e) \sum_{\mathbf{k}} n(\mathbf{k}) |\phi_{\mathbf{k}}(\mathbf{r})|^2$$

with $n(\mathbf{k})$ the Fermi occupation numbers. From first order perturbation theory (see Appendix B), the change in wavefunction of state $\mathbf{k}$ due to a perturbation represented by the potential $V^{int}(\mathbf{r})$ is given by

$$|\delta\phi_{\mathbf{k}}\rangle = \sum_{\mathbf{k}'} \frac{\langle \phi^{(0)}_{\mathbf{k}'} | V^{int} | \phi^{(0)}_{\mathbf{k}} \rangle}{\epsilon^{(0)}_{\mathbf{k}} - \epsilon^{(0)}_{\mathbf{k}'}} |\phi^{(0)}_{\mathbf{k}'}\rangle$$

with $|\phi_{\mathbf{k}}^{(0)}\rangle$ the unperturbed wavefunctions and $\epsilon_{\mathbf{k}}^{(0)}$ the corresponding energies. These changes in the wavefunctions give rise to the induced charge density $\rho^{ind}(\mathbf{r})$ to first order in V^{int}.

(a) Derive the expression for the Lindhard dielectric response function, given in Eq. (2.107), for free electrons with energy $\epsilon_{\mathbf{k}}^{(0)} = \hbar^2|\mathbf{k}|^2/2m_e$ and with Fermi energy ϵ_{F}, by keeping only first order terms in V^{int} in the perturbation expansion.
(b) Evaluate the zero-temperature Lindhard response function, Eq. (2.108), at $k = 2k_{\mathrm{F}}$, and the corresponding dielectric constant $\varepsilon = 1 - 4\pi\chi/k^2$; interpret their behavior in terms of the single-particle picture.

9. Show that at zero temperature the Thomas–Fermi inverse screening length l_s^{TF}, defined in Eq. (2.110), with the total occupation n_T given by Eq. (2.111), takes the form

$$l_s^{TF}(T=0) = \frac{2}{\sqrt{\pi}}\sqrt{\frac{k_{\mathrm{F}}}{a_0}}$$

with k_{F} the Fermi momentum and a_0 the Bohr radius.

10. Consider a fictitious atom which has a harmonic potential for the radial equation:

$$\left[-\frac{\hbar^2}{2m_e}\frac{d^2}{dr^2} + \frac{1}{2}m_e\omega^2 r^2\right]\phi_i(r) = \epsilon_i\phi_i(r) \qquad (2.129)$$

and has nine electrons. The harmonic oscillator potential is discussed in detail in Appendix B. The first four states, $\phi_0(r), \phi_1(r), \phi_2(r), \phi_3(r)$ are fully occupied core states, and the last state $\phi_4(r)$ is a valence state with one electron in it. We want to construct a pseudopotential which gives a state $\psi_4(r)$ that is smooth and nodeless in the core region. Choose as the cutoff radius r_c the position of the last extremum of $\phi_4(r)$, and use the simple expressions

$$\begin{aligned} \psi_4(r) &= Az^2 e^{-Bz^2} \quad & r \le r_c \\ \psi_4(r) &= \phi_4(r) \quad & r > r_c \end{aligned} \qquad (2.130)$$

for the pseudo-wavefunction, where $z = r\sqrt{m_e\omega/\hbar}$. Determine the parameters A, B so that the pseudo-wavefunction $\psi_4(r)$ and its derivative are continuous at r_c. Then invert the radial Schrödinger equation to obtain the pseudopotential which has $\psi_4(r)$ as its solution. Plot the pseudopotential you obtained as a function of r. Does this procedure produce a physically acceptable pseudopotential?

3

Electrons in crystal potential

In chapter 2 we provided the justification for the single-particle picture of electrons in solids. We saw that the proper interpretation of single particles involves the notion of quasiparticles: these are fermions which resemble real electrons, but are not identical to them since they also embody the effects of the presence of all other electrons, as in the exchange-correlation hole. Here we begin to develop the quantitative description of the properties of solids in terms of quasiparticles and collective excitations for the case of a perfectly periodic solid, i.e., an ideal crystal.

3.1 Periodicity – Bloch states

A crystal is described in real space in terms of the primitive lattice vectors $\mathbf{a}_1, \mathbf{a}_2, \mathbf{a}_3$ and the positions of atoms inside a primitive unit cell (PUC). The lattice vectors $\mathbf{R}$ are formed by all the possible combinations of primitive lattice vectors, multiplied by integers:

$$\mathbf{R} = n_1\mathbf{a}_1 + n_2\mathbf{a}_2 + n_3\mathbf{a}_3, \qquad n_1, n_2, n_3 : \text{integers} \tag{3.1}$$

The lattice vectors connect all equivalent points in space; this set of points is referred to as the "Bravais lattice". The PUC is defined as the volume enclosed by the three primitive lattice vectors:

$$\Omega_{PUC} = \mid \mathbf{a}_1 \cdot (\mathbf{a}_2 \times \mathbf{a}_3) \mid \tag{3.2}$$

This is a useful definition: we only need to know all relevant real-space functions for $\mathbf{r}$ *within* the PUC since, due to the periodicity of the crystal, these functions have the same value at an equivalent point of any other unit cell related to the PUC by a translation $\mathbf{R}$. There can be one or many atoms inside the primitive unit cell, and the origin of the coordinate system can be located at any position in space; for convenience, it is often chosen to be the position of one of the atoms in the PUC.

The foundation for describing the behavior of electrons in a crystal is the reciprocal lattice, which is the inverse space of the real lattice. The reciprocal primitive lattice vectors are defined by

$$\mathbf{b}_1 = \frac{2\pi(\mathbf{a}_2 \times \mathbf{a}_3)}{\mathbf{a}_1 \cdot (\mathbf{a}_2 \times \mathbf{a}_3)}, \quad \mathbf{b}_2 = \frac{2\pi(\mathbf{a}_3 \times \mathbf{a}_1)}{\mathbf{a}_2 \cdot (\mathbf{a}_3 \times \mathbf{a}_1)}, \quad \mathbf{b}_3 = \frac{2\pi(\mathbf{a}_1 \times \mathbf{a}_2)}{\mathbf{a}_3 \cdot (\mathbf{a}_1 \times \mathbf{a}_2)} \tag{3.3}$$

with the obvious consequence

$$\mathbf{a}_i \cdot \mathbf{b}_j = 2\pi\delta_{ij} \tag{3.4}$$

The vectors $\mathbf{b}_i, i = 1, 2, 3$ define a cell in reciprocal space which also has useful consequences, as we describe below. The volume of that cell in reciprocal space is given by

$$| \mathbf{b}_1 \cdot (\mathbf{b}_2 \times \mathbf{b}_3) | = \frac{(2\pi)^3}{| \mathbf{a}_1 \cdot (\mathbf{a}_2 \times \mathbf{a}_3) |} = \frac{(2\pi)^3}{\Omega_{PUC}} \tag{3.5}$$

We can construct vectors which connect all equivalent points in reciprocal space, which we call $\mathbf{G}$, by analogy to the Bravais lattice vectors defined in Eq. (3.1):

$$\mathbf{G} = m_1\mathbf{b}_1 + m_2\mathbf{b}_2 + m_3\mathbf{b}_3, \quad m_1, m_2, m_3 : \text{integers} \tag{3.6}$$

By construction, the dot product of any $\mathbf{R}$ vector with any $\mathbf{G}$ vector gives

$$\mathbf{R} \cdot \mathbf{G} = 2\pi l, \quad l = n_1 m_1 + n_2 m_2 + n_3 m_3 \tag{3.7}$$

where l is always an integer. This relationship can serve to define one set of vectors in terms of the other set. This also gives

$$\mathrm{e}^{\mathrm{i}\mathbf{G}\cdot\mathbf{R}} = 1 \tag{3.8}$$

for all $\mathbf{R}$ and $\mathbf{G}$ vectors defined by Eqs. (3.1) and (3.6). Any function that has the periodicity of the Bravais lattice can be written as

$$f(\mathbf{r}) = \sum_{\mathbf{G}} \mathrm{e}^{\mathrm{i}\mathbf{G}\cdot\mathbf{r}} f(\mathbf{G}) \tag{3.9}$$

with $f(\mathbf{G})$ the Fourier Transform (FT) components. Due to the periodicity of the lattice, any such function need only be studied for $\mathbf{r}$ within the PUC. This statement applied to the single-particle wavefunctions is known as "Bloch's theorem".

Bloch's theorem: When the potential in the single-particle hamiltonian has the translational periodicity of the Bravais lattice

$$V^{sp}(\mathbf{r} + \mathbf{R}) = V^{sp}(\mathbf{r}) \tag{3.10}$$

Table 3.1. *Vectors* $\mathbf{a}_1, \mathbf{a}_2, \mathbf{a}_3$ *that define the primitive unit cell of simple crystals.*

Only crystals with one or two atoms per unit cell are considered; the position of one atom in the PUC is always assumed to be at the origin; when there are two atoms in the PUC, the position of the second atom $\mathbf{t}_2$ is given with respect to the origin. All vectors are given in cartesian coordinates and in terms of the standard lattice parameter a, the side of the conventional cube or parallelpiped. For the HCP lattice, a second parameter is required, namely the c/a ratio. For graphite, only the two-dimensional honeycomb lattice of a single graphitic plane is defined. d_{NN} is the distance between nearest neighbors in terms of the lattice constant a. These crystals are illustrated in Fig. 3.1.

Lattice	$\mathbf{a}_1$	$\mathbf{a}_2$	$\mathbf{a}_3$	$\mathbf{t}_2$	c/a	d_{NN}
Cubic	$(a, 0, 0)$	$(0, a, 0)$	$(0, 0, a)$			a
BCC	$(\frac{a}{2}, -\frac{a}{2}, -\frac{a}{2})$	$(\frac{a}{2}, \frac{a}{2}, -\frac{a}{2})$	$(\frac{a}{2}, \frac{a}{2}, \frac{a}{2})$			$\frac{a\sqrt{3}}{2}$
FCC	$(\frac{a}{2}, \frac{a}{2}, 0)$	$(\frac{a}{2}, 0, \frac{a}{2})$	$(0, \frac{a}{2}, \frac{a}{2})$			$\frac{a}{\sqrt{2}}$
Diamond	$(\frac{a}{2}, \frac{a}{2}, 0)$	$(\frac{a}{2}, 0, \frac{a}{2})$	$(0, \frac{a}{2}, \frac{a}{2})$	$(\frac{a}{4}, \frac{a}{4}, \frac{a}{4})$		$\frac{a}{4\sqrt{3}}$
HCP	$(\frac{a}{2}, \frac{\sqrt{3}a}{2}, 0)$	$(\frac{a}{2}, -\frac{\sqrt{3}a}{2}, 0)$	$(0, 0, c)$	$(\frac{a}{2}, \frac{a}{2\sqrt{3}}, \frac{c}{2})$	$\sqrt{\frac{8}{3}}$	$\frac{a}{\sqrt{3}}$
Graphite	$(\frac{a}{2}, \frac{\sqrt{3}a}{2}, 0)$	$(\frac{a}{2}, -\frac{\sqrt{3}a}{2}, 0)$		$(\frac{a}{2}, \frac{a}{2\sqrt{3}}, 0)$		$\frac{a}{\sqrt{3}}$

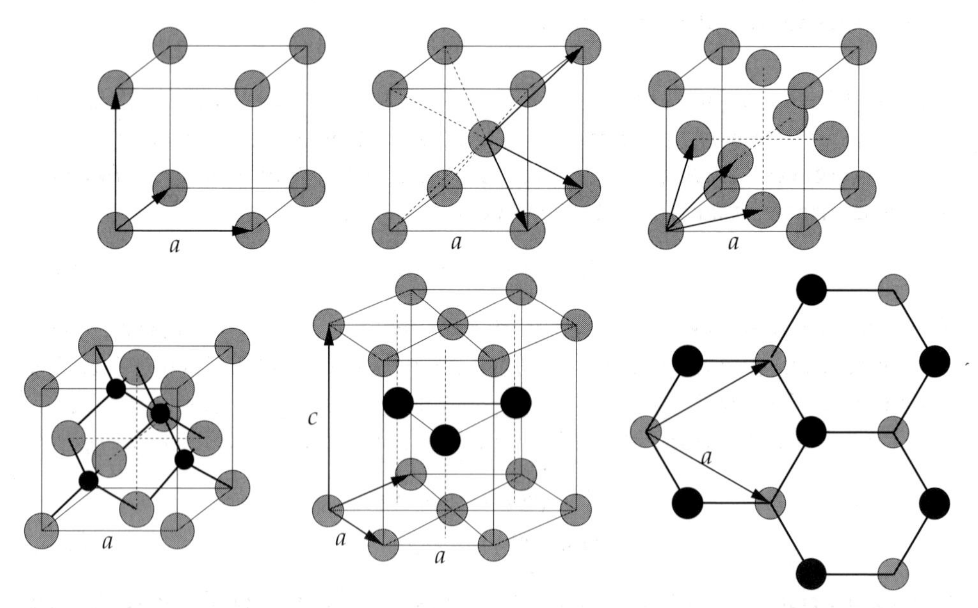

Figure 3.1. The crystals defined in Table 3.1. **Top**: simple cubic, BCC, FCC. **Bottom**: diamond, HCP, graphite (single plane). In all cases the lattice vectors are indicated by arrows (the lattice vectors for the diamond lattice are identical to those for the FCC lattice). For the diamond, HCP and graphite lattices two different symbols, gray and black circles, are used to denote the two atoms in the unit cell. For the diamond and graphite lattices, the bonds between nearest neighbors are also shown as thicker lines.

the single-particle wavefunctions have the same symmetry, up to a phase factor:

$$\psi_{\mathbf{k}}(\mathbf{r} + \mathbf{R}) = e^{i\mathbf{k}\cdot\mathbf{R}}\psi_{\mathbf{k}}(\mathbf{r}) \tag{3.11}$$

A different formulation of Bloch's theorem is that the single-particle wavefunctions must have the form

$$\psi_{\mathbf{k}}(\mathbf{r}) = e^{i\mathbf{k}\cdot\mathbf{r}}u_{\mathbf{k}}(\mathbf{r}), \quad u_{\mathbf{k}}(\mathbf{r} + \mathbf{R}) = u_{\mathbf{k}}(\mathbf{r}) \tag{3.12}$$

that is, the wavefunctions $\psi_{\mathbf{k}}(\mathbf{r})$ can be expressed as the product of the phase factor $\exp(i\mathbf{k}\cdot\mathbf{r})$ multiplied by the functions $u_{\mathbf{k}}(\mathbf{r})$ which have the full translational periodicity of the Bravais lattice.

The two formulations of Bloch's theorem are equivalent. For any wavefunction $\psi_{\mathbf{k}}(\mathbf{r})$ that can be put in the form of Eq. (3.12), the relation of Eq. (3.11) must obviously hold. Conversely, if Eq. (3.11) holds, we can factor out of $\psi_{\mathbf{k}}(\mathbf{r})$ the phase factor $\exp(i\mathbf{k}\cdot\mathbf{r})$, in which case the remainder

$$u_{\mathbf{k}}(\mathbf{r}) = \frac{\psi_{\mathbf{k}}(\mathbf{r})}{e^{i\mathbf{k}\cdot\mathbf{r}}}$$

must have the translational periodicity of the Bravais lattice by virtue of Eq. (3.11). The states $\psi_{\mathbf{k}}(\mathbf{r})$ are referred to as Bloch states. At this point $\mathbf{k}$ is just a subscript index for identifying the wavefunctions.

Proof of Bloch's theorem: A convenient way to prove Bloch's theorem is through the definition of translation operators, whose eigenvalues and eigenfunctions can be easily determined. We define the translation operator $T_{\mathbf{R}}$ which acts on any function $f(\mathbf{r})$ and changes its argument by a lattice vector $-\mathbf{R}$:

$$T_{\mathbf{R}}f(\mathbf{r}) = f(\mathbf{r} - \mathbf{R}) \tag{3.13}$$

This operator commutes with the hamiltonian $\mathcal{H}^{sp}$: it obviously commutes with the kinetic energy operator, and it leaves the potential energy unaffected since this potential has the translational periodicity of the Bravais lattice. Consequently, we can choose all eigenfunctions of $\mathcal{H}^{sp}$ to be simultaneous eigenfunctions of $T_{\mathbf{R}}$:

$$\begin{aligned} \mathcal{H}^{sp}\psi_{\mathbf{k}}(\mathbf{r}) &= \epsilon_{\mathbf{k}}\psi_{\mathbf{k}}(\mathbf{r}) \\ T_{\mathbf{R}}\psi_{\mathbf{k}}(\mathbf{r}) &= c_{\mathbf{R}}\psi_{\mathbf{k}}(\mathbf{r}) \end{aligned} \tag{3.14}$$

with $c_{\mathbf{R}}$ the eigenvalue corresponding to the operator $T_{\mathbf{R}}$. Our goal is to determine the eigenfunctions of $T_{\mathbf{R}}$ so that we can use them as the basis to express the eigenfunctions of $\mathcal{H}^{sp}$. To this end, we will first determine the eigenvalues of $T_{\mathbf{R}}$. We notice that

$$T_{\mathbf{R}}T_{\mathbf{R}'} = T_{\mathbf{R}'}T_{\mathbf{R}} = T_{\mathbf{R}+\mathbf{R}'} \Rightarrow c_{\mathbf{R}+\mathbf{R}'} = c_{\mathbf{R}}c_{\mathbf{R}'} \tag{3.15}$$

Considering $c_{\mathbf{R}}$ as a function of $\mathbf{R}$, we conclude that it must be an exponential in $\mathbf{R}$, which is the only function that satisfies the above relation. Without loss of generality, we define

$$c_{\mathbf{a}_j} = \mathrm{e}^{\mathrm{i}2\pi\kappa_j} \quad (j = 1, 2, 3) \tag{3.16}$$

where κ_j is an unspecified complex number, so that $c_{\mathbf{a}_j}$ can take any complex value. By virtue of Eq. (3.4), the definition of $c_{\mathbf{a}_j}$ produces for the eigenvalue $c_{\mathbf{R}}$:

$$c_{\mathbf{R}} = \mathrm{e}^{-\mathrm{i}\mathbf{k}\cdot\mathbf{R}}$$
$$\mathbf{k} = \kappa_1\mathbf{b}_1 + \kappa_2\mathbf{b}_2 + \kappa_3\mathbf{b}_3 \tag{3.17}$$

where now the index $\mathbf{k}$, introduced earlier to label the wavefunctions, is expressed in terms of the reciprocal lattice vectors $\mathbf{b}_j$ and the complex constants κ_j. Having established that the eigenvalues of the operator $T_{\mathbf{R}}$ are $c_{\mathbf{R}} = \exp(-\mathrm{i}\mathbf{k}\cdot\mathbf{R})$, we find by inspection that the eigenfunctions of this operator are $\exp(\mathrm{i}(\mathbf{k}+\mathbf{G})\cdot\mathbf{r})$, since

$$T_{\mathbf{R}}\mathrm{e}^{\mathrm{i}(\mathbf{k}+\mathbf{G})\cdot\mathbf{r}} = \mathrm{e}^{\mathrm{i}(\mathbf{k}+\mathbf{G})\cdot(\mathbf{r}-\mathbf{R})} = \mathrm{e}^{-\mathrm{i}\mathbf{k}\cdot\mathbf{R}}\mathrm{e}^{\mathrm{i}(\mathbf{k}+\mathbf{G})\cdot\mathbf{r}} = c_{\mathbf{k}}\mathrm{e}^{\mathrm{i}(\mathbf{k}+\mathbf{G})\cdot\mathbf{r}} \tag{3.18}$$

because $\exp(-\mathrm{i}\mathbf{G}\cdot\mathbf{R}) = 1$. Then we can write the eigenfunctions of $\mathcal{H}^{sp}$ as an expansion over all eigenfunctions of $T_{\mathbf{R}}$ corresponding to the same eigenvalue of $T_{\mathbf{R}}$:

$$\psi_{\mathbf{k}}(\mathbf{r}) = \sum_{\mathbf{G}} \alpha_{\mathbf{k}}(\mathbf{G})\mathrm{e}^{\mathrm{i}(\mathbf{k}+\mathbf{G})\cdot\mathbf{r}} = \mathrm{e}^{\mathrm{i}\mathbf{k}\cdot\mathbf{r}}u_{\mathbf{k}}(\mathbf{r})$$
$$u_{\mathbf{k}}(\mathbf{r}) = \sum_{\mathbf{G}} \alpha_{\mathbf{k}}(\mathbf{G})\mathrm{e}^{\mathrm{i}\mathbf{G}\cdot\mathbf{r}} \tag{3.19}$$

which proves Bloch's theorem, since $u_{\mathbf{k}}(\mathbf{r}+\mathbf{R}) = u_{\mathbf{k}}(\mathbf{r})$ for $u_{\mathbf{k}}(\mathbf{r})$ defined in Eq. (3.19).

When this form of the wavefunction is inserted in the single-particle Schrödinger equation, we obtain the equation for $u_{\mathbf{k}}(\mathbf{r})$:

$$\left[\frac{1}{2m_e}\left(\frac{\hbar\nabla_{\mathbf{r}}}{\mathrm{i}} + \hbar\mathbf{k}\right)^2 + V^{sp}(\mathbf{r})\right]u_{\mathbf{k}}(\mathbf{r}) = \epsilon_{\mathbf{k}}u_{\mathbf{k}}(\mathbf{r}) \tag{3.20}$$

Solving this last equation determines $u_{\mathbf{k}}(\mathbf{r})$, which with the factor $\exp(\mathrm{i}\mathbf{k}\cdot\mathbf{r})$ makes up the solution to the original single-particle equation. The great advantage is that we only need to solve this equation for $\mathbf{r}$ within a PUC of the crystal, since $u_{\mathbf{k}}(\mathbf{r}+\mathbf{R}) = u_{\mathbf{k}}(\mathbf{r})$, where $\mathbf{R}$ is any vector connecting equivalent Bravais lattice points. This result can also be thought of as equivalent to changing the momentum operator in the hamiltonian $\mathcal{H}(\mathbf{p}, \mathbf{r})$ by $+\hbar\mathbf{k}$, when dealing with the states $u_{\mathbf{k}}$ instead of the states $\psi_{\mathbf{k}}$:

$$\mathcal{H}(\mathbf{p}, \mathbf{r})\psi_{\mathbf{k}}(\mathbf{r}) = \epsilon_{\mathbf{k}}\psi_{\mathbf{k}}(\mathbf{r}) \Longrightarrow \mathcal{H}(\mathbf{p}+\hbar\mathbf{k}, \mathbf{r})u_{\mathbf{k}}(\mathbf{r}) = \epsilon_{\mathbf{k}}u_{\mathbf{k}}(\mathbf{r}) \tag{3.21}$$

For future use, we derive another relation between the two forms of the single-particle hamiltonian; multiplying the first expression from the left by $\exp(-i\mathbf{k}\cdot\mathbf{r})$ we get

$$e^{-i\mathbf{k}\cdot\mathbf{r}}\mathcal{H}(\mathbf{p},\mathbf{r})\psi_{\mathbf{k}}(\mathbf{r}) = e^{-i\mathbf{k}\cdot\mathbf{r}}\epsilon_{\mathbf{k}}e^{i\mathbf{k}\cdot\mathbf{r}}u_{\mathbf{k}}(\mathbf{r}) = \epsilon_{\mathbf{k}}u_{\mathbf{k}}(\mathbf{r}) = \mathcal{H}(\mathbf{p}+\hbar\mathbf{k},\mathbf{r})u_{\mathbf{k}}(\mathbf{r})$$

and comparing the first and last term, we conclude that

$$e^{-i\mathbf{k}\cdot\mathbf{r}}\mathcal{H}(\mathbf{p},\mathbf{r})e^{i\mathbf{k}\cdot\mathbf{r}} = \mathcal{H}(\mathbf{p}+\hbar\mathbf{k},\mathbf{r}) \tag{3.22}$$

This last expression will prove useful in describing the motion of crystal electrons under the influence of an external electric field.

3.2 k-space – Brillouin zones

In the previous section we introduced $\mathbf{k} = \kappa_1\mathbf{b}_1 + \kappa_2\mathbf{b}_2 + \kappa_3\mathbf{b}_3$ as a convenient index to label the wavefunctions. Here we will show that this index actually has physical meaning. Consider that the crystal is composed of N_j unit cells in the direction of vector $\mathbf{a}_j$ ($j = 1, 2, 3$), where we think of the values of N_j as macroscopically large. $N = N_1N_2N_3$ is equal to the total number of unit cells in the crystal (of order Avogadro's number, 6.023×10^{23}). We need to specify the proper boundary conditions for the single-particle states within this crystal. Consistent with the idea that we are dealing with an infinite solid, we can choose periodic boundary conditions, also known as the Born–von Karman boundary conditions,

$$\psi_{\mathbf{k}}(\mathbf{r}) = \psi_{\mathbf{k}}(\mathbf{r} + N_j\mathbf{a}_j) \tag{3.23}$$

with $\mathbf{r}$ lying within the first PUC. Bloch's theorem and Eq. (3.23) imply that

$$e^{i\mathbf{k}\cdot(N_j\mathbf{a}_j)} = 1 \Rightarrow e^{i2\pi\kappa_jN_j} = 1 \Rightarrow \kappa_j = \frac{n_j}{N_j} \tag{3.24}$$

where n_j is any integer. This shows two important things.

(1) The vector $\mathbf{k}$ is real because the parameters κ_j are real. Since $\mathbf{k}$ is defined in terms of the reciprocal lattice vectors $\mathbf{b}_j$, it can be thought of as a wave-vector; $\exp(i\mathbf{k}\cdot\mathbf{r})$ represents a plane wave of wave-vector $\mathbf{k}$. The physical meaning of this result is that the wavefunction does not decay within the crystal but rather extends throughout the crystal like a wave modified by the periodic function $u_{\mathbf{k}}(\mathbf{r})$. This fact was first introduced in chapter 1.

(2) The number of distinct values that $\mathbf{k}$ may take is $N = N_1N_2N_3$, because n_j can take N_j inequivalent values that satisfy Eq. (3.24), which can be any N_j consecutive integer values. Values of n_j beyond this range are equivalent to values within this range, because they correspond to adding integer multiples of $2\pi i$ to the argument of the exponential in Eq. (3.24). Values of $\mathbf{k}$ that differ by a reciprocal lattice vector $\mathbf{G}$ are equivalent,

since adding a vector **G** to **k** corresponds to a difference of an integer multiple of $2\pi\mathrm{i}$ in the argument of the exponential in Eq. (3.24). This statement is valid even in the limit when $N_j \to \infty$, that is, in the case of an infinite crystal when the values of **k** become continuous.

The second statement has important consequences: it restricts the inequivalent values of **k** to a volume in reciprocal space, which is the analog of the PUC in real space. This volume in reciprocal space is known as the first Brillouin Zone (BZ in the following). By convention, we choose the first BZ to correspond to the following N_j consecutive values of the index n_j:

$$n_j = -\frac{N_j}{2}, \ldots, 0, \ldots, \frac{N_j}{2} - 1 \quad (j = 1, 2, 3) \tag{3.25}$$

where we assume N_j to be an even integer (since we are interested in the limit $N_j \to \infty$ this assumption does not impose any significant restrictions).

To generalize the concept of the BZ, we first introduce the notion of Bragg planes. Consider a plane wave of incident radiation and wave-vector **q**, which is scattered by the planes of atoms in a crystal to a wave-vector $\mathbf{q}'$. For elastic scattering $|\mathbf{q}| = |\mathbf{q}'|$. As the schematic representation of Fig. 3.2 shows, the difference in paths along

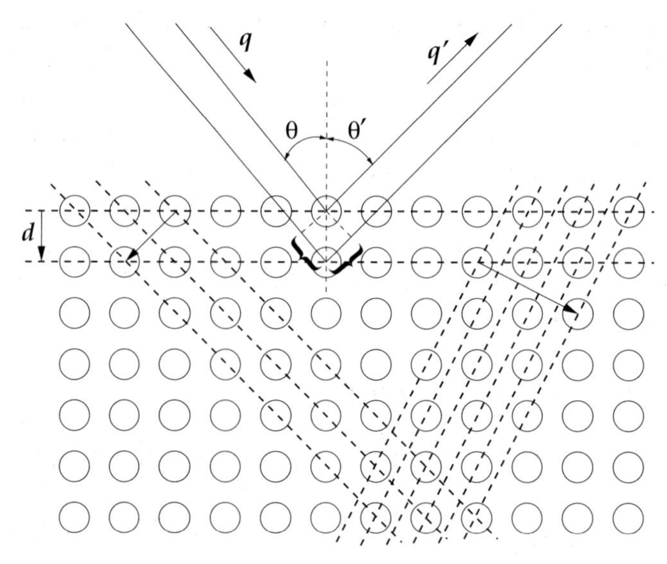

Figure 3.2. Schematic representation of Bragg scattering from atoms on successive atomic planes. Some families of parallel atomic planes are identified by sets of parallel dashed lines, together with the lattice vectors that are perpendicular to the planes and join equivalent atoms; notice that the closer together the planes are spaced the longer is the corresponding perpendicular lattice vector. The difference in path for two scattered rays from the horizontal family of planes is indicated by the two inclined curly brackets ({,}).

incident and reflected radiation from two consecutive planes is

$$d\cos\theta + d\cos\theta' = \mathbf{d}\cdot\hat{\mathbf{q}} - \mathbf{d}\cdot\hat{\mathbf{q}}' \tag{3.26}$$

with $\hat{\mathbf{q}}$ the unit vector along $\mathbf{q}$ ($\hat{\mathbf{q}} = \mathbf{q}/|\mathbf{q}|$) and $d = |\mathbf{d}|$, $\mathbf{d}$ being a vector that connects equivalent lattice points. For constructive interference between incident and reflected waves, this difference must be equal to $l\lambda$, where l is an integer and λ is the wavelength. Using $\mathbf{q} = (2\pi/\lambda)\hat{\mathbf{q}}$, we obtain the condition for constructive interference,

$$\mathbf{R}\cdot(\mathbf{q}-\mathbf{q}') = 2\pi l \Rightarrow \mathbf{q}-\mathbf{q}' = \mathbf{G} \tag{3.27}$$

where we have made use of two facts: first, that $\mathbf{d} = \mathbf{R}$ since $\mathbf{d}$ represents a distance between equivalent lattice points in neighboring atomic planes; and second, that the reciprocal lattice vectors are defined through the relation $\mathbf{G}\cdot\mathbf{R} = 2\pi l$, as shown in Eq. (3.7).

From the above equation we find $\mathbf{q}' = \mathbf{q} - \mathbf{G}$. By squaring both sides of this equation and using the fact that for elastic scattering $|\mathbf{q}| = |\mathbf{q}'|$, we obtain

$$\mathbf{q}\cdot\hat{\mathbf{G}} = \frac{1}{2}|\mathbf{G}| \tag{3.28}$$

This is the definition of the Bragg plane: it is formed by the tips of all the vectors $\mathbf{q}$ which satisfy Eq. (3.28) for a given $\mathbf{G}$. This relation determines all vectors $\mathbf{q}$ that lead to constructive interference. Since the angle of incidence and the magnitude of the wave-vector $\mathbf{q}$ can be varied arbitrarily, Eq. (3.28) serves to identify all the families of planes that can reflect radiation constructively. Therefore, by scanning the values of the angle of incidence and the magnitude of $\mathbf{q}$, we can determine all the $\mathbf{G}$ vectors, and from those all the $\mathbf{R}$ vectors, i.e. the Bravais lattice of the crystal.

For a crystal with N_j unit cells in the direction $\mathbf{a}_j$ ($j = 1, 2, 3$), the differential volume change in $\mathbf{k}$ is

$$\begin{aligned}\Delta^3\mathbf{k} &= \Delta\mathbf{k}_1\cdot(\Delta\mathbf{k}_2\times\Delta\mathbf{k}_3)\\ &= \frac{\Delta n_1\mathbf{b}_1}{N_1}\cdot\left(\frac{\Delta n_2\mathbf{b}_2}{N_2}\times\frac{\Delta n_3\mathbf{b}_3}{N_3}\right) \Rightarrow |\mathrm{d}\mathbf{k}| = \frac{(2\pi)^3}{N\Omega_{PUC}}\end{aligned} \tag{3.29}$$

where we have used $\Delta n_j = 1$, and $N = N_1N_2N_3$ is the total number of unit cells in the crystal; we have also made use of Eq. (3.5) for the volume of the basic cell in reciprocal space. For an infinite crystal $N\to\infty$, so that the spacing of $\mathbf{k}$ values becomes infinitesimal and $\mathbf{k}$ becomes a continuous variable.

Now consider the origin of reciprocal space and around it all the points that can be reached without crossing a Bragg plane. This corresponds to the first BZ. The condition in Eq. (3.28) means that the projection of $\mathbf{q}$ on $\mathbf{G}$ is equal to half the length of $\mathbf{G}$, indicating that the tip of the vector $\mathbf{q}$ must lie on a plane perpendicular to $\mathbf{G}$ that passes through its midpoint. This gives a convenient recipe for defining the first BZ: draw all reciprocal lattice vectors $\mathbf{G}$ and the planes that are perpendicular to them at their midpoints, which by the above arguments are identified as the Bragg planes; the volume enclosed by the first such set of Bragg planes around the origin is the first BZ. It also provides a convenient definition for the second, third, ..., BZs: the second BZ is the volume enclosed between the first set of Bragg planes and the second set of Bragg planes, going outward from the origin. A more rigorous definition is that the first BZ is the set of points that can be reached from the origin without crossing any Bragg planes; the second BZ is the set of points that can be reached from the origin by crossing only one Bragg plane, excluding the points in the first BZ, etc. The construction of the first three BZs for a two-dimensional square lattice, a case that is particularly easy to visualize, is shown in Fig. 3.3.

The usefulness of BZs is that they play an analogous role in reciprocal space as the primitive unit cells do in real space. We saw above that due to crystal periodicity, we only need to solve the single-particle equations inside the PUC. We also saw that values of $\mathbf{k}$ are equivalent if one adds to them any vector $\mathbf{G}$. Thus, we only need to solve the single-particle equations for values of $\mathbf{k}$ within the first BZ, or within any single BZ: points in other BZs are related by $\mathbf{G}$ vectors which make

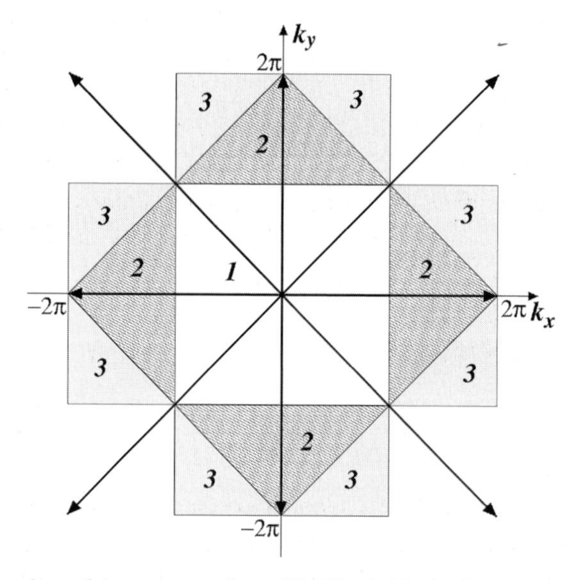

Figure 3.3. Illustration of the construction of Brillouin Zones in a two-dimensional crystal with $\mathbf{a}_1 = \hat{\mathbf{x}}$, $\mathbf{a}_2 = \hat{\mathbf{y}}$. The first two sets of reciprocal lattice vectors ($\mathbf{G} = \pm 2\pi\hat{\mathbf{x}}, \pm 2\pi\hat{\mathbf{y}}$ and $\mathbf{G} = 2\pi(\pm\hat{\mathbf{x}} \pm \hat{\mathbf{y}})$) are shown, along with the Bragg planes that bisect them. The first BZ, shown in white and labeled 1, is the central square; the second BZ, shown hatched and labeled 2, is composed of the four triangles around the central square; the third BZ, shown in lighter shade and labeled 3, is composed of the eight smaller triangles around the second BZ.

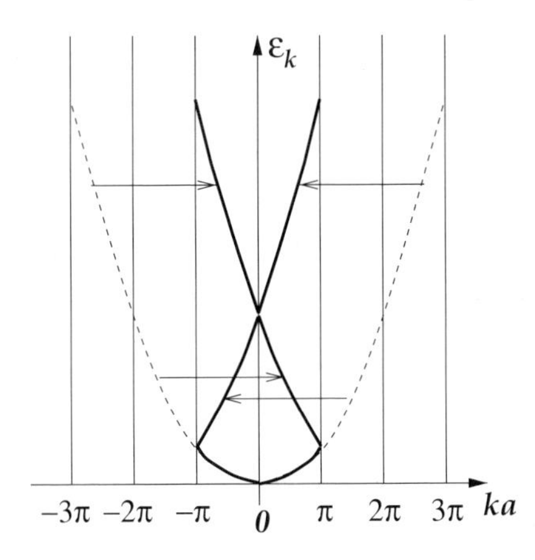

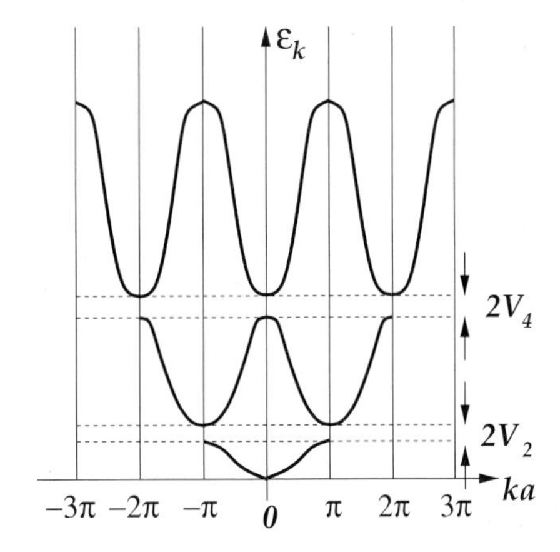

Figure 3.4. **Left:** the one-dimensional band structure of free electrons illustrating the reduced zone and extended zone schemes. **Right:** the one-dimensional band structure of electrons in a weak potential in the reduced and extended zone schemes, with splitting of the energy levels at BZ boundaries (Bragg planes): $V_2 = |V(2\pi/a)|$, $V_4 = |V(4\pi/a)|$.

them equivalent. Of course, within any single BZ and for the same value of $\mathbf{k}$ there may be several solutions of the single-particle equation, corresponding to the various allowed states for the quasiparticles. Therefore we need a second index to identify fully the solutions to the single-particle equations, which we denote by a superscript: $\psi_{\mathbf{k}}^{(n)}(\mathbf{r})$. The superscript index is discrete (it takes integer values), whereas, as argued above, for an infinite crystal the subscript index $\mathbf{k}$ is a continuous variable. The corresponding eigenvalues, also identified by two indices $\epsilon_{\mathbf{k}}^{(n)}$, are referred to as "energy bands". A plot of the energy bands is called the band structure. Keeping only the first BZ is referred to as the reduced zone scheme, keeping all BZs is referred to as the extended zone scheme. The eigenvalues and eigenfunctions in the two schemes are related by relabeling of the superscript indices. For a system of free electrons in one dimension, with $\epsilon_{\mathbf{k}} = \hbar^2\mathbf{k}^2/2m_e$ ($\mathbf{a} = a\hat{\mathbf{x}}$, $\mathbf{b} = 2\pi\hat{\mathbf{x}}/a$, $\mathbf{G}_m = m\mathbf{b}$, where m is an integer) the band structure in the reduced and the extended zone schemes is shown in Fig. 3.4.

It turns out that every BZ has the same volume, given by

$$\Omega_{BZ} = |\mathbf{b}_1 \cdot (\mathbf{b}_2 \times \mathbf{b}_3)| = \frac{(2\pi)^3}{|\mathbf{a}_1 \cdot (\mathbf{a}_2 \times \mathbf{a}_3)|} = \frac{(2\pi)^3}{\Omega_{PUC}} \tag{3.30}$$

By comparing this with Eq. (3.29) we conclude that in each BZ there are N distinct values of $\mathbf{k}$, where N is the total number of PUCs in the crystal. This is a very useful observation: if there are n electrons in the PUC (that is, nN electrons in the crystal), then we need exactly $nN/2$ different $\psi_{\mathbf{k}}(\mathbf{r})$ states to accommodate them, taking into account spin degeneracy (two electrons with opposite spins can coexist in state $\psi_{\mathbf{k}}(\mathbf{r})$). Since the first BZ contains N distinct values of $\mathbf{k}$, it can accommodate up to $2N$ electrons. Similarly, each subsequent BZ can accommodate $2N$ electrons because it has the same volume in reciprocal space. For n electrons per unit cell, we need to fill completely the states that occupy a volume in **k**-space equivalent to $n/2$ BZs. Which states will be filled is determined by their energy: in order to minimize the total energy of the system the lowest energy states must be occupied first. In the extended zone scheme, we need to occupy states that correspond to the lowest energy band and take up the equivalent of $n/2$ BZs. In the reduced zone scheme, we need to occupy a number of states that corresponds to a total of $n/2$ bands per **k**-point inside the first BZ. The Fermi level is defined as the value of the energy below which all single-particle states are occupied. For $n = 2$ electrons per PUC in the one-dimensional free-electron model discussed above, the Fermi level must be such that the first band in the first BZ is completely full. In the free-electron case this corresponds to the value of the energy at the first BZ boundary. The case of electrons in the weak periodic potential poses a more interesting problem which is discussed in detail in the next section: in this case, the first and the second bands are split in energy at the BZ boundary. For $n = 2$ electrons per PUC, there will be a gap between the highest energy of occupied states (the top of the first band) and the lowest energy of unoccupied states (the bottom of the second band); this gap, denoted by $2V_2$ in Fig. 3.4, is referred to as the "band gap". Given the above definition of the Fermi level, its position could be anywhere within the band gap. A more detailed examination of the problem reveals that actually the Fermi level is at the middle of the gap (see chapter 9).

We consider next another important property of the energy bands.

Theorem Since the hamiltonian is real, that is, the system is time-reversal invariant, we must have:

$$\epsilon_{\mathbf{k}}^{(n)} = \epsilon_{-\mathbf{k}}^{(n)} \tag{3.31}$$

for any state; this is known as Kramers' theorem.

Proof: We take the complex conjugate of the single-particle Schrödinger equation (in the following we drop the band index (n) for simplicity):

$$\mathcal{H}^{sp}\psi_{\mathbf{k}}(\mathbf{r}) = \epsilon_{\mathbf{k}}\psi_{\mathbf{k}}(\mathbf{r}) \Rightarrow \mathcal{H}^{sp}\psi_{\mathbf{k}}^*(\mathbf{r}) = \epsilon_{\mathbf{k}}\psi_{\mathbf{k}}^*(\mathbf{r}) \tag{3.32}$$

that is, the wavefunctions $\psi_{\mathbf{k}}$ and $\psi_{\mathbf{k}}^*$ have the same (real) eigenvalue $\epsilon_{\mathbf{k}}$. However, we can identify $\psi_{\mathbf{k}}^*(\mathbf{r})$ with $\psi_{-\mathbf{k}}(\mathbf{r})$ because

$$\begin{aligned}\psi_{-\mathbf{k}}(\mathbf{r}) &= \mathrm{e}^{-\mathrm{i}\mathbf{k}\cdot\mathbf{r}}\sum_{\mathbf{G}}\alpha_{-\mathbf{k}}(\mathbf{G})\mathrm{e}^{\mathrm{i}\mathbf{G}\cdot\mathbf{r}}\\ \psi_{\mathbf{k}}^*(\mathbf{r}) &= \mathrm{e}^{-\mathrm{i}\mathbf{k}\cdot\mathbf{r}}\sum_{\mathbf{G}}\alpha_{\mathbf{k}}^*(\mathbf{G})\mathrm{e}^{-\mathrm{i}\mathbf{G}\cdot\mathbf{r}}\end{aligned} \tag{3.33}$$

and the only requirement for these two wavefunctions to be the same is: $\alpha_{-\mathbf{k}}(\mathbf{G}) = \alpha^*_{\mathbf{k}}(-\mathbf{G})$, which we take as the definition of the $\alpha_{-\mathbf{k}}(\mathbf{G})$'s. Then the wavefunction $\psi_{-\mathbf{k}}(\mathbf{r})$ is a solution of the single-particle equation, with the proper behavior $\psi_{-\mathbf{k}}(\mathbf{r}+\mathbf{R}) = \exp(-\mathrm{i}\mathbf{k}\cdot\mathbf{R})u_{-\mathbf{k}}(\mathbf{r})$, and the eigenvalue $\epsilon_{-\mathbf{k}} = \epsilon_{\mathbf{k}}$.

A more detailed analysis which takes into account spin states explicitly reveals that, for spin 1/2 particles Kramers' theorem becomes

$$\epsilon_{-\mathbf{k},\uparrow} = \epsilon_{\mathbf{k},\downarrow}, \quad \psi_{-\mathbf{k},\uparrow}(\mathbf{r}) = \mathrm{i}\sigma_y \psi^*_{\mathbf{k},\downarrow}(\mathbf{r}) \tag{3.34}$$

where σ_y is a Pauli matrix (see Problem 3). For systems with equal numbers of up and down spins, Kramers' theorem amounts to inversion symmetry in reciprocal space.

A simple and useful generalization of the free-electron model is to consider that the crystal potential is not exactly vanishing but very weak. Using the Fourier expansion of the potential and the Bloch states,

$$\begin{aligned} V(\mathbf{r}) &= \sum_{\mathbf{G}} V(\mathbf{G})\mathrm{e}^{\mathrm{i}\mathbf{G}\cdot\mathbf{r}} \\ \psi_{\mathbf{k}}(\mathbf{r}) &= \mathrm{e}^{\mathrm{i}\mathbf{k}\cdot\mathbf{r}} \sum_{\mathbf{G}} \alpha_{\mathbf{k}}(\mathbf{G})\mathrm{e}^{\mathrm{i}\mathbf{G}\cdot\mathbf{r}} \end{aligned} \tag{3.35}$$

in the single-particle Schrödinger equation, we obtain the following equation:

$$\sum_{\mathbf{G}} \left[\frac{\hbar^2}{2m_e}(\mathbf{k}+\mathbf{G})^2 - \epsilon_{\mathbf{k}} + \sum_{\mathbf{G}'} V(\mathbf{G}')\mathrm{e}^{\mathrm{i}\mathbf{G}'\cdot\mathbf{r}} \right] \alpha_{\mathbf{k}}(\mathbf{G})\mathrm{e}^{\mathrm{i}(\mathbf{k}+\mathbf{G})\cdot\mathbf{r}} = 0 \tag{3.36}$$

Multiplying by $\exp(-\mathrm{i}(\mathbf{G}''+\mathbf{k})\cdot\mathbf{r})$ and integrating over $\mathbf{r}$ gives

$$\left[\frac{\hbar^2}{2m_e}(\mathbf{k}+\mathbf{G})^2 - \epsilon_{\mathbf{k}} \right] \alpha_{\mathbf{k}}(\mathbf{G}) + \sum_{\mathbf{G}'} V(\mathbf{G}-\mathbf{G}')\alpha_{\mathbf{k}}(\mathbf{G}') = 0 \tag{3.37}$$

where we have used the relation $\int \exp(\mathrm{i}\mathbf{G}\cdot\mathbf{r})\mathrm{d}\mathbf{r} = \Omega_{PUC}\delta(\mathbf{G})$. This is a linear system of equations in the unknowns $\alpha_{\mathbf{k}}(\mathbf{G})$, which can be solved to determine the values of these unknowns and hence find the eigenfunctions $\psi_{\mathbf{k}}(\mathbf{r})$. Now if the potential is very weak, $V(\mathbf{G}) \approx 0$ for all $\mathbf{G}$, which means that the wavefunction cannot have any components $\alpha_{\mathbf{k}}(\mathbf{G})$ for $\mathbf{G} \neq 0$, since these components can only arise from corresponding features in the potential. In this case we take $\alpha_{\mathbf{k}}(0) = 1$ and obtain

$$\psi_{\mathbf{k}}(\mathbf{r}) = \frac{1}{\sqrt{\Omega}}\mathrm{e}^{\mathrm{i}\mathbf{k}\cdot\mathbf{r}}, \quad \epsilon_{\mathbf{k}} = \frac{\hbar^2\mathbf{k}^2}{2m_e} \tag{3.38}$$

as we would expect for free electrons (here we are neglecting electron–electron interactions for simplicity).

Now suppose that all components of the potential are negligible except for one, $V(\mathbf{G}_0)$, which is small but not negligible, and consequently all coefficients $\alpha_{\mathbf{k}}(\mathbf{G})$ are negligible, except for $\alpha_{\mathbf{k}}(\mathbf{G}_0)$, and we take as before $\alpha_{\mathbf{k}}(0) = 1$, assuming $\alpha_{\mathbf{k}}(\mathbf{G}_0)$ to be much smaller. Then Eq. (3.37) reduces to

$$\alpha_{\mathbf{k}}(\mathbf{G}_0) = \frac{V(\mathbf{G}_0)}{\frac{\hbar^2}{2m_e}\left[\mathbf{k}^2 - (\mathbf{k}+\mathbf{G}_0)^2\right]} \tag{3.39}$$

where we have used the zeroth order approximation for $\epsilon_{\mathbf{k}}$. Given that $V(\mathbf{G}_0)$ is itself small, $\alpha_{\mathbf{k}}(\mathbf{G}_0)$ is indeed very small as long as the denominator is finite. The only chance for the coefficient $\alpha_{\mathbf{k}}(\mathbf{G}_0)$ to be large is if the denominator is vanishingly small, which happens for

$$(\mathbf{k}+\mathbf{G}_0)^2 = \mathbf{k}^2 \Rightarrow \mathbf{k}\cdot\hat{\mathbf{G}}_0 = -\frac{1}{2}|\mathbf{G}_0| \tag{3.40}$$

and this is the condition for Bragg planes! In this case, in order to obtain the correct solution we have to consider both $\alpha_{\mathbf{k}}(0)$ and $\alpha_{\mathbf{k}}(\mathbf{G}_0)$, without setting $\alpha_{\mathbf{k}}(0) = 1$. Since all $V(\mathbf{G}) = 0$ except for $\mathbf{G}_0$, we obtain the following linear system of equations:

$$\left[\frac{\hbar^2\mathbf{k}^2}{2m_e} - \epsilon_{\mathbf{k}}\right]\alpha_{\mathbf{k}}(0) + V^*(\mathbf{G}_0)\alpha_{\mathbf{k}}(\mathbf{G}_0) = 0$$

$$\left[\frac{\hbar^2(\mathbf{k}+\mathbf{G}_0)^2}{2m_e} - \epsilon_{\mathbf{k}}\right]\alpha_{\mathbf{k}}(\mathbf{G}_0) + V(\mathbf{G}_0)\alpha_{\mathbf{k}}(0) = 0 \tag{3.41}$$

where we have used $V(-\mathbf{G}) = V^*(\mathbf{G})$. Solving this system, we obtain

$$\epsilon_{\mathbf{k}} = \frac{\hbar^2\mathbf{k}^2}{2m_e} \pm |V(\mathbf{G}_0)| \tag{3.42}$$

for the two possible solutions. Thus, at Bragg planes (i.e. at the boundaries of the BZ), the energy of the free electrons is modified by the terms $\pm|V(\mathbf{G})|$ for the non-vanishing components of $V(\mathbf{G})$. This is illustrated for the one-dimensional case in Fig. 3.4.

3.3 Dynamics of crystal electrons

It can be shown straightforwardly, using second order perturbation theory, that if we know the energy $\epsilon_{\mathbf{k}}^{(n)}$ for all n at some point $\mathbf{k}$ in the BZ, we can obtain the energy at nearby points. The result is

$$\epsilon_{\mathbf{k}+\mathbf{q}}^{(n)} = \epsilon_{\mathbf{k}}^{(n)} + \frac{\hbar}{m_e}\mathbf{q}\cdot\mathbf{p}^{(nn)}(\mathbf{k}) + \frac{\hbar^2\mathbf{q}^2}{2m_e} + \frac{\hbar^2}{m_e^2}\sum_{n'\neq n}\frac{|\mathbf{q}\cdot\mathbf{p}^{(nn')}(\mathbf{k})|^2}{\epsilon_{\mathbf{k}}^{(n)} - \epsilon_{\mathbf{k}}^{(n')}} \tag{3.43}$$

where the quantities $\mathbf{p}^{(nn')}(\mathbf{k})$ are defined as

$$\mathbf{p}^{(nn')}(\mathbf{k}) = \frac{\hbar}{i}\langle\psi_{\mathbf{k}}^{(n')}|\nabla_{\mathbf{r}}|\psi_{\mathbf{k}}^{(n)}\rangle \tag{3.44}$$

Because of the appearance of terms $\mathbf{q}\cdot\mathbf{p}^{(nn')}(\mathbf{k})$ in the above expressions, this approach is known as $\mathbf{q}\cdot\mathbf{p}$ perturbation theory. The quantities defined in Eq. (3.44) are elements of a two-index matrix (n and n'); the diagonal matrix elements are simply the expectation value of the momentum operator in state $\psi_{\mathbf{k}}^{(n)}(\mathbf{r})$. We can also calculate the same quantity from

$$\nabla_{\mathbf{k}}\epsilon_{\mathbf{k}}^{(n)} = \lim_{\mathbf{q}\to 0}\frac{\partial\epsilon_{\mathbf{k}+\mathbf{q}}^{(n)}}{\partial\mathbf{q}} = \frac{\hbar}{m_e}\mathbf{p}^{(nn)}(\mathbf{k}) \tag{3.45}$$

which shows that the gradient of $\epsilon_{\mathbf{k}}^{(n)}$ with respect to $\mathbf{k}$ (multiplied by the factor $m_e/\hbar$) gives the expectation value of the momentum for the crystal states.

Let us consider the second derivative of $\epsilon_{\mathbf{k}}^{(n)}$ with respect to components of the vector $\mathbf{k}$, denoted by k_i, k_j:

$$\begin{aligned}\frac{1}{\overline{m}_{ij}^{(n)}(\mathbf{k})} &\equiv \frac{1}{\hbar^2}\frac{\partial^2\epsilon_{\mathbf{k}}^{(n)}}{\partial k_i\partial k_j} = \lim_{\mathbf{q}\to 0}\frac{1}{\hbar^2}\frac{\partial^2\epsilon_{\mathbf{k}+\mathbf{q}}^{(n)}}{\partial q_i\partial q_j} \\ &= \frac{1}{m_e}\delta_{ij} + \frac{1}{m_e^2}\sum_{n'\neq n}\frac{p_i^{(nn')}p_j^{(n'n)} + p_j^{(nn')}p_i^{(n'n)}}{\epsilon_{\mathbf{k}}^{(n)} - \epsilon_{\mathbf{k}}^{(n')}}\end{aligned} \tag{3.46}$$

The dimensions of this expression are 1/mass. This can then be directly identified as the inverse effective mass of the quasiparticles, which is no longer a simple scalar quantity but a second rank tensor.

It is important to recognize that, as the expression derived above demonstrates, the effective mass of a crystal electron depends on the wave-vector $\mathbf{k}$ and band index n of its wavefunction, as well as on the wavefunctions and energies of all other crystal electrons with the same $\mathbf{k}$-vector. This is a demonstration of the quasiparticle nature of electrons in a crystal. Since the effective mass involves complicated dependence on the direction of the $\mathbf{k}$-vector and the momenta and energies of many states, it can have different magnitude and even different signs along different crystallographic directions!

We wish next to derive expressions for the evolution with time of the position and velocity of a crystal electron in the state $\psi_{\mathbf{k}}^{(n)}$, that is, figure out its dynamics. To this end, we will need to allow the crystal momentum to acquire a time dependence, $\mathbf{k} = \mathbf{k}(t)$, and include its time derivatives where appropriate. Since we are dealing with a particular band, we will omit the band index n for simplicity.

Considering the time-dependent position of a crystal electron $\mathbf{r}(t)$ as a quantum mechanical operator, we have from the usual formulation in the Heisenberg picture (see Appendix B):

$$\frac{\mathrm{d}\mathbf{r}(t)}{\mathrm{d}t} = \frac{\mathrm{i}}{\hbar}[\mathcal{H}, \mathbf{r}] \Longrightarrow \langle\psi_{\mathbf{k}}|\frac{\mathrm{d}\mathbf{r}(t)}{\mathrm{d}t}|\psi_{\mathbf{k}}\rangle = \frac{\mathrm{i}}{\hbar}\langle\psi_{\mathbf{k}}|[\mathcal{H}, \mathbf{r}]|\psi_{\mathbf{k}}\rangle \tag{3.47}$$

with $[\mathcal{H}, \mathbf{r}]$ the commutator of the hamiltonian with the position operator. Now we can take advantage of the following identity:

$$\nabla_{\mathbf{k}}\left(\mathrm{e}^{-\mathrm{i}\mathbf{k}\cdot\mathbf{r}}\mathcal{H}\mathrm{e}^{\mathrm{i}\mathbf{k}\cdot\mathbf{r}}\right) = \mathrm{i}\mathrm{e}^{-\mathrm{i}\mathbf{k}\cdot\mathbf{r}}[\mathcal{H}, \mathbf{r}]\mathrm{e}^{\mathrm{i}\mathbf{k}\cdot\mathbf{r}} \tag{3.48}$$

whose proof involves simple differentiations of the exponentials with respect to $\mathbf{k}$, and of Eq. (3.22), to rewrite the right-hand side of the previous equation as

$$\begin{aligned}
\frac{\mathrm{i}}{\hbar}\langle\psi_{\mathbf{k}}|[\mathcal{H}, \mathbf{r}]|\psi_{\mathbf{k}}\rangle &\equiv \frac{\mathrm{i}}{\hbar}\int u^*_{\mathbf{k}}(\mathbf{r})\mathrm{e}^{-\mathrm{i}\mathbf{k}\cdot\mathbf{r}}[\mathcal{H}, \mathbf{r}]\mathrm{e}^{\mathrm{i}\mathbf{k}\cdot\mathbf{r}}u_{\mathbf{k}}(\mathbf{r})\mathrm{d}\mathbf{r} \\
&= \frac{1}{\hbar}\int u^*_{\mathbf{k}}(\mathbf{r})\left[\nabla_{\mathbf{k}}\left(\mathrm{e}^{-\mathrm{i}\mathbf{k}\cdot\mathbf{r}}\mathcal{H}\mathrm{e}^{\mathrm{i}\mathbf{k}\cdot\mathbf{r}}\right)\right]u_{\mathbf{k}}(\mathbf{r})\mathrm{d}\mathbf{r} \\
&= \frac{1}{\hbar}\int u^*_{\mathbf{k}}(\mathbf{r})\left[\nabla_{\mathbf{k}}\mathcal{H}(\mathbf{p}+\hbar\mathbf{k}, \mathbf{r})\right]u_{\mathbf{k}}(\mathbf{r})\mathrm{d}\mathbf{r}
\end{aligned}$$

Next, we move the differentiation with respect to $\mathbf{k}$, $\nabla_{\mathbf{k}}$, outside the integral and subtract the additional terms produced by this change, which leads to

$$\begin{aligned}
\frac{\mathrm{i}}{\hbar}\langle\psi_{\mathbf{k}}|[\mathcal{H}, \mathbf{r}]|\psi_{\mathbf{k}}\rangle &= \frac{1}{\hbar}\nabla_{\mathbf{k}}\int u^*_{\mathbf{k}}(\mathbf{r})\mathcal{H}(\mathbf{p}+\hbar\mathbf{k}, \mathbf{r})u_{\mathbf{k}}(\mathbf{r})\mathrm{d}\mathbf{r} \\
-\frac{1}{\hbar}\Big[\int\left(\nabla_{\mathbf{k}}u^*_{\mathbf{k}}(\mathbf{r})\right)\mathcal{H}(\mathbf{p}+\hbar\mathbf{k}, \mathbf{r})u_{\mathbf{k}}(\mathbf{r})\mathrm{d}\mathbf{r} &+ \int u^*_{\mathbf{k}}(\mathbf{r})\mathcal{H}(\mathbf{p}+\hbar\mathbf{k}, \mathbf{r})\left(\nabla_{\mathbf{k}}u_{\mathbf{k}}(\mathbf{r})\right)\mathrm{d}\mathbf{r}\Big]
\end{aligned}$$

We deal with the last two terms in the above expression separately: recalling that $u_{\mathbf{k}}(\mathbf{r})$ is an eigenfunction of the hamiltonian $\mathcal{H}(\mathbf{p}+\hbar\mathbf{k}, \mathbf{r})$ with eigenvalue $\epsilon_{\mathbf{k}}$, we obtain for these two terms:

$$\begin{aligned}
&\langle\nabla_{\mathbf{k}}u_{\mathbf{k}}|\mathcal{H}(\mathbf{p}+\hbar\mathbf{k})|u_{\mathbf{k}}\rangle + \langle u_{\mathbf{k}}|\mathcal{H}(\mathbf{p}+\hbar\mathbf{k})|\nabla_{\mathbf{k}}u_{\mathbf{k}}\rangle \\
&= \epsilon_{\mathbf{k}}\left(\langle\nabla_{\mathbf{k}}u_{\mathbf{k}}|u_{\mathbf{k}}\rangle + \langle u_{\mathbf{k}}|\nabla_{\mathbf{k}}u_{\mathbf{k}}\rangle\right) = \epsilon_{\mathbf{k}}\nabla_{\mathbf{k}}\langle u_{\mathbf{k}}|u_{\mathbf{k}}\rangle = 0
\end{aligned} \tag{3.49}$$

since $\langle u_{\mathbf{k}}|u_{\mathbf{k}}\rangle = 1$ for properly normalized wavefunctions.[1] This leaves the following result:

$$\langle\psi_{\mathbf{k}}|\frac{\mathrm{d}\mathbf{r}(t)}{\mathrm{d}t}|\psi_{\mathbf{k}}\rangle = \frac{1}{\hbar}\nabla_{\mathbf{k}}\int\psi^*_{\mathbf{k}}(\mathbf{r})\mathcal{H}\psi_{\mathbf{k}}(\mathbf{r})\mathrm{d}\mathbf{r} \Longrightarrow \langle\mathbf{v}_{\mathbf{k}}\rangle = \frac{1}{\hbar}\nabla_{\mathbf{k}}\epsilon_{\mathbf{k}} \tag{3.50}$$

where we have identified the velocity $\langle\mathbf{v}_{\mathbf{k}}\rangle$ of a crystal electron in state $\psi_{\mathbf{k}}$ with the expectation value of the time derivative of the operator $\mathbf{r}(t)$ in that state. This result

[1] This is a special case of the more general Hellmann–Feynman theorem, which we will encounter again in chapter 5.

is equivalent to Eq. (3.45) which we derived above for the momentum of a crystal electron in state $\psi_{\mathbf{k}}$.

Taking a derivative of the velocity in state $\psi_{\mathbf{k}}$ with respect to time, and using the chain rule for differentiation with respect to $\mathbf{k}$, we find

$$\frac{\mathrm{d}\langle \mathbf{v}_{\mathbf{k}} \rangle}{\mathrm{d}t} = \frac{1}{\hbar} \left(\frac{\mathrm{d}\mathbf{k}}{\mathrm{d}t} \cdot \nabla_{\mathbf{k}} \right) \nabla_{\mathbf{k}} \epsilon_{\mathbf{k}} \tag{3.51}$$

which we can write in terms of cartesian components $(i, j = x, y, z)$ as

$$\frac{\mathrm{d}\langle v^i_{\mathbf{k}} \rangle}{\mathrm{d}t} = \sum_j \left(\hbar \frac{\mathrm{d}k_j}{\mathrm{d}t} \right) \left[\frac{1}{\hbar^2} \frac{\partial^2 \epsilon_{\mathbf{k}}}{\partial k_j \partial k_i} \right] = \sum_j \left(\hbar \frac{\mathrm{d}k_j}{\mathrm{d}t} \right) \frac{1}{\overline{m}_{ji}(\mathbf{k})} \tag{3.52}$$

where we have identified the term in the square brackets as the inverse effective mass tensor derived in Eq. (3.46). With this identification, this equation has the form of *acceleration* = *force/mass*, as might be expected, but the mass is not a simple scalar quantity as in the case of free electrons; the mass is now a second rank tensor corresponding to the behavior of crystal electrons. The form of Eq. (3.52) compels us to identify the quantities in parentheses on the right-hand side as the components of the external force acting on the crystal electron:

$$\hbar \frac{\mathrm{d}\mathbf{k}}{\mathrm{d}t} = \mathbf{F} \tag{3.53}$$

We should note two important things. First, the identification of the time derivative of $\hbar\mathbf{k}$ with the external force at this point cannot be considered a proper proof; it is only an inference from the dimensions of the different quantities that enter into Eq. (3.52). In particular, so far we have discussed the dynamics of electrons in the crystal *without* the presence of any external forces, which could potentially change the wavefunctions and the eigenvalues of the single-particle hamiltonian; we examine this issue in detail in the next section. Second, to the extend that the above relation actually holds, it is a state-independent equation, so the wave-vectors evolve in the same manner for all states!

3.4 Crystal electrons in an electric field

We consider next what happens to crystal electrons when they are subjected to a constant external electric field $\mathbf{E}$, which gives rise to an electrostatic potential $\Phi(\mathbf{r}) = -\mathbf{E} \cdot \mathbf{r}$. External electric fields are typically used to induce electron transport, for instance in electronic devices. The periodicity of the crystal has profound effects on the transport properties of electrons, and influences strongly their response to external electric fields.

For the present discussion we will include the effects of the external field from the beginning, starting with the new hamiltonian in the presence of the external electric field, which is

$$\mathcal{H}_{\mathbf{E}} = \mathcal{H}_0 + q\Phi(\mathbf{r}) = \mathcal{H}_0 - q\mathbf{E} \cdot \mathbf{r}$$

where $\mathcal{H}_0$ is the hamiltonian of the crystal in zero field and q the charge of the particles (for electrons $q = -e$). To make a connection to the results of the previous section, we will need to construct the proper states, characterized by wave-vectors $\mathbf{k}$, that are relevant to the new hamiltonian: as we saw at the end of the previous section, we expect the wave-vectors themselves to acquire a time dependence. From the relation

$$\begin{aligned} \nabla_{\mathbf{k}}\psi_{\mathbf{k}}(\mathbf{r}) &= \nabla_{\mathbf{k}} \mathrm{e}^{\mathrm{i}\mathbf{k}\cdot\mathbf{r}} u_{\mathbf{k}}(\mathbf{r}) = \mathrm{i}\mathbf{r}\psi_{\mathbf{k}}(\mathbf{r}) + \mathrm{e}^{\mathrm{i}\mathbf{k}\cdot\mathbf{r}} \nabla_{\mathbf{k}} u_{\mathbf{k}}(\mathbf{r}) \\ \Longrightarrow \mathbf{r}\psi_{\mathbf{k}}(\mathbf{r}) &= -\mathrm{i}\nabla_{\mathbf{k}}\psi_{\mathbf{k}}(\mathbf{r}) + \mathrm{i}\mathrm{e}^{\mathrm{i}\mathbf{k}\cdot\mathbf{r}} \nabla_{\mathbf{k}} \mathrm{e}^{-\mathrm{i}\mathbf{k}\cdot\mathbf{r}} \psi_{\mathbf{k}}(\mathbf{r}) \end{aligned} \tag{3.54}$$

we conclude that the action of the hamiltonian $\mathcal{H}_{\mathbf{E}}$ on state $\psi_{\mathbf{k}}$ is equivalent to the action of

$$\mathcal{H}_0 - \mathrm{i}q\mathrm{e}^{\mathrm{i}\mathbf{k}\cdot\mathbf{r}}\mathbf{E} \cdot \nabla_{\mathbf{k}} \mathrm{e}^{-\mathrm{i}\mathbf{k}\cdot\mathbf{r}} + \mathrm{i}q\mathbf{E} \cdot \nabla_{\mathbf{k}}$$

The first new term in this hamiltonian has non-vanishing matrix elements between states of the same $\mathbf{k}$ only. To show this, consider the matrix element of this term between two states of different $\mathbf{k}$:

$$\begin{aligned} \langle\psi_{\mathbf{k}'}| \left(\mathrm{e}^{\mathrm{i}\mathbf{k}\cdot\mathbf{r}}\mathbf{E} \cdot \nabla_{\mathbf{k}} \mathrm{e}^{-\mathrm{i}\mathbf{k}\cdot\mathbf{r}}\right) |\psi_{\mathbf{k}}\rangle &= \mathbf{E} \cdot \int u^*_{\mathbf{k}'}(\mathbf{r}) \mathrm{e}^{-\mathrm{i}\mathbf{k}'\cdot\mathbf{r}} \mathrm{e}^{\mathrm{i}\mathbf{k}\cdot\mathbf{r}} \nabla_{\mathbf{k}} \mathrm{e}^{-\mathrm{i}\mathbf{k}\cdot\mathbf{r}} \mathrm{e}^{\mathrm{i}\mathbf{k}\cdot\mathbf{r}} u_{\mathbf{k}}(\mathbf{r}) \mathrm{d}\mathbf{r} \\ &= \mathbf{E} \cdot \int \mathrm{e}^{\mathrm{i}(\mathbf{k}-\mathbf{k}')\cdot\mathbf{r}} \left[u^*_{\mathbf{k}'}(\mathbf{r}) \nabla_{\mathbf{k}} u_{\mathbf{k}}(\mathbf{r})\right] \mathrm{d}\mathbf{r} \end{aligned}$$

Now, in this last expression, the square bracket includes only terms which have the full periodicity of the lattice, as we have discussed about the functions $u_{\mathbf{k}}(\mathbf{r})$:

$$\left[u^*_{\mathbf{k}'}(\mathbf{r}) \nabla_{\mathbf{k}} u_{\mathbf{k}}(\mathbf{r})\right] = f(\mathbf{r}) \to f(\mathbf{r} + \mathbf{R}) = f(\mathbf{r})$$

Therefore, this term can be expressed through its Fourier transform, which will take the form

$$f(\mathbf{r}) = \sum_{\mathbf{G}} f(\mathbf{G}) \mathrm{e}^{-\mathrm{i}\mathbf{G}\cdot\mathbf{r}}$$

with $\mathbf{G}$ the reciprocal lattice vectors; inserting this into the previous equation, after integrating over $\mathbf{r}$, we obtain terms which reduce to

$$f(\mathbf{G})\delta((\mathbf{k} - \mathbf{k}') - \mathbf{G}) \to \mathbf{k} - \mathbf{k}' = \mathbf{G}$$

for non-vanishing matrix elements, or, if we choose to work with $\mathbf{k}$, $\mathbf{k}'$ in the first BZ, we must have $\mathbf{k} = \mathbf{k}'$. This establishes that the first new term in the hamiltonian has non-vanishing matrix elements only between states of the same $\mathbf{k}$. Consequently, we can choose to work with basis states that are eigenfunctions of the hamiltonian

$$\tilde{\mathcal{H}}_{\mathbf{E}} = \mathcal{H}_0 - \left(\mathrm{i}q\mathrm{e}^{\mathrm{i}\mathbf{k}\cdot\mathbf{r}}\mathbf{E}\cdot\nabla_{\mathbf{k}}\mathrm{e}^{-\mathrm{i}\mathbf{k}\cdot\mathbf{r}}\right)$$

and are characterized by wave-vectors $\mathbf{k}$. These states, which we call $\tilde{\psi}_{\mathbf{k}}$, must have the form

$$\tilde{\psi}_{\mathbf{k}}(\mathbf{r}) = \mathrm{e}^{-(\mathrm{i}/\hbar)\int \tilde{\epsilon}_{\mathbf{k}}\mathrm{d}t}\mathrm{e}^{\mathrm{i}\mathbf{k}\cdot\mathbf{r}}\tilde{u}_{\mathbf{k}}(\mathbf{r})$$

where we have introduced the eigenvalues $\tilde{\epsilon}_{\mathbf{k}}$ of the hamiltonian $\tilde{\mathcal{H}}_{\mathbf{E}}$ explicitly, with $\mathbf{k}$ a time-dependent quantity. The $\tilde{\psi}_{\mathbf{k}}$ states will form the basis for the solution of the time-dependent Schrödinger equation of the full hamiltonian therefore:

$$\mathrm{i}\hbar\frac{\partial\tilde{\psi}_{\mathbf{k}}}{\partial t} = \left[\tilde{\mathcal{H}}_{\mathbf{E}} + \mathrm{i}q\mathbf{E}\cdot\nabla_{\mathbf{k}}\right]\tilde{\psi}_{\mathbf{k}} = \left[\tilde{\epsilon}_{\mathbf{k}} + \mathrm{i}q\mathbf{E}\cdot\nabla_{\mathbf{k}}\right]\tilde{\psi}_{\mathbf{k}} \tag{3.55}$$

while from the expression we wrote above for $\tilde{\psi}_{\mathbf{k}}$ we find for its time derivative

$$\frac{\partial\tilde{\psi}_{\mathbf{k}}}{\partial t} = \left[-\frac{\mathrm{i}}{\hbar}\tilde{\epsilon}_{\mathbf{k}} + \frac{\mathrm{d}\mathbf{k}(t)}{\mathrm{d}t}\cdot\nabla_{\mathbf{k}}\right]\tilde{\psi}_{\mathbf{k}} \tag{3.56}$$

Comparing the right-hand side of these two last equations term by term we find

$$\hbar\frac{\mathrm{d}\mathbf{k}}{\mathrm{d}t} = -e\mathbf{E} \tag{3.57}$$

This result gives the time evolution of the wave-vector $\mathbf{k}$ for the state $\tilde{\psi}_{\mathbf{k}}$, and is consistent with what we expected from Eq. (3.53).

Let us consider a simple example to illustrate this behavior: suppose we have a one-dimensional crystal with a single energy band,

$$\epsilon_k = \epsilon + 2t\cos(ka) \tag{3.58}$$

with ϵ a reference energy and t a negative constant.[2] This is shown in Fig. 3.5: the lattice constant is a, so that the first BZ for this crystal extends from $-\pi/a$ to π/a, and the energy ranges from a minimum of $\epsilon + 2t$ (at $k = 0$) to a maximum of $\epsilon - 2t$ (at $k = \pm\pi/a$). The momentum for this state is given by

$$p(k) = \frac{m_e}{\hbar}\frac{\mathrm{d}\epsilon_k}{\mathrm{d}k} = -\frac{2m_e ta}{\hbar}\sin(ka) \tag{3.59}$$

[2] In chapter 4 we discuss the physical system that can give rise to such an expression for the single-particle energy ϵ_k.

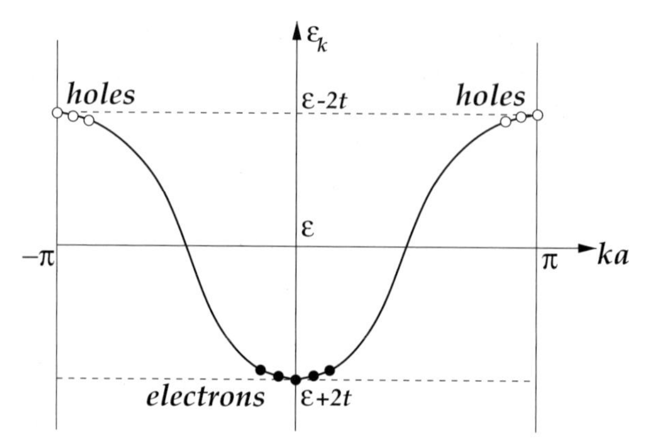

Figure 3.5. Single-particle energy eigenvalues ϵ_k for the simple example that illustrates the dynamics of electrons and holes in a one-band, one-dimensional crystal.

When the system is in an external electric field $\mathbf{E} = -E_0\hat{\mathbf{x}}$ with E_0 a positive constant, the time evolution of the wave-vector will be

$$\frac{\mathrm{d}\mathbf{k}}{\mathrm{d}t} = -\frac{e}{\hbar}\mathbf{E} \Rightarrow \frac{\mathrm{d}k}{\mathrm{d}t} = \frac{e}{\hbar}E_0 \tag{3.60}$$

We will consider some limiting cases for this idealized system.

(1) A single electron in this band would start at $k = 0$ at $t = 0$ (before the application of the external field it occupies the lowest energy state), then its k would increase at a constant rate $(eE_0/\hbar)$ until the value π/a is reached, and then it would re-enter the first BZ at $k = -\pi/a$ and continue this cycle. The same picture would hold for a few electrons initially occupying the bottom of the band.
(2) If the band were completely full, then all the wave-vectors would be changing in the same way, and all states would remain completely full. Since this creates no change in the system, we conclude that no current would flow: a full band cannot contribute to current flow in an ideal crystal!
(3) If the band were mostly full, with only a few states empty at wave-vectors near the boundaries of the first BZ (at $k = \pm\pi/a$) where the energy is a maximum, then the total current would be

$$\mathbf{I} = -e\sum_{k\leq k_{\mathrm{F}}}\mathbf{v}(k) = -e\sum_{k\leq k_{\mathrm{F}}}\frac{p(k)}{m_e}\hat{\mathbf{x}} = \left[-e\sum_{k\in\mathrm{BZ}}\frac{p(k)}{m_e} + e\sum_{k>k_{\mathrm{F}}}\frac{p(k)}{m_e}\right]\hat{\mathbf{x}}$$
$$= e\sum_{k>k_{\mathrm{F}}}\frac{p(k)}{m_e}\hat{\mathbf{x}} \tag{3.61}$$

where in the first two sums the summation is over k values in the first BZ, restricted to $k \leq k_{\mathrm{F}}$ that is, over occupied states only, whereas in the the last sum it is restricted over $k > k_{\mathrm{F}}$, that is, over unoccupied states only. The last equality follows from the fact that the sum over all values of $k \in \mathrm{BZ}$ corresponds to the full band, which as explained above does not contribute to the current. Thus, in this case the system behaves like a set of positively charged particles, referred to as holes (unoccupied states). In our simple one-dimensional example we can use the general expression for the effective mass derived in Eq. (3.46), to find near the top of the band:

$$\frac{1}{\overline{m}} = \frac{1}{\hbar^2}\frac{\mathrm{d}^2\epsilon_k}{\mathrm{d}k^2} \Longrightarrow \overline{m} = \hbar^2\left(\left[\frac{\mathrm{d}^2\epsilon_k}{\mathrm{d}k^2}\right]_{k=\pm\pi/a}\right)^{-1} = \frac{\hbar^2}{2ta^2} \tag{3.62}$$

which is a *negative* quantity (recall that $t < 0$). Using the Taylor expansion of cos near $k_0 = \pm\pi/a$, we can write the energy near the top of the band as

$$\begin{aligned}\epsilon_{\pm(\pi/a)+k} &= \epsilon + 2t\left[-1 + \frac{1}{2}(ka)^2 + \cdots\right] \\ &= (\epsilon - 2t) + ta^2k^2 = (\epsilon - 2t) + \frac{\hbar^2k^2}{2\overline{m}}\end{aligned} \tag{3.63}$$

with the effective mass $\overline{m}$ being the negative quantity found in Eq. (3.62). From the general expression Eq. (3.61), we find the time derivative of the current in our one-dimensional example to be

$$\frac{\mathrm{d}\mathbf{I}}{\mathrm{d}t} = e\sum_{k>k_\mathrm{F}}\frac{1}{m_e}\frac{\mathrm{d}p(k)}{\mathrm{d}t}\hat{\mathbf{x}} = e\sum_{k>k_\mathrm{F}}\frac{\mathrm{d}}{\mathrm{d}t}\left(\frac{1}{\hbar}\frac{\mathrm{d}}{\mathrm{d}k}\epsilon_k\right)\hat{\mathbf{x}} = \frac{e}{\overline{m}}\sum_{k>k_\mathrm{F}}\hbar\frac{\mathrm{d}k}{\mathrm{d}t}\hat{\mathbf{x}} \tag{3.64}$$

where we have used Eq. (3.45) to obtain the second equality and Eq. (3.63) to obtain the third equality. Now, assuming that we are dealing with a single hole state at the top of the band, and using the general result of Eq. (3.57), we obtain

$$\frac{\mathrm{d}\mathbf{I}}{\mathrm{d}t} = -\frac{e^2}{\overline{m}}\mathbf{E} = \frac{e^2}{|\overline{m}|}\mathbf{E} \tag{3.65}$$

which describes the response of a *positively* charged particle of charge $+e$ and *positive* mass $|\overline{m}|$ to an external field $\mathbf{E}$, as we expect for holes.

3.5 Crystal symmetries beyond periodicity

In the previous sections we discussed the effects of lattice periodicity on the single-particle wavefunctions and the energy eigenvalues. The one-dimensional examples we presented there can have only this type of symmetry. In two- and three-dimensional cases, a crystal can also have symmetries beyond the translational periodicity, such as rotations around axes, reflections on planes, and combinations of these operations among themselves and with translations by vectors that are *not* lattice vectors. All these symmetry operations are useful in calculating and analyzing the physical properties of a crystal. There are two basic advantages to using the symmetry operations of a crystal in describing its properties. First, the volume in reciprocal space for which solutions need to be calculated is further reduced, usually to a small fraction of the first Brillouin Zone, called the irreducible part; for example, in the FCC crystals with one atom per unit cell, the irreducible part is 1/48 of the full BZ. Second, certain selection rules and compatibility relations are dictated by symmetry alone, leading to a deeper understanding of the physical properties of the crystal as well as to simpler ways of calculating these properties in the context of the single-particle picture; for example, using symmetry arguments it is possible to identify the allowed optical transitions in a crystal, which involve excitation or de-excitation of electrons by absorption or emission of photons, thereby elucidating its optical properties.

Taking full advantage of the crystal symmetries requires the use of group theory. This very interesting branch of mathematics is particularly well suited to reduce the amount of work by effectively using group representations in the description of single-particle eigenfunctions. Although conceptually straightforward, the theory of group representations requires a significant amount of discussion, which is beyond the scope of the present treatment. Here, we will develop some of the basic concepts of group theory and employ them in simple illustrative examples.

To illustrate the importance of taking into account all the crystal symmetries we discuss a simple example. Consider a two-dimensional square lattice with lattice constant a, and one atom per unit cell. Assume that this atom has Z electrons so that there are Z/a^2 electrons per unit cell, that is, a total of NZ electrons in the crystal of volume Na^2, where N is the number of unit cells. The simplest case is that of free electrons. We want to find the behavior of energy eigenvalues as a function of the reciprocal lattice vector $\mathbf{k}$, for the various possible solutions. Let us first try to find the Fermi momentum and Fermi energy for this system. The Fermi momentum is obtained by integrating over all $\mathbf{k}$-vectors until we have enough states to accommodate all the electrons. Taking into account a factor of 2 for spin, the total number of states we need in reciprocal space to accommodate all the electrons of the crystal is given by

$$\sum_{\mathbf{k},|\mathbf{k}|<k_{\mathrm{F}}} 2 \rightarrow \frac{2}{(2\pi)^2}(Na^2)\int_{|\mathbf{k}|<k_{\mathrm{F}}} \mathrm{d}\mathbf{k} = NZ \Rightarrow \frac{2}{(2\pi)^2}\int_{|\mathbf{k}|<k_{\mathrm{F}}} \mathrm{d}\mathbf{k} = \frac{Z}{a^2} \tag{3.66}$$

where we have used $\mathrm{d}\mathbf{k} = (2\pi)^2/(Na^2)$ for the two-dimensional square lattice, by analogy to the general result for the three-dimensional lattice $\mathrm{d}\mathbf{k} = (2\pi)^3/(N\Omega_{PUC})$ (see Appendix D). Using spherical coordinates for the integration over $\mathbf{k}$, we obtain:

$$\frac{2}{(2\pi)^2}2\pi\int_0^{k_{\mathrm{F}}} k\mathrm{d}k = \frac{1}{2\pi}k_{\mathrm{F}}^2 = \frac{Z}{a^2} \Rightarrow k_{\mathrm{F}} = \left(\frac{2\pi Z}{a^2}\right)^{1/2} \tag{3.67}$$

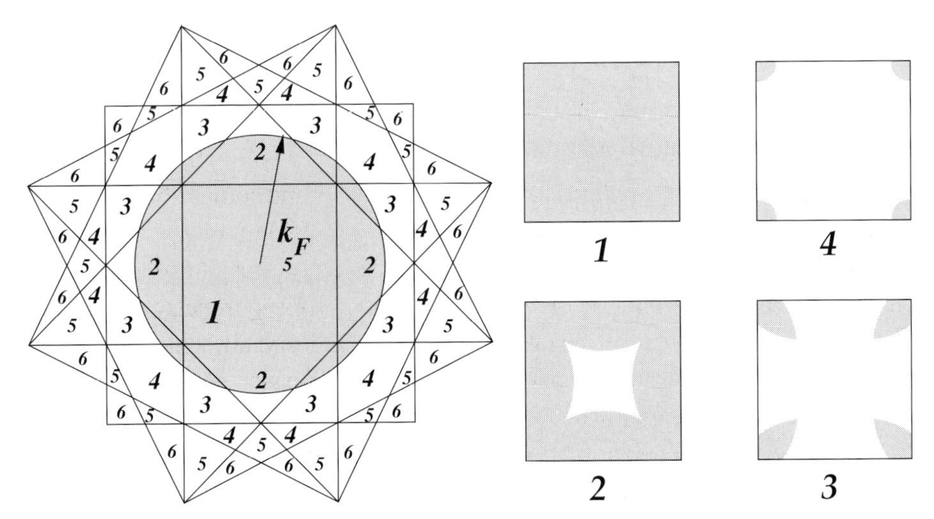

Figure 3.6. **Left:** the first six Brillouin Zones of the two-dimensional square lattice. The position of the Fermi sphere (Fermi surface) for a crystal with $Z = 4$ electrons per unit cell is indicated by a shaded circle. **Right:** the shape of occupied portions of the various Brillouin Zones for the 2D square lattice with $Z = 4$ electrons per unit cell.

and from this we can obtain the Fermi energy $\epsilon_{\rm F}$:

$$\epsilon_{\rm F} = \frac{\hbar^2 k_{\rm F}^2}{2m_e} = \frac{\hbar^2}{2m_e}\frac{2\pi Z}{a^2} \tag{3.68}$$

The value of $k_{\rm F}$ determines the so called "Fermi sphere" in reciprocal space, which contains all the occupied states for the electrons. This Fermi sphere corresponds to a certain number of BZs with all states occupied by electrons, and a certain number of partially occupied BZs with interestingly shaped regions for the occupied portions. The number of full BZs and the shapes of occupied regions in partially filled BZs depend on $k_{\rm F}$ through Z; an example for $Z = 4$ is shown in Fig. 3.6.

Suppose that we want to plot the energy levels as a function of the **k**-vector as we did in the one-dimensional case: what do these look like, and what does the occupation of states in the various BZs mean? The answer to these questions will help us visualize and understand the behavior of the physical system. First, we realize that the **k**-vectors are two-dimensional so we would need a three-dimensional plot to plot the energy eigenvalues, two axes for the k_x, k_y components and one for the energy. Moreover, in the extended zone scheme each component of **k** will span different ranges of values in different BZs, making the plot very complicated. Using the reduced Zone scheme is equally complicated now that we are dealing with a two-dimensional system. With the tools we have developed so far, there is no simple solution to the problem. We will discuss next how we can take advantage of symmetry to reduce the problem to something manageable.

3.6 Groups and symmetry operators

We first define the notion of a group: it is a finite or infinite set of operations, which satisfy the following four conditions:

1. **Closure**: if A and B are elements of the group so is $C = AB$.
2. **Associativity**: if A, B, C are members of a group then $(AB)C = A(BC)$.
3. **Unit element**: there is one element E of the group for which $EA = AE = A$, where A is every other element of the group.
4. **Inverse element**: for every element A of the group, there exists another element called its inverse A^{-1}, for which $AA^{-1} = A^{-1}A = E$.

A *subgroup* is a subset of operations of a group which form a group by themselves (they satisfy the above four conditions). Finally, two groups A and B are called *isomorphic* when there is a one-to-one correspondence between their elements $A_1, A_2, ..., A_n$ and $B_1, B_2, ..., B_n$, such that

$$A_1 A_2 = A_3 \Longrightarrow B_1 B_2 = B_3 \tag{3.69}$$

For a crystal we can define the following groups:

(a) S : the space group, which contains all symmetry operations that leave the crystal invariant.
(b) T : the translation group, which contains all the translations that leave the crystal invariant, that is, all the lattice vectors $\mathbf{R}$. T is a subgroup of S.
(c) P : the point group, which contains all space group operations with the translational part set equal to zero. P is not necessarily a subgroup of S, because when the translational part is set to zero, some operations may no longer leave the crystal invariant if they involved a translation by a vector not equal to any of the lattice vectors.

The space group can contain the following types of operations:

- lattice translations (all the Bravais lattice vectors $\mathbf{R}$), which form T;
- proper rotations by π, $2\pi/3$, $\pi/2$, $\pi/3$, around symmetry axes;
- improper rotations, like inversion I, or reflection on a plane σ;
- screw axes, which involve a proper rotation and a translation by $\mathbf{t} \neq \mathbf{R}$;
- glide planes, which involve a reflection and a translation by $\mathbf{t} \neq \mathbf{R}$.

To apply symmetry operations on various functions, we introduce the operators corresponding to the symmetries. These will be symbolized by $\{U|\mathbf{t}\}$, where U corresponds to a proper or improper rotation and $\mathbf{t}$ to a translation. These operators act on real space vectors according to the rule:

$$\{U|\mathbf{t}\}\mathbf{r} = U\mathbf{r} + \mathbf{t} \tag{3.70}$$

In terms of these operators, the translations can be described as $\{E|\mathbf{R}\}$ (where E is the identity, or unit element), while the pure rotations (proper or improper) are described as $\{U|0\}$. The first set corresponds to the translation group T, the second to the point group P. For any two symmetry operators belonging to the space group,

$$\{U_1|\mathbf{t}_1\}\{U_2|\mathbf{t}_2\}\mathbf{r} = \{U_1|\mathbf{t}_1\}[U_2\mathbf{r} + \mathbf{t}_2] = U_1U_2\mathbf{r} + U_1\mathbf{t}_2 + \mathbf{t}_1 \tag{3.71}$$

which means that the operator $\{U_1U_2|U_1\mathbf{t}_2 + \mathbf{t}_1\}$ corresponds to an element of the space group. Using this rule for multiplication of operators, we can easily show that the inverse of $\{U|\mathbf{t}\}$ is

$$\{U|\mathbf{t}\}^{-1} = \{U^{-1}| - U^{-1}\mathbf{t}\} \tag{3.72}$$

The validity of this can be verified by applying $\{U^{-1}| - U^{-1}\mathbf{t}\}\{U|\mathbf{t}\}$ on $\mathbf{r}$, whereupon we get back $\mathbf{r}$.

Some general remarks on group theory applications to crystals are in order. We saw that P consists of all space group operators with the translational part set equal to zero. Therefore generally $\mathsf{P} \times \mathsf{T} \neq \mathsf{S}$, because several of the group symmetries may involve non-lattice vector translations $\mathbf{t} \neq \mathbf{R}$. The point group leaves the Bravais lattice invariant but not the crystal invariant (recall that the crystal is defined by the Bravais lattice and the atomic basis in each unit cell). If the crystal has glide planes and screw rotations, that is, symmetry operations that involve $\{U|\mathbf{t}\}$ with $\mathbf{t} \neq 0$, $\{E|\mathbf{t}\} \notin \mathsf{T}$, then P does not leave the crystal invariant. Groups that include such symmetry operations are referred to as non-symmorphic groups. A symmorphic group is one in which a proper choice of the origin of coordinates eliminates all non-lattice translations, in which case P is a true subgroup of S. Finally, all symmetry operators in the space group commute with the hamiltonian, since they leave the crystal, and therefore the external potential, invariant.

In three dimensions there are 14 different types of Bravais lattices, 32 different types of point groups, and a total of 230 different space groups, of which 73 are symmorphic and 157 are non-symmorphic. We will give a more detailed account of symmetry groups in three dimensions after we have considered a simple example in two dimensions.

3.7 Symmetries of the band structure

In the following we will need to apply symmetry operations to functions of the space variable $\mathbf{r}$, so we define a new set of operators whose effect is to change $\mathbf{r}$:

$$\mathcal{O}_{\{U|\mathbf{t}\}} f(\mathbf{r}) = f(\{U|\mathbf{t}\}^{-1}\mathbf{r}) \tag{3.73}$$

In this definition of $\mathcal{O}_{\{U|\mathbf{t}\}}$, the action is always on the vector $\mathbf{r}$ itself. We will prove that the group formed by the operators $\mathcal{O}_{\{U|\mathbf{t}\}}$ is isomorphic to the group of operators $\{U|\mathbf{t}\}$:

$$\begin{aligned}
\mathcal{O}_{\{U_1|\mathbf{t}_1\}}\mathcal{O}_{\{U_2|\mathbf{t}_2\}}f(\mathbf{r}) &= \mathcal{O}_{\{U_1|\mathbf{t}_1\}}f(\{U_2|\mathbf{t}_2\}^{-1}\mathbf{r}) = \mathcal{O}_{\{U_1|\mathbf{t}_1\}}g(\mathbf{r}) \\
&= g(\{U_1|\mathbf{t}_1\}^{-1}\mathbf{r}) = f(\{U_2|\mathbf{t}_2\}^{-1}\{U_1|\mathbf{t}_1\}^{-1}\mathbf{r}) \\
&= f([\{U_1|\mathbf{t}_1\}\{U_2|\mathbf{t}_2\}]^{-1}\mathbf{r}) = \mathcal{O}_{\{U_1|\mathbf{t}_1\}\{U_2|\mathbf{t}_2\}}f(\mathbf{r}) \\
&\Rightarrow \mathcal{O}_{\{U_1|\mathbf{t}_1\}}\mathcal{O}_{\{U_2|\mathbf{t}_2\}} = \mathcal{O}_{\{U_1|\mathbf{t}_1\}\{U_2|\mathbf{t}_2\}}
\end{aligned} \tag{3.74}$$

where we have defined the intermediate function $g(\mathbf{r}) = f(\{U_2|\mathbf{t}_2\}^{-1}\mathbf{r})$ to facilitate the proof. We have also used the relation

$$\{U_2|\mathbf{t}_2\}^{-1}\{U_1|\mathbf{t}_1\}^{-1} = [\{U_1|\mathbf{t}_1\}\{U_2|\mathbf{t}_2\}]^{-1}$$

which can be easily proved from the definition of the inverse element Eq. (3.72). The last equality in Eq. (3.74) proves the isomorphism between the groups $\{U|\mathbf{t}\}$ and $\mathcal{O}_{\{U|\mathbf{t}\}}$, since their elements satisfy the condition of Eq. (3.69) in an obvious correspondence between elements of the two groups.

Having defined the basic formalism for taking advantage of the crystal symmetries, we will now apply it to simplify the description of the eigenfunctions and eigenvalues of the single-particle hamiltonian. We will prove that for any element of the space group $\{U|\mathbf{t}\} \in \mathsf{S}$, we have

$$\epsilon_{U\mathbf{k}}^{(n)} = \epsilon_{\mathbf{k}}^{(n)} \tag{3.75}$$

$$\psi_{U\mathbf{k}}^{(n)}(\mathbf{r}) = \mathcal{O}_{\{U|\mathbf{t}\}}\psi_{\mathbf{k}}^{(n)}(\mathbf{r}) \tag{3.76}$$

To show this we need to prove: first, that $\psi_{\mathbf{k}}^{(n)}(\mathbf{r})$ and $\mathcal{O}_{\{U|\mathbf{t}\}}\psi_{\mathbf{k}}^{(n)}(\mathbf{r})$ are eigenstates of the hamiltonian with the same eigenvalue; and second, that $\mathcal{O}_{\{U|\mathbf{t}\}}\psi_{\mathbf{k}}^{(n)}(\mathbf{r})$ is a Bloch state of wave-vector $U\mathbf{k}$. We first note that the symmetry operators of the space group commute with the hamiltonian, by their definition (they leave the crystal, hence the total potential invariant):

$$\mathcal{O}_{\{U|\mathbf{t}\}}[\mathcal{H}\psi_{\mathbf{k}}^{(n)}(\mathbf{r})] = \mathcal{O}_{\{U|\mathbf{t}\}}[\epsilon_{\mathbf{k}}^{(n)}\psi_{\mathbf{k}}^{(n)}(\mathbf{r})] \Rightarrow \mathcal{H}[\mathcal{O}_{\{U|\mathbf{t}\}}\psi_{\mathbf{k}}^{(n)}(\mathbf{r})] = \epsilon_{\mathbf{k}}^{(n)}[\mathcal{O}_{\{U|\mathbf{t}\}}\psi_{\mathbf{k}}^{(n)}(\mathbf{r})]$$

which shows that $\psi_{\mathbf{k}}^{(n)}(\mathbf{r})$ and $\mathcal{O}_{\{U|\mathbf{t}\}}\psi_{\mathbf{k}}^{(n)}(\mathbf{r})$ are eigenfunctions with the same eigenvalue, proving the first statement. To prove the second statement, we apply the operator $\mathcal{O}_{\{E|-\mathbf{R}\}}$ to the function $\mathcal{O}_{\{U|\mathbf{t}\}}\psi_{\mathbf{k}}^{(n)}(\mathbf{r})$: this is simply a translation operator (E is the identity), and therefore changes the argument of a function by $+\mathbf{R}$. We expect to find that this function is then changed by a phase factor $\exp(\mathrm{i}U\mathbf{k}\cdot\mathbf{R})$, if indeed it is a Bloch state of wave-vector $U\mathbf{k}$. To prove this we note that

$$\begin{aligned}
\{U|\mathbf{t}\}^{-1}\{E|-\mathbf{R}\}\{U|\mathbf{t}\} &= \{U|\mathbf{t}\}^{-1}\{U|\mathbf{t}-\mathbf{R}\} = \{U^{-1}|-U^{-1}\mathbf{t}\}\{U|\mathbf{t}-\mathbf{R}\} \\
&= \{E|U^{-1}\mathbf{t}-U^{-1}\mathbf{R}-U^{-1}\mathbf{t}\} = \{E|-U^{-1}\mathbf{R}\}
\end{aligned}$$

and multiplying the initial and final parts of this equation by $\{U|\mathbf{t}\}$ from the left, we obtain

$$\{E| - \mathbf{R}\}\{U|\mathbf{t}\} = \{U|\mathbf{t}\}\{E| - U^{-1}\mathbf{R}\} \Rightarrow \mathcal{O}_{\{E|-\mathbf{R}\}\{U|\mathbf{t}\}} = \mathcal{O}_{\{U|\mathbf{t}\}\{E|-U^{-1}\mathbf{R}\}}$$

due to isomorphism of the groups $\{U|\mathbf{t}\}$ and $\mathcal{O}_{\{U|\mathbf{t}\}}$. When we apply the operators in this last equation to $\psi_{\mathbf{k}}^{(n)}(\mathbf{r})$, we find

$$\mathcal{O}_{\{E|-\mathbf{R}\}}\mathcal{O}_{\{U|\mathbf{t}\}}\psi_{\mathbf{k}}^{(n)}(\mathbf{r}) = \mathcal{O}_{\{U|\mathbf{t}\}}\mathcal{O}_{\{E|-U^{-1}\mathbf{R}\}}\psi_{\mathbf{k}}^{(n)}(\mathbf{r}) = \mathcal{O}_{\{U|\mathbf{t}\}}\psi_{\mathbf{k}}^{(n)}(\mathbf{r} + U^{-1}\mathbf{R})$$

Using the fact that $\psi_{\mathbf{k}}^{(n)}(\mathbf{r})$ is a Bloch state, we obtain

$$\psi_{\mathbf{k}}^{(n)}(\mathbf{r} + U^{-1}\mathbf{R}) = \mathrm{e}^{\mathrm{i}\mathbf{k}\cdot U^{-1}\mathbf{R}}\psi_{\mathbf{k}}^{(n)}(\mathbf{r}) = \mathrm{e}^{\mathrm{i}U\mathbf{k}\cdot\mathbf{R}}\psi_{\mathbf{k}}^{(n)}(\mathbf{r})$$

where we have used the fact that $\mathbf{k} \cdot U^{-1}\mathbf{R} = U\mathbf{k} \cdot \mathbf{R}$; that is, a dot product of two vectors does not change if we rotate both vectors by U. When this last result is substituted into the previous equation, we get:

$$\mathcal{O}_{\{E|-\mathbf{R}\}}[\mathcal{O}_{\{U|\mathbf{t}\}}\psi_{\mathbf{k}}^{(n)}(\mathbf{r})] = \mathrm{e}^{\mathrm{i}U\mathbf{k}\cdot\mathbf{R}}[\mathcal{O}_{\{U|\mathbf{t}\}}\psi_{\mathbf{k}}^{(n)}(\mathbf{r})]$$

which proves that $\mathcal{O}_{\{U|\mathbf{t}\}}\psi_{\mathbf{k}}^{(n)}(\mathbf{r})$ is indeed a Bloch state of wave-vector $U\mathbf{k}$. Therefore, the corresponding eigenvalue must be $\epsilon_{U\mathbf{k}}^{(n)}$, which is equal to $\epsilon_{\mathbf{k}}^{(n)}$ since we have already proven that the two Bloch states $\psi_{\mathbf{k}}^{(n)}(\mathbf{r})$ and $\mathcal{O}_{\{U|\mathbf{t}\}}\psi_{\mathbf{k}}^{(n)}(\mathbf{r})$ have the same eigenvalue. We have then proven the two symmetry statements of Eq. (3.76). These results indicate that the energy spectrum $\epsilon_{\mathbf{k}}^{(n)}$ has the full symmetry of the point group $\mathsf{P} = [\{U|0\}]$. The set of vectors $U\mathbf{k}$ for $\{U|0\} \in \mathsf{P}$ is called the star of $\mathbf{k}$. Furthermore, we have shown that the two Bloch states are related by

$$\psi_{U\mathbf{k}}^{(n)}(\mathbf{r}) = \mathcal{O}_{\{U|\mathbf{t}\}}\psi_{\mathbf{k}}^{(n)}(\mathbf{r}) = \psi_{\mathbf{k}}^{(n)}(\{U|\mathbf{t}\}^{-1}\mathbf{r}) = \psi_{\mathbf{k}}^{(n)}(U^{-1}\mathbf{r} - U^{-1}\mathbf{t})$$

so that if we knew the eigenvalues $\epsilon_{\mathbf{k}}^{(n)}$ and eigenfunctions $\psi_{\mathbf{k}}^{(n)}(\mathbf{r})$ at a point $\mathbf{k}$ we could immediately find all the eigenvalues and eigenfunctions at points $U\mathbf{k}$, where U is any symmetry in the point group P. This offers great savings in computation, since we need only solve the single-particle equations in a small portion of the Brillouin Zone, which can then be used to obtain all the remaining solutions. This portion of the Brillouin Zone, which can be "unfolded" by the symmetry operations of the point group to give us all the solutions in the entire Brillouin Zone, is called the Irreducible Brillouin Zone (IBZ).

To summarize the discussion so far, we have found that the symmetry properties of the energy spectrum and the single-particle wavefunctions simplify the task of obtaining the band structure considerably, by requiring that we solve the single-particle Schrödinger equations in the IBZ only. Furthermore, both the energy and the wavefunctions are smooth functions of $\mathbf{k}$ which in the limit of an infinite crystal is a continuous variable, so we can understand the behavior of the energy bands by

Table 3.2. *Group multiplication table for symmetries of the 2D square lattice.*

E:	(x, y)	$\to$	(x, y)	E	C_4	C_2	C_4^3	σ_x	σ_y	σ_1	σ_3
C_4:	(x, y)	$\to$	$(-y, x)$	C_4	C_2	C_4^3	E	σ_1	σ_3	σ_y	σ_x
C_2:	(x, y)	$\to$	$(-x, -y)$	C_2	C_4^3	E	C_4	σ_y	σ_x	σ_3	σ_1
C_4^3:	(x, y)	$\to$	$(y, -x)$	C_4^3	E	C_4	C_2	σ_3	σ_1	σ_x	σ_y
σ_x:	(x, y)	$\to$	$(x, -y)$	σ_x	σ_3	σ_y	σ_1	E	C_2	C_4^3	C_4
σ_y:	(x, y)	$\to$	$(-x, y)$	σ_y	σ_1	σ_x	σ_3	C_2	E	C_4	C_4^3
σ_1:	(x, y)	$\to$	(y, x)	σ_1	σ_x	σ_3	σ_y	C_4	C_4^3	E	C_2
σ_3:	(x, y)	$\to$	$(-y, -x)$	σ_3	σ_y	σ_1	σ_x	C_4^3	C_4	C_2	E

considering their values along a few high-symmetry directions: the rest interpolate in a smooth fashion between these directions. We retake our simple example of the free-electron model in the 2D square lattice to illustrate these concepts.

The symmetries for this model, consisting of one atom per PUC, and having lattice vectors $\mathbf{a}_1 = a\hat{\mathbf{x}}$, $\mathbf{a}_2 = a\hat{\mathbf{y}}$, are:

- the identity E;
- rotation by $(2\pi)/4$ around an axis perpendicular to the plane, C_4;
- rotation by $(2\pi)/2$ around an axis perpendicular to the plane, C_2;
- rotation by $3(2\pi)/4$ around an axis perpendicular to the plane, C_4^3;
- reflection on the x axis, σ_x;
- reflection on the y axis, σ_x;
- reflection on the axis at $\theta = \pi/4$ from the x axis, σ_1;
- reflection on the axis at $\theta = 3\pi/4$ from the x axis, σ_3.

These symmetries constitute the point group P for this physical system. We assume that the origin is at the position of the atom, so that the group is symmorphic. This point group has the group multiplication table given in Table 3.2: an entry in this Table is the result of multiplying the element at the top of its column with the element at the left end of its row. The definition of each element in terms of its action on an arbitrary vector (x, y) on the plane is also given on the left half of Table 3.2. This group multiplication table can be used to prove that all the group properties are satisfied.

For this system the point group is a subgroup of the space group because it is a symmorphic group, and therefore the space group S is given by $\mathsf{S} = \mathsf{P} \times \mathsf{T}$ where T is the translation group. Using these symmetries we deduce that the symmetries of the point group give an IBZ which is one-eighth that of the first BZ, shown in Fig. 3.7.

There are a few high-symmetry points in this IBZ (in (k_x, k_y) notation):

(1) $\Gamma = (0, 0)$, which has the full symmetry of the point group;
(2) $M = (1, 1)(\pi/a)$, which also has the full symmetry of the point group;

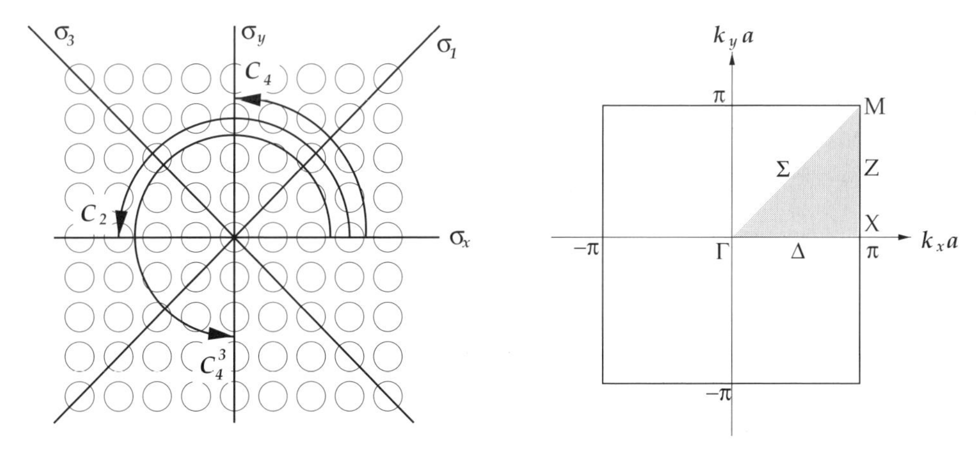

Figure 3.7. **Left:** the symmetry operations of the two-dimensional square lattice; the thick straight lines indicate the reflections (labeled $\sigma_x, \sigma_y, \sigma_1, \sigma_3$) and the curves with arrows the rotations (labeled C_4, C_2, C_4^3). **Right:** the Irreducible Brillouin Zone for the two-dimensional square lattice with full symmetry; the special points are labeled Γ, Σ, Δ, M, Z, X.

(3) $X = (1, 0)(\pi/a)$, which has the symmetries $E, C_2, \sigma_x, \sigma_y$;
(4) $\Delta = (k, 0)(\pi/a), 0 < k < 1$, which has the symmetries E, σ_x;
(5) $\Sigma = (k, k)(\pi/a), 0 < k < 1$, which has the symmetries E, σ_1;
(6) $Z = (1, k)(\pi/a), 0 < k < 1$, which has the symmetries E, σ_y.

Any point inside the IBZ that is not one of the high-symmetry points mentioned above does not have any symmetry, other than the identity E.

One important point to keep in mind is that Kramers' theorem always applies, which for systems with equal numbers of spin-up and spin-down electrons implies inversion symmetry in *reciprocal space*. Thus, even if the crystal does not have inversion symmetry as one of the point group elements, we can always use inversion symmetry, in addition to all the other point group symmetries imposed by the lattice, to reduce the size of the irreducible portion of the Brillouin Zone. In the example we have been discussing, C_2 is actually equivalent to inversion symmetry, so we do not have to add it explicitly in the list of symmetry operations when deriving the size of the IBZ.

As already described above, since both $\epsilon_{\mathbf{k}}$ and $\psi_{\mathbf{k}}(\mathbf{r})$ are smooth functions of $\mathbf{k}$, we need only calculate their values for the high-symmetry points mentioned, which will then provide essentially a full description of the physics. For the case of the free-electron system in the two-dimensional square lattice, the energy for the first nine bands at these high-symmetry points is shown in Fig. 3.8. In this simple model, the energy of single-particle levels is given by

$$\epsilon_{\mathbf{k}}^{(n)} = \frac{\hbar^2}{2m_e}|\mathbf{k}|^2$$

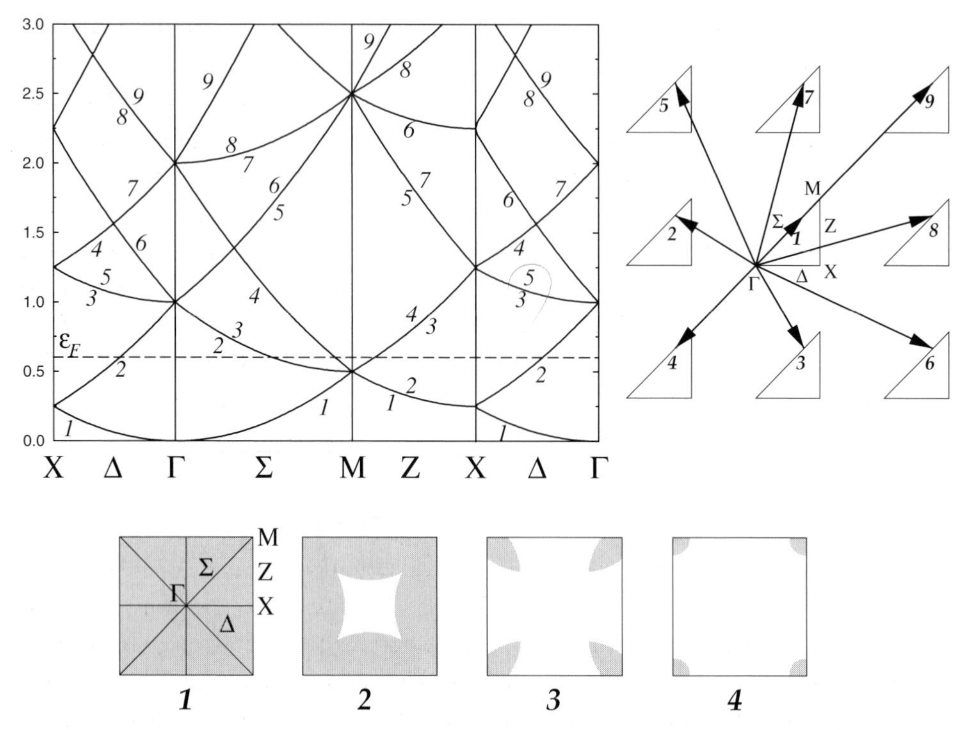

Figure 3.8. The band structure for the two-dimensional free-electron model at the high-symmetry points in the IBZ, for the first nine bands coming from the first nine BZs, which are shown on the right. The position of the Fermi level ϵ_F for $Z = 4$ electrons per unit cell is indicated. At the bottom we reproduce the filling of the first four BZs, as shown in Fig. 3.6.

that is, it depends only on the magnitude of the wave-vector **k**. The bands are obtained by scanning the values of the wave-vector **k** along the various directions in the first, second, third, etc. BZ, and then folding them within the first BZ. To illustrate the origin of the bands along the high-symmetry directions in the BZ, we show in Fig. 3.8 the wave-vectors that correspond to equivalent Σ points in the first nine BZs, along the $\Gamma - M$ direction. All these points differ by reciprocal lattice vectors, so in the reduced zone scheme they are mapped to the same Σ point in the first BZ. This makes it evident why the bands labeled 2 and 3 are degenerate along the $\Gamma - M$ line: the wave-vectors corresponding to these points in the BZs with labels 2 and 3 have equal magnitude. For the same reason, the pairs of bands labeled 5 and 6, or 7 and 8, are also degenerate along the $\Gamma - M$ line. Analogous degeneracies are found along the $\Gamma - X$ and $M - X$ lines, but they involve different pairs of bands. The labeling of the bands is of course arbitrary and serves only to identify electronic states with different energy at the same wave-vector **k**; the only requirement is that the label be kept consistent for the various parts of the BZ, as is done in Fig. 3.8. In this case, it is convenient to choose band labels that make the connection to the various BZs transparent.

Using this band structure, we can now interpret the filling of the different BZs in Fig. 3.6. The first band is completely full being entirely below the Fermi level, and so is the first BZ. The second band is full along the direction MX, but only partially full in the directions ΓX and ΓM, giving rise to the almost full second BZ with an empty region in the middle, which corresponds to the Fermi surface shape depicted. The third band has only a small portion occupied, in the MX and $M\Gamma$ directions, and the third BZ has occupied slivers near the four corners. Finally, the fourth band has a tiny portion occupied around the point M, corresponding to the so called "pockets" of occupied states at the corners. As this discussion exemplifies, the behavior of the energy bands along the high-symmetry directions of the BZ provides essentially a complete picture of the eigenvalues of the single-particle hamiltonian throughout reciprocal space in the reduced zone scheme.

3.8 Symmetries of 3D crystals

We revisit next the symmetry groups associated with three-dimensional solids. In three dimensions there are 14 different Bravais lattices. They are grouped in six crystal systems called triclinic, monoclinic, orthorhombic, tetragonal, hexagonal and cubic, in order of increasing symmetry. The hexagonal system is often split into two parts, called the trigonal and hexagonal subsystems. The definition of these systems in terms of the relations between the sides of the unit cell (a, b, c) and the angles between them (with $\alpha =$ the angle between sides a, c, $\beta =$ the angle between sides b, c, and $\gamma =$ the angle between sides a, b) are given in Table 3.3. Fig. 3.9 shows the conventional unit cells for the 14 Bravais lattices. Each corner of the cell is occupied by sites that are equivalent by translational symmetry.

Table 3.3. *The six crystal systems and the associated 32 point groups for crystals in 3D.*

The relations between the cell sides (a, b, c) and cell angles (α, β, γ) are shown and the corresponding lattices are labeled P = primitive, I = body-centered, C = side-centered, F = face-centered, R = rhombohedral.

System	Cell sides	Cell angles	Lattices	Point groups
Triclinic	$a \neq b \neq c$	$\alpha \neq \beta \neq \gamma$	P	C_1, C_i
Monoclinic	$a \neq b \neq c$	$\alpha = \beta = \frac{\pi}{2} \neq \gamma$	P, C	C_2, C_s, C_{2h}
Orthorhombic	$a \neq b \neq c$	$\alpha = \beta = \gamma = \frac{\pi}{2}$	P, I, C, F	D_2, D_{2h}, C_{2v}
Tetragonal	$a = b \neq c$	$\alpha = \beta = \gamma = \frac{\pi}{2}$	P, I	C_4, S_4, C_{4h}, D_4
				C_{4v}, D_{2d}, D_{4h}
Trigonal	$a = b \neq c$	$\alpha = \beta = \frac{\pi}{2}$	P	C_3, C_{3i}, C_{3v}, D_3
		$\gamma = \frac{2\pi}{3}$		$D_{3d}, C_6, C_{3h}, C_{6h}$
Hexagonal	$a = b \neq c$		R	$C_{6v}, D_{3h}, D_6, D_{6h}$
Cubic	$a = b = c$	$\alpha = \beta = \gamma = \frac{\pi}{2}$	P, I, F	T, T_h, O, T_d, O_h

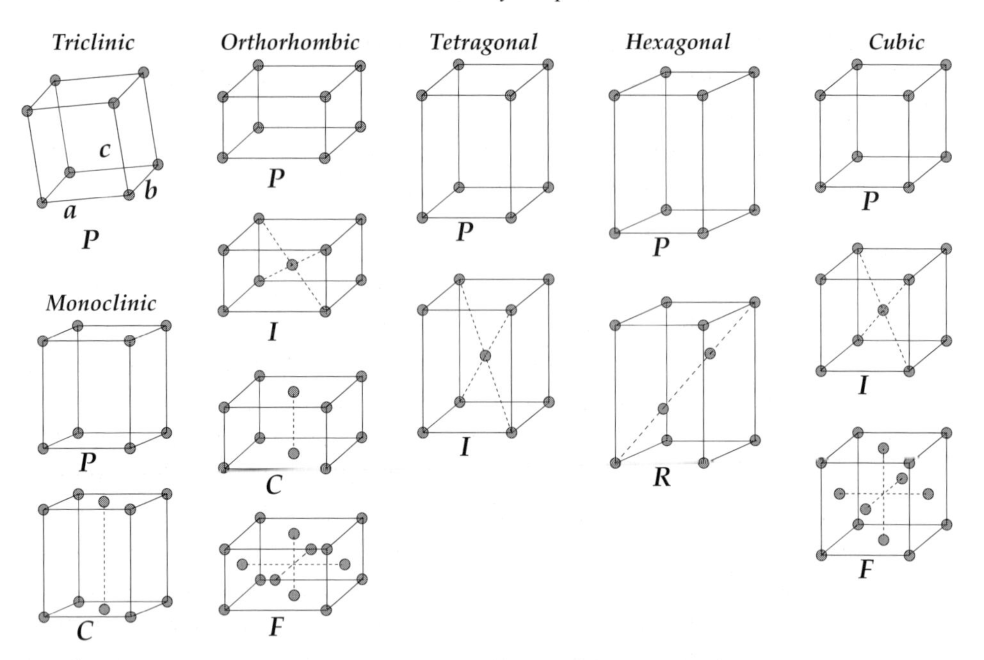

Figure 3.9. The conventional unit cells of the 14 Bravais lattices in three dimensions. The small gray circles indicate equivalent sites in the unit cell: those at the corners are equivalent by translational symmetry, the others indicate the presence of additional symmetries, implying a primitive unit cell smaller than the conventional.

When there are no other sites equivalent to the corners, the conventional cell is the same as the primitive cell for this lattice and it is designated P. When the cell has an equivalent site at its geometric center (a body-centered cell) it is designated I for the implied inversion symmetry; the primitive unit cell in this case is one-half the conventional cell. When the cell has an equivalent site at the center of one face it is designated C (by translational symmetry it must also have an equivalent site at the opposite face); the primitive unit cell in this case is one-half the conventional cell. When the cell has an equivalent site at the center of each face (a face-centered cell) it is designated F; the primitive unit cell in this case is one-quarter the conventional cell. Finally, in the hexagonal system there exists a lattice which has two more equivalent sites inside the conventional cell, at height $c/3$ and $2c/3$ along the main diagonal (see Fig. 3.9), and it is designated R for rhombohedral; the primitive unit cell in this case is one-third the conventional cell.

A number of possible point groups are associated with each of the 14 Bravais lattices, depending on the symmetry of the basis. There are 32 point groups in all, denoted by the following names:

- C for "cyclic", when there is a single axis of rotation – a number subscript m indicates the m-fold symmetry around this axis (m can be 2, 3, 4, 6);

- D for "dihedral", when there are two-fold axes at right angles to another axis;
- T for "tetrahedral", when there are four sets of rotation axes of three-fold symmetry, as in a tetrahedron (see Fig. 3.11);
- O for "octahedral", when there are four-fold rotation axes combined with perpendicular two-fold rotation axes, as in an octahedron (see Fig. 3.11).

In all of these cases, the existence of additional mirror plane symmetries is denoted by a second subscript which can be h for "horizontal", v for "vertical", or d for "diagonal" planes relative to the rotation axes. Inversion symmetry is denoted by the letter subscript i. In the case of a one-fold rotation axis, when inversion symmetry is present the notation C_i is adopted (instead of what would normally be called C_{1i}), while when a mirror plane symmetry is present the notation C_s is used (instead of what would normally be called C_{1h}). Finally, the group generated by the symmetry operation $\pi/2$-rotation followed by reflection on a vertical plane, the so called "roto-reflection" group, is denoted by S_4 (this is different than C_{4v} in which the $\pi/2$-rotation and the reflection on a vertical plane are independently symmetry operations). This set of conventions for describing the 32 crystallographic groups is referred to as the Schoenflies notation; the names of the groups in this notation are given in Table 3.3. There is a somewhat more rational set of conventions for naming crystallographic point groups described in the *International Tables for Crystallography*. This scheme is more complicated; we refer the interested reader to these Tables for further details.

There exists a useful way of visualizing the symmetry operations of the various point groups, referred to as "stereograms". These consist of two-dimensional projections of a point on the surface of a sphere and its images generated by acting on the sphere with the various symmetry operations of the point group. In order to illustrate the action of the symmetry operations, the initial point is chosen as some arbitrary point on the sphere, that is, it does not belong to any special axis or plane of symmetry. The convention for the equivalent points on the sphere is that the ones in the northern hemisphere are drawn as open circles, while the ones in the southern hemisphere are shown as solid dots and the sphere is viewed from the northern pole. In Fig. 3.10 we show the stereograms for 30 point groups belonging to five crystal systems; the stereograms for the triclinic system (point groups C_1 and C_i) are not shown because they are trivial: they contain only one point each, since these groups do not have any symmetry operations other than the trivial ones (identity and inversion).

As an illustration, we discuss briefly the symmetries of the cube, which has the highest degree of symmetry compatible with three-dimensional periodicity. When only two-fold, three-fold and four-fold rotations are allowed, as required by three-dimensional periodicity (see problem 1), there are 24 images of an arbitrary point

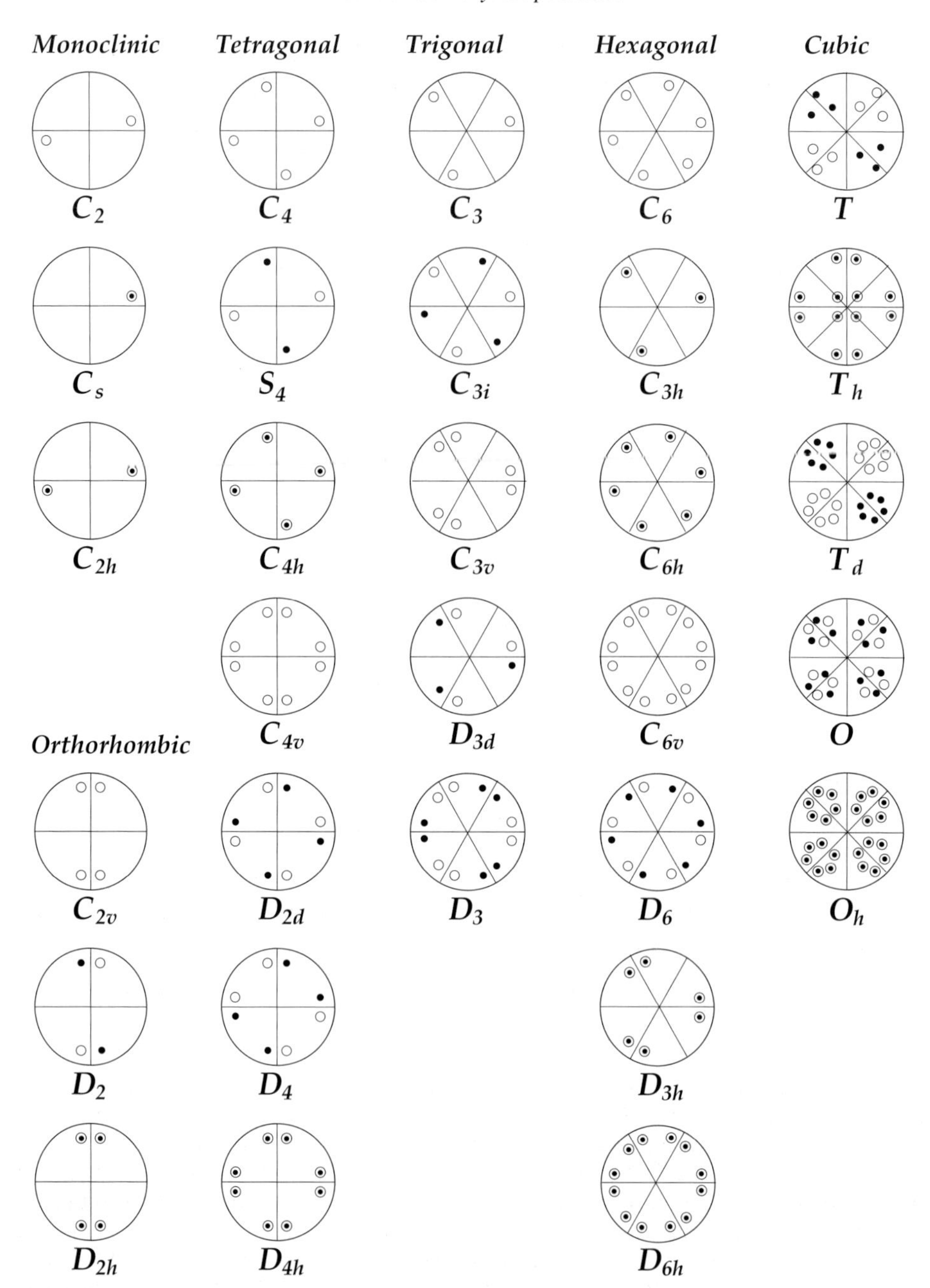

Figure 3.10. Stereograms for 30 point groups in 3D. The lines within the circles are visual aides, which in several cases correspond to reflection planes. The trigonal and hexagonal subsystems are shown separately, and the triclinic system (groups C_1 and C_i) is not shown at all since it has trivial representations.

Table 3.4. *The 24 symmetry operations of the cube.*

These operations involve rotations by π (the classes C_4^2 and C_2), $\pi/3$ (the class C_3), $\pi/2$ (the class C_4) and 2π (the identity E). Twenty-four more symmetry operations can be generated by combining each of the rotations with inversion I which corresponds to $(-x, -y, -z)$, i.e., a change of all signs.

Class	Operation				Class	Operation			
E		x	y	z		$R_{x/4}$	x	z	$-y$
	$R_{x/2}$	x	$-y$	$-z$		$\overline{R}_{x/4}$	x	$-z$	y
C_4^2	$R_{y/2}$	$-x$	y	$-z$	C_4	$R_{y/4}$	$-z$	y	x
	$R_{z/2}$	$-x$	$-y$	z		$\overline{R}_{y/4}$	z	y	$-x$
	R_1	y	z	x		$R_{z/4}$	y	$-x$	z
	$\overline{R}_1$	z	x	y		$\overline{R}_{z/4}$	$-y$	x	z
	R_2	z	$-x$	$-y$		$R_{\bar{x}/2}$	$-x$	z	y
C_3	$\overline{R}_2$	$-y$	$-z$	x		$R'_{\bar{x}/2}$	$-x$	$-z$	$-y$
	R_3	$-z$	x	$-y$	C_2	$R_{\bar{y}/2}$	z	$-y$	x
	$\overline{R}_3$	y	$-z$	$-x$		$R'_{\bar{y}/2}$	$-z$	$-y$	$-x$
	R_4	$-y$	z	$-x$		$R_{\bar{z}/2}$	y	x	$-z$
	$\overline{R}_4$	$-z$	$-x$	y		$R'_{\bar{z}/2}$	$-y$	$-x$	$-z$

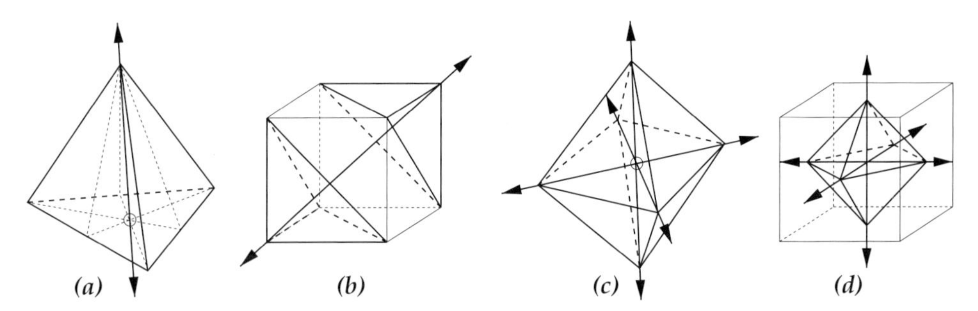

Figure 3.11. Illustration of symmetry axes of the tetrahedron and the octahedron. (a) A three-fold rotation axis of the *regular* tetrahedron, denoted by a double-headed arrow, passes through one of its corners and the geometric center of the equilateral triangle directly across it; there are four such axes in the regular tetrahedron, each passing through one of its corners. (b) Two tetrahedra (in this case *not* regular ones) with a common axis of three-fold rotational symmetry in the cube are identified by thicker lines; there are four pairs of such tetrahedra in the cube, whose three-fold rotational symmetry axes are the main diagonals of the cube. (c) Three different symmetry axes, the vertical and horizontal double-headed arrows, of an octahedron; when the edges of the horizontal square are not of the same length as the other edges, the vertical axis corresponds to four-fold rotation, while the two horizontal axes correspond to two-fold rotations. When all edges are of equal length all three axes correspond to four-fold rotational symmetry. (d) One such octahedron, with all edges equal and three four-fold rotation axes, is identified within the cube by thicker lines.

in space (x, y, z): these are given in Table 3.4. The associated 24 operations are separated into five classes:

1. the identity E;
2. two-fold (π) rotations around the axes x, y and z denoted as a class by C_4^2 and as individual operations by $R_{\nu/2}(\nu = x, y, z)$;
3. four-fold ($2\pi/4$) rotations around the axes x, y and z denoted as a class by C_4 and as individual operations by $R_{\nu/4}(\nu = x, y, z)$ for counter-clockwise or by $\overline{R}_{\nu/4}$ for clockwise rotation;
4. three-fold ($2\pi/3$) rotations around the main diagonals of the cube denoted as a class by C_3 and as individual operations by $R_n(n = 1, 2, 3, 4)$ for counter-clockwise or by $\overline{R}_n$ for clockwise rotation;
5. two-fold (π) rotations around axes that are perpendicular to the x, y, z axes and bisect the in-plane angles between the cartesian axes, denoted as a class by C_2 and as individual operations by $R_{\overline{\nu}/2}(\nu = x, y, z)$ or $R'_{\overline{\nu}/2}$ with the subscript indicating the cartesian axis perpendicular to the rotation axis.

Of these classes of operations the first three are trivial to visualize, while the last two are represented schematically in Fig. 3.12. When inversion is added to these operations, the total number of images of the arbitrary point becomes 48, since inversion can be combined with any rotation to produce a new operation. It is easy to rationalize this result: there are 48 different ways of rearranging the three cartesian components of the arbitrary point, including changes of sign: the first coordinate can be any of six possibilities $(\pm x, \pm y, \pm z)$, the second coordinate can be any of four possibilities, and the last coordinate any of two possibilities.

If these 48 operations are applied with respect to the center of the cube, the cube remains invariant. Therefore, in a crystal with a cubic Bravais lattice and a basis that does not break the cubic symmetry (such as a single atom per unit cell referred to as simple cubic, or an FCC or a BCC lattice), all 48 symmetry operations of the cube

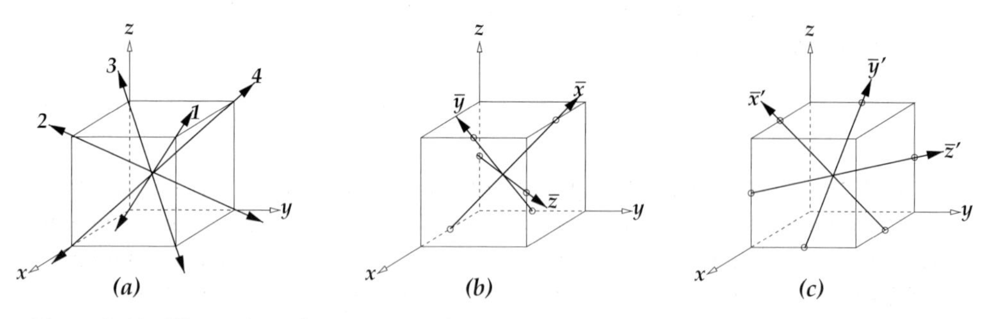

Figure 3.12. Illustration of two classes of symmetry operations of the cube. (a) The C_3 class consisting of three-fold rotation axes which correspond to the main diagonals of the cube, indicated by double-headed arrows and labeled 1–4. (b) and (c) The C_2 class consisting of two-fold rotation axes that are perpendicular to, and bisect the angles between, the cartesian axes; the rotation axes are shown as single-headed arrows and the points at which they intersect the sides of the cube are indicated by small circles.

will leave the crystal invariant, and therefore all these operations form the point group of the crystal. This is the largest point group compatible with three-dimensional periodicity. There are other larger point groups in three dimensions but they are not compatible with three-dimensional periodicity. For example, the icosahedral group has a total of 120 operations, but its five-fold symmetry is not compatible with translational periodicity in three dimensions.

3.9 Special k-points

Another useful application of group theory arguments is in the simplification of reciprocal-space integrations. Very often we need to calculate quantities of the type

$$\overline{g} = \frac{1}{N} \sum_{\mathbf{k} \in \mathrm{BZ}} g(\mathbf{k}) = \Omega_{PUC} \int \frac{\mathrm{d}\mathbf{k}}{(2\pi)^3} g(\mathbf{k}) \tag{3.77}$$

where $\mathbf{k} \in$ BZ stands for all values of the wave-vector $\mathbf{k}$ inside a Brillouin Zone (it is convenient to assume that we are dealing with the reduced zone scheme and the first BZ). For example, when calculating the electronic density in the single-particle picture we have to calculate the quantities

$$\rho_{\mathbf{k}}(\mathbf{r}) = \sum_{n, \epsilon_{\mathbf{k}}^{(n)} < \epsilon_{\mathrm{F}}} |\psi_{\mathbf{k}}^{(n)}(\mathbf{r})|^2, \quad \rho(\mathbf{r}) = \int \frac{\mathrm{d}\mathbf{k}}{(2\pi)^3} \rho_{\mathbf{k}}(\mathbf{r}) \tag{3.78}$$

with the single-particle wavefunctions $\psi_{\mathbf{k}}^{(n)}(\mathbf{r})$ normalized to unity in the PUC. We can take advantage of the crystal symmetry to reduce the summation over $\mathbf{k}$ values inside the IBZ. In doing this, we need to keep track of the multiplicity of each point in the IBZ, that is, the number of equivalent points to which a particular $\mathbf{k}$-point in the IBZ is mapped to when the different symmetry operations are applied to it. While this significantly simplifies our task, there is an even greater simplification: with the use of a very few $\mathbf{k}$ values we can obtain an excellent approximation to the sum in Eq. (3.77). These are called special $\mathbf{k}$-points [33–35], and we discuss them next.

We begin by defining the function

$$f(\mathbf{k}) = \frac{1}{N_P} \sum_{P \in \mathsf{P}} g(P\mathbf{k}) \tag{3.79}$$

where P is the point group, P is an operation in the point group, and N_P is the total number of operations in P. Then we will have $f(\mathbf{k}) = f(P\mathbf{k})$ since applying P to any operation in P gives another operation in P due to closure, and the definition of $f(\mathbf{k})$ includes already a summation over all the operations in P. We also can deduce that the sum of $f(\mathbf{k})$ over all $\mathbf{k}$-points in the BZ is equal to the sum of $g(\mathbf{k})$, because

$$\sum_{\mathbf{k} \in \mathrm{BZ}} f(\mathbf{k}) = \sum_{\mathbf{k} \in \mathrm{BZ}} \frac{1}{N_P} \sum_{P \in \mathsf{P}} g(P\mathbf{k}) = \frac{1}{N_P} \sum_{P \in \mathsf{P}} \sum_{P\mathbf{k} \in \mathrm{BZ}} g(P\mathbf{k}) \tag{3.80}$$

where we have used the fact that summation over $\mathbf{k} \in \mathrm{BZ}$ is the same as summation over $P\mathbf{k} \in \mathrm{BZ}$. Now, summation over $P\mathbf{k} \in \mathrm{BZ}$ of $g(P\mathbf{k})$ is the same for all P, so that doing this summation for all $P \in \mathsf{P}$ gives N_P times the same result:

$$\frac{1}{N_P} \sum_{P \in \mathsf{P}} \left[\sum_{P\mathbf{k} \in \mathrm{BZ}} g(P\mathbf{k}) \right] = \sum_{P\mathbf{k} \in \mathrm{BZ}} g(P\mathbf{k}) = \sum_{\mathbf{k} \in \mathrm{BZ}} g(\mathbf{k}) \tag{3.81}$$

Combining the results of the previous two equations, we obtain the desired result for the sums $\overline{f}, \overline{g}$:

$$\overline{f} = \frac{1}{N} \sum_{\mathbf{k} \in \mathrm{BZ}} f(\mathbf{k}) = \frac{1}{N} \sum_{\mathbf{k} \in \mathrm{BZ}} g(\mathbf{k}) = \overline{g} \tag{3.82}$$

We can expand the function $f(\mathbf{k})$ in a Fourier expansion with coefficients $\tilde{f}(\mathbf{R})$ as follows:

$$f(\mathbf{k}) = \sum_{\mathbf{R}} \tilde{f}(\mathbf{R}) \mathrm{e}^{-\mathrm{i}\mathbf{k}\cdot\mathbf{R}} \rightarrow \tilde{f}(\mathbf{R}) = \sum_{\mathbf{k} \in \mathrm{BZ}} f(\mathbf{k}) \mathrm{e}^{\mathrm{i}\mathbf{k}\cdot(\mathbf{R})} \tag{3.83}$$

which gives for the sum $\overline{f}$

$$\overline{f} = \frac{1}{N} \sum_{\mathbf{R}} \sum_{\mathbf{k} \in \mathrm{BZ}} \tilde{f}(\mathbf{R}) \mathrm{e}^{-\mathrm{i}\mathbf{k}\cdot\mathbf{R}} = \frac{1}{N} \sum_{\mathbf{R}} \tilde{f}(\mathbf{R}) \sum_{\mathbf{k} \in \mathrm{BZ}} \mathrm{e}^{-\mathrm{i}\mathbf{k}\cdot\mathbf{R}} \tag{3.84}$$

Now we can use the δ-function relations that result from summing the complex exponential $\exp(\pm \mathrm{i}\mathbf{k} \cdot \mathbf{R})$ over real-space lattice vectors or reciprocal-space vectors within the BZ, which are proven in Appendix G, to simplify Eq. (3.84). With these relations, $\overline{f}$ takes the form

$$\overline{f} = \frac{1}{N} \sum_{\mathbf{R}} \tilde{f}(\mathbf{R}) \delta(\mathbf{R}) = \tilde{f}(0) = \overline{g} = \frac{1}{N} \sum_{\mathbf{k} \in \mathrm{BZ}} g(\mathbf{k}) \tag{3.85}$$

so that we have found the sum of $g(\mathbf{k})$ in the BZ to be equal to the Fourier coefficients $\tilde{f}(\mathbf{R})$ of $f(\mathbf{k})$ evaluated at $\mathbf{R} = 0$. But we also have

$$f(\mathbf{k}) = \sum_{\mathbf{R}} \tilde{f}(\mathbf{R}) \mathrm{e}^{-\mathrm{i}\mathbf{k}\cdot\mathbf{R}} = \tilde{f}(0) + \sum_{\mathbf{R} \neq 0} \tilde{f}(\mathbf{R}) \mathrm{e}^{-\mathrm{i}\mathbf{k}\cdot\mathbf{R}} \tag{3.86}$$

We would like to find a value of $\mathbf{k}_0$ that makes the second term on the right side of this equation vanishingly small, because then

$$\sum_{\mathbf{R} \neq 0} \tilde{f}(\mathbf{R}) \mathrm{e}^{-\mathrm{i}\mathbf{k}_0\cdot\mathbf{R}} \approx 0 \Rightarrow \overline{g} = \tilde{f}(0) \approx f(\mathbf{k}_0) \tag{3.87}$$

The value $\mathbf{k}_0$ is a special point which allows us to approximate the sum of $g(\mathbf{k})$ over the entire BZ by calculating a single value of $f(\mathbf{k})$!

In practice it is not possible to make the second term in Eq. (3.86) identically zero, so we must find a reasonable way of determining values of the special **k**-point. To this end we notice that

$$\begin{aligned}\tilde{f}(P\mathbf{R}) &= \sum_{\mathbf{k}\in\mathrm{BZ}} f(\mathbf{k})\mathrm{e}^{\mathrm{i}\mathbf{k}\cdot(P\mathbf{R})} = \sum_{\mathbf{k}\in\mathrm{BZ}} f(\mathbf{k})\mathrm{e}^{\mathrm{i}(P^{-1}\mathbf{k})\cdot\mathbf{R}} \\ &= \sum_{P^{-1}\mathbf{k}\in\mathrm{BZ}} f(P^{-1}\mathbf{k})\mathrm{e}^{\mathrm{i}(P^{-1}\mathbf{k})\cdot\mathbf{R}} = \tilde{f}(\mathbf{R})\end{aligned} \tag{3.88}$$

where we have taken advantage of the fact that $f(\mathbf{k}) = f(P\mathbf{k})$ for any operation $P \in \mathsf{P}$; we have also used the expression for the inverse Fourier transform $\tilde{f}(\mathbf{R})$ from Eq. (3.83) (see also the discussion of Fourier transforms in Appendix G). Then for all **R** of the same magnitude which are connected by P operations we will have $\tilde{f}(\mathbf{R}) = \tilde{f}(|\mathbf{R}|)$, and consequently we can break the sum over **R** values to a sum over $|\mathbf{R}|$ and a sum over $P \in \mathsf{P}$ which connect **R** values of the same magnitude:

$$f(\mathbf{k}_0) = \tilde{f}(0) + \sum_{|\mathbf{R}|} \tilde{f}(|\mathbf{R}|)\left[\sum_{P\in\mathsf{P}} \mathrm{e}^{-\mathrm{i}\mathbf{k}_0\cdot(P\mathbf{R})}\right]_{|\mathbf{R}|} \tag{3.89}$$

where the summation inside the square brackets is done at constant $|\mathbf{R}|$. These sums are called "shells of **R**", and their definition depends on the Bravais lattice. This last equation gives us a practical means of determining $\mathbf{k}_0$: it must make as many shells of **R** vanish as possible. Typically $\tilde{f}(|\mathbf{R}|)$ falls fast as its argument increases, so it is only necessary to make sure that the first few shells of **R** vanish for a given $\mathbf{k}_0$, to make it a reasonable candidate for a special **k**-point. A generalization of this is to consider a set of special **k**-points for which the first few shells of **R** vanish, in which case we need to evaluate $f(\mathbf{k})$ at all of these points to obtain a good approximation for $\overline{g}$ (for details see Refs. [35, 36]).

Further reading

1. *Group Theory and Quantum Mechanics*, M. Tinkham (McGraw-Hill, New York, 1964).
2. *Group Theory in Quantum Mechanics*, V. Heine (Pergamon Press, New York, 1960).
3. *The Structure of Materials*, S.M. Allen and E.L. Thomas (J. Wiley, New York, 1998).
   This book contains a detailed discussion of crystal symmetries as they apply to different materials.

Problems

1. (a) Calculate the reciprocal lattice vectors for the crystal lattices given in Table 3.1.
   (b) Determine the lattice vectors, the positions of atoms in the primitive unit cell, and the reciprocal lattice vectors, for the NaCl, CsCl, and zincblende crystal structures discussed in chapter 1.
2. (a) Show that the single-particle equation obeyed by $u_{\mathbf{k}}(\mathbf{r})$, the part of the wavefunction which has full translational symmetry, is Eq. (3.20).
   (b) Show that the coefficients $\alpha_{\mathbf{k}}(\mathbf{G})$ in the Fourier expansion of $u_{\mathbf{k}}(\mathbf{r})$ obey Eq. (3.37).
3. (a) Show that the time-reversal operator for a spin-$\frac{1}{2}$ particle can be chosen to be $\mathcal{T} = \mathrm{i}\sigma_y C$, where C is complex conjugation and σ_y is a Pauli spin matrix (defined in Appendix B).
   (b) Kramers degeneracy : show that if the hamiltonian is time-reversal invariant, then the following relations hold for the energy and wavefunction of single-particle states:

$$\epsilon^{(n)}_{\downarrow -\mathbf{k}} = \epsilon^{(n')}_{\uparrow \mathbf{k}}, \quad \psi^{(n)}_{\downarrow -\mathbf{k}}(\mathbf{r}) = \mathrm{i}\sigma_y C \psi^{(n')}_{\uparrow \mathbf{k}}(\mathbf{r}) \tag{3.90}$$

4. Use second order perturbation theory to derive the expression for the energy at wave-vector $\mathbf{k} + \mathbf{q}$ in terms of the energy and wavefunctions at wave-vector $\mathbf{k}$, as given in Eq. (3.43). From that, derive the expression for the inverse effective mass, given in Eq. (3.46).
5. Show that the only rotations compatible with 3D periodicity are multiples of $2\pi/4$ and $2\pi/6$.
6. Find the symmetries and construct the group multiplication table for a 2D square lattice model, with two atoms per unit cell, at positions $\mathbf{t}_1 = 0$, $\mathbf{t}_2 = 0.5a\hat{\mathbf{x}} + 0.3a\hat{\mathbf{y}}$, where a is the lattice constant. Is this group symmorphic or non-symmorphic?
7. Draw the occupied Brillouin Zones for the 2D square lattice with $n = 2$ electrons per unit cell, by analogy to Fig. 3.6.
8. Consider the 2D honeycomb lattice with two atoms per unit cell, defined by the lattice vectors:

$$\mathbf{a}_1 = \frac{\sqrt{3}}{2}a\hat{\mathbf{x}} - \frac{1}{2}a\hat{\mathbf{y}}, \quad \mathbf{a}_2 = \frac{\sqrt{3}}{2}a\hat{\mathbf{x}} + \frac{1}{2}a\hat{\mathbf{y}} \tag{3.91}$$

(these form the 2D hexagonal lattice), and atomic positions $\mathbf{t}_1 = 0$, $\mathbf{t}_2 = 1/\sqrt{3}(a\hat{\mathbf{x}})$. Determine all the symmetries of this lattice, and construct its group multiplication table. Draw the Irreducible Brillouin Zone and indicate the high-symmetry points and the symmetry operations for each. Draw the first seven bands for the high-symmetry points and find the Fermi level for a system in which every atom has two valence electrons.

9. Find a special $\mathbf{k}$-point for the 2D square lattice and the 2D hexagonal lattice. How many shells can you make vanish with a single special $\mathbf{k}$-point in each case?

4

Band structure of crystals

In the previous two chapters we examined in detail the effects of crystal periodicity and crystal symmetry on the eigenvalues and wavefunctions of the single-particle equations. The models we used to illustrate these effects were artificial free-electron models, where the only effect of the presence of the lattice is to impose the symmetry restrictions on the eigenvalues and eigenfunctions. We also saw how a weak periodic potential can split the degeneracies of certain eigenvalues at the Bragg planes (the BZ edges). In realistic situations the potential is certainly not zero, as in the free-electron model, nor is it necessarily weak. Our task here is to develop methods for determining the solutions to the single-particle equations for realistic systems. We will do this by discussing first the so called tight-binding approximation, which takes us in the most natural way from electronic states that are characteristic of atoms (atomic orbitals) to states that correspond to crystalline solids. We will then discuss briefly more general methods for obtaining the band structure of solids, whose application typically involves a large computational effort. Finally, we will conclude the chapter by discussing the electronic structure of several representative crystals, as obtained by elaborate computational methods; we will also attempt to interpret these results in the context of the tight-binding approximation.

4.1 The tight-binding approximation

The simplest method for calculating band structures, both conceptually and computationally, is the so called Tight-Binding Approximation (TBA), also referred to as Linear Combination of Atomic Orbitals (LCAO). The latter term is actually used in a wider sense, as we will explain below. The basic assumption in the TBA is that we can use orbitals that are very similar to atomic states (i.e. wavefunctions tightly bound to the atoms, hence the term "tight-binding") as a basis for expanding the crystal wavefunctions. We will deal with the general theory of the TBA first, and then we will illustrate how it is applied through a couple of examples.

Suppose then that we start with a set of atomic wavefunctions

$$\phi_l(\mathbf{r} - \mathbf{t}_i) \tag{4.1}$$

where $\mathbf{t}_i$ is the position of atom with label i in the PUC, and $\phi_l(\mathbf{r})$ is one of the atomic states associated with this atom. The index l can take the usual values for an atom, that is, the angular momentum character $s, p, d, \ldots$. The state $\phi_l(\mathbf{r} - \mathbf{t}_i)$ is centered at the position of the atom with index i. It is assumed that we need as many orbitals as the number of valence states in the atom (this is referred to as the "minimal basis").

Our first task is to construct states which can be used as the basis for expansion of the crystal wavefunctions. These states must obey Bloch's theorem, and we call them $\chi_{\mathbf{k}li}(\mathbf{r})$:

$$\chi_{\mathbf{k}li}(\mathbf{r}) = \frac{1}{\sqrt{N}} \sum_{\mathbf{R}'} \mathrm{e}^{\mathrm{i}\mathbf{k}\cdot\mathbf{R}'} \phi_l(\mathbf{r} - \mathbf{t}_i - \mathbf{R}') \tag{4.2}$$

with the summation running over all the N unit cells in the crystal (the vectors $\mathbf{R}'$), for a given pair of indices i (used to denote the position $\mathbf{t}_i$ of the atom in the PUC) and l (used for the type of orbital). We first verify that these states have Bloch character:

$$\begin{aligned}
\chi_{\mathbf{k}li}(\mathbf{r} + \mathbf{R}) &= \frac{1}{\sqrt{N}} \sum_{\mathbf{R}'} \mathrm{e}^{\mathrm{i}\mathbf{k}\cdot(\mathbf{R}'-\mathbf{R})} \mathrm{e}^{\mathrm{i}\mathbf{k}\cdot\mathbf{R}} \phi_l((\mathbf{r} + \mathbf{R}) - \mathbf{t}_i - \mathbf{R}') \\
&= \mathrm{e}^{\mathrm{i}\mathbf{k}\cdot\mathbf{R}} \frac{1}{\sqrt{N}} \sum_{\mathbf{R}'} \mathrm{e}^{\mathrm{i}\mathbf{k}\cdot(\mathbf{R}'-\mathbf{R})} \phi_l(\mathbf{r} - \mathbf{t}_i - (\mathbf{R}' - \mathbf{R})) \\
&= \mathrm{e}^{\mathrm{i}\mathbf{k}\cdot\mathbf{R}} \frac{1}{\sqrt{N}} \sum_{\mathbf{R}''} \mathrm{e}^{\mathrm{i}\mathbf{k}\cdot\mathbf{R}''} \phi_l(\mathbf{r} - \mathbf{t}_i - \mathbf{R}'') = \mathrm{e}^{\mathrm{i}\mathbf{k}\cdot\mathbf{R}} \chi_{\mathbf{k}li}(\mathbf{r})
\end{aligned} \tag{4.3}$$

that is, Bloch's theorem is satisfied for our choice of $\chi_{\mathbf{k}li}(\mathbf{r})$, with the obvious definition $\mathbf{R}'' = \mathbf{R}' - \mathbf{R}$, which is another lattice vector. Now we can expand the crystal single-particle eigenstates in this basis:

$$\psi_{\mathbf{k}}^{(n)}(\mathbf{r}) = \sum_{l,i} c_{\mathbf{k}li}^{(n)} \chi_{\mathbf{k}li}(\mathbf{r}) \tag{4.4}$$

and all that remains to do is determine the coefficients $c_{\mathbf{k}li}^{(n)}$, assuming that the $\psi_{\mathbf{k}}^{(n)}(\mathbf{r})$ are solutions to the appropriate single-particle equation:

$$\mathcal{H}^{sp} \psi_{\mathbf{k}}^{(n)}(\mathbf{r}) = \epsilon_{\mathbf{k}} \psi_{\mathbf{k}}^{(n)}(\mathbf{r}) \Rightarrow \sum_{l,i} \left[\langle \chi_{\mathbf{k}mj} | \mathcal{H}^{sp} | \chi_{\mathbf{k}li} \rangle - \epsilon_{\mathbf{k}}^{(n)} \langle \chi_{\mathbf{k}mj} | \chi_{\mathbf{k}li} \rangle \right] c_{\mathbf{k}li}^{(n)} = 0 \tag{4.5}$$

In the above equation we only need to consider matrix elements of states with the same **k** index, because

$$\left\langle \psi_{\mathbf{k}}^{(n)} \middle| \psi_{\mathbf{k}'}^{(n')} \right\rangle \sim \delta(\mathbf{k} - \mathbf{k}') \tag{4.6}$$

where we are restricting the values of **k**, **k**′ to the IBZ. In Eq. (4.5) we have a secular equation of size equal to the total number of atomic orbitals in the PUC: the sum is over the number of different types of atoms and the number of orbitals associated with each type of atom. This is exactly the number of solutions (bands) that we can expect at each **k**-point. In order to solve this linear system we need to be able to evaluate the following integrals:

$$\begin{aligned}
\langle \chi_{\mathbf{k}mj} | \chi_{\mathbf{k}li} \rangle &= \frac{1}{N} \sum_{\mathbf{R}',\mathbf{R}''} e^{i\mathbf{k}\cdot(\mathbf{R}'-\mathbf{R}'')} \langle \phi_m(\mathbf{r} - \mathbf{t}_j - \mathbf{R}'') | \phi_l(\mathbf{r} - \mathbf{t}_i - \mathbf{R}') \rangle \\
&= \frac{1}{N} \sum_{\mathbf{R},\mathbf{R}'} e^{i\mathbf{k}\cdot\mathbf{R}} \langle \phi_m(\mathbf{r} - \mathbf{t}_j) | \phi_l(\mathbf{r} - \mathbf{t}_i - \mathbf{R}) \rangle \\
&= \sum_{\mathbf{R}} e^{i\mathbf{k}\cdot\mathbf{R}} \langle \phi_m(\mathbf{r} - \mathbf{t}_j) | \phi_l(\mathbf{r} - \mathbf{t}_i - \mathbf{R}) \rangle
\end{aligned} \tag{4.7}$$

where we have used the obvious definition $\mathbf{R} = \mathbf{R}' - \mathbf{R}''$, and we have eliminated one of the sums over the lattice vectors with the factor $1/N$, since in the last line of Eq. (4.7) there is no explicit dependence on $\mathbf{R}'$. We call the brackets in the last expression the "overlap matrix elements" between atomic states. In a similar fashion we obtain:

$$\langle \chi_{\mathbf{k}mj} | \mathcal{H}^{sp} | \chi_{\mathbf{k}li} \rangle = \sum_{\mathbf{R}} e^{i\mathbf{k}\cdot\mathbf{R}} \langle \phi_m(\mathbf{r} - \mathbf{t}_j) | \mathcal{H}^{sp} | \phi_l(\mathbf{r} - \mathbf{t}_i - \mathbf{R}) \rangle \tag{4.8}$$

and we call the brackets on the right-hand side of Eq. (4.8) the "hamiltonian matrix elements" between atomic states.

At this point we introduce an important approximation: in the spirit of the TBA, we take the overlap matrix elements in Eq. (4.7) to be non-zero only for the same orbitals on the same atom, i.e. only for $m = l$, $j = i$, $\mathbf{R} = 0$, which is expressed by the relation

$$\langle \phi_m(\mathbf{r} - \mathbf{t}_j) | \phi_l(\mathbf{r} - \mathbf{t}_i - \mathbf{R}) \rangle = \delta_{lm} \delta_{ij} \delta(\mathbf{R}) \tag{4.9}$$

This is referred to as an "orthogonal basis", since any overlap between different orbitals on the same atom or orbitals on different atoms is taken to be zero.[1]

[1] If the overlap between the $\phi_m(\mathbf{r})$ orbitals was strictly zero, there would be no interactions between nearest neighbors; this is only a convenient approximation.

Table 4.1. *Equations that define the TBA model.*

The first three equations are general, based on the atomic orbitals $\phi_l(\mathbf{r} - \mathbf{t}_i - \mathbf{R})$ of type l centered at an atom situated at the position $\mathbf{t}_i$ of the unit cell with lattice vector $\mathbf{R}$. The last three correspond to an orthogonal basis of orbitals and nearest neighbor interactions only, which define the on-site and hopping matrix elements of the hamiltonian.

$\chi_{\mathbf{k}li}(\mathbf{r}) = \frac{1}{\sqrt{N}} \sum_{\mathbf{R}} \mathrm{e}^{\mathrm{i}\mathbf{k}\cdot\mathbf{R}} \phi_l(\mathbf{r} - \mathbf{t}_i - \mathbf{R})$	Bloch basis
$\psi_{\mathbf{k}}^{(n)}(\mathbf{r}) = \sum_{l,i} c_{\mathbf{k}li}^{(n)} \chi_{\mathbf{k}li}(\mathbf{r})$	crystal states
$\sum_{l,i} \left[\langle \chi_{\mathbf{k}mj} \vert \mathcal{H}^{sp} \vert \chi_{\mathbf{k}li} \rangle - \epsilon_{\mathbf{k}}^{(n)} \langle \chi_{\mathbf{k}mj} \vert \chi_{\mathbf{k}li} \rangle \right] c_{\mathbf{k}li}^{(n)} = 0$	secular equation
$\langle \phi_m(\mathbf{r} - \mathbf{t}_j) \vert \phi_l(\mathbf{r} - \mathbf{t}_i - \mathbf{R}) \rangle = \delta_{lm} \delta_{ij} \delta(\mathbf{R})$	orthogonal orbitals
$\langle \phi_m(\mathbf{r} - \mathbf{t}_j) \vert \mathcal{H}^{sp} \vert \phi_l(\mathbf{r} - \mathbf{t}_i - \mathbf{R}) \rangle = \delta_{lm} \delta_{ij} \delta(\mathbf{R}) \epsilon_l$	on-site elements
$\langle \phi_m(\mathbf{r} - \mathbf{t}_j) \vert \mathcal{H}^{sp} \vert \phi_l(\mathbf{r} - \mathbf{t}_i - \mathbf{R}) \rangle = \delta((\mathbf{t}_j - \mathbf{t}_i - \mathbf{R}) - \mathbf{d}_{nn}) V_{lm,ij}$	hopping elements

Similarly, we will take the hamiltonian matrix elements in Eq. (4.8) to be non-zero only if the orbitals are on the same atom, i.e. for $j = i$, $\mathbf{R} = 0$, which are referred to as the "on-site energies":

$$\langle \phi_m(\mathbf{r} - \mathbf{t}_j) \vert \mathcal{H}^{sp} \vert \phi_l(\mathbf{r} - \mathbf{t}_i - \mathbf{R}) \rangle = \delta_{lm} \delta_{ij} \delta(\mathbf{R}) \epsilon_l \tag{4.10}$$

or, if the orbitals are on different atoms but situated at nearest neighbor sites, denoted in general as $\mathbf{d}_{nn}$:

$$\langle \phi_m(\mathbf{r} - \mathbf{t}_j) \vert \mathcal{H}^{sp} \vert \phi_l(\mathbf{r} - \mathbf{t}_i - \mathbf{R}) \rangle = \delta((\mathbf{t}_j - \mathbf{t}_i - \mathbf{R}) - \mathbf{d}_{nn}) V_{lm,ij} \tag{4.11}$$

The $V_{lm,ij}$ are also referred to as "hopping" matrix elements. When the nearest neighbors are in the same unit cell, $\mathbf{R}$ can be zero; when they are across unit cells $\mathbf{R}$ can be one of the primitive lattice vectors. The equations that define the TBA model, with the approximation of an orthogonal basis and nearest neighbor interactions only, are summarized in Table 4.1. Even with this drastic approximation, we still need to calculate the values of the matrix elements that we have kept. The parametrization of the hamiltonian matrix in an effort to produce a method with quantitative capabilities has a long history, starting with the work of Harrison [37] (see the Further reading section), and continues to be an active area of research. In principle, these matrix elements can be calculated using one of the single-particle hamiltonians we have discussed in chapter 2 (this approach is being actively pursued as a means of performing fast and reliable electronic structure calculations [38]). However, it is often more convenient to consider these matrix elements as parameters, which are fitted to reproduce certain properties and can then be used to calculate other properties of the solid (see, for instance, Refs. [39, 40]). We illustrate these concepts through two simple examples, the first concerning a 1D lattice, the second a 2D lattice of atoms.

4.1.1 Example: 1D linear chain with s or p orbitals

We consider first the simplest possible case, a linear periodic chain of atoms. Our system has only one type of atom and only one orbital associated with each atom. The first task is to construct the basis for the crystal wavefunctions using the atomic wavefunctions, as was done for the general case in Eq. (4.2). We notice that because of the simplicity of the model, there are no summations over the indices l (there is only one type of orbital for each atom) and i (there is only one atom per unit cell). We keep the index l to identify different types of orbitals in our simple model. Therefore, the basis for the crystal wavefunctions in this case will be simply

$$\chi_{kl}(x) = \sum_{n=-\infty}^{\infty} e^{ikx} \phi_l(x - na) \tag{4.12}$$

where we have further simplified the notation since we are dealing with a 1D example, with the position vector $\mathbf{r}$ set equal to the position x on the 1D axis and the reciprocal-space vector $\mathbf{k}$ set equal to k, while the lattice vectors $\mathbf{R}$ are given by na, with a the lattice constant and n an integer. We will consider atomic wavefunctions $\phi_l(x)$ which have either s-like or p-like character. The real parts of the wavefunction $\chi_{kl}(x)$, $l = s, p$ for a few values of k are shown in Fig. 4.1.

With these states, we can now attempt to calculate the band structure for this model. The TBA with an orthogonal basis and nearest neighbor interactions only implies that the overlap matrix elements are non-zero only for orbitals $\phi_l(x)$ on the same atom, that is,

$$\langle \phi_l(x) | \phi_l(x - na) \rangle = \delta_{n0} \tag{4.13}$$

Similarly, nearest neighbor interactions require that the hamiltonian matrix elements are non-zero only for orbitals that are on the same or neighboring atoms. If the orbitals are on the same atom, then we define the hamiltonian matrix element to be

$$\langle \phi_l(x) | \mathcal{H}^{sp} | \phi_l(x - na) \rangle = \epsilon_l \delta_{n0} \tag{4.14}$$

while if they are on neighboring atoms, that is $n = \pm 1$, we define the hamiltonian matrix element to be

$$\langle \phi_l(x) | \mathcal{H}^{sp} | \phi_l(x - na) \rangle = t_l \delta_{n\pm 1} \tag{4.15}$$

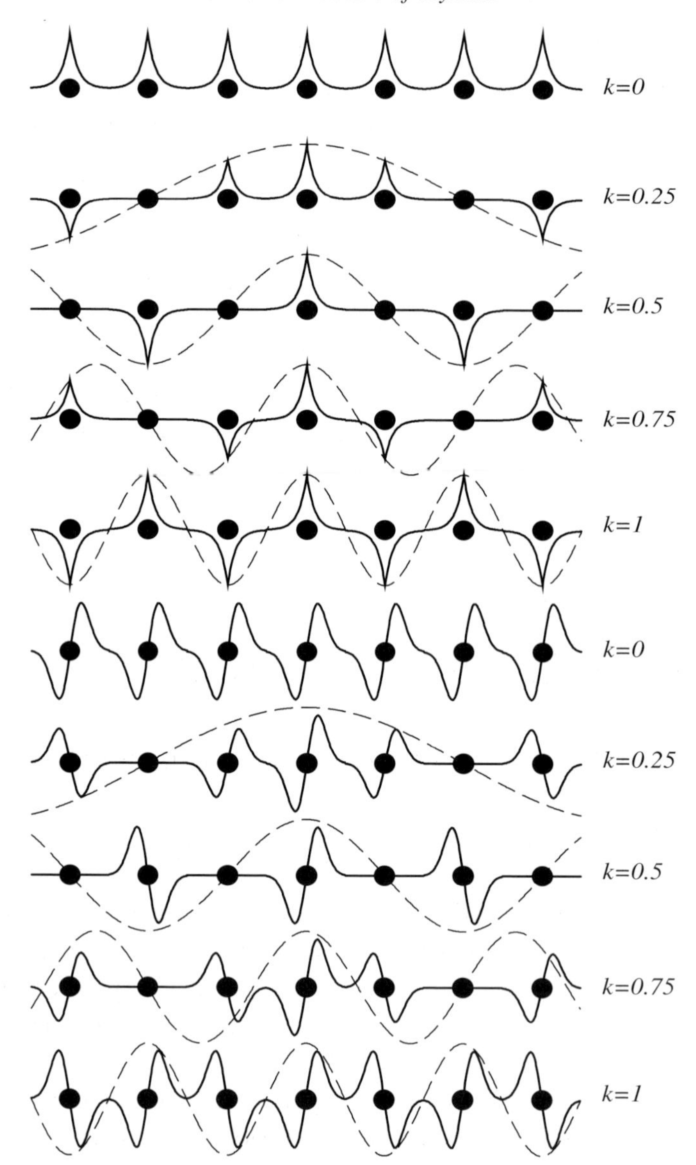

Figure 4.1. Real parts of the crystal wavefunctions $\chi_{kl}(x)$ for $k = 0, 0.25, 0.5, 0.75$ and 1 (in units of π/a); the dashed lines represent the term $\cos(kx)$, which determines the value of the phase factor when evaluated at the atomic sites $x = na$. Top five: s-like state; bottom five: p-like state. The solid dots represent the atoms in the one-dimensional chain.

where ϵ_l is the on-site hamiltonian matrix element and t_l is the hopping matrix element. We expect this interaction between orbitals on neighboring atoms to contribute to the cohesion of the solid, which implies that $t_s < 0$ for s-like orbitals and $t_p > 0$ for p-like orbitals, as we explain in more detail below.

We are now ready to use the $\chi_{kl}(x)$ functions as the basis to construct crystal wavefunctions and with these calculate the single-particle energy eigenvalues, that is, the band structure of the model. The crystal wavefunctions are obtained from the general expression Eq. (4.4):

$$\psi_k(x) = c_k \chi_{kl}(x) \tag{4.16}$$

where only the index k has survived due to the simplicity of the model (the index l simply denotes the character of the atomic orbitals). Inserting these wavefunctions into the secular equation, Eq. (4.5), we find that we have to solve a 1×1 matrix, because we have only one orbital per atom and one atom per unit cell. With the above definitions of the hamiltonian matrix elements between the atomic orbitals ϕ_l's, we obtain

$$[\langle \chi_{kl}(x)|\mathcal{H}^{sp}|\chi_{kl}(x)\rangle - \epsilon_k \langle \chi_{kl}(x)|\chi_{kl}(x)\rangle]c_k = 0$$
$$\Rightarrow \sum_n e^{ikna} \langle \phi_l(x)|\mathcal{H}^{sp}|\phi_l(x - na)\rangle = \epsilon_k \sum_n e^{ikna} \langle \phi_l(x)|\phi_l(x - na)\rangle$$
$$\Rightarrow \sum_n e^{ikna} [\epsilon_l \delta_{n0} + t_l \delta_{n\pm1}] = \epsilon_k \sum_n e^{ikna} \delta_{n0}$$

The solution to the last equation is straightforward, giving the energy band for this simple model:

$$\text{1D chain}: \quad \epsilon_k = \epsilon_l + 2t_l \cos(ka) \tag{4.17}$$

The behavior of the energy in the first BZ of the model, that is, for $-\pi/a \leq k \leq \pi/a$, is shown in Fig. 4.2 for the s and p orbitals. Since the coefficient c_k is undefined by the secular equation, we can take it to be unity, in which case the crystal wavefunctions $\psi_k(x)$ are the same as the basis functions $\chi_{kl}(x)$, which we have already discussed above (see Fig. 4.1).

We elaborate briefly on the sign of the hopping matrix elements and the dispersion of the bands. It is assumed that the single-particle hamiltonian is spherically symmetric. The s orbitals are spherically symmetric and have everywhere the same sign,[2] so that the overlap between s orbitals situated at nearest neighbor sites is positive. In order to produce an attractive interaction between these orbitals, the hopping matrix element must be negative:

$$t_s \equiv \int \phi_s^*(x)\mathcal{H}^{sp}(x)\phi_s(x - a)\mathrm{d}x < 0.$$

[2] We are concerned here with the sign implied only by the angular momentum character of the wavefunction and not by the radial part; of the latter part, only the tail beyond the core is involved, as discussed in chapter 2.

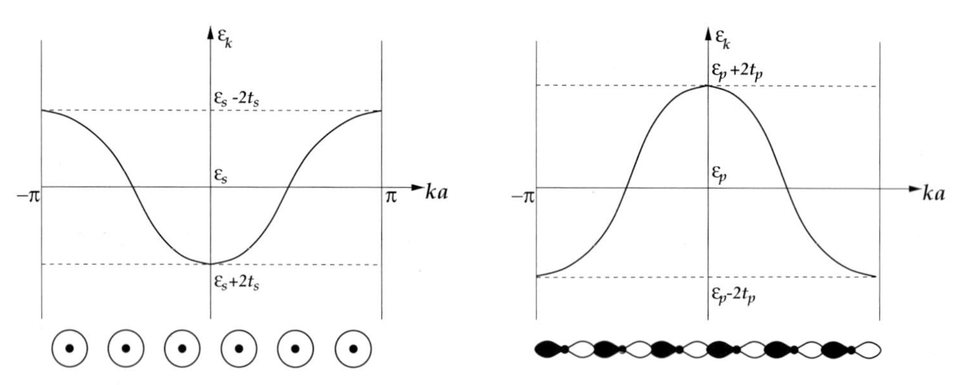

Figure 4.2. Single-particle energy eigenvalues ϵ_k in the first BZ ($-\pi \leq ka \leq \pi$), for the 1D infinite chain model, with one atom per unit cell and one orbital per atom, in the tight-binding approximation with nearest neighbor interactions. **Left:** s-like state; **right:** p-like state. ϵ_s, ϵ_p are the on-site hamiltonian matrix elements and $t_s < 0$ and $t_p > 0$ are the hopping matrix elements. The sketches at the bottom of each panel illustrate the arrangement of the orbitals in the 1D lattice with the positions of the atoms shown as small black dots and the positive and negative lobes of the p orbitals shown in white and black. Due to larger overlap we expect $|t_s| < |t_p|$, which leads to larger dispersion for the p band.

We conclude that the negative sign of this matrix element is due to the hamiltonian, since the product of the wavefunctions is positive. On the other hand, the p orbitals have a positive and a negative lobe (see Appendix B); consequently, the overlap between p orbitals situated at nearest neighbor sites and oriented in the same sense as required by translational periodicity, is negative, because the positive lobe of one is closest to the negative lobe of the next. Therefore, in order to produce an attractive interaction between these orbitals, and since the hamiltonian is the same as in the previous case, the hopping matrix element must be positive

$$t_p \equiv \int \phi_p^*(x)\mathcal{H}^{sp}(x)\phi_p(x-a)\mathrm{d}x > 0.$$

Thus, the band structure for a 1D model with one s-like orbital per unit cell will have a maximum at $k = \pm\pi/a$ and a minimum at $k = 0$, while that of the p-like orbital will have the positions of the extrema reversed, as shown in Fig. 4.2. Moreover, we expect that in general there will be larger overlap between the neighboring p orbitals than between the s orbitals, due to the directed lobes of the former, and therefore $|t_p| > |t_s|$, leading to larger dispersion for the p bands.

The generalization of the model to a two-dimensional square lattice with either one s-like orbital or one p-like orbital per atom and one atom per unit cell is straightforward; the energy eigenvalues are given by

$$\text{2D square}: \quad \epsilon_{\mathbf{k}} = \epsilon_l + 2t_l[\cos(k_x a) + \cos(k_y a)] \tag{4.18}$$

with the two-dimensional reciprocal-space vector defined as $\mathbf{k} = k_x\hat{\mathbf{x}} + k_y\hat{\mathbf{y}}$. Similarly, the generalization to the three-dimensional cubic lattice with either one s-like orbital or one p-like orbital per atom and one atom per unit cell leads to the energy eigenvalues:

$$\text{3D cube}: \quad \epsilon_{\mathbf{k}} = \epsilon_l + 2t_l[\cos(k_x a) + \cos(k_y a) + \cos(k_z a)] \tag{4.19}$$

where $\mathbf{k} = k_x\hat{\mathbf{x}} + k_y\hat{\mathbf{y}} + k_z\hat{\mathbf{z}}$ is the three-dimensional reciprocal-space vector. From these expressions, we can immediately deduce that for this simple model the band width of the energy eigenvalues is given by

$$W = 4dt_l = 2zt_l \tag{4.20}$$

where d is the dimensionality of the model ($d = 1, 2, 3$ in the above examples), or, equivalently, z is the number of nearest neighbors ($z = 2, 4, 6$ in the above examples). We will use this fact in chapter 12, in relation to disorder-induced localization of electronic states.

4.1.2 Example: 2D square lattice with s and p orbitals

We next consider a slightly more complex case, the two-dimensional square lattice with one atom per unit cell. We assume that there are four atomic orbitals per atom, one s-type and three p-type (p_x, p_y, p_z). We work again within the orthogonal basis of orbitals and nearest neighbor interactions only, as described by the equations of Table 4.1. The overlap matrix elements in this case are

$$\begin{aligned}
\langle\phi_m(\mathbf{r})|\phi_l(\mathbf{r}-\mathbf{R})\rangle &= \delta_{lm}\delta(\mathbf{R}) \\
\Rightarrow \langle\chi_{\mathbf{k}m}|\chi_{\mathbf{k}l}\rangle &= \sum_{\mathbf{R}} \mathrm{e}^{\mathrm{i}\mathbf{k}\cdot\mathbf{R}}\langle\phi_m(\mathbf{r})|\phi_l(\mathbf{r}-\mathbf{R})\rangle = \sum_{\mathbf{R}} \mathrm{e}^{\mathrm{i}\mathbf{k}\cdot\mathbf{R}}\delta_{lm}\delta(\mathbf{R}) \\
&= \delta_{lm}
\end{aligned} \tag{4.21}$$

while the hamiltonian matrix elements are

$$\begin{aligned}
\langle\phi_m(\mathbf{r})|\mathcal{H}^{sp}|\phi_l(\mathbf{r}-\mathbf{R})\rangle &\neq 0 \ \text{ only for } \ [\mathbf{R} = \pm a\hat{\mathbf{x}}, \pm a\hat{\mathbf{y}}, 0] \\
\Rightarrow \langle\chi_{\mathbf{k}m}|\mathcal{H}^{sp}|\chi_{\mathbf{k}l}\rangle &= \sum_{\mathbf{R}} \mathrm{e}^{\mathrm{i}\mathbf{k}\cdot\mathbf{R}}\langle\phi_m(\mathbf{r})|\mathcal{H}^{sp}|\phi_l(\mathbf{r}-\mathbf{R})\rangle \\
&\neq 0 \ \text{ only for } \ [\mathbf{R} = \pm a\hat{\mathbf{x}}, \pm a\hat{\mathbf{y}}, 0]
\end{aligned} \tag{4.22}$$

There are a number of different on-site and hopping matrix elements that are generated from all the possible combinations of $\phi_m(\mathbf{r})$ and $\phi_l(\mathbf{r})$ in Eq. (4.22), which we define as follows:

$$
\begin{aligned}
\epsilon_s &= \langle \phi_s(\mathbf{r})|\mathcal{H}^{sp}|\phi_s(\mathbf{r})\rangle \\
\epsilon_p &= \langle \phi_{p_x}(\mathbf{r})|\mathcal{H}^{sp}|\phi_{p_x}(\mathbf{r})\rangle = \langle \phi_{p_y}(\mathbf{r})|\mathcal{H}^{sp}|\phi_{p_y}(\mathbf{r})\rangle = \langle \phi_{p_z}(\mathbf{r})|\mathcal{H}^{sp}|\phi_{p_z}(\mathbf{r})\rangle \\
V_{ss} &= \langle \phi_s(\mathbf{r})|\mathcal{H}^{sp}|\phi_s(\mathbf{r} \pm a\hat{\mathbf{x}})\rangle = \langle \phi_s(\mathbf{r})|\mathcal{H}^{sp}|\phi_s(\mathbf{r} \pm a\hat{\mathbf{y}})\rangle \\
V_{sp} &= \langle \phi_s(\mathbf{r})|\mathcal{H}^{sp}|\phi_{p_x}(\mathbf{r} - a\hat{\mathbf{x}})\rangle = -\langle \phi_s(\mathbf{r})|\mathcal{H}^{sp}|\phi_{p_x}(\mathbf{r} + a\hat{\mathbf{x}})\rangle \\
V_{sp} &= \langle \phi_s(\mathbf{r})|\mathcal{H}^{sp}|\phi_{p_y}(\mathbf{r} - a\hat{\mathbf{y}})\rangle = -\langle \phi_s(\mathbf{r})|\mathcal{H}^{sp}|\phi_{p_y}(\mathbf{r} + a\hat{\mathbf{y}})\rangle \\
V_{pp\sigma} &= \langle \phi_{p_x}(\mathbf{r})|\mathcal{H}^{sp}|\phi_{p_x}(\mathbf{r} \pm a\hat{\mathbf{x}})\rangle = \langle \phi_{p_y}(\mathbf{r})|\mathcal{H}^{sp}|\phi_{p_y}(\mathbf{r} \pm a\hat{\mathbf{y}})\rangle \\
V_{pp\pi} &= \langle \phi_{p_y}(\mathbf{r})|\mathcal{H}^{sp}|\phi_{p_y}(\mathbf{r} \pm a\hat{\mathbf{x}})\rangle = \langle \phi_{p_x}(\mathbf{r})|\mathcal{H}^{sp}|\phi_{p_x}(\mathbf{r} \pm a\hat{\mathbf{y}})\rangle \\
V_{pp\pi} &= \langle \phi_{p_z}(\mathbf{r})|\mathcal{H}^{sp}|\phi_{p_z}(\mathbf{r} \pm a\hat{\mathbf{x}})\rangle = \langle \phi_{p_z}(\mathbf{r})|\mathcal{H}^{sp}|\phi_{p_z}(\mathbf{r} \pm a\hat{\mathbf{y}})\rangle
\end{aligned} \tag{4.23}
$$

The hopping matrix elements are shown schematically in Fig. 4.3. By the symmetry of the atomic orbitals we can deduce:

$$
\begin{aligned}
\langle \phi_s(\mathbf{r})|\mathcal{H}^{sp}|\phi_{p_\alpha}(\mathbf{r})\rangle &= 0 \quad (\alpha = x, y, z) \\
\langle \phi_s(\mathbf{r})|\mathcal{H}^{sp}|\phi_{p_\alpha}(\mathbf{r} \pm a\hat{\mathbf{x}})\rangle &= 0 \quad (\alpha = y, z) \\
\langle \phi_{p_\alpha}(\mathbf{r})|\mathcal{H}^{sp}|\phi_{p_\beta}(\mathbf{r} \pm a\hat{\mathbf{x}})\rangle &= 0 \quad (\alpha, \beta = x, y, z; \alpha \neq \beta) \\
\langle \phi_{p_\alpha}(\mathbf{r})|\mathcal{H}^{sp}|\phi_{p_\beta}(\mathbf{r})\rangle &= 0 \quad (\alpha, \beta = x, y, z; \alpha \neq \beta)
\end{aligned} \tag{4.24}
$$

as can be seen by the diagrams in Fig. 4.3, with the single-particle hamiltonian $\mathcal{H}^{sp}$ assumed to contain only spherically symmetric terms.

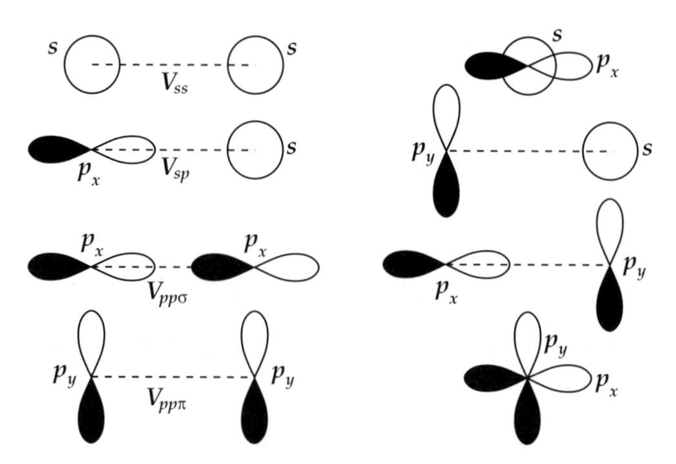

Figure 4.3. Schematic representation of hamiltonian matrix elements between s and p states. **Left:** elements that do not vanish; **right:** elements that vanish due to symmetry. The two lobes of opposite sign of the p_x, p_y orbitals are shaded black and white.

Having defined all these matrix elements, we can calculate the matrix elements between crystal states that enter in the secular equation; we find for our example

$$\begin{aligned}
\langle \chi_{\mathbf{k}s}(\mathbf{r})|\mathcal{H}^{sp}|\chi_{\mathbf{k}s}(\mathbf{r})\rangle &= \langle \phi_s(\mathbf{r})|\mathcal{H}^{sp}|\phi_s(\mathbf{r})\rangle \\
&+ \langle \phi_s(\mathbf{r})|\mathcal{H}^{sp}|\phi_s(\mathbf{r}-a\hat{\mathbf{x}})\rangle \mathrm{e}^{\mathrm{i}\mathbf{k}\cdot a\hat{\mathbf{x}}} \\
&+ \langle \phi_s(\mathbf{r})|\mathcal{H}^{sp}|\phi_s(\mathbf{r}+a\hat{\mathbf{x}})\rangle \mathrm{e}^{-\mathrm{i}\mathbf{k}\cdot a\hat{\mathbf{x}}} \\
&+ \langle \phi_s(\mathbf{r})|\mathcal{H}^{sp}|\phi_s(\mathbf{r}-a\hat{\mathbf{y}})\rangle \mathrm{e}^{\mathrm{i}\mathbf{k}\cdot a\hat{\mathbf{y}}} \\
&+ \langle \phi_s(\mathbf{r})|\mathcal{H}^{sp}|\phi_s(\mathbf{r}+a\hat{\mathbf{y}})\rangle \mathrm{e}^{-\mathrm{i}\mathbf{k}\cdot a\hat{\mathbf{y}}} \\
&= \epsilon_s + 2V_{ss}\big[\cos(k_x a) + \cos(k_y a)\big]
\end{aligned} \tag{4.25}$$

and similarly for the rest of the matrix elements

$$\begin{aligned}
\langle \chi_{\mathbf{k}s}(\mathbf{r})|\mathcal{H}^{sp}|\chi_{\mathbf{k}p_x}(\mathbf{r})\rangle &= 2i\,V_{sp}\sin(k_x a) \\
\langle \chi_{\mathbf{k}s}(\mathbf{r})|\mathcal{H}^{sp}|\chi_{\mathbf{k}p_y}(\mathbf{r})\rangle &= 2i\,V_{sp}\sin(k_y a) \\
\langle \chi_{\mathbf{k}p_z}(\mathbf{r})|\mathcal{H}^{sp}|\chi_{\mathbf{k}p_z}(\mathbf{r})\rangle &= \epsilon_p + 2V_{pp\pi}\big[\cos(k_x a) + \cos(k_y a)\big] \\
\langle \chi_{\mathbf{k}p_x}(\mathbf{r})|\mathcal{H}^{sp}|\chi_{\mathbf{k}p_x}(\mathbf{r})\rangle &= \epsilon_p + 2V_{pp\sigma}\cos(k_x a) + 2V_{pp\pi}\cos(k_y a) \\
\langle \chi_{\mathbf{k}p_y}(\mathbf{r})|\mathcal{H}^{sp}|\chi_{\mathbf{k}p_y}(\mathbf{r})\rangle &= \epsilon_p + 2V_{pp\pi}\cos(k_x a) + 2V_{pp\sigma}\cos(k_y a)
\end{aligned} \tag{4.26}$$

With these we can now construct the hamiltonian matrix for each value of $\mathbf{k}$, and obtain the eigenvalues and eigenfunctions by diagonalizing the secular equation.

For a quantitative discussion of the energy bands we will concentrate on certain portions of the BZ, which correspond to high-symmetry points or directions in the IBZ. Using the results of chapter 3 for the IBZ for the high-symmetry points for this lattice, we conclude that we need to calculate the band structure along $\Gamma - \Delta - X - Z - M - \Sigma - \Gamma$. We find that at $\Gamma = (k_x, k_y) = (0, 0)$, the matrix is already diagonal and the eigenvalues are given by

$$\epsilon_\Gamma^{(1)} = \epsilon_s + 4V_{ss},\ \epsilon_\Gamma^{(2)} = \epsilon_p + 4V_{pp\pi},\ \epsilon_\Gamma^{(3)} = \epsilon_\Gamma^{(4)} = \epsilon_p + 2V_{pp\pi} + 2V_{pp\sigma} \tag{4.27}$$

The same is true for the point $M = (1, 1)(\pi/a)$, where we get

$$\epsilon_M^{(1)} = \epsilon_M^{(3)} = \epsilon_p - 2V_{pp\pi} - 2V_{pp\sigma},\ \epsilon_M^{(2)} = \epsilon_p - 4V_{pp\pi},\ \epsilon_M^{(4)} = \epsilon_s - 4V_{ss} \tag{4.28}$$

Finally, at the point $X = (1, 0)(\pi/a)$ we have another diagonal matrix with eigenvalues

$$\epsilon_X^{(1)} = \epsilon_p + 2V_{pp\pi} - 2V_{pp\sigma},\ \epsilon_X^{(2)} = \epsilon_p, \epsilon_X^{(3)} = \epsilon_s, \epsilon_X^{(4)} = \epsilon_p - 2V_{pp\pi} + 2V_{pp\sigma} \tag{4.29}$$

We have chosen the labels of those energy levels to match the band labels as displayed on p. 135 in Fig. 4.4(a). Notice that there are doubly degenerate states at

Table 4.2. *Matrix elements for the 2D square lattice with s, p_x, p_y, p_z orbitals at the high-symmetry points Δ, Z, Σ.*

In all cases, $0 < k < 1$.

$\mathbf{k}$	$\Delta = (k, 0)(\pi/a)$	$Z = (1, k)(\pi/a)$	$\Sigma = (k, k)(\pi/a)$
$A_{\mathbf{k}}$	$2V_{ss}(\cos(k\pi) + 1)$	$2V_{ss}(\cos(k\pi) - 1)$	$4V_{ss}\cos(k\pi)$
$B_{\mathbf{k}}$	$2\mathrm{i}V_{sp}\sin(k\pi)$	$2\mathrm{i}V_{sp}\sin(k\pi)$	$2\sqrt{2}\mathrm{i}V_{sp}\sin(k\pi)$
$C_{\mathbf{k}}$	$2V_{pp\sigma}\cos(k\pi) + 2V_{pp\pi}$	$2V_{pp\sigma}\cos(k\pi) - 2V_{pp\pi}$	$2(V_{pp\sigma} + V_{pp\pi})\cos(k\pi)$
$D_{\mathbf{k}}$	$2V_{pp\sigma} + 2V_{pp\pi}\cos(k\pi)$	$2V_{pp\pi}\cos(k\pi) - 2V_{pp\sigma}$	$2(V_{pp\sigma} + V_{pp\pi})\cos(k\pi)$
$E_{\mathbf{k}}$	$2V_{pp\pi}(\cos(k\pi) + 1)$	$2V_{pp\pi}(\cos(k\pi) - 1)$	$4V_{pp\pi}\cos(k\pi)$

Γ and at M, dictated by symmetry, that is, by the values of $\mathbf{k}$ at those points and the form of the hopping matrix elements within the nearest neighbor approximation. For the three other high-symmetry points, Δ, Z, Σ, we obtain matrices of the type

$$\begin{bmatrix} A_{\mathbf{k}} & B_{\mathbf{k}} & 0 & 0 \\ B_{\mathbf{k}}^* & C_{\mathbf{k}} & 0 & 0 \\ 0 & 0 & D_{\mathbf{k}} & 0 \\ 0 & 0 & 0 & E_{\mathbf{k}} \end{bmatrix} \tag{4.30}$$

The matrices for Δ and Z can be put in this form straightforwardly, while the matrix for Σ requires a change of basis in order to be brought into this form, namely

$$\chi_{\mathbf{k}1}(\mathbf{r}) = \frac{1}{\sqrt{2}}\left[\chi_{\mathbf{k}p_x}(\mathbf{r}) + \chi_{\mathbf{k}p_y}(\mathbf{r})\right]$$

$$\chi_{\mathbf{k}2}(\mathbf{r}) = \frac{1}{\sqrt{2}}\left[\chi_{\mathbf{k}p_x}(\mathbf{r}) - \chi_{\mathbf{k}p_y}(\mathbf{r})\right] \tag{4.31}$$

with the other two functions, $\chi_{\mathbf{k}s}(\mathbf{r})$ and $\chi_{\mathbf{k}p_z}(\mathbf{r})$, the same as before. The different high-symmetry $\mathbf{k}$-points result in the matrix elements tabulated in Table 4.2. These matrices are then easily solved for the eigenvalues, giving:

$$\epsilon_{\mathbf{k}}^{(1,2)} = \frac{1}{2}\left[(A_{\mathbf{k}} + C_{\mathbf{k}}) \pm \sqrt{(A_{\mathbf{k}} - C_{\mathbf{k}})^2 + 4|B_{\mathbf{k}}|^2}\right], \; \epsilon_{\mathbf{k}}^{(3)} = D_{\mathbf{k}}, \; \epsilon_{\mathbf{k}}^{(4)} = E_{\mathbf{k}} \tag{4.32}$$

We have then obtained the eigenvalues for all the high-symmetry points in the IBZ. All that remains to be done is to determine the numerical values of the hamiltonian matrix elements.

In principle, one can imagine calculating the values of the hamiltonian matrix elements using one of the single-particle hamiltonians we discussed in chapter 2. There is a question as to what exactly the appropriate atomic basis functions $\phi_l(\mathbf{r})$ should be. States associated with free atoms are not a good choice, because in the solid the corresponding single-particle states are more compressed due to the presence of other electrons nearby. One possibility then is to solve for atomic-like states in fictitious atoms where the single-particle wavefunctions are compressed, by imposing for instance a constraining potential (typically a harmonic well) in addition to the Coulomb potential of the nucleus.

Alternatively, one can try to guess the values of the hamiltonian matrix so that they reproduce some important features of the band structure, which can be determined independently from experiment. Let us try to predict at least the sign and relative magnitude of the hamiltonian matrix elements, in an attempt to guess a set of reasonable values. First, the diagonal matrix elements ϵ_s, ϵ_p should have a difference approximately equal to the energy difference of the corresponding eigenvalues in the free atom. Notice that if we think of the atomic-like functions $\phi_l(\mathbf{r})$ as corresponding to compressed wavefunctions then the corresponding eigenvalues ϵ_l are not identical to those of the free atom, but we could expect the compression of eigenfunctions to have similar effects on the different eigenvalues. Since the energy scale is arbitrary, we can choose ϵ_p to be the zero of energy and ϵ_s to be lower in energy by approximately the energy difference of the corresponding free-atom states. The choice $\epsilon_s = -8$ eV is representative of this energy difference for several second row elements in the Periodic Table.

The matrix element V_{ss} represents the interaction of two $\phi_s(\mathbf{r})$ states at a distance a, the lattice constant of our model crystal. We expect this interaction to be attractive, that is, to contribute to the cohesion of the solid. Therefore, by analogy to our earlier analysis for the 1D model, we expect V_{ss} to be negative. The choice $V_{ss} = -2$ eV for this interaction would be consistent with our choice of the difference between ϵ_s and ϵ_p. Similarly, we expect the interaction of two p states to be attractive in general. In the case of $V_{pp\sigma}$ we are assuming the neighboring $\phi_{p_x}(\mathbf{r})$ states to be oriented along the x axis in the same sense, that is, with positive lobes pointing in the positive direction as required by translational periodicity. This implies that the negative lobe of the state to the right is closest to the positive lobe of the state to the left, so that the overlap between the two states will be negative. Because of this negative overlap, $V_{pp\sigma}$ should be positive so that the net effect is an attractive interaction, by analogy to what we discussed earlier for the 1D model. We expect this matrix element to be roughly of the same magnitude as V_{ss} and a little larger in magnitude, to reflect the larger overlap between the directed lobes of p states. A reasonable choice is $V_{pp\sigma} = +2.2$ eV. In the case of $V_{pp\pi}$, the two p states are parallel to each other at a distance a, so we expect the attractive interaction to be a little weaker than in the previous case, when the orbitals were pointing toward each other. A reasonable choice is $V_{pp\pi} = -1.8$ eV. Finally, we define V_{sp} to be the matrix element with $\phi_{p_x}(\mathbf{r})$ to the left of $\phi_s(\mathbf{r})$, so that the positive lobe of the p orbital is closer to the s orbital and their overlap is positive. As a consequence

Table 4.3. *Values of the on-site and hopping matrix elements for the band structure of the 2D square lattice with an orthogonal s and p basis and nearest neighbor interactions.*

ϵ_p is taken to be zero in all cases. All values are in electronvolts. (a)–(f) refer to parts in Fig. 4.4.

	(a)	(b)	(c)	(d)	(e)	(f)
ϵ_s	−8.0	−16.0	−8.0	−8.0	−8.0	−8.0
V_{ss}	−2.0	−2.0	−4.0	−2.0	−2.0	−2.0
$V_{pp\sigma}$	+2.2	+2.2	+2.2	+4.4	+2.2	+2.2
$V_{pp\pi}$	−1.8	−1.8	−1.8	−1.8	−3.6	−1.8
V_{sp}	−2.1	−2.1	−2.1	−2.1	−2.1	−4.2

of this definition, this matrix element, which also contributes to attraction, must be negative; we expect its magnitude to be somewhere between the V_{ss} and $V_{pp\sigma}$ matrix elements. A reasonable choice is $V_{sp} = -2.1$ eV. With these choices, the model yields the band structure shown in Fig. 4.4(a). Notice that in addition to the doubly degenerate states at Γ and M which are expected from symmetry, there is also a doubly degenerate state at X; this is purely accidental, due to our choice of parameters, as the following discussion also illustrates.

In order to elucidate the influence of the various matrix elements on the band structure we also show in Fig. 4.4 a number of other choices for their values. To keep the comparisons simple, in each of the other choices we increase one of the matrix elements by a factor of 2 relative to its value in the original set and keep all other values the same; the values for each case are given explicitly in Table 4.3. The corresponding Figs. 4.4 (b)–(f) provide insight into the origin of the bands. To facilitate the comparison we label the bands 1–4, according to their order in energy near Γ.

Comparing Figs. 4.4 (a) and (b) we conclude that band 1 arises from interaction of the s orbitals in neighboring atoms: a decrease of the corresponding eigenvalue ϵ_s from -8 to -16 eV splits this band off from the rest, by lowering its energy throughout the BZ by 8 eV, without affecting the other three bands, except for some minor changes in the neighborhood of M where bands 1 and 3 were originally degenerate. Since in plot (b) band 1 has split from the rest, now bands 3 and 4 have become degenerate at M, because there must be a doubly degenerate eigenvalue at M independent of the values of the parameters, as we found in Eq. (4.28). An increase of the magnitude of V_{ss} by a factor of 2, which leads to the band structure of plot (c), has as a major effect the increase of the dispersion of band 1; this confirms that band 1 is primarily due to the interaction between s orbitals. There are also some changes in band 4, which at M depends on the value of V_{ss}, as found in Eq. (4.28).

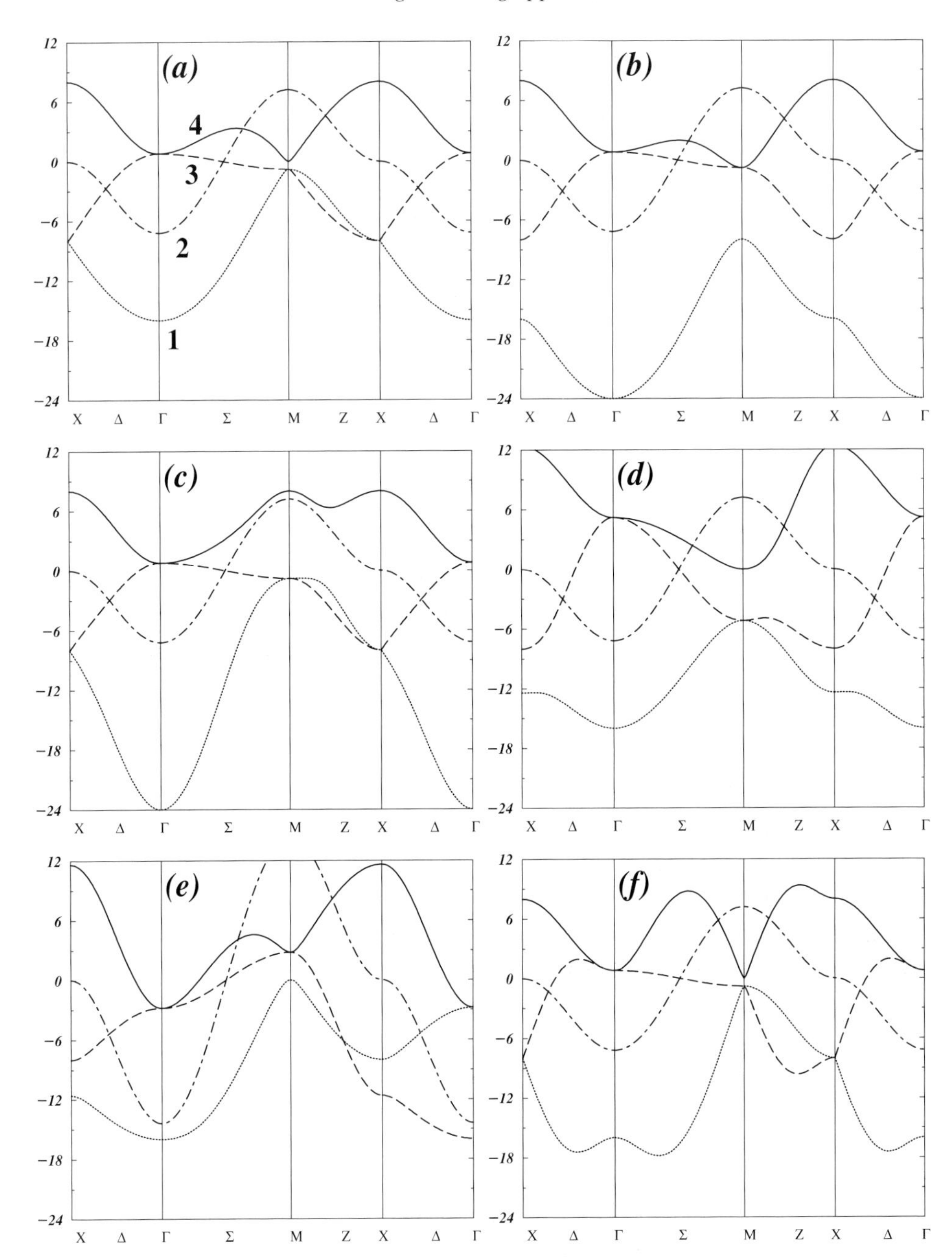

Figure 4.4. The band structure of the 2D square lattice with one atom per unit cell and an orthogonal basis consisting of s, p_x, p_y, p_z orbitals with nearest neighbor interactions. The values of the parameters for the six different plots are given in Table 4.3.

Increasing the magnitude of $V_{pp\sigma}$ by a factor of 2 affects significantly bands 3 and 4, somewhat less band 1, and not at all band 2, as seen from the comparison between plots (a) and (d). This indicates that bands 3 and 4 are essentially related to σ interactions between the p_x and p_y orbitals on neighboring atoms. This is also supported by plot (e), in which increasing the magnitude of $V_{pp\pi}$ by a factor of 2 has as a major effect the dramatic increase of the dispersion of band 2; this leads to the conclusion that band 2 arises from π-bonding interactions between p_z orbitals. The other bands are also affected by this change in the value of $V_{pp\pi}$, because they contain π-bonding interactions between p_x and p_y orbitals, but the effect is not as dramatic, since in the other bands there are also contributions from σ-bonding interactions, which lessen the importance of the $V_{pp\pi}$ matrix element. Finally, increasing the magnitude of V_{sp} by a factor of 2 affects all bands except band 2, as seen in plot (f); this is because all other bands except band 2 involve orbitals s and p interacting through σ bonds.

Two other features of the band structure are also worth mentioning: First, that bands 1 and 3 in Figs. 4.4(a) and (b) are nearly parallel to each other throughout the BZ. This is an accident related to our choice of parameters for these two plots, as the other four plots prove. This type of behavior has important consequences for the optical properties, as discussed in chapter 5, particularly when the lower band is occupied (it lies entirely below the Fermi level) and the upper band is empty (it lies entirely above the Fermi level). The second interesting feature is that the lowest band is parabolic near Γ, in all plots of Fig. 4.4 except for (f). The parabolic nature of the lowest band near the minimum is also a feature of the simple 1D model discussed in section 4.1.1, as well as of the free-electron model discussed in chapter 3. In all these cases, the lowest band near the minimum has essentially pure s character, and its dispersion is dictated by the periodicity of the lattice rather than interaction with other bands. Only for the choice of parameters in plot (f) is the parabolic behavior near the minimum altered; in this case the interaction between s and p orbitals (V_{sp}) is much larger than the interaction between s orbitals, so that the nature of the band near the minimum is not pure s any longer but involves also the p states. This last situation is unusual. Far more common is the behavior exemplified by plots (a) – (d), where the nature of the lowest band is clearly associated with the atomic orbitals with the lowest energy. This is demonstrated in more realistic examples later in this chapter.

4.1.3 Generalizations of the TBA

The examples we have discussed above are the simplest version of the TBA, with only orthogonal basis functions and nearest neighbor interactions, as defined in Eq. (4.21) and Eq. (4.22), respectively. We also encountered matrix elements in which the p wavefunctions are either parallel or point toward one another along the line that separates them; the case when they are perpendicular results in zero matrix elements by symmetry; see Fig. 4.3. It is easy to generalize all this to a more flexible model, as we discuss next. A comprehensive treatment of the tight-binding method and its application to elemental solids is given in the book by Papaconstantopoulos [41]. It is also worth mentioning that the TBA methods are increasingly employed to calculate the total energy of a solid. This practice is motivated by the desire to have a reasonably fast method for total-energy and force calculations while maintaining the flexibility of a quantum mechanical treatment as opposed to resorting to effective interatomic potentials (for details on how TBA methods are used for total-energy calculations see the original papers in the literature [42], [43]).

(1) Arbitrary orientation of orbitals

First, it is straightforward to include configurations in which the p orbitals are not just parallel or lie on the line that joins the atomic positions. We can consider each p orbital to be composed of a linear combination of two perpendicular p orbitals, one lying along the line that joins the atomic positions, the other perpendicular to it. This then leads to the general description of the interaction between two p-type orbitals oriented in random directions θ_1 and θ_2 relative to the line that joins the atomic positions where they are centered, as shown in Fig. 4.5:

$$
\begin{aligned}
\phi_{p_1}(\mathbf{r}) &= \phi_{p_{1x}}(\mathbf{r})\cos\theta_1 + \phi_{p_{1y}}(\mathbf{r})\sin\theta_1 \\
\phi_{p_2}(\mathbf{r}) &= \phi_{p_{2x}}(\mathbf{r})\cos\theta_2 + \phi_{p_{2y}}(\mathbf{r})\sin\theta_2 \\
\langle\phi_{p_1}|\mathcal{H}^{sp}|\phi_{p_2}\rangle &= \langle\phi_{p_{1x}}|\mathcal{H}^{sp}|\phi_{p_{2x}}\rangle\cos\theta_1\cos\theta_2 + \langle\phi_{p_{1y}}|\mathcal{H}^{sp}|\phi_{p_{2y}}\rangle\sin\theta_1\sin\theta_2 \\
&= V_{pp\sigma}\cos\theta_1\cos\theta_2 + V_{pp\pi}\sin\theta_1\sin\theta_2
\end{aligned}
\tag{4.33}
$$

where the line joining the atomic centers is taken to be the x axis, the direction perpendicular to it the y axis, and from symmetry we have $\langle\phi_{p_{1x}}|\mathcal{H}^{sp}|\phi_{p_{2y}}\rangle = 0$ and $\langle\phi_{p_{1y}}|\mathcal{H}^{sp}|\phi_{p_{2x}}\rangle = 0$. The matrix elements between an s and a p orbital with arbitrary orientation relative to the line joining their centers is handled by the same

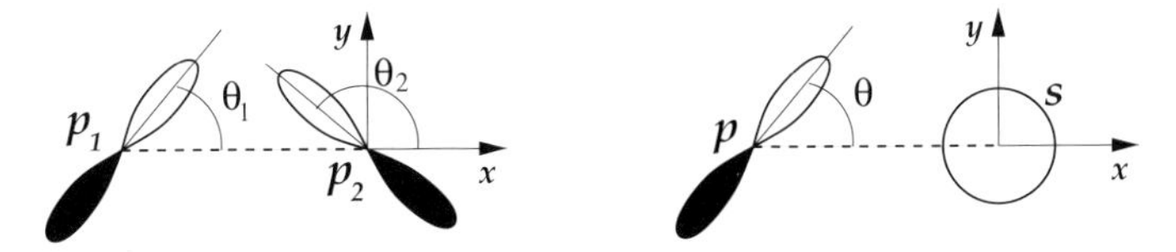

Figure 4.5. **Left:** two p orbitals oriented at arbitrary directions θ_1, θ_2, relative to the line that joins their centers. **Right:** an s orbital and a p orbital which lies at an angle θ relative to the line that join their centers.

procedure, leading to

$$\begin{aligned}\phi_p(\mathbf{r}) &= \phi_{p_x}(\mathbf{r})\cos\theta + \phi_{p_y}(\mathbf{r})\sin\theta \\ \langle\phi_p|\mathcal{H}^{sp}|s\rangle &= \langle\phi_{p_x}|\mathcal{H}^{sp}|s\rangle\cos\theta + \langle\phi_{p_y}|\mathcal{H}^{sp}|s\rangle\sin\theta \\ &= V_{sp}\cos\theta \end{aligned} \tag{4.34}$$

for the relative orientation of the p and s orbitals shown in Fig. 4.5.

(2) Non-orthogonal overlap matrix

A second generalization is to consider that the overlap matrix is not orthogonal. This is especially meaningful when we are considering contracted orbitals that are not true atomic orbitals, which is more appropriate in describing the wavefunctions of the solid. Then we will have

$$\langle\phi_m(\mathbf{r}-\mathbf{R}'-\mathbf{t}_j)|\phi_l(\mathbf{r}-\mathbf{R}-\mathbf{t}_i)\rangle = S_{\mu'\mu} \tag{4.35}$$

where we use the index μ to denote all three indices associated with each atomic orbital, that is, $\mu \to (li\mathbf{R})$ and $\mu' \to (mj\mathbf{R}')$. This new matrix is no longer diagonal, $S_{\mu'\mu} \neq \delta_{ml}\delta_{ji}\delta(\mathbf{R}-\mathbf{R}')$, as we had assumed earlier, Eq. (4.21). Then we need to solve the general secular equation (Eq. (4.5)) with the general definitions of the hamiltonian (Eq. (4.8)) and overlap (Eq. (4.7)) matrix elements. A common approximation is to take

$$S_{\mu'\mu} = f(|\mathbf{R}-\mathbf{R}'|)S_{mj,li} \tag{4.36}$$

where the function $f(r)$ falls fast with the magnitude of the argument or is cut off to zero beyond some distance $r > r_c$. In this case, consistency requires that the hamiltonian matrix elements are also cut off for $r > r_c$, where r is the distance between the atomic positions where the atomic-like orbitals are centered. Of course the larger r_c, the more matrix elements we will need to calculate (or fit), and the approximation becomes more computationally demanding.

(3) Multi-center integrals

The formulation of the TBA up to this point has assumed that the TBA hamiltonian matrix elements depend only on two single-particle wavefunctions centered at two different atomic sites. For example, we assumed that the hamiltonian matrix elements depend only on the relative distance and orientation of the two atomic-like orbitals between which we calculate the expectation value of the single-particle hamiltonian. This is referred to as the two-center approximation, but it is obviously another implicit approximation, on top of restricting the basis to the atomic-like wavefunctions. In fact, it is plausible that, in the environment of the solid, the presence of other electrons nearby will affect the interaction of any two given atomic-like wavefunctions. In principle we should consider all such interactions. An example of such terms is a three-center matrix element of the hamiltonian in which one orbital is centered at some atomic site, a second orbital is centered at a different atomic site, and a term in the hamiltonian (the ionic potential) includes the position of a third atomic site. One way of taking into account these types of interactions is to make the hamiltonian matrix elements environment dependent. In this case, the value of a two-center hamiltonian matrix element, involving explicitly the positions of only two atoms, will depend on the position of all other atoms around it and on the type of atomic orbitals on these other atoms. To accomplish this, we need to introduce more parameters to allow for the flexibility of having several possible environments around each two-center matrix element, making the approach much more complicated. The increase in realistic representation of physical systems is always accompanied by an increase in complexity and computational cost.

(4) Excited-state orbitals in basis

Finally, we can consider our basis as consisting not only of the valence states of the atoms, but including unoccupied (excited) atomic states. This is referred to as going beyond the minimal basis. The advantages of this generalization are obvious since including more basis functions always gives a better approximation (by the variational principle). This, however, presents certain difficulties: the excited states tend to be more diffuse in space, with tails extending farther away from the atomic core. This implies that the overlap between such states will not fall off fast with distance between their centers, and it will be difficult to truncate the non-orthogonal overlap matrix and the hamiltonian matrix at a reasonable distance. To avoid this problem, we perform the following operations. First we orthogonalize the states in the minimal basis. This is accomplished by diagonalizing the non-orthogonal overlap matrix and using as the new basis the linear combination of states that corresponds to the eigenvectors of the non-orthogonal overlap matrix. Next, we orthogonalize the excited states to the states in the orthogonal minimal basis, and finally we orthogonalize the new excited states among themselves. Each orthogonalization involves the diagonalization of the corresponding overlap matrix. The advantage of this procedure is that with each diagonalization, the energy of the new states is raised (since they are orthogonal to all previous states), and the overlap between them is reduced. In this way we create a basis that gives rise to a hamiltonian which can be truncated at a reasonable cutoff distance. Nevertheless, the increase in variational freedom that comes with the inclusion of excited states increases computational complexity, since we will have a larger basis and a correspondingly larger number of matrix elements that we need to calculate or obtain from fitting to known results.

This extension is suggestive of a more general approach: we can use an arbitrary set of functions centered at atomic sites to express the hamiltonian matrix elements. A popular set is composed of normalized gaussian functions (see Appendix 1) multiplied by the appropriate spherical harmonics to resemble a set of atomic-like orbitals. We can then calculate the hamiltonian and overlap matrix elements using this set of functions and diagonalize the resulting secular equation to obtain the desired eigenvalues and eigenfunctions. To the extent that these functions represent accurately all possible electronic states (if the original set does not satisfy this requirement we can simply add more basis functions), we can then consider that we have a variationally correct description of the system. The number of basis functions is no longer determined by the number of valence states in the constituent atoms, but rather by variational requirements. This is the more general Linear Combination of Atomic Orbitals (LCAO) method. It is customary in this case to use an explicit form for the single-particle hamiltonian and calculate the hamiltonian and overlap matrix elements exactly, either analytically, if the choice of basis functions permits it, or numerically.

4.2 General band-structure methods

Since the early days of solid state theory, a number of approaches have been introduced to solve the single-particle hamiltonian and obtain the eigenvalues (band structure) and eigenfunctions. These methods were the foundation on which modern approaches for electronic structure calculations have been developed. We review the basic ideas of these methods next.

Cellular or Linearized Muffin-Tin Orbital (LMTO) method This approach, originally developed by Wigner and Seitz [44], considers the solid as made up of cells (the Wigner–Seitz or WS cells), which are the analog of the Brillouin Zones in real space. In each cell, the potential felt by the electrons is the atomic potential, which is spherically symmetric around the atomic nucleus, but its boundaries are those of the WS cell, whose shape is dictated by the crystal. Due to the Bloch character of wavefunctions, the following boundary conditions must be obeyed at the boundary of the WS cell, denoted by $\mathbf{r}_b$:

$$\begin{aligned}\psi_{\mathbf{k}}(\mathbf{r}_b) &= \mathrm{e}^{-\mathrm{i}\mathbf{k}\cdot\mathbf{R}}\psi_{\mathbf{k}}(\mathbf{r}_b+\mathbf{R})\\ \hat{\mathbf{n}}(\mathbf{r}_b)\cdot\nabla\psi_{\mathbf{k}}(\mathbf{r}_b) &= -\mathrm{e}^{-\mathrm{i}\mathbf{k}\cdot\mathbf{R}}\hat{\mathbf{n}}(\mathbf{r}_b+\mathbf{R})\cdot\nabla\psi_{\mathbf{k}}(\mathbf{r}_b+\mathbf{R})\end{aligned} \tag{4.37}$$

where $\hat{\mathbf{n}}(\mathbf{r}_b)$ is the vector normal to the surface of the WS cell. Since the potential inside the WS cell is assumed to be spherical, we can use the standard expansion in spherical harmonics $Y_{lm}(\hat{\mathbf{r}})$ and radial wavefunctions $\rho_{\mathbf{k}l}(r)$ (for details see

Appendix B) which obey the following equation:

$$\left[\frac{\mathrm{d}^2}{\mathrm{d}r^2} + \frac{2}{r}\frac{\mathrm{d}}{\mathrm{d}r} + \left(\frac{\hbar^2}{2m_e}\right)^{-1}\left(\epsilon_{\mathbf{k}} - V(r) - \frac{\hbar^2}{2m_e}\frac{l(l+1)}{r^2}\right)\right]\rho_{\mathbf{k}l}(r) = 0 \quad (4.38)$$

where the dependence of the radial wavefunction on $\mathbf{k}$ enters through the eigenvalue $\epsilon_{\mathbf{k}}$. In terms of these functions the crystal wavefunctions become:

$$\psi_{\mathbf{k}}(\mathbf{r}) = \sum_{lm} \alpha_{\mathbf{k}lm} Y_{lm}(\hat{\mathbf{r}})\rho_{\mathbf{k}l}(r) \quad (4.39)$$

Taking matrix elements of the hamiltonian between such states creates a secular equation which can be solved to produce the desired eigenvalues. Since the potential cannot be truly spherical throughout the WS cell, it is reasonable to consider it to be spherical within a sphere which lies entirely within the WS, and to be zero outside that sphere. This gives rise to a potential that looks like a muffin-tin, hence the name of the method Linearized Muffin-Tin Orbitals (LMTO). This method is in use for calculations of the band structure of complex solids. The basic assumption of the method is that a spherical potential around the nuclei is a reasonable approximation to the true potential experienced by the electrons in the solid.

Augmented Plane Waves (APW) This method, introduced by Slater [45], consists of expanding the wavefunctions in plane waves in the regions between the atomic spheres, and in functions with spherical symmetry within the spheres. Then the two expressions must be matched at the sphere boundary so that the wavefunctions and their first and second derivatives are continuous. For core states, the wavefunctions are essentially unchanged within the spheres. It is only valence states that have significant weight in the regions outside the atomic spheres.

Both the LMTO and the APW methods treat all the electrons in the solid, that is, valence as well as core electrons. Accordingly, they are referred to as "all-electron" methods. The two methods share the basic concept of separating space in the solid into the spherical regions around the nuclei and the interstitial regions between these spheres. In the APW method, the spheres are touching while in the LMTO method they are overlapping. In many cases, especially in crystal structures other than the close-packed ones (FCC, HCP, BCC), this separation of space leads to inaccuracies, which can be corrected by elaborate extensions described as "full potential" treatment. There is also an all-electron electronic structure method based on multiple scattering theory known as the Korringa–Kohn–Rostocker (KKR) method. A detailed exposition of these methods falls beyond the scope of this book; they are discussed in specialized articles or books (see for example the book by Singh [46] often accompanied by descriptions of computer codes which are necessary for their application. In the remaining of this section we will examine other band structure methods in which the underlying concept is a separation between the electronic core and valence states. This separation makes it possible to treat a larger number of valence states, with a relatively small sacrifice in accuracy. The advantage is that structures with many more atoms in the unit cell can then be studied efficiently.

Orthogonalized Plane Waves (OPW) This method, due to Herring [47], is an elaboration on the APW approach. The trial valence wavefunctions are written at the outset as a combination of plane waves and core-derived states:

$$\phi_{\mathbf{k}}^{(v)}(\mathbf{r}) = \frac{1}{\Omega}\mathrm{e}^{\mathrm{i}\mathbf{k}\cdot\mathbf{r}} + \sum_{c'} \beta^{(c')}\psi_{\mathbf{k}}^{(c')}(\mathbf{r}) \tag{4.40}$$

where the $\psi_{\mathbf{k}}^{(c)}(\mathbf{r})$ are Bloch states formed out of atomic core states:

$$\psi_{\mathbf{k}}^{(c)}(\mathbf{r}) = \sum_{\mathbf{R}} \mathrm{e}^{\mathrm{i}\mathbf{k}\cdot\mathbf{R}}\phi^{(c)}(\mathbf{r}-\mathbf{R}) \tag{4.41}$$

With the choice of the parameters

$$\beta^{(c)} = -\langle\psi_{\mathbf{k}}^{(c)}|\mathbf{k}\rangle, \quad \langle\mathbf{r}|\mathbf{k}\rangle = \frac{1}{\Omega}\mathrm{e}^{\mathrm{i}\mathbf{k}\cdot\mathbf{r}}$$

we make sure that the wavefunctions $\phi_{\mathbf{k}}^{(v)}(\mathbf{r})$ are orthogonal to core Bloch states:

$$\langle\phi_{\mathbf{k}}^{(v)}|\psi_{\mathbf{k}}^{(c)}\rangle = \langle\mathbf{k}|\psi_{\mathbf{k}}^{(c)}\rangle - \sum_{c'}\langle\mathbf{k}|\psi_{\mathbf{k}}^{(c')}\rangle\langle\psi_{\mathbf{k}}^{(c')}|\psi_{\mathbf{k}}^{(c)}\rangle = 0 \tag{4.42}$$

where we have used $\langle\psi_{\mathbf{k}}^{(c')}|\psi_{\mathbf{k}}^{(c)}\rangle = \delta_{cc'}$ to reduce the sum to a single term. We can then use these trial valence states as the basis for the expansion of the true valence states:

$$\psi_{\mathbf{k}}^{(v)}(\mathbf{r}) = \sum_{\mathbf{G}} \alpha_{\mathbf{k}}(\mathbf{G})\phi_{\mathbf{k}+\mathbf{G}}^{(v)}(\mathbf{r}) \tag{4.43}$$

Taking matrix elements between such states produces a secular equation which can be diagonalized to obtain the eigenvalues of the energy.

Pseudopotential Plane Wave (PPW) method We can manipulate the expression in Eq. (4.43) to obtain something more familiar. First notice that

$$\psi_{\mathbf{k}+\mathbf{G}}^{(c)}(\mathbf{r}) = \sum_{\mathbf{R}} \mathrm{e}^{\mathrm{i}(\mathbf{k}+\mathbf{G})\cdot\mathbf{R}}\phi^{(c)}(\mathbf{r}-\mathbf{R}) = \psi_{\mathbf{k}}^{(c)}(\mathbf{r}) \tag{4.44}$$

and with this we obtain

$$\psi_{\mathbf{k}}^{(v)}(\mathbf{r}) = \sum_{\mathbf{G}} \alpha_{\mathbf{k}}(\mathbf{G})\left[\frac{1}{\Omega}\mathrm{e}^{\mathrm{i}(\mathbf{k}+\mathbf{G})\cdot\mathbf{r}} - \sum_{c}\langle\psi_{\mathbf{k}+\mathbf{G}}^{(c)}|(\mathbf{k}+\mathbf{G})\rangle\psi_{\mathbf{k}+\mathbf{G}}^{(c)}(\mathbf{r})\right]$$

$$= \left(\frac{1}{\Omega}\sum_{\mathbf{G}}\alpha_{\mathbf{k}}(\mathbf{G})\mathrm{e}^{\mathrm{i}(\mathbf{k}+\mathbf{G})\cdot\mathbf{r}}\right) - \sum_{c}\langle\psi_{\mathbf{k}}^{(c)}\left(\sum_{\mathbf{G}}\alpha_{\mathbf{k}}(\mathbf{G})|(\mathbf{k}+\mathbf{G})\rangle\right)\psi_{\mathbf{k}}^{(c)}(\mathbf{r})$$

which, with the definition

$$\tilde{\phi}_{\mathbf{k}}^{(v)}(\mathbf{r}) = \sum_{\mathbf{G}} \alpha_{\mathbf{k}}(\mathbf{G}) \frac{1}{\Omega} \mathrm{e}^{\mathrm{i}(\mathbf{k}+\mathbf{G})\cdot\mathbf{r}} = \frac{1}{\Omega} \mathrm{e}^{\mathrm{i}\mathbf{k}\cdot\mathbf{r}} \sum_{\mathbf{G}} \alpha_{\mathbf{k}}(\mathbf{G}) \mathrm{e}^{\mathrm{i}\mathbf{G}\cdot\mathbf{r}} \tag{4.45}$$

can be rewritten as

$$\psi_{\mathbf{k}}^{(v)}(\mathbf{r}) = \tilde{\phi}_{\mathbf{k}}^{(v)}(\mathbf{r}) - \sum_{c} \langle \psi_{\mathbf{k}}^{(c)} | \tilde{\phi}_{\mathbf{k}}^{(v)} \rangle \psi_{\mathbf{k}}^{(c)}(\mathbf{r}) \tag{4.46}$$

This is precisely the type of expression we saw in chapter 2, for the states we called pseudo-wavefunctions, Eq. (2.114), which contain a sum that projects out the core part. So the construction of the orthogonalized plane waves has led us to consider valence states from which the core part is projected out, which in turn leads to the idea of pseudopotentials, discussed in detail in chapter 2.

The crystal potential that the pseudo-wavefunctions experience is then given by

$$V_{cr}^{ps}(\mathbf{r}) = \sum_{\mathbf{R},i} V_{at}^{ps}(\mathbf{r} - \mathbf{t}_i - \mathbf{R}) \tag{4.47}$$

where $V_{at}^{ps}(\mathbf{r} - \mathbf{t}_i)$ is the pseudopotential of a particular atom in the unit cell, at position $\mathbf{t}_i$. We can expand the pseudopotential in the plane wave basis of the reciprocal lattice vectors:

$$V_{cr}^{ps}(\mathbf{r}) = \sum_{\mathbf{G}} V_{cr}^{ps}(\mathbf{G}) \mathrm{e}^{\mathrm{i}\mathbf{G}\cdot\mathbf{r}} \tag{4.48}$$

As we have argued before, the pseudopotential is much smoother than the true Coulomb potential of the ions, and therefore we expect its Fourier components $V_{cr}^{ps}(\mathbf{G})$ to fall fast with the magnitude of $\mathbf{G}$. To simplify the situation, we will assume that we are dealing with a solid that has several atoms of the same type in each unit cell; this is easily generalized to the case of several types of atoms. Using the expression from above in terms of the atomic pseudopotentials, the Fourier components take the form (with $N\Omega_{PUC}$ the volume of the crystal)

$$\begin{aligned} V_{cr}^{ps}(\mathbf{G}) &= \int \frac{\mathrm{d}\mathbf{r}}{N\Omega_{PUC}} V_{cr}^{ps}(\mathbf{r}) \mathrm{e}^{-\mathrm{i}\mathbf{G}\cdot\mathbf{r}} = \sum_{\mathbf{R},i} \int \frac{\mathrm{d}\mathbf{r}}{N\Omega_{PUC}} V_{at}^{ps}(\mathbf{r} - \mathbf{t}_i - \mathbf{R}) \mathrm{e}^{-\mathrm{i}\mathbf{G}\cdot\mathbf{r}} \\ &= \frac{1}{N} \sum_{\mathbf{R},i} \left[\int \frac{\mathrm{d}\mathbf{r}}{\Omega_{PUC}} V_{at}^{ps}(\mathbf{r}) \mathrm{e}^{-\mathrm{i}\mathbf{G}\cdot\mathbf{r}} \right] \mathrm{e}^{\mathrm{i}\mathbf{G}\cdot\mathbf{t}_i} = V_{at}^{ps}(\mathbf{G}) \sum_{i} \mathrm{e}^{\mathrm{i}\mathbf{G}\cdot\mathbf{t}_i} \end{aligned} \tag{4.49}$$

where we have eliminated a factor of N (the number of PUCs in the crystal) with a summation $\sum_{\mathbf{R}}$, since the summand at the end does not involve an explicit dependence on $\mathbf{R}$. We have also defined the Fourier transform of the atomic pseudopotential $V_{at}^{ps}(\mathbf{G})$ as the content of the square brackets in the next to last expression in the above equation. The sum appearing in the last step of this equation,

$$S(\mathbf{G}) = \sum_i e^{i\mathbf{G}\cdot\mathbf{t}_i} \tag{4.50}$$

is called the "structure factor". Depending on the positions of atoms in the unit cell, this summation can vanish for several values of the vector **G**. This means that the values of the crystal pseudopotential for these values of **G** are not needed for a band-structure calculation.

From the above analysis, we conclude that relatively few Fourier components of the pseudopotential survive, since those corresponding to large $|\mathbf{G}|$ are negligible because of the smoothness of the pseudopotential, whereas among those with small $|\mathbf{G}|$, several may be eliminated due to vanishing values of the structure factor $S(\mathbf{G})$. The idea then is to use a basis of plane waves $\exp(i\mathbf{G}\cdot\mathbf{r})$ to expand both the pseudo-wavefunctions and the pseudopotential, which will lead to a secular equation with a relatively small number of non-vanishing elements. Solving this secular equation produces the eigenvalues (band structure) and eigenfunctions for a given system.

To put these arguments in quantitative form, consider the single-particle equations which involve a pseudopotential (we neglect for the moment all the electron interaction terms, which are anyway isotropic; the full problem is considered in more detail in chapter 5):

$$\left[-\frac{\hbar^2}{2m_e}\nabla^2 + V_{cr}^{ps}(\mathbf{r})\right]\tilde{\phi}_{\mathbf{k}}^{(n)}(\mathbf{r}) = \epsilon_{\mathbf{k}}^{(n)}\tilde{\phi}_{\mathbf{k}}^{(n)}(\mathbf{r}) \tag{4.51}$$

These must be solved by considering the expansion for the pseudo-wavefunction in terms of plane waves, Eq. (4.45). Taking matrix elments of the hamiltonian with respect to plane wave states, we arrive at the following secular equation:

$$\sum_{\mathbf{G}'} \mathcal{H}_{\mathbf{k}}^{sp}(\mathbf{G}, \mathbf{G}')\alpha_{\mathbf{k}}^{(n)}(\mathbf{G}') = \epsilon_{\mathbf{k}}^{(n)}\alpha_{\mathbf{k}}^{(n)}(\mathbf{G}') \tag{4.52}$$

where the hamiltonian matrix elements are given by

$$\mathcal{H}_{\mathbf{k}}^{sp}(\mathbf{G}, \mathbf{G}') = \frac{\hbar^2(\mathbf{k}+\mathbf{G})^2}{2m_e}\delta(\mathbf{G}-\mathbf{G}') + V_{at}^{ps}(\mathbf{G}-\mathbf{G}')S(\mathbf{G}-\mathbf{G}') \tag{4.53}$$

Diagonalization of the hamiltonian matrix gives the eigenvalues of the energy $\epsilon_{\mathbf{k}}^{(n)}$ and corresponding eigenfunctions $\tilde{\phi}_{\mathbf{k}}^{(n)}(\mathbf{r})$. Obviously, Fourier components of $V_{at}^{ps}(\mathbf{r})$ which are multiplied by vanishing values of the structure factor $S(\mathbf{G})$ will not be of use in the above equation.

4.3 Band structure of representative solids

Having established the general methodology for the calculation of the band structure, we will now apply it to examine the electronic properties of several representative solids. In the following we will rely on the PPW method to do the actual calculations, since it has proven one of the most versatile and efficient approaches for calculating electronic properties. In the actual calculations we will employ the density functional theory single-particle hamiltonian, which we diagonalize numerically using a plane wave basis. We should emphasize that this approach does not give accurate results for semiconductor band gaps. This is due to the fact that the spectrum of the single-particle hamiltonian cannot describe accurately the true eigenvalue spectrum of the many-body hamiltonian. Much theoretical work has been devoted to develop accurate calculations of the eigenvalue spectrum with the use of many-body techniques; this is known as the GW approach [48, 49]. A simpler approach, based on extensions of DFT, can often give quite reasonable results [50]; it is this latter approach that has been used in the calculations described below (for details see Ref. [51]). We will also rely heavily on ideas from the TBA to interpret the results.

4.3.1 A 2D solid: graphite – a semimetal

As we have mentioned in chapter 3, graphite consists of stacked sheets of three-fold coordinated carbon atoms. On a plane of graphite the C atoms form a honeycomb lattice, that is, a hexagonal Bravais lattice with a two-atom basis. The interaction between planes is rather weak, of the van der Waals type, and the overlap of wavefunctions on different planes is essentially non-existent. We present in Fig. 4.6 the band structure for a single, periodic, infinite graphitic sheet.

In this plot, we easily recognize the lowest band as arising from a bonding state of s character; this is band $n = 1$ at Γ counting from the bottom, and corresponds to σ bonds between C atoms. The next three bands intersect each other at several points in the BZ. The two bands that are degenerate at Γ, labeled $n = 3, 4$, represent a p-like bonding state. There are two p states participating in this type of bonding, the two p orbitals that combine to form the sp^2 hybrids involved in the σ bonds on the plane. The single band intersecting the other two is a state with p character, arising from the p_z orbitals that contribute to the π-bonding; it is the symmetric (bonding) combination of these two p_z orbitals, labeled $n = 2$. The antisymmetric (antibonding) combination, labeled $n = 8$, has the reverse dispersion and lies higher in energy; it is almost the mirror image of the π-bonding state with respect to the Fermi level.

All these features are identified in the plots of the total charge densities and eigenfunction magnitudes shown in Fig. 4.7. Notice that because we are dealing with pseudo-wavefunctions, which vanish at the position of the ion, the charge is

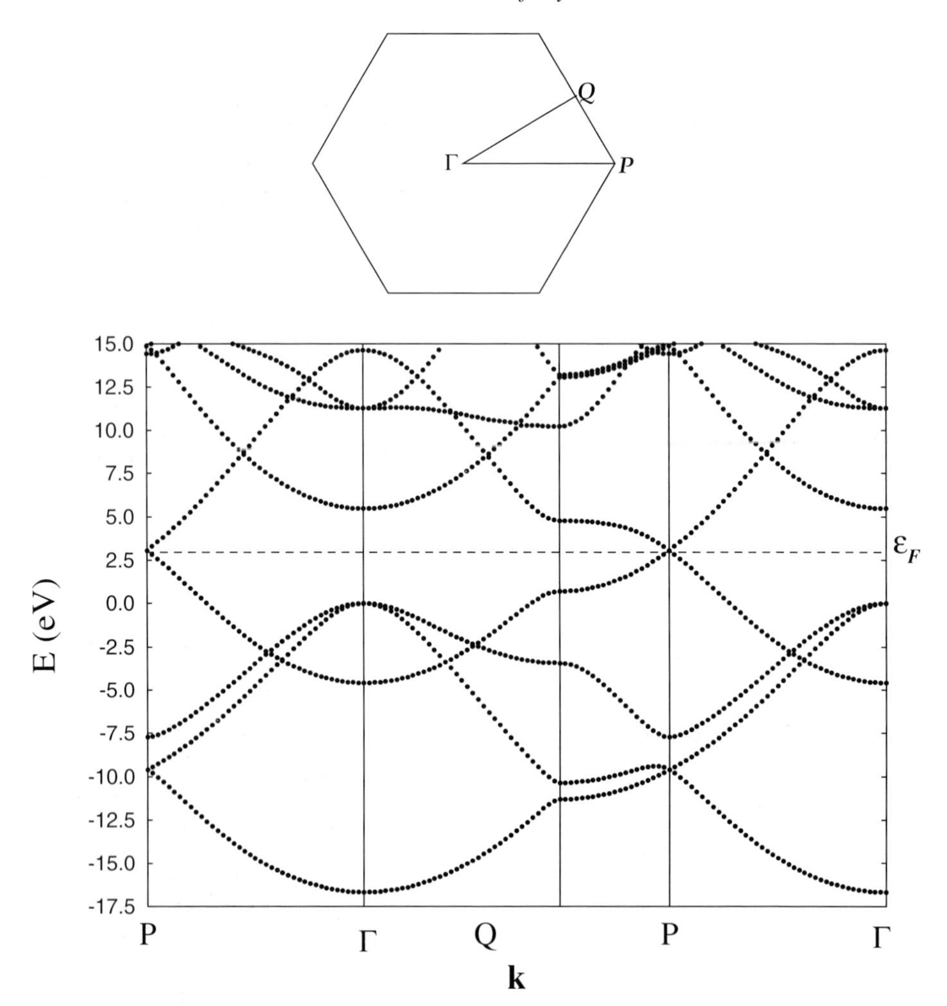

Figure 4.6. Band structure of a graphite sheet, calculated with the PPW method. The zero of the energy scale is set arbitrarily to the value of the highest σ-bonding state at Γ. The dashed line indicates the position of the Fermi level. The small diagram above the band structure indicates the corresponding Brillouin Zone with the special **k** points Γ, P, Q identified. (Based on calculations by I.N. Remediakis.)

mostly concentrated in the region between atoms. This is a manifestation of the expulsion of the valence states from the core region, which is taken into account by the pseudopotential. The s-like part ($n = 1$) of the σ-bonding state has uniform distribution in the region between the atoms; the center of the bond corresponds to the largest charge density. The p-like part ($n = 3, 4$) of the σ-bonding state has two pronounced lobes, and a dip in the middle of the bond. The π-bonding state ($n = 2$) shows the positive overlap between the two p_z orbitals. The π-antibonding state ($n = 8$) has a node in the region between the atoms.

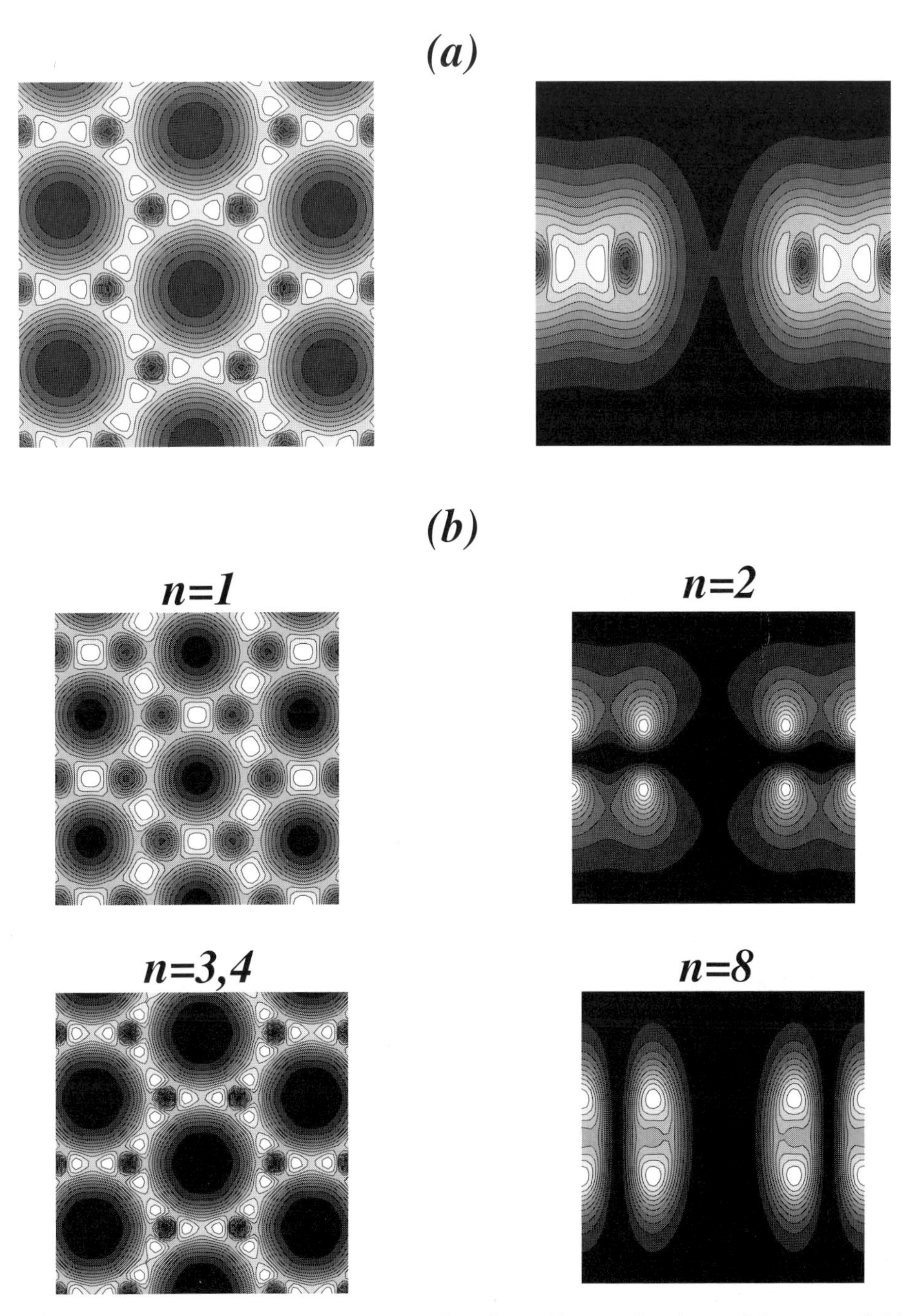

Figure 4.7. (a) Total electronic charge density of graphite, on the plane of the atoms (left) and on a plane perpendicular to it (right). (b) The wavefunction magnitude of states at Γ. (Based on calculations by I.N. Remediakis.)

Since in this system there are two atoms per unit cell with four valence electrons each, that is, a total of eight valence electrons per unit cell, we need four completely filled bands in the BZ to accommodate them. Indeed, the Fermi level must be at a position which makes the three σ-bonding states (one s-like and two p-like) and the π-bonding state completely full, while the π-antibonding state is completely empty. Similarly, antibonding states arising from antisymmetric combinations of s and p_x, p_y orbitals, lie even higher in energy and are completely empty.

The bonding and antibonding combinations of the p_z states are degenerate at the P high-symmetry point of the BZ. At zero temperature, electrons obey a Fermi distribution with an abrupt step cutoff at the Fermi level ϵ_F. At finite temperature T, the distribution will be smoother around the Fermi level, with a width at the step of order $k_B T$. This means that some states below the Fermi level will be unoccupied and some above will be occupied. This is the hallmark of metallic behavior, that is, the availability of states immediately below and immediately above the Fermi level, which makes it possible to excite electrons thermally. Placing electrons in unoccupied states at the bottom of empty bands allows them to move freely in these bands, as we discussed for the free-electron model.

In the case of graphite, the number of states immediately below and above the Fermi level is actually very small: the π-bonding and antibonding bands do not overlap but simply touch each other at the P point of the BZ. Accordingly, graphite is considered a semimetal, barely exhibiting the characteristics of metallic behavior, even though, strictly speaking according to the definition given above, it cannot be described as anything else.

4.3.2 3D covalent solids: semiconductors and insulators

Using the same concepts we can discuss the band structure of more complicated crystals. We consider first four crystals that have the following related structures: the diamond crystal, which consists of two interpenetrating FCC lattices and a two-atom basis in the PUC, with the two atoms being of the same kind; and the zincblende crystal in which the lattice is similar but the two atoms in the PUC are different. The crystals we will consider are Si, C, SiC and GaAs. The first two are elemental solids, the third has two atoms of the same valence (four valence electrons each) and the last consists of a group-III (Ga) and a group-V (As) atom. Thus, there are eight valence electrons per PUC in each of these crystals. The first two crystals are characteristic examples of covalent bonding, whereas the third and fourth have partially ionic and partially covalent bonding character (see also the discussion in chapter 1).

The band structure of these four crystals is shown in Fig. 4.8. The energy of the highest occupied band is taken to define the zero of the energy scale. In all four

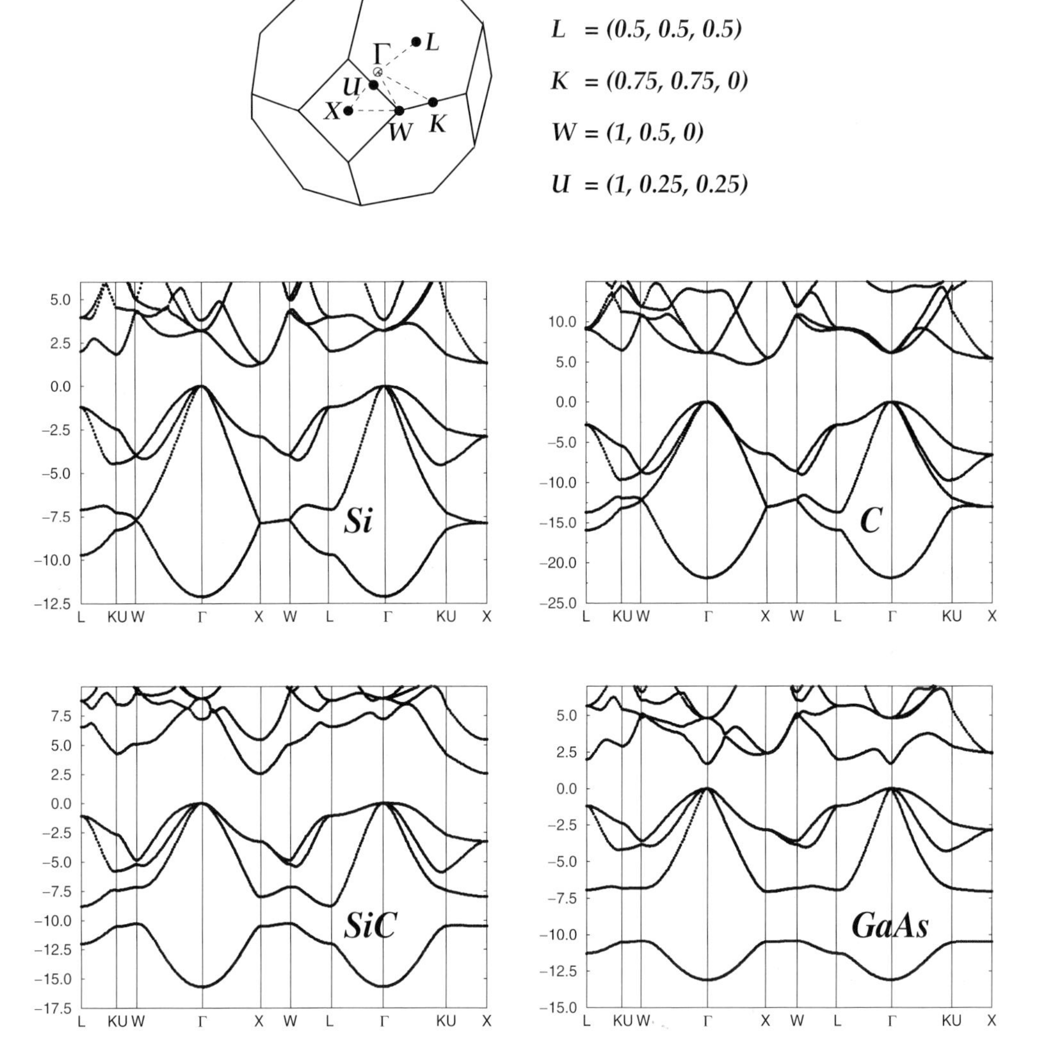

Figure 4.8. Band structure of four representative covalent solids: Si, C, SiC, GaAs. The first and the last are semiconductors, the other two are insulators. The small diagram above the band structure indicates the Brillouin Zone for the FCC lattice, with the special **k**-points X, L, K, W, U identified and expressed in units of $2\pi/a$, where a is the lattice constant; Γ is the center of the BZ. The energy scale is in electronvolts and the zero is set at the Valence Band Maximum. (Based on calculations by I.N. Remediakis.)

cases there is an important characteristic of the band structure, namely there is a range of energies where there are no electronic states across the entire BZ; this is the band gap, denoted by ϵ_{gap}. The ramifications of this feature are very important. We notice first that there are four bands below zero, which means that all four bands are fully occupied, since there are eight valence electrons in the PUC of each of these solids. Naively, we might expect that the Fermi level can be placed anywhere within the band gap, since for any such position all states below it remain occupied and all states above remain unoccupied. A more detailed analysis (see chapter 9) reveals that for an ideal crystal, that is, one in which there are no intrinsic defects or impurities, the Fermi level is at the middle of the gap. This means that there are no states immediately above or below the Fermi level, for an energy range of $\pm\epsilon_{gap}/2$. Thus, it will not be possible to excite thermally appreciable numbers of electrons from occupied to unoccupied bands, until the temperature reaches $\sim \epsilon_{gap}/2$. Since the band gap is typically of order 1 eV for semiconductors (1.2 eV for Si, 1.5 eV for GaAs, see Fig. 4.8), and 1 eV = 11 604 K, we conclude that for all practical purposes the states above the Fermi level remain unoccupied (these solids melt well below 5800 K). For insulators the band gap is even higher (2.5 eV for SiC, 5 eV for C-diamond, see Fig. 4.8). This is the hallmark of semiconducting and insulating behavior, that is, the absence of any states above the Fermi level to which electrons can be thermally excited. This makes it difficult for these solids to respond to external electric fields, since as we discussed in chapter 3 a filled band cannot carry current. In a perfect crystal it would take the excitation of electrons from occupied to unoccupied bands to create current-carrying electronic states. Accordingly, the states below the Fermi level are called "valence bands" while those above the Fermi level are called "conduction bands". Only when imperfections (defects) or impurities (dopants) are introduced into these crystals do they acquire the ability to respond to external electric fields; all semiconductors in use in electronic devices are of this type, that is, crystals with impurities (see detailed discussion in chapter 9).

The specific features of the band structure are also of interest. First we note that the highest occupied state (Valence Band Maximum, VBM) is always at the Γ point. The lowest unoccupied state (Conduction Band Minimum, CBM) can be at different positions in the BZ. For Si and C it is somewhere between the Γ and X points, while in SiC it is at the X point. Only for GaAs is the CBM at Γ: this is referred to as a direct gap; all the other cases discussed here are indirect gap semiconductors or insulators. The nature of the gap has important consequences for optical properties, as discussed in chapter 5.

It is also of interest to consider the "band width", that is, the range of energies covered by the valence states. In Si and GaAs it is about 12.5 eV, in SiC it is about 16 eV, and in C it is considerably larger, about 23 eV. There are two factors that influence the band width: the relative energy difference between the s and p atomic valence states, and the interaction between the hybrid orbitals in the solid. For instance, in C, where we are dealing with $2s$ and $2p$ atomic states, both their energy difference and the interaction of the hybrid orbitals is large, giving a large band width, almost twice that of Si.

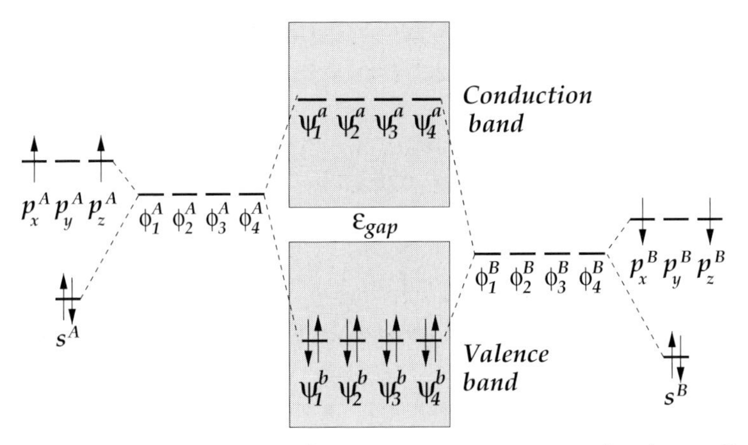

Figure 4.9. Origin of the bands in sp^3 bonded solids: the levels s^A, p^A and s^B, p^B correspond to the atoms A, B; the sp^3 hybrids in each case are denoted by ϕ_i^A, $\phi_i^B (i = 1, 2, 3, 4)$, and the bonding and antibonding states are denoted by ψ_i^b, $\psi_i^a (i = 1, 2, 3, 4)$ respectively; in the crystal, the bonding and antibonding states acquire dispersion, which leads to formation of the valence and conduction energy bands. The gap between the two manifolds of states, ϵ_{gap}, is indicated.

In all the examples considered here it is easy to identify the lowest band at Γ as the s-like state of the bonding orbitals, which arise from the interaction of sp^3 hybrids in nearest neighbor atoms. Since the sp^3 orbitals involve one s and three p states, the corresponding p-like states of the bonding orbitals are at the top of the valence manifold at Γ. This is illustrated in Fig. 4.9: in this example we show the relative energy of atomic-like s and p orbitals for two different tetravalent elements (for example Si and C) which combine to form a solid in the zincblende lattice. A similar diagram, but with different occupation of the atomic orbitals, would apply to GaAs, which has three electrons in the Ga orbitals and five electrons in the As orbitals. This illustration makes it clear why the states near the VBM have p bonding character and are associated with the more electronegative element in the solid, while those near the CBM have p antibonding character and are associated with the less electronegative element in the solid: the character derives from the hybrid states which are closest in energy to the corresponding bands. In the case of a homopolar solid, the two sets of atomic-like orbitals are the same and the character of bonding and antibonding states near the VBM and CBM is not differentiated among the two atoms in the unit cell. The VBM is in all cases three-fold degenerate at Γ. These states disperse and their s and p character becomes less clear away from Γ. It is also interesting that the bottom s-like band is split off from the other three valence bands in the solids with two types of atoms, SiC and GaAs. In these cases the s-like state bears more resemblance to the corresponding atomic state in

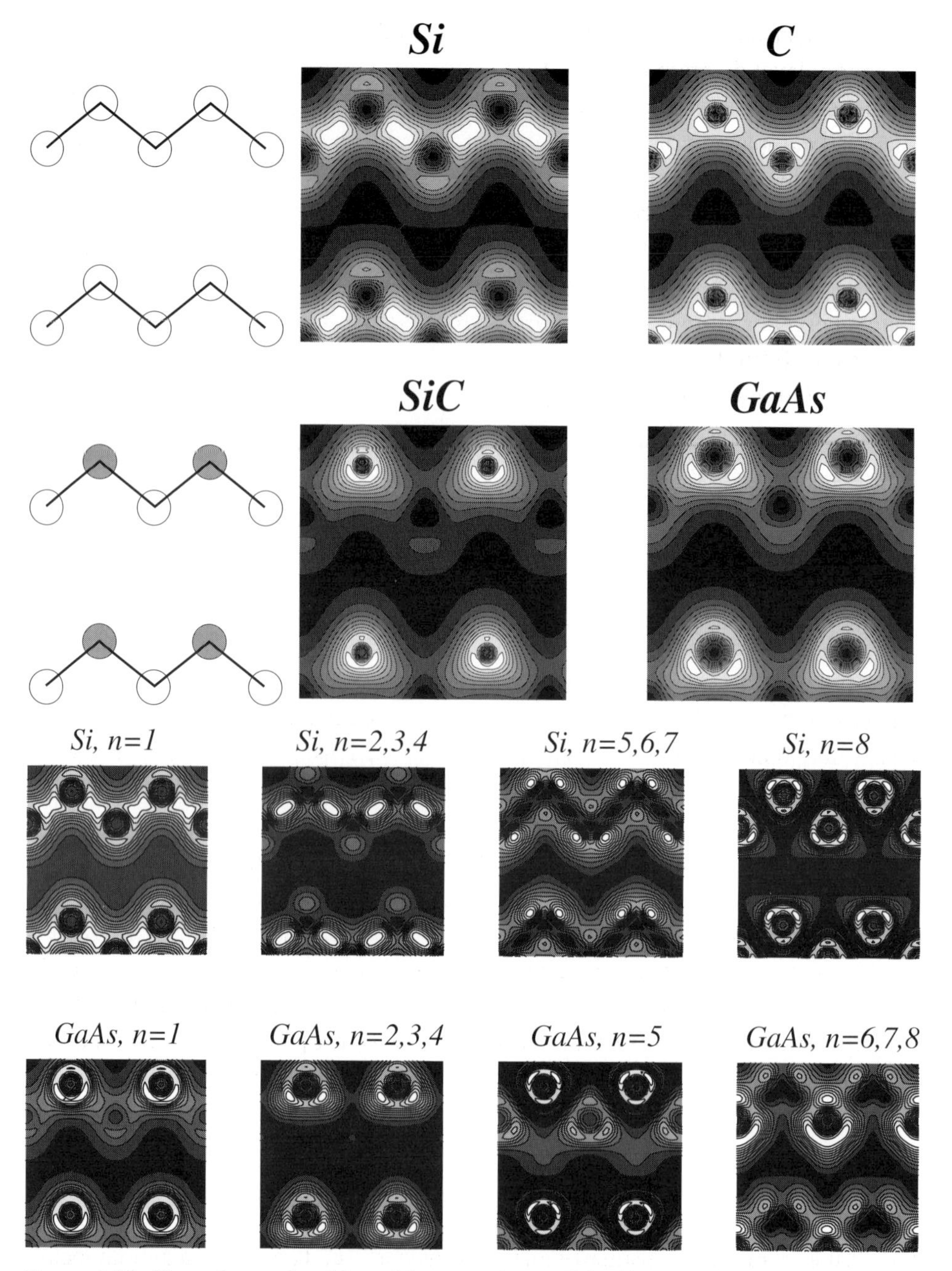

Figure 4.10. **Top:** charge densities of four covalent solids, Si, C, SiC, GaAs, on the (110) plane of the diamond or zincblende lattice. The diagrams on the left indicate the atomic positions on the (110) plane in the diamond and zincblende lattices. **Bottom:** wavefunction magnitude for the eight lowest states (four below and four above the Fermi level) for Si and GaAs, on the same plane as the total charge density. (Based on calculations by I.N. Remediakis.)

the more electronegative element (C in the case of SiC, As in the case of GaAs), as the energy level diagram of Fig. 4.9 suggests.

All these features can be identified in the charge density plots, shown in Fig. 4.10; a detailed comparison of such plots to experiment can be found in Ref. [52]. In all cases the valence electronic charge is concentrated between atomic positions. In the case of Si and C this distribution is the same relative to all atomic positions, whereas in SiC and GaAs it is polarized closer to the more electronegative atoms (C and As, respectively). The high concentration of electrons in the regions between the nearest neighbor atomic sites represents the covalent bonds between these atoms. Regions far from the bonds are completely devoid of charge. Moreover, there are just enough of these bonds to accommodate all the valence electrons. Specifically, there are four covalent bonds emanating from each atom (only two can be seen on the plane of Fig. 4.10), and since each bond is shared by two atoms, there are two covalent bonds per atom. Since each bonding state can accommodate two electrons due to spin degeneracy, these covalent bonds take up all the valence electrons in the solid.

The magnitude of the wavefunctions at Γ for Si and GaAs, shown in Fig. 4.10, reveals features that we would expect from the preceding analysis: the lowest state ($n = 1$) has s-like bonding character around all atoms in Si and around mostly the As atoms in GaAs. The next three states which are degenerate ($n = 2, 3, 4$), have p-like bonding character. The next three unoccupied degenerate states in Si ($n = 5, 6, 7$) have p antibonding character, with pronounced lobes pointing away from the direction of the nearest neighbors and nodes in the middle of the bonds; the next unoccupied state in Si ($n = 8$) has s antibonding character, also pointing away from the direction of the nearest neighbors. In GaAs, the states $n = 2, 3, 4$ have As p-like bonding character, with lobes pointing in the direction of the nearest neighbors. The next state ($n = 5$) has clearly antibonding character, with a node in the middle of the bond and significant weight at both the As and Ga sites. Finally, the next three unoccupied degenerate states ($n = 6, 7, 8$) have clearly antibonding character with nodes in the middle of the bonds and have large weight at the Ga atomic sites.

4.3.3 3D metallic solids

As a last example we consider two metals, Al and Ag. The first is a simple metal, in the sense that only s and p orbitals are involved: the corresponding atomic states are $3s, 3p$. Its band structure, shown in Fig. 4.11, has the characteristics of the free-electron band structure in the corresponding 3D FCC lattice. Indeed, Al is the prototypical solid with behavior close to that of free electrons. The dispersion near the bottom of the lowest band is nearly a perfect parabola, as would be expected for

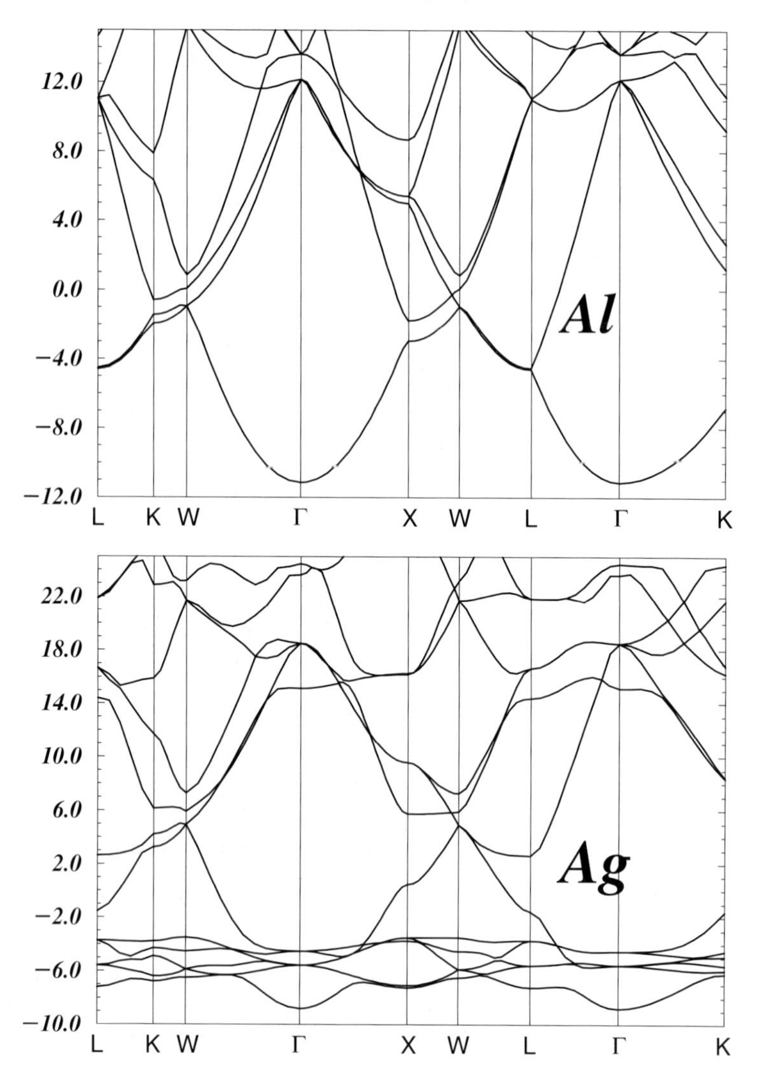

Figure 4.11. Band structure of two representative metallic solids: Al, a free-electron metal, and Ag, a d-electron metal. The zero of energy denotes the Fermi level. (Based on calculations by I.N. Remediakis.)

free electrons. Since there are only three valence electrons per atom and one atom per unit cell, we expect that on average 1.5 bands will be occupied throughout the BZ. As seen in Fig. 4.11 the Fermi level is at a position which makes the lowest band completely full throughout the BZ, and small portions of the second band full, especially along the X–W–L high symmetry lines.

The total charge density and magnitude of the wavefunctions for the lowest states at Γ, shown in Fig. 4.12, reveal features expected from the analysis above. Specifically, the total charge density is evenly distributed throughout the crystal.

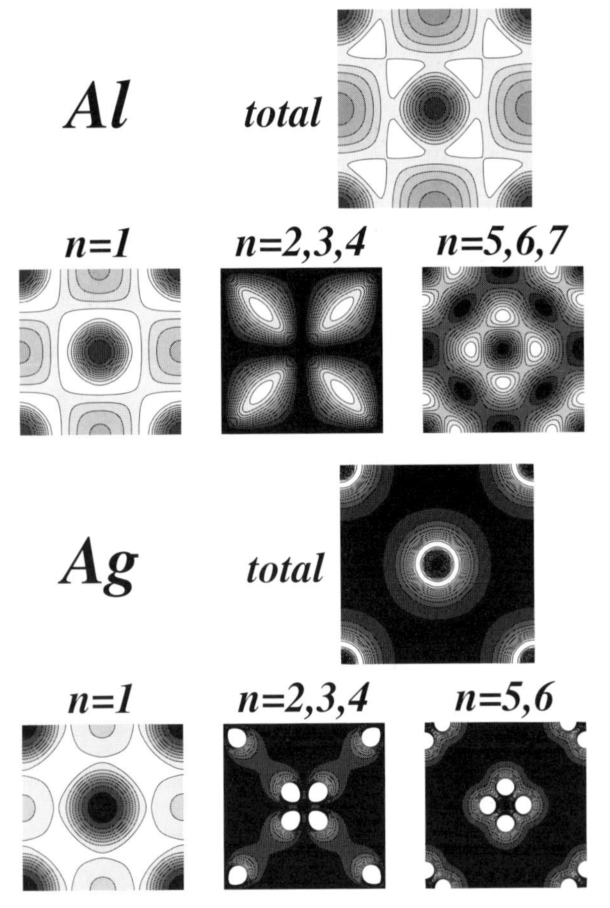

Figure 4.12. Electronic charge densities on a (100) plane of the FCC lattice for Al and Ag. The total charge density and the magnitude of the wavefunctions for the three lowest states (not counting degeneracies) at Γ are shown. The atomic positions are at the center and the four corners of the square. (Based on calculations by I.N. Remediakis.)

Although there is more concentration of electrons in the regions between nearest neighbor atomic positions, this charge concentration cannot be interpreted as a covalent bond, because there are too many neighbors (12) sharing the three valence electrons. As far as individual states are concerned, the lowest band ($n = 1$) is clearly of s character, uniformly distributed around each atomic site, and with large weight throughout the crystal. This corresponds to the metallic bonding state. The next three degenerate states ($n = 2, 3, 4$) are clearly of p character, with pronounced lobes pointing toward nearest neighbors. At Γ these states are unoccupied, but at other points in the BZ they are occupied, thus contributing to bonding. These states give rise to the features of the total charge density that appear like directed bonds. Finally, the next set of unoccupied states ($n = 5, 6, 7$) are clearly of antibonding character with nodes in the direction of the nearest neighbors; their energy is above the Fermi level throughout the BZ.

The second example, Ag, is more complicated because it involves s and d electrons: the corresponding atomic states are $4d$, $5s$. In this case we have 11 valence electrons, and we expect 5.5 bands to be filled on average in the BZ. Indeed, we see five bands with little dispersion near the bottom of the energy range, all of which are filled states below the Fermi level. There is also one band with large dispersion, which intersects the Fermi level at several points, and is on average half-filled. The five low-energy occupied bands are essentially bands arising from the $4d$ states. Their low dispersion is indicative of weak interactions among these orbitals. The next band can be identified with the s-like bonding band. This band interacts and hybridizes with the d bands, as the mixing of the spectrum near Γ suggests. In fact, if we were to neglect the five d bands, the rest of the band structure looks remarkably similar to that of Al. In both cases, the Fermi level intersects bands with high dispersion at several points, and thus there are plenty of states immediately below and above the Fermi level for thermal excitation of electrons. This indicates that both solids will act as good metals, being able to carry current when placed in an external electric field.

The total valence charge density and the magnitude of wavefunctions for the few lowest energy states at Γ, shown in Fig. 4.12, reveal some interesting features. First, notice how the total charge density is mostly concentrated around the atoms, and there seems to be little interaction between these atoms. This is consistent with the picture we had discussed of the noble metals, namely that they have an essentially full electronic shell (the $4d$ shell in Ag), and one additional s electron which is shared among all the atoms in the crystal. Indeed the wavefunction of the lowest energy state at Γ ($n = 1$) clearly exhibits this character: it is uniformly distributed in the regions between atoms and thus contributes to bonding. The distribution of charge corresponding to this state is remarkably similar to the corresponding state in Al. This state is strongly repelled from the atomic cores, leaving large holes at the positions where the atoms are. The next five states are in two groups, one three-fold degenerate, and one two-fold degenerate. All these states have clearly d character, with the characteristic four lobes emanating from the atomic sites. They seem to be very tightly bound to the atoms, with very small interaction across neighboring sites, as one would expect for core-like completely filled states. When the charge of these states is added up, it produces the completely spherically symmetric distribution shown in the total-density panel, as expected for non-interacting atoms. The lack of interaction among these states is reflected in the lack of dispersion of their energies in the band structure plot, Fig. 4.11. Only the s state, which is shared among all atoms, shows significant dispersion and contributes to the metallic bonding in this solid.

Further reading

1. *Electronic Structure and the Properties of Solids*, W.A. Harrison (W.H. Freeman, San Francisco, 1980).
   This book contains a general discussion of the properties of solids based on the tight-binding approximation.
2. *Handbook of the Band Structure of Elemental Solids*, D.A. Papaconstantopoulos (Plenum Press, New York, 1989).
   This is a comprehensive account of the tight-binding approximation and its application to the electronic structure of elemental solids.
3. *Planewaves, Pseudopotentials and the LAPW Method*, D. Singh (Kluwer Academic, Boston, 1994).
   This book contains a detailed account of the augmented plane wave approach and the plane wave approach and their applications to the electronic structure of solids.
4. *Electronic Structure and Optical Properties of Semiconductors*, M.L. Cohen and J.R. Chelikowsky (Springer-Verlag, Berlin, 1988).
   This book is a thorough compilation of information relevant to semiconductor crystals, a subfield of great importance to the theory of solids.
5. *Electronic States and Optical Transitions in Solids*, F. Bassani and G. Pastori Parravicini (Pergamon Press, Oxford, 1975).
6. *Calculated Electronic Properties of Metals*, V.L. Moruzzi, J.F. Janak and A.R. Williams (Pergamon Press, New York, 1978).
   This book is a thorough compilation of the electronic structure of elemental metallic solids, as obtained with the APW method.

Problems

1. Prove the orthogonality relation, Eq. (4.6).
2. The relationship between the band width and the number of nearest neighbors, Eq. (4.20), was derived for the simple chain, square and cubic lattices in one, two and three dimensions, using the simplest tight-binding model with one atom per unit cell and one s-like orbital per atom. For these lattices, the number of neighbors z is always $2d$, where d is the dimensionality. Consider the same simple tight-binding model for the close-packed lattices in two and three dimensions, that is, the simple hexagonal lattice in 2D and the FCC lattice in 3D, and derive the corresponding relation between the band width and the number of nearest neighbors.
3. Consider a single plane of the graphite lattice, defined by the lattice vectors $\mathbf{a}_1$, $\mathbf{a}_2$ and atomic positions $\mathbf{t}_1$, $\mathbf{t}_2$:

$$\mathbf{a}_1 = \frac{\sqrt{3}}{2}a\hat{\mathbf{x}} + \frac{1}{2}a\hat{\mathbf{y}},\ \mathbf{a}_2 = \frac{\sqrt{3}}{2}a\hat{\mathbf{x}} - \frac{1}{2}a\hat{\mathbf{y}},\ \mathbf{t}_1 = 0,\ \mathbf{t}_2 = \frac{a}{\sqrt{3}}\hat{\mathbf{x}} \tag{4.54}$$

   where a is the lattice constant of the 2D hexagonal plane; this plane is referred to as graphene.

   (a) Take a basis for each atom which consists of four orbitals, s, p_x, p_y, p_z. Determine the hamiltonian matrix for this system at each high-symmetry point of the IBZ, with nearest neighbor interactions in the tight-binding approximation, assuming

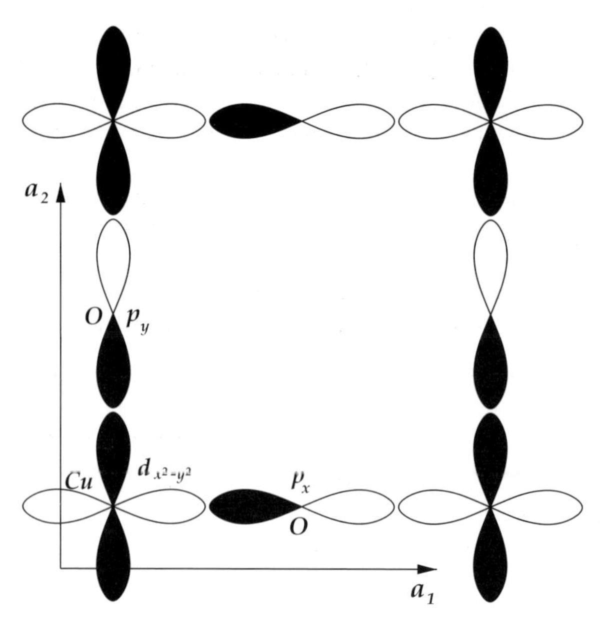

Figure 4.13. Model for the CuO_2 plane band structure: the Cu $d_{x^2-y^2}$ and the O p_x and p_y orbitals are shown, with positive lobes in white and negative lobes in black. The lattice vectors $\mathbf{a}_1$ and $\mathbf{a}_2$ are offset from the atomic positions for clarity.

an orthogonal overlap matrix. Use proper combinations of the atomic orbitals to take advantage of the symmetry of the problem (see also chapter 1).

(b) Choose the parameters that enter into the hamiltonian by the same arguments that were used in the 2D square lattice, scaled appropriately to reflect the interactions between carbon orbitals. How well does your choice of parameters reproduce the important features of the true band structure, shown in Fig. 4.6?

(c) Show that the π bands can be described reasonably well by a model consisting of a single orbital per atom, with on-site energy ϵ_0 and nearest neighbor hamiltonian matrix element t, which yields the following expression for the bands:

$$\epsilon_{\mathbf{k}}^{(\pm)} = \epsilon_0 \pm t \left[1 + 4 \cos\left(\frac{\sqrt{3}a}{2} k_x\right) \cos\left(\frac{a}{2} k_y\right) + 4 \cos^2\left(\frac{a}{2} k_y\right) \right]^{1/2} \tag{4.55}$$

Choose the values of the parameters in this simple model to obtain as close an approximation as possible to the true bands of Fig. 4.6. Comment on the differences between these model bands and the true π bands of graphite.

4. Consider the following two-dimensional model: the lattice is a square of side a with lattice vectors $\mathbf{a}_1 = a\hat{\mathbf{x}}$ and $\mathbf{a}_2 = a\hat{\mathbf{y}}$; there are three atoms per unit cell, one Cu atom and two O atoms at distances $\mathbf{a}_1/2$ and $\mathbf{a}_2/2$, as illustrated in Fig. 4.13. We will assume that the atomic orbitals associated with these atoms are orthogonal.

(a) Although there are five d orbitals associated with the Cu atom and three p orbitals associated with each O atom, only one of the Cu orbitals (the $d_{x^2-y^2}$ one) and two of the O orbitals (the p_x one on the O atom at $\mathbf{a}_1/2$ and the p_y one on the O atom at $\mathbf{a}_2/2$) are relevant, because of the geometry; the remaining Cu and O orbitals do not interact with their neighbors. Explain why this is a reasonable approximation, and in particular why the Cu–$d_{3z^2-r^2}$ orbitals do not interact with their nearest neighbor O–p_z orbitals.

(b) Define the hamiltonian matrix elements between the relevant Cu–d and O–p nearest neighbor orbitals and take the on-site energies to be ϵ_d and ϵ_p, with $\epsilon_p < \epsilon_d$. Use these matrix elements to calculate the band structure for this model.

(c) Discuss the position of the Fermi level for the case when there is one electron in each O orbital and one or two electrons in the Cu orbital.

Historical note: even though this model may appear as an artificial one, it has been used extensively to describe the basic electronic structure of the copper-oxide–rare earth materials which are high-temperature superconductors [53].

5. Calculate the free-electron band structure in 3D for an FCC lattice and compare it with the band structure of Al given in Fig. 4.11. To what extent are the claims made in the text, about the resemblance of the two band structures, valid?

5

Applications of band theory

In the previous chapter we examined in detail methods for solving the single-particle equations for electrons in solids. The resulting energy eigenvalues (band structure) and corresponding eigenfunctions provide insight into how electrons are arranged, both from an energetic and from a spatial perspective, to produce the cohesion between atoms in the solid. The results of such calculations can be useful in several other ways. The band structure of the solid can elucidate the way in which the electrons will respond to external perturbations, such as absorption or emission of light. This response is directly related to the optical and electrical properties of the solid. For example, using the band structure one can determine the possible optical excitations which in turn determine the color, reflectivity, and dielectric response of the solid. A related effect is the creation of excitons, that is, bound pairs of electrons and holes, which are also important in determining optical and electrical properties. Finally, the band structure can be used to calculate the total energy of the solid, from which one can determine a wide variety of thermodynamic and mechanical properties. In the present chapter we examine the theoretical tools for calculating all these aspects of a solid's behavior.

5.1 Density of states

A useful concept in analyzing the band structure of solids is the density of states as a function of the energy. To illustrate this concept we consider first the free-electron model. The density of states $g(\epsilon)\mathrm{d}\epsilon$ for energies in the range $[\epsilon, \epsilon + \mathrm{d}\epsilon]$ is given by a sum over all states with energy in that range. Since the states are characterized by their wave-vector $\mathbf{k}$, we simply need to add up all states with energy in the interval of interest. Taking into account the usual factor of 2 for spin degeneracy and normalizing by the volume of the solid Ω, we get

$$g(\epsilon)\mathrm{d}\epsilon = \frac{1}{\Omega} \sum_{\mathbf{k}, \epsilon_{\mathbf{k}} \in [\epsilon, \epsilon+\mathrm{d}\epsilon]} 2 = \frac{2}{(2\pi)^3} \int_{\epsilon_{\mathbf{k}} \in [\epsilon, \epsilon+\mathrm{d}\epsilon]} \mathrm{d}\mathbf{k} = \frac{1}{\pi^2} k^2 \mathrm{d}k \qquad (5.1)$$

where we have used spherical coordinates in $\mathbf{k}$-space to obtain the last result as well as the fact that in the free-electron model the energy does not depend on the angular orientation of the wave-vector:

$$\epsilon_{\mathbf{k}} = \frac{\hbar^2|\mathbf{k}|^2}{2m_e} \Longrightarrow k\mathrm{d}k = \frac{m_e}{\hbar^2}\mathrm{d}\epsilon, \quad k = \left(\frac{2m_e\epsilon}{\hbar^2}\right)^{1/2} \tag{5.2}$$

for $\epsilon_{\mathbf{k}} \in [\epsilon, \epsilon + \mathrm{d}\epsilon]$. These relations give for the density of states in this simple model in 3D:

$$g(\epsilon) = \frac{1}{2\pi^2}\left(\frac{2m_e}{\hbar^2}\right)^{3/2}\sqrt{\epsilon} \tag{5.3}$$

In a crystal, instead of this simple relationship between energy and momentum which applies to free electrons, we have to use the band-structure calculation for $\epsilon_{\mathbf{k}}$. Then the expression for the density of states becomes

$$\begin{aligned} g(\epsilon) &= \frac{1}{\Omega}\sum_{n,\mathbf{k}} 2\delta(\epsilon - \epsilon_{\mathbf{k}}^{(n)}) = \frac{2}{(2\pi)^3}\sum_n \int \delta(\epsilon - \epsilon_{\mathbf{k}}^{(n)})\mathrm{d}\mathbf{k} \\ &= \frac{2}{(2\pi)^3}\sum_n \int_{\epsilon_{\mathbf{k}}^{(n)}=\epsilon} \frac{1}{|\nabla_{\mathbf{k}}\epsilon_{\mathbf{k}}^{(n)}|}\mathrm{d}S_{\mathbf{k}} \end{aligned} \tag{5.4}$$

where the last integral is over a surface in $\mathbf{k}$-space on which $\epsilon_{\mathbf{k}}^{(n)}$ is constant and equal to ϵ. In this final expression, the roots of the denominator are of first order and therefore contribute a finite quantity to the integration over a smooth two-dimensional surface represented by $S_{\mathbf{k}}$; these roots introduce sharp features in the function $g(\epsilon)$, which are called "van Hove singularities". For the values $\mathbf{k}_0$ where $\nabla_{\mathbf{k}}\epsilon_{\mathbf{k}} = 0$, we can expand the energy in a Taylor expansion (from here on we consider the contribution of a single band and drop the band index for simplicity):

$$[\nabla_{\mathbf{k}}\epsilon_{\mathbf{k}}]_{\mathbf{k}=\mathbf{k}_0} = 0 \Rightarrow \epsilon_{\mathbf{k}} = \epsilon_{\mathbf{k}_0} + \sum_{i=1}^{d} \alpha_i(k_i - k_{0,i})^2 \tag{5.5}$$

The expansion is over as many principal axes as the dimensionality of our system: in three dimensions ($d = 3$) there are three principal axes, characterized by the symbols α_i, $i = 1, 2, 3$. Depending on the signs of these coefficients, the extremum can be a minimum (zero negative coefficients, referred to as "type 0 critical point"), two types of saddle point ("type 1 and 2 critical points") or a maximum ("type 3 critical point"). There is a useful theorem that tells us exactly how many critical points of each type we can expect.

Theorem Given a function of d variables periodic in all of them, there are $d!/l!(d-l)!$ critical points of type l, where l is the number of negative coefficients in the Taylor expansion of the energy, Eq. (5.5).

Table 5.1. *Critical points in* d $= 3$ *dimensions.* Symbols (l, M_l), multiplicity $= d!/l!(d-l)!$, type, and characteristic behavior of the coefficients $\alpha_i, i = 1, 2, 3$ along the principal axes.

l	Symbol	Multiplicity	Type	Coefficients
0	M_0	$3!/0!3! = 1$	minimum	$\alpha_1, \alpha_2, \alpha_3 > 0$
1	M_1	$3!/1!2! = 3$	saddle point	$\alpha_1, \alpha_2 > 0, \alpha_3 < 0$
2	M_2	$3!/2!1! = 3$	saddle point	$\alpha_1 > 0, \alpha_2, \alpha_3 < 0$
3	M_3	$3!/3!0! = 1$	maximum	$\alpha_1, \alpha_2, \alpha_3 < 0$

With the help of this theorem, we can obtain the number of each type of critical point in $d = 3$ dimensions, which are given in Table 5.1.

Next we want to extract the behavior of the density of states (DOS) explicitly near each type of critical point. Let us first consider the critical point of type 0, M_0, in which case $\alpha_i > 0, i = 1, 2, 3$. In order to perform the **k**-space integrals involved in the DOS we first make the following changes of variables: in the neighborhood of the critical point at $\mathbf{k}_0$,

$$\epsilon_{\mathbf{k}} - \epsilon_{\mathbf{k}_0} = \alpha_1 k_1^2 + \alpha_2 k_2^2 + \alpha_3 k_3^2 \tag{5.6}$$

where $\mathbf{k}$ is measured relative to $\mathbf{k}_0$. We can choose the principal axes so that α_1, α_2 have the same sign; we can always do this since there are always at least two coefficients with the same sign (see Table 5.1). We rescale these axes so that $\alpha_1 = \alpha_2 = \beta$ after the scaling and introduce cylindrical coordinates for the rescaled variables k_1, k_2:

$$q^2 = k_1^2 + k_2^2, \quad \theta = \tan^{-1}(k_2/k_1) \tag{5.7}$$

With these changes, the DOS function takes the form

$$g(\epsilon) = \frac{\lambda}{(2\pi)^3} \int \delta(\epsilon - \epsilon_{\mathbf{k}_0} - \beta q^2 - \alpha_3 k_3^2) \mathrm{d}\mathbf{k} \tag{5.8}$$

where the factor λ comes from rescaling the principal axes 1 and 2. We can rescale the variables q, k_3 so that their coefficients become unity, to obtain

$$g(\epsilon) = \frac{\lambda}{(2\pi)^2 \beta \alpha_3^{1/2}} \int \delta(\epsilon - \epsilon_{\mathbf{k}_0} - q^2 - k_3^2) q \mathrm{d}q \mathrm{d}k_3 \tag{5.9}$$

Now we can consider the expression in the argument of the δ-function as a function of k_3:

$$f(k_3) = \epsilon - \epsilon_{\mathbf{k}_0} - q^2 - k_3^2 \Rightarrow f'(k_3) = -2k_3 \tag{5.10}$$

and we can integrate over k_3 with the help of the expression for δ-function integration Eq. (G.60) (derived in Appendix G), which gives

$$g(\epsilon) = \lambda_0 \int_0^Q \frac{1}{(\epsilon - \epsilon_{\mathbf{k}_0} - q^2)^{1/2}} q \mathrm{d}q \tag{5.11}$$

where the factor λ_0 embodies all the constants in front of the integral from rescaling and integration over k_3. The upper limit of integration Q for the variable q is determined by the condition

$$k_3^2 = \epsilon - \epsilon_{\mathbf{k}_0} - q^2 \geq 0 \Rightarrow Q = (\epsilon - \epsilon_{\mathbf{k}_0})^{1/2} \tag{5.12}$$

and with this we can now perform the final integration to obtain

$$g(\epsilon) = -\lambda_0 \left[(\epsilon - \epsilon_{\mathbf{k}_0} - q^2)^{1/2} \right]_0^{(\epsilon - \epsilon_{\mathbf{k}_0})^{1/2}} = \lambda_0 (\epsilon - \epsilon_{\mathbf{k}_0})^{1/2} \tag{5.13}$$

This result holds for $\epsilon > \epsilon_{\mathbf{k}_0}$. For $\epsilon < \epsilon_{\mathbf{k}_0}$, the δ-function cannot be satisfied for the case we are investigating, with $\alpha_i > 0, i = 1, 2, 3$, so the DOS must be zero.

By an exactly analogous calculation, we find that for the maximum M_3, with $\alpha_i < 0, i = 1, 2, 3$, the DOS behaves as

$$g(\epsilon) = \lambda_3 (\epsilon_{\mathbf{k}_0} - \epsilon)^{1/2} \tag{5.14}$$

for $\epsilon < \epsilon_{\mathbf{k}_0}$ and it is zero for $\epsilon > \epsilon_{\mathbf{k}_0}$.

For the other two cases, M_1, M_2, we can perform a similar analysis. We outline briefly the calculation for M_1: in this case, we have after rescaling $\alpha_1 = \alpha_2 = \beta > 0$ and $\alpha_3 < 0$, which leads to

$$\begin{aligned} g(\epsilon) &= \frac{\lambda}{(2\pi)^2 \beta \alpha_3^{1/2}} \int \delta(\epsilon - \epsilon_{\mathbf{k}_0} - q^2 + k_3^2) q \mathrm{d}q \mathrm{d}k_3 \\ &\rightarrow \lambda_1 \int_{Q_1}^{Q_2} \frac{1}{(q^2 + \epsilon_{\mathbf{k}_0} - \epsilon)^{1/2}} q \mathrm{d}q \end{aligned} \tag{5.15}$$

and we need to specify the limits of the last integral from the requirement that $q^2 + \epsilon_{\mathbf{k}_0} - \epsilon \geq 0$. There are two possible situations:

(i) For $\epsilon < \epsilon_{\mathbf{k}_0}$ the condition $q^2 + \epsilon_{\mathbf{k}_0} - \epsilon \geq 0$ is always satisfied, so that the lower limit of q is $Q_1 = 0$, and the upper limit is any positive value $Q_2 = Q > 0$. Then the DOS becomes

$$g(\epsilon) = \lambda_1 \left[(q^2 + \epsilon_{\mathbf{k}_0} - \epsilon)^{1/2} \right]_0^Q = \lambda_1 (Q^2 + \epsilon_{\mathbf{k}_0} - \epsilon)^{1/2} - \lambda_1 (\epsilon_{\mathbf{k}_0} - \epsilon)^{1/2} \tag{5.16}$$

For $\epsilon \rightarrow \epsilon_{\mathbf{k}_0}$, expanding in powers of the small quantity $(\epsilon_{\mathbf{k}_0} - \epsilon)$ gives

$$g(\epsilon) = \lambda_1 Q - \lambda_1 (\epsilon_{\mathbf{k}_0} - \epsilon)^{1/2} + \mathcal{O}(\epsilon_{\mathbf{k}_0} - \epsilon) \tag{5.17}$$

(ii) For $\epsilon > \epsilon_{\mathbf{k}_0}$ the condition $q^2 + \epsilon_{\mathbf{k}_0} - \epsilon \geq 0$ is satisfied for a lower limit $Q_1 = (\epsilon - \epsilon_{\mathbf{k}_0})^{1/2}$ and an upper limit being any positive value $Q_2 = Q > (\epsilon - \epsilon_{\mathbf{k}_0})^{1/2} > 0$. Then the DOS becomes

$$g(\epsilon) = \lambda_1 \left[(q^2 + \epsilon_{\mathbf{k}_0} - \epsilon)^{1/2}\right]^{Q}_{(\epsilon-\epsilon_{\mathbf{k}_0})^{1/2}} = \lambda_1 (Q^2 + \epsilon_{\mathbf{k}_0} - \epsilon)^{1/2} \tag{5.18}$$

For $\epsilon \to \epsilon_{\mathbf{k}_0}$, expanding in powers of the small quantity $(\epsilon_{\mathbf{k}_0} - \epsilon)$ gives

$$g(\epsilon) = \lambda_1 Q + \frac{\lambda_1}{2Q}(\epsilon_{\mathbf{k}_0} - \epsilon) + \mathcal{O}[(\epsilon_{\mathbf{k}_0} - \epsilon)^2] \tag{5.19}$$

By an exactly analogous calculation, we find that for the other saddle point M_2, with $\alpha_1, \alpha_2 < 0, \alpha_3 > 0$, the DOS behaves as

$$g(\epsilon) = \lambda_2 Q + \frac{\lambda_2}{2Q}(\epsilon - \epsilon_{\mathbf{k}_0}) + \mathcal{O}[(\epsilon - \epsilon_{\mathbf{k}_0})^2] \tag{5.20}$$

for $\epsilon < \epsilon_{\mathbf{k}_0}$ and

$$g(\epsilon) = \lambda_2 Q - \lambda_2 (\epsilon - \epsilon_{\mathbf{k}_0})^{1/2} + \mathcal{O}(\epsilon - \epsilon_{\mathbf{k}_0}) \tag{5.21}$$

for $\epsilon > \epsilon_{\mathbf{k}_0}$. The behavior of the DOS for all the critical points is summarized graphically in Fig. 5.1. We should caution the reader that very detailed calculations are required to resolve these critical points. In particular, methods must be developed that allow the inclusion of eigenvalues at a very large number of sampling points in reciprocal space, usually by interpolation between points at which electronic eigenvalues are actually calculated (for detailed treatments of such methods see Refs. [54, 55]). As an example, we show in Fig. 5.2 the DOS for three real solids: a typical semiconductor (Si), a free-electron metal (Al) and a transition (d-electron) metal (Ag). In each case the DOS shows the characteristic features we would expect from the detailed discussion of the band structure of these solids, presented in chapter 4. For instance, the valence bands in Si show a low-energy hump associated with the s-like states and a broader set of features associated with the p-like states which have larger dispersion; the DOS also reflects the presence of the band gap, with valence and conduction bands clearly separated. Al has an almost featureless DOS, corresponding to the behavior of free electrons with the characteristic $\epsilon^{1/2}$

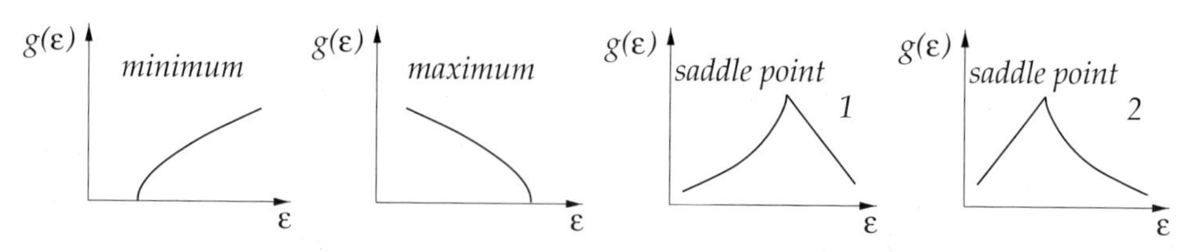

Figure 5.1. The behavior of the DOS near critical points of different type in three dimensions.

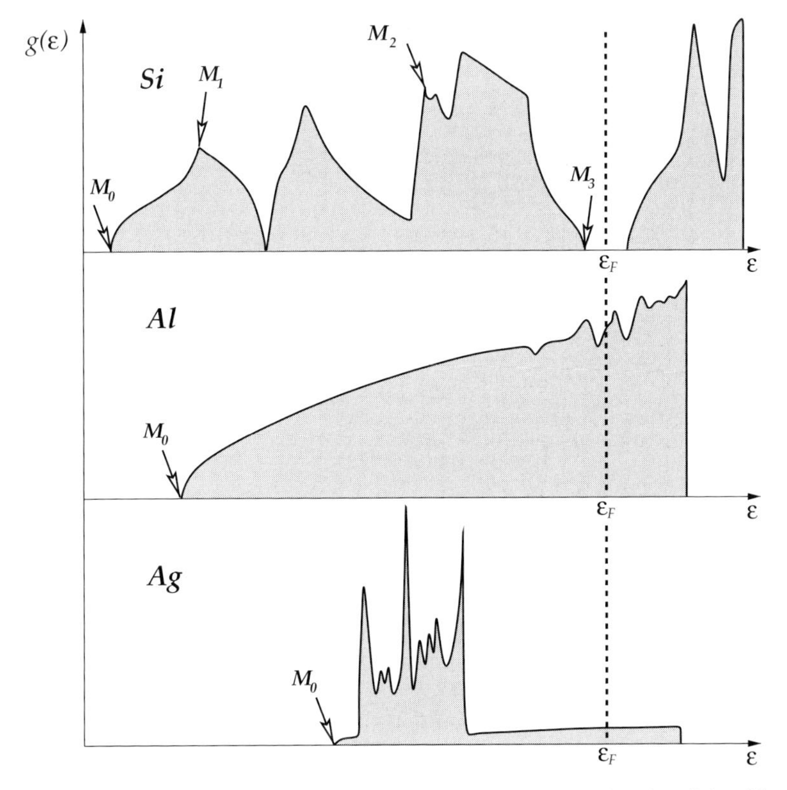

Figure 5.2. Examples of calculated electronic density of states of real solids: silicon (Si), a semiconductor with the diamond crystal structure, aluminum (Al), a free-electron metal with the FCC crystal structure, and silver (Ag), a transition (d-electron) metal also with the FCC crystal structure. The Fermi level is denoted in each case by ϵ_F and by a vertical dashed line. Several critical points are identified by arrows for Si (a minimum, a maximum and a saddle point of each kind) and for the metals (a minimum in each case). The density of states scale is not the same for the three cases, in order to bring out the important features.

dependence at the bottom of the energy range. Finally, Ag has a large DOS with significant structure in the range of energies where the d-like states lie, but has a very low and featureless DOS beyond that range, corresponding to the s-like state which has free-electron behavior.

5.2 Tunneling at metal–semiconductor contact

We consider next a contact between a metal and a semiconductor which has a band gap ϵ_{gap}. For an intrinsic semiconductor without any impurities, the Fermi level will be in the middle of the band gap (see chapter 9). At equilibrium the Fermi level on both sides of the contact must be the same. We will assume that the gap is sufficiently small to consider the density of states in the metal as being constant over a range of energies at least equal to ϵ_{gap} around the Fermi level. When a voltage bias V is applied to the metal side, all its states are shifted by $+eV$ so that at a given energy ϵ the density of states sampled is $g^{(m)}(\epsilon - eV)$. In general, the current flowing from side 2 to side 1 of a contact will be given by

$$I_{2\to 1} = -\sum_{\mathbf{kk}'} |\mathcal{T}_{\mathbf{kk}'}|^2 (1 - n_{\mathbf{k}}^{(1)}) n_{\mathbf{k}'}^{(2)} \tag{5.22}$$

where $\mathcal{T}_{\mathbf{kk}'}$ is the tunneling matrix element between the relevant single-particle states and $n_{\mathbf{k}}^{(1)}$, $n_{\mathbf{k}'}^{(2)}$ are the Fermi occupation numbers on each side; the overall minus sign comes from the fact that the expression under the summation accounts for electrons transferred from side 2 to 1, whereas the standard definition of the current involves transfer of positive charge. We will assume that, to a good approximation, the tunneling matrix elements are independent of $\mathbf{k}$, $\mathbf{k}'$. We can then turn the summations over $\mathbf{k}$ and $\mathbf{k}'$ into integrals over the energy by introducing the density of states associated with each side. We will also take the metal side to be under a bias voltage V. With these assumptions, applying the above expression for the current to the semiconductor-metal contact (with each side now identified by the corresponding superscript) we obtain

$$I_{m\to s} = -|\mathcal{T}|^2 \int_{-\infty}^{\infty} g^{(s)}(\epsilon) \left[1 - n^{(s)}(\epsilon)\right] g^{(m)}(\epsilon - eV) n^{(m)}(\epsilon - eV) \mathrm{d}\epsilon$$

$$I_{s\to m} = -|\mathcal{T}|^2 \int_{-\infty}^{\infty} g^{(s)}(\epsilon) n^{(s)}(\epsilon) g^{(m)}(\epsilon - eV) \left[1 - n^{(m)}(\epsilon - eV)\right] \mathrm{d}\epsilon$$

Subtracting the two expressions to obtain the total current flowing through the contact we find

$$I = -|\mathcal{T}|^2 \int_{-\infty}^{\infty} g^{(s)}(\epsilon) g^{(m)}(\epsilon - eV) \left[n^{(s)}(\epsilon) - n^{(m)}(\epsilon - eV)\right] \mathrm{d}\epsilon \tag{5.23}$$

What is usually measured experimentally is the so called differential conductance, given by the derivative of the current with respect to applied voltage:

$$\frac{\mathrm{d}I}{\mathrm{d}V} = |\mathcal{T}|^2 \int_{-\infty}^{\infty} g^{(s)}(\epsilon) g^{(m)}(\epsilon - eV) \left[\frac{\partial n^{(m)}(\epsilon - eV)}{\partial V}\right] \mathrm{d}\epsilon \tag{5.24}$$

where we have taken $g^{(m)}(\epsilon - eV)$ to be independent of V in the range of order ϵ_{gap} around the Fermi level, consistent with our assumption about the behavior of the metal density of states. At zero temperature the Fermi occupation number has the form

$$n^{(m)}(\epsilon) = 1 - \theta(\epsilon - \epsilon_{\mathrm{F}})$$

where $\theta(\epsilon)$ is the Heavyside step-function, and therefore its derivative is a δ-function (see Appendix G):

$$\frac{\partial n^{(m)}(\epsilon - eV)}{\partial V} = -e\frac{\partial n^{(m)}(\epsilon')}{\partial \epsilon'} = e\delta(\epsilon' - \epsilon_{\mathrm{F}})$$

where we have introduced the auxiliary variable $\epsilon' = \epsilon - eV$. Using this result in the expression for the differential conductance we find

$$\left[\frac{\mathrm{d}I}{\mathrm{d}V}\right]_{T=0} = e|\mathcal{T}|^2 g^{(s)}(\epsilon_{\mathrm{F}} + eV)g^{(m)}(\epsilon_{\mathrm{F}}) \tag{5.25}$$

which shows that by scanning the voltage V one samples the density of states of the semiconductor. Thus, the measured differential conductance will reflect all the features of the semiconductor density of states, including the gap.

5.3 Optical excitations

Let us consider what can happen when an electron in a crystalline solid absorbs a photon and jumps to an excited state. The transition rate for such an excitation is given by Fermi's golden rule (see Appendix B, Eq. (B.63)):

$$P_{i\to f}(\omega) = \frac{2\pi}{\hbar}\left|\langle\psi_{\mathbf{k}'}^{(n')}|\mathcal{H}^{int}|\psi_{\mathbf{k}}^{(n)}\rangle\right|^2 \delta(\epsilon_{\mathbf{k}'}^{(n')} - \epsilon_{\mathbf{k}}^{(n)} - \hbar\omega) \tag{5.26}$$

where $\langle f| = \langle\psi_{\mathbf{k}'}^{(n')}|$, $|i\rangle = |\psi_{\mathbf{k}}^{(n)}\rangle$ are the final and initial states of the electron in the crystal with the corresponding eigenvalues $\epsilon_{\mathbf{k}}^{(n)}$, $\epsilon_{\mathbf{k}'}^{(n')}$, and the interaction hamiltonian is given by (see Appendix B):

$$\mathcal{H}^{int}(\mathbf{r}, t) = \frac{e}{m_e c}\left[\mathrm{e}^{\mathrm{i}(\mathbf{q}\cdot\mathbf{r}-\omega t)}\mathbf{A}_0 \cdot \mathbf{p} + \mathrm{c.c.}\right] \tag{5.27}$$

with $\hbar\mathbf{q}$ the momentum, ω the frequency and $\mathbf{A}_0$ the vector potential of the photon radiation field; $\mathbf{p} = (\hbar/\mathrm{i})\nabla_{\mathbf{r}}$ is the momentum operator, and c.c. stands for complex conjugate. Of the two terms in Eq. (5.27), the one with $-\mathrm{i}\omega t$ in the exponential corresponds to absorption, while the other, with $+\mathrm{i}\omega t$, corresponds to emission of radiation. With the expression for the interaction hamiltonian of Eq. (5.27), the probability for absorption of light becomes:

$$P_{i\to f} = \frac{2\pi}{\hbar}\left(\frac{e}{m_e c}\right)^2 \left|\langle\psi_{\mathbf{k}'}^{(n')}|\mathrm{e}^{\mathrm{i}\mathbf{q}\cdot\mathbf{r}}\mathbf{A}_0 \cdot \mathbf{p}|\psi_{\mathbf{k}}^{(n)}\rangle\right|^2 \delta(\epsilon_{\mathbf{k}'}^{(n')} - \epsilon_{\mathbf{k}}^{(n)} - \hbar\omega) \tag{5.28}$$

In order to have non-vanishing matrix elements in this expression, the initial state $|\psi_{\mathbf{k}}^{(n)}\rangle$ must be occupied (a valence state), while the final state $\langle\psi_{\mathbf{k}'}^{(n')}|$ must be unoccupied (conduction state). Since all electronic states in the crystal can be expressed in our familiar Bloch form,

$$\psi_{\mathbf{k}}^{(n)}(\mathbf{r}) = \mathrm{e}^{\mathrm{i}\mathbf{k}\cdot\mathbf{r}} \sum_{\mathbf{G}} \alpha_{\mathbf{k}}^{(n)}(\mathbf{G}) \mathrm{e}^{\mathrm{i}\mathbf{G}\cdot\mathbf{r}} \tag{5.29}$$

we find the following expression for the matrix element:

$$\langle \psi_{\mathbf{k}'}^{(n')} | \mathrm{e}^{\mathrm{i}\mathbf{q}\cdot\mathbf{r}} \mathbf{A}_0 \cdot \mathbf{p} | \psi_{\mathbf{k}}^{(n)} \rangle = \sum_{\mathbf{G}\mathbf{G}'} \left[\alpha_{\mathbf{k}'}^{(n')}(\mathbf{G}') \right]^* \alpha_{\mathbf{k}}^{(n)}(\mathbf{G}) \left[\hbar \mathbf{A}_0 \cdot (\mathbf{k} + \mathbf{G}) \right]$$
$$\times \int \mathrm{e}^{\mathrm{i}(\mathbf{k}-\mathbf{k}'+\mathbf{q}+\mathbf{G}-\mathbf{G}')\cdot\mathbf{r}} \mathrm{d}\mathbf{r}$$

The integral over real space produces a δ-function in reciprocal-space vectors (see Appendix G):

$$\int \mathrm{e}^{\mathrm{i}(\mathbf{k}-\mathbf{k}'+\mathbf{q}+\mathbf{G})\cdot\mathbf{r}} \mathrm{d}\mathbf{r} \sim \delta(\mathbf{k} - \mathbf{k}' + \mathbf{q} + \mathbf{G}) \Rightarrow \mathbf{k}' = \mathbf{k} + \mathbf{q}$$

where we have set $\mathbf{G} = 0$ in the argument of the δ-function because we consider only values of $\mathbf{k}, \mathbf{k}'$ within the first BZ. However, taking into account the relative magnitudes of the three wave-vectors involved in this condition reveals that it boils down to $\mathbf{k}' = \mathbf{k}$: the momentum of radiation for optical transitions is $|\mathbf{q}| = (2\pi/\lambda)$ with a wavelength $\lambda \sim 10^4$ Å, while the crystal wave-vectors have typical wavelength values of order the interatomic distances, that is, ~ 1 Å. Consequently, the difference between the wave-vectors of the initial and final states due to the photon momentum is negligible. Taking $\mathbf{q} = 0$ in the above equation leads to so called "direct" transitions, that is, transitions at the same value of $\mathbf{k}$ in the BZ. These are the only allowed optical transitions when no other excitations are present. When other excitations that can carry crystal momentum are present, such as phonons (see chapter 6), the energy and momentum conservation conditions can be independently satisfied even for indirect transitions, in which case the initial and final photon states can have different momenta. The direct and indirect transitions are illustrated in Fig. 5.3.

Now suppose we are interested in calculating the transition probability for absorption of radiation of frequency ω. Then we would have to sum over all the possible pairs of states that have an energy difference $\epsilon_{\mathbf{k}}^{(n')} - \epsilon_{\mathbf{k}}^{(n)} = \hbar\omega$, where we are assuming that the wave-vector is the same for the two states, as argued above. We have also argued that state $\langle \psi_{\mathbf{k}_2}^{(n_2)} |$ is a conduction state (for which we will use the symbol c for the band index) and state $| \psi_{\mathbf{k}_1}^{(n_1)} \rangle$ is a valence state (for which we will use the symbol v for the band index). The matrix elements involved in the transition probability can be approximated as nearly constant, independent of the wave-vector $\mathbf{k}$ and the details of the initial and final state wavefunctions. With this

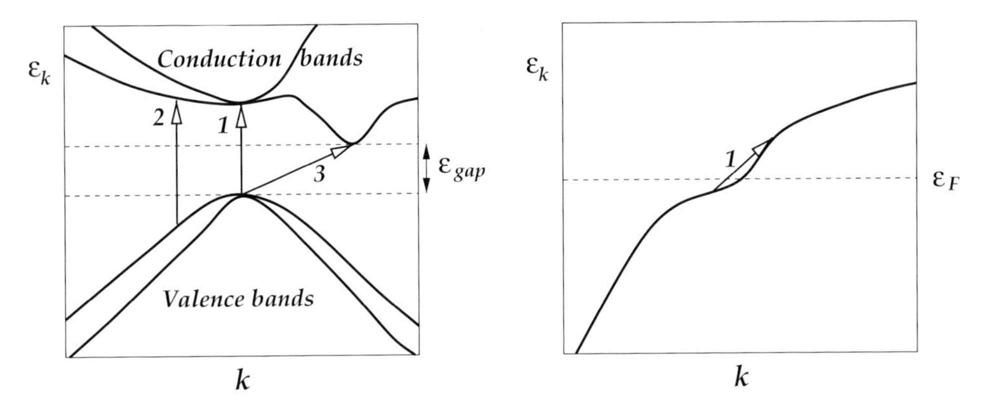

Figure 5.3. Illustration of optical transitions. **Left:** interband transitions in a semiconductor, between valence and conduction states: 1 is a direct transition at the minimal *direct* gap, 2 is another direct transition at a larger energy, and 3 is an indirect transition across the minimal gap ϵ_{gap}. **Right:** intraband transitions in a metal across the Fermi level ϵ_{F}

approximation, we obtain for the total transition probability

$$P(\omega) = P_0 \sum_{\mathbf{k},c,v} \delta(\epsilon_{\mathbf{k}}^{(c)} - \epsilon_{\mathbf{k}}^{(v)} - \hbar\omega) \tag{5.30}$$

where P_0 contains the constant matrix elements and the other constants that appear in Eq. (5.28). This expression is very similar to the expressions that we saw earlier for the DOS. The main difference here is that we are interested in the density of pairs of states that have energy difference $\hbar\omega$, rather than the density of states at given energy ϵ. This new quantity is called the "joint density of states" (JDOS), and its calculation is exactly analogous to that of the DOS. The JDOS appears in expressions of experimentally measurable quantities such as the dielectric function of a crystal, as we discuss in the next section.

5.4 Conductivity and dielectric function

The response of a crystal to an external electric field $\mathbf{E}$ is described in terms of the conductivity σ through the relation $\mathbf{J} = \sigma\mathbf{E}$, where $\mathbf{J}$ is the induced current (for details see Appendix A). Thus, the conductivity is the appropriate response function which relates the intrinsic properties of the solid to the effect of the external perturbation that the electromagnetic field represents. The conductivity is actually inferred from experimental measurements of the dielectric function. Accordingly, our first task is to establish the relationship between these two quantities. Before we do this, we discuss briefly how the dielectric function is measured experimentally.

The fraction R of reflected power for normally incident radiation on a solid with dielectric constant ε is given by the classical theory of electrodynamics (see Appendix A, Eq. (A.41)) as

$$R = \left| \frac{1 - \sqrt{\varepsilon}}{1 + \sqrt{\varepsilon}} \right|^2 \tag{5.31}$$

The real and imaginary parts of the dielectric function $\varepsilon = \varepsilon_1 + \mathrm{i}\varepsilon_2$, are related by

$$\varepsilon_1(\omega) = 1 + P \int_{-\infty}^{\infty} \frac{\mathrm{d}w}{\pi} \frac{\varepsilon_2(w)}{w - \omega}, \quad \varepsilon_2(\omega) = -P \int_{-\infty}^{\infty} \frac{\mathrm{d}w}{\pi} \frac{\varepsilon_1(w) - 1}{w - \omega} \tag{5.32}$$

known as the Kramers–Kronig relations. In these expressions the P in front of the integrals stands for the principal value. In essence, this implies that there is only one unknown function (either ε_1 or ε_2). Assuming the reflectivity R can be measured over a wide range of frequencies, and using the Kramers–Kronig relations, both ε_1 and ε_2 can then be determined.

We next derive the relation between the dielectric function and the conductivity. Using the plane wave expressions for the charge and current densities and the electric field,

$$\mathbf{J}(\mathbf{r}, t) = \mathbf{J}(\mathbf{q}, \omega)\mathrm{e}^{\mathrm{i}(\mathbf{q}\cdot\mathbf{r} - \omega t)}, \quad \rho^{ind}(\mathbf{r}, t) = \rho^{ind}(\mathbf{q}, \omega)\mathrm{e}^{\mathrm{i}(\mathbf{q}\cdot\mathbf{r} - \omega t)},$$
$$\mathbf{E}(\mathbf{r}, t) = \mathbf{E}(\mathbf{q}, \omega)\mathrm{e}^{\mathrm{i}(\mathbf{q}\cdot\mathbf{r} - \omega t)}$$

the definition of the conductivity in the frequency domain takes the form[1]

$$\mathbf{J}(\mathbf{q}, \omega) = \sigma(\mathbf{q}, \omega)\mathbf{E}(\mathbf{q}, \omega) \tag{5.33}$$

With the use of this relation, the continuity equation connecting the current $\mathbf{J}$ to the induced charge ρ^{ind} gives

$$\nabla_{\mathbf{r}} \cdot \mathbf{J} + \frac{\partial \rho^{ind}}{\partial t} = 0 \Rightarrow \mathbf{q} \cdot \mathbf{J}(\mathbf{q}, \omega) = \sigma(\mathbf{q}, \omega)\mathbf{q} \cdot \mathbf{E}(\mathbf{q}, \omega) = \omega\rho^{ind}(\mathbf{q}, \omega) \tag{5.34}$$

We can separate out the longitudinal and transverse parts of the current $(\mathbf{J}_l, \mathbf{J}_t)$ and the electric field $(\mathbf{E}_l, \mathbf{E}_t)$, with the longitudinal component of the electric field given by

$$\mathbf{E}_l(\mathbf{r}, t) = -\nabla_{\mathbf{r}}\Phi(\mathbf{r}, t) \Rightarrow \mathbf{E}_l(\mathbf{q}, \omega) = -i\mathbf{q}\Phi(\mathbf{q}, \omega)$$

where $\Phi(\mathbf{r}, t)$ is the scalar potential (see Appendix A, Eqs. (A.42)–(A.53) and accompanying discussion). This gives for the induced charge

$$\rho^{ind}(\mathbf{q}, \omega) = -\frac{\mathrm{i}q^2\sigma(\mathbf{q}, \omega)}{\omega}\Phi(\mathbf{q}, \omega) \tag{5.35}$$

[1] In the time domain, that is, when the explicit variable in $\mathbf{J}$ is the time t rather than the frequency ω, the right-hand side of this relation is replaced by a convolution integral.

As we have discussed in chapter 2, for weak external fields we can use the linear response expression: $\rho^{ind}(\mathbf{q}, \omega) = \chi(\mathbf{q}, \omega)\Phi(\mathbf{q}, \omega)$, where $\chi(\mathbf{q}, \omega)$ is the susceptibility or response function. This gives for the conductivity

$$\sigma(\mathbf{q}, \omega) = \frac{\mathrm{i}\omega}{q^2}\chi(\mathbf{q}, \omega) \tag{5.36}$$

Comparing this result with the general relation between the response function and the dielectric function

$$\varepsilon(\mathbf{q}, \omega) = 1 - \frac{4\pi}{q^2}\chi(\mathbf{q}, \omega) \tag{5.37}$$

we obtain the desired relation between the conductivity and the dielectric function:

$$\sigma(\mathbf{q}, \omega) = \frac{\mathrm{i}\omega}{4\pi}\left[1 - \varepsilon(\mathbf{q}, \omega)\right] \Longrightarrow \varepsilon(\mathbf{q}, \omega) = 1 - \frac{4\pi}{\mathrm{i}\omega}\sigma(\mathbf{q}, \omega) \tag{5.38}$$

Having established the connection between conductivity and dielectric function, we will next express the conductivity in terms of microscopic properties of the solid (the electronic wavefunctions and their energies and occupation numbers); as a final step, we will use the connection between conductivity and dielectric function to obtain an expression of the latter in terms of the microscopic properties of the solid. This will provide the direct link between the electronic structure at the microscopic level and the experimentally measured dielectric function, which captures the macroscopic response of the solid to an external electromagnetic field.

Before proceeding with the detailed derivations, it is worth mentioning two simple expressions which capture much of the physics. The first, known as the Drude model, refers to the frequency-dependent dielectric function for the case where only transitions within a single band, intersected by the Fermi level, are allowed:

$$\varepsilon(\omega) = 1 - \frac{\omega_p^2}{\omega(\omega + \mathrm{i}/\tau)} \tag{5.39}$$

where ω_p and τ are constants (known as the plasma frequency and relaxation time, respectively). The second, known as the Lorentz model, refers to the opposite extreme, that is, the case where the only allowed transitions are between occupied and unoccupied bands separated by a band gap:

$$\varepsilon(\omega) = 1 - \frac{\omega_p^2}{(\omega^2 - \omega_0^2) + \mathrm{i}\eta\omega} \tag{5.40}$$

with ω_p the same constant as in the previous expression and ω_0, η two additional constants. The expressions we will derive below can be ultimately reduced to these simple expressions; this exercise provides some insight to the meaning of the constants involved in the Drude and Lorentz models.

For the calculation of the conductivity we will rely on the general result for the expectation value of a many-body operator $\mathcal{O}(\{\mathbf{r}_i\})$, which can be expressed as a sum of single-particle operators $o(\mathbf{r}_i)$,

$$\mathcal{O}(\{\mathbf{r}_i\}) = \sum_i o(\mathbf{r}_i)$$

in terms of the matrix elements in the single-particle states, as derived in Appendix B. For simplicity we will use a single index, the subscript $\mathbf{k}$, to identify the single-particle states in the crystal, with the understanding that this index represents in a short-hand notation both the band index and the wave-vector index. From Eq. (B.21) we find that the expectation value of the many-body operator $\mathcal{O}$ in the single-particle states labeled by $\mathbf{k}$ is

$$\langle \mathcal{O} \rangle = \sum_{\mathbf{k},\mathbf{k}'} o_{\mathbf{k},\mathbf{k}'} \gamma_{\mathbf{k}',\mathbf{k}} \tag{5.41}$$

where $o_{\mathbf{k},\mathbf{k}'}$ and $\gamma_{\mathbf{k},\mathbf{k}'}$ are the matrix elements of the operator $o(\mathbf{r})$ and the single-particle density matrix $\gamma(\mathbf{r},\mathbf{r}')$. We must therefore identify the appropriate single-particle density matrix and single-particle operator $o(\mathbf{r})$ for the calculation of the conductivity. As derived by Ehrenreich and Cohen [56], to first order perturbation theory the interaction term of the hamiltonian gives for the single-particle density matrix

$$\gamma^{int}_{\mathbf{k}',\mathbf{k}} = \frac{n_0(\epsilon_{\mathbf{k}'}) - n_0(\epsilon_{\mathbf{k}})}{\epsilon_{\mathbf{k}'} - \epsilon_{\mathbf{k}} - \hbar\omega - i\hbar\eta} \mathcal{H}^{int}_{\mathbf{k}',\mathbf{k}} \tag{5.42}$$

where $n_0(\epsilon_{\mathbf{k}})$ is the Fermi occupation number for the state with energy $\epsilon_{\mathbf{k}}$ in the unperturbed system, and η is an infinitesimal positive quantity with the dimensions of frequency. For the calculation of the conductivity, the relevant interaction term is that for absorption or emission of photons, the carriers of the electromagnetic field, which was discussed earlier, Eq. (5.27) (see Appendix B for more details). In the usual Coulomb gauge we have the following relation between the transverse electric field $\mathbf{E}_t$ ($\nabla_{\mathbf{r}} \cdot \mathbf{E}_t = 0$), and the vector potential $\mathbf{A}$:

$$\mathbf{E}_t(\mathbf{r}, t) = -\frac{1}{c}\frac{\partial \mathbf{A}}{\partial t} \Rightarrow \mathbf{E}_t(\mathbf{r}, t) = \frac{i\omega}{c}\mathbf{A}(\mathbf{r}, t) \Rightarrow \mathbf{A}(\mathbf{r}, t) = \frac{c}{i\omega}\mathbf{E}_t(\mathbf{r}, t) \tag{5.43}$$

With this, the interaction term takes the form

$$\mathcal{H}^{int}(\mathbf{r}, t) = -\frac{e\hbar}{m_e\omega}\mathbf{E}_t \cdot \nabla_{\mathbf{r}} \Rightarrow \mathcal{H}^{int}_{\mathbf{k}',\mathbf{k}} = -\frac{e\hbar}{m_e\omega}\mathbf{E}_t \cdot \langle \psi_{\mathbf{k}'}|\nabla_{\mathbf{r}}|\psi_{\mathbf{k}}\rangle \tag{5.44}$$

which is the expression to be used in $\gamma^{int}_{\mathbf{k}',\mathbf{k}}$.

As far as the relevant single-particle operator $o(\mathbf{r})$ is concerned, it must describe the response of the physical system to the external potential, which is of course the induced current. The single-particle current operator is

$$\mathbf{j}(\mathbf{r}) = \frac{-e}{\Omega}\mathbf{v} = \frac{-e}{\Omega}\frac{\mathbf{p}}{m_e} = \frac{-e\hbar}{im_e\Omega}\nabla_{\mathbf{r}}.$$

Using this expression as the single-particle operator $o(\mathbf{r})$, and combining it with the expression for the single-particle density matrix $\gamma^{int}_{\mathbf{k}',\mathbf{k}}$ derived above, we obtain for the expectation value of the total current:

$$\begin{aligned}\langle\mathbf{J}\rangle &= \sum_{\mathbf{k},\mathbf{k}'} \mathbf{j}_{\mathbf{k},\mathbf{k}'}\gamma^{int}_{\mathbf{k}',\mathbf{k}} = \sum_{\mathbf{k},\mathbf{k}'} \langle\psi_{\mathbf{k}}| - \frac{e\hbar}{im_e\Omega}\nabla_{\mathbf{r}}|\psi_{\mathbf{k}'}\rangle \frac{n_0(\epsilon_{\mathbf{k}'}) - n_0(\epsilon_{\mathbf{k}})}{\epsilon_{\mathbf{k}'} - \epsilon_{\mathbf{k}} - \hbar\omega - i\hbar\eta}\mathcal{H}^{int}_{\mathbf{k}',\mathbf{k}} \\ &= -\frac{ie^2\hbar^2}{m_e^2\Omega}\sum_{\mathbf{k},\mathbf{k}'}\frac{1}{\omega}\frac{n_0(\epsilon_{\mathbf{k}'}) - n_0(\epsilon_{\mathbf{k}})}{\epsilon_{\mathbf{k}'} - \epsilon_{\mathbf{k}} - \hbar\omega - i\hbar\eta}\langle\psi_{\mathbf{k}}|\nabla_{\mathbf{r}}|\psi_{\mathbf{k}'}\rangle\langle\psi_{\mathbf{k}'}|\nabla_{\mathbf{r}}|\psi_{\mathbf{k}}\rangle \cdot \mathbf{E}_t \quad (5.45)\end{aligned}$$

From the relation between current and electric field, Eq. (5.33), expressed in tensor notation

$$\mathbf{J} = \sigma\mathbf{E} \rightarrow J_\alpha = \sum_\beta \sigma_{\alpha\beta}E_\beta$$

we obtain, for the real and imaginary parts of the conductivity,

$$\begin{aligned}\mathrm{Re}[\sigma_{\alpha\beta}] &= \frac{\pi e^2\hbar^2}{m_e^2\Omega}\sum_{\mathbf{k},\mathbf{k}'}\frac{1}{\omega}\left[n_0(\epsilon_{\mathbf{k}'}) - n_0(\epsilon_{\mathbf{k}})\right]\langle\psi_{\mathbf{k}}|\frac{\partial}{\partial x_\alpha}|\psi_{\mathbf{k}'}\rangle\langle\psi_{\mathbf{k}'}|\frac{\partial}{\partial x_\beta}|\psi_{\mathbf{k}}\rangle \\ &\quad \times \delta(\epsilon_{\mathbf{k}'} - \epsilon_{\mathbf{k}} - \hbar\omega) \\ \mathrm{Im}[\sigma_{\alpha\beta}] &= -\frac{e^2\hbar^2}{m_e^2\Omega}\sum_{\mathbf{k},\mathbf{k}'}\frac{1}{\omega}[n_0(\epsilon_{\mathbf{k}'}) - n_0(\epsilon_{\mathbf{k}})]\langle\psi_{\mathbf{k}}|\frac{\partial}{\partial x_\alpha}|\psi_{\mathbf{k}'}\rangle\langle\psi_{\mathbf{k}'}|\frac{\partial}{\partial x_\beta}|\psi_{\mathbf{k}}\rangle \\ &\quad \times \frac{1}{\epsilon_{\mathbf{k}'} - \epsilon_{\mathbf{k}} - \hbar\omega} \quad (5.46)\end{aligned}$$

where we have used the mathematical identity Eq. (G.55) (proven in Appendix G) to obtain the real and imaginary parts of σ. This expression for the conductivity is known as the Kubo–Greenwood formula [57]. These expressions can then be used to obtain the real and imaginary parts of the dielectric function, from Eq. (5.38). The result is precisely the type of expression we mentioned above: of the two states involved in the excitation $|\psi_{\mathbf{k}}\rangle$, $|\psi_{\mathbf{k}'}\rangle$, one must be occupied and the other empty, otherwise the difference of the Fermi occupation numbers $[n_0(\epsilon_{\mathbf{k}'}) - n_0(\epsilon_{\mathbf{k}})]$ will lead to a vanishing contribution. Notice that if $\epsilon_{\mathbf{k}'} - \epsilon_{\mathbf{k}} = \hbar\omega$ and $n_0(\epsilon_{\mathbf{k}'}) = 0$, $n_0(\epsilon_{\mathbf{k}}) = 1$, then we have absorption of a photon with energy $\hbar\omega$, whereas if $n_0(\epsilon_{\mathbf{k}'}) = 1$, $n_0(\epsilon_{\mathbf{k}}) = 0$, then we have emission of a photon with energy $\hbar\omega$; the two processes give contributions of opposite sign to the conductivity. In the case of the real part of the conductivity which is related to the imaginary part of the dielectric function, if we assume that the matrix elements are approximately independent of $\mathbf{k}, \mathbf{k}'$, then we obtain a sum over δ-functions that ensure conservation of energy upon absorption or emission of radiation with frequency ω, which leads to the JDOS, precisely as derived in the previous section, Eq. (5.30).

In order to obtain the dielectric function from the conductivity, we will first assume an isotropic solid, in which case $\sigma_{\alpha\beta} = \sigma\delta_{\alpha\beta}$, and then use the relation we derived in Eq. (5.38),

$$\varepsilon(\mathbf{q}, \omega) = 1 + \frac{4\pi e^2}{m_e^2 \Omega} \sum_{\mathbf{k}\mathbf{k}', nn'} \frac{1}{\omega^2} \left| \langle \psi_{\mathbf{k}'}^{(n')} | (-i\hbar\nabla_{\mathbf{r}}) | \psi_{\mathbf{k}}^{(n)} \rangle \right|^2 \frac{n_0(\epsilon_{\mathbf{k}'}^{(n')}) - n_0(\epsilon_{\mathbf{k}}^{(n)})}{\epsilon_{\mathbf{k}'}^{(n')} - \epsilon_{\mathbf{k}}^{(n)} - \hbar\omega - i\hbar\eta} \tag{5.47}$$

with $\mathbf{q} = \mathbf{k}' - \mathbf{k}$. In the above expression we have introduced again explicitly the band indices n, n'. To analyze the behavior of the dielectric function, we will consider two different situations: we will examine first transitions within the same band, $n = n'$, but at slightly different values of the wave-vector, $\mathbf{k}' = \mathbf{k} + \mathbf{q}$, in the limit $\mathbf{q} \to 0$; and second, transitions at the same wave-vector $\mathbf{k}' = \mathbf{k}$ but between different bands $n \neq n'$. The first kind are called *intraband* or *free-electron* transitions: they correspond to situations where an electron makes a transition by absorption or emission of a photon across the Fermi level by changing slightly its wave-vector, as illustrated in Fig. 5.3. The second kind correspond to *interband* or *bound-electron* transitions: they correspond to situations where an electron makes a direct transition by absorption or emission of a photon acros the band gap in insulators or semiconductors, also as illustrated in Fig. 5.3. Since in both cases we are considering the limit of $\mathbf{k}' \to \mathbf{k}$, we will omit from now on the dependence of the dielectric function on the wave-vector difference $\mathbf{q} = \mathbf{k}' - \mathbf{k}$ with the understanding that all the expressions we derive concern the limit $\mathbf{q} \to 0$.

We begin with the intraband transitions. To simplify the notation, we define

$$\hbar\omega_{\mathbf{k}}^{(n)}(\mathbf{q}) = \epsilon_{\mathbf{k}+\mathbf{q}}^{(n)} - \epsilon_{\mathbf{k}}^{(n)} = \frac{\hbar}{m_e}\mathbf{q} \cdot \mathbf{p}^{(nn)}(\mathbf{k}) + \frac{\hbar^2}{2\overline{m}^{(n)}(\mathbf{k})}\mathbf{q}^2 \tag{5.48}$$

where we have used the expressions of Eqs. (3.43) and (3.46) for the effective mass of electrons and, consistent with our assumption of an isotropic solid, we have taken the effective mass to be a scalar rather than a tensor. We will also use the result of Eq. (3.45), to relate the matrix elements of the momentum operator $\mathbf{p}^{(nn)}(\mathbf{k}) = \langle \psi_{\mathbf{k}}^{(n)} | (-i\hbar\nabla_{\mathbf{r}}) | \psi_{\mathbf{k}}^{(n)} \rangle$ to the gradient of the band energy $\nabla_{\mathbf{k}}\epsilon_{\mathbf{k}}^{(n)}$. With these considerations, we obtain

$$\varepsilon(\omega) = 1 - \frac{4\pi e^2}{\hbar^3 \Omega} \sum_{\mathbf{k}\mathbf{q},n} \frac{1}{\omega^2} \left[\frac{2}{\omega - \omega_{\mathbf{k}}^{(n)}(\mathbf{q}) + \mathrm{i}\eta} - \frac{2}{\omega + \omega_{\mathbf{k}}^{(n)}(\mathbf{q}) + \mathrm{i}\eta} \right] \left| \nabla_{\mathbf{k}} \epsilon_{\mathbf{k}}^{(n)} \right|^2$$

$$= 1 - \frac{4\pi e^2}{\hbar^2 \Omega} \sum_{\mathbf{k}\mathbf{q},n} \left[\frac{2\mathbf{q} \cdot \nabla_{\mathbf{k}} \epsilon_{\mathbf{k}}^{(n)}}{\omega^2} + \frac{\mathbf{q}^2}{\omega^2} \frac{1}{\overline{m}^{(n)}(\mathbf{k})} \right] \frac{2 \left| \nabla_{\mathbf{k}} \epsilon_{\mathbf{k}}^{(n)} \right|^2}{(\omega + \mathrm{i}\eta)^2 - \left(\omega_{\mathbf{k}}^{(n)}(\mathbf{q}) \right)^2}$$

where in the first equation we have written explicitly the two contributions from the absorption and emission terms, averaging over the two spin states, and in the second equation we have made use of the expression introduced in Eq. (5.48). At this point, we can take advantage of the fact that $\mathbf{q}$ is the momentum of the photon to write $\omega^2 = \mathbf{q}^2 c^2$ in the second term in the square brackets and, in the limit $\mathbf{q} \to 0$, we will neglect the first term, as well as the term $\omega_{\mathbf{k}}^{(n)}(\mathbf{q})$ compared to $(\omega + \mathrm{i}\eta)$ in the denominator, to obtain

$$\varepsilon(\omega) = 1 - \left(\frac{4\pi e^2 N}{m_e \Omega} \right) \left[\frac{2m_e}{\hbar^2 c^2} \sum_{\mathbf{k},n} \frac{1}{\overline{m}^{(n)}(\mathbf{k})} |\nabla_{\mathbf{k}} \epsilon_{\mathbf{k}}^{(n)}|^2 \right] \frac{1}{(\omega + \mathrm{i}\eta)^2} \tag{5.49}$$

with N the total number of unit cells in the crystal, which we take to be equal to the number of valence electrons for simplicity (the two numbers are always related by an integer factor). The quantity in the first parenthesis has the dimensions of a frequency squared, called the *plasma frequency*, ω_p:

$$\omega_p \equiv \left(\frac{4\pi e^2}{m_e} \frac{N}{\Omega} \right)^{1/2} \tag{5.50}$$

This is the characteristic frequency of the response of a uniform electron gas of density N/Ω in a uniform background of compensating positive charge (see problem 5); the modes describing this response are called *plasmons* (mentioned in chapter 2). Using the standard expressions for the radius r_s of the sphere corresponding to the average volume per electron, we can express the plasma frequency as

$$\omega_p = 0.716 \left(\frac{1}{r_s/a_0} \right)^{3/2} \times 10^{17}\ \mathrm{Hz}$$

and since typical values of r_s/a_0 in metals are 2–6, we find that the frequencies of plasmon oscillations are in the range of 10^3–10^4 THz.

Returning to Eq. (5.49), the quantity in the square brackets depends exclusively on the band structure and fundamental constants; once we have summed over all the relevant values of $\mathbf{k}$ and n, this quantity takes real positive values which we denote by Λ^2, with Λ a real positive constant. We note that the relevant values of the wave-vector and band index are those that correspond to crossings of the Fermi level, since we have assumed transitions across the Fermi level within the same band.[2] The remaining term in the expression for the dielectric function is the one which determines its dependence on the frequency. Extracting the real and imaginary parts from this term we obtain

$$\varepsilon_1(\omega) = 1 - \frac{(\Lambda\omega_p)^2}{\omega^2 + \eta^2} \Rightarrow \lim_{\eta \to 0} [\varepsilon_1(\omega)] = 1 - \frac{(\Lambda\omega_p)^2}{\omega^2} \tag{5.51}$$

$$\varepsilon_2(\omega) = \frac{(\Lambda\omega_p)^2}{\omega} \frac{\eta}{\omega^2 + \eta^2} \Rightarrow \lim_{\eta \to 0} [\varepsilon_2(\omega)] = \pi \frac{(\Lambda\omega_p)^2}{\omega} \delta(\omega) \tag{5.52}$$

These expressions describe the behavior of the dielectric function for a material with one band only, which is intersected by the Fermi level, that is, a one-band metal.

We next consider the effect of direct interband transitions, that is, transitions for which $\mathbf{k} = \mathbf{k}'$ and $n \neq n'$. To ensure direct transitions we introduce a factor $(2\pi)^3\delta(\mathbf{k} - \mathbf{k}')$ in the double summation over wave-vectors in the general expression for the dielectric function, Eq. (5.47), from which we obtain

$$\varepsilon(\omega) = 1 + \frac{4\pi e^2}{\hbar} \sum_{\mathbf{k}, n \neq n'} \frac{1}{\omega^2} \left| \langle \psi_{\mathbf{k}}^{(n')} | \frac{\mathbf{p}}{m_e} | \psi_{\mathbf{k}}^{(n)} \rangle \right|^2 \frac{n_0(\epsilon_{\mathbf{k}}^{(n')}) - n_0(\epsilon_{\mathbf{k}}^{(n)})}{\omega_{\mathbf{k}}^{(nn')} - \omega - i\eta} \tag{5.53}$$

where we have defined

$$\hbar\omega_{\mathbf{k}}^{(nn')} = \epsilon_{\mathbf{k}}^{(n')} - \epsilon_{\mathbf{k}}^{(n)}$$

We notice that the matrix elements of the operator $\mathbf{p}/m_e$ can be put in the following form:

$$\frac{\mathbf{p}}{m_e} = \mathbf{v} = \frac{\mathrm{d}}{\mathrm{d}t}\mathbf{r} \Rightarrow \left| \langle \psi_{\mathbf{k}}^{(n')} | \frac{\mathbf{p}}{m_e} | \psi_{\mathbf{k}}^{(n)} \rangle \right|^2 = \left(\omega_{\mathbf{k}}^{(nn')} \right)^2 \left| \langle \psi_{\mathbf{k}}^{(n')} | \mathbf{r} | \psi_{\mathbf{k}}^{(n)} \rangle \right|^2$$

where we have used Eq. (3.47) to obtain the last expression. With this result, writing out explicitly the contributions from the absorption and emission terms and averaging over the two spin states, we obtain for the real and imaginary parts of the dielectric function in the limit $\eta \to 0$

$$\varepsilon_1(\omega) = 1 - \frac{8\pi e^2}{\hbar} \sum_{\mathbf{k}, v, c} \frac{2\omega_{\mathbf{k}}^{(vc)}}{\omega^2 - \left(\omega_{\mathbf{k}}^{(vc)}\right)^2} \left| \langle \psi_{\mathbf{k}}^{(c)} | \mathbf{r} | \psi_{\mathbf{k}}^{(v)} \rangle \right|^2 \tag{5.54}$$

$$\varepsilon_2(\omega) = \frac{8\pi^2 e^2}{\hbar} \sum_{\mathbf{k}, v, c} \left[\delta(\omega - \omega_{\mathbf{k}}^{(vc)}) - \delta(\omega + \omega_{\mathbf{k}}^{(cv)}) \right] \left| \langle \psi_{\mathbf{k}}^{(c)} | \mathbf{r} | \psi_{\mathbf{k}}^{(v)} \rangle \right|^2 \tag{5.55}$$

[2] At non-zero temperatures, these values must be extended to capture all states which have partial occupation across the Fermi level.

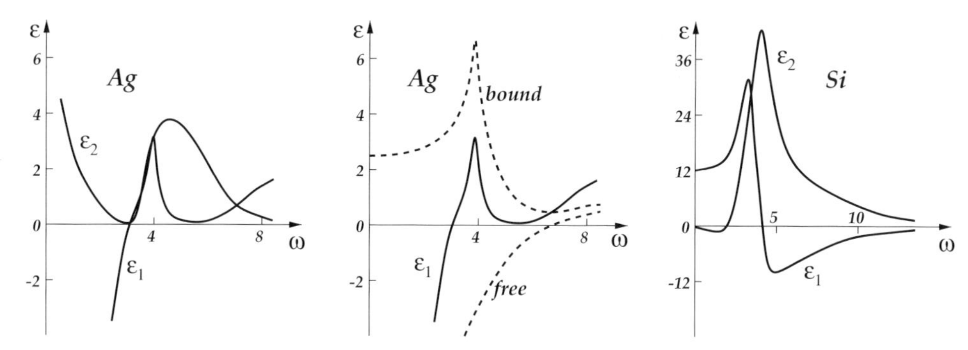

Figure 5.4. Examples of dielectric functions $\varepsilon(\omega)$ as a function of the frequency ω (in eV). **Left:** the real (ε_1) and imaginary (ε_2) parts of the dielectric function of Ag; **center:** an analysis of the real part into its bound-electron (interband) and free-electron (intraband) components. **Right:** the real and imaginary parts of the dielectric function of Si. Source: Refs. [57–59].

where we have made explicit the requirement that one of the indices must run over valence bands (v) and the other over conduction bands (c), since one state is occupied and the other empty. This would be the behavior of the dielectric function of a material in which only interband transitions are allowed, that is, a material with a band gap.

From the preceding derivations we conclude that the dielectric function of a semiconductor or insulator will derive from interband contributions only, as given by Eqs. (5.54) and (5.55) while that of a metal with several bands will have interband contributions as well as intraband contributions, described by Eqs. (5.51) and (5.52) (see also Ref. [61]). A typical example for a multi-band d-electron metal, Ag, is shown in Fig. 5.4. For more details the reader is referred to the review articles and books mentioned in the Further reading section. For semiconductors, it is often easy to identify the features of the band structure which are responsible for the major features of the dielectric function. The latter are typically related to transitions between occupied and unoccupied bands which happen to be parallel, that is, they have a constant energy difference over a large portion of the BZ, because this produces a large joint density of states (see problem 8).

5.5 Excitons

Up to this point we have been discussing excitations of electrons from an occupied to an unoccupied state. We have developed the tools to calculate the response of solids to this kind of external perturbation, which can apply to any situation. Often we are interested in applying these tools to situations where there is a gap between occupied and unoccupied states, as in semiconductors and insulators, in which case the difference in energy between the initial and final states must be larger than or equal to the band gap. This implies a lower cutoff in the energy of the photons that

Table 5.2. *Examples of Frenkel excitons in ionic solids and Mott-Wannier excitons in semiconductors.*

Binding energies of the excitons are given in units of electronvolts

Frenkel excitons		Mott–Wannier excitons	
KI	0.48	Si	0.015
KCl	0.40	Ge	0.004
KBr	0.40	GaAs	0.004
RbCl	0.44	CdS	0.029
LiF	1.00	CdSe	0.015

Source: C. Kittel [63].

can be absorbed or emitted, equal to the band gap energy. However, there are cases where optical excitations can occur for energies smaller than the band gap, because they involve the creation of an electron-hole pair in which the electron and hole are bound by their Coulomb attraction. These are called excitons.

There are two types of excitons. The first consists of an electron and a hole that are tightly bound with a large binding energy of order 1 eV. This is common in insulators, such as ionic solids (for example, SiO_2). These excitons are referred to as Frenkel excitons [62]. The second type consists of weakly bound excitons, delocalized over a range of several angstroms, and with a small binding energy of order 0.001–0.01 eV. This is common in small band gap systems, especially semiconductors. These excitons are referred to as Mott–Wannier excitons [63]. These two limiting cases are well understood, while intermediate cases are more difficult to treat. In Table 5.2 we give some examples of materials that have Frenkel and Mott–Wannier excitons and the corresponding binding energies.

The presence of excitons is a genuine many-body effect, so we need to invoke the many-body hamiltonian to describe the physics. For the purposes of the following treatment, we will find it convenient to cast the problem in the language of Slater determinants composed of single-particle states, so that solving it becomes an exercise in dealing with single-particle wavefunctions. We begin by writing the total many-body hamiltonian $\mathcal{H}$ in the form

$$\mathcal{H}(\{\mathbf{r}_i\}) = \sum_i h(\mathbf{r}_i) + \frac{1}{2}\sum_{i \neq j} \frac{e^2}{\mid \mathbf{r}_i - \mathbf{r}_j \mid},$$
$$h(\mathbf{r}) = -\frac{\hbar^2}{2m_e}\nabla^2_{\mathbf{r}} + \sum_{\mathbf{R},I} \frac{-Z_I e^2}{\mid \mathbf{r} - \mathbf{R} - \mathbf{t}_I \mid} \quad (5.56)$$

where $\mathbf{R}$ are the lattice vectors and $\mathbf{t}_I$ are the positions of the ions in the unit cell.

In this manner, we separate the part that can be dealt with strictly in the single-particle framework, namely $h(\mathbf{r})$, which is the non-interacting part, and the part that contains all the complications of electron-electron interactions. For simplicity, in the following we will discuss the case where there is only one atom per unit cell, thus eliminating the index I for the positions of atoms within the unit cell. The many-body wavefunction will have Bloch-like symmetry:

$$\Psi_{\mathbf{K}}(\mathbf{r}_1 + \mathbf{R}, \mathbf{r}_2 + \mathbf{R}, \ldots) = \mathrm{e}^{\mathrm{i}\mathbf{K}\cdot\mathbf{R}} \Psi_{\mathbf{K}}(\mathbf{r}_1, \mathbf{r}_2, \ldots) \tag{5.57}$$

We will discuss in some detail only the case of Frenkel excitons. We will assume that we are dealing with atoms that have two valence electrons, giving rise to a simple band structure consisting of a fully occupied valence band and an empty conduction band; generalization to more bands is straightforward. The many-body wavefunction will be taken as a Slater determinant in the positions $\mathbf{r}_1, \mathbf{r}_2, \ldots$ and spins $s_1, s_2, \ldots$ of the electrons: there are N atoms in the solid, hence $2N$ electrons in the full valence band, with one spin-up and one spin-down electron in each state, giving the many-body wavefunction

$$\Psi_{\mathbf{K}}(\mathbf{r}_1, s_1, \mathbf{r}_2, s_2, \ldots) = \frac{1}{\sqrt{(2N)!}} \begin{vmatrix} \psi^{(v)}_{\mathbf{k}_1\uparrow}(\mathbf{r}_1) & \psi^{(v)}_{\mathbf{k}_1\uparrow}(\mathbf{r}_2) & \cdots & \psi^{(v)}_{\mathbf{k}_1\uparrow}(\mathbf{r}_{2N}) \\ \psi^{(v)}_{\mathbf{k}_1\downarrow}(\mathbf{r}_1) & \psi^{(v)}_{\mathbf{k}_1\downarrow}(\mathbf{r}_2) & \cdots & \psi^{(v)}_{\mathbf{k}_1\downarrow}(\mathbf{r}_{2N}) \\ \cdot & \cdot & & \cdot \\ \cdot & \cdot & & \cdot \\ \cdot & \cdot & & \cdot \\ \psi^{(v)}_{\mathbf{k}_N\uparrow}(\mathbf{r}_1) & \psi^{(v)}_{\mathbf{k}_N\uparrow}(\mathbf{r}_2) & \cdots & \psi^{(v)}_{\mathbf{k}_N\uparrow}(\mathbf{r}_{2N}) \\ \psi^{(v)}_{\mathbf{k}_N\downarrow}(\mathbf{r}_1) & \psi^{(v)}_{\mathbf{k}_N\downarrow}(\mathbf{r}_2) & \cdots & \psi^{(v)}_{\mathbf{k}_N\downarrow}(\mathbf{r}_{2N}) \end{vmatrix} \tag{5.58}$$

In the ground state, all the states corresponding to $\mathbf{k}$ values in the first BZ will be occupied, and the total wave-vector will be equal to zero, since for every occupied $\mathbf{k}$-state there is a corresponding $-\mathbf{k}$-state with the same energy. Similarly the total spin will be zero, because of the equal occupation of single-particle states with up and down spins.

For localized Frenkel excitons, it is convenient to use a unitary transformation to a new set of basis functions which are localized at the positions of the ions in each unit cell of the lattice. These so called Wannier functions are defined in terms of the usual band states $\psi^{(v)}_{\mathbf{k},s}(\mathbf{r})$ through the following relation:

$$\phi^{(v)}_s(\mathbf{r} - \mathbf{R}) = \frac{1}{\sqrt{N}} \sum_{\mathbf{k}} \mathrm{e}^{-\mathrm{i}\mathbf{k}\cdot\mathbf{R}} \psi^{(v)}_{\mathbf{k},s}(\mathbf{r}) \tag{5.59}$$

Using this new basis we can express the many-body wavefunction as

$$\Psi_0(\mathbf{r}_1, s_1, \mathbf{r}_2, s_2, ...) = \frac{1}{\sqrt{(2N)!}} \begin{vmatrix} \phi_\uparrow^{(v)}(\mathbf{r}_1 - \mathbf{R}_1) & \cdots & \phi_\uparrow^{(v)}(\mathbf{r}_{2N} - \mathbf{R}_1) \\ \phi_\downarrow^{(v)}(\mathbf{r}_1 - \mathbf{R}_1) & \cdots & \phi_\downarrow^{(v)}(\mathbf{r}_{2N} - \mathbf{R}_1) \\ \cdot & & \cdot \\ \cdot & & \cdot \\ \cdot & & \cdot \\ \phi_\uparrow^{(v)}(\mathbf{r}_1 - \mathbf{R}_N) & \cdots & \phi_\uparrow^{(v)}(\mathbf{r}_{2N} - \mathbf{R}_N) \\ \phi_\downarrow^{(v)}(\mathbf{r}_1 - \mathbf{R}_N) & \cdots & \phi_\downarrow^{(v)}(\mathbf{r}_{2N} - \mathbf{R}_N) \end{vmatrix} \quad (5.60)$$

In order to create an exciton wavefunction we remove a single electron from state $\phi_{s_h}^{(v)}(\mathbf{r} - \mathbf{R}_h)$ (the subscript h standing for "hole") and put it in state $\phi_{s_p}^{(c)}(\mathbf{r} - \mathbf{R}_p)$ (the subscript p standing for "particle"). This is the expected excitation, from an occupied valence to an unoccupied conduction state, by the absorption of a photon as discussed earlier in this chapter. When this is done, the total momentum and total spin of the many-body wavefunction must be preserved, because there are no terms in the interaction hamiltonian to change these values (we are assuming as before that the wave-vector of the incident radiation is negligible compared with the wave-vectors of electrons).

Because of the difference in the nature of holes and particles, we need to pay special attention to the possible spin states of the entire system. When the electron is removed from state $\phi_{s_h}^{(v)}(\mathbf{r} - \mathbf{R}_h)$, the many-body state has a total spin z-component $S_z = -s_h$, since the original ground state had spin 0. Therefore, the new state created by adding a particle in state $\phi_{s_p}^{(c)}(\mathbf{r} - \mathbf{R}_p)$ produces a many-body state with total spin z-component $S_z = s_p - s_h$. This reveals that when we deal with hole states, we must take their contribution to the spin as the opposite of what a normal particle would contribute. Taking into consideration the fact that the hole has opposite wave-vector of a particle in the same state, we conclude that the hole corresponds to the time-reversed particle state, since the effect of the time-reversal operator $\mathcal{T}$ on the energy and the wavefunction is:

$$\mathcal{T}\epsilon_{\mathbf{k}\uparrow}^{(n)} = \epsilon_{-\mathbf{k}\downarrow}^{(n)}, \quad \mathcal{T}|\psi_{\mathbf{k}\uparrow}^{(n)}\rangle = |\psi_{-\mathbf{k}\downarrow}^{(n)}\rangle$$

as discussed in chapter 3. Notice that for the particle–hole system, if we start as usual with the highest S_z state, which in this case is $\uparrow\downarrow$, and proceed to create the rest by applying spin-lowering operators (see Appendix B), there will be an overall minus sign associated with the hole spin-lowering operator due to complex conjugation implied by time reversal. The resulting spin states for the particle–hole system, as identified by the total spin S and its z-component S_z, are given in Table 5.3 and contrasted against the particle–particle spin states.

Table 5.3. *Spin configurations for a particle–particle pair and a particle–hole pair for spin-1/2 particles.*

In the latter case the first spin refers to the particle, the second to the hole.

Spin state	Particle–particle	Particle–hole
$S = 1, S_z = 1$	$\uparrow\uparrow$	$\uparrow\downarrow$
$S = 1, S_z = 0$	$\frac{1}{\sqrt{2}}(\uparrow\downarrow + \downarrow\uparrow)$	$\frac{1}{\sqrt{2}}(\uparrow\uparrow - \downarrow\downarrow)$
$S = 1, S_z = -1$	$\downarrow\downarrow$	$\downarrow\uparrow$
$S = 0, S_z = 0$	$\frac{1}{\sqrt{2}}(\uparrow\downarrow - \downarrow\uparrow)$	$\frac{1}{\sqrt{2}}(\uparrow\uparrow + \downarrow\downarrow)$

Now the first task is to construct Bloch states from the proper basis. Let us denote by

$$\Phi^{(S,S_z)}([\mathbf{r}_p, s_p; \mathbf{r}_h, s_h], \mathbf{r}_2, s_2, \ldots, \mathbf{r}_{2N}, s_{2N})$$

the many-body wavefunction produced by exciting one particle from $\phi^{(v)}_{s_h}(\mathbf{r}_h)$ to $\phi^{(c)}_{s_p}(\mathbf{r}_p)$. This many-body state has total spin and z-projection (S, S_z), produced by the combination of s_p, s_h, in the manner discussed above. The real-space variable associated with the particle and the hole states is the same, but we use different symbols ($\mathbf{r}_p$ and $\mathbf{r}_h$) to denote the two different states associated with this excitation. Then the Bloch state obtained by appropriately combining such states is

$$\Psi^{(S,S_z)}_{\mathbf{K}}(\mathbf{r}_1, s_1, \ldots, \mathbf{r}_{2N}, s_{2N}) = \frac{1}{\sqrt{N}} \sum_{\mathbf{R},\mathbf{R}'} F_{\mathbf{R}}(\mathbf{R}') \mathrm{e}^{\mathrm{i}\mathbf{K}\cdot\mathbf{R}} \Phi^{(S,S_z)}_{\mathbf{R},\mathbf{R}'}(\mathbf{r}_1, s_1, \ldots, \mathbf{r}_{2N}, s_{2N}) \tag{5.61}$$

where the wavefunction $\Phi^{(S,S_z)}_{\mathbf{R},\mathbf{R}'}(\mathbf{r}_1, s_1, \ldots, \mathbf{r}_{2N}, s_{2N})$ is defined by the following relation:

$$\begin{aligned} &\Phi^{(S,S_z)}_{\mathbf{R},\mathbf{R}'}(\mathbf{r}_1, s_1, \ldots, \mathbf{r}_{2N}, s_{2N}) \\ &\quad = \Phi^{(S,S_z)}([\mathbf{r}_p - \mathbf{R}, s_p; \mathbf{r}_h - \mathbf{R} - \mathbf{R}', s_h], \ldots, \mathbf{r}_{2N} - \mathbf{R}, s_{2N}) \end{aligned} \tag{5.62}$$

that is, all particle variables in this many-body wavefunction, which is multiplied by the phase factor $\exp(i\mathbf{K}\cdot\mathbf{R})$, are shifted by $\mathbf{R}$ but the hole variable is shifted by any other lattice vector $\mathbf{R}'' = \mathbf{R} + \mathbf{R}'$, since it is not explicitly involved in the new many-body wavefunction. We need to invoke a set of coefficients $F_{\mathbf{R}}(\mathbf{R}')$ for this possible difference in shifts, as implemented in Eq. (5.61). The values of these coefficients will determine the wavefunction for this excited many-body state.

Let us consider a simple example, in which $F_{\mathbf{R}}(\mathbf{R}') = \delta(\mathbf{R} - \mathbf{R}')$. The physical meaning of this choice of coefficients is that only when the electron and the hole are localized at the same lattice site that the wavefunction does not vanish, which represents the extreme case of a localized Frenkel exciton. The energy of the state corresponding to this choice of coefficients is

$$E_{\mathbf{K}}^{(S,S_z)} = \langle \Psi_{\mathbf{K}}^{(S,S_z)} | \mathcal{H} | \Psi_{\mathbf{K}}^{(S,S_z)} \rangle = \frac{1}{N} \sum_{\mathbf{R},\mathbf{R}'} \mathrm{e}^{\mathrm{i}\mathbf{K}\cdot(\mathbf{R}-\mathbf{R}')} \langle \Phi_{\mathbf{R}',\mathbf{R}'}^{(S,S_z)} | \mathcal{H} | \Phi_{\mathbf{R},\mathbf{R}}^{(S,S_z)} \rangle \tag{5.63}$$

Now we can define the last expectation value as $E_{\mathbf{R}',\mathbf{R}}^{(S,S_z)}$, and obtain for the energy:

$$\begin{aligned} E_{\mathbf{K}}^{(S,S_z)} &= \frac{1}{N} \sum_{\mathbf{R},\mathbf{R}'} \mathrm{e}^{\mathrm{i}\mathbf{K}\cdot(\mathbf{R}-\mathbf{R}')} E_{\mathbf{R}',\mathbf{R}}^{(S,S_z)} \\ &= \frac{1}{N} \sum_{\mathbf{R}} E_{\mathbf{R},\mathbf{R}}^{(S,S_z)} + \frac{1}{N} \sum_{\mathbf{R}} \sum_{\mathbf{R}' \neq \mathbf{R}} \mathrm{e}^{\mathrm{i}\mathbf{K}\cdot(\mathbf{R}-\mathbf{R}')} E_{\mathbf{R}',\mathbf{R}}^{(S,S_z)} \\ &= E_{0,0}^{(S,S_z)} + \sum_{\mathbf{R}\neq 0} \mathrm{e}^{-\mathrm{i}\mathbf{K}\cdot\mathbf{R}} E_{\mathbf{R},0}^{(S,S_z)} \end{aligned} \tag{5.64}$$

where we have taken advantage of the translational symmetry of the hamiltonian to write $E_{\mathbf{R}',\mathbf{R}}^{(S,S_z)} = E_{\mathbf{R}'-\mathbf{R},0}^{(S,S_z)}$; we have also eliminated summations over $\mathbf{R}$ when the summand does not depend explicitly on $\mathbf{R}$, together with a factor of $(1/N)$.

We can express the quantities $E_{0,0}^{(S,S_z)}$, $E_{\mathbf{R},0}^{(S,S_z)}$ in terms of the single-particle states $\phi^{(v)}(\mathbf{r})$ and $\phi^{(c)}(\mathbf{r})$ as follows:

$$\begin{aligned} E_{0,0}^{(S,S_z)} &= E_0 + \epsilon^{(c)} - \epsilon^{(v)} + V^{(c)} - V^{(v)} + U^{(S)} \\ E_0 &= \langle \Psi_0 | \mathcal{H} | \Psi_0 \rangle \\ \epsilon^{(c)} &= \langle \phi^{(c)} | h | \phi^{(c)} \rangle \\ \epsilon^{(v)} &= \langle \phi^{(v)} | h | \phi^{(v)} \rangle \\ V^{(c)} &= \sum_{\mathbf{R}} \left[2\langle \phi^{(c)} \phi_{\mathbf{R}}^{(v)} | \frac{e^2}{\mid \mathbf{r}-\mathbf{r}' \mid} | \phi^{(c)} \phi_{\mathbf{R}}^{(v)} \rangle - \langle \phi^{(c)} \phi_{\mathbf{R}}^{(v)} | \frac{e^2}{\mid \mathbf{r}-\mathbf{r}' \mid} | \phi_{\mathbf{R}}^{(v)} \phi^{(c)} \rangle \right] \\ V^{(v)} &= \sum_{\mathbf{R}} \left[2\langle \phi^{(v)} \phi_{\mathbf{R}}^{(v)} | \frac{e^2}{\mid \mathbf{r}-\mathbf{r}' \mid} | \phi^{(v)} \phi_{\mathbf{R}}^{(v)} \rangle - \langle \phi^{(v)} \phi_{\mathbf{R}}^{(v)} | \frac{e^2}{\mid \mathbf{r}-\mathbf{r}' \mid} | \phi_{\mathbf{R}}^{(v)} \phi^{(v)} \rangle \right] \\ U^{(S)} &= 2\delta_{S,0} \langle \phi^{(c)} \phi^{(v)} | \frac{e^2}{\mid \mathbf{r}-\mathbf{r}' \mid} | \phi^{(v)} \phi^{(c)} \rangle - \langle \phi^{(c)} \phi^{(v)} | \frac{e^2}{\mid \mathbf{r}-\mathbf{r}' \mid} | \phi^{(c)} \phi^{(v)} \rangle \end{aligned} \tag{5.65}$$

where we have used the short-hand notation $\langle \mathbf{r} | \phi_{\mathbf{R}}^{(n)} \rangle = \phi^{(n)}(\mathbf{r} - \mathbf{R})$ with $n = v$ or c for the valence or conduction states; the states with no subscript correspond to $\mathbf{R} = 0$. In the above equations we do not include spin labels in the single-particle states $\phi^{(n)}(\mathbf{r})$ since the spin degrees of freedom have explicitly been taken into account to arrive at these expressions. $\mathcal{H}(\{\mathbf{r}_i\})$, $h(\mathbf{r})$ are the many-body and single-particle hamiltonians defined in Eq. (5.56). The interpretation of these terms is straightforward: the energy of the system with the exciton is given, relative to the energy of the ground state E_0, by adding the particle energy $\epsilon^{(c)}$, subtracting the hole energy $\epsilon^{(v)}$ and taking into account all the Coulomb interactions, consisting of the interaction energy of the particle $V^{(c)}$, which is added, the interaction energy of the hole $V^{(v)}$, which is subtracted, and the particle–hole interaction $U^{(S)}$, which depends on the total spin S. The interaction terms are at the Hartree–Fock level, including the direct and exchange contribution, which is a natural consequence of our choice of a Slater determinant for the many-body wavefunction, Eq. (5.60). With the same conventions, the last term appearing under the summation over $\mathbf{R}$ in Eq. (5.64) takes the form:

$$\begin{aligned} E_{\mathbf{R},0}^{(S,S_z)} &= \langle \Phi_{\mathbf{R},\mathbf{R}}^{(S,S_z)} | \mathcal{H} | \Phi_{0,0}^{(S,S_z)} \rangle \\ &= 2\delta_{S,0} \langle \phi_{\mathbf{R}}^{(v)} \phi^{(c)} | \frac{e^2}{\mid \mathbf{r} - \mathbf{r}' \mid} | \phi_{\mathbf{R}}^{(c)} \phi^{(v)} \rangle - \langle \phi_{\mathbf{R}}^{(v)} \phi^{(c)} | \frac{e^2}{\mid \mathbf{r} - \mathbf{r}' \mid} | \phi^{(v)} \phi_{\mathbf{R}}^{(c)} \rangle \end{aligned}$$

This term describes an interaction between the particle and hole states, in addition to their local Coulomb interaction $U^{(S)}$, which arises from the periodicity of the system.

How do these results change the interaction of the solid with light? The absorption probability will be given as before by matrix elements of the interaction hamiltonian $\mathcal{H}^{int}$ with eigenfunctions of the many-body system:

$$\begin{aligned} P(\omega) &= \frac{2\pi}{\hbar} \left| \langle \Psi_0 | \mathcal{H}^{int} | \Psi_{\mathbf{K}} \rangle \right|^2 \delta(E_{\mathbf{K}}^{(S,S_z)} - E_0 - \hbar\omega) \\ &= \frac{2\pi}{\hbar} \left(\frac{e}{m_e c} \right)^2 \left| \langle \phi^{(v)} | \mathbf{A}_0 \cdot \mathbf{p} | \phi^{(c)} \rangle \right|^2 \delta_{S,0} \delta(\mathbf{K}) \delta(E_{\mathbf{K}}^{(S,S_z)} - E_0 - \hbar\omega) \end{aligned}$$

where we have used the appropriate generalization of the interaction hamiltonian to a many-body system,

$$\mathcal{H}^{int}(\{\mathbf{r}_i\}, t) = \sum_i \frac{e}{m_e c} \left[\mathrm{e}^{\mathrm{i}(\mathbf{q}\cdot\mathbf{r}_i - \omega t)} \mathbf{A}_0 \cdot \mathbf{p}_i + \mathrm{c.c.} \right] \tag{5.66}$$

with the summation running over all electron coordinates. The total-spin conserving δ-function is introduced because there are no terms in the interaction hamiltonian which can change the spin, and the wave-vector conserving δ-function is introduced by arguments similar to what we discussed earlier for the conservation of $\mathbf{k}$ upon absorption or emission of optical photons. In this expression we can now use the results from above, giving for the argument of the energy-conserving δ-function,

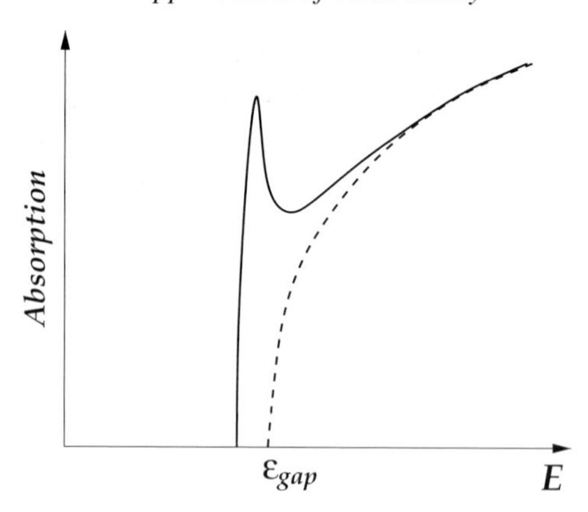

Figure 5.5. Modification of the absorption spectrum in the presence of excitons (solid line) relative to the spectrum in the absence of excitons (dashed line); in the latter case absorption begins at exactly the value of the band gap, ϵ_{gap}.

after we take into account that only terms with $\mathbf{K} = 0$, $S = S_z = 0$ survive:

$$\hbar\omega = E_0^{(0,0)} - E_0 = [\epsilon^{(c)} + V^{(c)}] - [\epsilon^{(v)} + V^{(v)}] + U^{(0)} + \sum_{\mathbf{R}\neq 0} E_{0,\mathbf{R}}^{(0,0)} \tag{5.67}$$

Taking into account the physical origin of the various terms as described above, we conclude that the two terms in square brackets on the right-hand side of Eq. (5.67) will give the band gap,

$$\epsilon_{gap} = \left[\epsilon^{(c)} + V^{(c)}\right] - \left[\epsilon^{(v)} + V^{(v)}\right]$$

Notice that the energies $\epsilon^{(c)}$, $\epsilon^{(v)}$ are eigenvalues of the single-particle hamiltonian $h(\mathbf{r})$ which does not include electron–electron interactions, and therefore their difference is not equal to the band gap; we must also include the contributions from electron–electron interactions in these single-particle energies, which are represented by the terms $V^{(c)}$, $V^{(v)}$ at the Hartree–Fock level, to obtain the proper quasiparticle energies whose difference equals the band gap. The last two terms in Eq. (5.67) give the particle–hole interaction, which is overall negative (attractive between two oppositely charged particles). This implies that there will be absorption of light for photon energies *smaller* than the band gap energy, as expected for an electron–hole pair that is bound by the Coulomb interaction. In our simple model with one occupied and one empty band, the absorption spectrum will have an extra peak at energies just below the band gap energy, corresponding to the presence of excitons, as illustrated in Fig. 5.5. In more realistic situations there will be several exciton peaks reflecting the band structure features of valence and conduction bands.

For Mott–Wannier excitons, the treatment is similar, only instead of the localized Wannier functions $\phi_{s_h}^{(v)}(\mathbf{r} - \mathbf{R}_h)$, $\phi_{s_p}^{(c)}(\mathbf{r} - \mathbf{R}_p)$ for the single-particle states we will need to use the Bloch states $\psi_{\mathbf{k}_h,s_h}^{(v)}(\mathbf{r})$, $\psi_{\mathbf{k}_p,s_p}^{(c)}(\mathbf{r})$, in terms of which we must express the many-body wavefunction. In this case, momentum conservation implies $\mathbf{k}_p - \mathbf{k}_h = \mathbf{K}$, so that the total momentum $\mathbf{K}$ is unchanged. This condition is implemented by choosing $\mathbf{k}_p = \mathbf{k} + \mathbf{K}/2$, $\mathbf{k}_h = \mathbf{k} - \mathbf{K}/2$ and allowing $\mathbf{k}$ to take all possible values in the BZ. Accordingly, we have to construct a many-body wavefunction which is a summation over all such possible states. The energy corresponding to this wavefunction can be determined by steps analogous to the discussion for the Frenkel excitons. The energy of these excitons exhibits dispersion just like electron states, and lies within the band gap of the crystal.

5.6 Energetics and dynamics

In the final section of the present chapter we discuss an application of band theory which is becoming a dominant component in the field of atomic and electronic structure of solids: it is the calculation of the total energy of the solid as a function of the arrangement of the atoms. The ability to obtain accurate values for the total energy as a function of atomic configuration is crucial in explaining a number of thermodynamic and mechanical properties of solids. For example,

- phase transitions as a function of pressure can be predicted if the total energy of different phases is known as a function of volume;
- alloy phase diagrams as a function of temperature and composition can be constructed by calculating the total energy of various structures with different types of elements;
- the relative stability of competing surface structures can be determined through their total energies, and from those one can predict the shape of solids;
- the dynamics of atoms in the interior and on the surface of a solid can be described by calculating the total energy of relevant configurations, which can elucidate complex phenomena like bulk diffusion and surface growth;
- the energetics of extended deformations, like shear or cleavage of a solid, are crucial in understanding its mechanical response, such as brittle or ductile behavior;
- the properties of defects of dimensionality zero, one and two, can be elucidated through total-energy calculations of model structures, which in turn provides insight into complex phenomena like fracture, catalysis, corrosion, adhesion, etc.

Calculations of this type have proliferated since the early 1980s, providing a wealth of useful information on the behavior of real materials. We will touch upon some topics where such calculations have proven particularly useful in chapters 9–11 of this book. It is impossible to provide a comprehensive review of such applications here, which are being expanded and refined at a very rapid pace by many practitioners worldwide. The contributions of M.L. Cohen, who pioneered this type of application and produced a large number of scientific descendants responsible for extensions of this field in many directions, deserve special mention. For a glimpse of the range of possible applications, the reader may consult the review article of Chelikowsky and Cohen [65].

5.6.1 The total energy

We will describe the calculation of the total energy and its relation to the band structure in the framework of Density Functional Theory (DFT in the following); see chapter 2. The reason is that this formulation has proven the most successful compromise between accuracy and efficiency for total-energy calculations in a very wide range of solids. In the following we will also adopt the pseudopotential method for describing the ionic cores, which, as we discussed in chapter 2, allows for an efficient treatment of the valence electrons only. Furthermore, we will assume that the ionic pseudopotentials are the same for all electronic states in the atom, an approximation known as the "local pseudopotential"; this will help keep the discussion simple. In realistic calculations the pseudopotential typically depends on the angular momentum of the atomic state it represents, which is known as a "non-local pseudopotential".

The total energy of a solid for a particular configuration of the ions $E^{tot}(\{\mathbf{R}\})$ is given by

$$E^{tot} = T + U^{ion-el} + U^{el-el} + U^{ion-ion} \tag{5.68}$$

where T is the kinetic energy of electrons, U^{ion-el} is the energy due to the ion–electron attractive interaction, U^{el-el} is the energy due to the electron–electron interaction including the Coulomb repulsion and exchange and correlation effects, and $U^{ion-ion}$ is the energy due to the ion–ion repulsive interaction. This last term is the Madelung energy:

$$U^{ion-ion} = \frac{1}{2}\sum_{I \neq J} \frac{Z_I Z_J e^2}{| \mathbf{R}_I \mathbf{R}_J |} \tag{5.69}$$

with Z_I the valence charge of ion at position $\mathbf{R}_I$, which can be calculated by the Ewald method, as discussed in Appendix F. Using the theory described in chapter 2, the rest of the terms take the form

$$T = \sum_{\mathbf{k}} \langle \psi_{\mathbf{k}} | -\frac{\hbar^2 \nabla_{\mathbf{r}}^2}{2m_e} | \psi_{\mathbf{k}} \rangle \tag{5.70}$$

$$U^{ion-el} = \sum_{\mathbf{k}} \langle \psi_{\mathbf{k}} | V^{ps}(\mathbf{r}) | \psi_{\mathbf{k}} \rangle \tag{5.71}$$

$$U^{el-el} = \frac{1}{2}\sum_{\mathbf{k}\mathbf{k}'} \langle \psi_{\mathbf{k}} \psi_{\mathbf{k}'} | \frac{e^2}{| \mathbf{r} - \mathbf{r}' |} | \psi_{\mathbf{k}} \psi_{\mathbf{k}'} \rangle + E^{XC}[n(\mathbf{r})] \tag{5.72}$$

where $| \psi_{\mathbf{k}} \rangle$ are the single-particle states obtained from a self-consistent solution of the set of single-particle Schrödinger equations and $n(\mathbf{r})$ is the electron density; for simplicity, we use a single index $\mathbf{k}$ to identify these single-particle wavefunctions, with the understanding that it encompasses both the wave-vector and the band index. In terms of these states, the density is given by

$$n(\mathbf{r}) = \sum_{\mathbf{k}(\epsilon_{\mathbf{k}}<\epsilon_{\mathrm{F}})} |\psi_{\mathbf{k}}(\mathbf{r})|^2$$

with the summation running over all occupied states with energy $\epsilon_{\mathbf{k}}$ below the Fermi level ϵ_{F}. $V^{ps}(\mathbf{r})$ is the external potential that each valence electron in the solid experiences due to the presence of ions described by pseudopotentials, and $E^{XC}[n(\mathbf{r})]$ is the exchange and correlation contribution to the total energy, which, in the framework of DFT, depends on the electron density only. We will adopt the local density approximation (LDA in the following) for the exchange correlation functional in terms of the electron density:

$$E^{XC}[n(\mathbf{r})] = \int \epsilon^{XC}[n(\mathbf{r})]n(\mathbf{r})\mathrm{d}\mathbf{r} \tag{5.73}$$

with $\epsilon^{XC}[n]$ the local function of the density that accounts for exchange and correlation effects (see also chapter 2). The potential due to exchange and correlation effects that appears in the single-particle equations is then

$$V^{XC}(\mathbf{r}) = \frac{\partial}{\partial n(\mathbf{r})}\left[\epsilon^{XC}[n(\mathbf{r})]n(\mathbf{r})\right] = \epsilon^{XC}[n(\mathbf{r})] + \frac{\partial \epsilon^{XC}[n(\mathbf{r})]}{\partial n(\mathbf{r})}n(\mathbf{r}) \tag{5.74}$$

and the single-particle equations take the form

$$\left[-\frac{\hbar^2\nabla^2_{\mathbf{r}}}{2m_e} + V^{ps}(\mathbf{r}) + \int \frac{e^2 n(\mathbf{r}')}{|\mathbf{r}-\mathbf{r}'|}d\mathbf{r}' + V^{XC}(\mathbf{r})\right] | \psi_{\mathbf{k}}\rangle = \epsilon_{\mathbf{k}} | \psi_{\mathbf{k}}\rangle \tag{5.75}$$

Multiplying this equation by $\langle\psi_{\mathbf{k}} |$ from the left and summing over all occupied states, we obtain

$$\sum_{\mathbf{k}(\epsilon_{\mathbf{k}}<\epsilon_{\mathrm{F}})}\left[\langle\psi_{\mathbf{k}} | -\frac{\hbar^2\nabla^2_{\mathbf{r}}}{2m_e} | \psi_{\mathbf{k}}\rangle + \langle\psi_{\mathbf{k}} | V^{ps}(\mathbf{r}) | \psi_{\mathbf{k}}\rangle\right]$$
$$+\int\int \frac{e^2 n(\mathbf{r})n(\mathbf{r}')}{|\mathbf{r}-\mathbf{r}'|}d\mathbf{r}d\mathbf{r}' + \int V^{XC}(\mathbf{r})n(\mathbf{r})d\mathbf{r} = \sum_{\mathbf{k}(\epsilon_{\mathbf{k}}<\epsilon_{\mathrm{F}})} \epsilon_{\mathbf{k}} \tag{5.76}$$

By comparing this with our earlier expression for the total energy, Eqs. (5.68) and (5.72), we find

$$E^{tot} = \sum_{\mathbf{k}(\epsilon_{\mathbf{k}}<\epsilon_{\mathrm{F}})} \epsilon_{\mathbf{k}} - \frac{1}{2}\int V^{Coul}(\mathbf{r})n(\mathbf{r})\mathrm{d}\mathbf{r} - \int \Delta V^{XC}(\mathbf{r})n(\mathbf{r})\mathrm{d}\mathbf{r} + U^{ion-ion} \tag{5.77}$$

where we have defined the Coulomb potential as

$$V^{Coul}(\mathbf{r}) = \int \frac{e^2 n(\mathbf{r}')}{|\mathbf{r}-\mathbf{r}'|} d\mathbf{r}' \tag{5.78}$$

and the difference in exchange and correlation potential as

$$\Delta V^{XC}(\mathbf{r}) = V^{XC}(\mathbf{r}) - \epsilon^{XC}[n(\mathbf{r})] \tag{5.79}$$

For a periodic solid, the Fourier transforms of the density and the potentials provide a particularly convenient platform for evaluating the various terms of the total energy. We follow here the original formulation of the total energy in terms of the plane waves $\exp(\mathrm{i}\mathbf{G}\cdot\mathbf{r})$, defined by the reciprocal-space lattice vectors $\mathbf{G}$, as derived by Ihm, Zunger and Cohen [66]. The Fourier transforms of the density and potentials are given by

$$\begin{aligned} n(\mathbf{r}) &= \sum_{\mathbf{G}} e^{\mathrm{i}\mathbf{G}\cdot\mathbf{r}} n(\mathbf{G}) \\ V^{Coul}(\mathbf{r}) &= \sum_{\mathbf{G}} e^{\mathrm{i}\mathbf{G}\cdot\mathbf{r}} V^{Coul}(\mathbf{G}) \\ \Delta V^{XC}(\mathbf{r}) &= \sum_{\mathbf{G}} e^{\mathrm{i}\mathbf{G}\cdot\mathbf{r}} \Delta V^{XC}(\mathbf{G}) \end{aligned} \tag{5.80}$$

In terms of these Fourier transforms, we obtain for the exchange and correlation term in Eq. (5.77)

$$\int \Delta V^{XC}(\mathbf{r}) n(\mathbf{r}) d\mathbf{r} = \Omega \sum_{\mathbf{G}} \Delta V^{XC}(\mathbf{G}) n(\mathbf{G}) \tag{5.81}$$

with Ω the total volume of the solid. For the Coulomb term, the calculation is a bit more tricky. First, we can use the identity

$$\int \frac{e^{\mathrm{i}\mathbf{q}\cdot\mathbf{r}}}{r} d\mathbf{r} = \frac{4\pi}{|\mathbf{q}|^2} \tag{5.82}$$

and the relation between electrostatic potential and electrical charge, that is, the Poisson equation

$$\nabla^2_{\mathbf{r}} V^{Coul}(\mathbf{r}) = -4\pi e^2 n(\mathbf{r}) \Rightarrow V^{Coul}(\mathbf{G}) = \frac{4\pi e^2}{|\mathbf{G}|^2} n(\mathbf{G}) \tag{5.83}$$

to rewrite the Coulomb term in Eq. (5.77) as

$$\int V^{Coul}(\mathbf{r}) n(\mathbf{r}) d\mathbf{r} = \Omega \sum_{\mathbf{G}} V^{Coul}(\mathbf{G}) n(\mathbf{G}) = 4\pi e^2 \Omega \sum_{\mathbf{G}} \frac{[n(\mathbf{G})]^2}{|\mathbf{G}|^2} \tag{5.84}$$

Here we need to be careful with handling of the infinite terms. We note that in a solid which is overall neutral, the total positive charge of the ions is canceled by the total negative charge of the electrons. This means that the average potential is zero. The $\mathbf{G} = 0$ term in the Fourier transform corresponds to the average over all space, which implies that due to charge neutrality we can omit the $\mathbf{G} = 0$ term from all the Coulomb contributions. However, we have introduced an alteration of the physical system by representing the ions with pseudopotentials in the solid: the pseudopotential matches exactly the Coulomb potential beyond the cutoff radius, but deviates from it inside the core. When accounting for the infinite terms, we have to compensate for the alteration introduced by the pseudopotential. This is done by adding to the total energy the following term:

$$\Delta U^{ps} = \sum_I Z_I \int \left[V_I^{ps}(\mathbf{r}) + \frac{Z_I e^2}{r} \right] d\mathbf{r} \tag{5.85}$$

where the summation is over all the ions and $V_I^{ps}(\mathbf{r})$ is the pseudopotential corresponding to ion I; the integrand in this term does not vanish inside the core region, where the pseudopotential is different from the Coulomb potential. With these contributions, the total energy of the solid takes the form

$$E^{tot} = \sum_{\mathbf{k}(\epsilon_\mathbf{k} < \epsilon_F)} \epsilon_\mathbf{k} - 2\pi e^2 \Omega \sum_{\mathbf{G} \neq 0} \frac{[n(\mathbf{G})]^2}{|\mathbf{G}|^2} - \Omega \sum_{\mathbf{G}} \Delta V^{XC}(\mathbf{G}) n(\mathbf{G}) + \Delta U^{ps} + U^{ion-ion} \tag{5.86}$$

This final expression has an appealing form: the total energy is essentially the sum of all occupied single-particle states, the first term in Eq. (5.86), corrected for double-counting the Coulomb and exchange-correlation contributions, second and third terms in Eq. (5.86), and adjusted for the infinite terms due to Coulomb interactions between the various charges, fourth term in Eq. (5.86). The very last term is of course the classical contribution of the ions as point charges, the Madelung energy, which is treated explicitly in Appendix F.

This expression for the total energy has inspired the derivation of semi-empirical schemes, which involve much smaller computational cost than fully self-consistent methods like DFT. A standard approach is to use a tight-binding hamiltonian which, when properly parametrized, can provide an accurate estimate of the first term in the total energy, the sum of single-particle energies. This is actually the dominant term in the total energy. The three other terms are then viewed as corrections which can be approximated by an empirical term in the form of a classical interatomic potential that depends only on the distance between atoms. This "correction potential" is fitted to reproduce the exact energies of a few possible structures, and can then be used to calculate the energy of a variety of other structures. Although not as accurate as a fully self-consistent approach, this method is quite useful because it makes feasible the application of total-energy calculations to systems involving large numbers of atoms. In particular, this method captures the essential aspects of bonding in solids because it preserves the quantum mechanical nature of bond formation and destruction through the electronic states with energies $\epsilon_{\mathbf{k}}$. There is certainly a loss in accuracy due to the very approximate treatment of the terms other than the sum of the electronic eigenvalues; this can be tolerated for certain classes of problems, where the exact changes in the energy are not crucial, but a reasonable estimate of the energy differences associated with various processes is important.

As an example of the use of total-energy calculations in elucidating the properties of solids, we discuss the behavior of the energy as a function of volume for different phases of the same element. This type of analysis using DFT calculations was pioneered by Yin and Cohen [67]. The classic case is silicon which, as we have mentioned before, has as its ground state the diamond lattice and is a semiconductor (see chapter 1). The question we want to address is: what is the energy of other possible crystalline phases of silicon relative to its ground state? The answer to this question is shown in Fig. 5.6, where we present the energy per atom of the simple cubic lattice with six nearest neighbors for each atom, the FCC and BCC high-coordination lattices with 12 and eight nearest neighbors, respectively, a body-centered tetragonal (BCT) structure with five nearest neighbors [68], the so called beta-tin structure which has six nearest neighbors, and the ground state diamond lattice which has four nearest neighbors. We must compare the energy of the various structures as a function of the volume per atom, since we do not know *a priori* at which volume the different crystalline phases will have their minimum value. The minimum in each curve corresponds to a different volume, as the variations of coordination in the different phases suggest. At their minimum, all other phases have higher energy than the diamond lattice. This comparison shows that it would not be possible to form any of these phases under equilibrium conditions. However, if the volume per atom of the solid were somehow reduced, for example by applying pressure, the results of Fig. 5.6 indicate that the lowest energy structure would become the beta-tin phase. It turns out that when pressure is applied to silicon, the structure indeed becomes beta-tin. A Maxwell (common tangent) construction between the energy-versus-volume curves of the two lowest energy phases gives the critical pressure at which the diamond phase begins to transform to the beta-tin phase; the value of this critical pressure predicted from the curves in Fig. 5.6 is in excellent agreement with experimental measurements (for more details see Ref. [67]).

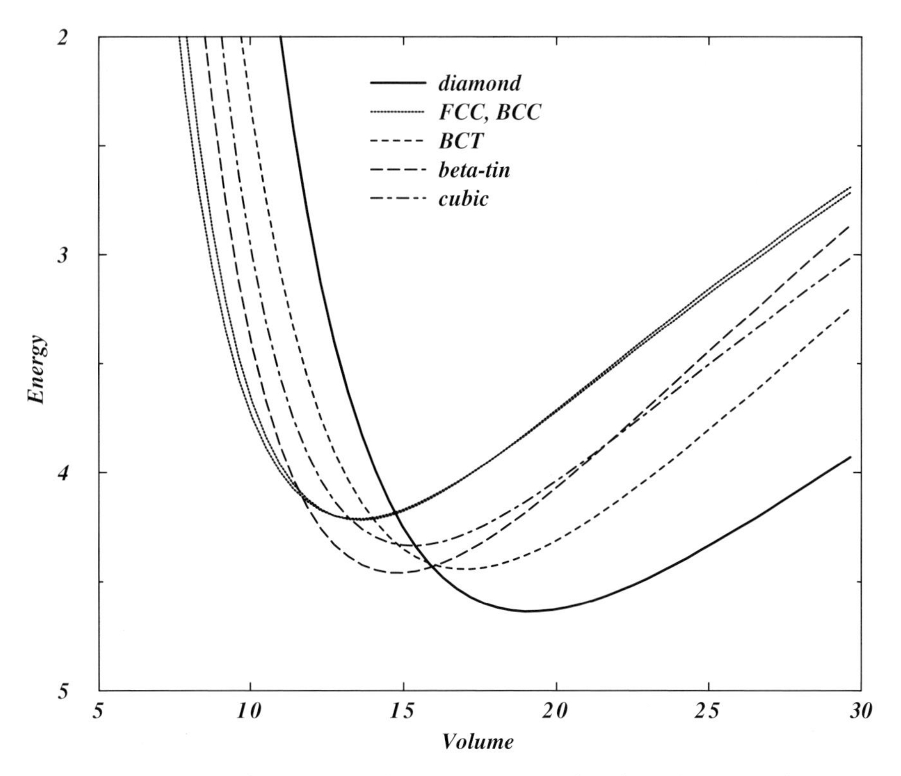

Figure 5.6. Total energy (in electronvolts per atom) as a function of volume (in angstoms cubed) for various crystal phases of silicon: diamond (a semiconducting phase with four nearest neighbors), BCC, FCC, BCT, beta-tin and cubic (all metallic phases with higher coordination). The zero of the energy scale corresponds to isolated silicon atoms.

As the example just discussed demonstrates, total-energy calculations allow the determination of the optimal lattice constant of any crystalline phase, its cohesive energy and its bulk modulus. Since these are all very basic properties of solids, we elaborate on this issue briefly. First, as is evident from the curves of Fig. 5.6, the behavior of the energy near the minimum of each curve is smooth and very nearly quadratic. A simple polynomial fit to the calculated energy as a function of volume near the minimum readily yields estimates for the volume, and hence the lattice constant, that corresponds to the lowest energy.[3] The bulk modulus B of a solid is defined as the change in the pressure P when the volume Ω changes, normalized by the volume (in the present calculation we are assuming zero temperature conditions):

$$B \equiv -\Omega \frac{\partial P}{\partial \Omega} \tag{5.87}$$

[3] The standard way to fit the total-energy curve near its minimum is by a polynomial in powers of $\Omega^{1/3}$, where Ω is the volume.

Using the relation between the pressure and the total energy, we obtain for the bulk modulus

$$P = -\frac{\partial E}{\partial \Omega} \Rightarrow B = \Omega \frac{\partial^2 E}{\partial \Omega^2} \tag{5.88}$$

which is easily calculated from the total energy versus volume curves, as the curvature at the minimum.

Finally, the cohesive energy of the solid is given by the difference between the total energy per atom at the minimum of the curve and the total energy of a free atom of the same element (for elemental solids). While this is a simple definition, it actually introduces significant errors because of certain subtleties: the calculation of the total energy of the free atom cannot be carried out within the framework developed earlier for solids, because in the case of the free atom it is not possible to define a real-space lattice and the corresponding reciprocal space of **G** vectors for expanding the wavefunctions and potentials, as was done for the solid. The calculation of the total energy of the free atom has to be carried out in a real-space approach, without involving reciprocal-space vectors and summations. The difference in computational methods for the solid and the free atom total energies introduces numerical errors which are not easy to eliminate. Even if both total-energy calculations were carried out with the same computational approach,[4] the underlying formalism does not give results of equal accuracy for the two cases. This is because the DFT/LDA formalism is well suited for solids, where the variations in the electron density are not too severe. In the case of a free atom the electron density goes to zero at a relatively short distance from the nucleus (the electronic wavefunctions decay exponentially), which presents a greater challenge to the formalism. Modifications of the formalism that include gradient corrections to the density functionals, referred to as Generalized Gradient Approximations (GGAs), can improve the situation considerably (see Ref. [27] and chapter 2).

In light of the above discussion, it is worthwhile mentioning two simple phenomenological theories, which predict with remarkable accuracy some important quantities related to the energetics of solids. The first, developed by M.L. Cohen [69] and based on the arguments of J.C. Phillips [70] about the origin of energy bands in semiconductors (see the Further reading section), deals with the bulk modulus of covalently bonded solids. The second, developed by Rose, Smith, Guinea and Ferrante [71], concerns the energy-versus-volume relation and asserts that, when scaled appropriately, it is a universal relation for all solids; this is referred to as the Universal Binding Energy Relation (UBER). We discuss these phenomenological theories in some detail, because they represent important tools, as they distil the results of elaborate calculations into simple, practical expressions.

[4] This can be done, for example, by placing an atom at the center of a large box and repeating this box periodically in 3D space, thus artificially creating a periodic solid that approximates the isolated atom.

Cohen's theory gives the bulk modulus of a solid which has average coordination $\langle N_c \rangle$, as:

$$B = \frac{\langle N_c \rangle}{4}(1971 - 220\lambda)d^{-3.5}\ \text{GPa} \tag{5.89}$$

where d is the bond-length in Å and λ is a dimensionless number which describes the ionicity: $\lambda = 0$ for the homopolar crystals C, Si, Ge, Sn; $\lambda = 1$ for III–V compounds (where the valence of each element differs by 1 from the average valence of 4); and $\lambda = 2$ for II–VI compounds (where the valence of each element differs by 2 from the average valence of 4). The basic physics behind this expression is that the bulk modulus is intimately related to the average electron density in a solid where the electrons are uniformly distributed, as was discussed in chapter 2, Problem 4. For covalently bonded solids, the electronic density is concentrated between nearest neighbor sites, forming the covalent bonds. Therefore, in this case the crucial connection is between the bulk modulus and the bond length, which provides a measure of the electron density relevant to bonding strength in the solid. Cohen's argument [69] produced the relation $B \sim d^{-3.5}$, which, dressed with the empirical constants that give good fits for a few representative solids, leads to Eq. (5.89). As can be checked from the values given in Table 5.5, actual values of bulk moduli are captured with an accuracy which is remarkable, given the simplicity of the expression in Eq. (5.89). In fact, this expression has been used to *predict* theoretically solids with very high bulk moduli [72], possibly exceeding those of naturally available hard materials such as diamond.

The UBER theory gives the energy-versus-volume relation as a two-parameter expression:

$$\begin{aligned} E(a) &= E^{coh}\overline{E}(\overline{a}) \\ \overline{E}(\overline{a}) &= -[1 + \overline{a} + 0.05\overline{a}^3]\exp(-\overline{a}) \\ \overline{a} &= (a - a_0)/l \end{aligned} \tag{5.90}$$

where a is the lattice constant (which has a very simple relation to the volume per atom, depending on the lattice), a_0 is its equilibrium value, l is a length scale and E^{coh} is the cohesive energy; the latter two quantities are the two parameters that are specific to each solid. This curve has a single minimum and goes asymptotically to zero for large distances; the value of its minimum corresponds to the cohesive energy. For example, we show in Fig. 5.7 the fit of this expression to DFT/GGA values for a number of representative elemental solids, including metals with s (Li, K, Ca) and sp (Al) electrons, sd electron metals with magnetic (Fe, Ni) or non-magnetic (Cu, Mo) ground states, and semiconductors (Si) and insulators (C). As seen from this example, the fit to this rather wide range of solids is quite satisfactory;

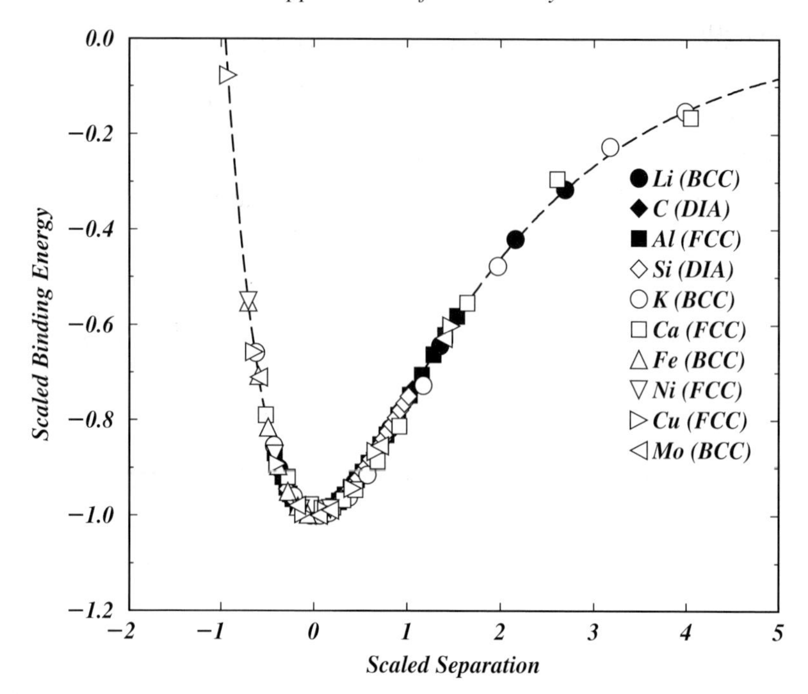

Figure 5.7. Fit of calculated total energies of various elemental solids to the Universal Binding Energy Relation; the calculations are based on DFT/GGA.

the values of the lattice constant and cohesive energy, a_0 and E^{coh}, extracted from this fit differ somewhat from the experimental values, but this is probably more related to limitations of the DFT/GGA formalism rather than the UBER theory. The degree to which this simple expression fits theoretically calculated values for a wide range of solids, as well as the energy of adhesion, molecular binding and chemisorption of gas molecules on surfaces of metals, is remarkable. The UBER expression is very useful for determining, with a few electronic structure calculations of the total energy, a number of important quantities such as the cohesive energy (or absorption, chemisorption, adhesion energies, when applied to surface phenomena), bulk modulus, critical stresses, etc.

5.6.2 Forces and dynamics

Having calculated the total energy of a solid, it should be feasible to calculate forces on the ions by simply taking the derivative of the total energy with respect to individual ionic positions. This type of calculation is indeed possible, and opens up a very broad field, that is, the study of the dynamics of atoms in the solid state. Here, by dynamics we refer to the behavior of ions as classical particles, obeying

Newton's equations of motion, with the calculated forces at each instant in time. We discuss next the details of the calculation of forces in the context of DFT. First, we note that there are two contributions, one from the ion–ion interaction energy, and one from the ion–electron interaction energy. For the contribution from the ion-ion interaction energy, the derivative of $U^{ion-ion}$, defined in Eq. (5.69), with respect to the position of a particular ion, $\mathbf{R}_I$, gives for the force on this ion

$$\mathbf{F}_I^{ion-ion} = -\frac{\partial}{\partial \mathbf{R}_I} U^{ion-ion} = \sum_{J \neq I} \frac{Z_I Z_J e^2}{\mid \mathbf{R}_I - \mathbf{R}_J \mid^3} (\mathbf{R}_I - \mathbf{R}_J) \qquad (5.91)$$

which can be evaluated by methods analogous to those used for the Madelung energy, discussed in Appendix F.

In order to calculate the contribution from the ion-electron interaction, we first prove a very useful general theorem, known as the Hellmann-Feynman theorem. This theorem states that the force on an atom I at position $\mathbf{R}_I$ is given by

$$\mathbf{F}_I^{ion-el} = -\langle \Psi_0 \mid \frac{\partial \mathcal{H}}{\partial \mathbf{R}_I} \mid \Psi_0 \rangle \qquad (5.92)$$

where $\mathcal{H}$ is the hamiltonian of the system and $\mid \Psi_0 \rangle$ is the normalized ground state wavefunction (we assume that the hamiltonian does not include ion-ion interactions, which have been taken into account explicitly with the $\mathbf{F}^{ion-ion}$ term). To prove this theorem, we begin with the standard definition of the force on ion I, as the derivative of the total energy $\langle \Psi_0 \mid \mathcal{H} \mid \Psi_0 \rangle$ with respect to $\mathbf{R}_I$:

$$\begin{aligned} \mathbf{F}_I^{ion-el} &= -\frac{\partial \langle \Psi_0 \mid \mathcal{H} \mid \Psi_0 \rangle}{\partial \mathbf{R}_I} \\ &= -\langle \frac{\partial \Psi_0}{\partial \mathbf{R}_I} \mid \mathcal{H} \mid \Psi_0 \rangle - \langle \Psi_0 \mid \frac{\partial \mathcal{H}}{\partial \mathbf{R}_I} \mid \Psi_0 \rangle - \langle \Psi_0 \mid \mathcal{H} \mid \frac{\partial \Psi_0}{\partial \mathbf{R}_I} \rangle \end{aligned}$$

Now using $\mathcal{H} \mid \Psi_0 \rangle = E_0 \mid \Psi_0 \rangle$ and $\langle \Psi_0 \mid \mathcal{H} = E_0 \langle \Psi_0 \mid$, where E_0 is the energy corresponding to the eigenstate $\mid \Psi_0 \rangle$, we can rewrite the second and third terms in the expression for the ion–electron contribution to the force as

$$\langle \frac{\partial \Psi_0}{\partial \mathbf{R}_I} \mid \Psi_0 \rangle E_0 + E_0 \langle \Psi_0 \mid \frac{\partial \Psi_0}{\partial \mathbf{R}_I} \rangle = E_0 \frac{\partial \langle \Psi_0 \mid \Psi_0 \rangle}{\partial \mathbf{R}_I} = 0 \qquad (5.93)$$

where the last result is due to the normalization of the wavefunction, $\langle \Psi_0 \mid \Psi_0 \rangle = 1$. This then proves the Hellmann–Feynman theorem, as stated in Eq. (5.92).

The importance of the Hellmann–Feynman theorem is the following: the only terms needed to calculate the contribution of ion–electron interactions to the force are those terms in the hamiltonian that depend *explicitly* on the atomic positions $\mathbf{R}$. From our analysis of the various terms that contribute to the total energy, it is clear that there is only one such term in the hamiltonian, namely U^{ion-el}, defined in Eq. (5.72). The expectation value of this term in the ground state wavefunction, expressed in terms of the single-particle wavefunctions $|\psi_{\mathbf{k}}\rangle$, is given by

$$\sum_{\mathbf{k},I} \langle \psi_{\mathbf{k}} \mid V_I^{ps}(\mathbf{r} - \mathbf{R}_I) \mid \psi_{\mathbf{k}} \rangle \tag{5.94}$$

Taking the derivative of this term with respect to $\mathbf{R}_I$, we obtain the force contribution $\mathbf{F}_I^{ion-el}$

$$\mathbf{F}_I^{ion-el} = -\frac{\partial}{\partial \mathbf{R}_I} \sum_{\mathbf{k},I} \langle \psi_{\mathbf{k}} \mid V_I^{ps}(\mathbf{r} - \mathbf{R}_I) \mid \psi_{\mathbf{k}} \rangle \tag{5.95}$$

We will derive the ion–electron contribution to the force explicitly for the case where there is only one atom per unit cell, and the origin of the coordinate system is chosen so that the atomic positions coincide with the lattice vectors. Following the derivation of the expression for the total energy, we use again the Fourier transforms of the potentials and single-particle wavefunctions

$$\psi_{\mathbf{k}}(\mathbf{r}) = \sum_{\mathbf{G}} \mathrm{e}^{\mathrm{i}(\mathbf{k}+\mathbf{G})\cdot\mathbf{r}} \alpha_{\mathbf{k}}(\mathbf{G})$$

$$V^{ps}(\mathbf{r}) = \sum_{\mathbf{G}} \mathrm{e}^{\mathrm{i}\mathbf{G}\cdot\mathbf{r}} V^{ps}(\mathbf{G})$$

where we have dropped the index I from the pseudopotential since there is only one type of ion in our example. With these definitions, we obtain for the ion–electron contribution to the force:

$$\begin{aligned} \mathbf{F}_I^{ion-el} &= -\frac{\partial}{\partial \mathbf{R}_I} \Omega \sum_{\mathbf{k},\mathbf{G},\mathbf{G}'} \alpha_{\mathbf{k}}^*(\mathbf{G}) \alpha_{\mathbf{k}}(\mathbf{G}') V^{ps}(\mathbf{G}' - \mathbf{G}) S(\mathbf{G}' - \mathbf{G}) \\ &= -\mathrm{i}\Omega_{at} \sum_{\mathbf{G}} \mathbf{G} \mathrm{e}^{\mathrm{i}\mathbf{G}\cdot\mathbf{R}_I} V^{ps}(\mathbf{G}) \rho(\mathbf{G}) \end{aligned} \tag{5.96}$$

where $S(\mathbf{G}) = (1/N) \sum_I \mathrm{e}^{\mathrm{i}\mathbf{G}\cdot\mathbf{R}_I}$ is the structure factor, Ω_{at} is the atomic volume $\Omega_{at} = \Omega/N$ (N being the total number of atoms in the solid), and $\rho(\mathbf{G})$ is defined as

$$\rho(\mathbf{G}) = \sum_{\mathbf{k},\mathbf{G}'} \alpha_{\mathbf{k}}^*(\mathbf{G}') \alpha_{\mathbf{k}}(\mathbf{G}' - \mathbf{G}) \tag{5.97}$$

We see from Eq. (5.96) that when the single-particle wavefunctions have been calculated (i.e. the coefficients $\alpha_{\mathbf{k}}(\mathbf{G})$ have been determined, when working with the basis of plane waves $\mathbf{G}$), we have all the necessary information to calculate the forces on the ions.

This last statement can be exploited to devise a scheme for simulating the dynamics of ions in a computationally efficient manner. This scheme was first proposed by Car and Parrinello [73], and consists of evolving simultaneously the electronic and ionic degrees of freedom. The big advantage of this method is that the single-particle electronic wavefunctions, from which the forces on the ions can be calculated, do not need to be obtained from scratch through a self-consistent calculation when the ions have been moved. This is achieved by coupling the electron and ion dynamics through an effective lagrangian, defined as

$$\mathcal{L} = \sum_i \frac{1}{2}\mu\langle\frac{\mathrm{d}\psi_i}{\mathrm{d}t} \mid \frac{\mathrm{d}\psi_i}{\mathrm{d}t}\rangle + \sum_I \frac{1}{2}M_I\left(\frac{\mathrm{d}\mathbf{R}_I}{\mathrm{d}t}\right)^2$$
$$-E[\{\psi_i\}, \{\mathbf{R}_I\}] + \sum_{ij} \Lambda_{ij}(\langle\psi_i \mid \psi_i\rangle - \delta_{ij}) \qquad (5.98)$$

Here the first two terms represent the kinetic energy of electrons and ions (M_I and $\mathbf{R}_I$ are the mass and position of ion I), the third term is the total energy of the electron–ion system which plays the role of the potential energy in the lagrangian, and the last term contains the Laplace multipliers Λ_{ij} which ensure the orthogonality of the wavefunctions $|\psi_i\rangle$.

It is important to note that the kinetic energy of the electron wavefunctions in the Car–Parrinello lagrangian is a fictitious term, introduced for the sole purpose of coupling the dynamics of ionic and electronic degrees of freedom; it has nothing to do with the true quantum-mechanical kinetic energy of electrons, which is of course part of the total energy $E[\{\psi_i\}, \{\mathbf{R}_I\}]$. Accordingly, the "mass" μ associated with the kinetic energy of the electron wavefunctions is a fictitious mass, and is essentially a free parameter that determines the coupling between electron and ion dynamics. When the velocities associated with both the electronic and ionic degrees of freedom are reduced to zero, the system is in its equilibrium state with $E[\{\psi_i\}, \{\mathbf{R}_I\}]$ a minimum at zero temperature. This method can be applied to the study of solids in two ways.

(a) It can be applied under conditions where the velocities of electrons and ions are steadily reduced so that the true ground state of the system is reached, in the sense that both ionic and electronic degrees of freedom have been optimized; in this case the temperature is reduced to zero.

(b) It can be applied at finite temperature to study the time evolution of the system, which is allowed to explore the phase space consistent with the externally imposed temperature and pressure conditions.

The value of the electron effective mass μ, together with the value of the time step $\mathrm{d}t$ used to update the electronic wavefunctions and the ionic positions, are adjusted to make the evolution of the system stable as well as computationally

Table 5.4. *Basic structural and electronic properties of common metals.*

a_0 = lattice constant in angstroms (and c/a ratio for HCP crystals); B = bulk modulus in gigapascals; E^{coh} = cohesive energy per atom in electronvolts; $g(\epsilon_F)$ = density of states /eV-atom at the Fermi energy ϵ_F (values marked by $*$ are for the FCC structure). All values are from experiment.

Solid	Crystal	a_0	(c/a)	B	E^{coh}	$g(\epsilon_F)$
Li	BCC	3.491		11.6	1.63	0.48
Na	BCC	4.225		6.8	1.11	0.45
K	BCC	5.225		3.2	0.93	0.73
Rb	BCC	5.585		3.1	0.85	0.90
Be	HCP	2.27	(1.5815)	100	3.32	0.054*
Mg	HCP	3.21	(1.6231)	35.4	1.51	0.45*
Ca	FCC	5.58		15.2	1.84	1.56
Sr	FCC	6.08		11.6	1.72	0.31
Sc	HCP	3.31	(1.5921)	43.5	3.90	1.73*
Y	HCP	3.65	(1.5699)	36.6	4.37	1.41*
Ti	HCP	2.95	(1.5864)	105	4.85	1.49*
Zr	HCP	3.23	(1.5944)	83.3	6.25	1.28*
V	BCC	3.03		162	5.31	1.64
Nb	BCC	3.30		170	7.57	1.40
Cr	BCC	2.88		190	4.10	0.70
Mo	BCC	3.15		273	6.82	0.65
Fe	BCC	2.87		168	4.28	3.06
Ru	HCP	2.71	(1.5793)	321	6.74	1.13*
Co	HCP	2.51	(1.6215)	191	4.39	2.01*
Rh	FCC	3.80		270	5.75	1.35
Ni	FCC	3.52		186	4.44	4.06
Pd	FCC	3.89		181	3.89	2.31
Cu	FCC	3.61		137	3.49	0.29
Ag	FCC	4.09		101	2.95	0.27
Zn	HCP	2.66	(1.8609)	59.8	1.35	0.30*
Cd	HCP	2.98	(1.8859)	46.7	1.16	0.36*
Al	FCC	4.05		72.2	3.39	0.41
Ga	FCC†	4.14		44.0	3.22	0.41
In	FCC†	4.74		35.0	2.90	0.50

Source: C. Kittel [64], except for $g(\epsilon_F)$ (*source:* Moruzzi, Janak and Williams [75]).
† This is not the lowest energy crystal structure, but one for which theoretical data are available.

efficient. Specifically, the time step dt should be as large as possible in order either to reach the true ground state of the system in the fewest possible steps, or to simulate the evolution of the system for the largest possible time interval at finite temperature, with the total number of steps in the simulation determined by the available computational resources. However, a large time step might throw the system out of balance, in the sense that the time-evolved wavefunctions may not correspond to the ground state of the ionic system. For small enough values of dt,

Table 5.5. *Basic structural and electronic properties of common semiconductors.* The crystal structures are denoted as DIA = diamond, ZBL = zincblende, WRZ = wurtzite; a_0 = lattice constant in angstroms (and c/a ratio for the wurtzite structures); B = bulk modulus in gigapascals; E^{coh} = cohesive energy per atom in electronvolts; ϵ_{gap} = minimal band gap in electronvolts and position of the CBM (the VBM is always at Γ, the center of the Brillouin Zone); ε = static dielectric function (ε_2 at $\omega = 0$), measured at room temperature. All values are from experiment.

Solid	Crystal	a_0	(c/a)	B	E^{coh}	ϵ_{gap}	CBM	ε
C	DIA	3.57		443	7.41	5.48	$0.76X$	5.9
Si	DIA	5.43		98.8	4.66	1.17	$0.85X$	12.1
Ge	DIA	5.66		77.2	3.86	0.74	L	15.8
SiC	ZBL	4.34		224	6.37	2.39	X	9.7
BN	ZBL	3.62			6.70	6.4	X	7.1
BP	ZBL	4.54				2.4	X	
AlN	WRZ	3.11	(1.6013)		5.81	6.28	Γ	9.1
AlP	ZBL	5.45		86.0	4.21	2.50	X	
AlAs	ZBL	5.62		77.0		2.32	$\sim X$	
AlSb	ZBL	6.13		58.2		1.65	$\sim X$	14.4
GaN	ZBL	4.69			4.43	3.52	Γ	
GaP	ZBL	5.45		88.7	3.51	2.39	$\sim X$	11.1
GaAs	ZBL	5.66		74.8	3.34	1.52	Γ	13.0
GaSb	ZBL	6.12		57.0	2.99	0.80	Γ	15.7
InN	WRZ	3.54	(1.6102)			1.89	Γ	
InP	ZBL	5.87		71.0	3.36	1.42	Γ	12.6
InAs	ZBL	6.04		60.0	3.13	0.41	Γ	14.6
InSb	ZBL	6.48		47.4	2.78	0.23	Γ	17.9
ZnS	ZBL	5.40		77.1	3.11	3.84	Γ	8.3
ZnSe	ZBL	5.66		62.4	2.70	2.83	Γ	9.1
ZnTe	ZBL	6.10		51.0		2.39	Γ	10.1
CdS	ZBL	5.82		62.0	2.85	2.58	Γ	9.7
CdSe	ZBL	6.08		53.0		1.73	Γ	9.7
CdSe	WRZ	4.30	(1.6302)			1.74	Γ	
CdTe	ZBL	6.48		42.4	2.08	1.61	Γ	10.2

Sources: [64, 74, 76].

the balance between the time-evolved wavefunctions and positions of the ions is maintained – this is referred to as the system remaining on the Born–Oppenheimer surface. In this way, a fully self-consistent solution of the electronic problem for a given ionic configuration, which is the most computationally demanding part of the calculation, is avoided except at the initial step. The ensuing dynamics of the ions are always governed by accurate quantum mechanical forces, obtained by the general method outlined above. This produces very realistic simulations of the dynamic evolution of complicated physical systems.

The total-energy and force calculations which are able to determine the lowest energy configuration of a system, taken together with the electronic structure information inherent in these calculations, provide a powerful method for a deeper understanding of the properties of solids. We have already mentioned above some examples where such calculations can prove very useful. The computational approach we have discussed, based on DFT/LDA and its extensions using gradient corrections to the exchange-correlation density functional, in conjunction with atomic pseudopotentials to eliminate core electrons, has certain important advantages. Probably its biggest advantage is that it does not rely on any empirical parameters: the only input to the calculations is the atomic masses of constituent atoms and the number of valence electrons, that is, the atomic number of each element. These calculations are not perfect: for example, calculated lattice constants differ from experimental ones by a few percent (typically 1–2%), while calculated phonon frequencies, bulk moduli and elastic constants may differ from experimental values by 5–10%. Moreover, in the simplest version of the computations, the calculated band gap of semiconductors and insulators can be off from the experimental value by as much as 50%, but this can be corrected, either within the same theoretical framework or by more elaborate calculations, as already mentioned in chapter 4. Nevertheless, the ability of these calculations to address a wide range of properties for systems composed of almost any element in the periodic table without involving any empirical parameters, makes them a truly powerful tool for the study of the properties of solids. For reference, we collect in Tables 5.4 and 5.5 the basic structural and electronic properties of a variety of common metals and semiconductors, as obtained either from experiment or theory.

Further reading

1. *Bonds and Bands in Semiconductors*, J.C. Phillips (Academic Press, New York, 1973).
   This is a classic work, containing a wealth of physical insight into the origin and nature of energy bands in semiconductors.
2. *Electromagnetic Transport in Solids: Optical Properties and Plasma Effects*, H. Ehrenreich, in *Optical Properties of Solids*, ed. J. Tauc (Course 34 of E. Fermi International Summer School, Varenna, Italy, 1966).
   This article presents a fundamental theory for solids and includes several examples, some of which were discussed in the text.
3. *Ultraviolet Optical Properties*, H.R. Philipp and H. Ehrenreich, in *Semiconductors and Semimetals*, vol. 3, eds. R.K. Willardson and A.C. Beer (Academic Press, 1967), p. 93.
   This is a general review article, including the theoretical background within the one-electron framework, but with emphasis on the experimental side.
4. *Optical Properties of Solids*, F. Wooten (Academic Press, 1972).
   This book contains a comprehensive treatment of optical properties of solids.
5. R.O. Jones and O. Gunnarsson, *Rev. Mod. Phys.* **61**, 689 (1989).
   This article is an excellent early review of density functional theory methods with emphasis on applications to molecules.
6. D.C. Allan, T.A. Arias, J.D. Joannopoulos, M.C. Payne and M.P. Teter, *Rev. Mod. Phys.* **64**, 1045 (1992).

This article is a concise review of total-energy methods based on density functional theory and their applications to the physics of solids.

7. H. Ehrenreich, *Science* **235**, 1029 (1987).
This article provides a penetrating perspective on the importance of electronic structure calculations for materials science.

Problems

1. Prove the last equality in Eq. (5.4).
2. Derive the density of states for the free-electron model in one and two dimensions.
3. Derive the type, multiplicity and behavior of the DOS at critical points in one and two dimensions.
4. Calculate the DOS for the 2D square lattice with one atom per unit cell, with s, p_x, p_y, p_z orbitals, using the TBA (the example problem solved explicitly in chapter 4). For this calculation it is *not* sufficient to obtain the bands along the high-symmetry points only, but the bands for points in the entire IBZ are needed. Make sure you include enough points to be able to see the characteristic behavior at the critical points.
5. (a) Prove that if a uniform electron gas is displaced by a small amount in one direction relative to the uniform background of compensating positive charge and then is allowed to relax, it will oscillate with the plasma frequency, Eq. (5.50).
   (b) Prove the expressions for the real and imaginary parts of the dielectric function as a function of frequency due to intraband transitions, given in Eqs. (5.51) and (5.52), starting with the expression of Eq. (5.49).
   (c) Prove the expressions for the real and imaginary parts of the dielectric function as a function of frequency due to interband transitions, given in Eqs. (5.54) and (5.55), starting with the expression of Eq. (5.53).
6. Compare the expression for the dielectric function due to intraband transitions, Eq. (5.49), to the Drude result, Eq. (5.39), and identify the constants in the latter in terms of fundamental constants and the microscopic properties of the solid. Provide a physical interpretation for the constants in the Drude expression.
7. Compare the expression for the dielectric function due to interband transitions, Eq. (5.53), to the Lorentz result, Eq. (5.40), and identify the constants in the latter in terms of fundamental constants and the microscopic properties of the solid. Provide a physical interpretation for the constants in the Lorentz expression. Apply this analysis to the case of Si, whose band structure is shown in Fig. 4.8.
8. Use the band structure of Si, Fig. 4.8, to interpret the main features of the dielectric function in Fig. 5.4, that is, the major peak of $\epsilon_2(\omega)$ and the node of $\epsilon_1(\omega)$, both appearing near $\hbar\omega \sim 4$ eV.
9. Prove the Bloch statement for the many-body wavefunction, Eq. (5.57), and relate the value of the wave-vector $\mathbf{K}$ to the single-particle wave-vectors.
10. (a) Prove that the transformation from the band states to the Wannier states, Eq. (5.59), is a unitary one, that is, the Wannier functions are orthonormal:

$$\psi_{\mathbf{k}}^{(n)}(\mathbf{r}) = \sum_{\mathbf{R}} \phi_{\mathbf{R}}^{(n)}(\mathbf{r}) e^{i\mathbf{k}\cdot\mathbf{R}} \Rightarrow \langle \phi_{\mathbf{R}'}^{(m)} | \phi_{\mathbf{R}}^{(n)} \rangle = \delta_{nm}\delta(\mathbf{R} - \mathbf{R}') \tag{5.99}$$

   Discuss the uniqueness of Wannier functions.

(b) Estimate the degree of localization of a Wannier function in an FCC cubic lattice: in order to do this assume that the BZ is almost spherical and take the periodic part of the Bloch state $u_{\mathbf{k}}^{(n)}(\mathbf{r})$ to be nearly independent of $\mathbf{k}$.
(c) Consider a perturbation $V(\mathbf{r}, t)$, not necessarily periodic, and use the Wannier functions as a basis. Derive the equation that the expansion coefficients obey, and comment on the physical situations where such an expansion is appropriate.

11. Prove the expressions for the various terms in the energy of a Frenkel exciton represented by a Slater determinant of Wannier functions, given in Eq. (5.65).
12. Derive the equations of motion for the electronic and ionic degrees of freedom from the Car–Parrinello lagrangian, Eq. (5.98).

6

Lattice vibrations

At a finite temperature the atoms that form a crystalline lattice vibrate about their equilibrium positions, with an amplitude that depends on the temperature. Because a crystalline solid has symmetries, these thermal vibrations can be analyzed in terms of collective modes of motion of the ions. These modes correspond to collective excitations, which can be excited and populated just like electronic states. These excitations are called phonons.[1] Unlike electrons, phonons are bosons: their total number is not fixed, nor is there a Pauli exclusion principle governing the occupation of any particular phonon state. This is easily rationalized, if we consider the real nature of phonons, that is, collective vibrations of the atoms in a crystalline solid which can be excited arbitrarily by heating (or hitting) the solid. In this chapter we discuss phonons and how they can be used to describe thermal properties of solids.

6.1 Phonon modes

The nature and physics of phonons are typically described in the so called harmonic approximation. Suppose that the positions of the ions in the crystalline solid at zero temperature are determined by the vectors

$$\mathbf{R}_{ni} = \mathbf{R}_n + \mathbf{t}_i \tag{6.1}$$

where the $\mathbf{R}_n$ are the Bravais lattice vectors and the $\mathbf{t}_i$ are the positions of ions in one PUC, with the convention that $|\mathbf{t}_i| < |\mathbf{R}_n|$ for all non-zero lattice vectors. Then the deviation of each ionic position at finite temperature from its zero-temperature position can be denoted as

$$\mathbf{S}_{ni} = \delta\mathbf{R}_{ni} \tag{6.2}$$

[1] The word phonon derives from the Greek noun $\phi\omega\nu\eta$, "phoni", for sound; its actual meaning is more restrictive, specifically a human-generated sound, i.e. voice.

with n running over all the PUCs of the crystal and i running over all ions in the PUC. In terms of these vectors the kinetic energy K of the ions will be

$$K = \sum_{n,i} \frac{1}{2} M_i \left[\frac{\mathrm{d}\mathbf{S}_{ni}}{\mathrm{d}t} \right]^2 = \sum_{ni\alpha} \frac{1}{2} M_i \left(\frac{\mathrm{d}S_{ni\alpha}}{\mathrm{d}t} \right)^2 \tag{6.3}$$

where M_i is the mass of ion i and α labels the cartesian coordinates of the vectors $\mathbf{S}_{ni}$ ($\alpha = x, y, z$ in 3D). The potential energy of the system can be written as a Taylor series expansion in powers of $\mathbf{S}_{ni}$. The harmonic approximation consists of keeping only the second order terms in this Taylor series, with the zeroth order term being an arbitrary constant (set to zero for convenience) and the first order terms taken to be zero since the system is expected to be in an equilibrium configuration at zero temperature, which represents a minimum in the total energy. Higher order terms in the expansion are considered negligible. In this approximation, the potential energy V is then given by

$$V = \frac{1}{2} \sum_{n,i,\alpha;m,j,\beta} \frac{\partial^2 E}{\partial R_{ni\alpha} \partial R_{mj\beta}} S_{ni\alpha} S_{mj\beta} \tag{6.4}$$

where the total energy E depends on all the atomic coordinates $\mathbf{R}_{ni}$. We define the so called force-constant matrix by

$$F_{ni\alpha,mj\beta} = \frac{\partial^2 E}{\partial R_{ni\alpha} \partial R_{mj\beta}} \tag{6.5}$$

in terms of which the potential energy becomes

$$V = \frac{1}{2} \sum_{n,i,\alpha;m,j,\beta} F_{ni\alpha,mj\beta} S_{ni\alpha} S_{mj\beta} \tag{6.6}$$

The size of the force-constant matrix is $d \times \nu \times N$, where d is the dimensionality of space (the number of values for α), ν is the number of ions in the PUC, and N is the number of PUCs in the crystal. We notice that the following relations hold:

$$\frac{\partial^2 V}{\partial S_{ni\alpha} \partial S_{mj\beta}} = F_{ni\alpha,mj\beta} = \frac{\partial^2 E}{\partial R_{ni\alpha} \partial R_{mj\beta}} \tag{6.7}$$

$$\frac{\partial V}{\partial S_{ni\alpha}} = \sum_{m,j,\beta} F_{ni\alpha,mj\beta} S_{mj\beta} = \frac{\partial E}{\partial R_{ni\alpha}} \tag{6.8}$$

where the first equation is a direct consequence of the definition of the force-constant matrix and the harmonic approximation for the energy, while the second equation is a consequence of Newton's third law, since the left-hand side represents the negative of the α component of the total force on ion i in the unit cell labeled by $\mathbf{R}_n$.

The motion of the ions will be governed by the following equations:

$$M_i \frac{\mathrm{d}^2 S_{ni\alpha}}{\mathrm{d}t^2} = -\frac{\partial E}{\partial R_{ni\alpha}} = -\sum_{m,j,\beta} F_{ni\alpha,mj\beta} S_{mj\beta} \tag{6.9}$$

where we have used Eq. (6.8). We can try to solve the equations of motion by assuming sinusoidal expressions for the time dependence of their displacements:

$$S_{ni\alpha}(t) = \frac{1}{\sqrt{M_i}} \tilde{u}_{ni\alpha} \mathrm{e}^{-\mathrm{i}\omega t} \tag{6.10}$$

where ω is the frequency of oscillation and we have explicitly introduced the mass of the ions in the definition of the new variables $\tilde{u}_{ni\alpha}$. This gives, when substituted into the equations of motion,

$$\omega^2 \tilde{u}_{ni\alpha} = \sum_{m,j,\beta} F_{ni\alpha,mj\beta} \frac{1}{\sqrt{M_i M_j}} \tilde{u}_{mj\beta} \tag{6.11}$$

We define a new matrix, which we will call the dynamical matrix, through

$$\tilde{D}_{ni\alpha,mj\beta} = \frac{1}{\sqrt{M_i M_j}} F_{ni\alpha,mj\beta} \tag{6.12}$$

In terms of this matrix, the equations of motion can be written as

$$\sum_{m,j,\beta} \tilde{D}_{ni\alpha,mj\beta} \tilde{u}_{mj\beta} = \omega^2 \tilde{u}_{ni\alpha} \Longrightarrow \tilde{\mathbf{D}} \cdot \tilde{\mathbf{u}} = \omega^2 \tilde{\mathbf{u}} \tag{6.13}$$

where we have used bold symbols for the dynamical matrix and the vector of ionic displacements in the last expression. This is an eigenvalue equation, the solution of which gives the values of the frequency and the vectors that describe the corresponding ionic displacements. The size of the dynamical matrix is the same as the size of the force-constant matrix, i.e. $d \times \nu \times N$. Obviously, it is impossible to diagonalize such a matrix for a crystal in order to find the eigenvalues and eigenfunctions when $N \to \infty$.

We need to reduce this eigenvalue equation to a manageable size, so that it can be solved. To this end, we note that from the definition of the dynamical matrix

$$\tilde{D}_{ni\alpha,mj\beta} = \frac{1}{\sqrt{M_i M_j}} F_{ni\alpha,mj\beta} = \frac{1}{\sqrt{M_i M_j}} \frac{\partial^2 E}{\partial R_{ni\alpha} \partial R_{mj\beta}} \tag{6.14}$$

If both positions $R_{ni\alpha}$, $R_{mj\beta}$ were to be shifted by the same lattice vector $\mathbf{R}'$, the result of differentiation of the energy with ionic positions must be the same because of the translational invariance of the hamiltonian. This leads to the conclusion that the dynamical matrix can only depend on the distance $\mathbf{R}_n - \mathbf{R}_m$ and not on the specific values of n and m, that is,

$$\tilde{D}_{ni\alpha,mj\beta} = \tilde{D}_{i\alpha,j\beta}(\mathbf{R}_n - \mathbf{R}_m) \tag{6.15}$$

Accordingly, we can define the ionic displacements as follows:

$$\tilde{u}_{ni\alpha} = u_{i\alpha} \mathrm{e}^{\mathrm{i}\mathbf{k}\cdot\mathbf{R}_n} \tag{6.16}$$

which gives for the eigenvalue equation

$$\sum_{j,\beta} \sum_{m} \tilde{D}_{i\alpha,j\beta}(\mathbf{R}_n - \mathbf{R}_m) \mathrm{e}^{-\mathrm{i}\mathbf{k}\cdot(\mathbf{R}_n - \mathbf{R}_m)} u_{j\beta} = \omega^2 u_{i\alpha} \tag{6.17}$$

and with the definition

$$D_{i\alpha,j\beta}(\mathbf{k}) = \sum_{\mathbf{R}} \tilde{D}_{i\alpha,j\beta}(\mathbf{R}) \mathrm{e}^{-\mathrm{i}\mathbf{k}\cdot\mathbf{R}} = \sum_{n} \mathrm{e}^{-\mathrm{i}\mathbf{k}\cdot\mathbf{R}_n} \frac{1}{\sqrt{M_i M_j}} \frac{\partial^2 V}{\partial S_{ni\alpha} \partial S_{0j\beta}} \tag{6.18}$$

where the last expression is obtained with the use of Eq. (6.7), the eigenvalue equation takes the form:

$$\sum_{j,\beta} D_{i\alpha,j\beta}(\mathbf{k}) u_{j\beta} = \omega^2 u_{i\alpha} \Longrightarrow \mathbf{D}(\mathbf{k}) \cdot \mathbf{u} = \omega^2 \mathbf{u} \tag{6.19}$$

Since $D_{i\alpha,j\beta}(\mathbf{k})$ has a dependence on the wave-vector $\mathbf{k}$, so will the eigenvalues ω and eigenvectors $u_{i\alpha}$. The size of the new matrix is $d \times \nu$, which is a manageable size for crystals that typically contain few atoms per PUC. In transforming the problem to a manageable size, we realize that we now have to solve this eigenvalue problem for all the allowed values of $\mathbf{k}$, which of course are all the values in the first BZ. There are N distinct values of $\mathbf{k}$ in the first BZ, where N is the number of PUCs in the crystal, that is, no information has been lost in the transformation. Arguments similar to those applied to the solution of the single-particle hamiltonian for the electronic states can also be used here, to reduce the reciprocal-space volume where we need to obtain a solution down to the IBZ only.

The solutions to the eigenvalue equation will need to be labeled by two indices, $\mathbf{k}$ for each value of the wave-vector in the BZ, and l, which takes $d \times \nu$ values, for all the different ions in the PUC and the cartesian coordinates. The solution for the displacement of ion j in the PUC at lattice vector $\mathbf{R}_n$ will then be given by

$$\mathbf{S}^{(l)}_{nj}(\mathbf{k}, t) = \frac{1}{\sqrt{M_j}} \hat{\mathbf{e}}^{(l)}_{\mathbf{k}j} \mathrm{e}^{\mathrm{i}(\mathbf{k}\cdot\mathbf{R}_n - \omega^{(l)}_{\mathbf{k}} t)} \tag{6.20}$$

where $\hat{\mathbf{e}}^{(l)}_{\mathbf{k}j}$ is the set of d components of the eigenvector that denote the displacement of ion j in d dimensions. The eigenvectors can be chosen to be orthonormal:

$$\sum_j \left[\hat{\mathbf{e}}^{(l)}_{\mathbf{k}j}\right]^* \cdot \hat{\mathbf{e}}^{(l')}_{\mathbf{k}j} = \delta_{ll'} \tag{6.21}$$

In terms of these displacements, the most general ionic motion of the crystal can be expressed as

$$\mathbf{S}_{n,j}(t) = \sum_{l,\mathbf{k}} c_{\mathbf{k}}^{(l)} \frac{1}{\sqrt{M_j}} \hat{\mathbf{e}}_{\mathbf{k}j}^{(l)} \mathrm{e}^{\mathrm{i}(\mathbf{k}\cdot\mathbf{R}_n - \omega_{\mathbf{k}}^{(l)} t)} \tag{6.22}$$

where the coefficients $c_{\mathbf{k}}^{(l)}$ correspond to the amplitude of oscillation of the mode with frequency $\omega_{\mathbf{k}}^{(l)}$. Finally, we note that the eigenvalues of the frequency $\omega_{\mathbf{k}}^{(l)}$ obey the symmetry $\omega_{\mathbf{k}+\mathbf{G}}^{(l)} = \omega_{\mathbf{k}}^{(l)}$, where $\mathbf{G}$ is any reciprocal lattice vector. This symmetry is a direct consequence of the property $D_{i\alpha,j\beta}(\mathbf{k}) = D_{i\alpha,j\beta}(\mathbf{k}+\mathbf{G})$, which is evident from the definition of the dynamical matrix, Eq. (6.18). The symmetry of the eigenvalue spectrum allows us to solve the eigenvalue equations for $\omega_{\mathbf{k}}^{(l)}$ in the first BZ only, just as we did for the electronic energies. In both cases the underlying symmetry of the crystal that leads to this simplification is the translational periodicity.

6.2 The force-constant model

We discuss next a simple model for calculating the dynamical matrix, which is the analog of the tight-binding approximation for electronic states. In the present case, we will use the bonds between the ions as the basic units, by analogy to the atomic-like orbitals for the electronic states. In its ideal, zero-temperature configuration, the crystal can be thought of as having bonds between each pair of nearest neighbor atoms. We assume that there are two types of distortions of the bonds relative to their ideal positions and orientations: there is bond stretching (elongation or contraction of the length of bonds) and bond bending (change of the orientation of bonds relative to their original position). For bond stretching, we will take the energy to be proportional to the square of the amount of stretching. For bond bending, we will take the energy to be proportional to the square of the angle that describes the change in bond orientation. These choices are consistent with the harmonic approximation.

To determine the contributions of bond stretching and bond bending to the potential energy, we will use the diagram shown in Fig. 6.1. In the following analysis we employ the lower case symbols $\mathbf{r}_i$, $\mathbf{s}_i$ for the ionic positions and their displacements and omit any reference to the lattice vectors for simplicity. We restore the full notation, including lattice vectors, to the final expressions when we apply them to a model periodic system. The bond stretching energy is given by

$$V_r = \frac{1}{2}\kappa_r \sum_{\langle ij\rangle} \left(\Delta|\mathbf{r}_{ij}|\right)^2 \tag{6.23}$$

where κ_r is the force constant for bond stretching, $\langle ij\rangle$ stands for summation over all the distinct nearest neighbor pairs, and $\Delta|\mathbf{r}_{ij}|$ is the change in the length of the

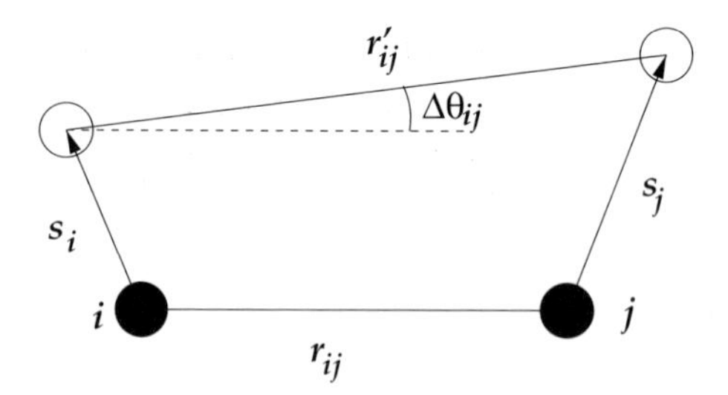

Figure 6.1. Schematic representation of bond stretching and bond bending in the force-constant model. The filled circles denote the positions of atoms in the ideal configuration and the open circles represent their displaced positions; s_i, s_j are the atomic displacements, r_{ij} is the original bond length, r'_{ij} the distorted bond length, and $\Delta\theta_{ij}$ the change in orientation of the bond (the angular distortion).

bond between atoms i and j. If we assume that all the bonds are equal to b (this can be easily generalized for systems with several types of bonds), then the change in bond length between atoms i and j is given by $\Delta|\mathbf{r}_{ij}| = |\mathbf{r}'_{ij}| - b$, where $\mathbf{r}'_{ij}$ is the new distance between atoms i and j. We will denote by $\mathbf{s}_i$ the departure of ion i from its equilibrium ideal position, which gives for the new distance between displaced atoms i and j:

$$\mathbf{r}'_{ij} = \mathbf{s}_j - \mathbf{s}_i + \mathbf{r}_{ij} \Longrightarrow |\mathbf{r}'_{ij}| = \left[\left(\mathbf{s}_j - \mathbf{s}_i + \mathbf{r}_{ij}\right)^2\right]^{1/2} \tag{6.24}$$

The displacements $\mathbf{s}_i$ will be taken to be always much smaller in magnitude than the bond distances $\mathbf{r}_{ij}$, consistent with the assumption of small deviations from equilibrium, in which case the harmonic approximation makes sense. Using this fact, we can expand in powers of the small quantity $|\mathbf{s}_i - \mathbf{s}_j|/b$, which gives to lowest order

$$|\mathbf{r}'_{ij}| = b\left(1 + \frac{\mathbf{r}_{ij}\cdot(\mathbf{s}_i - \mathbf{s}_j)}{b^2}\right) \Longrightarrow \Delta|\mathbf{r}_{ij}| = |\mathbf{r}'_{ij}| - b = \hat{\mathbf{r}}_{ij}\cdot(\mathbf{s}_i - \mathbf{s}_j) \tag{6.25}$$

which in turn gives for the bond stretching energy

$$V_r = \frac{1}{2}\kappa_r \sum_{\langle ij\rangle} \left[(\mathbf{s}_j - \mathbf{s}_i)\cdot\hat{\mathbf{r}}_{ij}\right]^2 \tag{6.26}$$

By similar arguments we can obtain an expression in terms of the variables $\mathbf{s}_i$, $\mathbf{s}_j$ for the bond bending energy, defined to be

$$V_\theta = \frac{1}{2}\kappa_\theta \sum_{\langle ij\rangle} b^2\left(\Delta\theta_{ij}\right)^2 \tag{6.27}$$

with κ_θ the bond bending force constant and $\Delta\theta_{ij}$ the change in the orientation of the bond between atoms i and j. Using the same notation for the new positions of the ions (see Fig. 6.1), and the assumption that the displacements $\mathbf{s}_i$ are much smaller than the bond distances $\mathbf{r}_{ij}$, we obtain for the bond bending term

$$\mathbf{r}'_{ij} \cdot \mathbf{r}_{ij} = |\mathbf{r}'_{ij}| b \cos(\Delta\theta_{ij}) \approx \left[1 - \frac{1}{2}(\Delta\theta_{ij})^2\right] |\mathbf{r}'_{ij}| b$$

$$\Longrightarrow \frac{\mathbf{r}'_{ij} \cdot \mathbf{r}_{ij}}{|\mathbf{r}'_{ij}| b} = 1 - \frac{1}{2}(\Delta\theta_{ij})^2 \tag{6.28}$$

For the left-hand side of this last equation, we find

$$\frac{\mathbf{r}'_{ij} \cdot \mathbf{r}_{ij}}{|\mathbf{r}'_{ij}| b} = \left[(\mathbf{s}_j - \mathbf{s}_i + \mathbf{r}_{ij}) \cdot \mathbf{r}_{ij}\right] \frac{1}{b^2} \left[1 + \frac{(\mathbf{s}_i - \mathbf{s}_j)^2}{b^2} - \frac{2\mathbf{r}_{ij} \cdot (\mathbf{s}_i - \mathbf{s}_j)}{b^2}\right]^{-1/2}$$

$$= \left[1 + \frac{\mathbf{r}_{ij} \cdot (\mathbf{s}_j - \mathbf{s}_i)}{b^2}\right] \left[1 + \frac{\mathbf{r}_{ij} \cdot (\mathbf{s}_i - \mathbf{s}_j)}{b^2} - \frac{1}{2}\frac{(\mathbf{s}_i - \mathbf{s}_j)^2}{b^2} + \frac{3}{2}\frac{|\mathbf{r}_{ij} \cdot (\mathbf{s}_i - \mathbf{s}_j)|^2}{b^4}\right]$$

$$= 1 - \frac{|\mathbf{s}_j - \mathbf{s}_i|^2}{2b^2} + \frac{\left[(\mathbf{s}_j - \mathbf{s}_i) \cdot \mathbf{r}_{ij}\right]^2}{2b^4}$$

where we have used the Taylor expansion of $(1 + x)^{-1/2}$ (see Appendix G) and kept only terms up to second order in the small quantities $|\mathbf{s}_i - \mathbf{s}_j|/b$. This result, when compared to the right-hand side of the previous equation, gives

$$(\Delta\theta_{ij})^2 b^2 = |\mathbf{s}_j - \mathbf{s}_i|^2 - \left[(\mathbf{s}_j - \mathbf{s}_i) \cdot \hat{\mathbf{r}}_{ij}\right]^2 \tag{6.29}$$

which leads to the following expression for the bond bending energy:

$$V_\theta = \frac{1}{2}\kappa_\theta \sum_{\langle ij \rangle} \left[|\mathbf{s}_j - \mathbf{s}_i|^2 - \left[(\mathbf{s}_j - \mathbf{s}_i) \cdot \hat{\mathbf{r}}_{ij}\right]^2\right] \tag{6.30}$$

Combining the two contributions, Eqs. (6.26) and (6.30), we obtain for the total potential energy

$$V = \frac{1}{2} \sum_{\langle ij \rangle} \left[(\kappa_r - \kappa_\theta)\left[(\mathbf{s}_j - \mathbf{s}_i) \cdot \hat{\mathbf{r}}_{ij}\right]^2 + \kappa_\theta |\mathbf{s}_j - \mathbf{s}_i|^2\right] \tag{6.31}$$

where the sum runs over all pairs of atoms in the crystal that are bonded, that is, over all pairs of nearest neighbors.

6.2.1 Example: phonons in 2D periodic chain

We next apply the force-constant model to a simple example that includes all the essential features needed to demonstrate the behavior of phonons. Our example is based on a system that is periodic in one dimension (with lattice vector $\mathbf{a}_1 = (a/\sqrt{2})[\hat{\mathbf{x}} + \hat{\mathbf{y}}]$), but exists in a two-dimensional space, and has two atoms per unit cell at positions $\mathbf{t}_1 = (a/2\sqrt{2})\hat{\mathbf{x}}$, $\mathbf{t}_2 = -(a/2\sqrt{2})\hat{\mathbf{x}}$. This type of atomic arrangement

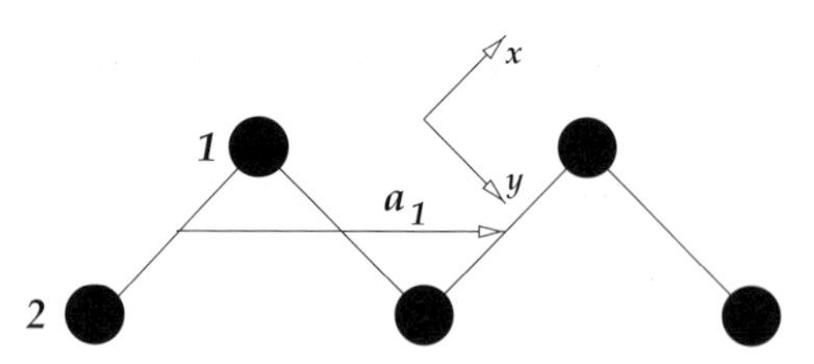

Figure 6.2. Definition of the model for the phonon calculation: it consists of a chain of atoms in 2D space.

will generate phonons of different nature, as we will see below. Since there are two atoms per unit cell ($\nu = 2$), and the system exists in two dimensions ($d = 2$), the size of the dynamical matrix will be $2 \times 2 = 4$.

The physical system is shown schematically in Fig. 6.2. To identify the atoms, we use two indices as in the general formulation developed above, the lattice vector index (n), and the index identifying the position of the atom inside the unit cell (i). By convention, we take the unit cell of interest to correspond to the zero lattice vector. We denote the potential energy corresponding to a pair of atoms by $V_{ni,mj}$. We will take into account nearest neighbor interactions only, in the spirit of using local force constants. Consequently, for each unit cell we only need to take into account the interaction between atoms within the unit cell and the interaction of these atoms with their nearest neighbors in the adjacent unit cells. In our example this means that we need only the following three terms in the expression for the potential energy:

$$V_{01,02} = \frac{1}{2}(S_{01x} - S_{02x})^2(\kappa_r - \kappa_\theta) + \frac{1}{2}\kappa_\theta[(S_{01x} - S_{02x})^2 + (S_{01y} - S_{02y})^2]$$

$$V_{01,12} = \frac{1}{2}(S_{01y} - S_{12y})^2(\kappa_r - \kappa_\theta) + \frac{1}{2}\kappa_\theta[(S_{01x} - S_{12x})^2 + (S_{01y} - S_{12y})^2]$$

$$V_{\bar{1}1,02} = \frac{1}{2}(S_{\bar{1}1y} - S_{02y})^2(\kappa_r - \kappa_\theta) + \frac{1}{2}\kappa_\theta[(S_{\bar{1}1x} - S_{02x})^2 + (S_{\bar{1}1y} - S_{02y})^2]$$

where we have reverted to the original notation for the displacement of ions $S_{ni\alpha}$, with n labeling the unit cell (in our case $n = 0, \pm 1$, with $\bar{1} = -1$), i labeling the atoms in the unit cell (in our case $i = 1, 2$) and α labeling the spatial coordinates (in our case $\alpha = x, y$). Using the definition of the dynamical matrix Eq. (6.18), and taking the masses of the two ions to be the same $M_1 = M_2 = M$, we obtain for the various matrix elements of the dynamical matrix

$$D_{i\alpha,j\beta}(\mathbf{k}) = \frac{1}{M}\sum_{\mathbf{R}} \mathrm{e}^{-\mathrm{i}\mathbf{k}\cdot\mathbf{R}} \frac{\partial^2 V}{\partial S_{\mathbf{R}i\alpha}\partial S_{0j\beta}} \qquad (6.32)$$

We notice that, since we have adopted the approximation of nearest neighbor interactions, for the diagonal elements $D_{i\alpha,i\alpha}$ ($i = 1, 2; \alpha = x, y$) we must only use

Table 6.1. *Dynamical matrix for the linear chain in 2D.*
All elements have been multiplied by M.

	$1, x$	$2, x$	$1, y$	$2, y$
$1, x$	$\kappa_r + \kappa_\theta$	$-\kappa_r - \kappa_\theta e^{-i\mathbf{k}\cdot\mathbf{a}_1}$	0	0
$2, x$	$-\kappa_r - \kappa_\theta e^{i\mathbf{k}\cdot\mathbf{a}_1}$	$\kappa_r + \kappa_\theta$	0	0
$1, y$	0	0	$\kappa_r + \kappa_\theta$	$-\kappa_\theta - \kappa_r e^{-i\mathbf{k}\cdot\mathbf{a}_1}$
$2, y$	0	0	$-\kappa_\theta - \kappa_r e^{i\mathbf{k}\cdot\mathbf{a}_1}$	$\kappa_r + \kappa_\theta$

the $\mathbf{R} = 0$ contribution because there are no nearest neighbor atoms with the same index i, whereas for the off-diagonal elements $D_{1x,2x}$, $D_{2x,1x}$, $D_{1y,2y}$, $D_{2y,1y}$ we need to take into account contributions from $\mathbf{R} = 0$, $\mathbf{R} = \pm\mathbf{a}_1$, as appropriate for the nearest neighbors. The remaining matrix elements are identically equal to zero because there are no contributions to the potential energy that involve the variables S_{nix} and S_{mjy} together. The explicit expressions for the matrix elements are tabulated in Table 6.1.

The solution of this matrix gives for the frequency eigenvalues

$$\omega_{\mathbf{k}}^2 = \frac{1}{\sqrt{M}}\left[(\kappa_r + \kappa_\theta) \pm \sqrt{\kappa_r^2 + \kappa_\theta^2 + 2\kappa_r\kappa_\theta \cos(\mathbf{k}\cdot\mathbf{a}_1)}\right] \tag{6.33}$$

with each value of $\omega_{\mathbf{k}}$ doubly degenerate (each value coming from one of the two submatrices).

A plot of the two different eigenvalues, denoted in the following by $\omega_{\mathbf{k}}^{(+)}, \omega_{\mathbf{k}}^{(-)}$ from the sign in front of the square root, is given in Fig. 6.3. We have used values of the force constants such that $\kappa_r > \kappa_\theta$ since the cost of stretching bonds is always greater than the cost of bending bonds.

It is instructive to analyze the behavior of the eigenvalues and the corresponding eigenvectors near the center ($\mathbf{k} = 0$) and near the edge of the BZ ($\mathbf{k} = (\pi/a)\hat{\mathbf{x}}$). This gives the following results:

$$\begin{aligned}
\omega_0^{(-)} &= \frac{1}{\sqrt{M}}\left[\frac{1}{2}\frac{\kappa_r\kappa_\theta}{\kappa_r + \kappa_\theta}\right]^{1/2} ak \\
\omega_0^{(+)} &= \frac{1}{\sqrt{M}}\left[2(\kappa_r + \kappa_\theta) - \frac{1}{2}\frac{\kappa_r\kappa_\theta}{\kappa_r + \kappa_\theta}a^2k^2\right]^{1/2} \\
\omega_1^{(-)} &= \frac{1}{\sqrt{M}}[2\kappa_\theta]^{1/2} \\
\omega_1^{(+)} &= \frac{1}{\sqrt{M}}[2\kappa_r]^{1/2}
\end{aligned} \tag{6.34}$$

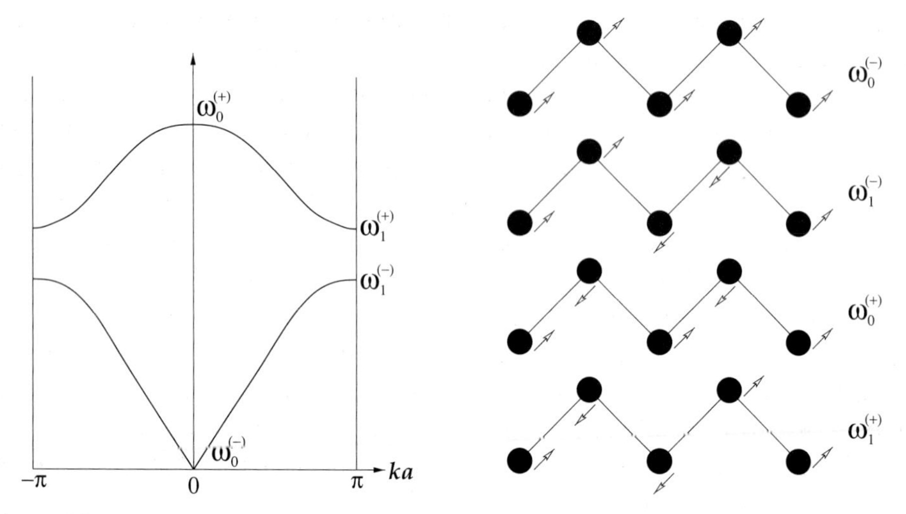

Figure 6.3. Solution of the chain model for phonons in 2D. **Left:** the frequency eigenvalues. **Right:** the motion of ions corresponding to the eigenvalues at the center and at the edge of the BZ.

where we have used the subscript 0 for values of $|\mathbf{k}| \approx 0$ and the subscript 1 for values of $|\mathbf{k}| \approx \pi/a$. We see that the lowest frequency at $\mathbf{k} \approx 0$, $\omega_0^{(-)}$, is linear in $k = |\mathbf{k}|$, in contrast to the standard behavior of the energy of electrons at the bottom of the valence band, which was quadratic in k. If we examine the eigenvector of this eigenvalue in the limit $k \to 0$, we find that it corresponds to a normal mode in which all the ions move in the same direction and by the same amount, that is, uniform translation of the crystal. This motion does not involve bond bending or bond stretching distortions. The eigenvector of the lower branch eigenvalue $\omega_1^{(-)}$ at $|\mathbf{k}| = (\pi/a)$ corresponds to motion of the two ions which is in phase within one unit cell, but π out of phase in neighboring unit cells. This motion involves bond bending distortions only, hence the frequency eigenvalue $\omega_1^{(-)} \sim \sqrt{\kappa_\theta}$. For the eigenvalue $\omega_0^{(+)}$, we find an eigenvector which corresponds to motion of the ions against each other within one unit cell, while the motion of equivalent ions in neighboring unit cells is in phase. This mode involves both bond stretching and bond bending, hence the frequency eigenvalue $\omega_0^{(+)} \sim \sqrt{\kappa_r + \kappa_\theta}$. Finally, for the eigenvalue $\omega_1^{(+)}$, we find an eigenvector which corresponds to motion of the ions against each other in the same unit cell, while the motion of equivalent ions in neighboring unit cells is π out of phase. This distortion involves bond stretching only, hence the frequency eigenvalue $\omega_1^{(+)} \sim \sqrt{\kappa_r}$. The motion of ions corresponding to these four modes is illustrated in Fig. 6.3. There are four other modes which are degenerate to these and involve similar motion of the ions in the $\hat{\mathbf{y}}$ direction.

The lower branch of the eigenvalues of the frequency, in which the motion of ions within one unit cell is in phase, is referred to as "acoustic", while the upper branch in which the motion of ions within one unit cell is π out of phase, is referred to as "optical". We note that in this example we get two branches because we have two atoms per unit cell, so the in-phase and out-of-phase motion of atoms in the unit cell produce two distinct modes. Moreover, there are doubly degenerate modes at each value of $\mathbf{k}$ ($\omega_{\mathbf{k}}^{(1)} = \omega_{\mathbf{k}}^{(2)} = \omega_{\mathbf{k}}^{(-)}$ and $\omega_{\mathbf{k}}^{(3)} = \omega_{\mathbf{k}}^{(4)} = \omega_{\mathbf{k}}^{(+)}$) because we are dealing with a 2D crystal which has symmetric couplings in the $\hat{\mathbf{x}}$ and $\hat{\mathbf{y}}$ directions. In general, the degeneracy will be broken if the couplings between ions are not symmetric in the different directions. In 3D there would be three acoustic and three optical modes for a crystal with two atoms per unit cell, and their frequencies would not be degenerate, except when required by symmetry. In cases where there is only one atom per unit cell there are only acoustic modes. For crystals with ν atoms per unit cell, there are $3 \times \nu$ modes for each value of $\mathbf{k}$.

6.2.2 Phonons in a 3D crystal

As an illustration of the above ideas in a more realistic example we discuss briefly the calculation of phonon modes in Si. Si is the prototypical covalent solid, where both bond bending and bond stretching terms are very important. If we wanted to obtain the phonon spectrum of Si by a force-constant model, we would need parameters that can describe accurately the relevant contributions of the bond bending and bond stretching forces. For example, fitting the values of κ_r and κ_θ to reproduce the experimental values for the highest frequency optical mode at Γ (labeled LTO(Γ))[2] and the lowest frequency acoustic mode at X (labeled TA(X)) gives the spectrum shown in Fig. 6.4 (see also Problem 1). Although through such a procedure it is possible to obtain estimates for the bond stretching and bond bending parameters, the model is not very accurate: for instance, the values of the two other phonon modes at X are off by $+5\%$ for the mode labeled TO(X) and -12% for the mode labeled LOA(X), compared with experimental values. Interactions to several neighbors beyond the nearest ones are required in order to obtain frequencies closer to the experimentally measured spectrum throughout the BZ. However, establishing the values of force constants for interactions beyond nearest neighbors is rather complicated and must rely on experimental input. Moreover, with such extensions the model loses its simplicity and the transparent physical meaning of the parameters.

Alternatively, one can use the formalism discussed in chapters 4 and 5 (DFT/LDA and its refinements) for calculating total energies or forces, as follows. The atomic displacements that correspond to various phonon modes can be established

[2] In phonon nomenclature "L" stands for longitudinal, "T" for transverse, "O" for optical and "A" for acoustic.

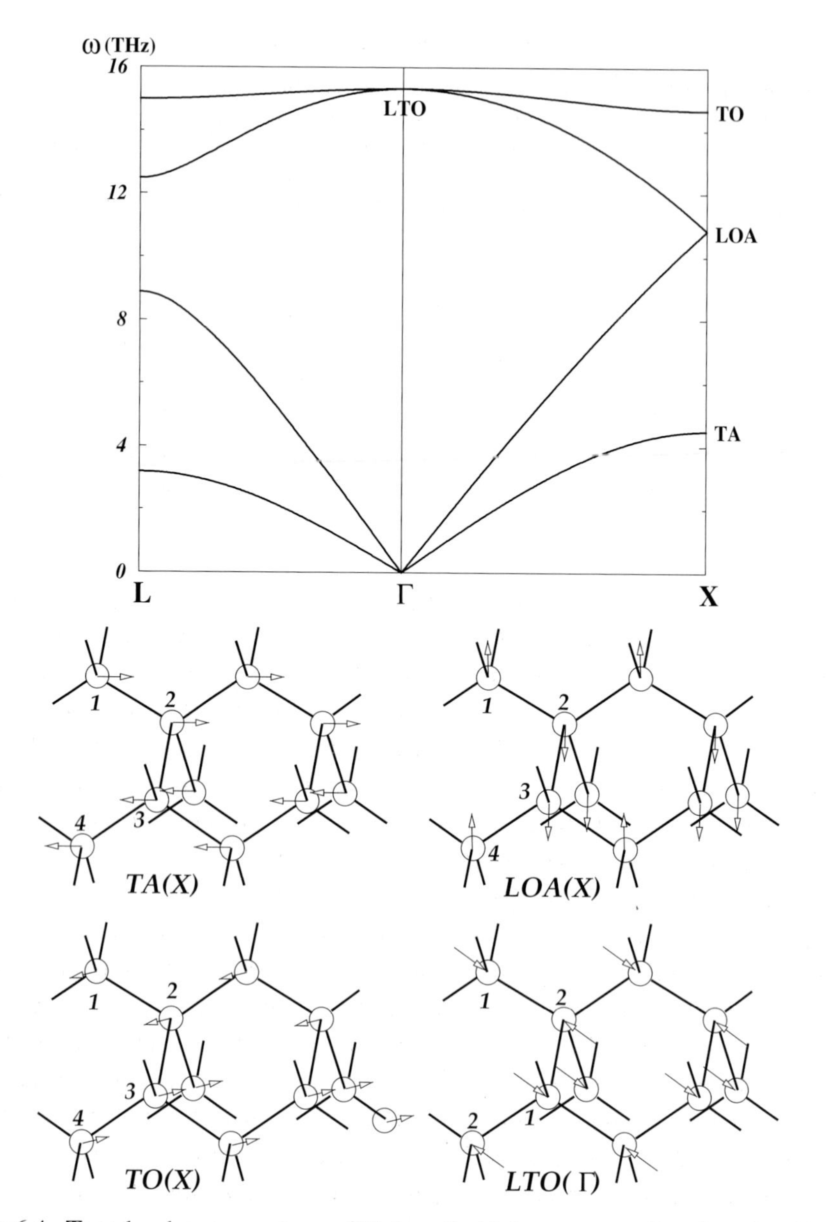

Figure 6.4. **Top:** the phonon spectrum of Si along the high-symmetry directions $L - \Gamma - X$, calculated within the force-constant model with $\sqrt{\kappa_r/M_{Si}} = 8.828$ THz, $\sqrt{\kappa_\theta/M_{Si}} = 2.245$ THz. **Bottom:** atomic displacements associated with the phonon modes in Si at the Γ and X points in the BZ; inequivalent atoms are labeled 1–4.

either by symmetry (for instance, using group theory arguments) or in conjunction with simple calculations based on a force-constant model, as in the example discussed above. For instance, we show in Fig. 6.4 the atomic displacements corresponding to certain high-symmetry phonon modes for Si, as obtained from the

Table 6.2. *Phonon modes in Si.*

Frequencies of four high-symmetry modes are given (in THz) at the center (Γ) and the boundary (X) of the Brillouin Zone, as obtained by theoretical calculations and by experimental measurements. DFT results are from Refs. [65, 67]. The force-constant model is based on nearest neighbor interactions only, with values of κ_r and κ_θ chosen to reproduce exactly the experimental frequencies marked by asterisks. The atomic displacements corresponding to these modes are shown in Fig. 6.4.

	LTO(Γ)	LOA(X)	TO(X)	TA(X)
DFT energy calculation	15.16	12.16	13.48	4.45
DFT force calculation	15.14	11.98	13.51	4.37
Force-constant model	15.53*	10.83	14.65	4.49*
Experiment	15.53	12.32	13.90	4.49

force-constant model. For a given atomic displacement, the energy of the system relative to its equilibrium structure, or the restoring forces on the atoms, can be calculated as a function of the phonon amplitude. These energy differences are then fitted by a second or higher order polynomial in the phonon amplitude, and the coefficient of the second order term gives the phonon frequency (in the case of a force calculation the fit starts at first order in the phonon amplitude). Higher order terms give the anharmonic contributions which correspond to phonon–phonon interactions. This straightforward method, called the "frozen phonon" approach, gives results that are remarkably close to experimental values. Its limitation is that only phonons of relatively high symmetry can be calculated, such as at the center, the boundaries and a few special points within the Brillouin Zone. The reason is that phonon modes with low symmetry involve the coherent motion of atoms across many unit cells of the ideal crystal; in order to represent this motion, a large supercell (a multiple of the ideal crystal unit cell) that contains all the inequivalent atoms in a particular phonon mode must be used. For instance, the unit cell for the LTO(Γ) mode in Fig. 6.4 involves only two atoms, the same as in the PUC of the perfect crystal, whereas the supercell for all the modes at X involves four atoms, that is, twice as many as in the PUC of the perfect crystal. The computational cost of the quantum mechanical calculations of energy and forces increases sharply with the size of the unit cell, making computations for large unit cells with many atoms intractable. However, the information from the few high-symmetry points can usually be interpolated to yield highly accurate phonon frequencies throughout the entire BZ. In Table 6.2 we compare the calculated frequencies for the phonon modes of Fig. 6.4 (from Yin and Cohen [77]), to experimentally measured ones; for a review and more details, see Ref. [65].

6.3 Phonons as harmonic oscillators

We next draw the analogy between the phonon hamiltonian and that of a collection of independent harmonic oscillators. We begin with the most general expression for the displacement of ions, Eq. (6.22). In that equation, we combine the amplitude $c_{\mathbf{k}}^{(l)}$ with the time-dependent part of the exponential $\exp(-\mathrm{i}\omega_{\mathbf{k}}^{(l)}t)$ into a new variable $Q_{\mathbf{k}}^{(l)}(t)$, and express the most general displacement of ions as:

$$\mathbf{S}_{nj}(t) = \sum_{l,\mathbf{k}} Q_{\mathbf{k}}^{(l)}(t)\frac{1}{\sqrt{M_j}}\hat{\mathbf{e}}_{\mathbf{k}j}^{(l)}\mathrm{e}^{\mathrm{i}\mathbf{k}\cdot\mathbf{R}_n} \tag{6.35}$$

These quantities must be real, since they describe ionic displacements in real space. Therefore, for the time dependence we should only consider the real part, that is $\cos(\omega_{\mathbf{k}}^{(l)}t)$. Moreover, the coefficients of factors $\exp(\mathrm{i}\mathbf{k}\cdot\mathbf{R}_n)$ and $\exp(\mathrm{i}(-\mathbf{k})\cdot\mathbf{R}_n)$ that appear in the expansion must be complex conjugate, leading to the relation

$$\left[Q_{\mathbf{k}}^{(l)}(t)\hat{\mathbf{e}}_{\mathbf{k}j}^{(l)}\right]^* = Q_{-\mathbf{k}}^{(l)}(t)\hat{\mathbf{e}}_{-\mathbf{k}j}^{(l)} \tag{6.36}$$

and since this relation must hold for any time t, we deduce that

$$\left[Q_{\mathbf{k}}^{(l)}(t)\right]^* = Q_{-\mathbf{k}}^{(l)}(t), \quad \left[\hat{\mathbf{e}}_{\mathbf{k}j}^{(l)}\right]^* = \hat{\mathbf{e}}_{-\mathbf{k}j}^{(l)} \tag{6.37}$$

The kinetic energy of the system of ions will be given in terms of the displacements $\mathbf{S}_{nj}(t)$ as

$$K = \sum_{n,j}\frac{M_j}{2}\left(\frac{\mathrm{d}\mathbf{S}_{nj}}{\mathrm{d}t}\right)^2 = \frac{1}{2}\sum_{n,j}\sum_{\mathbf{k}l,\mathbf{k}'l'}\frac{\mathrm{d}Q_{\mathbf{k}}^{(l)}}{\mathrm{d}t}\frac{\mathrm{d}Q_{\mathbf{k}'}^{(l')}}{\mathrm{d}t}\hat{\mathbf{e}}_{\mathbf{k}j}^{(l)}\cdot\hat{\mathbf{e}}_{\mathbf{k}'j}^{(l')}\mathrm{e}^{\mathrm{i}(\mathbf{k}+\mathbf{k}')\cdot\mathbf{R}_n} \tag{6.38}$$

Using the relations of Eq. (6.37) and $(1/N)\sum_n \exp(\mathrm{i}\mathbf{k}\cdot\mathbf{R}_n) = \delta(\mathbf{k}-\mathbf{G})$ from Eq. (G.69), we obtain for the kinetic energy

$$K = \frac{1}{2}\sum_{\mathbf{k},l,l',j}\frac{\mathrm{d}Q_{\mathbf{k}}^{(l)}}{\mathrm{d}t}\frac{\mathrm{d}Q_{\mathbf{k}}^{(l')*}}{\mathrm{d}t}\hat{\mathbf{e}}_{\mathbf{k}j}^{(l)}\cdot\hat{\mathbf{e}}_{\mathbf{k}j}^{(l')*} = \frac{1}{2}\sum_{\mathbf{k},l}\frac{\mathrm{d}Q_{\mathbf{k}}^{(l)}}{\mathrm{d}t}\frac{\mathrm{d}Q_{\mathbf{k}}^{(l)*}}{\mathrm{d}t} \tag{6.39}$$

where we have set $\mathbf{G} = 0$ in the argument of the δ-function since the wave-vectors $\mathbf{k}$, $\mathbf{k}'$ lie in the first BZ, and the last equality was obtained with the help of Eq. (6.21). The potential energy can also be expressed in terms of the atomic displacements and the force-constant matrix:

$$V = \sum_{n,i,\alpha;m,j,\beta}\frac{1}{2\sqrt{M_iM_j}}\sum_{\mathbf{k},l;\mathbf{k}',l'} Q_{\mathbf{k}}^{(l)}\hat{\mathbf{e}}_{\mathbf{k}i\alpha}^{(l)}\mathrm{e}^{\mathrm{i}\mathbf{k}\cdot\mathbf{R}_n}F_{ni\alpha,mj\beta}Q_{\mathbf{k}'}^{(l')}\hat{\mathbf{e}}_{\mathbf{k}'j\beta}^{(l')}\mathrm{e}^{\mathrm{i}\mathbf{k}'\cdot\mathbf{R}_m} \tag{6.40}$$

which, with the help of the same relations that were used in the derivation of the kinetic energy, becomes

$$V = \frac{1}{2} \sum_{\mathbf{k},l} Q_{\mathbf{k}}^{(l)} Q_{\mathbf{k}}^{(l)*} \left(\omega_{\mathbf{k}}^{(l)}\right)^2 \tag{6.41}$$

To obtain the last expression we have also used Eqs. (6.10) and (6.11) to relate the displacements to the force-constant matrix and the frequency eigenvalues $\omega_{\mathbf{k}}^{(l)}$. Combining the expressions for the kinetic and potential energies, we obtain the total energy of a system of phonons:

$$E^{phon} = \frac{1}{2} \sum_{\mathbf{k},l} \left[\left| \frac{\mathrm{d}Q_{\mathbf{k}}^{(l)}}{\mathrm{d}t} \right|^2 + \left(\omega_{\mathbf{k}}^{(l)}\right)^2 \left| Q_{\mathbf{k}}^{(l)} \right|^2 \right] \tag{6.42}$$

which is formally identical to the total energy of a collection of independent harmonic oscillators with frequencies $\omega_{\mathbf{k}}^{(l)}$, where $Q_{\mathbf{k}}^{(l)}$ is the free variable describing the motion of the harmonic oscillator. This expression also makes it easy to show that the kinetic and potential contributions to the total energy are equal, as expected for harmonic oscillators (see Problem 2).

This analysis shows that a solid in which atoms are moving is equivalent to a number of phonon modes that are excited, with the atomic motion given as a superposition of the harmonic modes corresponding to the excited phonons. The total energy of the excitation is that of a collection of independent harmonic oscillators with the proper phonon frequencies. Notice that the $Q_{\mathbf{k}}^{(l)}$ contain the amplitude of the vibration, as seen from Eq. (6.35), and consequently, the total energy involves the absolute value squared of the amplitude of every phonon mode that is excited. Since the atomic motion in the solid must be quantized, the harmonic oscillators describing this motion should be treated as quantum mechanical ones. Indeed, in Eq. (6.42), $Q_{\mathbf{k}}^{(l)}$ can be thought of as the quantum mechanical position variable and $\mathrm{d}Q_{\mathbf{k}}^{(l)}/\mathrm{d}t$ as the conjugate momentum variable of a harmonic oscillator identified by index l and wave-vector $\mathbf{k}$. The amplitude of the vibration, contained in $Q_{\mathbf{k}}^{(l)}$, can be interpreted as the number of excited phonons of this particular mode. In principle, there is no limit for this amplitude, therefore an arbitrary number of phonons of each frequency can be excited; all these phonons contribute to the internal energy of the solid. The fact that an arbitrary number of phonons of each mode can be excited indicates that phonons must be treated as bosons. With this interpretation, $Q_{\mathbf{k}}^{(l)}$ and its conjugate variable $\mathrm{d}Q_{\mathbf{k}}^{(l)}/\mathrm{d}t$ obey the proper commutation relations for bosons. This implies that the total energy due to the phonon excitations must be given by

$$E_s^{phon} = \sum_{\mathbf{k},l} \left(n_{\mathbf{k}s}^{(l)} + \frac{1}{2} \right) \hbar\omega_{\mathbf{k}}^{(l)} \tag{6.43}$$

where $n_{\mathbf{k}s}^{(l)}$ is the number of phonons of frequency $\omega_{\mathbf{k}}^{(l)}$ that have been excited in a particular state (denoted by s) of the system. This expression is appropriate for quantum harmonic oscillators, with $n_{\mathbf{k}s}^{(l)}$ allowed to take any non-negative integer value. One interesting aspect of this expression is that, even in the ground state of the system when none of the phonon modes is excited, that is, $n_{\mathbf{k}0}^{(l)} = 0$, there is a certain amount of energy in the system due to the so called zero-point motion associated with quantum harmonic oscillators; this arises from the factors of $\frac{1}{2}$ which are added to the phonon occupation numbers in Eq. (6.43); we will see below that this has measurable consequences (see section 6.5).

In practice, if the atomic displacements are not too large, the harmonic approximation to phonon excitations is reasonable. For large displacements anharmonic terms become increasingly important, and a more elaborate description is necessary which takes into account phonon–phonon interactions arising from the anharmonic terms. Evidently, this places a limit on the number of phonons that can be excited before the harmonic approximation breaks down.

6.4 Application: the specific heat of crystals

We can use the concept of phonons to determine the thermal properties of crystals, and in particular their specific heat. This is especially interesting at low temperatures, where the quantum nature of excitations becomes important, and gives behavior drastically different from the classical result. We discuss this topic next, beginning with a brief review of the result of the classical theory.

6.4.1 The classical picture

The internal energy per unit volume of a collection of classical particles at inverse temperature $\beta = 1/k_{\mathrm{B}}T$ (k_{B} is Boltzmann's constant) contained in a volume Ω is given by

$$\frac{E}{\Omega} = \frac{1}{\Omega} \frac{\int \{d\mathbf{r}\}\{d\mathbf{p}\} E(\{\mathbf{r}\},\{\mathbf{p}\}) e^{-\beta E(\{\mathbf{r}\},\{\mathbf{p}\})}}{\int \{d\mathbf{r}\}\{d\mathbf{p}\} e^{-\beta E(\{\mathbf{r}\},\{\mathbf{p}\})}}$$

$$= -\frac{1}{\Omega}\frac{\partial}{\partial \beta} \ln \left[\int \{d\mathbf{r}\}\{d\mathbf{p}\} e^{-\beta E(\{\mathbf{r}\},\{\mathbf{p}\})} \right] \tag{6.44}$$

where $\{\mathbf{r}\}$ and $\{\mathbf{p}\}$ are the coordinates and momenta of the particles. In the harmonic approximation, the energy is given in terms of the displacements $\delta\mathbf{r}_n$ from ideal positions as

$$E = E_0 + \frac{1}{2}\sum_{n,n'} \delta\mathbf{r}_n \cdot \mathbf{F}(\mathbf{r}_n - \mathbf{r}_{n'}) \cdot \delta\mathbf{r}_{n'} + \frac{1}{2}\sum_n \frac{\mathbf{p}_n^2}{M_n} \tag{6.45}$$

where E_0 is the energy of the ground state and $\mathbf{F}(\mathbf{r}_n - \mathbf{r}_{n'})$ is the force-constant matrix, defined in Eq. (6.5); the indices n, n' run over all the particles in the system. Since both the potential and kinetic parts involve quadratic expressions, we can rescale all coordinates by $\beta^{1/2}$, in which case the integral in the last expression in Eq. (6.44) is multiplied by a factor of $\beta^{-3N_{at}}$ where N_{at} is the total number of atoms in 3D space. What remains is independent of β, giving for the internal energy per unit volume:

$$\frac{E}{\Omega} = \frac{E_0}{\Omega} + \frac{3N_{at}}{\Omega} k_B T \tag{6.46}$$

from which the specific heat per unit volume can be calculated:

$$c(T) = \frac{\partial}{\partial T}\left[\frac{E}{\Omega}\right] = 3k_B(N_{at}/\Omega) \tag{6.47}$$

and turns out to be a constant independent of temperature. This behavior is referred to as the Dulong–Petit law, which is valid at high temperatures.

6.4.2 The quantum mechanical picture

The quantum mechanical calculation of the internal energy in terms of phonons gives the following expression:

$$\frac{E}{\Omega} = \frac{1}{\Omega}\frac{\sum_s E_s e^{-\beta E_s}}{\sum_s e^{-\beta E_s}} \tag{6.48}$$

where E_s is the energy corresponding to a particular state of the system, which involves a certain number of excited phonons. We will take E_s from Eq. (6.43) that we derived earlier. Just as in the classical discussion, we express the total internal energy as

$$\frac{E}{\Omega} = -\frac{1}{\Omega}\frac{\partial}{\partial \beta}\ln\left(\sum_s e^{-\beta E_s}\right) \tag{6.49}$$

There is a neat mathematical trick that allows us to express this in a more convenient form. Consider the expression

$$\sum_s e^{-\beta E_s} = \sum_s \exp\left[-\sum_{\mathbf{k},l}\left(n_{\mathbf{k}s}^{(l)} + \frac{1}{2}\right)\hbar\omega_{\mathbf{k}}^{(l)}\beta\right] \tag{6.50}$$

which involves a sum of all exponentials containing terms $(n_{\mathbf{k}}^{(l)} + \frac{1}{2})\beta\hbar\omega_{\mathbf{k}}^{(l)}$ with all possible non-negative integer values of $n_{\mathbf{k}}^{(l)}$. Now consider the expression

$$\prod_{\mathbf{k},l} \sum_{n=0}^{\infty} \mathrm{e}^{-\beta(n+\frac{1}{2})\hbar\omega_{\mathbf{k}}^{(l)}} = (\mathrm{e}^{-\beta\hbar\frac{1}{2}\omega_{\mathbf{k}_1}^{(l_1)}} + \mathrm{e}^{-\beta\hbar\frac{3}{2}\omega_{\mathbf{k}_1}^{(l_1)}} + \cdots)(\mathrm{e}^{-\beta\hbar\frac{1}{2}\omega_{\mathbf{k}_2}^{(l_2)}} + \mathrm{e}^{-\beta\hbar\frac{3}{2}\omega_{\mathbf{k}_2}^{(l_2)}} + \cdots) \tag{6.51}$$

It is evident that when these products are expanded out we obtain exponentials with exponents $(n + \frac{1}{2})\beta\hbar\omega_{\mathbf{k}}^{(l)}$ with all possible non-negative values of n. Therefore the expressions in these two equations are the same, and we can substitute the second expression for the first one in Eq. (6.49). Notice further that the geometric series summation gives

$$\sum_{n=0}^{\infty} \mathrm{e}^{-\beta(n+\frac{1}{2})\hbar\omega_{\mathbf{k}}^{(l)}} = \frac{\mathrm{e}^{-\beta\hbar\omega_{\mathbf{k}}^{(l)}/2}}{1 - \mathrm{e}^{-\beta\hbar\omega_{\mathbf{k}}^{(l)}}} \tag{6.52}$$

which gives for the total internal energy per unit volume

$$\frac{E}{\Omega} = -\frac{1}{\Omega}\frac{\partial}{\partial\beta} \ln\left[\prod_{\mathbf{k},l} \frac{\mathrm{e}^{-\beta\hbar\omega_{\mathbf{k}}^{(l)}/2}}{1 - \mathrm{e}^{-\beta\hbar\omega_{\mathbf{k}}^{(l)}}}\right] = \frac{1}{\Omega}\sum_{\mathbf{k},l} \hbar\omega_{\mathbf{k}}^{(l)}\left(\bar{n}_{\mathbf{k}}^{(l)} + \frac{1}{2}\right) \tag{6.53}$$

where we have defined $\bar{n}_{\mathbf{k}}^{(l)}$ as

$$\bar{n}_{\mathbf{k}}^{(l)}(T) = \frac{1}{\mathrm{e}^{\beta\hbar\omega_{\mathbf{k}}^{(l)}} - 1} \tag{6.54}$$

This quantity represents the average occupation of the phonon state with frequency $\omega_{\mathbf{k}}^{(l)}$, at temperature T, and is appropriate for bosons (see Appendix D). From Eq. (6.53) we can now calculate the specific heat per unit volume:

$$c(T) = \frac{\partial}{\partial T}\left[\frac{E}{\Omega}\right] = \frac{1}{\Omega}\sum_{\mathbf{k},l} \hbar\omega_{\mathbf{k}}^{(l)} \frac{\partial}{\partial T}\bar{n}_{\mathbf{k}}^{(l)}(T) \tag{6.55}$$

We examine first the behavior of this expression in the low-temperature limit. In this limit, $\beta\hbar\omega_{\mathbf{k}}^{(l)}$ becomes very large, and therefore $\bar{n}_{\mathbf{k}}^{(l)}(T)$ becomes negligibly small, except when $\omega_{\mathbf{k}}^{(l)}$ happens to be very small. We saw in an earlier discussion that $\omega_{\mathbf{k}}^{(l)}$ goes to zero linearly near the center of the BZ ($\mathbf{k} \approx 0$), for the acoustic branches. We can then write that near the center of the BZ $\omega_{\mathbf{k}}^{(l)} = v_{\hat{\mathbf{k}}}^{(l)}k$, where $k = |\mathbf{k}|$ is the wave-vector magnitude and $v_{\hat{\mathbf{k}}}^{(l)}$ is the sound velocity which depends on the direction of the wave-vector $\hat{\mathbf{k}}$ and the acoustic branch label l. For large β we can then approximate all frequencies as $\omega_{\mathbf{k}}^{(l)} = v_{\hat{\mathbf{k}}}^{(l)}k$ over the entire BZ, since for large values of k the contributions are negligible because of the factor $\exp(\hbar v_{\hat{\mathbf{k}}}^{(l)}k\beta)$

which appears in the denominator. Turning the sum into an integral as usual, we obtain for the specific heat at low temperature

$$c(T) = \frac{\partial}{\partial T} \sum_l \int \frac{d\mathbf{k}}{(2\pi)^3} \frac{\hbar v_{\hat{\mathbf{k}}}^{(l)} k}{e^{\hbar v_{\hat{\mathbf{k}}}^{(l)} k\beta} - 1} \tag{6.56}$$

We will change variables as in the case of the classical calculation to make the integrand independent of inverse temperature β, by taking $\hbar v_{\hat{\mathbf{k}}}^{(l)} k\beta = t_{\hat{\mathbf{k}}}^{(l)}$, which gives a factor of β^{-4} in front of the integral, while the integral now has become independent of temperature. In fact, we can use the same argument about the negligible contributions of large values of k to extend the integration to infinity in the variable k, so that the result is independent of the BZ shape. Denoting by c_0 the value of the constant which is obtained by performing the integration and encompasses all other constants in front of the integral, we then find that

$$c(T) = c_0 \frac{\partial}{\partial T} T^4 = 4c_0 T^3 \tag{6.57}$$

that is, the behavior of the specific heat at low temperature is cubic in the temperature and not constant as the Dulong-Petit law suggests from the classical calculation.

In the high-temperature limit, i.e. when $k_B T \gg \hbar\omega_{\mathbf{k}}^{(l)}$, the Bose occupation factors take the form

$$\frac{1}{e^{\beta\hbar\omega_{\mathbf{k}}^{(l)}} - 1} \approx \frac{k_B T}{\hbar\omega_{\mathbf{k}}^{(l)}} \tag{6.58}$$

and with this, the expression for the specific heat becomes

$$c(T) = \frac{1}{\Omega} \sum_{\mathbf{k},l} k_B = 3k_B(N_{at}/\Omega) \tag{6.59}$$

which is the same result as in the classical calculation. Thus, at sufficiently high temperatures the Dulong-Petit law is recovered.

6.4.3 The Debye model

A somewhat more quantitative discussion of the behavior of the specific heat as a function of the temperature is afforded by the Debye model. In this model we assume all frequencies to be linear in the wave-vector magnitude, as acoustic modes near $\mathbf{k} \approx 0$ are. For simplicity we assume we are dealing with an isotropic solid, so we can take $\omega = vk$ for all the different acoustic branches, with v, the sound velocity, being the same in all directions. The total number of phonon modes is equal to $3N_{at}$, where N_{at} is the total number of atoms in the crystal, and for each value of $\mathbf{k}$ there are 3ν normal modes in 3D (ν being the number of atoms in the PUC). Strictly speaking, the Debye model makes sense only for crystals with one atom per PUC, because it assumes all modes to be acoustic in nature (they behave like $\omega \sim k$ for $k \to 0$). Just as in the case of electronic states, we can use a normalization argument to relate the density of the crystal, $n = N_{at}/\Omega$, to the highest value of the wave-vector that phonons can assume (denoted by $k_{\rm D}$):

$$\sum_{\mathbf{k}} 3\nu = 3N_{at} \Longrightarrow 3\Omega \int_{k=0}^{k=k_{\rm D}} \frac{d\mathbf{k}}{(2\pi)^3} = 3N_{at} \Longrightarrow n = \frac{N_{at}}{\Omega} = \frac{k_{\rm D}^3}{6\pi^2} \tag{6.60}$$

from which the value of $k_{\rm D}$ is determined in terms of n. Notice that $k_{\rm D}$ is determined by considering the total number of *phonon modes*, not the actual number of phonons present in some excited state of the system, which of course can be anything since phonons are bosons. We can also define the Debye frequency $\omega_{\rm D}$ and Debye temperature $\Theta_{\rm D}$, through

$$\omega_{\rm D} = vk_{\rm D}, \qquad \Theta_{\rm D} = \frac{\hbar\omega_{\rm D}}{k_{\rm B}} \tag{6.61}$$

With these definitions, we obtain the following expression for the specific heat at any temperature:

$$c(T) = 9nk_{\rm B}\left(\frac{T}{\Theta_{\rm D}}\right)^3 \int_0^{\Theta_{\rm D}/T} \frac{t^4 e^t}{(e^t - 1)^2} dt \tag{6.62}$$

In the low-temperature limit, with $\Theta_{\rm D}$ much larger than T, the upper limit of the integral is approximated by ∞, and there is no temperature dependence left in the integral which gives a constant when evaluated explicitly. This reduces to the expression we discussed above, Eq. (6.57). What happens for higher temperatures? Notice first that the integrand is always a positive quantity. As the temperature increases, the upper limit of the integral becomes smaller and smaller, and the value obtained from integration is lower than the value obtained in the low-temperature limit. Therefore, as the temperature increases the value of the specific heat increases slower than T^3, as shown in Fig. 6.5. As we saw above, at sufficiently high temperature the specific heat eventually approaches the constant value given by the Dulong–Petit law.

The physical meaning of the Debye temperature is somewhat analogous to that of the Fermi level for electrons. For T above the Debye temperature all phonon modes are excited, that is, $\bar{n}_{\mathbf{k}}^{(l)}(T) > 0$ for all frequencies, while for T below the Debye temperature, the high-frequency phonon modes are frozen, that is, $\bar{n}_{\mathbf{k}}^{(l)}(T) \approx 0$ for those modes. The Debye model is oversimplified, because it treats the frequency spectrum as linearly dependent on the wave-vectors, $\omega = vk$ for all modes. For a realistic calculation we need to include the actual frequency spectrum, which produces a density of states with features that depend on the structure of the solid,

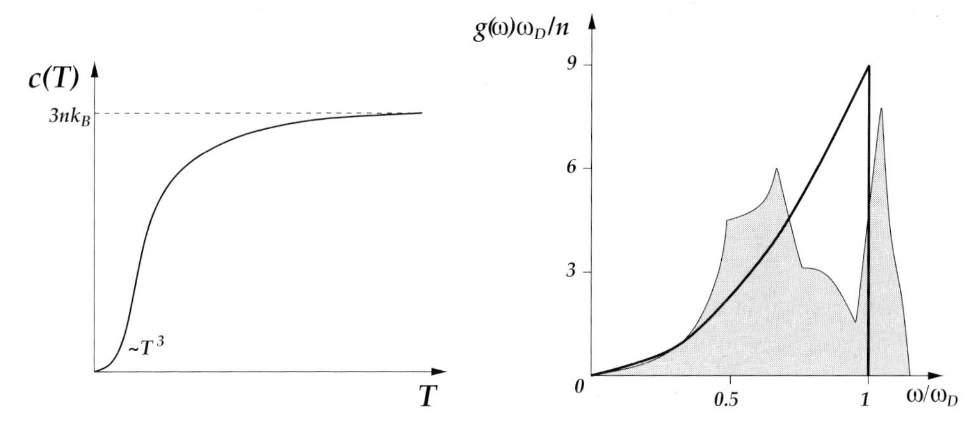

Figure 6.5. **Left:** behavior of the specific heat c as a function of temperature T in the Debye model. The asymptotic value for large T corresponds to the Dulong–Petit law. **Right:** comparison of the density of states in the Debye model as given by Eq. (6.65), shown by the thick solid line, and a realistic calculation (adapted from Ref. [78]), shown by the shaded curve, for Al.

analogous to what we had discussed for the electronic energy spectrum. An example (Al) is shown in Fig. 6.5.

Within the Debye model, the density of phonon modes at frequency ω per unit volume of the crystal, $g(\omega)$, can be easily obtained starting with the general expression

$$g(\omega) = \frac{1}{\Omega}\sum_{\mathbf{k},l}\delta(\omega - \omega_{\mathbf{k}}^{(l)}) = \sum_{l}\int\frac{d\mathbf{k}}{(2\pi)^3}\delta(\omega - \omega_{\mathbf{k}}^{(l)}) = \sum_{l}\int_{\omega=\omega_{\mathbf{k}}^{(l)}}\frac{dS_{\mathbf{k}}}{(2\pi)^3}\frac{1}{|\nabla_{\mathbf{k}}\omega_{\mathbf{k}}^{(l)}|}$$

where the last expression involves an integral over a surface in $\mathbf{k}$-space on which $\omega_{\mathbf{k}}^{(l)} = \omega$, the argument of the density of states. In the Debye model we have $\omega_{\mathbf{k}}^{(l)} = v|\mathbf{k}|$, which gives $|\nabla_{\mathbf{k}}\omega_{\mathbf{k}}^{(l)}| = v$, with v the average sound velocity. For fixed value of $\omega = \omega_{\mathbf{k}}^{(l)} = v|\mathbf{k}|$, the integral over a surface in $\mathbf{k}$-space that corresponds to this value of ω is equal to the surface of a sphere of radius $|\mathbf{k}| = \omega/v$, and since there are three phonon branches in 3D space (all having the same average sound velocity within the Debye model), we find for the density of states

$$g(\omega) = \frac{3}{(2\pi)^3}\frac{1}{v}4\pi\left(\frac{\omega}{v}\right)^2 = \frac{3}{2\pi^2}\frac{\omega^2}{v^3} \tag{6.63}$$

that is, a simple quadratic expression in ω. If we calculate the total number of available phonon modes per unit volume N_{ph}/Ω up to the Debye frequency ω_{D} we find

$$\frac{N_{ph}}{\Omega} = \int_0^{\omega_{\mathrm{D}}} g(\omega)\mathrm{d}\omega = \frac{1}{2\pi^2}\frac{\omega_{\mathrm{D}}^3}{v^3} = 3\frac{1}{6\pi^2}k_{\mathrm{D}}^3 \tag{6.64}$$

which from Eq. (6.60) is equal to $3(N_{at}/\Omega)$, exactly as we would expect for a 3D crystal containing N_{at} atoms. Using the definition of the Debye frequency we can express the density of states as

$$\frac{g(\omega)\omega_{\rm D}}{n} = 9\left(\frac{\omega}{\omega_{\rm D}}\right)^2 \tag{6.65}$$

which is a convenient relation between dimensionless quantities. In Fig. 6.5 we show a comparison of the density of states in the Debye model as given by Eq. (6.65), and as obtained by a realistic calculation [78], for Al. This comparison illustrates how simple the Debye model is relative to real phonon modes. Nevertheless, the model is useful in that it provides a simple justification for the behavior of the specific heat as a function of temperature as well as a rough measure of the highest phonon frequency, $\omega_{\rm D}$. In fact, the form of the specific heat predicted by the Debye model, Eq. (6.62), is often used to determine the Debye temperature, and through it the Debye frequency. Since this relation involves temperature dependence, a common practice is to fit the observed specific heat at given T to one-half the value of the Dulong–Petit law as obtained from Eq. (6.62) [79], which determines $\Theta_{\rm D}$. Results of this fitting approach for the Debye temperatures of several elemental solids are given in Table 6.3.

Table 6.3. *Debye temperatures $\Theta_{\rm D}$ and frequencies $\omega_{\rm D}$ for elemental solids.*

Values of $\Theta_{\rm D}$ (in K) and $\omega_{\rm D}$ (in THz) were determined by fitting the observed value of the specific heat at a certain temperature to half the value of the Dulong–Petit law through Eq. (6.62).

Element	$\Theta_{\rm D}$	$\omega_{\rm D}$	Element	$\Theta_{\rm D}$	$\omega_{\rm D}$	Element	$\Theta_{\rm D}$	$\omega_{\rm D}$
Li	400	8.33	Na	150	3.12	K	100	2.08
Be	1000	20.84	Mg	318	6.63	Ca	230	4.79
B	1250	26.05	Al	394	8.21	Ga	240	5.00
C[a]	1860	38.76	Si	625	13.02	Ge	350	7.29
As	285	5.94	Sb	200	4.17	Bi	120	2.50
Cu	315	6.56	Ag	215	4.48	Au	170	3.54
Zn	234	4.88	Cd	120	2.50	Hg	100	2.08
Cr	460	9.58	Mo	380	7.92	W	310	6.46
Mn	400	8.33	Fe	420	8.75	Co	385	8.02
Ni	375	7.81	Pd	275	5.73	Pt	230	4.79

[a] In the diamond phase. Source: Ref. [78].

6.4.4 Thermal expansion coefficient

One physical effect which is related to phonon modes and to the specific heat is the thermal expansion of a solid. The thermal expansion coefficient α is defined as the rate of change of the linear dimension L of the solid with temperature T, normalized by this linear dimension, at constant pressure P:

$$\alpha \equiv \frac{1}{L}\left(\frac{\partial L}{\partial T}\right)_P \tag{6.66}$$

The linear dimension of the solid L is related to its volume Ω through $L = \Omega^{1/3}$, which gives for the thermal expansion coefficient in terms of the volume

$$\alpha = \frac{1}{3\Omega}\left(\frac{\partial \Omega}{\partial T}\right)_P \tag{6.67}$$

This can also be expressed in terms of the bulk modulus B, the negative inverse of which is the rate of change of the volume with pressure, normalized by the volume, at constant temperature:

$$B^{-1} \equiv -\frac{1}{\Omega}\left(\frac{\partial \Omega}{\partial P}\right)_T \tag{6.68}$$

(see also the equivalent definition given in chapter 5, Eq. (5.87)). With the help of standard thermodynamic relations (see Appendix C) we obtain for the thermal expansion coefficient, in terms of the bulk modulus and the pressure

$$\alpha = \frac{1}{3B}\left(\frac{\partial P}{\partial T}\right)_\Omega \tag{6.69}$$

The pressure is given by the negative rate of change of the internal energy E with volume, at constant temperature

$$P = -\left(\frac{\partial E}{\partial \Omega}\right)_T \tag{6.70}$$

When changes in the internal energy E are exclusively due to phonon excitations, we can use E from Eq. (6.53) to obtain for the thermal expansion coefficient

$$\alpha = \frac{1}{3B}\sum_{\mathbf{k},l}\left(-\frac{\partial \hbar\omega_{\mathbf{k}}^{(l)}}{\partial \Omega}\right)_T\left(\frac{\partial \bar{n}_{\mathbf{k}}^{(l)}}{\partial T}\right)_\Omega \tag{6.71}$$

which involves the same quantities as the expression for the specific heat c, Eq. (6.55), that is, the phonon frequencies $\omega_{\mathbf{k}}^{(l)}$ and the corresponding average occupation numbers $\bar{n}_{\mathbf{k}}^{(l)}$ at temperature T. In fact, it is customary to express the thermal expansion coefficient in terms of the specific heat as

$$\alpha = \frac{\gamma c}{3B} \tag{6.72}$$

where the coefficient γ is known as the Grüneisen parameter. This quantity is given by

$$\gamma = \frac{1}{c}\sum_{\mathbf{k},l} -\frac{\partial \hbar\omega_{\mathbf{k}}^{(l)}}{\partial \Omega}\frac{\partial \bar{n}_{\mathbf{k}}^{(l)}}{\partial T} = -\Omega\left[\sum_{\mathbf{k},l}\frac{\partial \hbar\omega_{\mathbf{k}}^{(l)}}{\partial \Omega}\frac{\partial \bar{n}_{\mathbf{k}}^{(l)}}{\partial T}\right]\left[\sum_{\mathbf{k},l}\hbar\omega_{\mathbf{k}}^{(l)}\frac{\partial \bar{n}_{\mathbf{k}}^{(l)}}{\partial T}\right]^{-1} \tag{6.73}$$

We can simplify the notation in this equation by defining the contributions to the specific heat from each phonon mode as

$$c_{\mathbf{k}}^{(l)} = \hbar\omega_{\mathbf{k}}^{(l)}\frac{\partial \bar{n}_{\mathbf{k}}^{(l)}}{\partial T} \tag{6.74}$$

and the mode-specific Grüneisen parameters as

$$\gamma_{\mathbf{k}}^{(l)} = -\frac{\Omega}{\omega_{\mathbf{k}}^{(l)}}\frac{\partial \omega_{\mathbf{k}}^{(l)}}{\partial \Omega} = -\frac{\partial(\ln \omega_{\mathbf{k}}^{(l)})}{\partial(\ln \Omega)} \tag{6.75}$$

in terms of which the Grüneisen parameter takes the form

$$\gamma = \left[\sum_{\mathbf{k},l}\gamma_{\mathbf{k}}^{(l)}c_{\mathbf{k}}^{(l)}\right]\left[\sum_{\mathbf{k},l}c_{\mathbf{k}}^{(l)}\right]^{-1} \tag{6.76}$$

The Grüneisen parameter is a quantity that can be measured directly by experiment.

Important warning We should alert the reader to the fact that so far we have taken into account only the excitation of phonons as contributing to changes in the internal energy of the solid. This is appropriate for semiconductors and insulators where electron excitation across the band gap is negligible for usual temperatures: For these solids, for temperatures as high as their melting point, the thermal energy is still much lower than the band gap energy, the minimum energy required to excite electrons from their ground state. For metals, on the other hand, the excitation of electrons is as important as that of phonons; when it is included in the picture, it gives different behavior of the specific heat and the thermal expansion coefficient at low temperature. These effects can be treated explicitly by modeling the electrons as a Fermi liquid, which requires a many-body picture; a detailed description of the thermodynamics of the Fermi liquid is given in Fetter and Walecka [17].

6.5 Application: phonon scattering

We consider next the effects of scattering of particles or waves incident on a crystal, in the presence of phonon excitations. In fact, phonon frequencies are measured experimentally by inelastic neutron scattering. Neutrons are typically scattered by atomic nuclei in the solid; an inelastic collision involves changes in the total energy and total momentum, which imply a change in the number of phonons that are excited in the solid. Specifically, the change in energy will be given by

$$\Delta E = \sum_{\mathbf{k},l} \hbar\omega_{\mathbf{k}}^{(l)} \Delta n_{\mathbf{k}}^{(l)} \tag{6.77}$$

where $n_{\mathbf{k}}^{(l)}$ is the number of phonons of frequency $\omega_{\mathbf{k}}^{(l)}$ that are excited. Similarly, the change in momentum is given by

$$\Delta \mathbf{p} = \sum_{\mathbf{k},l} \hbar(\mathbf{k} + \mathbf{G}) \Delta n_{\mathbf{k}}^{(l)} \tag{6.78}$$

For processes that involve a single phonon we will have for the energy and momentum before $(E, \mathbf{p})$ and after $(E', \mathbf{p}')$ a collision:

$$\begin{aligned} E' &= E + \hbar\omega_{\mathbf{k}}^{(l)}, \quad \mathbf{p}' = \mathbf{p} + \hbar(\mathbf{k} + \mathbf{G}) \quad \text{(absorption)} \\ E' &= E - \hbar\omega_{\mathbf{k}}^{(l)}, \quad \mathbf{p}' = \mathbf{p} - \hbar(\mathbf{k} + \mathbf{G}) \quad \text{(emission)} \end{aligned} \tag{6.79}$$

Using the fact that $\omega_{\mathbf{k}\pm\mathbf{G}}^{(l)} = \omega_{\mathbf{k}}^{(l)}$, we obtain the following relation from conservation of energy:

$$\frac{\mathbf{p}'^2}{2M_n} = \frac{\mathbf{p}^2}{2M_n} \pm \hbar\omega_{(\mathbf{p}'-\mathbf{p})/\hbar}^{(l)} \tag{6.80}$$

with (+) corresponding to absorption and (−) to emission of a phonon, and M_n the mass of the neutron. In experiment $\mathbf{p}$ is the momentum of the incident neutron beam and $\mathbf{p}'$ is the momentum of the scattered beam. Eq. (6.80) has solutions only if $\omega_{(\mathbf{p}-\mathbf{p}')/\hbar}^{(l)}$ corresponds to a phonon frequency $\omega_{\mathbf{k}}^{(l)}$. In reality it is impossible to separate the single-phonon events from those that involve many phonons, for which we cannot apply the energy and momentum conservation equations separately. Thus, for every value of the energy and momentum of neutrons there will be a broad background, corresponding to multiphonon processes. However, the flux of scattered neutrons as a function of their energy (which is given by $\mathbf{p}'^2/2M_n$) will exhibit sharp peaks in certain directions $\hat{\mathbf{p}}'$ which correspond to phonon frequencies that satisfy Eq. (6.80). From these peaks, one determines the phonon spectrum by scanning the energy of the scattered neutron beam along different directions. The type of signal obtained from such experiments is shown schematically in Fig. 6.6.

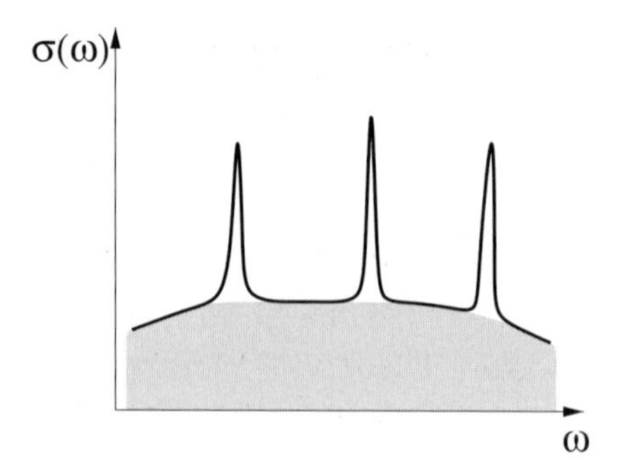

Figure 6.6. Schematic representation of the cross-section of inelastic neutron scattering experiments. The broad background is due to multiphonon processes; the individual peaks correspond to single-phonon events from which the phonon frequencies are determined. The width of the single-phonon peaks are due to anharmonic processes.

6.5.1 Phonon scattering processes

In order to provide a more quantitative picture of phonon scattering we consider the scattering of an electron in a plane wave state from a solid in the presence of phonons. We will denote the incident wave-vector of the electron by $\mathbf{k}$ and the scattered wave-vector by $\mathbf{k}'$. The matrix element $M_{\mathbf{kk}'}$ for the scattering process will be given by the expectation value of the scattering potential, that is, the potential of all the ions in the crystal, $V_{cr}(\mathbf{r})$, between initial and final states:

$$
\begin{aligned}
M_{\mathbf{kk}'} &= \frac{1}{N\Omega_{PUC}} \int \mathrm{e}^{-\mathrm{i}\mathbf{k}'\cdot\mathbf{r}} V_{cr}(\mathbf{r}) \mathrm{e}^{\mathrm{i}\mathbf{k}\cdot\mathbf{r}} \mathrm{d}\mathbf{r} \\
V_{cr}(\mathbf{r}) &= \sum_{nj} V_{at}(\mathbf{r} - \mathbf{t}_j - \mathbf{R}_n)
\end{aligned}
\tag{6.81}
$$

where $V_{at}(\mathbf{r} - \mathbf{t}_j - \mathbf{R}_n)$ is the ionic potential of an atom situated at $\mathbf{t}_j$ in the unit cell of the crystal at the lattice vector $\mathbf{R}_n$; N is the total number of unit cells in the crystal, each of volume Ω_{PUC}. For simplicity, we will assume that the solid contains only one type of atoms, so there is only one type of ionic potential $V_{at}(\mathbf{r})$. Inserting the expression for the crystal potential in the scattering matrix element, we obtain

$$
M_{\mathbf{kk}'} = \frac{1}{N} \sum_{n} V_{at}(\mathbf{q}) \sum_{j} \mathrm{e}^{-\mathrm{i}\mathbf{q}\cdot(\mathbf{t}_j + \mathbf{R}_n)} \tag{6.82}
$$

$$
V_{at}(\mathbf{q}) = \frac{1}{\Omega_{PUC}} \int V_{at}(\mathbf{r}) \mathrm{e}^{-\mathrm{i}\mathbf{q}\cdot\mathbf{r}} \mathrm{d}\mathbf{r} \tag{6.83}
$$

with $V_{at}(\mathbf{q})$ the Fourier transform of the ionic potential $V_{at}(\mathbf{r})$. In the above expressions we have introduced the vector $\mathbf{q} = \mathbf{k}' - \mathbf{k}$. If this vector happens to be equal to a reciprocal-lattice vector $\mathbf{G}$, the scattering matrix element takes the form

$$M_{\mathbf{kk}'} = \frac{1}{N}\sum_n V_{at}(\mathbf{G})\sum_j e^{-i\mathbf{G}\cdot\mathbf{t}_j} = V_{at}(\mathbf{G})S(\mathbf{G}) \tag{6.84}$$

where we have identified the sum over the atoms in the unit cell j as the structure factor $S(\mathbf{G})$, the quantity we had defined in chapter 4, Eq. (4.50); the other sum over the crystal cells n is canceled in this case by the factor $1/N$. If $\mathbf{q} \neq \mathbf{G}$, we can generalize the definition of the structure factor as follows:

$$S(\mathbf{q}) \equiv \frac{1}{N}\sum_{nj} e^{-i\mathbf{q}\cdot(\mathbf{t}_j+\mathbf{R}_n)} \tag{6.85}$$

in which case the scattering matrix element takes the form

$$M_{\mathbf{kk}'} = V_{at}(\mathbf{q})S(\mathbf{q}) \tag{6.86}$$

We are now interested in determining the behavior of this matrix element when, due to thermal motion, the ions are not at their ideal crystalline positions. Obviously, only the structure factor $S(\mathbf{q})$ is affected by this departure of the ions from their ideal positions, so we examine its behavior. The thermal motion leads to deviations from the ideal crystal positions, which we denote by $\mathbf{s}_{nj}$. With these deviations, the structure factor takes the form

$$S(\mathbf{q}) = \frac{1}{N}\sum_{n,j} e^{-i\mathbf{q}\cdot(\mathbf{t}_j+\mathbf{R}_n)}e^{-i\mathbf{q}\cdot\mathbf{S}_{nj}} \tag{6.87}$$

For simplicity, we will assume from now on that there is only one atom in each unit cell, which allows us to eliminate the summation over the index j. We can express the deviations of the ions from their crystalline positions $\mathbf{S}_n$ in terms of the amplitudes of phonons $Q^{(l)}_{\mathbf{q}'}$ and the corresponding eigenvectors $\hat{\mathbf{e}}^{(l)}_{\mathbf{q}'}$, with l denoting the phonon branch:

$$\mathbf{S}_n = \sum_{l,\mathbf{q}'} Q^{(l)}_{\mathbf{q}'}\hat{\mathbf{e}}^{(l)}_{\mathbf{q}'}e^{i\mathbf{q}'\cdot\mathbf{R}_n} \tag{6.88}$$

and with this expression the structure factor becomes

$$S(\mathbf{q}) = \frac{1}{N}\sum_n e^{-i\mathbf{q}\cdot\mathbf{R}_n}\exp\left[-i\mathbf{q}\cdot\sum_{l,\mathbf{q}'}(Q^{(l)}_{\mathbf{q}'}\hat{\mathbf{e}}^{(l)}_{\mathbf{q}'})e^{i\mathbf{q}'\cdot\mathbf{R}_n}\right] \tag{6.89}$$

We will take the deviations from ideal positions to be small quantities and the corresponding phonon amplitudes to be small. To simplify the calculations, we define the vectors $\mathbf{f}^{(l)}_{\mathbf{q}'n}$ as

$$\mathbf{f}^{(l)}_{n\mathbf{q}'} = Q^{(l)}_{\mathbf{q}'}\hat{\mathbf{e}}^{(l)}_{\mathbf{q}'}e^{i\mathbf{q}'\cdot\mathbf{R}_n} \tag{6.90}$$

which, by our assumption above, will have small magnitude. Using this fact, we can expand the exponential with square brackets in Eq. (6.89) as follows:

$$\exp\left[-\mathrm{i}\mathbf{q}\cdot\sum_{l,\mathbf{q}'}\mathbf{f}^{(l)}_{n\mathbf{q}'}\right]=\prod_{l,\mathbf{q}'}\exp\left[-\mathrm{i}\mathbf{q}\cdot\mathbf{f}^{(l)}_{n\mathbf{q}'}\right]$$
$$=\prod_{l,\mathbf{q}'}\left[1-\mathrm{i}\mathbf{q}\cdot\mathbf{f}^{(l)}_{n\mathbf{q}'}-\frac{1}{2}\left(\mathbf{q}\cdot\mathbf{f}^{(l)}_{n\mathbf{q}'}\right)^2+\cdots\right] \quad (6.91)$$

Keeping only terms up to second order in $|\mathbf{f}^{(l)}_{n\mathbf{q}'}|$ in the last expression gives the following result:

$$\left[1-\mathrm{i}\mathbf{q}\cdot\sum_{l,\mathbf{q}'}\mathbf{f}^{(l)}_{n\mathbf{q}'}-\frac{1}{2}\sum_{l,\mathbf{q}'}\left(\mathbf{q}\cdot\mathbf{f}^{(l)}_{n\mathbf{q}'}\right)^2-\sum_{l\mathbf{q}'<l'\mathbf{q}''}\left(\mathbf{q}\cdot\mathbf{f}^{(l)}_{n\mathbf{q}'}\right)\left(\mathbf{q}\cdot\mathbf{f}^{(l')}_{n\mathbf{q}''}\right)\right]$$
$$=\left[1-\mathrm{i}\mathbf{q}\cdot\sum_{l,\mathbf{q}'}\mathbf{f}^{(l)}_{n\mathbf{q}'}-\frac{1}{2}\sum_{ll',\mathbf{q}'\mathbf{q}''}\left(\mathbf{q}\cdot\mathbf{f}^{(l)}_{n\mathbf{q}'}\right)\left(\mathbf{q}\cdot\mathbf{f}^{(l')}_{n\mathbf{q}''}\right)\right] \quad (6.92)$$

Let us consider the physical meaning of these terms by order, when the above expression is substituted into Eq. (6.89)

(0) The zeroth order term in Eq. (6.92) gives

$$S_0(\mathbf{q})=\frac{1}{N}\sum_n \mathrm{e}^{-\mathrm{i}\mathbf{q}\cdot\mathbf{R}_n} \quad (6.93)$$

which is the structure factor for a crystal with the atomic positions frozen at the ideal crystal sites, that is, in the absence of any phonon excitations.

(1) The first order term in Eq. (6.92) gives

$$-\mathrm{i}\mathbf{q}\cdot\sum_{\mathbf{q}'l}\frac{1}{N}\sum_n\mathbf{f}^{(l)}_{n\mathbf{q}'}\mathrm{e}^{-\mathrm{i}\mathbf{q}\cdot\mathbf{R}_n}=-\mathrm{i}\mathbf{q}\cdot\sum_{\mathbf{q}'l}\frac{1}{N}\sum_n Q^{(l)}_{\mathbf{q}'}\hat{\mathbf{e}}^{(l)}_{\mathbf{q}'}\mathrm{e}^{\mathrm{i}(\mathbf{q}'-\mathbf{q})\cdot\mathbf{R}_n}$$
$$=-\mathrm{i}\mathbf{q}\cdot\sum_{\mathbf{q}'l}Q^{(l)}_{\mathbf{q}'}\hat{\mathbf{e}}^{(l)}_{\mathbf{q}'}\delta(\mathbf{q}'-\mathbf{q}-\mathbf{G})=-\mathrm{i}\sum_l(\mathbf{q}\cdot\hat{\mathbf{e}}^{(l)}_{\mathbf{q}+\mathbf{G}})Q^{(l)}_{\mathbf{q}+\mathbf{G}} \quad (6.94)$$

This is an interesting result: it corresponds to the scattering from a single-phonon mode, with wave-vector $\mathbf{q}$ and amplitude $Q^{(l)}_{\mathbf{q}+\mathbf{G}}$, and involves the projection of the scattering wave-vector $\mathbf{q}$ onto the polarization $\hat{\mathbf{e}}^{(l)}_{\mathbf{q}+\mathbf{G}}$ of this phonon. If $\mathbf{G}=0$, these processes are called normal single-phonon scattering processes; if $\mathbf{G}\neq 0$ they are called Umklapp processes. The latter usually contribute less to scattering, so we will ignore them in the following.

(2) The second order term in Eq. (6.92) gives

$$-\frac{1}{2}\sum_{ll'\mathbf{q}'\mathbf{q}''}\frac{1}{N}\sum_{n}(\mathbf{q}\cdot\mathbf{f}^{(l)}_{n\mathbf{q}'})(\mathbf{q}\cdot\mathbf{f}^{(l')}_{n\mathbf{q}''})\mathrm{e}^{-\mathrm{i}\mathbf{q}\cdot\mathbf{R}_n}$$

$$= -\frac{1}{2}\sum_{ll'\mathbf{q}'\mathbf{q}''}(\mathbf{q}\cdot Q^{(l)}_{\mathbf{q}'}\hat{\mathbf{e}}^{(l)}_{\mathbf{q}'})(\mathbf{q}\cdot Q^{(l')}_{\mathbf{q}''}\hat{\mathbf{e}}^{(l')}_{\mathbf{q}''})\frac{1}{N}\sum_{n}\mathrm{e}^{\mathrm{i}(\mathbf{q}'+\mathbf{q}''-\mathbf{q})\cdot\mathbf{R}_n}$$

$$= -\frac{1}{2}\sum_{ll'\mathbf{q}'\mathbf{q}''}(\mathbf{q}\cdot Q^{(l)}_{\mathbf{q}'}\hat{\mathbf{e}}^{(l)}_{\mathbf{q}'})(\mathbf{q}\cdot Q^{(l')}_{\mathbf{q}''}\hat{\mathbf{e}}^{(l')}_{\mathbf{q}''})\delta(\mathbf{q}'+\mathbf{q}''-\mathbf{q})$$

$$= -\frac{1}{2}\sum_{ll'\mathbf{q}'}(\mathbf{q}\cdot Q^{(l)}_{\mathbf{q}'}\hat{\mathbf{e}}^{(l)}_{\mathbf{q}'})(\mathbf{q}\cdot Q^{(l')}_{\mathbf{q}-\mathbf{q}'}\hat{\mathbf{e}}^{(l')}_{\mathbf{q}-\mathbf{q}'}) \tag{6.95}$$

This expression can be interpreted as the scattering from two phonons, one of wave-vector $\mathbf{q}'$, the other of wave-vector $\mathbf{q}-\mathbf{q}'$; the corresponding phonon amplitudes and projections of the scattering wave-vector onto the phonon polarizations are also involved.

This set of processes can be represented in a more graphical manner in terms of diagrams, as illustrated in Fig. 6.7. The incident and scattered wave-vectors are represented by normal vectors, while the wave-vectors of the phonons are represented by wavy lines of the proper direction and magnitude to satisfy momentum conservation by vector addition. Simple rules can be devised to make the connection between such diagrams and the expressions in the equations above that give their contribution to the structure factor, and hence the scattering cross-section. For example, the following simple rules would generate the terms we calculated above from the diagrams shown in Fig. 6.7.

(i) Every vertex where a phonon wave-vector $\mathbf{q}'$ intersects other wave-vectors introduces a factor of $(-\mathrm{i})\sum_l Q^{(l)}_{\mathbf{q}'}\mathbf{q}\cdot\hat{\mathbf{e}}^{(l)}_{\mathbf{q}'}$, with $\mathbf{q}$ the total scattering wave-vector.

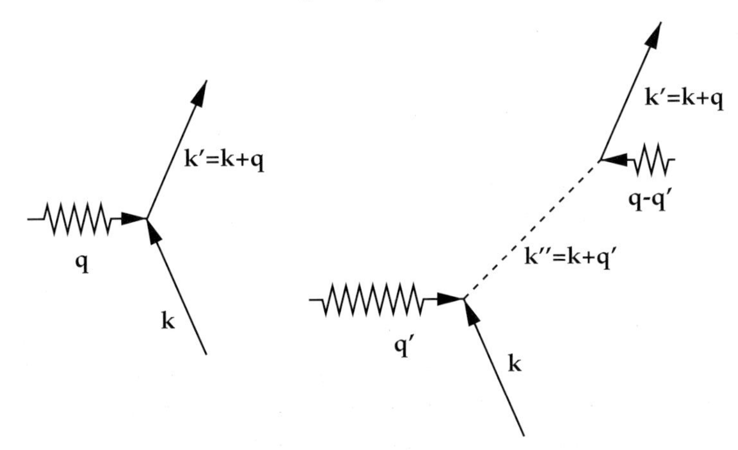

Figure 6.7. Diagrams for the one- and two-phonon scattering processes.

(ii) A factor of $(1/m!)$ accompanies the term of order m in $\mathbf{q}$.
(iii) When an intermediate phonon wave-vector $\mathbf{q}'$, other than the final scattering phonon wave-vector $\mathbf{q}$, appears in a diagram, it is accompanied by a sum over all allowed values.

These rules can be applied to generate diagrams of higher orders, which we have not discussed here. The use of diagrammatic techniques simplifies many calculations of the type outlined above. For perturbative calculations in many-body interacting systems such techniques are indispensable.

6.5.2 The Debye–Waller factor

For an arbitrary state of the system, we are not interested in individual phonon scattering processes but rather in the thermal average over such events. We are also typically interested in the absolute value squared of the scattering matrix element, which enters in the expression for the cross-section, the latter being the experimentally measurable quantity. When we average over the sums that appear in the absolute value squared of the structure factor, the contribution that survives is

$$\langle |S(\mathbf{q})|^2 \rangle = |S_0(\mathbf{q})|^2 \left[1 - \frac{1}{N} \sum_{l\mathbf{q}'} \left\langle \left| \mathbf{q} \cdot \mathbf{f}^{(l)}_{n\mathbf{q}'} \right|^2 \right\rangle \right] \tag{6.96}$$

Taking advantage again of the fact that the magnitude of $\mathbf{f}^{(l)}_{n\mathbf{q}}$ vectors is small, we can approximate the factor in square brackets by

$$\mathrm{e}^{-2W} \approx \left[1 - \frac{1}{N} \sum_{l\mathbf{q}'} \left\langle \left| \mathbf{q} \cdot \mathbf{f}^{(l)}_{n\mathbf{q}'} \right|^2 \right\rangle \right] \Longrightarrow W = \frac{1}{2N} \sum_{l\mathbf{q}'} \left\langle \left| \mathbf{q} \cdot \mathbf{f}^{(l)}_{n\mathbf{q}'} \right|^2 \right\rangle \tag{6.97}$$

The quantity W is called the Debye–Waller factor. Note that in the expression for W the index n of $\mathbf{f}^{(l)}_{n\mathbf{q}'}$ is irrelevant since it only appears in the complex exponential $\exp(-\mathrm{i}\mathbf{q}' \cdot \mathbf{R}_n)$ which is eliminated by the absolute value; thus, W contains no dependence on n. Substituting the expression of $\mathbf{f}^{(l)}_{n\mathbf{q}}$ from Eq. (6.90) we find

$$W = \frac{1}{2N} \sum_{l\mathbf{q}'} \left\langle \left| \mathbf{q} \cdot Q^{(l)}_{\mathbf{q}'} \hat{\mathbf{e}}^{(l)}_{\mathbf{q}'} \right|^2 \right\rangle \tag{6.98}$$

We will assume next that the amplitudes $|Q^{(l)}_{\mathbf{q}'}|^2$ are independent of the phonon mode l, which leads to

$$\sum_{l\mathbf{q}'} \left\langle \left| \mathbf{q} \cdot Q^{(l)}_{\mathbf{q}'} \hat{\mathbf{e}}^{(l)}_{\mathbf{q}'} \right|^2 \right\rangle = \sum_{\mathbf{q}'} \langle |Q_{\mathbf{q}'}|^2 \rangle \sum_{l} |\mathbf{q} \cdot \hat{\mathbf{e}}^{(l)}_{\mathbf{q}'}|^2 = |\mathbf{q}|^2 \sum_{\mathbf{q}'} \langle |Q_{\mathbf{q}'}|^2 \rangle \tag{6.99}$$

because in our simple example of one atom per unit cell the polarization vectors $\hat{\mathbf{e}}_{\mathbf{k}}^{(l)}$ cover the same space as the cartesian coordinates; therefore,

$$\sum_l |\mathbf{q} \cdot \hat{\mathbf{e}}_{\mathbf{k}}^{(l)}|^2 = |\mathbf{q}|^2$$

With these simplifications, W takes the form

$$W = \frac{1}{2N} |\mathbf{q}|^2 \sum_{\mathbf{q}'} \langle |Q_{\mathbf{q}'}|^2 \rangle \tag{6.100}$$

As we discussed earlier, phonons can be viewed as harmonic oscillators, for which the average kinetic and potential energies are equal and each is one-half of the total energy. This gives

$$M\omega_{\mathbf{q}'}^2 \langle |Q_{\mathbf{q}'}|^2 \rangle = \left(n_{\mathbf{q}'} + \frac{1}{2} \right) \hbar\omega_{\mathbf{q}'} \Longrightarrow \langle |Q_{\mathbf{q}'}|^2 \rangle = \frac{(n_{\mathbf{q}'} + \frac{1}{2})\hbar}{M\omega_{\mathbf{q}'}} \tag{6.101}$$

with $\omega_{\mathbf{q}}$ the phonon frequencies, M the mass of the atoms and $n_{\mathbf{q}}$ the phonon occupation numbers. When this is substituted in the expression for W it leads to

$$W = \frac{1}{2N} |\mathbf{q}|^2 \sum_{\mathbf{q}'} \frac{(n_{\mathbf{q}'} + \frac{1}{2})\hbar}{M\omega_{\mathbf{q}'}} = \frac{\hbar^2 |\mathbf{q}|^2}{2MN} \sum_{\mathbf{q}'} \left(n_{\mathbf{q}'} + \frac{1}{2} \right) \frac{1}{\hbar\omega_{\mathbf{q}'}} \tag{6.102}$$

As usual, we will turn the sum over $\mathbf{q}'$ into an integral over the first BZ. We will also employ the Debye approximation in which all phonon frequencies are given by $\omega_{\mathbf{q}} = vq$ with the same average v; there is a maximum frequency ω_{D}, and related to it is the Debye temperature $k_{\mathrm{B}}\Theta_{\mathrm{D}} = \hbar\omega_{\mathrm{D}}$. This approximation leads to the following expression for the Debye–Waller factor:

$$W = \frac{3\hbar^2 |\mathbf{q}|^2 T^2}{2Mk_{\mathrm{B}}\Theta_{\mathrm{D}}^3} \int_0^{\Theta_{\mathrm{D}}/T} \left(\frac{1}{e^t - 1} + \frac{1}{2} \right) t \, \mathrm{d}t \tag{6.103}$$

with T the temperature. The limits of high temperature and low temperature are interesting. At high temperature, the upper limit of the integral in Eq. (6.103) is a very small quantity, so that we can expand the exponential in the integrand in powers of t and evaluate the integral to obtain

$$W_\infty \equiv \lim_{T \gg \Theta_{\mathrm{D}}} W = \frac{3\hbar^2 |\mathbf{q}|^2 T}{2Mk_{\mathrm{B}}\Theta_{\mathrm{D}}^2} \tag{6.104}$$

which has a strong linear dependence on the temperature. Thus, at high enough temperature when all the phonon modes are excited, the effect of the thermal motion will be manifested strongly in the structure factor which, in absolute value squared, will be multiplied by a factor

$$e^{-2W_\infty} = \exp\left[-\frac{3\hbar^2|\mathbf{q}|^2 T}{M k_B \Theta_D^2}\right] \tag{6.105}$$

For low temperatures, the upper limit in the integral in Eq. (6.103) is a very large number; the exact result of the integration contains a complicated dependence on temperature, but we see that the term $\frac{1}{2}$ inside the bracket in the integrand is now significant. For $T \to 0$, this term provides the dominant contribution to the integral,

$$W_0 \equiv \lim_{T\to 0} W = \frac{3\hbar^2|\mathbf{q}|^2}{8 M k_B \Theta_D} \tag{6.106}$$

which is comparable to the value of W_∞ at $T = \Theta_D$. This suggests that even in the limit of zero temperature there will be a significant modification of the structure factor by $\exp(-2W_0)$ relative to its value for a frozen lattice of ions, due to the phonon degrees of freedom; this modification arises from the zero-point motion associated with phonons, i.e. the $\frac{1}{2}$ term added to the phonon occupation numbers. This effect is another manifestation of the quantum mechanical nature of phonon excitations in a crystal.

6.5.3 The Mössbauer effect

Another interesting application of phonon scattering is the emission of recoil-less radiation from crystals, known as the Mössbauer effect. Consider a nucleus that can emit γ-rays due to nuclear transitions. In free space, the energy and momentum conservation requirements produce a shift in the γ-ray frequency and a broadening of the spectrum. Both of these changes obscure the nature of the nuclear transition. We can estimate the amount of broadening by the following argument: the momentum of the emitted photon of wave-vector $\mathbf{K}$ is $\hbar\mathbf{K}$, which must be equal to the recoil of the nucleus $M\mathbf{v}$, with M the mass of the nucleus (assuming that the nucleus is initially at rest). The recoil energy of the nucleus is $E_R = M\mathbf{v}^2/2 = \hbar^2\mathbf{K}^2/2M$, and the photon frequency is given by $\hbar\omega_0/c = \hbar|\mathbf{K}|$, so that $E_R = \hbar^2\omega_0^2/2Mc^2$. The broadening is then obtained by $E_R = \hbar\Delta\omega$, which gives

$$\frac{\Delta\omega}{\omega_0} = \left(\frac{E_R}{2Mc^2}\right)^{1/2} \tag{6.107}$$

This can be much greater than the natural broadening which comes from the finite lifetime of nuclear processes that give rise to the γ-ray emission in the first place. When the emission of γ-rays takes place within a solid, the presence of phonons can change things significantly. Specifically, much of the photon momentum can be carried by phonons with low energy, near the center of the BZ, and the emission can be recoil-less, in which case the γ-rays will have a frequency and width equal to their natural values. We examine this process in some detail next.

The transition probability between initial ($|i\rangle$) and final ($\langle f|$) states is given by

$$P_{i\to f} = |\langle f|\mathcal{O}(\mathbf{K}, \{p\})|i\rangle|^2 \tag{6.108}$$

where $\mathcal{O}(\mathbf{K}, \{p\})$ is the operator for the transition, $\mathbf{K}$ is the wave-vector of the radiation and $\{p\}$ are nuclear variables that describe the internal state of the nucleus. Due to translational and Galilean invariance, we can write the operator for the transition as $\mathcal{O}(\mathbf{K}, \{p\}) = \exp(-\mathrm{i}\mathbf{K}\cdot\mathbf{r})\mathcal{O}(\{p\})$, where the last operator involves nuclear degrees of freedom only. We exploit this separation to write

$$P_{i\to f} = |\langle f|\mathcal{O}(\{p\})|i\rangle|^2 \, P_{0\to 0} \tag{6.109}$$

where the states $\langle f|, |i\rangle$ now refer to nuclear states, and $P_{0\to 0}$ is the probability that the crystal will be left in its ground state, with only phonons of low frequency $\omega_{\mathbf{q}\to 0}$ emitted. The hamiltonian for the crystal degrees of freedom involved in the transition is simply the phonon energy, as derived above, Eq. (6.42). For simplicity, in the following we assume that there is only one type of atom in the crystal and one atom per unit cell, so that all atoms have the same mass $M_i = M$. This allows us to exclude the factor of $1/\sqrt{M}$ from the definition of the coordinates $Q^{(l)}_{\mathbf{q}}(t)$ (since the mass now is the same for all phonon modes), in distinction to what we had done above, see Eq. (6.35); with this new definition, the quantities Q_s have the dimension of length. To simplify the notation, we also combine the $\mathbf{q}$ and l indices into a single index s. With these definitions, we obtain the following expression for the phonon hamiltonian:

$$\mathcal{H}^{phon} = \frac{1}{2}\sum_s \left[M\omega_s^2 Q_s^2 + M\left(\frac{\mathrm{d}Q_s}{\mathrm{d}t}\right)^2 \right] \tag{6.110}$$

which, as discussed before, corresponds to a set of independent harmonic oscillators. The harmonic oscillator coordinates Q_s are related to the atomic displacements $\mathbf{S}$ by

$$\mathbf{S} = \sum_s Q_s \hat{\mathbf{e}}_s \tag{6.111}$$

where $\hat{\mathbf{e}}_s$ is the eigenvector corresponding to the phonon mode with frequency ω_s. The wavefunction of each independent harmonic oscillator in its ground state is

$$\phi_0^{(s)}(Q_s) = \left(\frac{a_s^2}{\pi}\right)^{1/4} \mathrm{e}^{-a_s^2 Q_s^2/2} \tag{6.112}$$

where the constant $a_s = M\omega_s/\hbar$.

We are now in a position to calculate the probability of recoil-less emission in the ground state of the crystal, at $T = 0$, so that all oscillators are in their ground state before and after the transition. This probability is given by

$$P_{0\to 0} \sim \left|\langle 0|\mathrm{e}^{\mathrm{i}\mathbf{K}\cdot\mathbf{S}}|0\rangle\right|^2 = \left|\prod_s \langle \phi_0^{(s)}|\mathrm{e}^{\mathrm{i}\mathbf{K}\cdot\hat{\mathbf{e}}_s Q_s}|\phi_0^{(s)}\rangle\right|^2$$
$$= \left|\prod_s \left(\frac{a_s^2}{\pi}\right)^{1/2} \int \mathrm{e}^{-a_s^2 Q_s^2}\mathrm{e}^{\mathrm{i}\mathbf{K}\cdot\hat{\mathbf{e}}_s Q_s}\mathrm{d}Q_s\right|^2 = \prod_s \mathrm{e}^{-(\mathbf{K}\cdot\hat{\mathbf{e}}_s)^2/2a_s^2} \quad (6.113)$$

But there is a simple relation between $1/2a_s^2$ and the average value of Q_s^2 in the ground state of the harmonic oscillator:

$$\frac{1}{2a_s^2} = \int \left(\frac{a_s^2}{\pi}\right)^{1/2} Q_s^2 \mathrm{e}^{-a_s^2 Q_s^2}\mathrm{d}Q_s = \langle \phi_0^{(s)}|Q_s^2|\phi_0^{(s)}\rangle = \langle Q_s^2\rangle \quad (6.114)$$

with the help of which we obtain

$$\left|\langle 0|\mathrm{e}^{\mathrm{i}\mathbf{K}\cdot\mathbf{S}}|0\rangle\right|^2 = \exp\left[-\sum_s (\mathbf{K}\cdot\hat{\mathbf{e}}_s)^2 \langle Q_s^2\rangle\right] \quad (6.115)$$

Now we will assume that all the phonon modes l have the same average $(Q_{\mathbf{q}}^{(l)})^2$, which, following the same steps as in Eq. (6.99), leads to

$$\left|\langle 0|\mathrm{e}^{\mathrm{i}\mathbf{K}\cdot\mathbf{S}}|0\rangle\right|^2 = \exp\left[-\sum_{\mathbf{q}} \mathbf{K}^2\langle Q_{\mathbf{q}}^2\rangle\right] \quad (6.116)$$

By a similar argument we obtain

$$\langle \mathbf{S}^2\rangle = \langle \left|\sum_{\mathbf{q},l} Q_{\mathbf{q}}^{(l)}\hat{\mathbf{e}}_{\mathbf{q}}^{(l)}\right|^2 \rangle = 3\sum_{\mathbf{q}}\langle Q_{\mathbf{q}}^2\rangle \quad (6.117)$$

and from this

$$P_{0\to 0} \sim \mathrm{e}^{-\mathbf{K}^2\langle \mathbf{S}^2\rangle/3} \quad (6.118)$$

We invoke the same argument as in Eq. (6.101) to express $\langle Q_s^2\rangle$ in terms of the corresponding occupation number n_s and frequency ω_s:

$$M\omega_s^2\langle Q_s^2\rangle = (n_s + \tfrac{1}{2})\hbar\omega_s \Longrightarrow \langle Q_s^2\rangle = \frac{(n_s+\frac{1}{2})\hbar}{M\omega_s} \quad (6.119)$$

This result holds in general for a canonical distribution, from which we can obtain

$$\frac{1}{3}\mathbf{K}^2\langle \mathbf{S}^2\rangle = \frac{\hbar^2\mathbf{K}^2}{2M}\frac{2}{3}\sum_s \frac{n_s+\frac{1}{2}}{\hbar\omega_s} = \frac{2}{3}E_R\sum_s \frac{n_s+\frac{1}{2}}{\hbar\omega_s} \quad (6.120)$$

with E_R the recoil energy as before, which is valid for a system with a canonical distribution at temperature T. At $T = 0$, n_s vanishes and if we use the Debye model with $\omega_s = vk$ (i.e. we take an average velocity $v_s = v$), we obtain

$$\frac{1}{3}\mathbf{K}^2\langle \mathbf{S}^2\rangle = \frac{1}{3}E_R \sum_{\mathbf{q}} \frac{1}{\hbar v k} \tag{6.121}$$

from which we can calculate explicitly the sum over $\mathbf{q}$ by turning it into an integral with upper limit k_{D} (the Debye wave-vector), to obtain

$$P_{0\to 0} \sim \exp\left[-\frac{3}{2}\frac{E_R}{k_{\mathrm{B}}\Theta_{\mathrm{D}}}\right] \tag{6.122}$$

with Θ_{D} the Debye temperature. As we have seen before, energy conservation gives $E_R = \hbar\omega_0^2/2Mc^2$, which is a fixed value since the value of ω_0 is the natural frequency of the emitted γ-rays. This last expression shows that if the Debye temperature is high enough ($k_{\mathrm{B}}\Theta_{\mathrm{D}} \gg E_R$) then $P_{0\to 0} \sim 1$; in other words, the probability of the transition involving *only* nuclear degrees of freedom, without having to balance the recoil energy, will be close to unity. This corresponds to recoil-less emission.

Problems

1. Use the Born force-constant model to calculate the phonon frequencies for silicon along the $L - \Gamma - X$ directions, where $L = \frac{\pi}{a}(1, 1, 1)$ and $X = \frac{2\pi}{a}(1, 0, 0)$. Take the ratio of the bond stretching and bond bending force constants to be $\kappa_r/\kappa_\theta = 16$. Fit the value of κ_r to reproduce the experimental value for the highest optical mode at Γ, which is 15.53 THz, and use this value to obtain the frequencies of the various modes at X; compare these with the values given in Table 6.2. Determine the atomic displacements for the normal modes at Γ, the lowest acoustic branch at X and L, and the highest optical branch at X and L.
2. Show that the kinetic and potential energy contributions to the energy of a simple harmonic oscillator are equal. Show explicitly (i.e., not by invoking the analogy to a set of independent harmonic oscillators) that the same holds for the kinetic and potential energy contributions to the energy of a collection of phonons, as given in Eq. (6.42).
3. Do phonons carry physical momentum? What is the meaning of the phonon wave-vector $\mathbf{q}$? How does the phonon wave-vector $\mathbf{q}$ enter selection rules in scattering processes, such as a photon of momentum $\mathbf{k}$ scattering off a crystal and emerging with momentum $\mathbf{k}'$ after creating a phonon of momentum $\mathbf{q}$?
4. Show that in the Debye model all mode-specific Grüneisen parameters $\gamma_{\mathbf{k}}^{(l)}$, defined in Eq. (6.75), are equal. Using this fact, find the behavior of the thermal expansion coefficient α as a function of temperature T, in the low- and high-temperature limits ($T \to 0$ and $T \gg \Theta_{\mathrm{D}}$, respectively); assume that the bulk modulus B has a very weak temperature dependence.
5. Provide the steps in the derivation of Eq. (6.103) from Eq. (6.102).

7

Magnetic behavior of solids

We begin with a brief overview of magnetic behavior in different types of solids. We first define the terms used to describe the various types of magnetic behavior. A system is called **paramagnetic** if it has no inherent magnetization, but when subjected to an external field it develops magnetization which is aligned with the field; this corresponds to situations where the microscopic magnetic moments (like the spins) tend to be oriented in the same direction as the external magnetic field. A system is called **diamagnetic** if it has no inherent magnetization, but when subjected to an external field it develops magnetization that is opposite to the field; this corresponds to situations where the induced microscopic magnetic moments tend to shield the external magnetic field. Finally, a system may exhibit magnetic order even in the absence of an external field. If the microscopic magnetic moments tend to be oriented in the same direction, the system is described as **ferromagnetic**. A variation on this theme is the situation in which microscopic magnetic moments tend to have parallel orientation but they are not necessarily equal at neighboring sites, which is described as **ferrimagnetic** behavior. If magnetic moments at neighboring sites tend to point in opposite directions, the system is described as **antiferromagnetic**. In the latter case there is inherent magnetic order due to the orientation of the microscopic magnetic moments, but the net macroscopic magnetization is zero.

Magnetic order, induced either by an external field or inherent in the system, may be destroyed by thermal effects, that is, the tendency to increase the entropy by randomizing the direction of microscopic magnetic moments. Thus, magnetic phenomena are usually observable at relatively low temperatures where the effect of entropy is not strong enough to destroy magnetic ordering.

A natural way to classify magnetic phenomena is by the origin of the microscopic magnetic moments. This origin can be ascribed to two factors: the intrinsic magnetic moment (spin) of quantum mechanical particles, or the magnetic moment arising from the motion of charged particles in an external electromagnetic field (orbital moment). In real systems the two sources of magnetic moment are actually coupled. In particular, the motion of electrons in the electric field of the nuclei leads to what is known as spin-orbit coupling. This coupling produces typically a net moment (total spin) which characterizes the entire atom or ion. In solids, there is another effect that leads to magnetic behavior, namely the motion in an external field of itinerant (metallic) electrons, which are not bound to any particular nucleus. In order to make the discussion of magnetic phenomena more coherent, we consider first the behavior of systems in which the microscopic magnetic moments are individual spins, whether these are due to single electrons (which have spin $\frac{1}{2}$) or to compound objects such as atoms or ions (which can have total spin of various values). Such phenomena include the behavior of magnetic atoms or ions as non-interacting spins in insulators, the behavior of non-interacting electron spins in metals, and the behavior of interacting quantum spins on a lattice; each of these topics is treated in a separate section. In the final section of this chapter we consider the behavior of crystal electrons in an external magnetic field, that is, orbital moments, which gives rise to very interesting classical and quantum phenomena.

7.1 Magnetic behavior of insulators

The magnetic behavior of insulators can usually be discussed in terms of the physics that applies to isolated atoms or ions which have a magnetic moment. In such situations the most important consideration is how the spin and angular momentum of electrons in the electronic shells combine to produce the total spin of the atom or ion. The rules that govern this behavior are the usual quantum mechanical rules for combining spin and angular momentum vectors, supplemented by Hund's rules that specify which states are preferred energetically. There are three quantum numbers that identify the state of an atom or ion, the total spin S, the orbital angular momentum L and the total angular momentum J (this assumes that spin-orbit coupling is weak, so that S and L can be independently used as good quantum numbers). Hund's rules determine the values of S, L and J for a given electronic shell and a given number of electrons. The first rule states that the spin state is such that the total spin S is maximized. The second rule states that the occupation of angular momentum states l_z in the shell (there are $2l + 1$ of them for a shell with nominal angular momentum l) is such that the orbital angular momentum L is maximized, for those situations consistent with the first rule. The third rule states that $J = |L - S|$ when the electronic shell is less than half filled ($n \leq 2l + 1$) and $J = L + S$ when it is more than half filled ($n \geq 2l + 1$), where n is the number of electrons in the shell; when the shell is exactly half filled the two expressions happen to give the same result for J. Of course, the application of all rules must also be consistent with the Pauli exclusion principle.

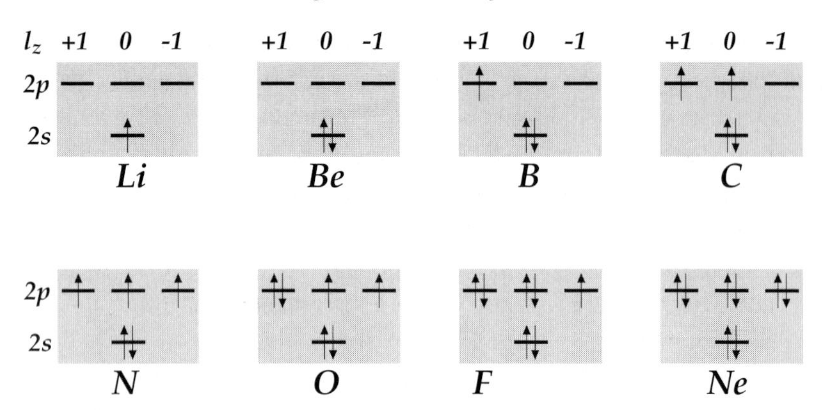

Figure 7.1. Schematic representation of the filling of the $2s$ and $2p$ electronic shells by electrons according to Hund's rules. The resulting total spin and orbital angular momentum values are: Li:($S = \frac{1}{2}$, $L = 0$), Be:($S = 0$, $L = 0$), B:($S = \frac{1}{2}$, $L = 1$), C:($S = 1$, $L = 1$), N:($S = \frac{3}{2}$, $L = 0$), O:($S = 1$, $L = 1$), F:($S = \frac{1}{2}$, $L = 1$), Ne:($S = 0$, $L = 0$).

To illustrate how Hund's rules work, consider the elements in the second row of the Periodic Table, columns I-A through to VII-A, as shown in Fig. 7.1. In Li, the one valence electron is in the $2s$ shell in a total spin $S = \frac{1}{2}$ state, this being the only possibility; the corresponding orbital angular momentum is $L = 0$ and the total angular momentum is $J = |L - S| = |L + S| = \frac{1}{2}$. In Be, the two valence electrons fill the $2s$ shell with opposite spins in a total spin $S = 0$ state; the corresponding orbital angular momentum is $L = 0$ and the total angular momentum is $J = 0$. In B, two of the three valence electrons are in a filled $2s$ shell and one is in the $2p$ shell, in a total-spin $S = \frac{1}{2}$ state, this being the only possibility for the spin; the angular momentum state $l_z = 1$ of the $2p$ shell is singly occupied, resulting in a state with $L = 1$ and $J = |L - S| = \frac{1}{2}$. In C, two of the four valence electrons are in a filled $2s$ shell and the other two are in the $2p$ shell, both with the same spin in a total spin $S = 1$ state; the angular momentum states $l_z = 1$ and $l_z = 0$ of the $2p$ shell are singly occupied, resulting in a state with $L = 1$ and $J = |L - S| = 0$. In N, two of the five valence electrons are in a filled $2s$ shell and the other three are in the $2p$ shell, all of them with the same spin in a total spin $S = \frac{3}{2}$ state; the angular momentum states $l_z = 1$, $l_z = 0$ and $l_z = -1$ of the $2p$ shell are singly occupied, resulting in a state with $L = 0$ and $J = |L - S| = L + S = \frac{3}{2}$. In O, two of the six valence electrons are in a filled $2s$ shell and the other four are in the $2p$ shell, a pair with opposite spins and the remaining two with the same spin in a total spin $S = 1$ state; the angular momentum states $l_z = 0$ and $l_z = -1$ of the $2p$ shell are singly occupied and the state $l_z = +1$ is doubly occupied, resulting in a state with $L = 1$ and $J = L + S = 2$. In F, two of the seven valence electrons are in a filled $2s$ shell and the other five are in the $2p$ shell, in two pairs with opposite spins and the remaining one giving rise to a total spin $S = \frac{1}{2}$ state; the angular momentum states $l_z = +1$ and $l_z = 0$ of the $2p$ shell are doubly occupied and the state $l_z = -1$ is singly occupied, resulting in a state with $L = 1$ and $J = L + S = \frac{3}{2}$. Ne, with closed $2s$ and $2p$ electronic shells has $S = 0$, $L = 0$ and therefore also $J = 0$. It is straightforward to derive the corresponding results for the d and f shells, which are given in Table 7.1. We also provide in this table the standard symbols for the various states, as traditionally denoted in spectroscopy: ${}^{(2S+1)}X_J$, with $X = S, P, D, F, G, H, I$ for $L = 0, 1, 2, 3, 4, 5, 6$, by analogy to the usual notation for the nominal angular momentum of the shell, s, p, d, f, g, h, i for $l = 0, 1, 2, 3, 4, 5, 6$.

From the above discussion we conclude that the B, C, N and O isolated atoms would have non-zero magnetic moment due to electronic spin states alone. However, when they are close to other atoms, their valence electrons form hybrid orbitals from which bonding and antibonding states are produced which are filled by pairs of electrons of opposite spin in the energetically favorable configurations (see the discussion in chapter 1 of how covalent bonds are formed between such atoms); Li can donate its sole valence electron to become a positive ion with closed electronic shell and F can gain an electron to become a negative ion with closed electronic shell, while Be is already in a closed electronic shell configuration. Thus, none of these atoms would lead to magnetic behavior in solids due to electronic spin alone.

Insulators in which atoms have completely filled electronic shells, such as the noble elements or the purely ionic alkali halides (composed of atoms from columns I and VII of the Periodic Table) are actually the simplest cases: the atoms or ions in these solids differ very little from isolated atoms or ions, because of the stability of the closed electronic shells. The presence of the crystal produces a very minor perturbation to the atomic configuration of electronic shells. The magnetic behavior of noble element solids and alkali halides predicted by the analysis at the individual atom or ion level is in excellent agreement with experimental measurements.

There also exist insulating solids that contain atoms with partially filled $4f$ electronic shells (the lanthanides series; see chapter 1). The electrons in these shells are largely unaffected by the presence of the crystal, because they are relatively tightly bound to the nucleus: for the lanthanides there are no core f electrons and therefore the wavefunctions of these electrons penetrate close to the nucleus. Consequently, insulators that contain such atoms or ions will exhibit the magnetic response of a collection of individual atoms or ions. It is tempting to extend this description to the atoms with partially filled $3d$ electronic shells (the fourth row of the Periodic Table, containing the magnetic elements Fe, Co and Ni); these d electrons also can penetrate close to the nucleus since there are no d electrons in the core. However, these electrons mix more strongly with the other valence electrons and therefore the presence of the crystal has a significant influence on their behavior. The predictions of the magnetic behavior of these solids based on

Table 7.1. *Atomic spin states according to Hund's rules.*

Total spin S, orbital angular momentum L and total angular momentum J numbers for the $l = 2$ (d shell) and $l = 3$ (f shell) as they are being filled by n electrons, where $1 \leq n \leq 2(2l + 1)$. Of the two expressions for J, the $-$ sign applies for $n \leq (2l + 1)$ and the $+$ sign applies for $n \geq (2l + 1)$. The standard spectroscopic symbols, ${}^{(2S+1)}X_J$, are also given with $X = S, P, D, F, G, H, I$ for $L = 0, 1, 2, 3, 4, 5, 6$. For the $3d$-shell transition metals and the $4f$-shell rare earth elements we give the calculated (p_{th}) and experimental (p_{exp}) values of the effective Bohr magneton number; note the ionization of the various elements (right superscript) which makes them correspond to the indicated state.

l	n	$S = \left\|\sum s_z\right\|$	$L = \left\|\sum l_z\right\|$	$J = \|L \mp S\|$	${}^{(2S+1)}X_J$		p_{th}	p_{exp}
2	1	1/2	2	3/2	${}^2D_{3/2}$	Ti^{3+}	1.73[a]	1.8
2	2	1	3	2	${}^3F_2$	V^{3+}	2.83[a]	2.8
2	3	3/2	3	3/2	${}^4F_{3/2}$	Cr^{3+}	3.87[a]	3.7
2	4	2	2	0	${}^5D_0$	Mn^{3+}	4.90[a]	5.0
2	5	5/2	0	5/2	${}^6S_{5/2}$	Fe^{3+}	5.92[a]	5.9
2	6	2	2	4	${}^5D_4$	Fe^{2+}	4.90[a]	5.4
2	7	3/2	3	9/2	${}^4F_{9/2}$	Co^{2+}	3.87[a]	4.8
2	8	1	3	4	${}^3F_4$	Ni^{2+}	2.83[a]	3.2
2	9	1/2	2	5/2	${}^2D_{5/2}$	Cu^{2+}	1.73[a]	1.9
2	10	0	0	0	${}^1S_0$	Zn^{2+}	0.00	
3	1	1/2	3	5/2	${}^2F_{5/2}$	Ce^{3+}	2.54	2.4
3	2	1	5	4	${}^3H_4$	Pr^{3+}	3.58	3.5
3	3	3/2	6	9/2	${}^4I_{9/2}$	Nd^{3+}	3.62	3.5
3	4	2	6	4	${}^5I_4$	Pm^{3+}	2.68	
3	5	5/2	5	5/2	${}^6H_{5/2}$	Sm^{3+}	0.84	1.5
3	6	3	3	0	${}^7F_0$	Eu^{3+}	0.00	3.4
3	7	7/2	0	7/2	${}^8S_{7/2}$	Gd^{3+}	7.94	8.0
3	8	3	3	6	${}^7F_6$	Tb^{3+}	9.72	9.5
3	9	5/2	5	15/2	${}^6H_{15/2}$	Dy^{3+}	10.63	10.6
3	10	2	6	8	${}^5I_8$	Ho^{3+}	10.60	10.4
3	11	3/2	6	15/2	${}^4I_{15/2}$	Er^{3+}	9.59	9.5
3	12	1	5	6	${}^3H_6$	Tm^{3+}	7.57	7.3
3	13	1/2	3	7/2	${}^2F_{7/2}$	Yb^{3+}	4.54	4.5
3	14	0	0	8	${}^1S_0$	Lu^{3+}	0.00	

[a] Values obtained from the quenched total angular momentum expression ($L = 0 \Rightarrow J = S$) which are in much better agreement with experiment than those from the general expression, Eq. (7.8). Source: [74].

the behavior of the constituent atoms is not very close to experimental observations. A more elaborate description, along the lines presented in the following sections, is required for these solids.

In insulating solids whose magnetic behavior arises from individual ions or atoms, there is a common feature in the response to an external magnetic field H, known as the Curie law. This response is measured through the magnetic susceptibility, which according to the Curie law is inversely proportional to the temperature. The magnetic susceptibility is defined as

$$\chi = \frac{\partial M}{\partial H}, \qquad M = -\frac{\partial F}{\partial H}$$

where M is the magnetization and F is the free energy (see Appendix D for a derivation of these expressions from statistical mechanics). In order to prove the Curie law we use a simple model. We assume that the ion or atom responsible for the magnetic behavior has a total angular momentum J, so that there are $(2J + 1)$ values of its J_z component, $J_z = -J, \ldots, +J$, with the axis z defined by the direction of the external magnetic field, which give rise to the following energy levels for the system:

$$E_{J_z} = m_0 J_z H, \qquad J_z = -J, \ldots, +J \tag{7.1}$$

with m_0 the magnetic moment of the atoms or ions. The canonical partition function for this system, consisting of N atoms or ions, is given by

$$Z_N = \left(\sum_{J_z=-J}^{J} \mathrm{e}^{-\beta m_0 J_z H} \right)^N = \left(\frac{\mathrm{e}^{\beta m_0 (J+\frac{1}{2})H} - \mathrm{e}^{-\beta m_0 (J+\frac{1}{2})H}}{\mathrm{e}^{\beta m_0 \frac{1}{2} H} - \mathrm{e}^{-\beta m_0 \frac{1}{2} H}} \right)^N$$

where $\beta = 1/k_{\mathrm{B}}T$ is the inverse temperature. We define the variable

$$w = \frac{\beta m_0 H}{2}$$

in terms of which the free energy becomes

$$F = -\frac{1}{\beta} \ln Z_N = -\frac{N}{\beta} \ln \left[\frac{\sinh(2J+1)w}{\sinh w} \right]$$

which gives for the magnetization M and the susceptibility χ per unit volume

$$M = \frac{N m_0}{2\Omega} \left[(2J+1)\coth(2J+1)w - \coth w \right]$$

$$\chi = \frac{N m_0^2}{4\Omega} \beta \left[\frac{1}{\sinh^2 w} - \frac{(2J+1)^2}{\sinh^2(2J+1)w} \right]$$

with Ω the total volume containing the N magnetic ions or atoms. In the limit of small w, which we justify below, the lowest few terms in the series expansions of $\coth x$ and $\sinh x$ (see Appendix G) give

$$M = \frac{N}{\Omega}\frac{J(J+1)m_0^2}{3}\frac{1}{k_{\mathrm{B}}T}H \tag{7.2}$$

$$\chi = \frac{N}{\Omega}\frac{J(J+1)m_0^2}{3}\frac{1}{k_{\mathrm{B}}T} \tag{7.3}$$

This is exactly the form of the Curie law, that is, the susceptibility is inversely proportional to the temperature. In order for this derivation to be valid, we have to make sure that w is much smaller than unity. From the definition of w we obtain the condition

$$m_0 H << 2k_{\mathrm{B}}T \tag{7.4}$$

Typical values of the magnetic moment m_0 (see following discussion) are of order the *Bohr magneton*, which is defined as

$$\mu_B \equiv \frac{e\hbar}{2m_e c} = 0.579 \times 10^{-8} \text{ eV/G}$$

The Bohr magneton may also be written as

$$\mu_B = \frac{e\hbar}{2m_e c} = \frac{e}{2\pi\hbar c}\frac{\pi\hbar^2}{m_e} = \frac{2\pi a_0^2}{\phi_0}\text{ Ry}, \quad \phi_0 \equiv \frac{hc}{e} \tag{7.5}$$

where $h = 2\pi\hbar$ is Planck's constant and ϕ_0 is the value of the flux quantum which involves only the fundamental constants h, c, e. From the units involved in the Bohr magneton we see that, even for very large magnetic fields of order 10^4 G, the product $m_0 H$ is of order 10^{-4} eV $\sim$ 1 K, so the condition of Eq. (7.4) is satisfied reasonably well except at very low temperatures (below 1 K) and very large magnetic fields (larger than 10^4 G).

The values of the total spin S, orbital angular momentum L and total angular momentum J for the state of the atom or ion determine the exact value of the magnetic moment m_0. The interaction energy of an electron in a state of total spin $\mathbf{S}$ and orbital angular momentum $\mathbf{L}$, with an external magnetic field $\mathbf{H}$, to lowest order in the field is given by

$$\langle \mu_B(\mathbf{L} + g_0\mathbf{S}) \cdot \mathbf{H} \rangle \tag{7.6}$$

where g_0 is the *electronic g-factor*

$$g_0 = 2\left(1 + \frac{\alpha}{2\pi} + \mathcal{O}(\alpha^2) + \cdots\right) = 2.0023$$

with $\alpha = e^2/\hbar c = 1/137$ the fine-structure constant. The angular brackets in Eq. (7.6) denote the expectation value of the operator in the electron state. Choosing the direction of the magnetic field to be the z axis, we conclude that, in order to evaluate the effect of the magnetic field, we must calculate the expectation values of the operator $(L_z + g_0 S_z)$, which, in the basis of states with definite J, L, S and J_z quantum numbers, are given by

$$\mu_B H \langle (L_z + g_0 S_z) \rangle = \mu_B H g(JLS) J_z \tag{7.7}$$

with $g(JLS)$ the Landé g-factors

$$g(JLS) = \frac{1}{2}(g_0 + 1) + \frac{1}{2}(g_0 - 1)\frac{S(S+1) - L(L+1)}{J(J+1)}$$

(see Appendix B, Eq. (B.45) with $\lambda = g_0$). Comparing Eq. (7.7) with Eq. (7.1) we conclude that the magnetic moment is given by

$$m_0 = g(JLS)\mu_B \approx \left[\frac{3}{2} + \frac{S(S+1) - L(L+1)}{2J(J+1)}\right]\mu_B \tag{7.8}$$

where we have used $g_0 \approx 2$, a very good approximation. When this expression is used for the theoretical prediction of the Curie susceptibility, Eq. (7.3), the results compare very well with experimental measurements for rare earth $4f$ ions. Typically, the comparison is made through the so called "effective Bohr magneton number", defined in terms of the magnetic moment m_0 as

$$p = \frac{m_0\sqrt{J(J+1)}}{\mu_B} = g(JLS)\sqrt{J(J+1)}$$

Values of p calculated from the expression of Eq. (7.8) and obtained from experiment by fitting the measured susceptibility to the expression of Eq. (7.3) are given in Table 7.1. There is only one blatant disagreement between theoretical and experimental values for these rare earth ions, namely Eu, for which $J = 0$; in this case, the vanishing value of the total angular momentum means that the linear (lowest order) contribution of the magnetic field to the energy assumed in Eq. (7.6) is not adequate, and higher order contributions are necessary to capture the physics.

When it comes to the effective Bohr magneton number for the $3d$ transition metal ions, the expression of Eq. (7.8) is not successful in reproducing the values measured experimentally. In that case, much better agreement between theory and experiment is obtained if instead we use for the magnetic moment the expression $m_0 = g_0\mu_B \approx 2\mu_B$, and take $J = S$ in the expression for the magnetic susceptibility Eq. (7.3), that is, if we set $L = 0$. The reason is that in the $3d$ transition metals the presence of the crystal environment strongly affects the valence electrons due to the nature of their wavefunctions, as discussed above, and consequently their total angular momentum quantum numbers are not those determined by Hund's rules, which apply to spherical atoms or ions. Apparently, the total spin quantum number is still good, that is, Hund's first rule still holds, and it can be used to describe the magnetic behavior of ions in the crystalline environment, but the orbital angular momentum number is no longer a good quantum number. This phenomenon is referred to as "quenching" of the orbital angular momentum due to crystal field splitting of the levels.

7.2 Magnetic behavior of metals

We consider next the magnetic behavior of metallic electrons. We provide first the general picture and then examine in some detail specific models that put this picture on a quantitative basis. In the simplest possible picture, we can view the electrons in metals as free fermions of spin $\frac{1}{2}$ in an external magnetic field. This leads to paramagnetic behavior which is referred to as **Pauli paramagnetism**, a phenomenon discussed in detail in Appendix D. All results from that analysis are directly applicable to the electron gas, taking the magnetic moment of each particle to be $m_0 = g_0 \mu_B$.

We quote here some basic results from the discussion of Appendix D: the high-temperature and low-temperature limits of the susceptibility per unit volume are given by

$$\chi(T \to \infty) = \frac{g_0^2 \mu_B^2}{k_B T} n, \qquad \chi(T = 0) = \frac{3}{2} \frac{g_0^2 \mu_B^2}{\epsilon_F} n$$

with $n = N/\Omega$ the density of particles. We see that in the high-temperature limit the susceptibility exhibits behavior consistent with the Curie law. In this case, the relevant scale for the temperature is the Fermi energy, which gives for the condition at which the Curie law should be observed in the free-electron gas,

$$\epsilon_F = \left(\frac{9\pi}{4}\right)^{2/3} \frac{1}{(r_s/a_0)^2} \text{ Ry} < k_B T$$

For typical values of r_s in metals, $(r_s/a_0) \sim 2$–6, we find that the temperature must be of order 10^4–10^5 K in order to satisfy this condition, which is too high to be relevant for any real solids. In the opposite extreme, at $T = 0$, it is natural to expect that the only important quantities are those related to the filling of the Fermi sphere, that is, the density n and Fermi energy ϵ_F. Using the relation between the density and the Fermi momentum $k_F^3 = 3\pi^2 n$, which applies to the uniform electron gas in 3D, we can rewrite the susceptibility of the free-electron gas at $T = 0$ as

$$\chi = g_0^2 \mu_B^2 \frac{m_e k_F}{\pi^2 \hbar^2} = g_0^2 \mu_B^2 g_F \tag{7.9}$$

where $g_F = g(\epsilon_F)$ is the density of states evaluated at the Fermi level (recall that the density of states in the case of free fermions of spin $\frac{1}{2}$ in 3D has the form

$$g(\epsilon) = \frac{\sqrt{2}}{\pi^2}\left(\frac{m_e}{\hbar^2}\right)^{3/2} \epsilon^{1/2},$$

as shown in chapter 5, Eq. (5.3)). Eq. (7.9) is a general form of the susceptibility, useful in several contexts (see, for example, the following discussion on magnetization of band electrons).

The orbital motion of electrons in an external electromagnetic field always leads to diamagnetic response because it tries to shield the external field. This behavior is referred to as **Landau diamagnetism**; for a treatment of this effect see Ref. [80]. Pauli paramagnetism and Landau diamagnetism are effects encountered in solids that are very good conductors, that is, the behavior of their valence electrons is close to that of the free-electron gas. The two effects are of similar magnitude but opposite sign, as mentioned above, and they are both much weaker than the response of individual atoms or ions. Thus, in order to measure these effects one must be able to separate the effect that comes from the ions in the solid, which is not always easy. In the following we discuss in more detail the theory of the magnetic response of metallic solids in certain situations where it is feasible to identify its source.

We specialize the discussion to models that provide a more detailed account of the magnetic behavior of metals. The first step in making the free-electron picture more realistic is to include exchange effects explicitly, that is, to invoke a Hartree–Fock picture, but without the added complications imposed by band-structure effects. We next analyze a model that takes into account band-structure effects in an approximate manner. These models are adequate to introduce the important physics of magnetic behavior in metals.

7.2.1 Magnetization in Hartree–Fock free-electron gas

In our discussion of the free-electron gas in chapter 2 we had assumed that there are equal numbers of up and down spins, and therefore we took the ground state of the system to have non-magnetic character. This is not necessarily the case. We show here that, depending on the density of the electron gas, the ground state can have magnetic character. This arises from considering explicitly the exchange interaction of electrons due to their fermionic nature. The exchange interaction is taken into account by assuming a Slater determinant of free-particle states for the many-body wavefunction, as we did in the Hartree–Fock approximation (see chapter 2). In that case, after averaging over the spin states, we had found for the ground state energy per particle, Eq. (2.43)

$$\frac{E^{HF}}{N} = \frac{3}{5}\frac{\hbar^2 k_{\mathrm{F}}^2}{2m_e} - \frac{3e^2}{4\pi}k_{\mathrm{F}} = \left[\frac{3}{5}(k_{\mathrm{F}}a_0)^2 - \frac{3}{2\pi}(k_{\mathrm{F}}a_0)\right] \mathrm{Ry} \tag{7.10}$$

where k_{F} is the Fermi momentum; in the last expression we have used the rydberg as the unit of energy and the Bohr radius a_0 as the unit of length. In the following discussion we will express all energies in units of rydbergs, and for simplicity we do not include that unit in the equations explicitly. The first term in the total energy E^{HF} is the kinetic energy, while the second represents the effect of the exchange interactions. When the two spin states of the electrons are averaged, the Fermi momentum is related to the total number of fermions N in volume Ω by

$$n = \frac{N}{\Omega} = \frac{k_{\mathrm{F}}^3}{3\pi^2}$$

as shown in Appendix D, Eq. (D.10). By a similar calculation as the one that led to the above result, we find that if we had $N_\uparrow$ ($N_\downarrow$) electrons with spin up (down), occupying states up to Fermi momentum $k_{F\uparrow}$ ($k_{F\downarrow}$), the relation between this Fermi momentum and the number of fermions in volume Ω would be

$$N_\uparrow = \Omega\frac{k_{F\uparrow}^3}{6\pi^2}, \quad N_\downarrow = \Omega\frac{k_{F\downarrow}^3}{6\pi^2}$$

Notice that for $N_\uparrow = N_\downarrow = N/2$, we would obtain:

$$k_{F\uparrow} = k_{F\downarrow} = k_{\mathrm{F}}$$

as expected for the non-magnetic case. For each set of electrons with spin up or spin down considered separately, we would have a total energy given by an equation analogous to the spin-averaged case:

$$\frac{E_\uparrow^{HF}}{N_\uparrow} = \frac{3}{5}(k_{F\uparrow}a_0)^2 - \frac{3}{2\pi}(k_{F\uparrow}a_0), \quad \frac{E_\downarrow^{HF}}{N_\downarrow} = \frac{3}{5}(k_{F\downarrow}a_0)^2 - \frac{3}{2\pi}(k_{F\downarrow}a_0)$$

The coefficients for the kinetic energy and exchange energy terms in the spin-polarized cases are exactly the same as for the spin-averaged case, because the factors of 2 that enter into the calculation of the spin-averaged results are compensated for by the factor of 2 difference in the definition of $k_{F\uparrow}$, $k_{F\downarrow}$, relative to that of k_{F}.

We next consider a combined system consisting of a total number of N electrons, with $N_\uparrow$ of them in the spin-up state and $N_\downarrow$ in the spin-down state. We define the magnetization number M and the magnetization per particle m as

$$M = N_\uparrow - N_\downarrow, \quad m = \frac{M}{N} \tag{7.11}$$

From these definitions, and from the fact that $N = N_\uparrow + N_\downarrow$, we find

$$N_\uparrow = (1+m)\frac{N}{2}, \quad N_\downarrow = (1-m)\frac{N}{2} \tag{7.12}$$

and using these expressions in the total energies $E^{HF}_\uparrow$, $E^{HF}_\downarrow$, we obtain for the total energy per particle

$$\frac{E^{HF}_\uparrow + E^{HF}_\downarrow}{N} = \frac{3a_0^2}{10}\left(\frac{3\pi^2 N}{\Omega}\right)^{2/3}\left[(1+m)^{5/3} + (1-m)^{5/3}\right]$$
$$-\frac{3a_0}{4\pi}\left(\frac{3\pi^2 N}{\Omega}\right)^{1/3}\left[(1+m)^{4/3} + (1-m)^{4/3}\right]$$

Notice that in the simple approximation we are considering, the free-electron gas, there are no other terms involved in the total energy since we neglect the Coulomb interactions and there is no exchange interaction between spin-up and spin-down electrons.

We recall the definition of the average volume per particle through the variable r_s, the radius of the sphere that encloses this average volume:

$$\frac{4\pi}{3}r_s^3 = \frac{N}{\Omega} \Longrightarrow \left(\frac{3\pi^2 N}{\Omega}\right)^{1/3} = \left(\frac{9\pi}{4}\right)^{1/3} r_s$$

which, when used in the expression for the total energy of the general case, gives

$$\frac{E^{HF}_\uparrow + E^{HF}_\downarrow}{N} = \frac{3}{10}\left(\frac{9\pi}{4}\right)^{2/3}\frac{1}{(r_s a_0)^2}\left[(1+m)^{5/3} + (1-m)^{5/3}\right]$$
$$-\frac{3}{4\pi}\left(\frac{9\pi}{4}\right)^{1/3}\frac{1}{(r_s a_0)}\left[(1+m)^{4/3} + (1-m)^{4/3}\right]$$

Using this expression for the total energy we must determine what value of the magnetization corresponds to the lowest energy state for a given value of r_s. From the powers of r_s involved in the two terms of the total energy, it is obvious that for low r_s the kinetic energy dominates while for high r_s the exchange energy dominates. We expect that when exchange dominates it is more likely to get a magnetized ground state. Indeed, if $m \neq 0$ there will be more than half of the particles in one of the two spin states, and that will give larger exchange energy than the $m = 0$ case with exactly half of the electrons in each spin state. The extreme case of this is $m = \pm 1$, when all of the electrons are in one spin state. Therefore, we expect that for low enough r_s (high-density limit) the energy will be dominated by the kinetic part and the magnetization will be zero, while for large enough r_s (low-density limit) the energy will be dominated by the exchange part and the magnetization will be $m = \pm 1$. This is exactly what is found by scanning the values

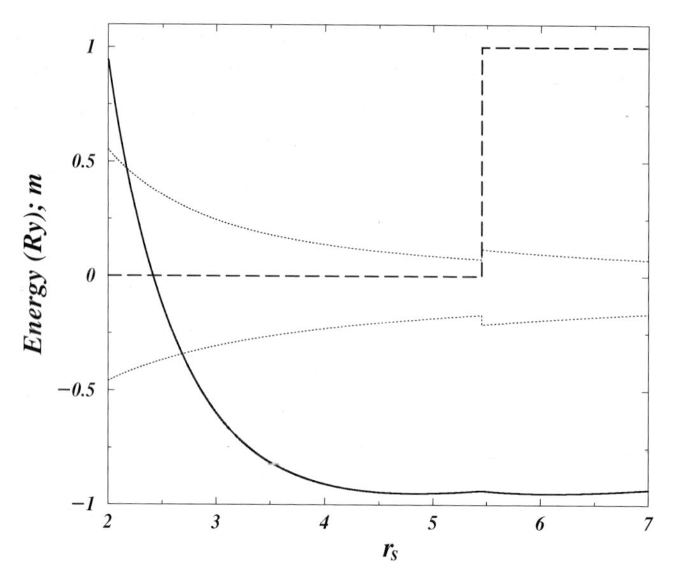

Figure 7.2. The total energy multiplied by a factor of 10 (solid line) and the magnetization m (dashed line) of the polarized electron gas, as functions of r_s (in units of a_0). The thinner dotted lines indicate the contributions of the kinetic part (positive) and the exchange part (negative). For low values of r_s the kinetic energy dominates and the total energy is positive ($r_s \leq 2.411$), while for larger values of r_s the exchange energy dominates and the total energy is negative. Notice that both contributions are discontinuous at $r_s = 5.450$, the transition point at which the magnetization jumps from a value of $m = 0$ to a value of $m = 1$, but that the total energy is continuous.

of r_s and determining for each value the lowest energy state as a function of m: the ground state has $m = 0$ for $r_s \leq 5.450$ and $m = \pm 1$ for $r_s > 5.450$, as shown in Fig. 7.2. It is interesting that the transition from the unpolarized to the polarized state is abrupt and complete (from zero polarization $m = 0$ to full polarization, $m = \pm 1$). It is also interesting that the kinetic energy and exchange energy contributions to the total energy are discontinuous at the transition point, but the total energy itself is continuous. The transition point is a local maximum in the total energy, with two local minima of equal energy on either side of it, the one on the left (at $r_s = 4.824$) coresponding to a non-magnetic ground state and the one on the right (at $r_s = 6.077$) to a ferromagnetic ground state. The presence of the two minima of equal energy is indicative of how difficult it would be to stabilize the ferromagnetic ground state based on exchange interactions, as we discuss in more detail next.

While the above discussion is instructive on how exchange interactions alone can lead to non-zero magnetization, this simple model is not adequate to describe realistic systems. There are several reasons for this, having to do with the limitations of both the Hartree–Fock approximation and the uniform electron gas approximation. First, the Hartree–Fock approximation is not realistic because of screening and correlation effects (see the discussion in chapter 2); attempts to fix the Hartree–Fock approximation destroy the transition to the non-zero magnetization state. Moreover, there are actually other solutions beyond the uniform density solution, which underlies the preceding discussion, that have lower energy; these are referred to as spin density waves. Finally, for very low densities where the state with non-zero magnetization is supposed to be stable, there is a different state in which electrons maximally avoid each other to minimize the total energy: this state is known as the "Wigner crystal", with the electrons localized at crystal sites to optimize the Coulomb repulsion. Thus, it is not surprising that a state with non-zero magnetization, arising purely from exchange interactions in a uniform electron gas, is not observed in real solids. The model is nevertheless useful in motivating the origin of interactions that can produce ferromagnetic order in metals: the essential idea is that, neglecting all other contributions, the exchange effect can lead to polarization of the spins at low enough density, resulting in a ferromagnetic state.

7.2.2 Magnetization of band electrons

We next develop a model, in some sense a direct extension of the preceding analysis, which includes the important aspects of the band structure in magnetic phenomena. The starting point for this theory is the density of states derived from the band structure, $g(\epsilon)$. We assume that the electronic states can be filled up to the Fermi level ϵ_F for a spin-averaged configuration with $N_\uparrow = N_\downarrow = N/2$. We then perturb this state of the system by moving a few electrons from the spin-down state to the spin-up state, so that $N_\uparrow > N/2$, $N_\downarrow < N/2$. The energy of the electrons that have been transferred to the spin-up state will be slightly higher than the Fermi level, while the highest energy of electrons in the depleted spin state will be somewhat lower than the Fermi energy, as shown in Fig. 7.3. We will take the number of electrons with spin up and spin down to be given by

$$N_\uparrow = \int_{-\infty}^{\epsilon_F+\Delta} g(\epsilon)d\epsilon, \quad N_\downarrow = \int_{-\infty}^{\epsilon_F-\Delta} g(\epsilon)d\epsilon \tag{7.13}$$

where we have introduced Δ to denote the deviation of the highest occupied state in each case from the Fermi level (a positive deviation in the case of spin-up electrons and a negative one in the case of spin-down electrons). Notice that, in the small Δ limit, in order to conserve the total number of electrons when we transfer some from the spin-down state to the spin-up state, the density of states must be symmetric around its value at ϵ_F, which means that $g(\epsilon)$ is constant in the neighborhood of ϵ_F or it is a local extremum at $\epsilon = \epsilon_F$; in either case, the derivative of the density of states $g'(\epsilon)$ must vanish at $\epsilon = \epsilon_F$.

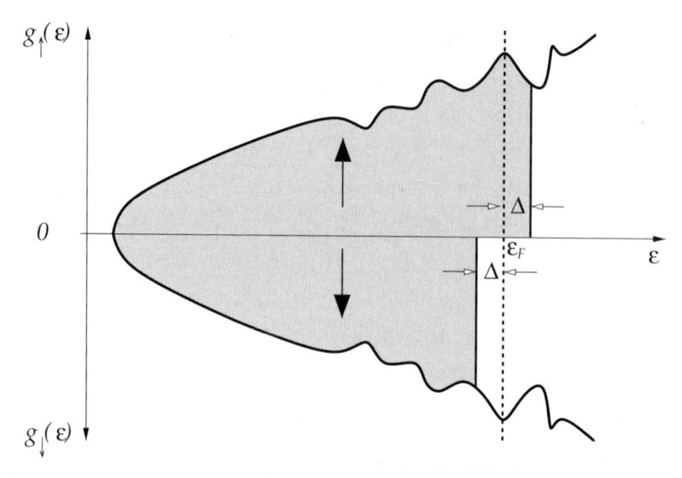

Figure 7.3. Illustration of the occupation of spin-up electron states $g_{\uparrow}(\epsilon)$ and spin-down electron states $g_{\downarrow}(\epsilon)$, with corresponding Fermi levels $\epsilon_{\mathrm{F}} + \Delta$ and $\epsilon_{\mathrm{F}} - \Delta$, where ϵ_{F} is the Fermi energy for the spin-averaged state. In this example we have chosen $g(\epsilon)$ to have a symmetric local maximum over a range of 2Δ around $\epsilon = \epsilon_{\mathrm{F}}$.

Expanding the integrals over the density of states as Taylor series in Δ through the expressions given in Eqs. (G.24) and (G.40), and using the definition of the magnetization number from Eq. (7.11), we obtain

$$M = 2\Delta g_F \left[1 + \mathcal{O}(\Delta^2)\right] \tag{7.14}$$

where we have used the symbol $g_F = g(\epsilon_{\mathrm{F}})$ for the density of states at the Fermi level. We can also define the band energy of the system, which in this case can be identified as the Hartree energy E^H, that is, the energy due to electron-electron interactions other than those from exchange and correlation; this contribution comes from summing the energy of electronic levels associated with the two spin states:

$$E^H = E_{\uparrow} + E_{\downarrow} = \int_{-\infty}^{\epsilon_{\mathrm{F}}+\Delta} \epsilon g(\epsilon) \mathrm{d}\epsilon + \int_{-\infty}^{\epsilon_{\mathrm{F}}-\Delta} \epsilon g(\epsilon) \mathrm{d}\epsilon \tag{7.15}$$

Through the same procedure of expanding the integrals as Taylor series in Δ, this expression gives for the band energy

$$E^H = 2\int_{-\infty}^{\epsilon_{\mathrm{F}}} \epsilon g(\epsilon) \mathrm{d}\epsilon + \Delta^2 \left[g_F + \epsilon_{\mathrm{F}} g'(\epsilon_{\mathrm{F}})\right] \tag{7.16}$$

where we have kept terms only up to second order in Δ. Consistent with our earlier assumptions about the behavior of the density of states near the Fermi level, we will take $g'(\epsilon_{\mathrm{F}}) = 0$. The integral in the above equation represents the usual band energy for the spin-averaged system, with the factor of 2 coming from the two spin states associated with each electronic state with energy ϵ. The remaining term is the change in the band energy due to spin polarization:

$$\delta E^H = \Delta^2 g_F \tag{7.17}$$

Our goal is to compare this term, which is always positive, with the change in the exchange energy due to spin polarization. We will define the exchange energy in terms of a parameter $-J$, which we refer to as the "exchange integral": this is the contribution to the energy due to exchange of a pair of particles. We will take $J > 0$, that is, we assume that the system gains energy due to exchange interactions, as we proved explicitly for the Hartree–Fock free-electron model in chapter 3. Since exchange applies only to particles of the same spin, the total contribution to the exchange energy from the spin-up and the spin-down sets of particles will be given by

$$E^X = -J\left[\frac{N_\uparrow(N_\uparrow - 1)}{2} + \frac{N_\downarrow(N_\downarrow - 1)}{2}\right] \tag{7.18}$$

Using the expressions for $N_\uparrow$ and $N_\downarrow$ in terms of the magnetization, Eqs. (7.11) and (7.12), we find that the exchange energy is

$$E^X = -\frac{J}{2}\left(\frac{N}{2}\right)^2\left[(1+m)\left(1+m-\frac{2}{N}\right) + (1-m)\left(1-m-\frac{2}{N}\right)\right]$$

If we subtract from this expression the value of the exchange energy for $m = 0$, corresponding to the spin-averaged state with $N_\uparrow = N_\downarrow = N/2$, we find that the change in the exchange energy due to spin polarization is

$$\delta E^X = -J\frac{M^2}{4} \tag{7.19}$$

which is always negative (recall that $J > 0$). The question then is, under what conditions will the gain in exchange energy due to spin polarization be larger than the cost in band energy, Eq. (7.17)? Using our result for the magnetization from Eq. (7.14), and keeping only terms to second order in Δ, we find that the two changes in the energy become equal for $J = 1/g_F$ and the gain in exchange energy dominates over the cost in band energy for

$$J > \frac{1}{g_F}$$

which is known as the "Stoner criterion" for spontaneous magnetization [81]. This treatment applies to zero temperature; for finite temperature, in addition to the density of states in Eq. (7.13), we must also take into account the Fermi occupation numbers (see Problem 3).

To complete the discussion, we examine the magnetic susceptibility χ of a system in which spontaneous magnetization arises when the Stoner criterion is met. We will assume that the magnetization develops adiabatically, producing an effective magnetic field H whose change is proportional to the magnetization change

$$\mathrm{d}H = \chi^{-1}\mathrm{d}M\mu_B$$

where we have also included the Bohr magneton μ_B to give the proper units for the field, and the constant of proportionality is, by definition, the inverse of the susceptibility. From the preceding discussion we find that the change in total energy due to the spontaneous magnetization is given by

$$\delta E_{tot} = \delta E^H + \delta E^X = \frac{1 - g_F J}{4g_F}M^2$$

and is a negative quantity for $J > 1/g_F$ (the system gains energy due to the magnetization). This change can also be calculated from the magnetization in the presence of the induced field H', which ranges from zero to some final value H when the magnetization M' ranges from zero to its final value M:

$$\delta E_{tot} = -\int_0^H (M'\mu_B)\mathrm{d}H' = -\chi^{-1}\mu_B^2 \int_0^M M'\mathrm{d}M' = -\frac{1}{2}\chi^{-1}M^2\mu_B^2$$

where we have included an overall minus sign to indicate that the spontaneous magnetization lowers the total energy. From the two expressions for the change in total energy we find

$$\chi = \frac{2g_F}{|1 - g_F J|}\mu_B^2 \tag{7.20}$$

which should be compared with the Pauli susceptibility, Eq. (7.9): the susceptibility of band electrons is enhanced by the factor $|1 - g_F J|^{-1}$, which involves explicitly the density of states at the Fermi energy g_F and the exchange integral J. For example, the value of this factor for Pd, extracted from measurements of the specific heat, is $|1 - g_F J| \sim 13$ (for more details see chapter 4, volume 1 of Jones and March [82]).

The preceding discussion of magnetic effects in band electrons is based on the assumption of a particularly simple behavior of the band structure near the Fermi level. When these assumptions are not valid, a more elaborate theory which takes into account the realistic band structure is needed (see, for example, Refs. [83]–[85]).

7.3 Heisenberg spin model

The discussion of the previous section concerned non-interacting microscopic magnetic moments. Many aspects of magnetic behavior involve magnetic moments that are strongly coupled. A model that captures the physics of such systems is the so called Heisenberg spin model. This model consists of interacting spins on lattice sites, $\mathbf{S}(\mathbf{R})$, with $\mathbf{S}$ denoting a quantum mechanical spin variable and $\mathbf{R}$ the lattice point where it resides. A pair of spins at lattice sites $\mathbf{R}$ and $\mathbf{R}'$ interact by the so called exchange term J, which depends on the relative distance $\mathbf{R} - \mathbf{R}'$ between the spins. The Heisenberg spin model hamiltonian is:

$$\mathcal{H}_H = -\frac{1}{2}\sum_{\mathbf{R}\neq\mathbf{R}'} J(\mathbf{R}-\mathbf{R}')\mathbf{S}(\mathbf{R})\cdot\mathbf{S}(\mathbf{R}'), \quad J(\mathbf{R}-\mathbf{R}') = J(\mathbf{R}'-\mathbf{R}) \tag{7.21}$$

If the exchange integral J is positive, then the model describes ferromagnetic order because the spins will tend to be oriented in the same direction to give a positive value for $\langle\mathbf{S}(\mathbf{R})\cdot\mathbf{S}(\mathbf{R}')\rangle$ as this minimizes the energy. In the opposite case, a negative value for J will lead to antiferromagnetic order, where nearest neighbor spins will tend to be oriented in opposite directions. We discuss first the physics of the Heisenberg ferromagnet, which is the simpler case.

7.3.1 Ground state of the Heisenberg ferromagnet

As a first step in the study of the Heisenberg ferromagnet, we express the dot product of two spin operators, with the help of the spin raising S_+ and lowering S_- operators (for details see Appendix B), as

$$S_+ \equiv S_x + \mathrm{i}S_y, \quad S_- \equiv S_x - \mathrm{i}S_y \Longrightarrow S_x = \frac{1}{2}(S_+ + S_-), \quad S_y = \frac{1}{2\mathrm{i}}(S_+ - S_-)$$

which give for a pair of spins situated at $\mathbf{R}$ and $\mathbf{R}'$

$$\begin{aligned}\mathbf{S}(\mathbf{R})\cdot\mathbf{S}(\mathbf{R}') &= S_x(\mathbf{R})S_x(\mathbf{R}') + S_y(\mathbf{R})S_y(\mathbf{R}') + S_z(\mathbf{R})S_z(\mathbf{R}')\\ &= \frac{1}{2}\left[S_+(\mathbf{R})S_-(\mathbf{R}') + S_-(\mathbf{R})S_+(\mathbf{R}')\right] + S_z(\mathbf{R})S_z(\mathbf{R}')\end{aligned}$$

When this result is substituted into the Heisenberg spin model, Eq. (7.21), taking into account that raising and lowering operators at different lattice sites commute with each other because they operate on different spins, we find

$$\mathcal{H}_H = -\frac{1}{2}\sum_{\mathbf{R}\neq\mathbf{R}'} J(\mathbf{R}-\mathbf{R}')\left[S_-(\mathbf{R})S_+(\mathbf{R}') + S_z(\mathbf{R})S_z(\mathbf{R}')\right] \tag{7.22}$$

where the presence of a sum over all lattice vectors and the relation $J(\mathbf{R}-\mathbf{R}') = J(\mathbf{R}'-\mathbf{R})$ conspire to eliminate the two separate appearances of the lowering–raising operator product at sites $\mathbf{R}$ and $\mathbf{R}'$. If the lowering–raising operator product were neglected, and only the $S_z(\mathbf{R})S_z(\mathbf{R}')$ part were retained, the model would be equivalent to the classical Ising spin model; the physics of the Ising model with nearest neighbor interactions is discussed in Appendix D.

Even though we developed this model with electrons in mind, we will allow the spin of the particles to be arbitrary. The z direction is determined by an external magnetic field which is taken to be vanishingly small, with its sole purpose being to break the symmetry and define a direction along which the spins tend to orient. If the particles were classical, the z component of the spin would be a continuous variable with values $S_z = S$ to $S_z = -S$. For quantum mechanical particles the S_z values are quantized: $S_z = S, S-1, \ldots, -S+1, -S$. In the classical case, it is easy to guess the ground state of this system: it would be the state with all spins pointing in the same direction and S_z for each spin assuming the highest value it can take, $S_z = S$. It turns out that this is also the ground state of the quantum mechanical system. We prove this statement in two stages. First, we will show that this state is a proper eigenfunction of the hamiltonian and calculate its energy; second, we will show that any other state we can construct cannot have lower energy than it.

We define the state with all $S_z = S$ as

$$|S_z(\mathbf{R}_1) = S, \ldots, S_z(\mathbf{R}_N) = S\rangle = |S, \ldots, S\rangle \equiv |S^{(N)}\rangle$$

and apply the hamiltonian $\mathcal{H}_H$ in the form of Eq. (7.22) to it: the product of operators $S_-(\mathbf{R})S_+(\mathbf{R}')$ applied to $|S^{(N)}\rangle$ gives zero, because the raising operator applied to the maximum S_z value gives zero. Thus, the only terms in the hamiltonian that give non-vanishing contributions are the $S_z(\mathbf{R})S_z(\mathbf{R}')$ products, which give S^2 for each pair of spins and leave the spins unchanged. Therefore, the state $|S^{(N)}\rangle$ is indeed an eigenfunction of the hamiltonian with eigenvalue:

$$E_0^{(N)} = -\frac{1}{2}S^2 \sum_{\mathbf{R}\neq\mathbf{R}'} J(\mathbf{R}-\mathbf{R}') \tag{7.23}$$

Next we try to construct eigenstates of the hamiltonian that have different energy than $E_0^{(N)}$. Since the hamiltonian contains only pair interactions, we can focus on the possible configurations of a pair of spins situated at the given sites $\mathbf{R}$, $\mathbf{R}'$ and consider the spins at all the other sites fixed. We are searching for the lowest energy states, so it makes sense to consider only changes of one unit in the S_z value of the spins, starting from the configuration

$$|S_z(\mathbf{R}) = S, S_z(\mathbf{R}') = S\rangle \equiv |S, S\rangle$$

There are two configurations that can be constructed with one S_z value lowered by one unit,

$$|S_z(\mathbf{R}) = S-1, S_z(\mathbf{R}') = S\rangle \equiv |S-1, S\rangle$$
$$|S_z(\mathbf{R}) = S, S_z(\mathbf{R}') = S-1\rangle \equiv |S, S-1\rangle$$

Since the hamiltonian contains a sum over all values of $\mathbf{R}, \mathbf{R}'$, both $S_-(\mathbf{R})S_+(\mathbf{R}')$ and $S_-(\mathbf{R}')S_+(\mathbf{R})$ will appear in it. When applied to the two configurations defined above, these operators will produce

$$S_-(\mathbf{R}')S_+(\mathbf{R})|S-1, S\rangle = 2S|S, S-1\rangle$$
$$S_-(\mathbf{R})S_+(\mathbf{R}')|S, S-1\rangle = 2S|S-1, S\rangle$$

where we have used the general expression for the action of raising or lowering operators on a spin state of total spin S and z component S_z,

$$S_{\pm}|S_z\rangle = \sqrt{(S \mp S_z)(S + 1 \pm S_z)}|S_z \pm 1\rangle \qquad (7.24)$$

as discussed in Appendix B. Thus, we see that the raising–lowering operators turn $|S, S-1\rangle$ to $|S-1, S\rangle$ and vice versa, which means that both of these configurations need to be included in the wavefunction in order to have an eigenstate of the hamiltonian. Moreover, the coefficients of the two configurations must have the same magnitude in order to produce an eigenfunction. Two simple choices are to take the coefficients of the two configurations to be equal, which, forgetting normalization for the moment, we can choose to be $+1$, or to be opposite, which we can choose to be ± 1. These choices lead to the following two states:

$$|S^{(+)}\rangle = [|S, S-1\rangle + |S, S-1\rangle], \quad |S^{(-)}\rangle = [|S, S-1\rangle - |S, S-1\rangle]$$

We also notice that these states are eigenfunctions of the product of operators $S_z(\mathbf{R})S_z(\mathbf{R}')$ or $S_z(\mathbf{R}')S_z(\mathbf{R})$, with eigenvalue $S(S-1)$. Ignoring all other spins which are fixed to $S_z = S$, we can then apply the hamiltonian to the states $|S^{(+)}\rangle$, $|S^{(-)}\rangle$, to find that they are both eigenstates with energies:

$$\epsilon^{(+)} = -S^2 J(\mathbf{R} - \mathbf{R}'), \quad \epsilon^{(-)} = -(S^2 - 2S)J(\mathbf{R} - \mathbf{R}')$$

These energies should be compared with the contribution to the ground state energy of the corresponding state $|S, S\rangle$, which from Eq. (7.23) is found to be

$$\epsilon^{(0)} = -S^2 J(\mathbf{R} - \mathbf{R}')$$

Thus, we conclude that of the two states we constructed, $|S^{(+)}\rangle$ has the same energy as the ground state, while $|S^{(-)}\rangle$ has higher energy by $2SJ(\mathbf{R} - \mathbf{R}')$ (recall that all $J(\mathbf{R} - \mathbf{R}')$ are positive in the ferromagnetic case). This analysis shows that when the S_z component of individual spins is reduced by one unit, only a special state has the same energy as the ground state; this state is characterized by equal coefficients for the spin pairs which involve one reduced spin. All other states which have unequal coefficients for pairs of spins, with one of the two spins reduced, have higher energy. Notice that both states $|S^{(+)}\rangle$, $|S^{(-)}\rangle$ have only a single spin with $S_z = S - 1$ and all other spins with $S_z = S$, so they represent the smallest possible change from state $|S^{(N)}\rangle$. Evidently, lowering the z component of spins by more than one unit or lowering the z component of several spins simultaneously will produce states of even higher energy. These arguments lead to the conclusion that, except for the special state which is degenerate with the state $|S^{(N)}\rangle$, all other states have higher energy than $E_0^{(N)}$. It turns out that the state with the same energy as the ground state corresponds to the infinite wavelength, or zero wave-vector ($\mathbf{k} = 0$), spin-wave state, while the states with higher energy are spin-wave states with wave-vector $\mathbf{k} \neq 0$.

7.3.2 Spin waves in the Heisenberg ferromagnet

The above discussion also leads us naturally to the low-lying excitations starting from the ground state $|S^{(N)}\rangle$. These excitations will consist of a linear superposition of configurations with the z component of only one spin reduced by one unit. In order to produce states that reflect the crystal periodicity, just as we did for the construction of states based on atomic orbitals in chapter 4, we must multiply the spin configuration which has the reduced spin at site $\mathbf{R}$ by the phase factor $\exp(\mathrm{i}\mathbf{k}\cdot\mathbf{R})$. The resulting states are

$$|S_{\mathbf{k}}^{(N-1)}\rangle \equiv \frac{1}{\sqrt{N}}\sum_{\mathbf{R}} \mathrm{e}^{\mathrm{i}\mathbf{k}\cdot\mathbf{R}} \frac{1}{\sqrt{2S}} S_-(\mathbf{R})|S^{(N)}\rangle \tag{7.25}$$

where the factors $1/\sqrt{N}$ and $1/\sqrt{2S}$ are needed to ensure proper normalization, assuming that $|S^{(N)}\rangle$ is a normalized state (recall the result of the action of S_- on a state with total spin S and z component $S_z = S$, Eq. (7.24)). These states are referred to as "spin waves" or "magnons". We will show that they are eigenstates of the Heisenberg hamiltonian and calculate their energy and their properties.

To apply the Heisenberg ferromagnetic hamiltonian to the spin-wave state $|S_{\mathbf{k}}^{(N-1)}\rangle$, we start with the action of the operator $S_z(\mathbf{R}')S_z(\mathbf{R}'')$ on configuration $S_-(\mathbf{R})|S^{(N)}\rangle$,

$$\begin{aligned}
&S_z(\mathbf{R}')S_z(\mathbf{R}'')S_-(\mathbf{R})|S^{(N)}\rangle \\
&\quad = \left[S(S-1)(\delta_{\mathbf{R}'\mathbf{R}} + \delta_{\mathbf{R}''\mathbf{R}}) + S^2(1 - \delta_{\mathbf{R}'\mathbf{R}} - \delta_{\mathbf{R}''\mathbf{R}})\right] S_-(\mathbf{R})|S^{(N)}\rangle \\
&\quad = \left[S^2 - S\delta_{\mathbf{R}\mathbf{R}'} - S\delta_{\mathbf{R}\mathbf{R}''}\right] S_-(\mathbf{R})|S^{(N)}\rangle
\end{aligned} \tag{7.26}$$

so that the $S_z(\mathbf{R}')S_z(\mathbf{R}'')$ part of the hamiltonian when applied to state $|S_{\mathbf{k}}^{(N-1)}\rangle$ gives

$$\begin{aligned}
&\left[-\frac{1}{2}\sum_{\mathbf{R}'\neq\mathbf{R}''} J(\mathbf{R}'-\mathbf{R}'')S_z(\mathbf{R}')S_z(\mathbf{R}'')\right]|S_{\mathbf{k}}^{(N-1)}\rangle \\
&\quad = \left[-\frac{1}{2}S^2\sum_{\mathbf{R}'\neq\mathbf{R}''} J(\mathbf{R}'-\mathbf{R}'') + S\sum_{\mathbf{R}\neq 0} J(\mathbf{R})\right]|S_{\mathbf{k}}^{(N-1)}\rangle
\end{aligned} \tag{7.27}$$

We next consider the action of the operator $S_-(\mathbf{R}')S_+(\mathbf{R}'')$ on configuration $S_-(\mathbf{R})|S^{(N)}\rangle$:

$$S_-(\mathbf{R}')S_+(\mathbf{R}'')S_-(\mathbf{R})|S^{(N)}\rangle = 2S\delta_{\mathbf{R}\mathbf{R}''}S_-(\mathbf{R}')|S^{(N)}\rangle \tag{7.28}$$

which gives for the action of the $S_-(\mathbf{R}')S_+(\mathbf{R}'')$ part of the hamiltonian on state $|S_{\mathbf{k}}^{(N-1)}\rangle$

$$\left[-\frac{1}{2}\sum_{\mathbf{R}'\neq\mathbf{R}''} J(\mathbf{R}'-\mathbf{R}'')S_-(\mathbf{R}')S_+(\mathbf{R}'')\right]|S_{\mathbf{k}}^{(N-1)}\rangle$$
$$= \left[-S\sum_{\mathbf{R}\neq 0} J(\mathbf{R})\mathrm{e}^{\mathrm{i}\mathbf{k}\cdot\mathbf{R}}\right]|S_{\mathbf{k}}^{(N-1)}\rangle \tag{7.29}$$

Combining the two results, Eqs. (7.27) and (7.29), we find that the state $|S_{\mathbf{k}}^{(N-1)}\rangle$ is an eigenfunction of the hamiltonian with energy

$$E_{\mathbf{k}}^{(N-1)} = E_0^{(N)} + S\sum_{\mathbf{R}\neq 0}\left(1-\mathrm{e}^{\mathrm{i}\mathbf{k}\cdot\mathbf{R}}\right)J(\mathbf{R}) \tag{7.30}$$

with the ground state energy $E_0^{(N)}$ defined in Eq. (7.23). Since we have assumed the $J(\mathbf{R})$'s to be positive, we see that the spin wave energies are higher than the ground state energy, except for $\mathbf{k} = 0$, which is degenerate with the ground state, as anticipated from our analysis based on a pair of spins.

We wish next to analyze the behavior of the system in a spin-wave state. To this end, we define the transverse spin–spin correlation operator which measures the correlation between the non-z components of two spins at sites $\mathbf{R}$, $\mathbf{R}'$; from its definition, this operator is

$$S_\perp(\mathbf{R})S_\perp(\mathbf{R}') \equiv S_x(\mathbf{R})S_x(\mathbf{R}') + S_y(\mathbf{R})S_y(\mathbf{R}')$$
$$= \frac{1}{2}\left[S_+(\mathbf{R})S_-(\mathbf{R}') + S_-(\mathbf{R})S_+(\mathbf{R}')\right]$$

The expectation value of this operator in state $|S_{\mathbf{k}}^{(N-1)}\rangle$ will involve the application of the lowering and raising operators $S_+(\mathbf{R})S_-(\mathbf{R}')$ or $S_-(\mathbf{R})S_+(\mathbf{R}')$ on configuration $S_-(\mathbf{R}'')|S^{(N)}\rangle$, which is similar to what we calculated above, Eq. (7.28), leading to

$$\langle S_{\mathbf{k}}^{(N-1)}|S_\perp(\mathbf{R})S_\perp(\mathbf{R}')|S_{\mathbf{k}}^{(N-1)}\rangle = \frac{2S}{N}\cos[\mathbf{k}\cdot(\mathbf{R}-\mathbf{R}')] \tag{7.31}$$

This shows that spins which are apart by $\mathbf{R}-\mathbf{R}'$ have a difference in transverse orientation of $\cos[\mathbf{k}\cdot(\mathbf{R}-\mathbf{R}')]$, which provides a picture of what the spin-wave state means: spins are mostly oriented in the z direction but have a small transverse component which changes orientation by an angle $\mathbf{k}\cdot(\mathbf{R}-\mathbf{R}')$ for spins separated by a distance $\mathbf{R}-\mathbf{R}'$. This angle obviously depends on the wave-vector

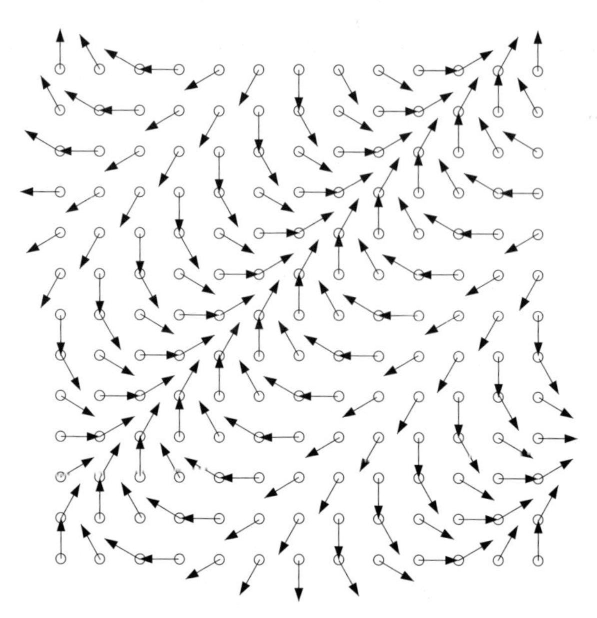

Figure 7.4. Illustration of a spin wave in the 2D square lattice with $\mathbf{k} = (\frac{1}{6}, \frac{1}{6})\frac{\pi}{a}$, where a is the lattice constant. Only the transverse component of the spins is shown, that is, the projection on the xy plane.

$\mathbf{k}$ of the spin wave. An illustration for spins on a 2D square lattice is shown in Fig. 7.4.

Finally, we calculate the magnetization in the system with spin waves. First, we notice that in a spin-wave state defined as $|S_-(\mathbf{R})S^{(N)}\rangle$ only one out of N S_z spin components has been lowered by one unit in each configuration, which means that the z component spin of each configuration and therefore of the entire $|S_{\mathbf{k}}^{(N-1)}\rangle$ state is $NS - 1$. Now suppose there are several spin-wave states excited at some temperature T. We will assume that we can treat the spin waves like phonons, so we can take a superposition of them to describe the state of the system, with boson occupation numbers

$$n_{\mathbf{k}}^{(N-1)} = \frac{1}{\exp[\epsilon_{\mathbf{k}}^{(N-1)}/k_{\mathrm{B}}T] - 1}$$

where the excitation energies are measured with respect to the ground state energy $E_0^{(N)}$:

$$\epsilon_{\mathbf{k}}^{(N-1)} = E_{\mathbf{k}}^{(N-1)} - E_0^{(N)} = 2S\sum_{\mathbf{R}\neq 0} J(\mathbf{R})\sin^2\left(\frac{1}{2}\mathbf{k}\cdot\mathbf{R}\right)$$

In the above expression we have used the fact that $J(\mathbf{R}) = J(-\mathbf{R})$ to retain only the real part of the complex exponential in the expression for $E_{\mathbf{k}}^{(N-1)}$, Eq. (7.30).

We must emphasize that treating the system as a superposition of spin waves is an approximation, because they cannot be described as independent bosons, as was the case for phonons (see chapter 6). Within this approximation, the total z component of the spin, which we identify with the magnetization of the system, will be given by

$$M(T) = NS - \sum_{\mathbf{k}} n_{\mathbf{k}}^{(N-1)} = NS \left[1 - \frac{1}{NS} \sum_{\mathbf{k}} \frac{1}{\exp[\epsilon_{\mathbf{k}}^{(N-1)}/k_{\mathrm{B}}T] - 1} \right] \quad (7.32)$$

since each state $|S_{\mathbf{k}}^{(N-1)}\rangle$ reduces the total z component of the ground state spin, NS, by one unit. In the small $|\mathbf{k}|$ limit, turning the sum over $\mathbf{k}$ into an integral as usual, and defining the new variable $\mathbf{q} = \mathbf{k}/\sqrt{2k_{\mathrm{B}}T/S}$, we obtain for the magnetization per particle $m(T) = M(T)/N$

$$m(T) = S - \left(\frac{2k_{\mathrm{B}}T}{n^{2/3}S} \right)^{3/2} \int \frac{\mathrm{d}\mathbf{q}}{(2\pi)^3} \left[\exp\left(\sum_{\mathbf{R}\neq 0} J(\mathbf{R})(\mathbf{q}\cdot\mathbf{R})^2 \right) - 1 \right]^{-1} \quad (7.33)$$

with n the density of particles. This expression gives the temperature dependence of the magnetization as being $\sim -T^{3/2}$, which is known as the Bloch $T^{3/2}$ law. This behavior is actually observed in experiments performed on isotropic ferromagnets, that is, systems in which $J(\mathbf{R}-\mathbf{R}')$ is the same for all operators appearing in the Heisenberg spin hamiltonian, as we have been assuming so far. There exist also anisotropic systems in which the value of $J_{\perp}(\mathbf{R}-\mathbf{R}')$ corresponding to the transverse operators $S_{\perp}(\mathbf{R})S_{\perp}(\mathbf{R}') = S_x(\mathbf{R})S_x(\mathbf{R}') + S_y(\mathbf{R})S_y(\mathbf{R}')$ is different than the value $J_z(\mathbf{R}-\mathbf{R}')$ corresponding to the operator $S_z(\mathbf{R})S_z(\mathbf{R}')$, in which case the Bloch $T^{3/2}$ law does not hold.

One last thing worth mentioning, which is a direct consequence of the form of the magnetization given in Eq. (7.33), is that for small enough $|\mathbf{q}| = q$ we can expand the exponential to obtain

$$\left[\exp\left(\sum_{\mathbf{R}\neq 0} J(\mathbf{R})(\mathbf{q}\cdot\mathbf{R})^2 \right) - 1 \right]^{-1} \approx \frac{1}{q^2} \left(\sum_{\mathbf{R}\neq 0} J(\mathbf{R})(\hat{\mathbf{q}}\cdot\mathbf{R})^2 \right)^{-1}$$

If we assume for the moment (we will analyze this assumption below) that only small values of $|\mathbf{q}|$ contribute to the integral, then the integrand contains the factor $1/q^2$, which in 3D is canceled by the factor q^2 in the infinitesimal $\mathrm{d}\mathbf{q} = q^2 \mathrm{d}q \mathrm{d}\hat{\mathbf{q}}$, but in 2D and 1D is not canceled, because in those cases $\mathrm{d}\mathbf{q} = q \mathrm{d}q \mathrm{d}\hat{\mathbf{q}}$ and $\mathrm{d}\mathbf{q} = \mathrm{d}q \mathrm{d}\hat{\mathbf{q}}$, respectively. Since the integration includes the value $\mathbf{q} = 0$, we conclude that in 3D the integral gives finite value but in 2D and 1D it gives an infinite value. The interpretation of this result is that in 2D and 1D the presence of excitations above the ground state is so disruptive to the magnetic order in the system that the spin-wave picture itself breaks down at finite temperature. Another way to express this is that there can be no magnetic order at finite temperature in 2D and 1D within the isotropic Heisenberg model, a statement known as the Hohenberg–Mermin–Wagner theorem [86]. To complete the argument, we examine the validity of its basic assumption, namely the conditions under which only small values of $|\mathbf{k}|$ are relevant: from Eq. (7.32) we see that for large values of $\epsilon_{\mathbf{k}}^{(N-1)}/k_{\mathrm{B}}T$ the exponential in the denominator of the integrand leads to vanishing contributions to the integral. This implies that for large T spin-wave states with large energies will contribute, but for small T only states with very small energies can contribute to the integral. We have seen earlier that in the limit $\mathbf{k} \to 0$ the energy $\epsilon_{\mathbf{k}}^{(N-1)} \to 0$, so that the states with the smallest $\mathbf{k}$ also have the lowest energy above the ground state. Therefore, at low T only the states with lowest $|\mathbf{k}|$, and consequently lowest $|\mathbf{q}|$, will contribute to the magnetization. Thus, the Bloch $T^{3/2}$ law will hold for low temperatures. The scale over which the temperature can be considered low is set by the factor

$$\frac{S}{2}\sum_{\mathbf{R}\neq 0} J(\mathbf{R})(\mathbf{k}\cdot\mathbf{R})^2$$

which obviously depends on the value of the exchange integral $J(\mathbf{R})$.

7.3.3 Heisenberg antiferromagnetic spin model

The Heisenberg spin model with negative exchange integral J can be used to study the physics of antiferromagnetism. A useful viewpoint for motivating the Heisenberg antiferromagnetic spin model is based on the Hubbard model [18–20], which we have discussed in the context of the hydrogen molecule in chapter 2, and we revisit next. The Hubbard model assumes that the system can be described by electrons which, in the single-particle language, are localized at particular atomic sites and are allowed to hop between such sites. Due to Coulomb repulsion the electrons will avoid being on the same site, which can only happen if they have opposite spins due to the Pauli exclusion principle. The model is defined by two parameters, the energy gain t due to hopping of the electrons between nearest neighbor sites, which in effect represents the kinetic energy, and the energy cost U to have two electrons at the same site, which is the dominant contribution to the Coulomb energy. For simplicity, we assume that there is only one type of atom in the crystal, forming a simple Bravais lattice with lattice vectors $\mathbf{R}$, so that all the hopping energies are the same and all the on-site Coulomb repulsion energies are the same. We also ignore the interaction between electrons and ions, assuming that its main effect is to produce the localized single-particle states $|\psi_{\mathbf{R}}\rangle$ centered at the lattice sites $\mathbf{R}$, which coincide with the atomic sites. To be specific, the hopping and

Coulomb repulsion energies can be defined in terms of the corresponding operators of the hamiltonian and the single-particle states $|\psi_{\mathbf{R}}\rangle$:

$$t \equiv -\langle \psi_{\mathbf{R}} | \hat{T} | \psi_{\mathbf{R} \pm \mathbf{a}_i} \rangle, \qquad U \equiv e^2 \langle \psi_{\mathbf{R}} \psi_{\mathbf{R}} | \frac{\exp(-k|\mathbf{r}_1 - \mathbf{r}_2|)}{|\mathbf{r}_1 - \mathbf{r}_2|} | \psi_{\mathbf{R}} \psi_{\mathbf{R}} \rangle \tag{7.34}$$

In these expressions, $\mathbf{a}_i$ stands for any of the three primitive lattice vectors, $\hat{T}$ is the kinetic energy operator, and we have assumed a screened Coulomb potential between the electrons; t and U are both taken to be positive constants.

In the limit $t \ll U$ the system, to the extent possible, will avoid having double occupancy of any site due to the very high cost in Coulomb energy. If we consider the hopping part of the hamiltonian as a small perturbation on the Coulomb interaction term, we can classify the unperturbed states according to how many sites are doubly occupied: the lowest state with zero energy will have no doubly occupied sites, the next state with energy U will have one doubly occupied site, etc. In this spirit, it is possible to show using perturbation theory [87–89] that, to second order in t, the original hamiltonian $\mathcal{H}$ is equivalent to an effective hamiltonian $\mathcal{H}_{eff}$ which has three terms:

$$\mathcal{H}_{eff} = \mathcal{H}_1 + \mathcal{H}_2 + \mathcal{H}_3$$

The first term $\mathcal{H}_1$, of order t, is simply the hopping term; it requires the presence of an unoccupied site next to the site where the electron is, into which the electron can hop. The second term $\mathcal{H}_2$, of order t^2/U, leads to spin exchange between electrons at neighboring sites when they happen to have opposite spins. The third term $\mathcal{H}_3$, also of order t^2/U, leads to electron-pair hopping by one lattice vector, which may or may not be accompanied by spin exchange; this last term applies only to situations where two electrons at nearest neighbor sites have opposite spins and there is an unoccupied site next to one of them. $\mathcal{H}_2$ and $\mathcal{H}_3$ involve the virtual occupation of a site by two electrons of opposite spin, which is the origin of the energy cost U; the physical meaning of these three terms is illustrated in Fig. 7.5. From this analysis, we can now have an estimate of the exchange integral J: in the context of the Hubbard model and in the limit of strong on-site Coulomb repulsion, J must be of order t^2/U. In fact, it can be rigorously shown [87, 90] that the second term in the effective hamiltonian discussed above, in the case of one electron per crystal site, can be expressed as

$$\mathcal{H}_2 = -\frac{J}{2} \sum_{\langle ij \rangle} \mathbf{S}(\mathbf{R}_i) \cdot \mathbf{S}(\mathbf{R}_j)$$

where the exchange integral is given by $J = -(8t^2/U)$, and $\mathbf{S}(\mathbf{R}_i)$ is the spin operator for site i; in this example, i, j must be nearest neighbor indices. Since $J < 0$, we have arrived at a Heisenberg antiferromagnetic hamiltonian. It should be

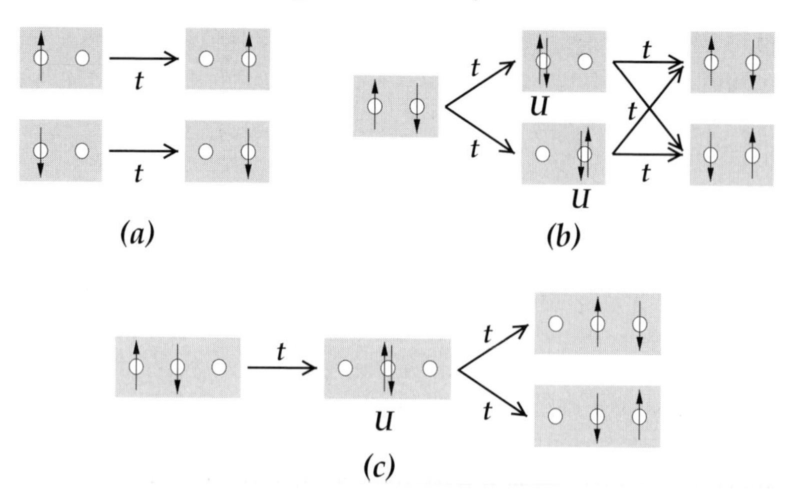

Figure 7.5. The three terms in the effective second order hamiltonian obtained from the Hubbard model in the limit $t \ll U$. (a) The hopping term, which allows spin-up or spin-down electrons to move by one site when the neighboring site is unoccupied. (b) The exchange term, which allows electrons to exchange spins or remain in the same configuration, at the cost of virtual occupation of a site by two electrons of opposite spin (intermediate configuration). (c) The pair hopping term, which allows a pair of electrons with opposite spins to move by one site, either with or without spin exchange, at the cost of virtual occupation of a site by two electrons of opposite spin (intermediate configuration). Adapted from Ref. [87].

emphasized that this derivation applies to a special case, an insulating crystal with exactly one electron per crystal site. However, there are several physical systems of interest which fall within, or can be considered small perturbations of, this special case; an example is the doped copper-oxide perovskites, which are high-temperature superconductors (see chapter 8).

The Heisenberg antiferromagnetic spin model is much more complicated than the ferromagnetic model. The basic problem is that it is considerably more difficult to find the ground state and the excitations of spins with antiferromagnetic interactions. In fact, in this case the model is only studied in its nearest neighbor interaction version, because it is obviously impossible to try to make all spins on a lattice oriented opposite from each other: two spins that are antiparallel to the same third spin must be parallel to each other. This is in stark contrast to the ferromagnetic case, where it is possible to make all spins on a lattice parallel, and this is actually the ground state of the model with $J > 0$. For nearest neighbor interactions only, it is possible to create a state in which every spin is surrounded by nearest neighbor spins pointing in the opposite direction, which is called the "Néel state". This is not true for every lattice, but only for lattices that can be split into two interpenetrating sublattices; these lattices are called *bipartite*. Some examples of bipartite lattices in 3D are the simple cubic lattice (with two interpenetrating cubic sublattices), the BCC lattice (also with two interpenetrating cubic sublattices), and the diamond

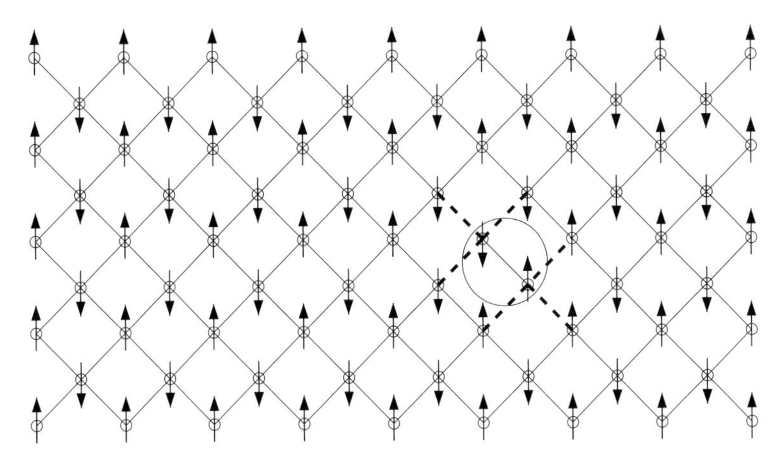

Figure 7.6. Illustration of the Néel state on a 2D square lattice. Two spins, shown by a circle, have been flipped from their ideal orientation, destroying locally the antiferromagnetic order: neighboring pairs of spins with wrong orientations are highlighted by thicker dashed lines.

lattice (with two interpenetrating FCC sublattices); in 2D examples of bipartite lattices are the square lattice (with two interpenetrating square sublattices) and the graphitic lattice (with two interpenetrating hexagonal sublattices). The lattices that cannot support a Néel state are called "frustrated lattices", some examples being the FCC lattice in 3D and the hexagonal lattice in 2D.

The difficulty with solving the Heisenberg antiferromagnet is that, even in cases where the lattice supports such a state, the Néel state is not an eigenstate of the hamiltonian. This can be easily seen from the fact that the operators $S_+(i)S_-(j)$ or $S_-(i)S_+(j)$ which appear in the hamiltonian, when applied to the pair of antiparallel spins at nearest neighbor sites $\langle ij \rangle$, flip both spins and as a result destroy the alternating spin pattern and create a different configuration. This is illustrated in Fig. 7.6. Much theoretical work has been directed toward understanding the physics of the Heisenberg antiferromagnet, especially after it was suggested by Anderson that this model may be relevant to the copper-oxide high-temperature superconductors [91]. Interestingly, spin waves may also be used to describe the physics of this model, at least in the limit of large spin S, assuming that deviations from the Néel state are small [92, 93].

7.4 Magnetic order in real materials

Having developed the theoretical background for describing magnetic order in ferromagnetic and antiferromagnetic ideal solids, we discuss next how this order is actually manifested in real materials. First, a basic consideration is the temperature

Table 7.2. *Examples of ferromagnets and antiferromagnets.* The Curie temperature T_C and Néel temperature T_N are given in kelvin; n_B is the number of Bohr magnetons per atom; M_0 is the saturation magnetization in gauss.

Elemental							
Ferromagnets				Antiferromagnets			
Solid	T_C	n_B	M_0	Solid	T_N	Solid	T_N
Fe	1043	2.2	1752	Cr	311	Sm	106
Co	1388	1.7	1446	Mn	100	Eu	91
Ni	627	0.6	510	Ce	13	Dy	176
Gd	293	7.0	1980	Nd	19	Ho	133

Compound							
Ferromagnets				Antiferromagnets			
Solid	T_C	Solid	T_C	Solid	T_N	Solid	T_N
MnB	152	Fe_3C	483	MnO	122	$FeKF_3$	115
MnAs	670	FeP	215	FeO	198	$CoKF_3$	125
MnBi	620	CrTe	339	CoO	291	MnF_2	67
MnSb	710	$CrBr_3$	37	NiO	600	$MnCl_2$	2
FeB	598	CrI_3	68	$MnRbF_3$	54	FeF_2	78
Fe_2B	1043	CrO_2	386	$MnKF_3$	88	CoF_2	38

Source: Ref. [73].

dependence of magnetic order. As we noted already in the introduction to this chapter, magnetic order is relatively weak and can be easily destroyed by thermal fluctuations. The critical temperature above which there is no magnetic order is called the Curie temperature (T_C) for ferromagnets and the Néel temperature (T_N) for antiferromagnets. Table 7.2 provides examples of elementary and compound magnetic solids and their Curie and Néel temperatures. In certain solids, these critical temperatures can be rather large, exceeding 1000 K, such as in Fe and Co. The Néel temperatures are generally much lower than the Curie temperatures, the antiferromagnetic state being more delicate than the ferromagnetic one. Another interesting characteristic feature of ferromagnets is the number of Bohr magnetons per atom, n_B. This quantity is a measure of the difference in occupation of spin-up and spin-down states in the solid; it is determined by the band structure of the spin-polarized system of electrons in the periodic solid (values of n_B can be readily obtained from band-structure calculations of the type described in chapter 4). The higher the value of n_B the more pronounced is the magnetic behavior of the solid. However, there is no direct correlation between n_B and the Curie temperature.

Finally, the behavior of ferromagnets is characterized by the saturation magnetization M_0, that is, the highest value that the magnetization can attain. This value corresponds to all the spins being aligned in the same direction; values of this quantity for elemental ferromagnets are given in Table 7.2. The saturation magnetization is not attained spontaneously in a real ferromagnet but requires the application of an external field. This is related to interesting physics: in a real ferromagnet the dominant magnetic interactions at the atomic scale tend to orient the magnetic moments at neighboring sites parallel to each other, but this does not correspond to a globally optimal state from the energetic point of view. Other dipoles placed in the field of a given magnetic dipole would be oriented at various angles relative to the original dipole in order to minimize the magnetic dipole–dipole interactions. The dipole–dipole interactions are much weaker than the short range interactions responsible for magnetic order. However, over macroscopically large distances the shear number of dipole–dipole interactions dictates that their contribution to the energy should also be optimized. The system accommodates this by breaking into regions of different magnetization called domains. Within each domain the spins are oriented parallel to each other, as dictated by the atomic scale interactions (the magnetization of a domain is the sum of all the spins in it). The magnetizations of different domains are oriented in a pattern that tries to minimize the magnetic dipole–dipole interactions over large distances. An illustration of this effect in a simple 2D case is shown in Fig. 7.7. The domains of different magnetization are separated by boundaries called domain walls. In a domain wall the orientation of spins is gradually changed from the one which dominates on one side of the wall to that of the other side. This

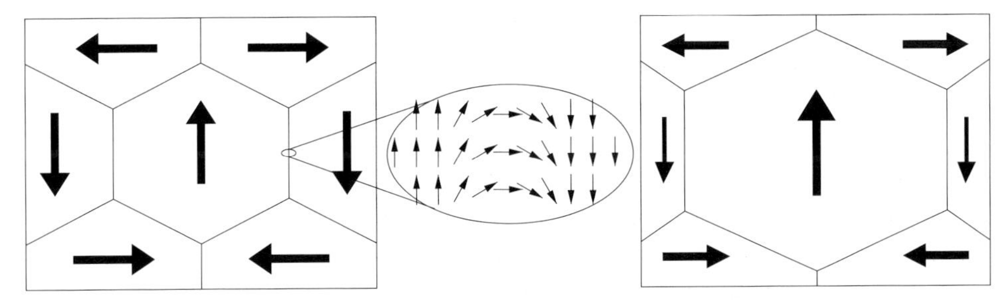

Figure 7.7. Illustration of magnetic domains in a ferromagnet. The large arrows indicate the magnetization in each domain. The pattern of domain magnetizations follows the field of a magnetic dipole situated at the center of each figure. **Left:** this corresponds to a zero external magnetic field, with the total magnetization averaging to zero. **Right:** this corresponds to a magnetic field in the same direction as the magnetization of the domain at the center, which has grown in size at the xpense of the other domains; domains with magnetization opposite to the external field have shrunk the most. **Inset:** the domain wall structure at microscopic scale in terms of individual spins.

change in spin orientation takes place over many interatomic distances in order to minimize the energy cost due to the disruption of order at the microscopic scale induced by the domain wall. If the extent of the domain wall were of order an interatomic distance the change in spin orientation would be drastic across the wall leading to high energy cost. Spreading the spin-orientation change over many interatomic distances minimizes this energy cost, very much like in the case of spin waves.

In an ideal solid the domain sizes and distribution would be such that the total magnetization is zero. In real materials the presence of a large number of defects (see Part II, especially chapters 9, 10 and 11), introduces limitations in the creation and placement of magnetic domain walls. The magnetization M can be increased by the application of an external magnetic field H. The external field will favor energetically the domains of magnetization parallel to it, and the domain walls will move to enlarge the size of these domains at the expense of domains with magnetization different from the external field. The more the magnetization of a certain domain deviates from the external field the higher its energy will be and consequently the more this domain will shrink, as illustrated in Fig. 7.7. For relatively small fields this process is reversible, since the domain walls move only by a small amount and they can revert to their original configuration upon removal of the field. For large fields this process becomes irreversible, because if the walls have to move far, eventually they will cross regions with defects which will make their return to the original configuration impossible, after the field is removed. Thus, when the field is reduced to zero after saturation, the magnetization does not return to its original zero value but has a positive value because of limitations in the mobility of domain walls due to defects in the solid. It is therefore necessary to apply a large field in the opposite direction, denoted as $-H_c$ in Fig. 7.8, to reduce the magnetization M back to zero; this behavior is referred to as "hysteresis".[1] Continuing to increase the field in the negative direction will again lead to a state with saturated magnetization, as indicated in Fig. 7.8. When the field is removed, the magnetization now has a non-zero value in the opposite direction, which requires the application of a positive field to reduce it to zero. This response of the system to the external field is called a "hysteresis loop".

7.5 Crystal electrons in an external magnetic field

In this section we consider the response of electrons in an external magnetic field. This response is particularly dramatic when the electrons are confined in two dimensions and are subjected to a very strong perpendicular magnetic field at

[1] From the Greek word υστερησις, which means lagging behind.

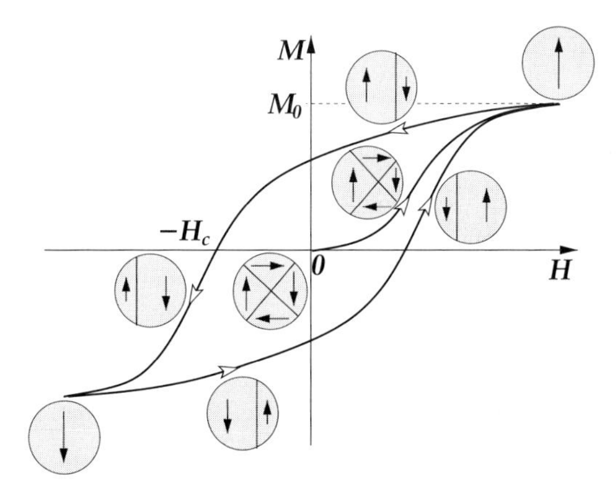

Figure 7.8. Hysteresis curve of the magnetization M in a ferromagnet upon application of an external field H. The original magnetization curve starts at $H = 0$, $M = 0$ and extends to the saturation point, $M = M_0$. When the field is reduced back to zero the magnetization has a finite value; zero magnetization is obtained again for the reversed field value $-H_c$. The circular insets next to the origin, the end-points and the arrows along the hysteresis curve indicate schematically the changes in magnetization due to domain wall motion.

very low temperature. We will therefore concentrate the discussion mostly to those conditions.

To begin, let us explore how electrons can be confined in two dimensions. In certain situations it is possible to form a flat interface between two crystals. We will take x, y to be the coordinate axes on the interface plane and z the direction perpendicular to it; the magnetic field will be $\mathbf{H} = H\hat{\mathbf{z}}$. A particular arrangement of an insulator and doped semiconductor produces a potential well in the direction perpendicular to the interface plane which quantizes the motion of electrons in this direction. This is illustrated in Fig. 7.9: the confining potential $V_{conf}(z)$ can be approximated as nearly linear on the doped semiconductor side and by a hard wall on the insulator side. The electron wavefunctions $\psi(\mathbf{r})$ can be factored into two parts:

$$\psi(\mathbf{r}) = \bar{\psi}(x, y) f(z) \tag{7.35}$$

where the first part $\bar{\psi}(x, y)$ describes the motion on the interface plane and the second part $f(z)$ describes motion perpendicular to it. The latter part experiences the confining potential $V_{conf}(z)$, which gives rise to discrete levels. When these levels are well separated in energy and the temperature is lower than the separation between levels, only the lowest level is occupied. When all electrons are at this lowest level corresponding to the wavefunction $f_0(z)$, their motion in the z direction is confined, making the system of electrons essentially a two-dimensional one. This arrangement is called a quantum well, and the system of electrons confined in two dimensions in the semiconductor is called an inversion layer. A more detailed

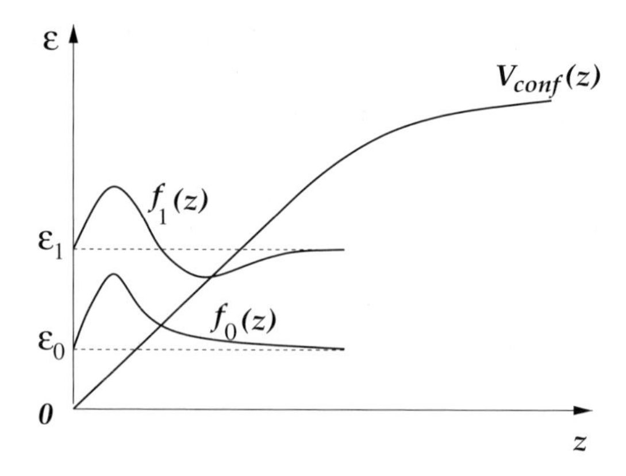

Figure 7.9. The confining potential in the z direction $V_{conf}(z)$, and the energies ϵ_0, ϵ_1, and wavefunctions $f_0(z)$, $f_1(z)$, of the two lowest states. The Fermi level lies between ϵ_0 and ϵ_1.

discussion of electronic states in doped semiconductors and the inversion layer is given in chapter 9.

In the following discussion we will ignore the crystal momentum **k** and band index normally associated with electronic states in the crystal, because the density of confined electrons in the inversion layer is such that only a very small fraction of one band is occupied. Specifically, the density of confined electrons is of order $n = 10^{12}$ cm^{-2}, whereas a full band corresponding to a crystal plane can accommodate of order $(10^{24}\ \text{cm}^{-3})^{2/3} = 10^{16}$ cm^{-2} states. Therefore, the confined electrons occupy a small fraction of the band near its minimum. It should be noted that the phenomena we will describe below can also be observed for a system of holes at similar density. For both the electron and the hole systems the presence of the crystal interface is crucial in achieving confinement in two dimensions, but is otherwise not important in the behavior of the charge carriers (electrons or holes) under the conditions considered here.

7.5.1 de Haas–van Alphen effect

Up to this point we have treated the electrons as a set of classical charge carriers, except for the confinement in the z direction which introduces quantization of the motion along z, and localization in the xy plane. Now we introduce the effects of quantization in the remaining two dimensions. The behavior of electrons in the plane exhibits quantization in two ways, which are actually related (see Problem 4). Here we give only a qualitative discussion of these two levels of quantization.

1. In the presence of a perpendicular magnetic field $\mathbf{H} = H\hat{\mathbf{z}}$ the electron orbits are quantized in units of the cyclotron frequency

$$\omega_c = \frac{eH}{mc}$$

The corresponding energy levels are equally spaced, separated by intervals $\hbar\omega_c$; these are called **Landau levels**. This is intuitively understood as the quantization of circular motion due to a harmonic oscillator potential.

2. The number of electronic states that each Landau level can accommodate is determined by the quantization of magnetic flux. This is intuitively understood as a consequence of the fact that there exist spatial limitations in the placement of the circular orbits.

The total magnetic flux through the plane where the electrons are confined is $H(WL)$, where W is the width and L the length of the plane; we define the area A of the plane as $A = WL$. The number of flux quanta corresponding to this total flux is

$$f = \frac{HA}{\phi_0},$$

with ϕ_0 the flux quantum defined in Eq. (7.5). Suppose that in order to accommodate the total number of electrons N we need to fill the Landau levels up to index l, that is, we need exactly $l + 1$ Landau levels starting with level 0. Then we will have

$$N = (l+1)f = (l+1)\frac{HA}{\phi_0} \Rightarrow l+1 = \frac{N}{A}\frac{\phi_0}{H} = n\frac{ch}{eH} \Rightarrow H = n\frac{ch}{e(l+1)} \tag{7.36}$$

where $n = N/A$ is the density of electrons in the plane. Thus, if the field H has precisely the value $H = nch/e(l+1)$ exactly $l+1$ Landau levels will be filled with index numbers $0, 1, \ldots, l$. If the electrons cannot be accommodated exactly by an integer number of Landau levels, then levels up to index l will be completely filled and the level with index $l+1$ will be partially filled, as illustrated in Fig. 7.10. In this case we will have

$$(l+1)f < N < (l+2)f \Longrightarrow \frac{1}{l+2} < b = \frac{AH}{N\phi_0} < \frac{1}{l+1}$$

where we have defined the new variable b for convenience. The total energy of the system in this situation will be given by

$$E(H) = \sum_{j=0}^{l} f\hbar\omega_c\left(j + \frac{1}{2}\right) + [N - (l+1)f]\hbar\omega_c\left(l + \frac{3}{2}\right)$$

Using the definition of the frequency ω_c, the Bohr magneton μ_B and the variable b, we obtain for the total energy per particle

$$\frac{E(H)}{N} = \mu_B\left(\frac{N\phi_0}{A}\right)\left[b(2l+3) - b^2(l^2+3l+2)\right], \quad \frac{1}{l+2} < b < \frac{1}{l+1}, l \geq 0$$

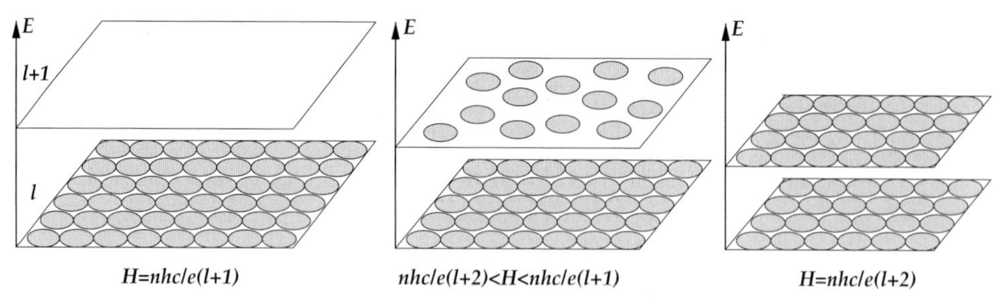

Figure 7.10. Schematic representation of filling of Landau levels by flux quanta ϕ_0 (shown as shaded circles): at some value of the magnetic field H each Landau level can accommodate HA/ϕ_0 quanta (with A the area of the plane) and the total number of electrons N fills all levels up to index l, that is, a total of $l+1$ levels; as the field decreases each level can accommodate fewer flux quanta and some electrons move to the next level, with index $l+1$; eventually, this level also becomes filled when the value of the field decreases enough. Between fillings of entire levels the Fermi energy is pinned at the energy of the partially filled level. Notice that the spacing of the levels, given by $\hbar\omega_c = \hbar(eH/mc)$, also decreases as the field decreases.

For large enough field, $H > N\phi_0/A$, only the lowest Landau level ($l = 0$) will be filled, and $b > 1$, giving for the energy per particle

$$\frac{E(H)}{N} = \mu_B \left(\frac{N\phi_0}{A}\right) b, \quad b > 1, l = 0$$

We can now calculate the magnetization M and susceptibility χ per unit volume, as

$$M \equiv -\frac{1}{\Omega}\frac{\partial E}{\partial H} = -\frac{N}{\Omega}\mu_B, \quad b > 1, l = 0$$

$$= \frac{N}{\Omega}\mu_B\left[2(l^2+3l+2)b - (2l+3)\right], \quad \frac{1}{l+2} < b < \frac{1}{l+1}, l \geq 0$$

$$\chi \equiv -\frac{1}{\Omega}\frac{\partial^2 E}{\partial H^2} = 0, \quad b > 1, l = 0$$

$$= \frac{1}{\Omega}\mu_B\left(\frac{2A}{\phi_0}\right)(l^2+3l+2), \quad \frac{1}{l+2} < b < \frac{1}{l+1}, l \geq 0$$

A plot of the magnetization in units of $(N/\Omega)\mu_B$ as a function of the variable b, which is the magnetic field in units of $N\phi_0/A$, is shown in Fig. 7.11: the magnetization shows oscillations between positive and negative values as a function of the magnetic field! This behavior is known as the de Haas–van Alphen effect. Although we have shown how it arises in a 2D system, the effect is observed in 3D systems as well, in which case the discontinuities in the magnetization are rounded off. A number of other physical quantities exhibit the same type of oscillations as a function of the magnetic field H; examples include the conductivity (this is known as the Shubnikov–de Haas effect and is usually easier to measure than the de Haas–van Alphen effect), the sound attenuation, the magnetoresistance and the magnetostriction (strain induced on the sample by the magnetic field).

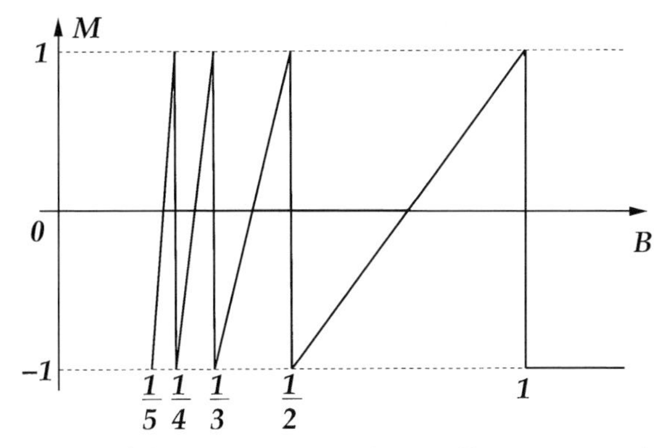

Figure 7.11. Illustration of the de Haas–van Alphen effect in a two-dimensional electron gas: the magnetization M, measured in units of $(N/\Omega)\mu_B$, exhibits oscillations as a function of the magnetic field H, measured in units of $N\phi_0/A$.

The oscillations of the magnetization as a function of the magnetic field can be used to map out the Fermi surface of metals, through the following argument, due to Onsager [94]. When a 3D metal is in an external magnetic field, the plane perpendicular to the field will intersect the Fermi surface of the metal producing a cross-sectional area that depends on the shape of the Fermi surface. Of course there are many cross-sections of a 3D Fermi surface along a certain direction, but the relevant ones are those with the largest and smallest areas (the extremal values). These cross-sectional areas will play a role analogous to the area of the plane on which the electrons were confined in the above example. Thus, our simple treatment of electrons confined in 2D becomes relevant to the behavior of electrons with wave-vectors lying on a certain plane that intersects the Fermi surface. Although additional complications arise from the band structure, in principle we can use the oscillations of the magnetization to determine the extremal cross-sectional areas enclosed by the Fermi surface on this plane. By changing the orientation of the metal with respect to the direction of the field, different cross-sections of the Fermi surface come into play, which makes it possible to determine the entire 3D Fermi surface by reconstructing it from its various cross-sections. Usually, a band-structure calculation of the type discussed in detail in chapter 4 is indispensable in reconstructing the exact Fermi surface, in conjunction with magnetic measurements which give the precise values of the extremal areas on selected planes.

7.5.2 Classical and quantum Hall effects

We consider next what happens if in addition to the perpendicular magnetic field we apply also an electric field in the x direction, E_x, as shown in Fig. 7.12. The applied electric field will induce a current $j_x = (-e)nv_x$ in the same direction, where v_x is the velocity and n is the density of electrons in the plane. Due to the presence of

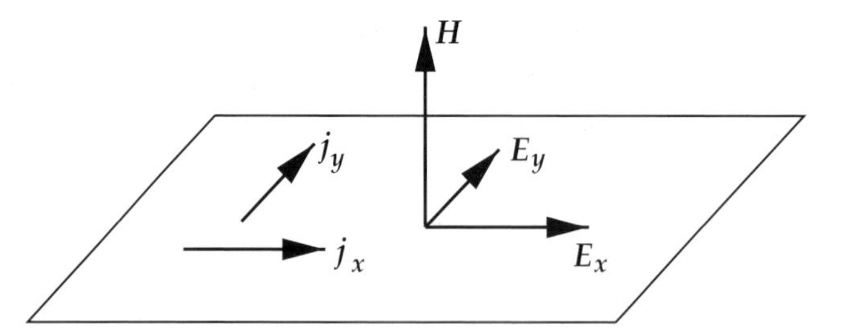

Figure 7.12. Geometry of the two-dimensional electrons in the Hall effect, with the external magnetic field **H** and the x, y components of the electric field E_x, E_y, and the current j_x, j_y.

the magnetic field, the moving electrons will experience a force in the y direction given by

$$F_y = (-e)\frac{v_x}{c}H \tag{7.37}$$

Using the expression for the current, and the fact that the motion in the y direction can be associated with an effective electric field E_y, we obtain

$$E_y = \frac{F_y}{(-e)} = \frac{j_x}{(-e)nc}H \Rightarrow \frac{j_x}{E_y} = \frac{(-e)nc}{H} \tag{7.38}$$

The expression derived in Eq. (7.38) is defined as the Hall conductivity σ_H: it is the ratio of the current in the x direction to the effective electric field E_y in the perpendicular direction, as usually defined in the classical Hall effect in electrodynamics. In more general terms, this ratio can be viewed as one of the off-diagonal components of the conductivity tensor $\sigma_{ij}(i, j = x, y)$, which relates the current to the electric field:

$$\begin{aligned} j_x &= \sigma_{xx}E_x + \sigma_{xy}E_y \\ j_y &= \sigma_{yx}E_x + \sigma_{yy}E_y \end{aligned} \tag{7.39}$$

The electric field can be expressed in terms of the current using the resistivity tensor $\rho_{ij}(i, j = x, y)$:

$$\begin{aligned} E_x &= \rho_{xx}j_x + \rho_{xy}j_y \\ E_y &= \rho_{yx}j_x + \rho_{yy}j_y \end{aligned} \tag{7.40}$$

which is the inverse of the conductivity tensor. Assuming that $\rho_{yx} = -\rho_{xy}$, $\rho_{xx} = \rho_{yy}$, that is, an isotropic solid, it is easy to find the relation between the two tensors:[2]

$$\begin{aligned} \sigma_{xx} &= \frac{\rho_{xx}}{\rho_{xy}^2 + \rho_{xx}^2} = \sigma_{yy} \\ \sigma_{xy} &= \frac{-\rho_{xy}}{\rho_{xy}^2 + \rho_{xx}^2} = -\sigma_{yx} \end{aligned} \tag{7.41}$$

[2] In three dimensions the two indices of the resistivity and conductivity tensors take three values each (x, y, z).

We assume next that the conditions in the system of confined electrons are such that $\sigma_{xx} = 0 \Rightarrow \rho_{xx} = 0$; the meaning of this assumption will become evident shortly. We then find for the Hall conductivity

$$\sigma_H = \sigma_{xy} = -\frac{1}{\rho_{xy}} = \frac{j_x}{E_y} \tag{7.42}$$

Combining this with the expression from Eq. (7.38), we obtain

$$\sigma_H = \frac{-enc}{H} = -\frac{e^2}{h}(l+1) \tag{7.43}$$

where $l + 1$ is the number of filled Landau levels (up to index l). Equivalently, we can obtain the Hall resistivity (the off-diagonal component of the resistivity tensor) as

$$\rho_{xy} = -\frac{1}{\sigma_{xy}} = \frac{h}{e^2}\frac{1}{l+1} \tag{7.44}$$

These relations suggest that the Hall conductivity and resistivity are quantized. This conclusion rests on the assumption that the value of the magnetic field and the density of electrons in the plane are such that exactly $l + 1$ Landau levels are filled. However, it is obvious that this need not be always the case, since for a given density of electrons n, a small change in the value of the magnetic field will violate the condition of Eq. (7.36) that $n(ch/eH)$ is an integer, as illustrated in Fig. 7.10. In 1980, experiments by von Klitzing and coworkers [95] showed that for very high magnetic fields and low temperatures, the Hall resistivity ρ_{xy} as a function of H has wide plateaus that correspond to integer values of $l + 1$ in Eq. (7.44). When ρ_{xy} has the quantized values suggested by Eq. (7.44), the diagonal resistivity ρ_{xx} vanishes. This is shown schematically in Fig. 7.13. This fascinating observation was called the integer quantum Hall effect (IQHE); for its discovery, K. von Klitzing was awarded the 1985 Nobel prize for Physics.

Why does the IQHE exist? To put it differently, what might we expect for values of n and H that *do not* satisfy the condition of an integer number of filled Landau levels? Suppose that we start the experiment with a value of the magnetic field that corresponds to exactly one filled Landau level. According to Eq. (7.36) this level can accommodate a density of electrons $n = eH/ch$. Let us consider the density n to be fixed by the number of electrons and the geometry of the system. We expect that in this case there is no current, since a filled Landau level corresponds to a saturated system analogous to the filled band discussed in chapter 3. Then we expect ρ_{xy} to have the value given by Eq. (7.44) with $l = 0$ and $\rho_{xx} = \sigma_{xx} = 0$, as we had to assume earlier in order to arrive at the quantized values of ρ_{xy}. As soon as the value of the magnetic field is decreased slightly, the total magnetic flux through the plane will decrease, and the first Landau level will not be able to accommodate

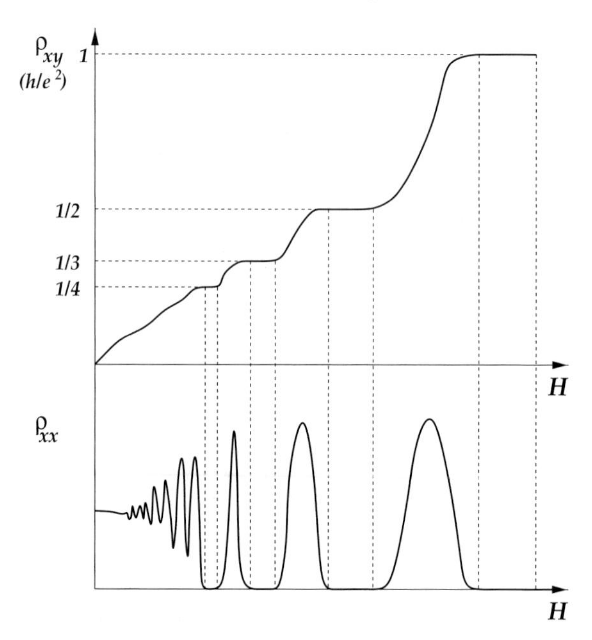

Figure 7.13. Schematic representation of the integral quantum Hall effect. The plateaus in the Hall conductivity ρ_{xy} as a function of the magnetic field H, where its values are given by $(h/e^2)/(l+1)$ with $l+1$ an integer, are identified; the diagonal conductivity ρ_{xx} vanishes for the values of H corresponding to the plateaus.

all the electrons in the system, so that a few electrons will have to go to the next Landau level. This next level will have only a very small occupation, so that it will be able to carry current, in which case $\rho_{xx} \neq 0$, and the value of ρ_{xy} does not have to remain the same as before. Indeed, for an ideal system we would expect ρ_{xy} to take the quantized values of Eq. (7.44) only for the specific values of H that satisfy this equation (with fixed n), and ρ_{xx} to vanish at exactly these values only. For the rest of the values of H, the system should behave like the classical system of electrons, with linear Hall resistivity in H. The Fermi energy will jump between the values of the Landau levels and will be pinned at these values, as successive levels are being filled one after the other with decreasing magnetic field, as illustrated in Fig. 7.14.

What leads to the existence of plateaus in the value of ρ_{xy} is the fact that not all states in the system are extended Bloch states. In a real system, the Landau levels are not δ-functions in energy but are broadened due to the presence of impurities, as indicated in Fig. 7.15. Only those states with energy very close to the ideal Landau levels are extended and can carry current, while the states lying between ideal Landau levels are localized and cannot carry current. As the magnetic field is varied and localized states with energy away from an ideal Landau level are being filled, there can be no change in the conductivity of the system which remains stuck to its quantized value corresponding to the ideal Landau level. This produces the

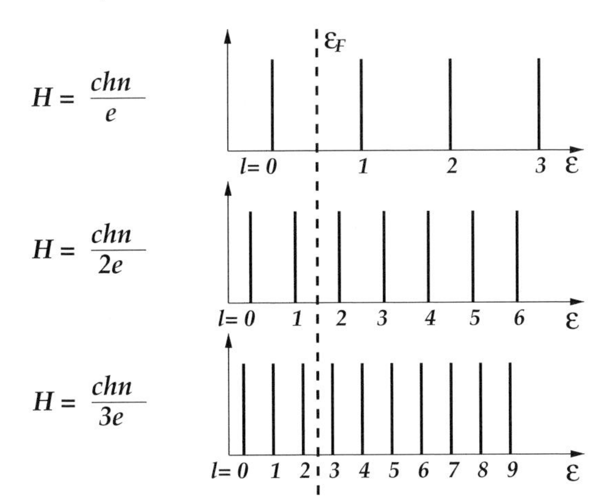

Figure 7.14. Position of the Fermi level E_F in the spectrum of Landau levels, as a function of the magnetic field H; the levels are shown as vertical lines representing infinitely sharp δ-functions. When the magnetic field takes the values $H = nch/e(l+1)$, with $l = 0, 1, 2, \ldots$, exactly $l+1$ Landau levels are filled and the Fermi level, denoted by a vertical dashed line, lies between levels with index l and $l+1$. The spacing between levels is $\hbar\omega_c = \hbar(eH/mc)$.

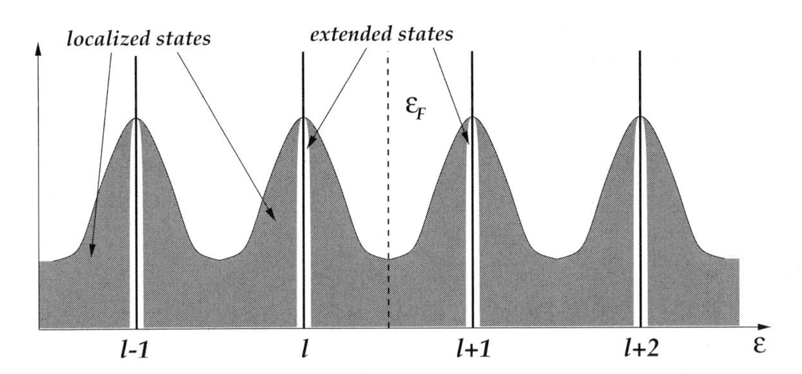

Figure 7.15. Landau levels in the two-dimensional electron gas: the solid lines represent the ideal levels (no impurities), the grey areas represent the localized states due to the presence of impurities, and the unshaded narrow regions around the ideal levels represent the extended states. When the Fermi level ϵ_F lies between levels l and $l+1$ only localized states are available and these do not contribute to conduction.

plateaus in the Hall resistivity ρ_{xy}. This fascinating phenomenon demonstrates the importance of defects in the crystal, which can dominate the electronic behavior.

We should emphasize that the above discussion is an oversimplified version of very careful theoretical analysis. For more detailed arguments we refer the reader to the original theoretical papers by Laughlin [96] and by Halperin [97]. We should also note that the plateaus in Hall resistivity can be measured with unprecedented precision, reaching one part in 10^7. This has led to advances in metrology, since the ratio e^2/h appears in several fundamental units. For example, the fine-structure constant is given by

$$\alpha = \frac{e^2}{\hbar c} = \left(\frac{e^2}{h}\right)\left(\frac{2\pi}{c}\right) \tag{7.45}$$

and the speed of light c is known to one part in 10^9 from independent measurements. Thus, using phenomena related to the behavior of electrons in a solid, one can determine the value of a fundamental constant that plays a central role in high energy physics!

Fascinating though the discovery of the IQHE may have seemed, an even more remarkable observation was in store: a couple of years after the experiments of von Klitzing and co-workers, experiments on samples of very high purity at high magnetic fields and low temperatures by Tsui, Störmer and Gossard [98] revealed that the plateaus in the Hall resistivity can also occur for values *larger* than (h/e^2). In particular, they occur at values 3, 3/2, 5, 5/2, 5/3, ..., in units of (h/e^2). According to Eq. (7.44), this implies that when specific fractions of the Landau levels are filled, corresponding to fillings of $\nu = 1/3, 2/3, 1/5, 2/5, 3/5, \ldots$, the system again behaves like a saturated band. This discovery was called the fractional quantum Hall effect (FQHE). The explanation of the FQHE involves very interesting many-body physics which lies beyond the scope of the present treatment. Suffice to say that the FQHE effect has generated enormous interest in the condensed matter physics community. A hierarchy of states at fractions

$$\nu = \frac{p}{2qp + 1} \tag{7.46}$$

with p, q integers, that lead to plateaus in the Hall conductivity has been predicted theoretically [99] and observed experimentally; all these fractions have odd denominators. The states corresponding to the partial fillings are referred to as "incompressible quantum fluids". Laughlin originally postulated a many-body wavefunction to describe the states with $\nu = 1/(2q + 1)$ where q is a positive integer [100]; this wavefunction for N electrons has the form

$$\Psi_q(\mathbf{r}_1, \ldots, \mathbf{r}_N) = \prod_{j<k}^{N} (z_j - z_k)^{(2q+1)} \exp\left(-\sum_{k=1}^{N} |z_k|^2\right), \quad z_k = x_k + \mathrm{i}y_k \tag{7.47}$$

where only the in-plane coordinates (x, y) of each electron are involved in a special linear combination that gives the complex coordinate $z = x + \mathrm{i}y$. Evidently, this is an antisymmetric wavefunction upon exchange of any pair of electron positions. The magnitude squared of this wavefunction can be written as

$$| \Psi_q(\mathbf{r}_1, \ldots, \mathbf{r}_N) |^2 = \exp\left[\Phi_q(z_1, \ldots, z_N)/(q + 1/2)\right]$$

$$\Phi_q(z_1, \ldots, z_N) = (2q + 1)^2 \sum_{j<k}^{N} \ln |z_j - z_k| - (2q + 1) \sum_{k=1}^{N} |z_k|^2$$

where $\Phi_q(z_1, \ldots, z_N)$ has the form of an electrostatic potential energy between charged particles in 2D: the first term represents the mutual repulsion between particles of charge $(2q + 1)$, and the second term the attraction to a uniform background of opposite charge. This analogy can be used to infer the relation of this wavefunction to a system of quasiparticles with fractional charge.

Many interesting theoretical ideas have been developed to account for the behavior of the incompressible quantum fluids which involves the highly correlated motion of all the electrons in the system. More recently, theorists have also addressed the properties of partially filled Landau levels with even denominator fractions (such as $\nu = 1/2$), at which there is no quantized Hall effect; it is intriguing that those states, despite the highly correlated motion of the underlying physical particles, seem to be described well in terms of *independent fermions*, each of which carries an even number of fictitious magnetic flux quanta [101]. R.B. Laughlin, D.C. Tsui and H.L. Störmer were awarder the 1998 Nobel prize for Physics for their work on the FQHE, a testament to the great excitement which these experimental discoveries and theoretical developments have created.

Further reading

1. For a thorough review of the Heisenberg antiferromagnetic model we refer the reader to the article by E. Manousakis, *Rev. Mod. Phys.* **63**, 1 (1991).
2. An extensive treatment of theoretical models for magnetic behavior can be found in *Theoretical Solid State Physics*, vol. 1, W. Jones and N.H. March (J. Wiley, London, 1973).
3. *The Quantum Hall Effect*, R.E. Prange and S.M. Girvin, eds. (2nd edn, Springer-Verlag, New York, 1990).
   This is a collection of interesting review articles on the quantum Hall effect.
4. 'The quantum Hall effect', T. Chakraborty in *Handbook on Semiconductors*, vol. 1, pp. 977–1038 (ed. P.T. Landsberg, North Holland, 1992).
5. *The Fractional Quantum Hall Effect*, T. Chakraborty and P. Pietiläinen (Springer-Verlag, Berlin, 1988).
6. *Perspectives in Quantum Hall Effects*, S. Das Sarma and A. Pinczuk, eds. (J. Wiley, New York, 1997).

Problems

1. Verify the assignment of the total spin state S, angular momentum state L and total angular momentum state J of the various configurations for the d and f electronic shells, according to Hund's rules, given in Table 7.1.
2. Prove the expressions for the magnetization Eq. (7.14) and band energy Eq. (7.16) using the spin-up and spin-down filling of the two spin states defined in Eq. (7.13).
3. In the Stoner theory of spontaneous magnetization of a system consisting of N electrons, the energy ϵ of an electron with magnetic moment $\pm\mu_B$ is modified by the term $\mp\mu_B(\mu_B M)$, where M is the magnetization number (the difference of spin-up and spin-down numbers of electrons). We introduce a characteristic temperature Θ_S

for the onset on spontaneous magnetization by defining the magnetic contribution to the energy as

$$E_{mag} = \bar{s} k_{\rm B} \Theta_S,$$

with $k_{\rm B}$ the Boltzmann constant and $\bar{s} = M/N$ the dimensionless magnetic moment per particle. At finite temperature T, the magnetization number M is given by

$$M = \Omega \int_{-\infty}^{\infty} \left[n(\epsilon - \bar{s} k_{\rm B} \Theta_S) - n(\epsilon + \bar{s} k_{\rm B} \Theta_S) \right] g(\epsilon) {\rm d}\epsilon$$

while the total number of electrons is given by

$$N = \Omega \int_{-\infty}^{\infty} \left[n(\epsilon - \bar{s} k_{\rm B} \Theta_S) + n(\epsilon + \bar{s} k_{\rm B} \Theta_S) \right] g(\epsilon) {\rm d}\epsilon$$

where Ω is the system volume, $n(\epsilon)$ is the Fermi occupation number,

$$n(\epsilon) = \frac{1}{{\rm e}^{(\epsilon - \mu)/k_{\rm B} T} + 1}$$

$g(\epsilon)$ is the density of states and μ is the chemical potential. For simplicity we will consider the free-electron model or, equivalently, a single band with parabolic energy dependence,

$$\epsilon_{\mathbf{k}} = \frac{\hbar^2}{2\overline{m}} \mathbf{k}^2$$

with $\overline{m}$ the effective mass of electrons. We also define the Fermi integral function:

$$F_{\frac{1}{2}}(y) = \int_0^{\infty} \frac{x^{\frac{1}{2}}}{{\rm e}^{x-y} + 1} {\rm d}x$$

Show that the number of electrons per unit volume $n = N/\Omega$ and the magnetic moment per unit volume $m = M/\Omega$ are given by

$$n = \frac{2\pi}{\lambda^3} \left[F_{\frac{1}{2}}(y + z) + F_{\frac{1}{2}}(y - z) \right]$$

$$m = \frac{2\pi}{\lambda^3} \left[F_{\frac{1}{2}}(y + z) - F_{\frac{1}{2}}(y - z) \right]$$

where the variables y, z, λ are defined as

$$y = \frac{\mu}{k_{\rm B} T}, \qquad z = \frac{\Theta_S}{T} \bar{s}, \qquad \lambda = \left(\frac{4\pi^2 \hbar^2}{2\overline{m} k_{\rm B} T} \right)^{1/2}$$

Next, show that the equations derived above lead to

$$2\bar{s} k_{\rm B} \Theta_S = \frac{\epsilon_{\rm F}}{2} \left[(1 + \bar{s})^{2/3} - (1 - \bar{s})^{2/3} \right]$$

with $\epsilon_{\rm F}$ the Fermi energy of the unpolarized electrons. From this result, derive the condition for the existence of spontaneous magnetization,

$$k_{\rm B} \Theta_S > \frac{2}{3} \epsilon_{\rm F}$$

Find the upper limit of Θ_S. (Hint: show that the Fermi integral function for large values of its argument can be approximated by

$$F_{\frac{1}{2}}(y) = \frac{2}{3}y^{3/2} + \frac{\pi^2}{12}y^{-1/2} + \cdots$$

The results quoted above are obtained by keeping only the leading order term in this expansion.)

4. We want to derive the quantization of energy and flux of the Landau levels in the quantum Hall effect. Assume that the magnetic field is generated by a vector potential $\mathbf{A} = (-yH, 0, 0)$, which is known as the Landau gauge. The Schrödinger equation for electrons confined in the xy plane is

$$\frac{1}{2m}\left(\mathbf{p} - \frac{e}{c}\mathbf{A}\right)^2 \bar{\psi}(x, y) = \epsilon \bar{\psi}(x, y) \tag{7.48}$$

(a) Take $\bar{\psi}(x, y) = \mathrm{e}^{\mathrm{i}kx}\tilde{\psi}(y)$ and show that $\tilde{\psi}(y)$ satisfies a Schrödinger equation with a harmonic oscillator potential, with frequency $\omega_c = eH/mc$ (the cyclotron frequency). The free variable in this equation is y, scaled by $1/\lambda$ and shifted by $k\lambda$, where $\lambda^2 = \hbar/(\omega_c m)$. Find the eigenvalues and eigenfunctions of this Schrödinger equation. These are the energies of the Landau levels and the corresponding wavefunctions.

(b) Assume that the extent of the system is L in the x direction and W in the y direction. The eigenfunctions of the Schrödinger equation we solved above are centered at $y_0 = -\lambda^2 k$, which implies $0 < |k| \leq W/\lambda^2$. Use Born–von Karman boundary conditions for x,

$$\bar{\psi}(x + L, y) = \bar{\psi}(x, y)$$

to find the allowed values for k, and from those determine the number of states per unit area that each Landau level can accommodate. The answer must be the same as the one suggested by flux quantization arguments, Eq. (7.36), with $l = 0$.

8
Superconductivity

Superconductivity was discovered in 1911 by Kamerling Onnes [102] and remains one of the most actively studied aspects of the behavior of solids. It is a truly remarkable phenomenon of purely quantum mechanical nature; it is also an essentially many-body phenomenon which cannot be described within the single-particle picture. Because of its fascinating nature and of its many applications, superconductivity has been the focus of intense theoretical and experimental investigations ever since its discovery. Studies of superconductivity have gained new vigor since the discovery of high-temperature superconductors in 1987.

8.1 Overview of superconducting behavior

Superconductivity is mostly characterized by a vanishing electrical resistance below a certain temperature T_c, called the critical temperature. Below T_c there is no measurable DC resistance in a superconductor and, if a current is set up in it, it will flow without dissipation practically forever: experiments trying to detect changes in the magnetic field associated with current in a superconductor give estimates that it is constant for $10^6 - 10^9$ years! Thus, the superconducting state is not a state of merely very low resistance, but one with a truly zero resistance. This is different than the case of very good conductors. In fact, materials which are very good conductors in their normal state typically do not exhibit superconductivity. The reason is that in very good conductors there is little coupling between phonons and electrons, since it is scattering by phonons which gives rise to resistance in a conductor, whereas electron-phonon coupling is crucial for superconductivity. The drop in resistance from its normal value above T_c to zero takes place over a range of temperatures of order $10^{-2} - 10^{-3} T_c$, as illustrated in Fig. 8.1(a), that is, the transition is rather sharp.

For typical superconductors T_c is in the range of a few degrees kelvin, which has made it difficult to take advantage of this extraordinary behavior in practical applications, because cooling the specimen to within a few degrees of absolute zero

Table 8.1. *Elemental conventional superconductors.*

Critical temperature T_c (in kelvin), critical field H_0 (in oersted), and Debye frequency $\hbar\omega_D$ (in milielectronvolts).

Element	T_c	H_0	$\hbar\omega_D$	Element	T_c	H_0	$\hbar\omega_D$
Al	1.175	105	36.2	Ru	0.49	69	50.0
Cd	0.517	28	18.0	Nb	9.25	2060	23.8
Gd	1.083	58	28.0	Ta	4.47	829	22.2
Hg	3.949	339	8.0	Tc	7.80	1410	35.4
In	3.408	282	9.4	Th	1.38	1.6	14.2
Ir	0.113	16	36.6	Ti	0.40	56	35.8
Mo	0.915	96	39.6	Tl	2.38	178	6.8
Sn	3.722	305	50.0	V	5.40	1408	33.0
W	0.015	1.2	33.0	Os	0.66	70	43.1
Pb	7.196	803	8.3	Zn	0.85	54	26.7
Re	1.697	200	0.4	Zr	0.61	47	25.0

Source: Ref. [74].

is quite difficult (it requires the use of liquid He, the only non-solid substance at such low temperatures, as the coolant, which is expensive and cumbersome). In 1986 a new class of superconducting materials was discovered by Bednorz and Müller, dubbed high-temperature superconductors, in which the T_c is much higher than in typical superconductors: in general it is in the range of $\sim$ 90 K, but in certain compounds it can exceed 130 K (see Table 8.2). This is well above the freezing point of N_2 (77 K), so that this much more abundant and cheap substance can be used as the coolant to bring the superconducting materials below their critical point. The new superconductors have complex crystal structures characterized by Cu–O octahedra arranged in various ways as in the perovskites (see discussion of these types of structures in chapter 1) and decorated by various other elements. In the following we will refer to the older, low-temperature superconductors as the conventional superconductors, to distinguish them from the high-temperature kind. We give examples of both conventional and high-temperature superconductors in Tables 8.1 and 8.2, respectively.

The discovery of high-temperature superconductors has opened the possibility of many practical applications, but these materials are ceramics and therefore more difficult to utilize than the classical superconductors, which are typically elemental metals or simple metallic alloys. It also re-invigorated theoretical interest in superconductivity, since it seemed doubtful that the microscopic mechanisms responsible for low-temperature superconductivity could also explain its occurrence at such high temperatures. J.G. Bednorz and K.A. Müller were awarded the 1987 Nobel prize for Physics for their discovery, which has sparked an extraordinary activity,

Table 8.2. *High-temperature superconductors.*

Critical temperature T_c (in kelvin). In several cases there is fractional occupation of the dopant atoms, denoted by x (with $0 < x < 1$), or there are equivalent structures with two different elements, denoted by (X,Y); in such cases the value of the highest T_c is given.

Material	T_c	Material	T_c
$La_2CuO_{4+\delta}$	39	$SmBaSrCu_3O_7$	84
$La_{2-x}(Sr,Ba)_xCuO_4$	35	$EuBaSrCu_3O_7$	88
$La_2Ca_{1-x}Sr_xCu_2O_6$	60	$GdBaSrCu_3O_7$	86
$YBa_2Cu_3O_7$	93	$DyBaSrCu_3O_7$	90
$YBa_2Cu_4O_8$	80	$HoBaSrCu_3O_7$	87
$Y_2Ba_4Cu_7O_{15}$	93	$YBaSrCu_3O_7$	84
$Tl_2Ba_2CuO_6$	92	$ErBaSrCu_3O_7$	82
$Tl_2CaBa_2Cu_2O_8$	119	$TmBaSrCu_3O_7$	88
$Tl_2Ca_2Ba_2Cu_3O_{10}$	128	$HgBa_2CuO_4$	94
$TlCaBa_2Cu_2O_7$	103	$HgBa_2CaCu_2O_6$	127
$TlSr_2Y_{0.5}Ca_{0.5}Cu_2O_7$	90	$HgBa_2Ca_2Cu_3O_8$	133
$TlCa_2Ba_2Cu_3O_8$	110	$HgBa_2Ca_3Cu_4O_{10}$	126
$Bi_2Sr_2CuO_6$	10	$Pb_2Sr_2La_{0.5}Ca_{0.5}Cu_3O_8$	70
$Bi_2CaSr_2Cu_2O_8$	92	$Pb_2(Sr,La)_2Cu_2O_6$	32
$Bi_2Ca_2Sr_2Cu_3O_{10}$	110	$(Pb,Cu)Sr_2(La,Ca)Cu_2O_7$	50

Source: Ref. [74].

both in experiment and in theory, to understand the physics of the high-temperature superconductors.

A number of other properties are also characteristic of typical superconductors. These are summarized in Fig. 8.1. The AC resistance in the superconducting state is also zero below a certain frequency ω_g, which is related to the critical temperature by $\hbar\omega_g \approx 3.5k_B T_c$. This is indicative of a gap in the excitation spectrum of the superconductor that can be overcome only above a sufficiently high frequency to produce a decrease in the superconductivity and consequently an increase in the resistance, which then approaches its normal state value as shown in Fig. 8.1(b). Superconductors can expel completely an external magnetic field H imposed on them, which is known as the Meissner effect [103]. The total magnetic field B inside the superconductor is given by $B = H + 4\pi m$, where m is the magnetic moment. B is zero inside the superconductor, that is, $m = -(1/4\pi)H$. In other words, the superconductor is a perfect diamagnet, a behavior that can be easily interpreted as perfect shielding of the external field by the dissipationless currents that can be set up inside the superconductor. This is true up to a certain critical value of the external field H_c, above which the field is too strong for the superconductor to resist: the field abruptly penetrates into the superconductor and the magnetic moment becomes zero, as shown in Fig. 8.1(c). Materials that exhibit this behavior

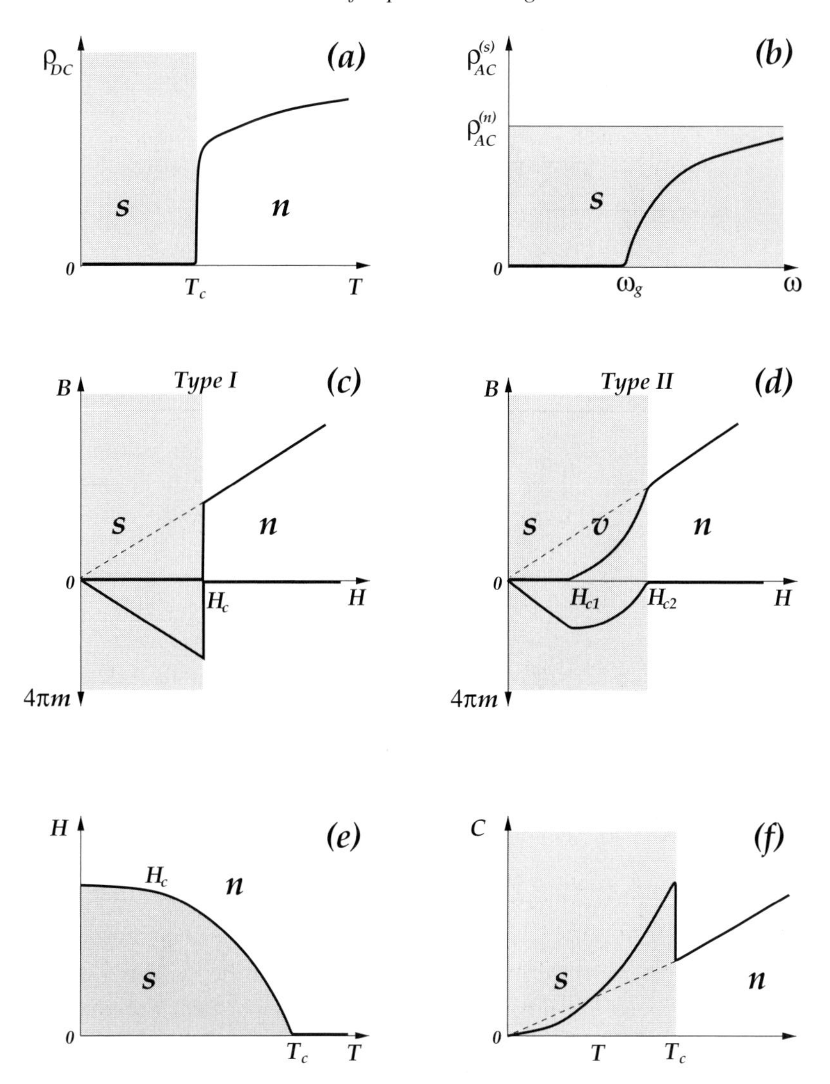

Figure 8.1. Experimental portrait of a typical superconductor: s denotes the superconducting (shaded), v the vortex and n the normal (unshaded) state. (a) The DC conductivity ρ_{DC} as a function of temperature. (b) The AC conductivity $\rho_{AC}^{(s)}$ in the superconducting state as a function of frequency, which approaches the normal state value $\rho_{AC}^{(n)}$ above ω_g. (c) and (d) The total magnetic field B and the magnetic moment $4\pi m$ as a function of the external field H for type I and type II superconductors. (e) The critical field H_c as a function of temperature. (f) The specific heat as a function of temperature.

are called "type I" superconductors. The expulsion of magnetic field costs energy because it bends the magnetic field lines around the superconductor. Depending on the shape of the specimen, even when the magnetic field is smaller than H_c it may still penetrate some regions which revert to the normal state, while the rest

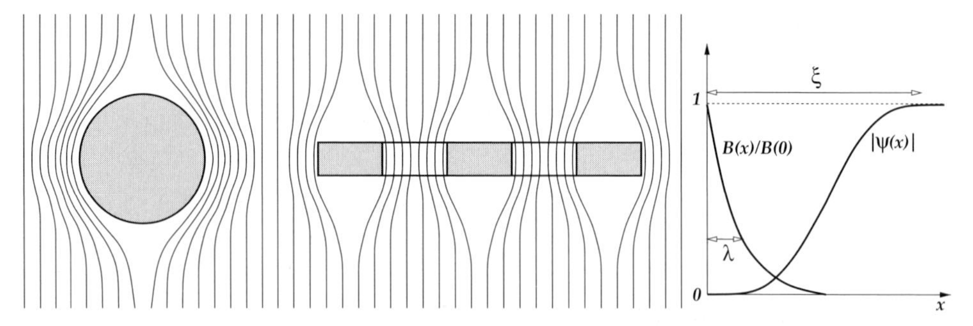

Figure 8.2. Illustration of expulsion of magnetic field (Meissner effect) in a type I superconductor. **Left:** a specimen of spherical shape. **Center:** a specimen of rectangular shape with its long dimension perpendicular to the direction of the field lines. It is shown in the intermediate state, with some regions in the superconducting (shaded) and others in the normal (unshaded) state. **Right:** magnetic field $B(x)$ and the magnitude of the order parameter $|\psi(x)|$ as functions of the distance x from the normal-superconducting interface, located at $x = 0$. The behavior of $B(x)$ and $\psi(x)$ determine the penetration length λ and coherence length ξ.

remains in the superconducting state; this is called the "intermediate" state. In type I superconductors, the normal or superconducting regions in the intermediate state are of *macroscopic* dimensions, as illustrated in Fig. 8.2.

There exists a different class of superconductors, in which the Meissner effect is also observed up to a critical field H_{c1} but then a gradual penetration of the field into the superconductor begins that is completed at a higher critical field H_{c2}, beyond which the magnetic moment is again zero as illustrated in Fig. 8.1(d). These materials are called "type II" superconductors. The phase in which the external field partially penetrates in the superconductor is called the "vortex" state and has very interesting physics in itself. A vortex consists of a cylindrical core where the external field has penetrated the specimen, that is, the material is in the normal state, and is surrounded by material in the superconducting state. The flux contained in the core of the vortex is equal to the fundamental flux quantum, Eq. (7.5). The core of the vortex has *microscopic* dimensions in cross-section and is not rigid but can change direction subject to line-tension forces. The ground state of the vortex state is a regular array of the vortex lines, which on a plane perpendicular to the lines forms a triangular lattice; this is called the Abrikosov lattice. Vortices can move in response to external forces or thermal fluctuations. The motion of vortices can be quite elaborate; for instance, vortices can form bundles, as first suggested by Anderson [104], the motion of which depends on the presence and distribution of defects in the crystal.

In type I superconductors, the critical field H_c is a function of temperature and vanishes at T_c, as shown in Fig. 8.1(e); the behavior of H_c with temperature is described well by the expression

$$H_c(T) = H_0 \left[1 - \left(\frac{T}{T_c} \right)^2 \right] \tag{8.1}$$

with H_0 a constant in the range of $10^2 - 10^3$ Oe. An interesting consequence of this behavior is that the slope of the critical field with temperature, $\mathrm{d}H_c/\mathrm{d}T$, vanishes at $T = 0$ but not at $T = T_c$, where it takes the value $-2H_0/T_c$; this is related to the nature of the phase transition between the superconducting and normal state, as discussed below. Finally, the electronic specific heat[1] of a superconductor has a discontinuity at T_c, falling by a finite amount from its superconducting value below T_c to its normal value above T_c, as shown in Fig. 8.1(f); this behavior is another indication of a gap in the excitation spectrum of the superconducting state. One last important feature of superconductivity, not illustrated in Fig. 8.1, is the so called "isotope effect": for superconductors made of different isotopes of the same element, it is found that the critical temperature T_c depends on the isotope mass M. In general this dependence is given by the relation

$$T_c \sim M^{\alpha} \tag{8.2}$$

with $\alpha = -0.5$ in the simplest cases. This fact led Fröhlich to suggest that there is a connection between electron–phonon interactions and superconductivity [105], since the phonon frequencies naturally introduce a mass dependence of the form $M^{-1/2}$ (see chapter 6).

The theoretical explanation of superconductivity remained a big puzzle for almost a half century after its discovery. Although extremely useful phenomenological theories were proposed to account for the various aspects of superconductivity, due to London and London [106], Pippard [107], Ginzburg and Landau [108], Abrikosov [109] and others, a theory based on microscopic considerations was lacking until 1957, when it was developed by J. Bardeen, L.N. Cooper and J.R. Schrieffer [110], who were awarded the 1972 Nobel prize for Physics for their contribution; this is referred to as BCS theory.

A common feature of the phenomenological theories is the presence of two important length scales, the so called "penetration length" denoted by λ, and the "coherence length", denoted by ξ. The penetration length gives the scale over which the magnetic field inside the superconductor is shielded by the supercurrents: if x measures the distance from the surface of the superconducting sample towards its interior, the total magnetic field behaves like

$$B(x) = B(0)\mathrm{e}^{-x/\lambda}$$

[1] In this entire section, by specific heat we mean the electronic contribution only.

that is, it decays exponentially over a distance of order λ. It is natural that such a length scale should exist, because the field cannot drop to zero immediately inside the superconductor as this would imply an infinite surface current. The coherence length determines the scale over which there exist strong correlations which stabilize the superconducting state, or, equivalently, it is a measure of the distance over which the superconducting state is affected by fluctuations in the external field or other variables. If the length scale of these fluctuations is much larger than ξ, the superconducting state is not affected (the correlations responsible for its stability are not destroyed); if it is of order ξ, it is affected. In phenomenological theories the superconducting state is typically described by a quantity called the "order parameter", $\psi(x)$, whose absolute value is between unity and zero: at these two extremes the superconducting state is either at full strength or has been completely destroyed. The typical behavior of $B(x)$ and $|\psi(x)|$ in a type I superconductor is shown in Fig. 8.2.

Both λ and ξ are temperature-dependent quantities. Their ratio, $\kappa = \lambda/\xi$, serves to distinguish type I from type II superconductors: $\kappa < 1/\sqrt{2}$ corresponds to type I and $\kappa > 1/\sqrt{2}$ corresponds to type II superconductors, as shown by Abrikosov [109]. The temperature dependence of λ and ξ is essentially the same, which produces a temperature-independent value for κ. The distinction between the two types of superconductivity in terms of κ has to do with the interface energy between the normal and superconducting parts. In type I superconductors the interface energy is positive and the sample tries to minimize it by splitting into macroscopic normal or superconducting domains with the smallest possible interface area. In type II superconductors the interface energy is negative and it is actually energetically favorable to create the microscopic regions which contain a quantum of flux, the vortices. This behavior can be understood from the point of view of phenomenological theories of superconductivity, and especially through the so called Ginzburg–Landau theory.

We present here a brief heuristic argument for the reasons behind these two types of behavior. Anticipating the discussion of the microscopic mechanism for superconductivity, we will assume that the superconducting state consists of bound pairs of electrons with opposite momenta and spins. We can use the distance between such a pair of electrons as a measure for ξ, the coherence length. The electron pairs that create the superconducting state cannot exist within a region of size λ near the boundary between the normal and superconducting states; within this region the superconducting state and the magnetic field both decay to zero, the first going from superconducting toward the normal region, the second going in the opposite direction, as illustrated in Fig. 8.2. If ξ is considerably larger than λ, the electron pairs are awkwardly rigid on the scale of λ and the superconducting state is too "inflexible" to shield the magnetic field as it tries to penetrate the sample. When the magnetic field becomes very strong and its expulsion from the sample is energetically costly, the only option for the system is to reduce the volume of the superconducting sample and yield some volume to the normal state within which the magnetic field can exist; this implies a positive interface energy. For sufficiently large magnetic field the superconducting state is abruptly destroyed altogether: this is the behavior characteristic of type I superconductors. On the other hand, if ξ is considerably smaller than λ, the electron pairs are quite nimble and the superconducting state is more "flexible" and able to shield small regions of normal state. This allows the creation of vortices of microscopic size where the magnetic field penetrates the sample. The generation of vortices can be continued to much larger values of the magnetic field, before their density becomes such that the superconducting state is destroyed; this is the behavior characteristic of type II superconductors. This picture is represented by Fig. 8.3. It is then reasonable to expect that the ratio $\kappa = \lambda/\xi$ can characterize the two types of behavior, with a critical value κ_c separating them. Abrikosov's theory established that $\kappa_c = 1/\sqrt{2}$.

In the following we provide first a simple thermodynamic treatment of the superconducting transition, which sheds some light onto the behavior of the physical system, and then we give an elementary account of the BCS theory of superconductivity. We conclude this section with a short discussion of the high-temperature superconductors and their differences from the conventional ones.

8.2 Thermodynamics of the superconducting transition

In order to discuss the thermodynamics of the superconducting transition as a function of the external magnetic field H and temperature T we consider a simplified description in terms of two components, the normal (denoted by the superscript (n)) and superconducting (denoted by the superscript (s)) parts. The system attains equilibrium under given H and T by transferring some amount from one to the other part. We will assume that the normal part occupies a volume Ω_n and the superconducting part occupies a volume Ω_s, which add up to the fixed total volume of the specimen $\Omega = \Omega_n + \Omega_s$. For simplicity we consider a type I superconductor in a homogeneous external field, in which case the magnetic moment in the superconducting part is $m = -(1/4\pi)H$ and vanishes in the normal part. The magnetization M is given as usual by

$$M = \int m(\mathbf{r})\mathrm{d}\mathbf{r} = m\Omega_s = -\frac{1}{4\pi}H\Omega_s$$

The total magnetic field is given by

$$B(\mathbf{r}) = H + 4\pi m(\mathbf{r})$$

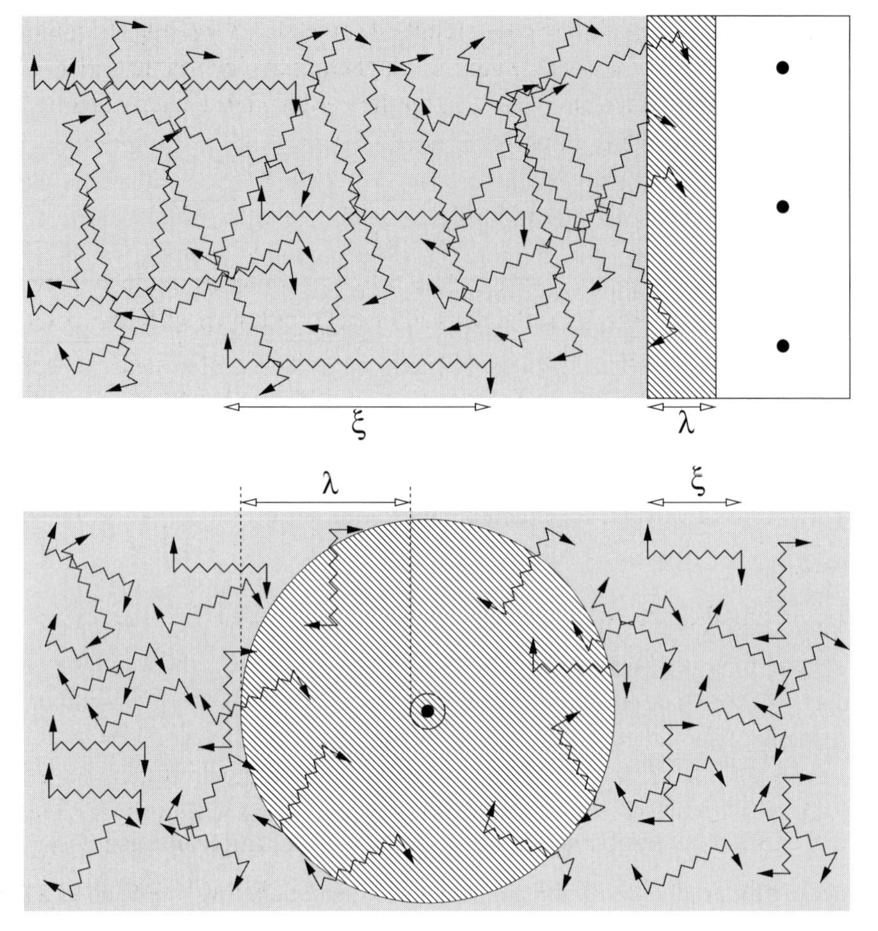

Figure 8.3. Schematic illustration of the behavior of type I **(top)** and type II **(bottom)** superconductors. The superconducting state, occupying the shaded areas, consists of pairs of electrons with opposite momenta and spins, shown as small arrows pointing in opposite directions and linked by a wavy line; the size of the electron pairs is a measure of the coherence length ξ. The normal state is shown as white and the black dots indicate the lines of magnetic field (perpendicular to the plane of the paper in this example) which penetrate the normal but not the superconducting state. The dashed area at the interface between normal and superconducting regions has length scale λ, over which the superconducting state and the field decay to zero from opposite directions. In type I superconductors the system responds to increasing magnetic field by shrinking the superconducting region and minimizing the interface. In type II superconductors it responds by creating microscopic vortices which contain one quantum of magnetic flux each (a white circle with a single magnetic line) which increases the interface area.

which vanishes in the superconducting part and is equal to H in the normal part, as shown in Fig. 8.1(c). The Gibbs free energy of the total system is given by

$$G = F_0 + \frac{1}{8\pi}\int_{\Omega}[B(\mathbf{r})]^2 \mathrm{d}\mathbf{r} - HM = F_0 + \frac{1}{8\pi}H^2\Omega_n - H\left(-\frac{1}{4\pi}H\Omega_s\right)$$

where we have defined F_0 to be the Helmholtz free energy arising from the non-magnetic contributions. We denote the non-magnetic contributions to the Helmholtz free energy per unit volume as $f_0^{(n)}$, $f_0^{(s)}$, for the normal and superconducting parts, respectively, with which the Gibbs free energy becomes

$$G = f_0^{(n)}\Omega_n + f_0^{(s)}\Omega_s + \frac{1}{8\pi}H^2\Omega + 2\pi m^2\Omega_s$$

where we have also used the fact that $H = -4\pi m$ in the superconducting part and $\Omega = \Omega_n + \Omega_s$. At equilibrium, the derivative of G with respect to the volume of either part will be zero, indicating a balance between the amount of superconducting and normal volumes consistent with the external conditions (the values of H and T). Since the total volume Ω of the specimen is fixed, we must have $\mathrm{d}\Omega_n = -\mathrm{d}\Omega_s$, which gives

$$\frac{\partial G}{\partial \Omega_s} = f_0^{(s)} - f_0^{(n)} + 2\pi m^2 = 0 \Longrightarrow f_0^{(s)} - f_0^{(n)} = -2\pi m^2 = -\frac{1}{8\pi}H^2$$

Differentiating the last expression with respect to T we obtain

$$\frac{\partial f_0^{(s)}}{\partial T} - \frac{\partial f_0^{(n)}}{\partial T} = -\frac{H}{4\pi}\frac{\mathrm{d}H}{\mathrm{d}T}$$

From the definition of $f_0^{(n)}$, $f_0^{(s)}$, we can view the two partial derivatives on the left-hand side of the above equation as derivatives of Helmholtz free energies at constant magnetization M (in this case zero), whereupon, using standard thermodynamic relations (see Appendix C), we can equate these derivatives to the entropies per unit volume of the normal and superconducting parts, $s^{(n)}$, $s^{(s)}$:

$$\frac{\partial f_0^{(n)}}{\partial T} = \left(\frac{\partial f^{(n)}}{\partial T}\right)_{M=0} = -s_0^{(n)}, \quad \frac{\partial f_0^{(s)}}{\partial T} = \left(\frac{\partial f^{(s)}}{\partial T}\right)_{M=0} = -s_0^{(s)}$$

with the subscripts 0 in the entropies serving as a reminder of the validity of these expressions only at zero magnetization. For the normal part, this is always the case, since the magnetic moment in the normal part is zero. For the superconducting part, this condition is satisfied at $H = H_c$, the critical field at which superconductivity is destroyed. Therefore, with this connection to the entropy, the earlier equation relating the partial derivatives of the Helmholtz free energies per unit volume to the external field becomes

$$s^{(n)} - s^{(s)} = -\frac{H_c}{4\pi}\frac{\mathrm{d}H_c}{\mathrm{d}T} \tag{8.3}$$

where we have dropped the subscripts 0 from the entropies and specialized the relation to the values of the critical field only. Eq. (8.3) is an interesting result: if we define the latent heat per unit volume $L = T\Delta s$, we can rewrite this equation in the following manner:

$$\frac{\mathrm{d}H_c}{\mathrm{d}T} = -\frac{4\pi}{H_c}\frac{L}{T}$$

which has the familiar form of the Clausius–Clapeyron equation for a first order phase transition (see Appendix C). We note that $\mathrm{d}H_c/\mathrm{d}T \leq 0$ as suggested by the plot in Fig. 8.1(e), which means that when going from the normal to the superconducting state the latent heat is a positive quantity, that is, the superconducting state is more ordered (has lower entropy) than the normal state. The above arguments are valid as long as $H_c > 0$, but break down at the critical temperature where $H_c = 0$. At this point, $\mathrm{d}H_c/\mathrm{d}T$ is not zero, as indicated in Fig. 8.1(e), which implies that the latent heat must vanish there since the ratio L/H_c must give a finite value equal to $-(T_c/4\pi)(\mathrm{d}H_c/\mathrm{d}T)_{T=T_c}$. From this argument we conclude that at T_c the transition is second order.

Differentiating both sides of Eq. (8.3) with respect to temperature, and using the standard definition of the specific heat $C = \mathrm{d}Q/\mathrm{d}T = T(\mathrm{d}S/\mathrm{d}T)$, we find for the difference in specific heats per unit volume:

$$c^{(n)}(T) - c^{(s)}(T) = -\frac{T}{4\pi}\left[\left(\frac{\mathrm{d}H_c}{\mathrm{d}T}\right)^2 + H_c\frac{\mathrm{d}^2H_c}{\mathrm{d}T^2}\right] \tag{8.4}$$

To explore the consequences of this result, we will assume that H_c as a function of temperature is given by Eq. (8.1), which leads to the following expression for the difference in specific heats between the two states:

$$c^{(s)}(T) - c^{(n)}(T) = \frac{H_0^2}{2\pi}\frac{T}{T_c^2}\left[3\left(\frac{T}{T_c}\right)^2 - 1\right]$$

At $T = T_c$ this expression reduces to

$$c^{(s)}(T_c) - c^{(n)}(T_c) = \frac{H_0^2}{\pi T_c}$$

that is, the specific heat has a discontinuity at the critical temperature, dropping by a finite amount from its value in the superconducting state to its value in the normal state, as the temperature is increased through the transition point. This behavior is indicated in Fig. 8.1(f). Eq. (8.4) also shows that the specific heat of the superconducting state is higher than the specific heat of the normal state for $T_c/\sqrt{3} < T < T_c$, which is also indicated schematically in Fig. 8.1(f).

8.3 BCS theory of superconductivity

There are two main ingredients in the microscopic theory of superconductivity developed by Bardeen, Cooper and Schrieffer. The first is an effective attractive interaction between two electrons that have opposite momenta (larger in magnitude than the Fermi momentum) and opposite spins, which leads to the formation of the so called "Cooper pairs". The second is the condensation of the Cooper pairs into a single coherent quantum state which is called the "superconducting condensate"; this is the state responsible for all the manifestations of superconducting behavior. We discuss both ingredients in some detail.

8.3.1 Cooper pairing

The Coulomb interaction between two electrons is of course repulsive. For two free electrons at a distance $\mathbf{r}$, the interaction potential in real space is

$$V(\mathbf{r}) = \frac{e^2}{|\mathbf{r}|}$$

We can think of the interaction between two free electrons as a scattering process corresponding to the exchange of photons, the carriers of the electromagnetic field: an electron with initial momentum $\hbar\mathbf{k}$ scatters off another electron by exchanging a photon of momentum $\hbar\mathbf{q}$. Due to momentum conservation, the final momentum of the electron will be $\hbar\mathbf{k}' = \hbar(\mathbf{k} - \mathbf{q})$. We can calculate the matrix element corresponding to this scattering of an electron, taken to be in a plane wave state $\langle \mathbf{r}|\mathbf{k}\rangle = (1/\sqrt{\Omega})\exp(\mathrm{i}\mathbf{k}\cdot\mathbf{r})$, as

$$\langle \mathbf{k}'|V|\mathbf{k}\rangle = \frac{1}{\Omega}\int \mathrm{e}^{-\mathrm{i}(\mathbf{k}-\mathbf{q})\cdot\mathbf{r}} V(\mathbf{r})\mathrm{e}^{\mathrm{i}\mathbf{k}\cdot\mathbf{r}}\mathrm{d}\mathbf{r} = \frac{4\pi e^2}{|\mathbf{q}|^2}$$

the last expression being simply the Fourier transform $V(\mathbf{q})$ of the bare Coulomb potential (see Appeindix G), with $\mathbf{q} = \mathbf{k} - \mathbf{k}'$. In the solid, this interaction is screened by all the other electrons, with the dielectric function describing the effect of screening. We have seen in chapter 3 that in the simplest description of metallic behavior, the Thomas–Fermi model, screening changes the bare Coulomb interaction to a Yukawa potential:

$$V(\mathbf{r}) = \frac{e^2 \mathrm{e}^{-k_s|\mathbf{r}|}}{|\mathbf{r}|}$$

with k_s the inverse screening length; the Fourier transform of this potential is (see Appendix G)

$$V(\mathbf{q}) = \frac{4\pi e^2}{|\mathbf{q}|^2 + k_s^2}$$

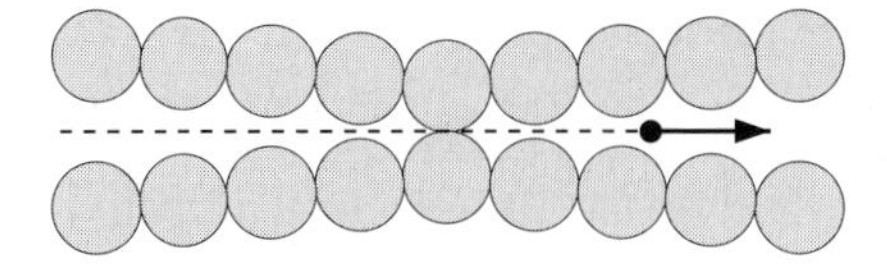

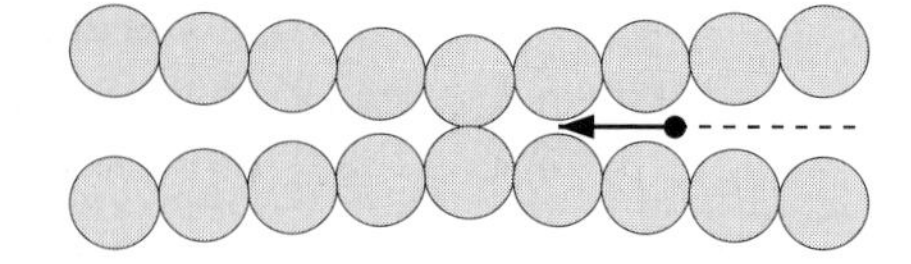

Figure 8.4. Illustration of attractive effective interaction between two electrons mediated by phonons. **Left:** the distortion that the first electron induces to the lattice of ions. **Right:** the second electron with opposite momentum at the same position but at a later time. The lattice distortion favors energetically the presence of the second electron in that position.

that is, the presence of the screening term $\exp(-k_s|\mathbf{r}|)$ eliminates the singularity at $|\mathbf{q}| \to 0$ of the bare Coulomb potential (this singularity is a reflection of the infinite range of the bare potential). The strength of the interaction is characterized by the constant e^2, the charge of the electrons. These considerations indicate that, when the interaction is viewed as a scattering process, the relevant physical quantity is the Fourier transform of the real-space interaction potential. Accordingly, in the following we will only consider the interaction potential in this form.

The preceding discussion concerned the interaction between two electrons in a solid assuming that the ions are fixed in space. When the ions are allowed to move, a new term in the interaction between electrons is introduced. The physical origin of this new term is shown schematically in Fig. 8.4: an electron moving through the solid attracts the positively charged ions which come closer to it as it approaches and then return slowly to their equilibrium positions once the electron has passed them by. We can describe this motion in terms of phonons emitted by the traveling electron. It is natural to assume that the other electrons will be affected by this distortion of the ionic positions; since the electrons themselves are attracted to the ions, the collective motion of ions toward one electron will translate into an effective attraction of other electrons toward the first one. Frölich [111] and Bardeen and Pines [112] showed that the effective interaction between electrons due to exchange of a phonon is given in reciprocal space by

$$V_{\mathbf{k}\mathbf{k}'}^{phon}(\mathbf{q}) = \frac{g^2\hbar\omega_{\mathbf{q}}}{(\epsilon_{\mathbf{k}'} - \epsilon_{\mathbf{k}})^2 - (\hbar\omega_{\mathbf{q}})^2} \tag{8.5}$$

where $\mathbf{k}$, $\mathbf{k}'$ and $\epsilon_{\mathbf{k}}$, $\epsilon_{\mathbf{k}'}$ are the incoming and outgoing wave-vectors and energies of the electrons, and $\mathbf{q}$, $\hbar\omega_{\mathbf{q}}$ is the wave-vector and energy of the exchanged phonon; g is a constant that describes the strength of the electron-phonon interaction. From this expression, it is evident that if the energy difference $(\epsilon_{\mathbf{k}'} - \epsilon_{\mathbf{k}})$ is smaller than the phonon energy $\hbar\omega_{\mathbf{q}}$ the effective interaction is attractive.

To show that an attractive interaction due to phonons can actually produce binding between electron pairs, we consider the following simple model [113]: the interaction potential in reciprocal space is taken to be a constant $V_{\mathbf{kk}'} = -V_0 < 0$, independent of $\mathbf{k}, \mathbf{k}'$, only when the single-particle energies $\epsilon_{\mathbf{k}'}, \epsilon_{\mathbf{k}}$ lie within a narrow shell of width t_0 above the Fermi level, and is zero otherwise. Moreover, we will take the pair of interacting electrons to have opposite wave-vectors $\pm\mathbf{k}$ in the initial and $\pm\mathbf{k}'$ in the final state, both larger in magnitude than the Fermi momentum k_{F}. We provide a rough argument to justify this choice; the detailed justification has to do with subtle issues related to the optimal choice for the superconducting ground state, which lie beyond the scope of the present treatment. As indicated in Fig. 8.4, the distortion of the lattice induced by an electron would lead to an attractive interaction with any other electron put in the same position at a later time. The delay is restricted to times of order $1/\omega_{\mathrm{D}}$, where ω_{D} is the Debye frequency, or the distortion will decay away. There is, however, no restriction on the momentum of the second electron from these considerations. It turns out that the way to maximize the effect of the interaction is to take the electrons in pairs with opposite momenta because this ensures that no single-particle state is doubly counted or left out of the ground state.

For simplicity, we will assume that the single-particle states are plane waves, as in the jellium model. We denote the interaction potential in real space as $V^{phon}(\mathbf{r})$, where $\mathbf{r}$ is the relative position of the two electrons (in an isotropic solid V^{phon} would be a function of the relative distance $r = |\mathbf{r}|$ only). Taking the center of mass of the two electrons to be fixed, consistent with the assumption that they have opposite momenta, we arrive at the following Schrödinger equation for their *relative* motion:

$$\left[-\frac{\hbar^2\nabla_{\mathbf{r}}^2}{2\mu} + V^{phon}(\mathbf{r})\right]\psi(\mathbf{r}) = E^{pair}\psi(\mathbf{r})$$

where $\mu = m_e/2$ is the reduced mass of the pair. We expand the wavefunction $\psi(\mathbf{r})$ of the relative motion in plane waves with momentum $\mathbf{k}'$ larger in magnitude than the Fermi momentum k_{F},

$$\psi(\mathbf{r}) = \frac{1}{\sqrt{\Omega}} \sum_{|\mathbf{k}'|>k_{\mathrm{F}}} \alpha_{\mathbf{k}'} \mathrm{e}^{-\mathrm{i}\mathbf{k}'\cdot\mathbf{r}}$$

and insert this expansion in the above equation to obtain, after multiplying through by $(1/\sqrt{\Omega})\exp(\mathrm{i}(\mathbf{k}\cdot\mathbf{r}))$ and integrating over $\mathbf{r}$, the following relation for the coefficients $\alpha_{\mathbf{k}}$:

$$(2\epsilon_{\mathbf{k}} - E^{pair})\alpha_{\mathbf{k}} + \sum_{|\mathbf{k}'|>k_{\mathrm{F}}} \alpha_{\mathbf{k}'} V_{\mathbf{kk}'}^{phon} = 0 \tag{8.6}$$

where $V^{phon}_{\mathbf{k}\mathbf{k}'}$ is the Fourier transform of the phonon interaction potential

$$V^{phon}_{\mathbf{k}\mathbf{k}'} = \frac{1}{\Omega} \int \mathrm{e}^{-\mathrm{i}\mathbf{k}'\cdot\mathbf{r}} V^{phon}(\mathbf{r}) \mathrm{e}^{\mathrm{i}\mathbf{k}\cdot\mathbf{r}} \mathrm{d}\mathbf{r}$$

and $\epsilon_{\mathbf{k}}$ is the single-particle energy of the electrons (in the present case equal to $\hbar^2\mathbf{k}^2/2m_e$). We now employ the features of the simple model outlined above, which gives for the summation in Eq. (8.6)

$$\sum_{|\mathbf{k}'|>k_\mathrm{F}} \alpha_{\mathbf{k}'} V^{phon}_{\mathbf{k}\mathbf{k}'} = -V_0 \left(\sum_{|\mathbf{k}'|>k_\mathrm{F}} \alpha_{\mathbf{k}'} \right) \theta(\epsilon_\mathrm{F} + t_0 - \epsilon_{\mathbf{k}})$$

where the step function is introduced to ensure that the single-particle energy lies within t_0 of the Fermi energy ϵ_F (recall that by assumption we also have $\epsilon_{\mathbf{k}} > \epsilon_\mathrm{F}$). The last equation leads to the following expression for $\alpha_{\mathbf{k}}$:

$$\alpha_{\mathbf{k}} = V_0 \left(\sum_{|\mathbf{k}'|>k_\mathrm{F}} \alpha_{\mathbf{k}'} \right) \theta(\epsilon_\mathrm{F} + t_0 - \epsilon_{\mathbf{k}}) \frac{1}{2\epsilon_{\mathbf{k}} - E^{pair}},$$

which, upon summing both sides over $\mathbf{k}$ with $|\mathbf{k}| > k_\mathrm{F}$, produces the following relation:

$$\left(\sum_{|\mathbf{k}|>k_\mathrm{F}} \alpha_{\mathbf{k}} \right) = V_0 \sum_{|\mathbf{k}|>k_\mathrm{F}} \left(\sum_{|\mathbf{k}'|>k_\mathrm{F}} \alpha_{\mathbf{k}'} \right) \theta(\epsilon_\mathrm{F} + t_0 - \epsilon_{\mathbf{k}}) \frac{1}{2\epsilon_{\mathbf{k}} - E^{pair}}$$

Assuming that the sum in the parentheses does not vanish identically, we obtain

$$1 = V_0 \sum_{|\mathbf{k}|>k_\mathrm{F}} \theta(\epsilon_\mathrm{F} + t_0 - \epsilon_{\mathbf{k}}) \frac{1}{2\epsilon_{\mathbf{k}} - E^{pair}}$$

The sum over $\mathbf{k}$ is equivalent to an integral over the energy ϵ that also includes the density of states $g(\epsilon)$; in addition, taking the range t_0 to be very narrow, we can approximate the density of states over such a narrow range by its value at the Fermi energy, $g_F = g(\epsilon_\mathrm{F})$. These considerations lead to

$$\sum_{|\mathbf{k}|>k_\mathrm{F}} \to g_F \int_{\epsilon_\mathrm{F}}^{\epsilon_\mathrm{F}+t_0} \mathrm{d}\epsilon$$

and with this identification, in which the range of the integral automatically satisfies the step function appearing in the summation, the previous equation yields

$$1 = V_0 g_F \int_{\epsilon_F}^{\epsilon_F + t_0} \frac{1}{2\epsilon - E^{pair}} d\epsilon$$

This is easily solved for E^{pair}:

$$1 = \frac{1}{2} V_0 g_F \ln \left(\frac{2(\epsilon_F + t_0) - E^{pair}}{2\epsilon_F - E^{pair}} \right) \Longrightarrow E^{pair} = 2\epsilon_F - \frac{2t_0}{\exp(2/V_0 g_F) - 1}$$

In the so called "weak coupling" limit $V_0 g_F \ll 1$, which is justified *a posteriori* since it is reasonably obeyed by classical superconductors, the above expression reduces to

$$E^{pair} = 2\epsilon_F - 2t_0 e^{-2/V_0 g_F}$$

This relation proves that the pair of electrons forms a bound state with binding energy E_b given by

$$E_b \equiv 2\epsilon_F - E^{pair} = 2t_0 e^{-2/V_0 g_F} \tag{8.7}$$

A natural choice for t_0 is $\hbar\omega_D$, where ω_D is the Debye frequency: this is an approximate upper limit of the frequency of phonons in the solid (see chapter 6), so with both $\epsilon_{\mathbf{k}'}$ and $\epsilon_{\mathbf{k}}$ within a shell of this thickness above the Fermi level, their difference is likely to be smaller than the phonon energy $\hbar\omega_{\mathbf{q}}$ and hence their interaction, given by Eq. (8.5), attractive, as assumed in the Cooper model. Cooper also showed that the radius R of the bound electron pair is

$$R \sim \frac{\hbar^2 k_F}{m_e E_b}$$

For typical values of k_F and E_b, this radius is $R \sim 10^4$ Å, which is a very large distance on the atomic scale. Eq. (8.7) was a seminal result: it showed that the effective interaction can be very weak and still lead to a bound pair of electrons; it established that electrons close to the Fermi level are the major participants in the pairing; and it indicated that their binding energy is a non-analytic function of the effective interaction V_0, which implied that a perturbative approach to the problem would not work.

8.3.2 BCS ground state

The fact that creation of Cooper pairs is energetically favorable (it has a positive binding energy) naturally leads to the following question: what is the preferred state of the system under conditions where Cooper pairs can be formed? A tempting answer would be to construct a state with the maximum possible number of Cooper pairs. However, this cannot be done for all available electrons since then the Fermi surface would collapse, removing the basis for the stability of Cooper pairs. A different scheme must be invented, in which the Fermi surface can survive, yet a sufficiently large number of Cooper pairs can be created to take maximum advantage of the benefits of pairing. To address these issues, BCS proposed that the many-body wavefunction for the superconducting state can be chosen from the restricted Hilbert space consisting of a direct product of four-dimensional vector spaces: each such space includes the states generated by a pair of electrons with opposite momenta, $\mathbf{k}\uparrow, \mathbf{k}\downarrow, -\mathbf{k}\uparrow, -\mathbf{k}\downarrow$. Each vector subspace contains a total of 16 states, depending on whether the individual single-particle states are occupied or not. The ground state is further restricted to only two of those states in each subspace, the ones with paired electrons of opposite spin in a singlet configuration:

$$\Psi_0^{(s)} = \prod_{\mathbf{k}} \left[u_{\mathbf{k}} |0_{\mathbf{k}} 0_{-\mathbf{k}}\rangle + v_{\mathbf{k}} |\psi_{\mathbf{k}} \psi_{-\mathbf{k}}\rangle \right] \tag{8.8}$$

where $|\psi_{\mathbf{k}}\psi_{-\mathbf{k}}\rangle$ represents the presence of a pair of electrons with opposite spins and momenta and $|0_{\mathbf{k}}0_{-\mathbf{k}}\rangle$ represents the absence of such a pair. We discuss first the physical meaning of this wavefunction. Each pair of electrons retains the antisymmetric nature of two fermions due to the spin component which is a singlet state, but as a whole it has zero total momentum and zero total spin. In this sense, each pair can be thought of as a composite particle with bosonic character, and the total wavefunction as a coherent state of all these bosons occupying a zero-momentum state. This is reminiscent of the Bose-Einstein condensation of bosons at low enough temperature (see Appendix D). This analogy cannot be taken literally, because of the special nature of the composite particles in the BCS wavefunction, but it is helpful in establishing that this wavefunction describes a coherent state of the entire system of electrons.

We explore next the implications of the BCS ground state wavefunction. For properly normalized pair states we require that

$$u_{\mathbf{k}} u_{\mathbf{k}}^* + v_{\mathbf{k}} v_{\mathbf{k}}^* = 1 \tag{8.9}$$

In Eq. (8.8), $|v_{\mathbf{k}}|^2$ is the probability that a Cooper pair of wave-vector $\mathbf{k}$ is present in the ground state and $|u_{\mathbf{k}}|^2$ the probability that it is not. We note that from the definition of these quantities, $v_{-\mathbf{k}} = v_{\mathbf{k}}$ and $u_{-\mathbf{k}} = u_{\mathbf{k}}$. Moreover, in terms of these quantities the normal (non-superconducting) ground state is described by $|v_{\mathbf{k}}| = 1, |u_{\mathbf{k}}| = 0$ for $|\mathbf{k}| < k_{\mathrm{F}}$ and $|u_{\mathbf{k}}| = 1, |v_{\mathbf{k}}| = 0$ for $|\mathbf{k}| > k_{\mathrm{F}}$, which is a consequence of the fact that at zero temperature all states with wave-vectors up to the Fermi momentum k_{F} are filled.

The hamiltonian for this system contains two terms,

$$\mathcal{H} = \mathcal{H}_0^{sp} + \mathcal{H}^{phon}$$

The first term describes the electron interactions in the single-particle picture with the ions frozen, and the second term describes the electron interactions mediated by the exchange of phonons, when the ionic motion is taken into account. In principle, we can write the first term as a sum over single-particle terms and the second as a sum over pair-wise interactions:

$$\mathcal{H}_0^{sp} = \sum_i h(\mathbf{r}_i), \quad \mathcal{H}^{phon} = \frac{1}{2}\sum_{ij} V^{phon}(\mathbf{r}_i - \mathbf{r}_j) \tag{8.10}$$

The single-particle wavefunctions $|\psi_{\mathbf{k}}\rangle$ are eigenfunctions of the first term, with eigenvalues $\epsilon_{\mathbf{k}}$,

$$\mathcal{H}_0^{sp}|\psi_{\mathbf{k}}\rangle = \epsilon_{\mathbf{k}}|\psi_{\mathbf{k}}\rangle$$

and form a complete orthonormal set. It is customary to refer to $\epsilon_{\mathbf{k}}$ as the "kinetic energy",[2] even though in a single-particle picture it also includes all the electron–ion as well as the electron-electron interactions, as discussed in chapter 2. Furthermore, it is convenient to measure all single-particle energies $\epsilon_{\mathbf{k}}$ relative to the Fermi energy ϵ_F, a convention we adopt in the following. This is equivalent to having a variable number of electrons in the system, with the chemical potential equal to the Fermi level. When we take matrix elements of the first term in the hamiltonian with respect to the ground state wavefunction defined in Eq. (8.8), we find that it generates the following types of terms:

$$u_{\mathbf{k}}u_{\mathbf{k}'}^*\langle 0_{\mathbf{k}'}0_{-\mathbf{k}'}|\mathcal{H}_0^{sp}|0_{\mathbf{k}}0_{-\mathbf{k}}\rangle, \quad u_{\mathbf{k}}v_{\mathbf{k}'}^*\langle \psi_{\mathbf{k}'}\psi_{-\mathbf{k}'}|\mathcal{H}_0^{sp}|0_{\mathbf{k}}0_{-\mathbf{k}}\rangle$$

$$v_{\mathbf{k}}u_{\mathbf{k}'}^*\langle 0_{\mathbf{k}'}0_{-\mathbf{k}'}|\mathcal{H}_0^{sp}|\psi_{\mathbf{k}}\psi_{-\mathbf{k}}\rangle, \quad v_{\mathbf{k}}v_{\mathbf{k}'}^*\langle \psi_{\mathbf{k}'}\psi_{-\mathbf{k}'}|\mathcal{H}_0^{sp}|\psi_{\mathbf{k}}\psi_{-\mathbf{k}}\rangle$$

Of all these terms, only the last one gives a non-vanishing contribution, because the terms that include a $|0_{\mathbf{k}}0_{-\mathbf{k}}\rangle$ state give identically zero when $\mathcal{H}_0^{sp}$ is applied to them. The last term gives

$$v_{\mathbf{k}}v_{\mathbf{k}'}^*\langle \psi_{\mathbf{k}'}\psi_{-\mathbf{k}'}|\mathcal{H}_0^{sp}|\psi_{\mathbf{k}}\psi_{-\mathbf{k}}\rangle = |v_{\mathbf{k}}|^2\delta(\mathbf{k}-\mathbf{k}')\epsilon_{\mathbf{k}} + |v_{\mathbf{k}}|^2\delta(\mathbf{k}+\mathbf{k}')\epsilon_{\mathbf{k}}$$

where we have used the facts that $v_{\mathbf{k}} = v_{-\mathbf{k}}$ and $\epsilon_{\mathbf{k}} = \epsilon_{-\mathbf{k}}$ (the latter from the discussion of electronic states in chapter 3). Summing over all values of $\mathbf{k}$, $\mathbf{k}'$, we

[2] The reason for this terminology is that in the simplest possible treatment of metallic electrons, they can be considered as free particles in the positive uniform background of the ions (the jellium model), in which case $\epsilon_{\mathbf{k}}$ is indeed the kinetic energy; in the following we adopt this terminology to conform to literature conventions.

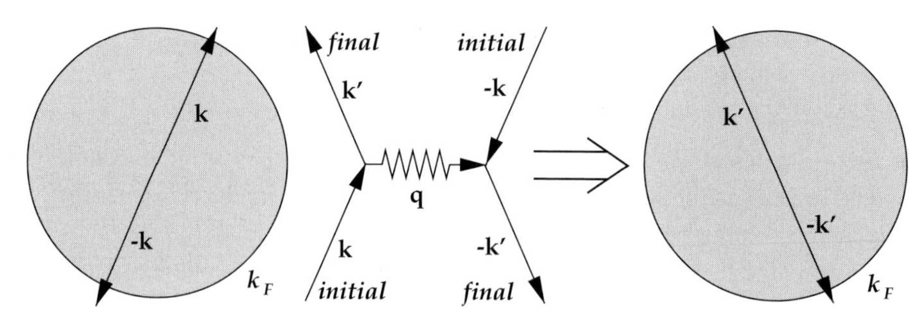

Figure 8.5. Scattering between electrons of wave-vectors $\pm\mathbf{k}$ (initial) and $\pm\mathbf{k}'$ (final) through the exchange of a phonon of wave-vector $\mathbf{q}$: in the initial state the Cooper pair of wave-vector $\mathbf{k}$ is occupied and that of wave-vector $\mathbf{k}'$ is not, while in the final state the reverse is true. The shaded region indicates the Fermi sphere of radius k_F.

find that the contribution of the first term in the hamiltonian to the total energy is given by

$$\langle \Psi_0^{(s)} | \mathcal{H}_0^{sp} | \Psi_0^{(s)} \rangle = \sum_{\mathbf{k}} 2\epsilon_{\mathbf{k}} |v_{\mathbf{k}}|^2$$

Turning our attention next to the second term in the hamiltonian, we see that it must describe the types of processes illustrated in Fig. 8.5. If a Cooper pair of wave-vector $\mathbf{k}$ is initially present in the ground state, the interaction of electrons through the exchange of phonons will lead to this pair being kicked out of the ground state and being replaced by another pair of wave-vector $\mathbf{k}'$ which was not initially part of the ground state. This indicates that the only non-vanishing matrix elements of the second term in the hamiltonian must be of the form

$$\left(v_{\mathbf{k}'}^* \langle \psi_{\mathbf{k}'} \psi_{-\mathbf{k}'} | u_{\mathbf{k}}^* \langle 0_{\mathbf{k}} 0_{-\mathbf{k}} | \right) \mathcal{H}^{phon} \left(v_{\mathbf{k}} | \psi_{\mathbf{k}} \psi_{-\mathbf{k}} | \rangle u_{\mathbf{k}'} | 0_{\mathbf{k}'} 0_{-\mathbf{k}'} \rangle \right)$$

In terms of the potential that appears in Eq. (8.10), which describes the exchange of phonons, these matrix elements will take the form

$$V_{\mathbf{k}\mathbf{k}'} = \langle \psi_{\mathbf{k}'} \psi_{-\mathbf{k}'} | V^{phon} | \psi_{\mathbf{k}} \psi_{-\mathbf{k}} \rangle$$

from which we conclude that the contribution of the phonon interaction hamiltonian to the total energy will be

$$\langle \Psi_0^{(s)} | \mathcal{H}^{phon} | \Psi_0^{(s)} \rangle = \sum_{\mathbf{k}\mathbf{k}'} V_{\mathbf{k}\mathbf{k}'} u_{\mathbf{k}}^* v_{\mathbf{k}'}^* u_{\mathbf{k}'} v_{\mathbf{k}}$$

Finally, putting together the two contributions, we find that the total energy of the ground state defined in Eq. (8.8) is

$$E_0^{(s)} = \langle \Psi_0^{(s)} | \mathcal{H}_0^{sp} + \mathcal{H}^{phon} | \Psi_0^{(s)} \rangle = \sum_{\mathbf{k}} 2\epsilon_{\mathbf{k}} |v_{\mathbf{k}}|^2 + \sum_{\mathbf{k}\mathbf{k}'} V_{\mathbf{k}\mathbf{k}'} u_{\mathbf{k}}^* v_{\mathbf{k}'}^* u_{\mathbf{k}'} v_{\mathbf{k}} \tag{8.11}$$

where in the final expression we have omitted for simplicity the phonon wave-vector dependence of the matrix elements $V_{\mathbf{k}\mathbf{k}'}$.

In order to determine the values of $u_{\mathbf{k}}$, $v_{\mathbf{k}}$, which are the only parameters in the problem, we will employ a variational argument: we will require that the ground state energy be a minimum with respect to variations in $v^*_{\mathbf{k}}$ while keeping $v_{\mathbf{k}}$ and $u_{\mathbf{k}}$ fixed. This is the usual procedure of varying only the bra of a single-particle state (in the present case a single-pair state) while keeping the ket fixed, as was done in chapter 2 for the derivation of the single-particle equations from the many-body ground state energy. This argument leads to

$$\frac{\partial E_0^{(s)}}{\partial v^*_{\mathbf{k}}} = 2\epsilon_{\mathbf{k}} v_{\mathbf{k}} + \sum_{\mathbf{k}'} V_{\mathbf{k}'\mathbf{k}} u^*_{\mathbf{k}'} u_{\mathbf{k}} v_{\mathbf{k}'} + \sum_{\mathbf{k}'} V_{\mathbf{k}\mathbf{k}'} \frac{\partial u^*_{\mathbf{k}}}{\partial v^*_{\mathbf{k}}} v^*_{\mathbf{k}'} u_{\mathbf{k}'} v_{\mathbf{k}} = 0$$

From the normalization condition, Eq. (8.9), and the assumptions we have made above, we find that

$$\frac{\partial u^*_{\mathbf{k}}}{\partial v^*_{\mathbf{k}}} = -\frac{v_{\mathbf{k}}}{u_{\mathbf{k}}}$$

which we can substitute into the above equation, and with the use of the definitions

$$\Delta_{\mathbf{k}} = \sum_{\mathbf{k}'} V_{\mathbf{k}'\mathbf{k}} u^*_{\mathbf{k}'} v_{\mathbf{k}'} \rightarrow \Delta^*_{\mathbf{k}} = \sum_{\mathbf{k}'} V_{\mathbf{k}\mathbf{k}'} u_{\mathbf{k}'} v^*_{\mathbf{k}'} \tag{8.12}$$

we arrive at the following relation:

$$2\epsilon_{\mathbf{k}} v_{\mathbf{k}} u_{\mathbf{k}} + \Delta_{\mathbf{k}} u^2_{\mathbf{k}} - \Delta^*_{\mathbf{k}} v^2_{\mathbf{k}} = 0$$

In the above derivation we have used the fact that $V^*_{\mathbf{k}'\mathbf{k}} = V_{\mathbf{k}\mathbf{k}'}$, since $V_{\mathbf{k}\mathbf{k}'}$ represents the Fourier transform of a real potential. Next, we will assume an explicit form for the parameters $u_{\mathbf{k}}$, $v_{\mathbf{k}}$ which automatically satisfies the normalization condition, Eq. (8.9), and allows for a phase difference $w_{\mathbf{k}}$ between them:

$$u_{\mathbf{k}} = \cos\frac{\theta_{\mathbf{k}}}{2} \mathrm{e}^{\mathrm{i}w_{\mathbf{k}}/2}, \qquad v_{\mathbf{k}} = \sin\frac{\theta_{\mathbf{k}}}{2} \mathrm{e}^{-\mathrm{i}w_{\mathbf{k}}/2} \tag{8.13}$$

When these expressions are substituted into the previous equation we find

$$2\epsilon_{\mathbf{k}} \sin\frac{\theta_{\mathbf{k}}}{2} \cos\frac{\theta_{\mathbf{k}}}{2} = \Delta^*_{\mathbf{k}} \sin^2\frac{\theta_{\mathbf{k}}}{2} \mathrm{e}^{-\mathrm{i}w_{\mathbf{k}}} - \Delta_{\mathbf{k}} \cos^2\frac{\theta_{\mathbf{k}}}{2} \mathrm{e}^{\mathrm{i}w_{\mathbf{k}}} \tag{8.14}$$

In this relation the left-hand side is a real quantity, therefore the right-hand side must also be real; if we express the complex number $\Delta_{\mathbf{k}}$ as its magnitude times a phase factor,

$$\Delta_{\mathbf{k}} = |\Delta_{\mathbf{k}}| \mathrm{e}^{-\mathrm{i}\tilde{w}_{\mathbf{k}}}$$

and substitute this expression into the previous equation, we find that the right-hand side becomes

$$|\Delta_{\mathbf{k}}| \sin^2 \frac{\theta_{\mathbf{k}}}{2} \mathrm{e}^{-\mathrm{i}(w_{\mathbf{k}} - \tilde{w}_{\mathbf{k}})} - |\Delta_{\mathbf{k}}| \cos^2 \frac{\theta_{\mathbf{k}}}{2} \mathrm{e}^{\mathrm{i}(w_{\mathbf{k}} - \tilde{w}_{\mathbf{k}})}$$

which must be real, and therefore its imaginary part must vanish:

$$\left(|\Delta_{\mathbf{k}}| \sin^2 \frac{\theta_{\mathbf{k}}}{2} + |\Delta_{\mathbf{k}}| \cos^2 \frac{\theta_{\mathbf{k}}}{2} \right) \sin(w_{\mathbf{k}} - \tilde{w}_{\mathbf{k}}) = 0 \Longrightarrow w_{\mathbf{k}} = \tilde{w}_{\mathbf{k}}$$

With this result, Eq. (8.14) simplifies to

$$\epsilon_{\mathbf{k}} \sin \theta_{\mathbf{k}} + |\Delta_{\mathbf{k}}| \cos \theta_{\mathbf{k}} = 0 \tag{8.15}$$

This equation has two possible solutions:

$$\sin \theta_{\mathbf{k}} = -\frac{|\Delta_{\mathbf{k}}|}{\zeta_{\mathbf{k}}}, \quad \cos \theta_{\mathbf{k}} = \frac{\epsilon_{\mathbf{k}}}{\zeta_{\mathbf{k}}} \tag{8.16}$$

$$\sin \theta_{\mathbf{k}} = \frac{|\Delta_{\mathbf{k}}|}{\zeta_{\mathbf{k}}}, \quad \cos \theta_{\mathbf{k}} = -\frac{\epsilon_{\mathbf{k}}}{\zeta_{\mathbf{k}}} \tag{8.17}$$

where we have defined the quantity

$$\zeta_{\mathbf{k}} \equiv \sqrt{\epsilon_{\mathbf{k}}^2 + |\Delta_{\mathbf{k}}|^2} \tag{8.18}$$

Of the two solutions, the first one leads to the lowest energy state while the second leads to an excited state (see Problem 1). In the following discussion, which is concerned with the ground state of the system, we will use the values of the parameters $u_{\mathbf{k}}$, $v_{\mathbf{k}}$ given in Eq. (8.16). From the definition of $\zeta_{\mathbf{k}}$, Eq. (8.18) and using Eq. (8.13), it is straightforward to derive the relations

$$|u_{\mathbf{k}}|^2 = \frac{1}{2}\left(1 + \frac{\epsilon_{\mathbf{k}}}{\zeta_{\mathbf{k}}}\right), \quad |v_{\mathbf{k}}|^2 = \frac{1}{2}\left(1 - \frac{\epsilon_{\mathbf{k}}}{\zeta_{\mathbf{k}}}\right)$$

$$|u_{\mathbf{k}}||v_{\mathbf{k}}| = \frac{1}{2}\left(1 - \frac{\epsilon_{\mathbf{k}}^2}{\zeta_{\mathbf{k}}^2}\right)^{1/2} = \frac{1}{2}\frac{|\Delta_{\mathbf{k}}|}{\zeta_{\mathbf{k}}} \tag{8.19}$$

The quantity $\zeta_{\mathbf{k}}$ is actually the excitation energy above the BCS ground state, associated with breaking the pair of wave-vector **k** (see Problem 2). For this reason, we refer to $\zeta_{\mathbf{k}}$ as the energy of quasiparticles associated with excitations above the superconducting ground state.

In order to gain some insight into the meaning of this solution, we consider a simplified picture in which we neglect the **k**-dependence of all quantities and use

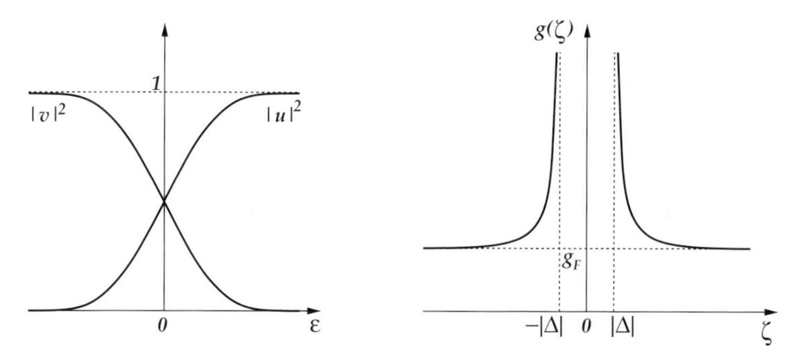

Figure 8.6. Features of the BCS model: **Left:** Cooper pair occupation variables $|u|^2$ and $|v|^2$ as a function of the energy ϵ. **Right:** superconducting density of states $g(\zeta)$ as a function of the energy ζ; g_F is the density of states in the normal state at the Fermi level.

ϵ as the only relevant variable. In this picture, the magnitudes of the parameters $u_{\mathbf{k}}$, $v_{\mathbf{k}}$ reduce to

$$|u|^2 = \frac{1}{2}\left(1 + \frac{\epsilon}{\sqrt{\epsilon^2 + |\Delta|^2}}\right), \quad |v|^2 = \frac{1}{2}\left(1 - \frac{\epsilon}{\sqrt{\epsilon^2 + |\Delta|^2}}\right)$$

which are plotted as functions of ϵ in Fig. 8.6. We see that the parameters u, v exhibit behavior similar to Fermi occupation numbers, with $|v|^2$ being close to unity for $\epsilon \ll 0$ and approaching zero for $\epsilon \gg 0$, and $|u|^2$ taking the reverse values. At $\epsilon = 0$, which coincides with ϵ_{F}, $|u|^2 = |v|^2 = 0.5$. The change in $|v|^2$, $|u|^2$ from one asymptotic value to the other takes place over a range in ϵ of order $|\Delta|$ around zero (the exact value of this range depends on the threshold below which we take $|u|$ and $|v|$ to be essentially equal to their asymptotic values). Thus, the occupation of Cooper pairs is significant only for energies around the Fermi level, within a range of order $|\Delta|$. It is instructive to contrast this result to the normal state which, as discussed earlier, corresponds to the occupations $|v| = 1$, $|u| = 0$ for $\epsilon < \epsilon_{\mathrm{F}}$ and $|u| = 1$, $|v| = 0$ for $\epsilon > \epsilon_{\mathrm{F}}$. We see that the formation of the superconducting state changes these step functions to functions similar to Fermi occupation numbers.

We can also use this simplified picture to obtain the density of states in the superconducting state. Since the total number of electronic states must be conserved in going from the normal to the superconducting state, we will have

$$g(\zeta)\mathrm{d}\zeta = g(\epsilon)\mathrm{d}\epsilon \Rightarrow g(\zeta) = g(\epsilon)\frac{\mathrm{d}\epsilon}{\mathrm{d}\zeta}$$

which, from the definition of $\zeta_{\mathbf{k}}$, Eq. (8.18), viewed here as a function of ϵ, gives

$$\begin{aligned} g(\zeta) &= g_F \frac{|\zeta|}{\sqrt{\zeta^2 - |\Delta|^2}}, \qquad &\text{for } |\zeta| > |\Delta| \\ &= 0, &\text{for } |\zeta| < |\Delta| \end{aligned}$$

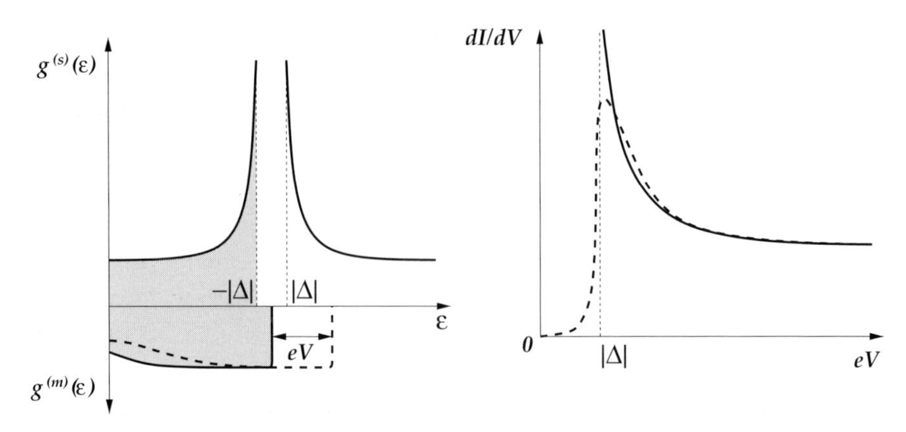

Figure 8.7. **Left:** density of states at a metal–superconductor contact: the shaded regions represent occupied states, with the Fermi level in the middle of the superconducting gap. When a voltage bias V is applied to the metal side the metal density of states is shifted in energy by eV (dashed line). **Right:** differential conductance $\mathrm{d}I/\mathrm{d}V$ as a function of the bias voltage eV at $T = 0$ (solid line): in this case the measured curve should follow exactly the features of the superconductor density of states $g^{(s)}(\epsilon)$. At $T > 0$ (dashed line) the measured curve is a smoothed version of $g^{(s)}(\epsilon)$.

In the last expression we have approximated $g(\epsilon)$ by its value at the Fermi level g_F. The function $g(\zeta)$ is also plotted in Fig. 8.6. This is an intriguing result: it shows that the superconducting state opens a gap in the density of states around the Fermi level equal to $2|\Delta|$, within which there are no quasiparticle states. This is known as the "superconducting gap" and is directly observable in tunneling experiments.

To illustrate how the superconducting gap is observed experimentally we consider a contact between a superconductor and a metal. At equilibrium the Fermi level on both sides must be the same, leading to the situation shown in Fig. 8.7. Typically, the superconducting gap is small enough to allow us to approximate the density of states in the metal as a constant over a range of energies at least equal to $2|\Delta|$ around the Fermi level. With this approximation, the situation at hand is equivalent to a metal-semiconductor contact, which was discussed in detail in chapter 5. We can therefore apply the results of that discussion directly to the metal–superconductor contact: the measured differential conductance at $T = 0$ will be given by Eq. (5.25),

$$\left[\frac{\mathrm{d}I}{\mathrm{d}V}\right]_{T=0} = e|\mathcal{T}|^2 g^{(s)}(\epsilon_\mathrm{F} + eV) g_F^{(m)}$$

with $\mathcal{T}$ the tunneling matrix element and $g^{(s)}(\epsilon)$, $g^{(m)}(\epsilon)$ the density of states of the superconductor and the metal, the latter evaluated at the Fermi level, $g_F^{(m)} = g^{(m)}(\epsilon_\mathrm{F})$. This result shows that by scanning the bias voltage V one samples the density of states of the superconductor. Thus, the measured differential conductance will reflect all the features of the superconductor density of states, including the superconducting gap, as illustrated in Fig. 8.7. However, making measurements at $T = 0$ is not physically possible. At finite temperature $T > 0$ the occupation number $n^{(m)}(\epsilon)$ is not a sharp step function but a smooth function whose derivative is an analytic representation of the δ-function (see Appendix G). In this case, when the voltage V is scanned the measured differential conductance is also a smooth function representing a smoothed version of the superconductor density of states, as shown in Fig. 8.7: the superconducting gap is still clearly evident, although not as an infinitely sharp feature.

The physical interpretation of the gap is that it takes at least this much energy to create excitations above the ground state; since an excitation is related to the creation of quasiparticles as a result of Cooper pair breaking, we may conclude that the lowest energy of each quasiparticle state is half the value of the gap, that is, $|\Delta|$. This can actually be derived easily in a simple picture where the single-particle energies $\epsilon_{\mathbf{k}}$ are taken to be the kinetic energy of electrons in a jellium model. Consistent with our earlier assumptions, we measure these energies relative to the Fermi level ϵ_{F}, which gives

$$\epsilon_{\mathbf{k}} = \frac{\hbar^2}{2m_e}(\mathbf{k}^2 - k_{\mathrm{F}}^2) = \frac{\hbar^2}{2m_e}(k - k_{\mathrm{F}})(k + k_{\mathrm{F}}) \approx \frac{\hbar^2 k_{\mathrm{F}}}{m_e}(k - k_{\mathrm{F}})$$

where in the last expression we have assumed that the single-particle energies $\epsilon_{\mathbf{k}}$ (and the wave-vectors $\mathbf{k}$) lie within a very narrow range of the Fermi energy ϵ_{F} (and the Fermi momentum k_{F}). With this expression, the quasiparticle energies $\zeta_{\mathbf{k}}$ take the form

$$\zeta_{\mathbf{k}} = \sqrt{\left(\frac{\hbar^2 k_{\mathrm{F}}}{m_e}\right)^2 (k - k_{\mathrm{F}})^2 + |\Delta|^2}$$

This result proves our assertion that the lowest quasiparticle energy is $|\Delta|$ (see also Problem 2).

These considerations reveal that the quantity $\Delta_{\mathbf{k}}$ plays a key role in the nature of the superconducting state. We therefore consider its behavior in more detail. From Eqs. (8.13) and (8.16) and the definition of $\Delta_{\mathbf{k}}$, Eq. (8.12), we find

$$\Delta_{\mathbf{k}} = \sum_{\mathbf{k}'} V_{\mathbf{k}'\mathbf{k}} u_{\mathbf{k}'}^* v_{\mathbf{k}'} = \sum_{\mathbf{k}'} V_{\mathbf{k}'\mathbf{k}} \cos\frac{\theta_{\mathbf{k}'}}{2} \sin\frac{\theta_{\mathbf{k}'}}{2} \mathrm{e}^{-\mathrm{i}w_{\mathbf{k}'}} = -\sum_{\mathbf{k}'} V_{\mathbf{k}'\mathbf{k}} \frac{|\Delta_{\mathbf{k}'}|}{2\zeta_{\mathbf{k}'}} \mathrm{e}^{-\mathrm{i}w_{\mathbf{k}'}}$$

$$\Rightarrow \Delta_{\mathbf{k}} = -\sum_{\mathbf{k}'} V_{\mathbf{k}'\mathbf{k}} \frac{\Delta_{\mathbf{k}'}}{2\zeta_{\mathbf{k}'}} \qquad (8.20)$$

which is known as the "BCS gap equation". Within the simple model defined in relation to Cooper pairs, we find

$$|\Delta| = \frac{\hbar\omega_{\rm D}}{\sinh(1/g_F V_0)}$$

(see Problem 3). In the weak coupling limit we find

$$|\Delta| = 2\hbar\omega_{\rm D}{\rm e}^{-1/g_F V_0} \tag{8.21}$$

which is an expression similar to the binding energy of a Cooper pair, Eq. (8.7). As was mentioned in that case, this type of expression is non-analytic in V_0, indicating that a perturbative approach in the effective interaction parameter would not be suitable for this problem. The expression we have found for $|\Delta|$ shows that the magnitude of this quantity is much smaller than $\hbar\omega_{\rm D}$, which in turn is much smaller than the Fermi energy, as discussed in chapter 6:

$$|\Delta| \ll \hbar\omega_{\rm D} \ll \epsilon_{\rm F}$$

Thus, the superconducting gap in the density of states is a minuscule fraction of the Fermi energy.

We next calculate the total energy of the superconducting ground state using the values of the parameters $u_{\mathbf{k}}$, $v_{\mathbf{k}}$ and $\Delta_{\mathbf{k}}$ that we have obtained in terms of $\epsilon_{\mathbf{k}}$, $\zeta_{\mathbf{k}}$. From Eq. (8.11) and the relevant expressions derived above we find

$$\begin{aligned} E_0^{(s)} &= \sum_{\mathbf{k}} 2\epsilon_{\mathbf{k}}\frac{1}{2}\left(1-\frac{\epsilon_{\mathbf{k}}}{\zeta_{\mathbf{k}}}\right) + \sum_{\mathbf{k}} |\Delta_{\mathbf{k}}|{\rm e}^{{\rm i}w_{\mathbf{k}}} \cos\frac{\theta_{\mathbf{k}}}{2}{\rm e}^{-{\rm i}w_{\mathbf{k}}/2} \sin\frac{\theta_{\mathbf{k}}}{2}{\rm e}^{-{\rm i}w_{\mathbf{k}}/2} \\ &= \sum_{\mathbf{k}} \epsilon_{\mathbf{k}}\left(1-\frac{\epsilon_{\mathbf{k}}}{\zeta_{\mathbf{k}}}\right) + \sum_{\mathbf{k}} \frac{1}{2}|\Delta_{\mathbf{k}}|\sin\theta_{\mathbf{k}} \\ &= \sum_{\mathbf{k}} \epsilon_{\mathbf{k}}\left(1-\frac{\epsilon_{\mathbf{k}}}{\zeta_{\mathbf{k}}}\right) - \sum_{\mathbf{k}} \frac{|\Delta_{\mathbf{k}}|^2}{2\zeta_{\mathbf{k}}} \end{aligned} \tag{8.22}$$

This last expression for the total energy has a transparent physical interpretation. The first term is the kinetic energy cost to create the Cooper pairs; this term is always positive since $\epsilon_{\mathbf{k}} \leq \zeta_{\mathbf{k}}$ and comes from the fact that we need to promote some electrons to energies higher than the Fermi energy in order to create Cooper pairs. The second term is the gain in potential energy due to the binding of the Cooper pairs, which is evidently always negative corresponding to an effective attractive interaction. In the BCS model it is possible to show that this energy is always lower than the corresponding energy of the normal, non-superconducting state (see Problem 3).

8.3.3 BCS theory at finite temperature

Finally, we consider the situation at finite temperature. In this case, we will take the temperature-dependent occupation number of the single-particle state $\mathbf{k}$ to be $n_{\mathbf{k}}$ for each spin state. This leads to the assignment of $(1 - 2n_{\mathbf{k}})$ as the Cooper pair occupation number of the same wave-vector. Consequently, each occurrence of a Cooper pair will be accompanied by a factor $(1 - 2n_{\mathbf{k}})$, while the occurrence of an electronic state that is not part of a Cooper pair will be accompanied by a factor $2n_{\mathbf{k}}$, taking into account spin degeneracy. With these considerations, the total energy of the ground state at finite temperature will be given by

$$E_0^{(s)} = \sum_{\mathbf{k}} \left(2n_{\mathbf{k}}\epsilon_{\mathbf{k}} + 2(1 - 2n_{\mathbf{k}})|v_{\mathbf{k}}|^2\epsilon_{\mathbf{k}}\right) + \sum_{\mathbf{k}\mathbf{k}'} V_{\mathbf{k}\mathbf{k}'} u_{\mathbf{k}}^* v_{\mathbf{k}'}^* u_{\mathbf{k}'} v_{\mathbf{k}} (1 - 2n_{\mathbf{k}})(1 - 2n_{\mathbf{k}'}) \tag{8.23}$$

To this energy we must add the entropy contribution, to arrive at the free energy. The entropy of a gas of fermions consisting of spin-up and spin-down particles with occupation numbers $n_{\mathbf{k}}$ is given by (see Appendix D)

$$S^{(s)} = -2k_{\mathrm{B}} \sum_{\mathbf{k}} \left[n_{\mathbf{k}} \ln n_{\mathbf{k}} + (1 - n_{\mathbf{k}}) \ln(1 - n_{\mathbf{k}})\right] \tag{8.24}$$

By analogy to our discussion of the zero-temperature case, we define the quantities $\Delta_{\mathbf{k}}, \Delta_{\mathbf{k}}^*$ as

$$\Delta_{\mathbf{k}} = \sum_{\mathbf{k}} V_{\mathbf{k}'\mathbf{k}} u_{\mathbf{k}'}^* v_{\mathbf{k}'} (1 - 2n_{\mathbf{k}'}) \rightarrow \Delta_{\mathbf{k}}^* = \sum_{\mathbf{k}} V_{\mathbf{k}\mathbf{k}'} u_{\mathbf{k}'} v_{\mathbf{k}'}^* (1 - 2n_{\mathbf{k}'}) \tag{8.25}$$

We can now apply a variational argument to the free energy of the ground state $F^{(s)} = E_0^{(s)} - TS^{(s)}$ which, following exactly the same steps as in the zero-temperature case, leads to

$$\Delta_{\mathbf{k}} = -\sum_{\mathbf{k}'} V_{\mathbf{k}'\mathbf{k}} \frac{\Delta_{\mathbf{k}'}}{2\zeta_{\mathbf{k}'}} (1 - 2n_{\mathbf{k}'}) \tag{8.26}$$

with $\zeta_{\mathbf{k}}$ defined by the same expression as before, Eq. (8.18), only now it is a temperature-dependent quantity through the dependence of $\Delta_{\mathbf{k}}$ on $n_{\mathbf{k}}$. Eq. (8.26) is the BCS gap equation at finite temperature. We can also require that the free energy $F^{(s)}$ is a minimum with respect to variations in the occupation numbers $n_{\mathbf{k}}$, which leads to

$$\frac{\partial F^{(s)}}{\partial n_{\mathbf{k}}} = 0 \Rightarrow 2\epsilon_{\mathbf{k}}(1 - 2|v_{\mathbf{k}}|^2) - 2\Delta_{\mathbf{k}}^* u_{\mathbf{k}}^* v_{\mathbf{k}} - 2\Delta_{\mathbf{k}} u_{\mathbf{k}} v_{\mathbf{k}}^* + 2k_{\mathrm{B}}T \ln \frac{n_{\mathbf{k}}}{1 - n_{\mathbf{k}}} = 0$$

All the relations derived for $u_{\mathbf{k}}, v_{\mathbf{k}}, \Delta_{\mathbf{k}}$ in terms of the variables $\theta_{\mathbf{k}}, w_{\mathbf{k}}$ are still valid with the new definition of $\Delta_{\mathbf{k}}$, Eq. (8.25); when these relations are substituted into the above equation, they lead to

$$n_{\mathbf{k}} = \frac{1}{1 + e^{\zeta_{\mathbf{k}}/k_{\mathrm{B}}T}} \tag{8.27}$$

This result reveals that the occupation numbers $n_{\mathbf{k}}$ have the familiar Fermi function form but with the energy $\zeta_{\mathbf{k}}$, rather than the single-particle energy $\epsilon_{\mathbf{k}}$, as the relevant variable. Using this expression for $n_{\mathbf{k}}$ in the BCS gap equation at finite temperature, Eq. (8.26), we find

$$\Delta_{\mathbf{k}} = -\sum_{\mathbf{k}'} V_{\mathbf{k}'\mathbf{k}} \frac{\Delta_{\mathbf{k}'}}{2\zeta_{\mathbf{k}'}} \tanh \frac{\zeta_{\mathbf{k}}}{2k_{\mathrm{B}}T}$$

If we analyze this equation in the context of the BCS model (see Problem 4), in the weak coupling limit we obtain

$$k_{\mathrm{B}}T_c = 1.14\hbar\omega_{\mathrm{D}} e^{-1/g_F V_0} \tag{8.28}$$

This relation provides an explanation for the isotope effect in its simplest form: from our discussion of phonons in chapter 6, we can take the Debye frequency to be

$$\omega_{\mathrm{D}} \sim \left(\frac{\kappa}{M}\right)^{1/2}$$

where κ is the relevant force constant and M the mass of the ions; this leads to $T_c \sim M^{-1/2}$, the relation we called the isotope effect, Eq. (8.2), with $\alpha = -0.5$. Moreover, combining our earlier result for $|\Delta|$ at zero temperature, Eq. (8.21), with Eq. (8.28), we find

$$2|\Delta_0| = 3.52 k_{\mathrm{B}} T_c \tag{8.29}$$

where we have used the notation $2|\Delta_0|$ for the zero-temperature value of the superconducting gap. This relation is referred to as the "law of corresponding states" and is obeyed quite accurately by a wide range of conventional superconductors, confirming the validity of BCS theory.

8.3.4 The McMillan formula for T_c

We close this section with a brief discussion of how the T_c in *conventional* superconductors, arising from electron-phonon coupling as described by the BCS theory, can be calculated. McMillan [114] proposed a formula to evaluate T_c as

$$T_c = \frac{\Theta_{\mathrm{D}}}{1.45} \exp\left[-\frac{1.04(1+\lambda)}{\lambda - \mu^* - 0.62\lambda\mu^*}\right] \tag{8.30}$$

where Θ_D is the Debye temperature, λ is a constant describing electron–phonon coupling strength[3] and μ^* is another constant describing the repulsive Coulomb interaction strength. This expression is valid for $\lambda < 1.25$. The value of μ^* is difficult to obtain from calculations, but in any case the value of this constant tends to be small. For *sp* metals like Pb, its value has been estimated from tunneling measurements to be $\mu^* \sim 0.1$; for other cases this value is scaled by the density of states at the Fermi level, g_F, taking the value of g_F for Pb as the norm. The exponential dependence of T_c on the other parameter, λ, necessitates a more accurate estimate of its value. It turns out that λ can actually be obtained from electronic structure calculations, a procedure which has even led to predictions of superconductivity [115]. We outline the calculation of λ following the treatment of Chelikowsky and Cohen [65]. λ is expressed as the average over all phonon modes of the constants $\lambda_{\mathbf{k}}^{(l)}$:

$$\lambda = \sum_l \int \lambda_{\mathbf{k}}^{(l)} \mathrm{d}\mathbf{k} \tag{8.31}$$

which describe the coupling of an electron to a particular phonon mode identified by the index l and the wave-vector $\mathbf{k}$ (see chapter 6):

$$\lambda_{\mathbf{k}}^{(l)} = \frac{2g_F}{\hbar\omega_{\mathbf{k}}^{(l)}} \langle |f(n, \mathbf{q}, n', \mathbf{q}'; l, \mathbf{k})|^2 \rangle_F \tag{8.32}$$

where $f(n, \mathbf{q}, n', \mathbf{q}'; l, \mathbf{k})$ is the electron-phonon matrix element and $\langle \cdots \rangle_F$ is an average over the Fermi surface. The electron-phonon matrix elements are given by

$$f(n, \mathbf{q}, n', \mathbf{q}'; l, \mathbf{k}) = \sum_j \sqrt{\frac{\hbar}{2M_j\omega_{\mathbf{k}}^{(l)}}} \langle \psi_{\mathbf{q}}^{(n)} \left| \hat{\mathbf{e}}_{\mathbf{k}j}^{(l)} \cdot \frac{\delta V}{\delta \mathbf{t}_j} \right| \psi_{\mathbf{q}'}^{(n')} \rangle \delta(\mathbf{q} - \mathbf{q}' - \mathbf{k}) \tag{8.33}$$

where the summation on j is over the ions of mass M_j at positions $\mathbf{t}_j$ in the unit cell, $\psi_{\mathbf{q}}^{(n)}, \psi_{\mathbf{q}'}^{(n')}$ are electronic wavefunctions in the ideal crystal, and $\hat{\mathbf{e}}_{\mathbf{k}j}^{(l)}$ is the phonon polarization vector (more precisely, the part of the polarization vector corresponding to the position of ion j). The term $\delta V / \delta \mathbf{t}_j$ is the change in the crystal potential due to the presence of the phonon. This term can be evaluated as

$$\sum_l \hat{\mathbf{e}}_{\mathbf{k}j}^{(l)} \cdot \frac{\delta V}{\delta \mathbf{t}_j} = \frac{V_{\mathbf{k}}^{(l)} - V_0}{u_{\mathbf{k}}^{(l)}} \tag{8.34}$$

where V_0 is the ideal crystal potential, $V_{\mathbf{k}}^{(l)}$ is the crystal potential in the presence of the phonon mode identified by $l, \mathbf{k}$, and $u_{\mathbf{k}}^{(l)}$ is the average atomic displacement corresponding to this phonon mode. The terms in Eq. (8.34) can be readily evaluated through the computational methods discussed in chapter 5, by introducing atomic

[3] It is unfortunate that the same symbol is used in the literature for the electron–phonon coupling constant and the penetration length, but we will conform to this convention.

displacements $\mathbf{u}_{\mathbf{k}j}^{(l)}$ corresponding to a phonon mode and evaluating the crystal potential difference resulting from this distortion, with

$$u_{\mathbf{k}}^{(l)} = \frac{1}{N_{at}} \left(\sum_j |\mathbf{u}_{\mathbf{k}j}^{(l)}|^2 \right)^{1/2}$$

the atomic displacement averaged over the N_{at} atoms in the unit cell. Using this formalism, it was predicted and later verified experimentally that Si under high pressure would be a superconductor [115].

8.4 High-temperature superconductors

In this final section we give a short discussion of the physics of high-temperature superconductors, mostly in order to bring out their differences from the conventional ones. We first review the main points of the theory that explains the physics of conventional superconductors, as derived in the previous section. These are as follows.

(a) Electrons form pairs, called Cooper pairs, due to an attractive interaction mediated by the exchange of phonons. In a Cooper pair the electrons have opposite momenta and opposite spins in a spin-singlet configuration.
(b) Cooper pairs combine to form a many-body wavefunction which has lower energy than the normal, non-superconducting state, and represents a coherent state of the entire electron gas. Excitations above this ground state are represented by quasiparticles which correspond to broken Cooper pairs.
(c) In order to derive simple expressions relating the superconducting gap $|\Delta_0|$ to the critical temperature (or other experimentally accessible quantities) we assumed that there is no dependence of the gap, the quasiparticle energies, and the attractive interaction potential, on the electron wave-vectors $\mathbf{k}$. This assumption implies an isotropic solid and a spherical Fermi surface. This model is referred to as "s-wave" pairing, due to the lack of any spatial features in the physical quantities of interest (similar to the behavior of an s-like atomic wavefunction, see Appendix B).
(d) We also assumed that we are in the limit in which the product of the density of states at the Fermi level with the strength of the interaction potential, $g_F V_0$, which is a dimensionless quantity, is much smaller than unity; we called this the weak-coupling limit.

The high-temperature superconductors based on copper oxides (see Table 8.2, p. 284) conform to some but not all of these points. Points (a) and (b) apparently apply to those systems as well, with one notable exception: there seems to exist ground for doubting that electron-phonon interactions alone can be responsible for the pairing of electrons. Magnetic order plays an important role in the physics of these materials, and the presence of strong antiferromagnetic interactions may be

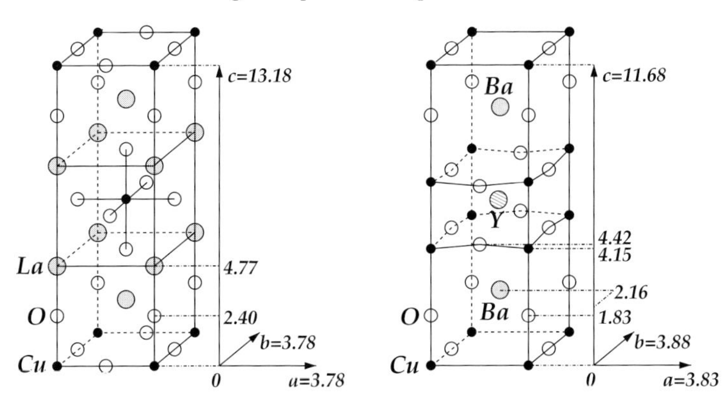

Figure 8.8. Representation of the structure of two typical high-temperature superconductors: La_2CuO_4 (**left**) is a crystal with tetragonal symmetry and $T_c = 39$ K, while $YBa_2Cu_3O_7$ (**right**) is a crystal with orthorhombic symmetry and $T_c = 93$ K. For both structures the conventional unit cell is shown, with the dimensions of the orthogonal axes and the positions of inequivalent atoms given in angstroms; in the case of La_2CuO_4 this cell is twice as large as the primitive unit cell. Notice the Cu–O octahedron, clearly visible at the center of the La_2CuO_4 conventional unit cell. Notice also the puckering of the Cu–O planes in the middle of the unit cell of $YBa_2Cu_3O_7$ and the absence of full Cu–O octahedra at its corners.

intimately related to the mechanism(s) of electron pairing. On the other hand, some variation of the basic theme of electron-phonon interaction, taking also into account the strong anisotropy in these systems (which is discussed next), may be able to capture the reason for electron pairing. Point (c) seems to be violated by these systems in several important ways. The copper-oxide superconductors are strongly anisotropic, having as a main structural feature planes of linked Cu–O octahedra which are decorated by other elements, as shown in Fig. 8.8 (see also chapter 1 for the structure of perovskites). In addition to this anisotropy of the crystal in real space, there exist strong indications that important physical quantities such as the superconducting gap are not featureless, but possess structure and hence have a strong dependence on the electron wave-vectors. These indications point to what is called "d-wave" pairing, that is, dependence of the gap on $\mathbf{k}$, similar to that exhibited by a d-like atomic wavefunction. Finally, point (d) also appears to be strongly violated by the high-temperature superconductors, which seem to be in the "strong-coupling" limit. Indications to this effect come from the very short coherence length in these systems, which is of order a few lattice constants as opposed to $\sim 10^4$ Å, and from the fact that the ratio $2|\Delta_0|/k_B T_c$ is not 3.52, as the weak-coupling limit of BCS theory, Eq. (8.29), predicts but is in the range $4-7$. These departures from the behavior of conventional superconductors have sparked much theoretical debate on what are the microscopic mechanisms responsible for this exotic behavior. At the present writing, the issue remains unresolved.

Further reading

1. *Theory of Superconductivity*, J.R. Schrieffer (Benjamin/Cummings, Reading, 1964).
   This is a classic account of the BCS theory of superconductivity.
2. *Superfluidity and Superconductivity*, D.R. Tilley and J. Tilley (Adam Hilger Ltd., Bristol, 1986).
   This book contains an interesting account of superconductivity as well as illuminating comparisons to superfluidity.
3. *Superconductivity*, C.P. Poole, H.A. Farach and R.J. Creswick (Academic Press, San Diego, 1995).
   This is a modern account of superconductivity, including extensive coverage of the high-temperature copper-oxide superconductors.
4. *Introduction to Superconductivity*, M. Tinkham (2nd edn, McGraw-Hill, New York, 1996).
   This is a standard reference, with comprehensive coverage of all aspects of superconductivity, including the high-temperature copper-oxide superconductors.
5. *Superconductivity in Metals and Alloys*, P.G. de Gennes (Addison-Wesley, Reading, MA, 1966).
6. "The structure of YBCO and its derivatives", R. Beyers and T.M. Shaw, in *Solid State Physics*, vol. 42, pp. 135–212 (eds. H. Ehrenreich and D. Turnbul, Academic Press, Boston, 1989).
   This article reviews the structure of representative high-temperature superconductor ceramic crystals.
7. "Electronic structure of copper-oxide semiconductors", K. C. Hass, in *Solid State Physics*, vol. 42, pp. 213–270 (eds. H. Ehrenreich and D. Turnbul, Academic Press, Boston, 1989).
   This article reviews the electronic properties of HTSC ceramic crystals.
8. "Electronic structure of the high-temperature oxide superconductors", W.E. Pickett, *Rev. Mod. Phys.*, **61**, p. 433 (1989).
   This is a thorough review of experimental and theoretical studies of the electronic properties of the oxide superconductors.

Problems

1. Show that the second solution to the BCS equation, Eq. (8.17), corresponds to an energy higher than the ground state energy, Eq. (8.22), by $2\zeta_{\mathbf{k}}$.
2. Show that the energy cost of removing the pair of wave-vector **k** from the BCS ground state is:

$$E_{\mathbf{k}}^{pair} = 2\epsilon_{\mathbf{k}}|v_{\mathbf{k}}|^2 + \Delta_{\mathbf{k}} u_{\mathbf{k}} v_{\mathbf{k}}^* + \Delta_{\mathbf{k}}^* u_{\mathbf{k}}^* v_{\mathbf{k}}$$

   where the first term comes from the kinetic energy loss and the two other terms come from the potential energy loss upon removal of this pair. Next, show that the excitation energy to break this pair, by having only one of the two single-particle states with wave-vectors $\pm\mathbf{k}$ occupied, is given by

$$\epsilon_{\mathbf{k}} - E_{\mathbf{k}}^{pair}$$

   where $\epsilon_{\mathbf{k}}$ is the energy associated with occupation of these single-particle states. Finally, using the expressions for $u_{\mathbf{k}}, v_{\mathbf{k}}, \Delta_{\mathbf{k}}$ in terms of $\theta_{\mathbf{k}}$ and $w_{\mathbf{k}}$, and the BCS solution Eq. (8.16), show that this excitation energy is equal to $\zeta_{\mathbf{k}}$, as defined in Eq. (8.18).

3. The BCS model consists of the following assumptions, in the context of BCS theory and Cooper pairs. A pair consists of two electrons with opposite spins and wave-vectors $\mathbf{k}, -\mathbf{k}$, with the single-particle energy $\epsilon_{\mathbf{k}}$ lying within a shell of $\pm\hbar\omega_{\rm D}$ around the Fermi level $\epsilon_{\rm F}$. The gap $\Delta_{\mathbf{k}}$ and the effective interaction potential $V_{\mathbf{kk}'}$ are considered independent of $\mathbf{k}$ and are set equal to the constant values Δ and $-V_0$ (with V_0 and $\Delta > 0$). The density of single-particle states is considered constant in the range $\epsilon_{\rm F} - \hbar\omega_{\rm D} < \epsilon_{\mathbf{k}} < \epsilon_{\rm F} + \hbar\omega_{\rm D}$ and is set equal to $g_F = g(\epsilon_{\rm F})$. Within these assumptions, and the convention that single-particle energies $\epsilon_{\mathbf{k}}$ are measured with respect to the Fermi level, the summation over wave-vectors $\mathbf{k}$ reduces to:

$$\sum_{\mathbf{k}} \to g_F \int_{-\hbar\omega_{\rm D}}^{\hbar\omega_{\rm D}} {\rm d}\epsilon$$

(a) Show that in this model the BCS gap equation, Eq. (8.20), gives

$$\Delta = \frac{\hbar\omega_{\rm D}}{\sinh(1/g_F V_0)}$$

and that, in the weak-coupling limit $g_F V_0 \ll 1$, this reduces to Eq. (8.21).

(b) Show that in this model the energy difference $E_0^{(s)} - E_0^{(n)}$ between the normal ground state energy given by

$$E_0^{(n)} = \sum_{|\mathbf{k}|<k_{\rm F}} 2\epsilon_{\mathbf{k}}$$

and the superconducting ground state energy $E_0^{(s)}$ given in Eq. (8.22), is always negative. Find the value of this energy difference in the weak-coupling limit and interpret its physical meaning.

4. Starting with the expressions for the energy and the entropy of the BCS superconducting state at finite temperature, Eqs. (8.23), (8.24), and the definition the temperature-dependent gap, Eq. (8.25), use a variational argument on the free energy to prove the finite-temperature gap equation, Eq. (8.26). Then use the BCS model defined in the previous problem, with the gap being a temperature-dependent quantity $\Delta(T)$, to show that:

(a) The gap equation at finite temperature gives

$$\int_0^{\hbar\omega_{\rm D}} \frac{1}{(\epsilon^2 + \Delta^2)^{1/2}} \tanh\left(\frac{(\epsilon^2 + \Delta^2)^{1/2}}{2k_{\rm B}T}\right) {\rm d}\epsilon = \frac{1}{g_F V_0}$$

From this expression prove that the gap $\Delta(T)$ is a monotonically decreasing function of T, which becomes zero at some temperature that we identify as the transition temperature T_c.

(b) Show that for $T = T_c$ in the weak-coupling limit the above equation yields Eq. (8.28). By comparing this with the result at zero temperature, Eq. (8.21), prove the law of corresponding states, Eq. (8.29).

Part II

Defects, non-crystalline solids and finite structures

The crystalline structure of solids forms the basis for our understanding of their properties. Most real systems, however, are not perfect crystals. To begin with, even in those real solids which come close to the definition of a perfect crystal there are a large number of defects. For instance, in the best quality crystals produced with great care for the exacting requirements of high-technology applications, such as Si wafers used to manufacture electronic devices, the concentration of defects is rarely below one per billion; this corresponds to roughly one defect per cube of only 1000 atomic distances on each side! These imperfections play an essential role in determining the electronic and mechanical properties of the real crystal. Moreover, there are many solids whose structure has none of the characteristic symmetries of crystals (amorphous solids or glasses) as well as many interesting finite systems whose structure has some resemblance to crystals but they are not of infinite extent. This second part of the book is concerned with all these types of solids and structures.

A natural way to classify defects in crystalline structure is according to their dimensionality:

(i) Zero-dimensional or "point" defects consist of single atomic sites, or complexes of very few atomic sites, which are not in the proper crystalline positions; examples of point defects are missing atoms called vacancies, or extra atoms called interstitials. Point defects may also consist of crystalline sites occupied by atoms foreign to the host crystal; these are called substitutional impurities.

(ii) One-dimensional defects consist of lines of atomic sites perturbed from their ideal positions; these can extend for distances comparable to the linear dimension of the solid. The linear defects are called dislocations.

(iii) Two-dimensional defects consist of a plane of atomic sites where the crystalline lattice is terminated; this is a true surface of a crystal. Alternatively, a two-dimensional defect may correspond to the intersection of two crystallites (grains); these are called grain boundaries and interfaces.

In the following three chapters we will take up the properties of point defects, line defects and surfaces and interfaces of crystals. Chapters 12 and 13 are devoted to solids that lack crystalline order and to structures which are finite (that is, not of macroscopic extent and therefore not representable as periodic) in one or more dimensions. These structures are becoming increasingly important in practical applications and give rise to a wide variety of interesting physical phenomena.

9

Defects I: point defects

Point defects can be either intrinsic, that is, atoms of the same type as the host crystal but in wrong positions, or extrinsic, that is, atoms of a different type than those of the ideal crystal. The most common intrinsic point defects are the vacancy (a missing atom from a crystalline site) and the interstitial (an extra atom in a non-crystalline site). The most common extrinsic defects are substitutional impurities, that is, atoms which are foreign to the crystal but are situated at crystalline sites, substituting for a regular crystal atom. The presence of defects often distorts significantly the atomic structure of the crystal in the immediate neighborhood of the defects. An illustration of these defects in a two-dimensional square lattice is given in Fig. 9.1; in this figure the types of distortions introduced by various defects are exaggerated to make them obvious. More realistic examples, consisting of the vacancy and interstitial defects in bulk Si, are shown in Fig. 9.2.

9.1 Intrinsic point defects

9.1.1 Energetics and electronic levels

Vacancies are quite common in many crystals, both in close-packed metallic structures as well as in open covalent structures. The creation of a vacancy costs some energy, due to the breaking of bonds, which is called the formation energy $\epsilon_f^{(vac)}$. This is defined as the energy of the crystal containing a vacancy, that is, one atom fewer than the ideal crystal which has N atoms, denoted as $E^{(vac)}(N-1)$, plus the energy of an atom in the ideal crystal, minus the energy of an equal number of atoms in a perfect crystal (with $N \to \infty$ for an infinite crystal):

$$\epsilon_f^{(vac)} = \lim_{N\to\infty} \left[\left(E^{(vac)}(N-1) + \frac{E_0(N)}{N} \right) - E_0(N) \right] \tag{9.1}$$

where $E_0(N)$ is the ground state energy of a crystal consisting of N atoms at ideal positions. Vacancies can move in the crystal, jumping from one crystalline

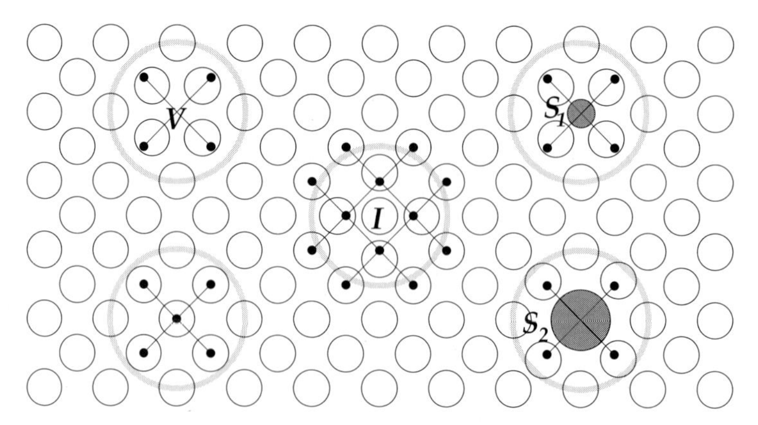

Figure 9.1. Illustration of point defects in a two-dimensional square lattice. V = vacancy, I = interstitial, S_1, S_2 = substitutional impurities of size smaller and larger than the host atoms, respectively. In all cases the defects are surrounded by a faintly drawn circle which includes their nearest neighbors. In the lower left corner an ideal atom and its nearest neighbors are shown for comparison, also surrounded by a faint circle of the same radius. The thin lines and small dots in the neighborhood of defects serve to identify the position of neighboring atoms in the ideal, undistorted crystal.

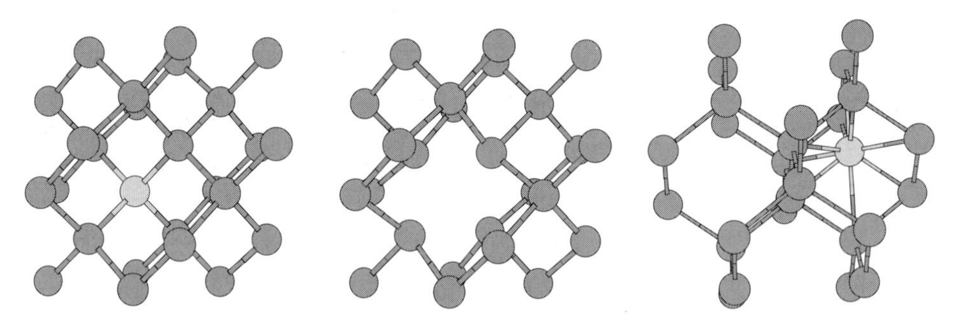

Figure 9.2. Examples of vacancy and interstitial defects in bulk Si. **Left:** the ideal bulk structure of Si, with one of the tetrahedrally bonded atoms highlighted. **Center:** the vacancy configuration, with the highlighted tetrahedrally bonded atom missing (note the relaxation of its four neighbors). **Right:** an interstitial atom placed in an empty region of the diamond lattice; the crystal is viewed at an angle that makes the open spaces evident. The actual lowest energy configuration of the Si interstitial is different and involves a more complicated arrangement of atoms.

site to the next – what actually happens is an atom jumps from one of the sites surrounding the vacancy, into the vacant site. This is a thermally activated process, that has to overcome an energy barrier, called the migration energy $\epsilon_m^{(vac)}$; this energy corresponds to the so called activated configuration, or saddle point, and is given relative to the energy of the equilibrium configuration of the defect, which corresponds to the formation energy $\epsilon_f^{(vac)}$. The concentration of vacancies per site in the crystal at temperature T under equilibrium conditions is determined by the vacancy formation energy, and is given by

$$c^{(vac)}(T) = \exp(-\epsilon_f^{(vac)}/k_B T) \tag{9.2}$$

as can be easily shown by standard statistical mechanics arguments (see Problem 1). Such arguments prove that the concentration of defects is determined by the balance between the entropy gain due to the various ways of arranging the defects in the crystal and the energy cost to introduce each defect.

Interstitial atoms are extra atoms that exist in the crystal at positions that do not coincide with regular crystalline sites. These defects are common in crystals with open structures since in close-packed crystals there simply is not enough room to accommodate extra atoms: it costs a great amount of energy to squeeze an extra atom into a close-packed crystal. The energy to create an interstitial, its formation energy $\epsilon_f^{(int)}$, by analogy to the formation energy of the vacancy, is given by

$$\epsilon_f^{(int)} = \lim_{N\to\infty}\left[\left(E^{(int)}(N+1) - \frac{E_0(N)}{N}\right) - E_0(N)\right] \tag{9.3}$$

where $E^{(int)}(N+1)$ is the energy of the solid containing a total of $N+1$ atoms, N of them at regular crystalline positions plus one at the interstitial position. Interstitials in crystals with open structures can also move, undergoing a thermally activated process by which they are displaced from one stable position to another, either through a direct jump, or by exchanging positions with an atom of the host lattice. By analogy to the vacancy, there is a thermal activation energy associated with the motion of interstitials, called the interstitial migration energy $\epsilon_m^{(int)}$, and the concentration of interstitials per site in the crystal at temperature T under equilibrium conditions is given by

$$c^{(int)}(T) = \exp(-\epsilon_f^{(int)}/k_B T) \tag{9.4}$$

The presence of intrinsic defects in a crystal has an important consequence: it introduces electronic states in the solid beyond those of the ideal crystal. If the energy of such defect-related states happens to be in the same range of energy as that of states of the perfect crystal, the presence of the defect does not make a big difference to the electronic properties of the solid. However, when the defect states happen to have energy different than that of the crystal states, for example, in the band gap of semiconductors, their effect can be significant. To illustrate this point, consider the vacancy in Si: the presence of the vacancy introduces four broken bonds (called "dangling bonds"), the energy of which is in the middle of the band gap, that is, outside the range of energy that corresponds to crystal states. By counting the available electrons, it becomes clear that each dangling bond state is occupied by one electron, since two electrons are required to form a covalent bond. If the

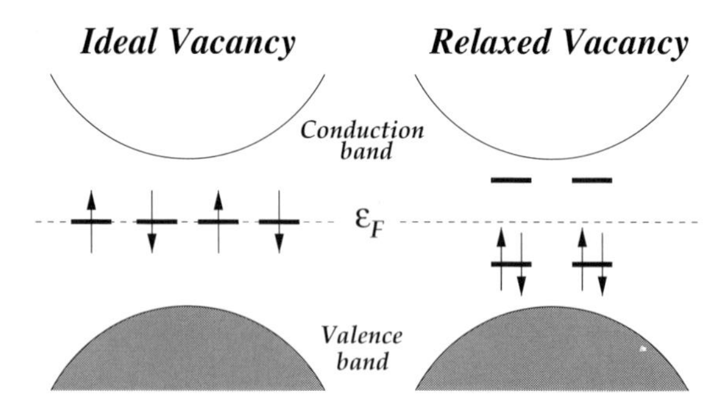

Figure 9.3. Schematic representation of electronic states associated with the Si vacancy. The shaded regions represent the conduction and valence states of the bulk. **Left:** the vacancy configuration before any relaxation of the neighbors; each broken bond contains one electron denoted by an up or a down arrow for the spin states, and all levels are degenerate, coincident with the Fermi level which is indicated by a dashed line. **Right:** the reconstructed vacancy, with pairs of broken bonds forming bonding and antibonding states; the former are fully occupied, the latter empty, and the two sets of states are separated by a small gap.

neighbors of the vacant site are not allowed to distort from their crystalline positions, symmetry requires that the four dangling bond states are degenerate in energy, and are each half filled (every electronic level can accommodate two electrons with opposite spin). The actual physical situation is interesting: the four dangling bonds combine in pairs, forming two bonding and two antibonding combinations. The four electrons end up occupying the two bonding states, leaving the antibonding ones empty. This is achieved by a slight distortion in the position of the immediate neighbors of the vacancy, which move in pairs closer to each other as shown in Fig. 9.2. The net effect is to produce a more stable structure, with lower energy and a new band gap between the occupied bonding and unoccupied antibonding states, as illustrated schematically in Fig. 9.3. This is called a Jahn–Teller distortion; it is common in defect configurations when the undistorted defect structure corresponds to partially occupied, degenerate electronic states.

9.1.2 Defect-mediated diffusion

One of the most important effects of intrinsic point defects is enhanced atomic diffusion. Vacancies and interstitials can move much more easily through the crystal than atoms residing at ideal lattice sites in the absence of any defects; the latter are fully bonded to their neighbors by strong bonds in the case of covalent or ionic solids, or have very little room to move in the case of close-packed metallic solids. An illustration of the motion of point defects in a two-dimensional square lattice is shown in Fig. 9.4. Diffusion is described in terms of the changes in the *local concentration* $c(\mathbf{r}, t)$ with respect to time t due to the current

$$\mathbf{j}(\mathbf{r}, t) = -\nabla_{\mathbf{r}} c(\mathbf{r}, t)$$

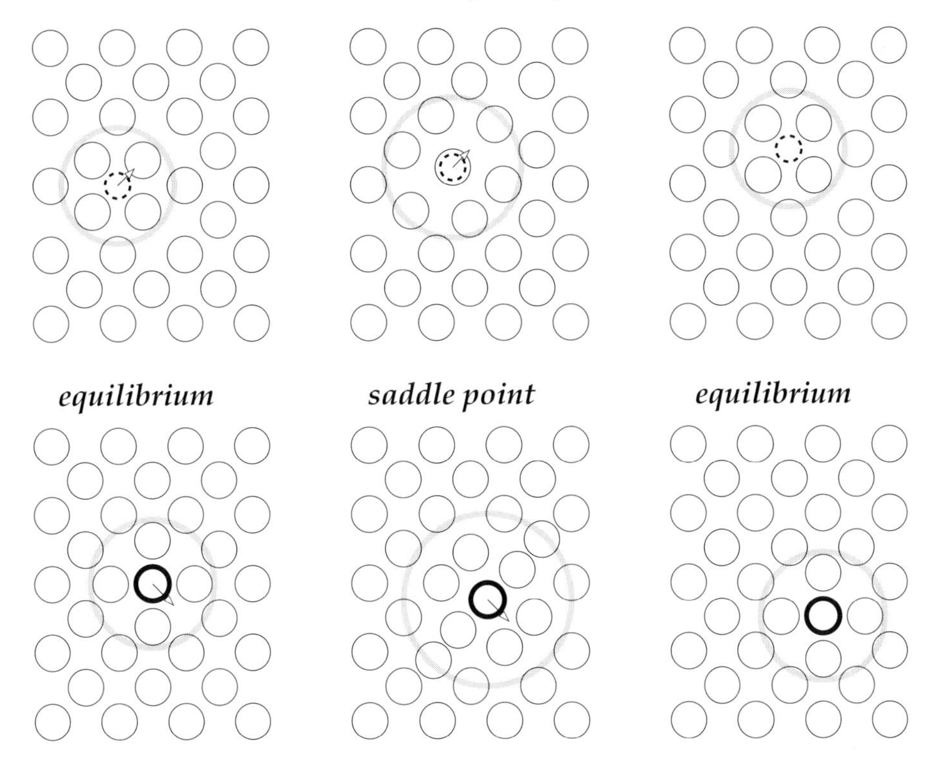

Figure 9.4. Illustration of motion of point defects in a two-dimensional square lattice. **Top:** the migration of the vacancy (indicated by a smaller circle in dashed line) between two equilibrium configurations. **Bottom:** the migration of the interstitial (indicated by a circle in thicker line) between two equilibrium configurations. Notice that in the case of the vacancy what is actually moving is one of its neighbors into the vacant site; the energy cost of breaking the bonds of this moving atom corresponds to the migration energy. The interstitial must displace some of its neighbors significantly in order to move; the energy cost of this distortion corresponds to the migration energy. The larger dashed circles indicate the region which is significantly distorted due to the defect; in both cases, the saddle point configuration induces larger distortions.

flowing into or out of a volume element. The mass conservation equation applied to this situation gives

$$\frac{\partial c(\mathbf{r}, t)}{\partial t} + D(T)\mathbf{j}(\mathbf{r}, t) = 0 \Longrightarrow \frac{\partial c(\mathbf{r}, t)}{\partial t} = D(T)\nabla^2_{\mathbf{r}} c(\mathbf{r}, t) \tag{9.5}$$

The last equation is known as Fick's law. $D(T)$ is the diffusion constant; it has dimensions $[\text{length}]^2[\text{time}]^{-1}$ and depends on the temperature T. If motion of vacancies and interstitials were the only possible means for diffusion in a solid, then the diffusion constant would be

$$D(T) = D^{(vac)}(T)\exp[-\epsilon_m^{(vac)}/k_{\mathrm{B}}T] + D^{(int)}(T)\exp[-\epsilon_m^{(int)}/k_{\mathrm{B}}T] \tag{9.6}$$

The pre-exponentials for the vacancy and the interstitial mechanisms contain the concentrations per site $c^{(vac)}(T)$ or $c^{(int)}(T)$, as well as three additional factors: the attempt frequency $\nu^{(vac)}$ or $\nu^{(int)}$, that is, the number of attempts to jump out of the stable configuration per unit time; the hopping length $a^{(vac)}$ or $a^{(int)}$, that is, the distance by which an atom moves in the elementary hop associated with the defect mechanism; and a geometrical factor, $f^{(vac)}$ or $f^{(int)}$, related to the defect structure which accounts for the different ways of moving from the intitial to the final configuration during an elementary hop. In the most general case, the diffusion constant is given by

$$D = \sum_n f^{(n)} \left(a^{(n)}\right)^2 \nu^{(n)} \mathrm{e}^{\Delta S^{(n)}/k_{\mathrm{B}}} \mathrm{e}^{-\epsilon_a^{(n)}/k_{\mathrm{B}}T} \tag{9.7}$$

where the sum runs over all contributing mechanisms n, $f^{(n)}$ is the geometric factor, $a^{(n)}$ is the hopping length, $\nu^{(n)}$ is the attempt frequency, and $\Delta S^{(n)}$, $\epsilon_a^{(n)}$ are the entropy change and activation energy associated with a particular mechanism. The activation energy is the sum of formation and migration energies:

$$\epsilon_a^{(n)} = \epsilon_f^{(n)} + \epsilon_m^{(n)} \tag{9.8}$$

The attempt frequency $\nu^{(n)}$ and entropy term $\exp(\Delta S^{(n)}/k_{\mathrm{B}})$ give the rate of the mechanism. Within classical thermodynamics this rate can be calculated through Vineyard's Transition State Theory (TST) [116]:

$$\Gamma^{(n)} \equiv \nu^{(n)} \mathrm{e}^{\Delta S^{(n)}/k_{\mathrm{B}}} = \sqrt{\frac{k_{\mathrm{B}}T}{2\pi}} \left[\int \mathrm{e}^{-\tilde{E}^{(n)}/k_{\mathrm{B}}T} \mathrm{d}A^{(n)}\right] \left[\int \mathrm{e}^{-E^{(n)}/k_{\mathrm{B}}T} \mathrm{d}\Omega^{(n)}\right]^{-1} \tag{9.9}$$

where the first integral is over a $(d-1)$-dimensional area $A^{(n)}$ called the "saddle point surface" and the second integral is over a d-dimensional volume $\Omega^{(n)}$ around the equilibrium configuration. In these two integrals, the spatial variables have been scaled by the square roots of the corresponding reduced masses:

$$\begin{aligned} \mathrm{d}\Omega^{(n)} &= \mathrm{d}\left(\sqrt{m_1^{(n)}}x_1\right) \mathrm{d}\left(\sqrt{m_2^{(n)}}x_2\right) \cdots \mathrm{d}\left(\sqrt{m_d^{(n)}}x_d\right) \\ \mathrm{d}A^{(n)} &= \mathrm{d}\left(\sqrt{\tilde{m}_1^{(n)}}\tilde{x}_1\right) \mathrm{d}\left(\sqrt{\tilde{m}_2^{(n)}}\tilde{x}_2\right) \cdots \mathrm{d}\left(\sqrt{\tilde{m}_{d-1}^{(n)}}\tilde{x}_{d-1}\right) \end{aligned} \tag{9.10}$$

The dimensionality of these two spaces is determined by the number of independent degrees of freedom involved in the diffusion mechanism; for example, if there are N atoms involved in the mechanism, then $d = 3N$. The energy $E^{(n)}$ is measured from the ground state energy, which is taken to define the zero of the energy scale, so that $E^{(n)}$ is always positive. The energy $\tilde{E}^{(n)}$ is measured from the activated state energy $\epsilon_a^{(n)}$, that is, $\tilde{E}^{(n)} = E^{(n)} - \epsilon_a^{(n)}$ (see Fig. 9.5); the saddle point surface passes through the saddle point configuration and separates the two stable configurations of the mechanism. This surface is locally perpendicular to constant energy contours,

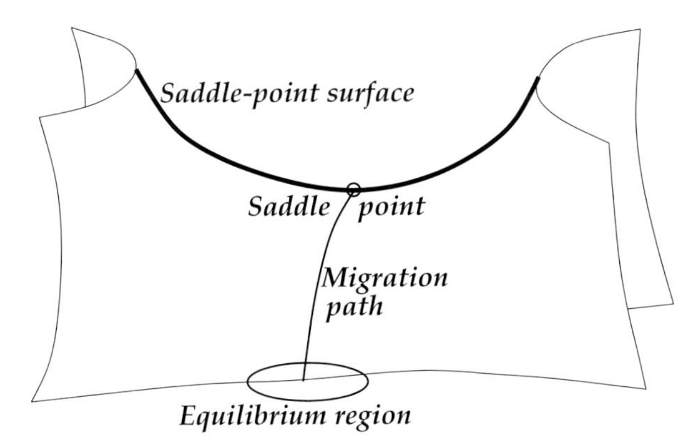

Figure 9.5. Energy surface corresponding to an activated diffusion mechanism. The equilibrium region, the saddle point configuration, the saddle point surface and the migration path are indicated.

so that the energy $\tilde{E}^{(n)}$ is always positive. The rate $\Gamma^{(n)}$ given by Eq. (9.9) is based on a calculation of the current of trajectories that depart from the equilibrium configuration and reach the saddle point with just enough velocity to cross to the other side of the saddle point surface. The first integral in the TST rate is a measure of the successful crossings of the saddle point surface, while the second is a normalization by the number of attempts of the system to leave the equilibrium configuration. One of the limitations of TST theory is that it does not take into account the backflow of trajectories, which in certain systems can be substantial.

In many situations it is convenient and adequately accurate to use the harmonic approximation to TST theory: this involves an expansion of the energy over all the normal modes of the system near the equilibrium and the saddle point configuration of the mechanism:

$$E^{(n)} \approx \frac{1}{2}\sum_{i=1}^{d} m_i^{(n)}(2\pi \nu_i^{(n)})^2 (x_i^{(n)})^2$$
$$\tilde{E}^{(n)} \approx \frac{1}{2}\sum_{i=1}^{d-1} \tilde{m}_i^{(n)}(2\pi \tilde{\nu}_i^{(n)})^2 (\tilde{x}_i^{(n)})^2 \tag{9.11}$$

With these expressions, the contribution of mechanism n to the diffusion constant becomes

$$D^{(n)} = f^{(n)}\left(a^{(n)}\right)^2 \left[\prod_{i=1}^{d} \nu_i^{(n)}\right]\left[\prod_{i=1}^{d-1} \tilde{\nu}_i^{(n)}\right]^{-1} \mathrm{e}^{-\epsilon_a^{(n)}/k_\mathrm{B}T} \tag{9.12}$$

This form of the diffusion constant helps elucidate the relevant importance of the various contributions. We note first that the geometric factor is typically a constant of order unity, while the hopping length is fixed by structural considerations (for example, in the case of the vacancy it is simply equal to the distance between two nearest neighbor sites in the ideal crystal). The remaining factors are related to the features of the energy surface associated with a particular mechanism. The dominant feature of the energy surface, which essentially determines its contribution, is evidently the activation energy $\epsilon_a^{(n)}$ which enters exponentially in the expression for the diffusion constant. To illustrate how the other features of the energy surface affect the diffusion rate, we consider a very simple example of a mechanism where there are only two relevant degrees of freedom; in this case, the expression for the diffusion constant reduces to

$$D = fa^2 \frac{\nu_1 \nu_2}{\tilde{\nu}_1} \mathrm{e}^{-\epsilon_a / k_{\mathrm{B}} T}$$

In this expression we identify ν_2 as the attempt frequency and we relate the ratio $\nu_1/\tilde{\nu}_1$ to the change in entropy through

$$\frac{\nu_1}{\tilde{\nu}_1} = \mathrm{e}^{\Delta S / k_{\mathrm{B}}} \Rightarrow \Delta S = k_{\mathrm{B}} \ln \left(\frac{\nu_1}{\tilde{\nu}_1} \right)$$

From this example we see that, for a given activation energy ϵ_a, diffusion is enhanced when the curvature of the energy at the saddle point is much smaller than the curvature in the corresponding direction around the equilibrium configuration, that is, when $\tilde{\nu}_1 \ll \nu_1$ which implies a large entropy ΔS. Equally important is the value of the attempt frequency ν_2: diffusion is enhanced if the curvature of the energy surface in the direction of the migration path is large, giving rise to a large attempt frequency.

When the measured diffusion constant is plotted on a logarithmic scale against $1/T$, the slope of the resulting line identifies the activation energy of the diffusion mechanism; this is known as an Arrhenius plot. Often, a single atomistic process has an activation energy $\epsilon_a^{(n)}$ compatible with experimental measurements and is considered the dominant diffusion mechanism. Thus, when trying to identify the microscopic processes responsible for diffusion, the first consideration is to determine whether or not a proposed mechanism has an activation energy compatible with experimental measurements. This issue can be established reasonably well in many cases of bulk and surface diffusion in solids, through the methods discussed in chapter 5 for calculating the total energy of various atomic configurations. If several hypothetical mechanisms have activation energies that cannot be distinguished within the experimental uncertainty, then the expression for the pre-exponential factor Eq. (9.12) can help identify the dominant processes: the ones with a very flat saddle point surface and a very steep equilibrium well, and hence low frequencies $\tilde{\nu}_i^{(n)}$ and high frequencies $\nu_i^{(n)}$, will give the largest contribution to diffusion (for an application of this approach see, for example, Ref. [117]). When the harmonic approximation is not appropriate, the entire energy landscape, that is, the energy variation with respect to defect motion near the equilibrium configuration and near the saddle point configuration, is needed to calculate the rate $\Gamma^{(n)}$ from Eq. (9.9). It is also possible that more sophisticated theories than TST are required, especially when backflow across the saddle point surface is important, but such situations tend to be rare in the context of diffusion in solids.

Vacancies, interstitials and point defects in general introduce significant distortions to the crystal positions in their neighborhood, both in the equilibrium configuration and in the activated configuration. These distortions can be revealed experimentally through the changes they induce on the volume of the crystal and their effect on the diffusion constant, using hydrostatic or non-hydrostatic pressure experiments. This necessitates the generalization of the above expressions for the diffusion constant to the Helmholtz free energy $F = E - TS + P\Omega$, with P the external pressure.[1] Experimental measurements of the effect of pressure on the diffusion rate can help identify the type of defect that contributes most to diffusion, and the manner in which it moves in the crystal through the distortions it induces to the lattice [118].

9.2 Extrinsic point defects

The most common type of extrinsic point defect is a substitutional impurity, that is, a foreign atom that takes the place of a regular crystal atom at a crystal site. When the natural size of the impurity is compatible with the volume it occupies in the host crystal (i.e. the distance between the impurity and its nearest neighbors in the crystal is not too different from its usual bond length), the substitution results in a stable structure with interesting physical properties. In certain cases, the impurity behaves essentially like an isolated atom in an external field possessing the point group symmetries of the lattice. This external field usually induces splitting of levels that would normally be degenerate in a free atom. The impurity levels are in general different than the host crystal levels. Electronic transitions between impurity levels, through absorption or emission of photons, reveal a signature that is characteristic of, and therefore can uniquely identify, the impurity. Such transitions are often responsible for the color of crystals that otherwise would be transparent. This type of impurity is called a color center.

9.2.1 Impurity states in semiconductors

A common application of extrinsic point defects is the doping of semiconductor crystals. This is a physical process of crucial importance to the operation of modern

[1] The pressure–volume term is replaced by a stress–strain term in the case of non-hydrostatic pressure.

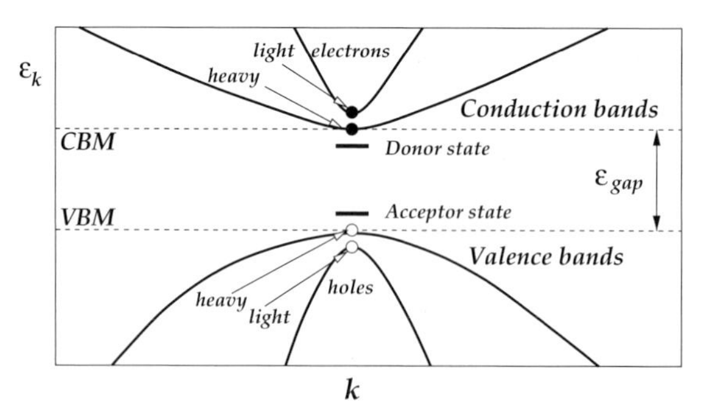

Figure 9.6. Schematic illustration of shallow donor and acceptor impurity states in a semiconductor with direct gap. The light and heavy electron and hole states are also indicated; the bands corresponding to the light and heavy masses are split in energy for clarity.

electronic devices, so we discuss its basic aspects next. We begin with a general discussion of the nature of impurity states in semiconductors. The energy of states introduced by the impurities, which are relevant to doping, lies within the band gap of the semiconductor. If the energy of the impurity-related states is near the middle of the band gap, these are called "deep" states; if it lies near the band extrema (Valence Band Maximum –VBM– or Conduction Band Minimum –CBM) they are called "shallow" states. It is the latter type of state that is extremely useful for practical applications, so we will assume from now on that the impurity-related states of interest have energies close to the band extrema. If the impurity atom has more valence electrons than the host atom, in which case it is called a donor, these extra electrons are in states near the CBM. In the opposite case, the impurity creates states near the VBM, which when occupied, leave empty states in the VBM of the crystal called "holes"; this type of impurity is called an acceptor. These features are illustrated in Fig. 9.6.

One important aspect of the impurity-related states is the effective mass of charge carriers (extra electrons or holes): due to band-structure effects, these can have an effective mass which is smaller or larger than the electron mass, described as "light" or "heavy" electrons or holes. We recall from our earlier discussion of the behavior of electrons in a periodic potential (chapter 3) that an electron in a crystal has an inverse effective mass which is a tensor, written as $[(\bar{m})^{-1}]_{\alpha\beta}$. The inverse effective mass depends on matrix elements of the momentum operator at the point of the BZ where it is calculated. At a given point in the BZ we can always identify the principal axes which make the inverse effective mass a diagonal tensor, i.e. $[(\bar{m})^{-1}]_{\alpha\beta} = \delta_{\alpha\beta}(1/\bar{m}_\alpha)$, as discussed in chapter 3. The energy of electronic states around this point will be given by quadratic expressions in $\mathbf{k}$, from $\mathbf{q}\cdot\mathbf{p}$ perturbation theory (see chapter 3). Applying this to the VBM and the CBM of a semiconductor yields, for the energy of states near the highest valence ($\epsilon^{(v)}_{\mathbf{k}}$) and the lowest conduction ($\epsilon_{\mathbf{k}}^{(c)}$) states, respectively,

$$\epsilon_{\mathbf{k}}^{(v)} = \epsilon_{vbm} + \frac{\hbar^2}{2}\left(\frac{k_1^2}{\bar{m}_{v1}} + \frac{k_2^2}{\bar{m}_{v2}} + \frac{k_3^2}{\bar{m}_{v3}}\right)$$

$$\epsilon_{\mathbf{k}}^{(c)} = \epsilon_{cbm} + \frac{\hbar^2}{2}\left(\frac{k_1^2}{\bar{m}_{c1}} + \frac{k_2^2}{\bar{m}_{c2}} + \frac{k_3^2}{\bar{m}_{c3}}\right) \tag{9.13}$$

where the effective masses for the valence states $\bar{m}_{v1}, \bar{m}_{v2}, \bar{m}_{v3}$ are negative, and those for the conduction states $\bar{m}_{c1}, \bar{m}_{c2}, \bar{m}_{c3}$ are positive.

We will assume the impurity has an effective charge $\bar{Z}^{(I)}$ which is the difference between the valence of the impurity and the valence of the crystal atom it replaces; with this definition $\bar{Z}^{(I)}$ can be positive or negative. The potential that the presence of the impurity creates is then

$$V^{(I)}(\mathbf{r}) = -\frac{\bar{Z}^{(I)}e^2}{\varepsilon r} \tag{9.14}$$

where ε is the dielectric constant of the solid. We expect that in such a potential, the impurity-related states will be similar to those of an atom with effective nuclear charge $\bar{Z}^{(I)}e/\varepsilon$. In order to examine the behavior of the extra electrons or the holes associated with the presence of donor or acceptor dopants in the crystal, we will employ what is referred to as "effective-mass theory" [119]. We will consider the potential introduced by the impurity to be a small perturbation to the crystal potential, and use the eigenfunctions of the perfect crystal as the basis for expanding the perturbed wavefunctions of the extra electrons or the holes. In order to keep the notation simple, we will assume that we are dealing with a situation where there is only one band in the host crystal, the generalization to many bands being straightforward. The wavefunction of the impurity state will then be given by

$$\psi^{(I)}(\mathbf{r}) = \sum_{\mathbf{k}} \beta_{\mathbf{k}} \psi_{\mathbf{k}}(\mathbf{r}) \tag{9.15}$$

where $\psi_{\mathbf{k}}(\mathbf{r})$ are the relevant crystal wavefunctions (conduction states for the extra electrons, valence states for the holes); we elaborate below on what are the relevant values of the index $\mathbf{k}$. This wavefunction will obey the single-particle equation

$$\left[\mathcal{H}_0^{sp} + V^{(I)}(\mathbf{r})\right] \psi^{(I)}(\mathbf{r}) = \epsilon^{(I)} \psi^{(I)}(\mathbf{r}) \tag{9.16}$$

with $\mathcal{H}_0^{sp}$ the single-particle hamiltonian for the ideal crystal, which of course, when applied to the crystal wavefunctions $\psi_{\mathbf{k}}(\mathbf{r})$, gives

$$\mathcal{H}_0^{sp} \psi_{\mathbf{k}}(\mathbf{r}) = \epsilon_{\mathbf{k}} \psi_{\mathbf{k}}(\mathbf{r}) \tag{9.17}$$

Substituting the expansion (9.15) into Eq. (9.16) and using the orthogonality of the crystal wavefunctions we obtain

$$\epsilon_{\mathbf{k}}\beta_{\mathbf{k}} + \sum_{\mathbf{k}'} \langle \psi_{\mathbf{k}} | V^{(I)} | \psi_{\mathbf{k}'} \rangle \beta_{\mathbf{k}'} = \epsilon^{(I)} \beta_{\mathbf{k}} \tag{9.18}$$

If we express the crystal wavefunctions in the familiar Bloch form,

$$\psi_{\mathbf{k}}(\mathbf{r}) = \mathrm{e}^{\mathrm{i}\mathbf{k}\cdot\mathbf{r}} \sum_{\mathbf{G}} \alpha_{\mathbf{k}}(\mathbf{G}) \mathrm{e}^{\mathrm{i}\mathbf{G}\cdot\mathbf{r}} \tag{9.19}$$

we can obtain the following expression for the matrix elements of the impurity potential that appear in Eq. (9.18):

$$\langle \psi_{\mathbf{k}} | V^{(I)} | \psi_{\mathbf{k}'} \rangle = \sum_{\mathbf{G}} \gamma_{\mathbf{k}\mathbf{k}'}(\mathbf{G}) V^{(I)}(\mathbf{G} - \mathbf{k} + \mathbf{k}') \tag{9.20}$$

where we have defined the quantity $\gamma_{\mathbf{k}\mathbf{k}'}(\mathbf{G})$ as

$$\gamma_{\mathbf{k}\mathbf{k}'}(\mathbf{G}) = \sum_{\mathbf{G}'} \alpha^*_{\mathbf{k}}(\mathbf{G}') \alpha_{\mathbf{k}'}(\mathbf{G}' + \mathbf{G}) \tag{9.21}$$

and $V^{(I)}(\mathbf{G})$ is the Fourier transform of the impurity potential,

$$V^{(I)}(\mathbf{G}) = \int V^{(I)}(\mathbf{r}) \mathrm{e}^{\mathrm{i}\mathbf{G}\cdot\mathbf{r}} \mathrm{d}\mathbf{r} \tag{9.22}$$

We will assume that the impurity potential is a very smooth one, which implies that its Fourier transform components fall very fast with the magnitude of the reciprocal-space vectors, and therefore the only relevant matrix elements are those for $\mathbf{G} = 0$ and $\mathbf{k} \approx \mathbf{k}'$. We will also assume that the expansion of the impurity wavefunction onto crystal wavefunctions needs only include wave-vectors near a band extremum at $\mathbf{k}_0$ (the VBM for holes or the CBM for extra electrons). But for $\mathbf{G} = 0$ and $\mathbf{k} = \mathbf{k}' = \mathbf{k}_0$, $\gamma_{\mathbf{k}\mathbf{k}'}(\mathbf{G})$ becomes

$$\gamma_{\mathbf{k}_0\mathbf{k}_0}(0) = \sum_{\mathbf{G}'} \alpha^*_{\mathbf{k}_0}(\mathbf{G}') \alpha_{\mathbf{k}_0}(\mathbf{G}') = \sum_{\mathbf{G}'} |\alpha_{\mathbf{k}_0}(\mathbf{G}')|^2 = 1 \tag{9.23}$$

for properly normalized crystal wavefunctions. Using these results, we arrive at the following equation for the coefficients $\beta_{\mathbf{k}}$:

$$\epsilon_{\mathbf{k}}\beta_{\mathbf{k}} + \sum_{\mathbf{k}'} V^{(I)}(\mathbf{k}' - \mathbf{k}) \beta_{\mathbf{k}'} = \epsilon^{(I)} \beta_{\mathbf{k}} \tag{9.24}$$

Rather than trying to solve this equation for each coefficient $\beta_{\mathbf{k}}$, we will try to construct a linear combination of these coefficients which embodies the important physics. To this end, we rewrite the wavefunction for the impurity state in Eq. (9.15) using the expressions for the crystal Bloch functions that involve the periodic functions $u_{\mathbf{k}}(\mathbf{r})$:

$$\psi^{(I)}(\mathbf{r}) = \sum_{\mathbf{k}} \beta_{\mathbf{k}} \mathrm{e}^{\mathrm{i}\mathbf{k}\cdot\mathbf{r}} u_{\mathbf{k}}(\mathbf{r})$$

Since the summation extends only over values of $\mathbf{k}$ close to $\mathbf{k}_0$, and for these values $u_{\mathbf{k}}(\mathbf{r})$ does not change significantly, we can take it as a common factor outside the summation; this leads to

$$\psi^{(I)}(\mathbf{r}) \approx u_{\mathbf{k}_0}(\mathbf{r}) \sum_{\mathbf{k}} \beta_{\mathbf{k}} \mathrm{e}^{\mathrm{i}\mathbf{k}\cdot\mathbf{r}} = \psi_{\mathbf{k}_0}(\mathbf{r}) \sum_{\mathbf{k}} \beta_{\mathbf{k}} \mathrm{e}^{\mathrm{i}(\mathbf{k}-\mathbf{k}_0)\cdot\mathbf{r}} = \psi_{\mathbf{k}_0}(\mathbf{r}) f(\mathbf{r})$$

where in the last expression we have defined the so called "envelope function",

$$f(\mathbf{r}) = \sum_{\mathbf{k}} \beta_{\mathbf{k}} \mathrm{e}^{\mathrm{i}(\mathbf{k}-\mathbf{k}_0)\cdot\mathbf{r}} \tag{9.25}$$

This expression is the linear combination of the coefficients $\beta_{\mathbf{k}}$ that we were seeking: since $\mathbf{k}$ is restricted to the neighborhood of $\mathbf{k}_0$, $f(\mathbf{r})$ is a smooth function because it involves only Fourier components of small magnitude $|\mathbf{k} - \mathbf{k}_0|$. We see from this analysis that the impurity state $\psi^{(I)}(\mathbf{r})$ is described well by the corresponding crystal state $\psi_{\mathbf{k}_0}(\mathbf{r})$, modified by the envelope function $f(\mathbf{r})$. Our goal then is to derive the equation which determines $f(\mathbf{r})$. If we multiply both sides of Eq. (9.24) by $\exp[\mathrm{i}(\mathbf{k} - \mathbf{k}_0) \cdot \mathbf{r}]$ and sum over $\mathbf{k}$ in order to create the expression for $f(\mathbf{r})$ on the right-hand side, we obtain

$$\sum_{\mathbf{k}} \epsilon_{\mathbf{k}} \beta_{\mathbf{k}} \mathrm{e}^{\mathrm{i}(\mathbf{k}-\mathbf{k}_0)\cdot\mathbf{r}} + \sum_{\mathbf{k},\mathbf{k}'} V^{(I)}(\mathbf{k}' - \mathbf{k}) \mathrm{e}^{\mathrm{i}(\mathbf{k}-\mathbf{k}_0)\cdot\mathbf{r}} \beta_{\mathbf{k}'} = \epsilon^{(I)} f(\mathbf{r}) \tag{9.26}$$

We rewrite the second term on the left-hand side of this equation as

$$\sum_{\mathbf{k}'} \left[\sum_{\mathbf{k}} V^{(I)}(\mathbf{k}' - \mathbf{k}) \mathrm{e}^{-\mathrm{i}(\mathbf{k}'-\mathbf{k})\cdot\mathbf{r}} \right] \mathrm{e}^{\mathrm{i}(\mathbf{k}'-\mathbf{k}_0)\cdot\mathbf{r}} \beta_{\mathbf{k}'}$$

In the square bracket of this expression we recognize the familiar sum of the inverse Fourier transform (see Appendix G). In order to have a proper inverse Fourier transform, the sum over $\mathbf{k}$ must extend over all values in the BZ, whereas in our discussion we have restricted $\mathbf{k}$ to the neighborhood of $\mathbf{k}_0$. However, since the potential $V^{(I)}(\mathbf{k})$ is assumed to vanish for large values of $|\mathbf{k}|$ we can extend the summation in the above equation to all values of $\mathbf{k}$ without any loss of accuracy. With this extension we find

$$\sum_{\mathbf{k}} V^{(I)}(\mathbf{k}' - \mathbf{k}) \mathrm{e}^{-\mathrm{i}(\mathbf{k}'-\mathbf{k})\cdot\mathbf{r}} = V^{(I)}(\mathbf{r}) \tag{9.27}$$

and with this, Eq. (9.26) takes the form

$$\sum_{\mathbf{k}} \epsilon_{\mathbf{k}} \beta_{\mathbf{k}} \mathrm{e}^{\mathrm{i}(\mathbf{k}-\mathbf{k}_0)\cdot\mathbf{r}} + V^{(I)}(\mathbf{r}) f(\mathbf{r}) = \epsilon^{(I)} f(\mathbf{r}) \tag{9.28}$$

We next use the expansion of Eq. (9.13) for the energy of crystal states near the band extremum,

$$\epsilon_{\mathbf{k}} = \epsilon_{\mathbf{k}_0} + \sum_i \frac{\hbar^2}{2\bar{m}_i}(k_i - k_{0,i})^2 \tag{9.29}$$

with i identifying the principal directions along which the effective mass tensor is diagonal with elements $\bar{m}_i$; with the help of this expansion, the equation for the coefficients $\beta_{\mathbf{k}}$ takes the form

$$\sum_{\mathbf{k}} \sum_i \frac{\hbar^2}{2\bar{m}_i}(k_i - k_{0,i})^2 \beta_{\mathbf{k}} \mathrm{e}^{\mathrm{i}(\mathbf{k}-\mathbf{k}_0)\cdot\mathbf{r}} + V^{(I)}(\mathbf{r}) f(\mathbf{r}) = \left[\epsilon^{(I)} - \epsilon_{\mathbf{k}_0}\right] f(\mathbf{r}) \tag{9.30}$$

Finally, using the identity

$$(k_i - k_{0,i})^2 \mathrm{e}^{\mathrm{i}(\mathbf{k}-\mathbf{k}_0)\cdot\mathbf{r}} = -\frac{\partial^2}{\partial x_i^2} \mathrm{e}^{\mathrm{i}(\mathbf{k}-\mathbf{k}_0)\cdot\mathbf{r}} \tag{9.31}$$

we arrive at the following equation for the function $f(\mathbf{r})$:

$$\left[\sum_i -\frac{\hbar^2}{2\bar{m}_i}\frac{\partial^2}{\partial x_i^2} + V^{(I)}(\mathbf{r})\right] f(\mathbf{r}) = \left[\epsilon^{(I)} - \epsilon_{\mathbf{k}_0}\right] f(\mathbf{r}) \tag{9.32}$$

This equation is equivalent to a Schrödinger equation for an atom with a nuclear potential $V^{(I)}(\mathbf{r})$, precisely the type of equation we were anticipating; in this case, not only is the effective charge of the nucleus modified by the dielectric constant of the crystal, but the effective mass of the particles $\bar{m}_i$ also bear the signature of the presence of the crystal.

With the impurity potential $V^{(I)}(\mathbf{r})$ given by Eq. (9.14), and assuming for simplicity an average effective mass $\bar{m}$ for all directions i (an isotropic crystal), the equation for the envelope function takes the form

$$\left[-\frac{\hbar^2 \nabla_{\mathbf{r}}^2}{2\bar{m}} - \frac{\bar{Z}^{(I)} e^2}{\varepsilon r}\right] f_n(\mathbf{r}) = \left[\epsilon_n^{(I)} - \epsilon_{\mathbf{k}_0}\right] f_n(\mathbf{r}) \tag{9.33}$$

where we have added the index n to allow for different possible solutions to this equation. This is identical to the equation for an electron in an atom with nuclear charge $\bar{Z}^{(I)} e/\varepsilon$. If the charge is negative, i.e. the impurity state is a hole, the effective mass is also negative, so the equation is formally the same as that of an electron associated with a positive ion. Typically the impurity atoms have an effective charge of $\bar{Z}^{(I)} = \pm 1$, so the solutions are hydrogen-like wavefunctions and energy eigenvalues, scaled appropriately. Specifically, the eigenvalues of the extra electron or the hole states are given by

$$\epsilon_n^{(I)} = \epsilon_{\mathbf{k}_0} - \frac{\bar{m}e^4}{2\hbar^2\varepsilon^2 n^2} \tag{9.34}$$

while the corresponding solutions for the envelope function are hydrogen-like wavefunctions with a length scale $\bar{a} = \hbar^2\varepsilon/\bar{m}e^2$; for instance, the wavefunction of the first state is

$$f_1(\mathbf{r}) = A_1 \mathrm{e}^{-r/\bar{a}} \tag{9.35}$$

with A_1 a normalization factor. Thus, the energy of impurity-related states is scaled by a factor

$$\frac{\bar{m}(e^2/\varepsilon)^2}{m_e e^4} = \frac{\bar{m}/m_e}{\varepsilon^2}$$

while the characteristic radius of the wavefunction is scaled by a factor

$$\frac{m_e e^2}{\bar{m}(e^2/\varepsilon)} = \frac{\varepsilon}{\bar{m}/m_e}$$

As is evident from Eq. (9.34), the energy of donor electron states relative to the CBM is negative ($\bar{m}$ being positive in this case), while the energy of holes relative to the VBM is positive ($\bar{m}$ being negative in this case); both of these energy differences, denoted as $\epsilon_n^{(I)}$ in Eq. (9.34), are referred to as the "binding energy" of the electron or hole related to the impurity. Typical values of $\bar{m}/m_e$ in semiconductors are of order 10^{-1} (see Table 9.1), and typical values of ε are of order 10. Using these values, we find that the binding energy of electrons or holes is scaled by a factor of $\sim 10^{-3}$ (see examples in Table 9.1), while the radius of the wavefunction is scaled by a factor of $\sim 10^2$. This indicates that in impurity-related states the electrons or the holes are loosely bound, both in energy and in wavefunction-spread around the impurity atom.

9.2.2 Effect of doping in semiconductors

To appreciate the importance of the results of the preceding discussion, we calculate first the number of conduction electrons or holes that exist in a semiconductor at finite temperature T, in the absence of any dopant impurities. The number of conduction electrons, denoted by $n_c(T)$, will be given by the sum over occupied states with energy above the Fermi level ϵ_F. In a semiconductor, this will be equal

Table 9.1. *Impurity states in semiconductors.*

The effective masses at the CBM and VBM of representative elemental (Si and Ge) and compound semiconductors (GaAs) are given in units of the electron mass m_e, and are distinguished in longitudinal $\bar{m}_L$ and transverse $\bar{m}_T$ states. The energy of the CBM relative to the VBM, given in parentheses, is the band gap. E_b is the binding energy of donor and acceptor states from the band extrema for various impurity atoms, in meV. In the case of compound semiconductors we also indicate which atom is substituted by the impurity.

		$\bar{m}_L/m_e$	$\bar{m}_T/m_e$	E_b	Impurity	$\bar{Z}^{(I)}$
Si:	CBM (1.153 eV)	0.98	0.19	45	P	+1
				54	As	+1
				39	Sb	+1
	VBM	0.52	0.16	46	B	−1
				67	Al	−1
				72	Ga	−1
Ge:	CBM (0.744 eV)	1.60	0.08	12.0	P	+1
				12.7	As	+1
				9.6	Sb	+1
	VBM	0.34	0.04	10.4	B	−1
				10.2	Al	−1
				10.8	Ga	−1
GaAs:	CBM (1.53 eV)	0.066	0.066	6	S/As	+1
				6	Se/As	+1
				30	Te/As	+1
				6	Si/Ga	+1
				6	Sn/Ga	+1
	VBM	0.80	0.12	28	Be/Ga	−1
				28	Mg/Ga	−1
				31	Zn/Ga	−1
				35	Cd/Ga	−1
				35	Si/As	−1

Source: Ref. [73].

to the integral over all conduction states of the density of conduction states $g_c(\epsilon)$ multiplied by the Fermi occupation number at temperature T:

$$n_c(T) = \int_{\epsilon_{cbm}}^{\infty} g_c(\epsilon) \frac{1}{e^{(\epsilon-\epsilon_F)/k_B T} + 1} d\epsilon \tag{9.36}$$

Similarly, the number of holes, denoted by $p_v(T)$, will be equal to the integral over all valence states of the density of valence states $g_v(\epsilon)$ multiplied by one minus the Fermi occupation number:

$$p_v(T) = \int_{-\infty}^{\epsilon_{vbm}} g_v(\epsilon) \left[1 - \frac{1}{e^{(\epsilon-\epsilon_F)/k_B T} + 1}\right] d\epsilon = \int_{-\infty}^{\epsilon_{vbm}} g_v(\epsilon) \frac{1}{e^{(\epsilon_F-\epsilon)/k_B T} + 1} d\epsilon \tag{9.37}$$

In the above equations we have used the traditional symbols n_c for electrons and p_v for holes, which come from the fact that the former represent *negatively* charged carriers (with the subscript c to indicate they are related to conduction bands) and the latter represent *positively* charged carriers (with the subscript v to indicate they are related to valence bands). We have mentioned before (see chapter 5) that in a perfect semiconductor the Fermi level lies in the middle of the band gap, a statement that can be easily proved from the arguments that follow. For the moment, we will use this fact to justify an approximation for the number of electrons or holes. Since the band gap of semiconductors is of order 1 eV and the temperatures of operation are of order 300 K, i.e. $\sim$ 1/40 eV, we can use the approximations

$$\frac{1}{\mathrm{e}^{(\epsilon_{cbm}-\epsilon_{\mathrm{F}})/k_{\mathrm{B}}T}+1} \approx \mathrm{e}^{-(\epsilon_{cbm}-\epsilon_{\mathrm{F}})/k_{\mathrm{B}}T}, \qquad \frac{1}{\mathrm{e}^{(\epsilon_{\mathrm{F}}-\epsilon_{vbm})/k_{\mathrm{B}}T}+1} \approx \mathrm{e}^{-(\epsilon_{\mathrm{F}}-\epsilon_{vbm})/k_{\mathrm{B}}T}$$

These approximations are very good for the values $\epsilon = \epsilon_{cbm}$ and $\epsilon = \epsilon_{vbm}$ in the expressions for the number of electrons and holes, Eqs. (9.36) and (9.37), respectively, and become even better for values larger than ϵ_{cbm} or lower than ϵ_{vbm}, so that we can use them for the entire range of values in each integral. This gives

$$n_c(T) = \bar{n}_c(T)\mathrm{e}^{-(\epsilon_{cbm}-\epsilon_{\mathrm{F}})/k_{\mathrm{B}}T}, \qquad \bar{n}_c(T) = \int_{\epsilon_{cbm}}^{\infty} g_c(\epsilon)\mathrm{e}^{-(\epsilon-\epsilon_{cbm})/k_{\mathrm{B}}T}\mathrm{d}\epsilon$$

$$p_v(T) = \bar{p}_v(T)\mathrm{e}^{-(\epsilon_{\mathrm{F}}-\epsilon_{vbm})/k_{\mathrm{B}}T}, \qquad \bar{p}_v(T) = \int_{-\infty}^{\epsilon_{vbm}} g_v(\epsilon)\mathrm{e}^{-(\epsilon_{vbm}-\epsilon)/k_{\mathrm{B}}T}\mathrm{d}\epsilon$$

where the quantities $\bar{n}_c(T)$ and $\bar{p}_v(T)$ are obtained by extracting from the integrand a factor independent of the energy ϵ; we will refer to these quantities as the reduced number of electrons or holes. The exponential factors in the integrals for $\bar{n}_c(T)$ and $\bar{p}_v(T)$ fall very fast when the values of ϵ are far from ϵ_{cbm} and ϵ_{vbm}; accordingly, only values of ϵ very near these end-points contribute significantly to each integral. We have seen above that near these values we can use $\mathbf{q} \cdot \mathbf{p}$ perturbation theory to express the energy near the CBM and VBM as quadratic expressions in $\mathbf{k}$. Moreover, from our analysis of the density of states near a minimum (such as the CBM) or a maximum (such as the VBM), Eqs. (5.13) and (5.14), we can use for $g_c(\epsilon)$ and $g_v(\epsilon)$ expressions which are proportional to $(\epsilon - \epsilon_{cbm})^{1/2}$ and $(\epsilon_{vbm} - \epsilon)^{1/2}$, respectively. These considerations make it possible to evaluate explicitly the integrals in the above expressions for $\bar{n}_c(T)$ and $\bar{p}_v(T)$, giving

$$\bar{n}_c(T) = \frac{1}{4}\left(\frac{2\bar{m}_c k_{\mathrm{B}}T}{\pi\hbar^2}\right)^{3/2}, \qquad \bar{p}_v(T) = \frac{1}{4}\left(\frac{2\bar{m}_v k_{\mathrm{B}}T}{\pi\hbar^2}\right)^{3/2} \tag{9.38}$$

where $\bar{m}_c$, $\bar{m}_v$ are the appropriate effective masses at the CBM and VBM.

Now we can derive simple expressions for the number of electrons or holes under equilibrium conditions at temperature T in a crystalline semiconductor without dopants. We note that the product of electrons and holes at temperature T is

$$n_c(T)p_v(T) = \frac{1}{16}\left(\frac{2k_{\rm B}T}{\pi\hbar^2}\right)^3 (\bar{m}_c\bar{m}_v)^{3/2}{\rm e}^{-(\epsilon_{cbm}-\epsilon_{vbm})/k_{\rm B}T} \tag{9.39}$$

and the two numbers must be equal because the number of electrons thermally excited to the conduction band is the same as the number of holes left behind in the valence band. Therefore, each number is equal to

$$n_c(T) = p_v(T) = \frac{1}{4}\left(\frac{2k_{\rm B}T}{\pi\hbar^2}\right)^{3/2} (\bar{m}_c\bar{m}_v)^{3/4}{\rm e}^{-\epsilon_{gap}/2k_{\rm B}T} \tag{9.40}$$

where $\epsilon_{gap} = \epsilon_{cbm} - \epsilon_{vbm}$. Thus, the number of electrons or holes is proportional to the factor $\exp(-\epsilon_{gap}/2k_{\rm B}T)$, which for temperatures of order room temperature is extremely small, the gap being of order 1 eV = 11 604 K.

It is now easy to explain why the presence of dopants is so crucial for the operation of real electronic devices. This operation depends on the presence of carriers that can be easily excited and made to flow in the direction of an external electric field (or opposite to it, depending on their sign). For electrons and holes which are due to the presence of impurities, we have seen that the binding energy is of order 10^{-3} of the binding energy of electrons in the hydrogen atom, that is, of order few meV (the binding energy of the $1s$ electron in hydrogen is 13.6 eV). The meaning of binding energy here is the amount by which the energy of the donor-related electron state is below the lowest unoccupied crystal state, the CBM, or the amount by which the energy of the acceptor-related electron state is above the highest occupied crystal state, the VBM. Excitation of electrons from the impurity state to a conduction state gives rise to a delocalized state that can carry current; this excitation is much easier than excitation of electrons from the VBM to the CBM across the entire band gap, in the undoped crystal. Thus, the presence of donor impurities in semiconductors makes it possible to have a reasonable number of thermally excited carriers at room temperature. Similarly, excitation of electrons from the top of the valence band into the acceptor-related state leaves behind a delocalized hole state that can carry current.

It is instructive to calculate the average occupation of donor and acceptor levels in thermal equilibrium. In the following discussion we assume for simplicity that donor and acceptor impurities have an effective charge of $|\bar{Z}^{(I)}| = 1$, i.e. they have one electron more or one electron fewer than the crystal atoms they replace. From statistical mechanics, the average number of electrons in a donor state is

$$\langle n_e\rangle = \frac{\sum_i n_i g_i \exp[-(E_i^{(d)} - n_i\epsilon_{\rm F})/k_{\rm B}T]}{\sum_i g_i \exp[-(E_i^{(d)} - n_i\epsilon_{\rm F})/k_{\rm B}T]} \tag{9.41}$$

where n_i is the number of electrons in state i whose degeneracy and energy are g_i and $E_i^{(d)}$; $\epsilon_{\rm F}$ is the Fermi level, which we take to be the chemical potential for electrons. There are three possible states associated with a donor impurity level: a state with no electrons (degeneracy 1), a state with one electron (degeneracy 2) and a state with two electrons (degeneracy 1). The energy for the no-electron state is $E_0^{(d)} = 0$, while the state with one electron has energy $E_1^{(d)} = \epsilon^{(d)}$, with $\epsilon^{(d)}$ the energy of the donor level. In the case of the two-electron state, an additional term needs to be included to account for the Coulomb repulsion between the two electrons of opposite spin which reside in the same localized state. We call this term $U^{(d)}$, and the energy of the two-electron state is given by $E_2^{(d)} = 2\epsilon^{(d)} + U^{(d)}$. Typically, the Coulomb repulsion is much greater than the difference $(\epsilon^{(d)} - \epsilon_F)$, so that the contribution of the two-electron state to the thermodynamic average is negligible. With these values the average number of electrons in the donor state takes the form

$$\langle n_e \rangle = \frac{2\exp[-(\epsilon^{(d)} - \epsilon_F)/k_B T]}{1 + 2\exp[-(\epsilon^{(d)} - \epsilon_F)/k_B T]} = \frac{1}{\frac{1}{2}\exp[(\epsilon^{(d)} - \epsilon_F)/k_B T] + 1} \tag{9.42}$$

The corresponding argument for an acceptor level indicates that the no-electron state has energy $E_0^{(a)} = U^{(a)}$ due to the repulsion between the two holes (equivalent to no electrons), the one-electron state has energy $E_1^{(a)} = \epsilon^{(a)}$, and the two-electron state has energy $E_2^{(a)} = 2\epsilon^{(a)}$, where $\epsilon^{(a)}$ is the energy of the acceptor level; the degeneracies of these states are the same as those of the donor states. We will again take the Coulomb repulsion energy between the holes to be much larger than the difference $(\epsilon^{(a)} - \epsilon_F)$, so that the no-electron state has negligible contribution to the thermodynamic averages. With these considerations, the average number of holes in the acceptor state takes the form

$$\begin{aligned}\langle p_h \rangle = 2 - \langle n_e \rangle &= 2 - \frac{2\exp[-(\epsilon^{(a)} - \epsilon_F)/k_B T] + 2\exp[-(2\epsilon^{(a)} - 2\epsilon_F)/k_B T]}{2\exp[-(\epsilon^{(a)} - \epsilon_F)/k_B T] + \exp[-(2\epsilon^{(a)} - 2\epsilon_F)/k_B T]} \\ &= \frac{1}{\frac{1}{2}\exp[(\epsilon_F - \epsilon^{(a)})/k_B T] + 1}\end{aligned} \tag{9.43}$$

This expression has a similar appearance to the one for the average number of electrons in the donor state.

Because the above assignments of energy may seem less than obvious, we present a more physical picture of the argument. For the donor state the situation is illustrated in Fig. 9.7: the presence of the donor impurity creates a localized state which lies just below the CBM. We define the zero of the energy scale to correspond to the state with no electrons in the donor state, in which case the valence bands are exactly full as in the perfect crystal. Taking an electron from the reservoir and placing it in the donor state costs energy of an amount $\epsilon^{(d)} - \epsilon_F$, while placing two electrons in it costs twice as much energy plus the Coulomb repulsion energy $U^{(d)}$ experienced

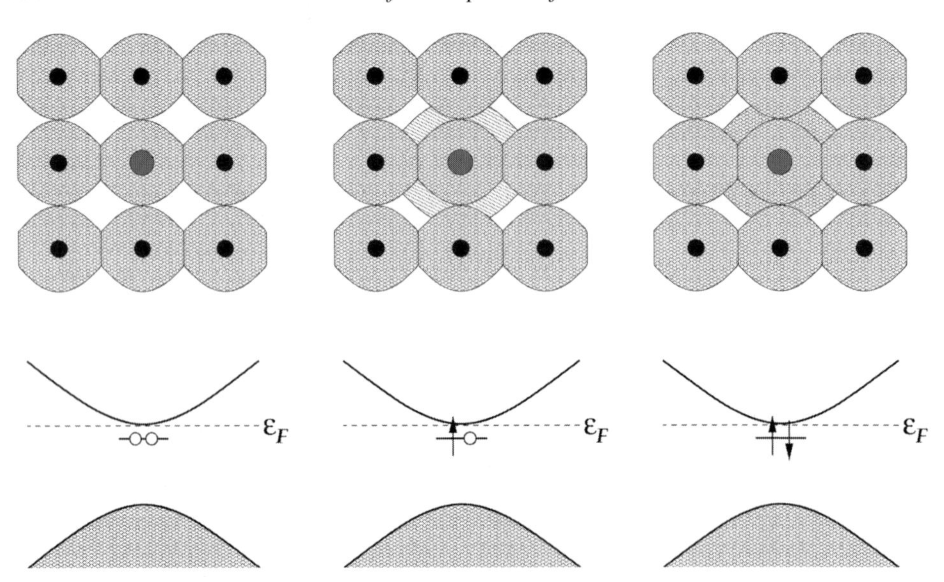

Figure 9.7. Electron occupation of of a donor-related state. **Top row:** this indicates schematically the occupation in real space, with regular crystal atoms shown as small black circles. The perfect crystal has two electrons per atomic site corresponding to $Z = 2$ and the donor has $Z^{(d)} = 3$ (that is, $\bar{Z}^{(d)} = +1$) and is shown as a slightly larger black circle. Electron states are shown as large shaded areas; singly hatched and cross-hatched regions represent occupation by one electron and two electrons, respectively. The defect-related states are shown as more extended than regular crystal states. **Bottom row:** this indicates the occupation of electronic levels in the band structure. The Fermi level is at the bottom of the conduction band and the defect state is slightly below it. The situation with no electrons in the defect state, which corresponds to a filled valence band, is shown in the left panel, that with one electron localized near the defect in the middle panel and that with two electrons in the right panel.

by the two electrons localized in the defect state. Notice that, if we consider an isolated system, the natural value of the chemical potential is $\epsilon_{\mathrm{F}} = \epsilon_{cbm}$, since this is the reservoir of electrons for occupation of the donor state; in this case, $\epsilon^{(d)} - \epsilon_{\mathrm{F}} < 0$, which means that the optimal situation corresponds to the donor state being occupied by one electron, assuming $U^{(d)} \gg |\epsilon^{(d)} - \epsilon_{\mathrm{F}}|$.

For the acceptor state the situation is illustrated in Fig. 9.8: the presence of the acceptor impurity creates a localized state which lies just above the VBM. Taking an electron from the reservoir and placing it in this state, equivalent to having one hole, costs energy of an amount $\epsilon^{(a)} - \epsilon_{\mathrm{F}}$, while placing two electrons in this state costs twice this energy, with no additional effect. However, if there are no electrons in this state, which is equivalent to having two holes, the localized holes experience a Coulomb repulsion $U^{(a)}$. This is another manifestation of the quasiparticle nature of electrons and holes in the solid. A hole is not a physical object, but a convenient way to describe the collective behavior of the electrons in the solid. The binding of the

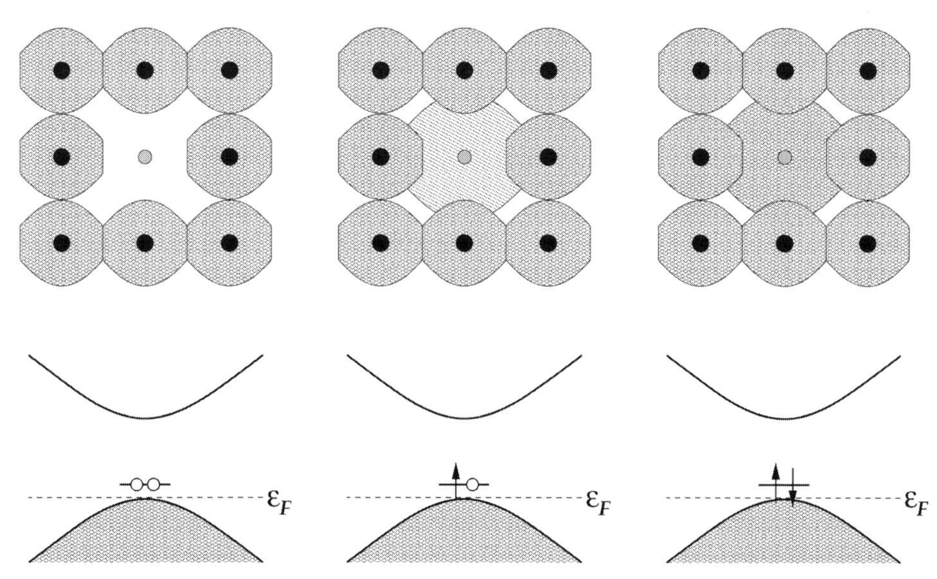

Figure 9.8. Electron occupation of an acceptor-related state; the symbols are the same as in Fig. 9.7, except that the acceptor has $Z^{(a)} = 1$ (that is, $\bar{Z}^{(a)} = -1$) and is shown as a smaller gray circle. The Fermi level is at the top of the valence band and the defect state is slightly above it. The situation with no electrons in the defect state, which corresponds to two electrons missing from the perfect crystal (equivalent to two holes), is shown in the left panel, that with one electron localized near the defect in the middle panel and that with two electrons in the right panel.

hole at the acceptor level means that electrons collectively avoid the neighborhood of the acceptor defect, which has a weaker ionic potential than a regular crystal atom at the same site (equivalent to a negative effective charge, $\bar{Z}^{(a)} = -1$). Thus, the occupation of this state by zero electrons is equivalent to having two localized holes, which repel each other by an energy $U^{(a)}$; this statement expresses the cost in energy to have the electrons completely expelled from the neighborhood of the acceptor defect. If we consider an isolated system, the natural value of the chemical potential is $\epsilon_F = \epsilon_{vbm}$, since this is the reservoir of electrons for occupation of the acceptor state; in this case, $\epsilon^{(a)} - \epsilon_F > 0$, which means that the optimal situation corresponds to the acceptor state being occupied by one electron, or equivalently by one hole, assuming $U^{(a)} \gg |\epsilon^{(a)} - \epsilon_F|$. These results are summarized in Table 9.2. In both cases, the optimal situation has one electron in the defect state which corresponds to a locally neutral system. For both the donor and the acceptor state, when the system is in contact with an external reservoir which determines the chemical potential of electrons (see, for instance, the following sections on the metal–semiconductor and metal–oxide–semiconductor junctions), the above arguments still hold with the position of the Fermi level not at the band extrema, but determined by the reservoir.

Table 9.2. *The energies of the donor and acceptor states for different occupations.*

$U^{(d)}$, $U^{(a)}$ is the Coulomb repulsion between two localized electrons or holes, respectively.

State i	Occupation n_i	Degeneracy g_i	Energy (donor) $E_i^{(d)} - n_i\epsilon_F$	Energy (acceptor) $E_i^{(a)} - n_i\epsilon_F$
1	no electrons	1	0	$U^{(a)}$
2	one electron	2	$(\epsilon^{(d)} - \epsilon_F)$	$(\epsilon^{(a)} - \epsilon_F)$
3	two electrons	1	$2(\epsilon^{(d)} - \epsilon_F) + U^{(d)}$	$2(\epsilon^{(a)} - \epsilon_F)$

9.2.3 The p–n junction

Finally, we will discuss in very broad terms the operation of electronic devices which are based on doped semiconductors (for details see the books mentioned in the Further reading section). The basic feature is the presence of two parts, one that is doped with donor impurities and has an excess of electrons, that is, it is negatively (n) doped, and one that is doped with acceptor impurities and has an excess of holes, that is, it is positively (p) doped. The two parts are in contact, as shown schematically in Fig. 9.9. Because electrons and holes are mobile and can diffuse in the system, some electrons will move from the n-doped side to the p-doped side leaving behind positively charged donor impurities. Similarly, some holes will diffuse from the p-doped side to the n-doped side leaving behind negatively charged acceptor impurities. An alternative way of describing this effect is that the electrons which have moved from the n-doped side to the p-doped side are captured by the acceptor impurities, which then lose their holes and become negatively charged; the reverse applies to the motion of holes from the p-doped to the n-doped side. In either case, the carriers that move to the opposite side are no longer mobile. Once enough holes have passed to the n-doped side and enough electrons to the p-doped side, an electric field is set up due to the imbalance of charge, which prohibits further diffusion of electric charges. The potential $\Phi(x)$ corresponding to this electric field is also shown in Fig. 9.9. The region near the interface from which holes and electrons have left to go the other side, and which is therefore depleted of carriers, is called the "depletion region". This arrangement is called a p–n junction.

The effect of the p–n junction is to rectify the current: when the junction is hooked to an external voltage bias with the plus pole connected to the p-doped side and the minus pole connected to the n-doped side, an arrangement called "forward bias", current flows because the holes are attracted to the negative pole and the electrons are attracted to the positive pole, freeing up the depletion region for additional charges to move into it. In this case, the external potential introduced by the bias

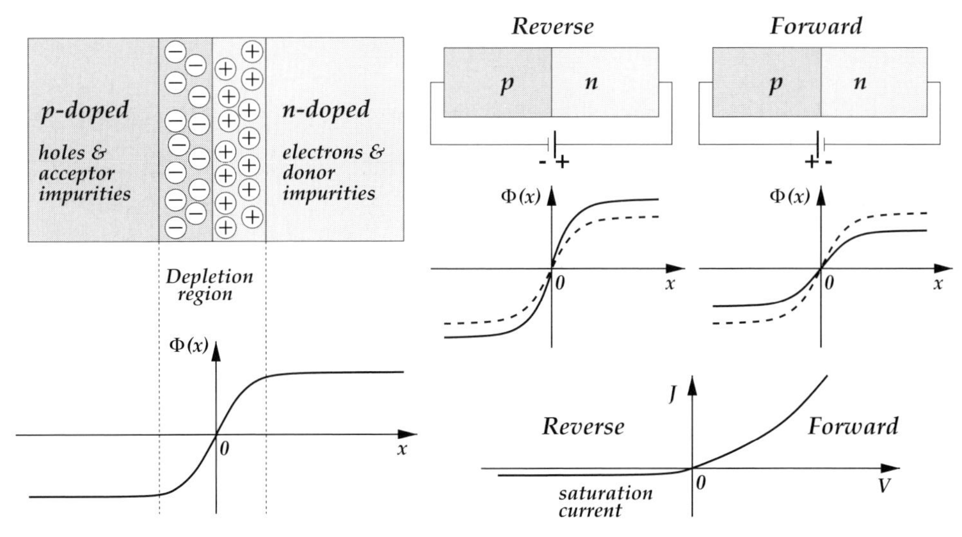

Figure 9.9. Schematic representation of p–n junction elements. **Left:** charge distribution, with the p-doped, n-doped and depletion regions identified. The positive and negative signs represent donor and acceptor impurities that have lost their charge carriers (electrons and holes, respectively) which have diffused to the opposite side of the junction; the resulting potential $\Phi(x)$ prohibits further diffusion of carriers. **Right:** operation of a p–n junction in the reverse bias and forward bias modes; the actual potential difference (solid lines) between the p-doped and n-doped regions relative to the zero-bias case (dashed lines) is enhanced in the reverse bias mode, which further restricts the motion of carriers, and is reduced in the forward bias mode, which makes it possible for current to flow. The rectifying behavior of the current J as a function of applied voltage V is also indicated; the small residual current for reverse bias is the saturation current.

voltage counteracts the potential due to the depletion region, as indicated in Fig. 9.9. If, on the other hand, the positive pole of the external voltage is connected to the the n-doped region and the negative pole to the p-doped region, an arrangement called "reverse bias", then current cannot flow because the motion of charges would be against the potential barrier. In this case, the external potential introduced by the bias voltage enhances the potential due to the depletion region, as indicated in Fig. 9.9. In reality, even in a reverse biased p–n junction there is a small amount of current that can flow due to thermal generation of carriers in the doped regions; this is called the saturation current. For forward bias, by convention taken as positive applied voltage, the current flow increases with applied voltage, while for reverse bias, by convention taken as negative applied voltage, the current is essentially constant and very small (equal to the saturation current). Thus, the p–n junction preferentially allows current flow in one bias direction, leading to rectification, as shown schematically in Fig. 9.9.

The formation of the p–n junction has interesting consequences on the electronic structure. We consider first the situation when the two parts, p-doped and n-doped,

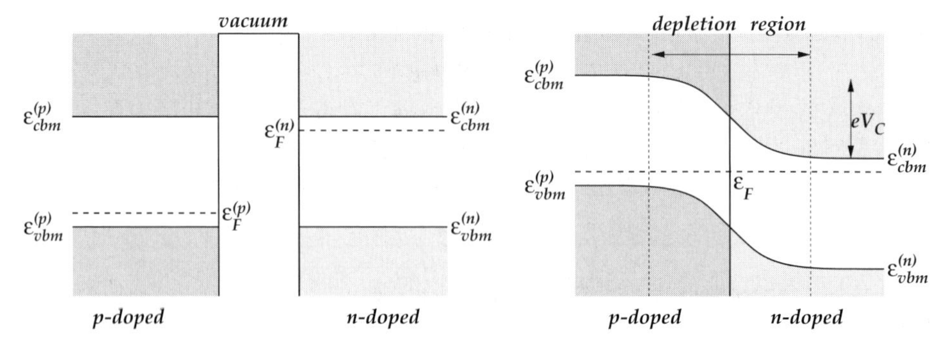

Figure 9.10. Band bending associated with a p–n junction. **Left:** the bands of the p-doped and n-doped parts when they are separated, with different Fermi levels $\epsilon_F^{(p)}$ and $\epsilon_F^{(n)}$. **Right:** the bands when the two sides are brought together, with a common Fermi level ϵ_F; the band bending in going from the p-doped side to the n-doped side is shown, with the energy change due to the contact potential eV_C.

are well separated. In this case, the band extrema (VBM and CBM) are at the same position for the two parts, but the Fermi levels are not: the Fermi level in the p-doped part, $\epsilon_F^{(p)}$, is near the VBM, while that of the n-doped part, $\epsilon_F^{(n)}$, is near the CBM (these assignments are explained below). When the two parts are brought into contact, the two Fermi levels must be aligned, since charge carriers move across the interface to establish a common Fermi level. When this happens, the bands on the two sides of the interface are distorted to accommodate the common Fermi level and maintain the same relation of the Fermi level to the band extrema on either side far from the interface, as shown in Fig. 9.10. This distortion of the bands is referred to as "band bending". The reason behind the band bending is the presence of the potential $\Phi(x)$ in the depletion region. In fact, the amount by which the bands are bent upon forming the contact between the p-doped and n-doped regions is exactly equal to the potential difference far from the interface,

$$V_C = \Phi(+\infty) - \Phi(-\infty)$$

which is called the "contact potential".[2] The difference in the band extrema on the two sides far from the interface is then equal to eV_C.

It is instructive to relate the width of the depletion region and the amount of band bending to the concentration of dopants on either side of the p–n junction and the energy levels they introduce in the band gap. From the discussion in the preceding subsections, we have seen that an acceptor impurity introduces a state whose energy $\epsilon^{(a)}$ lies just above the VBM and a donor impurity introduces a state whose energy $\epsilon^{(d)}$ lies just below the CBM (see Fig. 9.6). Moreover, in equilibrium these states are occupied by single electrons, assuming that the effective charge of

[2] Sometimes this is also referred to as the "bias potential", but we avoid this term here in order to prevent any confusion with the externally applied bias potential.

the impurities is $\bar{Z}^{(I)} = \pm 1$. If the concentration of the dopant impurities, denoted here as $N^{(a)}$ or $N^{(d)}$ for acceptors and donors, respectively, is significant then the presence of electrons in the impurity-related states actually determines the position of the Fermi level in the band gap. Thus, in n-doped material the Fermi level will coincide with $\epsilon^{(d)}$ and in p-doped material it will coincide with $\epsilon^{(a)}$:

$$\epsilon_F^{(n)} = \epsilon^{(d)}, \qquad \epsilon_F^{(p)} = \epsilon^{(a)}$$

From these considerations, and using the diagram of Fig. 9.10, we immediately deduce that

$$\begin{aligned} eV_C = \epsilon_F^{(n)} - \epsilon_F^{(p)} &= \epsilon_{gap} - \left(\epsilon_F^{(p)} - \epsilon_{vbm}^{(p)}\right) - \left(\epsilon_{cbm}^{(n)} - \epsilon_F^{(n)}\right) \\ &= \epsilon_{gap} - \left(\epsilon^{(a)} - \epsilon_{vbm}^{(p)}\right) - \left(\epsilon_{cbm}^{(n)} - \epsilon^{(d)}\right) \end{aligned}$$

since this is the amount by which the position of the Fermi levels in the p-doped and n-doped parts differ before contact.

We can also determine the lengths over which the depletion region extends into each side, l_p and l_n, respectively, by assuming uniform charge densities ρ_p and ρ_n in the p-doped and n-doped sides. In terms of the dopant concentrations, these charge densities will be given by

$$\rho_n = eN^{(d)}, \qquad \rho_p = -eN^{(a)}$$

assuming that within the depletion region all the dopants have been stripped of their carriers. The assumption of uniform charge densities is rather simplistic, but leads to correct results which are consistent with more realistic assumptions (see Problem 6). We define the direction perpendicular to the interface as the x axis and take the origin to be at the interface, as indicated in Fig. 9.9. We also define the zero of the potential $\Phi(x)$ to be at the interface, $\Phi(0) = 0$. We then use Poisson's equation, which for this one-dimensional problem gives

$$\begin{aligned} \frac{\mathrm{d}^2\Phi(x)}{\mathrm{d}x^2} &= -\frac{4\pi}{\varepsilon}\rho_n, \quad x > 0 \\ &= -\frac{4\pi}{\varepsilon}\rho_p, \quad x < 0 \end{aligned} \tag{9.44}$$

where ε is the dielectric constant of the material. Integrating once and requiring that $\mathrm{d}\Phi/\mathrm{d}x$ vanishes at $x = +l_n$ and at $x = -l_p$, the edges of the depletion region where the potential has reached its asymptotic value and becomes constant, we find:

$$\begin{aligned} \frac{\mathrm{d}\Phi(x)}{\mathrm{d}x} &= -\frac{4\pi}{\varepsilon}\rho_n(x - l_n), \quad x > 0 \\ &= -\frac{4\pi}{\varepsilon}\rho_p(x + l_p), \quad x < 0 \end{aligned} \tag{9.45}$$

The derivative of the potential, which is related to the electric field, must be continuous at the interface since there is no charge build up there and hence no discontinuity in the electric field (see Appendix A). This condition gives

$$N^{(d)} l_n = N^{(a)} l_p \tag{9.46}$$

where we have also used the relation between the charge densities and the dopant concentrations mentioned above. Integrating the Poisson equation once again, and requiring the potential to vanish at the interface, leads to

$$\begin{aligned} \frac{d\Phi(x)}{dx} &= -\frac{2\pi}{\varepsilon}\rho_n(x - l_n)^2 + \frac{2\pi}{\varepsilon}\rho_n l_n^2, \quad x > 0 \\ &= -\frac{2\pi}{\varepsilon}\rho_p(x + l_p)^2 + \frac{2\pi}{\varepsilon}\rho_p l_p^2, \quad x < 0 \end{aligned} \tag{9.47}$$

From this expression we can calculate the contact potential as

$$V_C = \Phi(l_n) - \Phi(-l_p) = \frac{2\pi e}{\varepsilon}\left[N^{(d)} l_n^2 + N^{(a)} l_p^2\right]$$

and using the relation of Eq. (9.46) we can solve for l_n and l_p:

$$l_n = \left[\left(\frac{\varepsilon V_C}{2\pi e}\right)\frac{N^{(a)}}{N^{(d)}}\frac{1}{N^{(a)} + N^{(d)}}\right]^{1/2}, \quad l_p = \left[\left(\frac{\varepsilon V_C}{2\pi e}\right)\frac{N^{(d)}}{N^{(a)}}\frac{1}{N^{(a)} + N^{(d)}}\right]^{1/2}$$

From these expressions we find that the total size of the depletion layer l_D is given by

$$l_D = l_p + l_n = \left[\left(\frac{\varepsilon V_C}{2\pi e}\right)\frac{N^{(a)} + N^{(d)}}{N^{(a)} N^{(d)}}\right]^{1/2} \tag{9.48}$$

It is interesting to consider the limiting cases in which one of the two dopant concentrations dominates:

$$N^{(a)} \gg N^{(d)} \Rightarrow l_D = \left(\frac{\varepsilon V_C}{2\pi e}\right)^{1/2}\left(N^{(d)}\right)^{-1/2}$$

$$N^{(d)} \gg N^{(a)} \Rightarrow l_D = \left(\frac{\varepsilon V_C}{2\pi e}\right)^{1/2}\left(N^{(a)}\right)^{-1/2}$$

which reveals that in either case the size of the depletion region is determined by the *lowest* dopant concentration.

Up to this point we have been discussing electronic features of semiconductor junctions in which the two doped parts consist of the same material. It is also possible to create p–n junctions in which the two parts consist of different semiconducting materials; these are called "heterojunctions". In these situations the band gap and position of the band extrema are different on each side of the junction. For doped semiconductors, the Fermi levels on the two sides of a heterojunction will also be

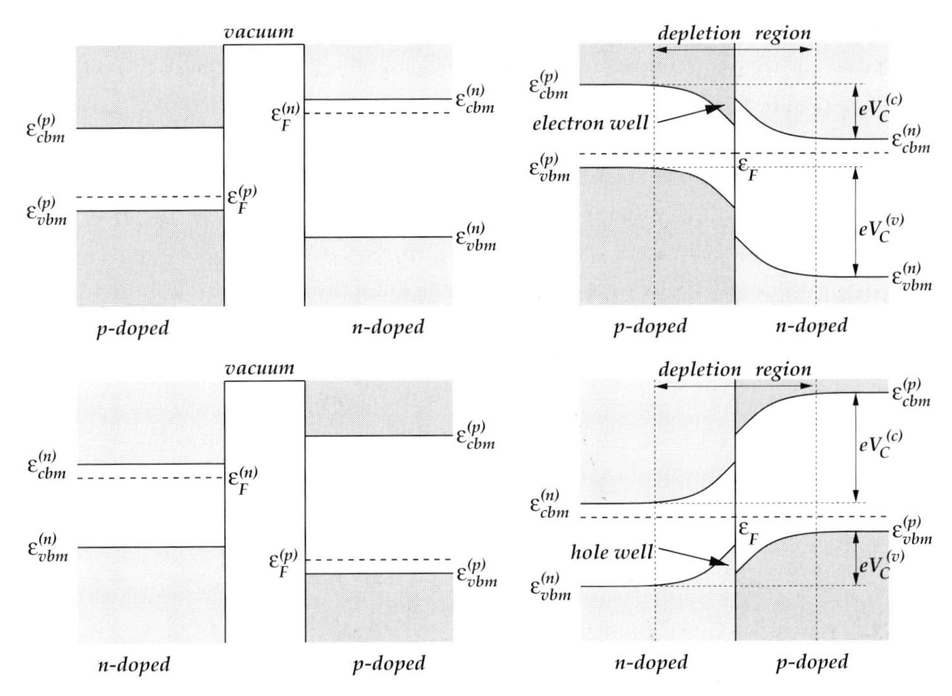

Figure 9.11. Illustration of band bending in doped heterojunctions. The left side in each case represents a material with smaller band gap than the right side. **Top:** a situation with p-doped small-gap material and n-doped large-gap material, before and after contact. **Bottom:** the situation with the reverse doping of the two sides. The band bending produces energy wells for electrons in the first case and for holes in the second case.

at different positions before contact. Two typical situations are shown in Fig. 9.11: in both cases the material on the left has a smaller band gap than the material on the right (this could represent, for example, a junction between GaAs on the left and $Al_xGa_{1-x}As$ on the right). When the Fermi levels of the two sides are aligned upon forming the contact, the bands are bent as usual to accommodate the common Fermi level. However, in these situations the contact potential is not the same for the electrons (conduction states) and the holes (valence states). As indicated in Fig. 9.11, in the case of a heterojunction with p-doping in the small-gap material and n-doping in the large-gap material, the contact potential for the conduction and valence states will be given by

$$eV_C^{(c)} = \Delta\epsilon_{\mathrm{F}} - \Delta\epsilon_{cbm}, \quad eV_C^{(v)} = \Delta\epsilon_{\mathrm{F}} + \Delta\epsilon_{vbm}$$

whereas in the reverse case of n-doping in the small-gap material and p-doping in the large-gap material, the two contact potentials will be given by

$$eV_C^{(c)} = \Delta\epsilon_{\mathrm{F}} + \Delta\epsilon_{cbm}, \quad eV_C^{(v)} = \Delta\epsilon_{\mathrm{F}} - \Delta\epsilon_{vbm}$$

where in both cases $\Delta\epsilon_F$ is the difference in Fermi level positions and $\Delta\epsilon_{cbm}$, $\Delta\epsilon_{vbm}$ are the differences in the positions of the band extrema before contact. It is evident from Fig. 9.11 that in both situations there are discontinuities in the potential across the junction due to the different band gaps on the two sides. Another interesting and very important feature is the presence of energy wells due to these discontinuities: such a well for electron states is created on the p-doped side in the first case and a similar well for hole states is created on the n-doped side in the second case. The states associated with these wells are discrete in the x direction, so if charge carriers are placed in the wells they will be localized in these discrete states and form a 2D gas parallel to the interface. Indeed, it is possible to populate these wells by additional dopant atoms far from the interface or by properly biasing the junction. The 2D gas of carriers can then be subjected to external magnetic fields, giving rise to very interesting quantum behavior. This phenomenon was discussed in more detail in chapter 7.

In real devices, the arrangement of n-doped and p-doped regions is more complicated. The most basic element of a device is the so called Metal–Oxide–Semiconductor-Field-Effect-Transistor (MOSFET). This element allows the operation of a rectifying channel with very little loss of power. A MOSFET is illustrated in Fig. 9.12. There are two n-doped regions buried in a larger p-doped region. The two n-doped regions act as source (S) and drain (D) of electrons. An external voltage is applied between the two n-doped regions with the two opposite poles attached to the source and drain through two metal electrodes which are separated by an insulating oxide layer. A different bias voltage is connected to an electrode placed

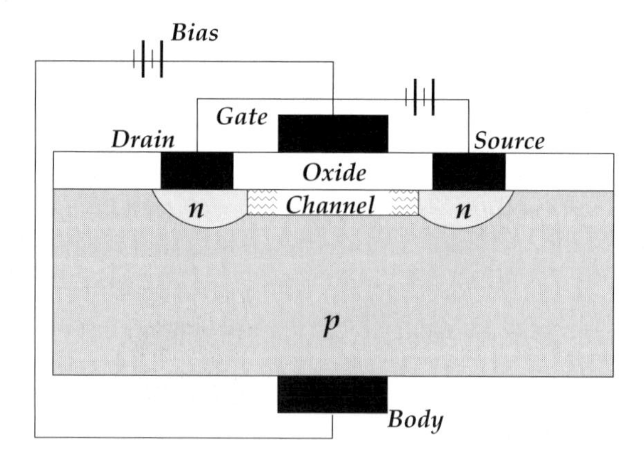

Figure 9.12. The basic features of a MOSFET: the source and drain, both n-doped regions, buried in a larger p-doped region and connected through two metal electrodes and an external voltage. The metal electrodes are separated by the oxide layer. Two additional metal electrodes, the gate and body, are attached to the oxide layer and to the bottom of the p-doped layer and are connected through the bias voltage. The conducting channel is between the two n-doped regions.

at the bottom of the p-doped layer, called the body (B), and to another electrode placed above the insulating oxide layer, called the gate (G). When a sufficiently large bias voltage is applied across the body and the gate electrodes, the holes in a region of the p-doped material below the gate are repelled, leaving a channel through which the electrons can travel from the source to the drain. The advantage of this arrangement is that no current flows between the body and the gate, even though it is this pair of electrodes to which the bias voltage is applied. Instead, the current flow is between the source and drain, which takes much less power to maintain, with correspondingly lower generation of heat. In modern devices there are several layers of this and more complicated arrangements of p-doped and n-doped regions interconnected by complex patterns of metal wires.

9.2.4 Metal–semiconductor junction

The connection between the metal electrodes and the semiconductor is of equal importance to the p–n junction for the operation of an electronic device. In particular, this connection affects the energy and occupation of electronic states on the semiconductor side, giving rise to effective barriers for electron transfer between the two sides. A particular model of this behavior, the formation of the so called Schottky barrier [120], is shown schematically in Fig. 9.13. When the metal and semiconductor are well separated, each has a well-defined Fermi level denoted by $\epsilon_F^{(m)}$ for the metal and $\epsilon_F^{(s)}$ for the semiconductor. For an *n-doped* semiconductor, the Fermi level lies close to the conduction band, as discussed above. The energy difference between the vacuum level, which is common to both systems, and the Fermi level is defined as the work function (the energy cost of removing electrons from the system); it is denoted by ϕ_m for the metal and ϕ_s for the semiconductor.

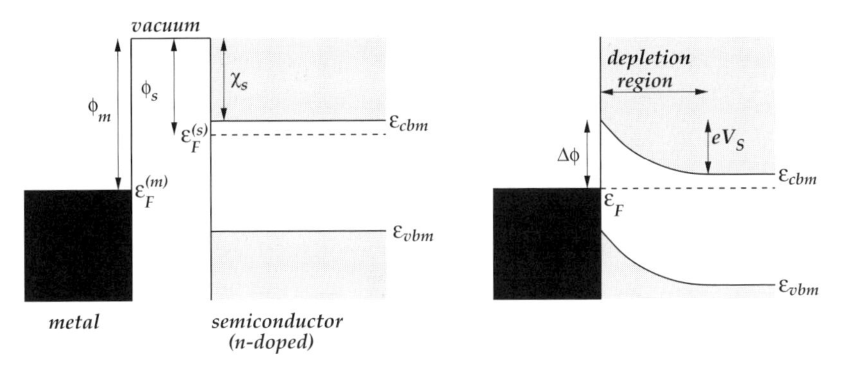

Figure 9.13. Band alignment in a metal–semiconductor junction: ϕ_m, ϕ_s are the work functions of the metal and the semiconductor, respectively; χ_s is the electron affinity; $\epsilon_{vbm}, \epsilon_{cbm}, \epsilon(s)_F, \epsilon_F^{(m)}$ represent the top of the valence band, the bottom of the conduction band, and the Fermi levels in the semiconductor and the metal, respectively; $\Delta\phi = \phi_m - \phi_s$ is the shift in work function; and V_S is the potential (Schottky) barrier.

The energy difference between the conduction band minimum, denoted by ϵ_{cbm}, and the vacuum level is called the electron affinity, denoted by χ_s.

When the metal and the semiconductor are brought into contact, the Fermi levels on the two sides have to be aligned. This is done by moving electrons from one side to the other, depending on the relative position of Fermi levels. For the case illustrated in Fig. 9.13, when the two Fermi levels are aligned, electrons have moved from the semiconductor (which originally had a higher Fermi level) to the metal. This creates a layer near the interface which has fewer electrons than usual on the semiconductor side, and more electrons than usual on the metal side, creating a charge depletion region on the semiconductor side. The presence of the depletion region makes it more difficult for electrons to flow across the interface. This corresponds to a potential barrier V_S, called the Schottky barrier. In the case of a junction between a metal and an n-type semiconductor, the Schottky barrier is given by

$$eV_S^{(n)} = \phi_m - \chi_s$$

as is evident from Fig. 9.13. Far away from the interface the relation of the semiconductor bands to the Fermi level should be the same as before the contact is formed. To achieve this, the electron energy bands of the semiconductor must bend, just like in the case of the p–n junction, since at the interface they must maintain their original relation to the metal bands. In the case of a junction between a metal and a p-type semiconductor, the band bending is in the opposite sense and the corresponding Schottky barrier is given by

$$eV_S^{(p)} = \epsilon_{gap} - (\phi_m - \chi_s)$$

Combining the two expressions for the metal/n-type and metal/p-type semiconductor contacts, we obtain

$$e\left[V_S^{(p)} + V_S^{(n)}\right] = \epsilon_{gap}$$

Two features of this picture of metal–semiconductor contact are worth emphasizing. First, it assumes there are no changes in the electronic structure of the metal or the semiconductor due to the presence of the interface between them, other than the band bending which comes from equilibrating the Fermi levels on both sides. Second, the Schottky barrier is proportional to the work function of the metal. Neither of these features is very realistic. The interface can induce dramatic changes in the electronic structure, as discussed in more detail in chapter 11, which alter the simple picture described above. Moreover, experiments indicate that measured Schottky barriers are indeed roughly proportional to the metal work function for large-gap semiconductors (ZnSe, ZnS), but they tend to be almost independent of the metal work function for small-gap semiconductors (Si, GaAs) [121].

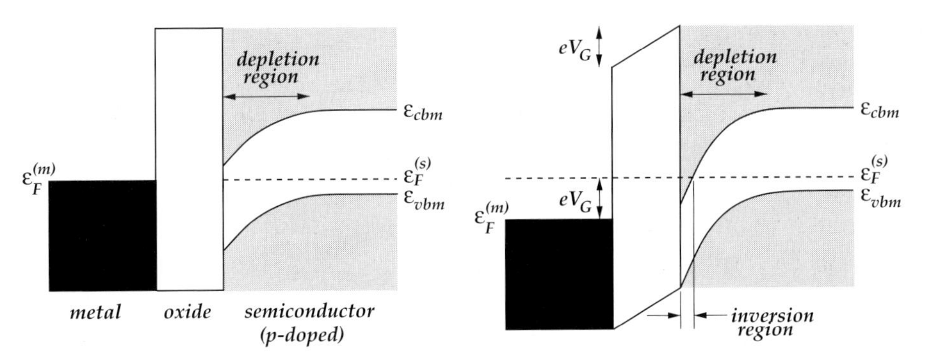

Figure 9.14. Band alignment in a metal–oxide–semiconductor junction, for a p-doped semiconductor. V_G is the gate (bias) voltage, which lowers the energy of electronic states in the metal by eV_G relative to the common Fermi level. Compare the band energy in the inversion region with the confining potential in Fig. 7.9.

The situation is further complicated by the presence of the insulating oxide layer between the metal and the semiconductor. Band bending occurs in this case as well, as shown for instance in Fig. 9.14 for a p-doped semiconductor. The interesting new feature is that the oxide layer can support an externally applied bias voltage, which we will refer to as the gate voltage V_G (see also Fig. 9.12). The gate voltage moves the electronic states on the metal side down in energy by eV_G relative to the common Fermi level. This produces additional band bending, which lowers both the valence and the conduction bands of the semiconductor in the immediate neighborhood of the interface. When the energy difference eV_G is sufficiently large, it can produce an "inversion region", that is, a narrow layer near the interface between the semiconductor and the oxide where the bands have been bent to the point where some conduction states of the semiconductor have moved below the Fermi level. When these states are occupied by electrons, current can flow from the source to the drain in the MOSFET. In the inversion layer the electrons in the occupied semiconductor conduction bands form a two-dimensional system of charge carriers, because the confining potential created by the distorted bands can support only one occupied level below the Fermi level. In such systems, interesting phenomena which are particular to two dimensions, such as the quantum Hall effect (integer and fractional) can be observed (see chapter 7).

Further reading

1. *Physics of Semiconductor Devices*, S.M. Sze (J. Wiley, New York, 1981).
   This is a standard reference with extensive discussion of all aspects of semiconductor physics from the point of view of application in electronic devices.
2. *The Physics of Semiconductors*, K.F. Brennan (Cambridge University Press, Cambridge, 1999).
   This is a modern account of the physics of semiconductors, with extensive discussion of the basic methods for studying solids in general.

3. *Interfaces in Crystalline Materials*, A.P. Sutton and R.W. Balluffi (Clarendon Press, Oxford, 1995).
This is a thorough and detailed discussion of all aspects of crystal interfaces, including a treatment of metal–semiconductor and semiconductor–semiconductor interfaces.

Problems

1. We wish to prove the relations for the point defect concentrations per site, given in Eqs. (9.2) and (9.4) for vacancies and interstitials. We consider a general point defect whose formation energy is ϵ_f relative to the ideal crystal, the energy of the latter being defined as the zero energy state. We assume that there are N atoms and N' defects in the crystal, occupying a total of $N + N'$ sites.[3] The number of atoms N will be considered fixed, while the number of defects N', and therefore the total number of crystal sites involved, will be varied to obtain the state of lowest free energy. We define the ratio of defects to atoms as $x = N'/N$.

   (a) Show that in the microcanonical ensemble the entropy of the system is given by

   $$S = k_{\mathrm{B}} N \left[(1+x)\ln(1+x) - x\ln x \right]$$

   (b) Show that the free energy at zero external pressure, $F = E - TS$, is minimized for

   $$x = \frac{1}{\mathrm{e}^{\epsilon_f/k_{\mathrm{B}}T} - 1}$$

   (c) Using this result, show that the concentration of the defect per site of the crystal is given by

   $$c(T) \equiv \frac{N'}{N + N'} = \mathrm{e}^{-\epsilon_f/k_{\mathrm{B}}T}$$

   as claimed for the point defects discussed in the text.

2. The formation energy of vacancies in Si is approximately 3.4 eV, and that of interstitials is approximately 3.7 eV. Determine the relative concentration of vacancies and interstitials in Si at room temperature and at a temperature of 100°C.
3. Prove the expressions for the reduced number of electrons $\bar{n}_c(T)$ or holes $\bar{p}_v(T)$ given by Eq. (9.38).
4. Using the fact that the number of electrons in the conduction band is equal to the number of holes in the valence band for an intrinsic semiconductor (containing no dopants), show that in the zero-temperature limit the Fermi level lies exactly at the middle of the band gap. Show that this result holds also for finite temperature if the densities of states at the VBM and CBM are the same.
5. Calculate the number of available carriers at room temperature in undoped Si and compare it with the number of carriers when it is doped with P donor impurities at a

[3] In this simple model we assume that the defect occupies a single crystal site, which is common for typical point defects, but can easily be extended to more general situations.

concentration of 10^{16} cm^{-3} or with As donor impurities at a concentration of 10^{18} cm^{-3}.

6. We will analyze the potential at a p–n junction employing a more realistic set of charge distributions than the uniform distributions assumed in the text. Our starting point will be the following expressions for the charge distributions in the n-doped and p-doped regions:

$$\rho_n(x) = \tanh\left(\frac{x}{l_n}\right)\left[1 - \tanh^2\left(\frac{x}{l_n}\right)\right] eN^{(d)}, \quad x > 0$$

$$\rho_p(x) = \tanh\left(\frac{x}{l_p}\right)\left[1 - \tanh^2\left(\frac{x}{l_p}\right)\right] eN^{(a)}, \quad x < 0$$

(a) Plot these functions and show that they correspond to smooth distributions with no charge build up at the interface, $x = 0$.
(b) Integrate Poisson's equation once to obtain the derivative of the potential $\mathrm{d}\Phi/\mathrm{d}x$ and determine the constants of integration by physical considerations. Show that from this result the relation of Eq. (9.46) follows.
(c) Integrate Poisson's equation again to obtain the potential $\Phi(x)$ and determine the constants of integration by physical considerations. Calculate the contact potential by setting a reasonable cutoff for the asymptotic values, for example, the point at which 99% of the charge distribution is included in either side. From this, derive expressions for the total size of the depletion region $l_D = l_p + l_n$, analogous to Eq. (9.48).

7. Describe in detail the nature of band bending at a metal–semiconductor interface for all possible situations: there are four possible cases, depending on whether the semiconductor is n-doped or p-doped and on whether $\epsilon_F^{(s)} > \epsilon_F^{(m)}$ or $\epsilon_F^{(s)} < \epsilon_F^{(m)}$; of these only one was discussed in the text, shown in Fig. 9.13.

10

Defects II: line defects

Line defects in crystals are called dislocations. Dislocations had been considered in the context of the elastic continuum theory of solids, beginning with the work of Volterra [122], as a one-dimensional mathematical cut in a solid. Although initially viewed as useful but abstract constructs, dislocations became indispensable in understanding the mechanical properties of solids and in particular the nature of plastic deformation. In 1934, Orowan [123], Polanyi [124] and Taylor [125], each independently, made the connection between the atomistic structure of crystalline solids and the nature of dislocations; this concerned what is now called an "edge dislocation". A few years later, Burgers [126] introduced the concept of a different type of dislocation, the "screw dislocation". The existence of dislocations in crystalline solids is confirmed experimentally by a variety of methods. The most direct observation of dislocations comes from transmission electron microscopy, in which electrons pass through a thin slice of the material and their scattering from atomic centers produces an image of the crystalline lattice and its defects (see, for example, Refs. [127, 128]). A striking manifestation of the presence of dislocations is the spiral growth pattern on a surface produced by a screw dislocation. The field of dislocation properties and their relation to the mechanical behavior of solids is enormous. Suggestions for comprehensive reviews of this field, as well as some classic treatments, are given in the Further reading section.

10.1 Nature of dislocations

The simplest way to visualize a dislocation is to consider a simple cubic crystal consisting of two halves that meet on a horizontal plane, with the upper half containing one more vertical plane of atoms than the lower half. This is called an edge dislocation and is shown in Fig. 10.1. The points on the horizontal plane where the extra vertical plane of atoms ends form the dislocation line. The region around the

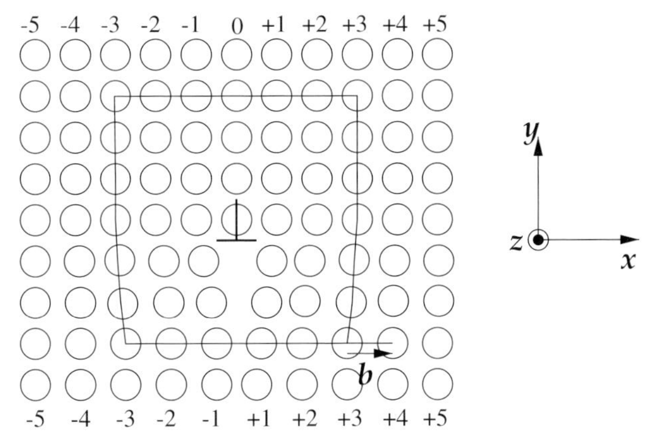

Figure 10.1. Illustration of an edge dislocation in a simple cubic crystal. The extra plane of atoms (labeled 0) is indicated by a vertical line terminated at an inverted T. The path shows the Burgers vector construction: starting at the lower right corner, we take six atomic-spacing steps in the $+y$ direction, six in the $-x$ direction, six in the $-y$ direction and six in the $+x$ direction; in this normally closed path, the end misses the beginning by the Burgers vector. The Burgers vector for this dislocation is along the x axis, indicated by the small arrow labeled **b**, and is perpendicular to the dislocation line which lies along the z axis.

dislocation line is called the dislocation core, and involves significant distortion of the atoms from their crystalline positions in order to accommodate the extra plane of atoms: atoms on the upper half are squeezed closer together while atoms on the lower half are spread farther apart than they would be in the ideal crystal. Far away from the dislocation core the vertical planes on either side of the horizontal plane match smoothly.

A dislocation is characterized by the Burgers vector and its angle with respect to the dislocation line. The Burgers vector is the vector by which the end misses the beginning when a path is formed around the dislocation core, consisting of steps that would have led to a closed path in the perfect crystal. The Burgers vector of the edge dislocation is perpendicular to its line, as illustrated in Fig. 10.1. The Burgers vector of a full dislocation is one of the Bravais lattice vectors. The energetically preferred dislocations have as a Burgers vector the shortest lattice vector, for reasons which will be discussed in detail below.

There are different types of dislocations, depending on the crystal structure and the Burgers vector. Another characteristic example is a screw dislocation, which has a Burgers vector parallel to its line, as shown in Fig. 10.2. Dislocations in which the Burgers vector lies between the two extremes (parallel or perpendicular to the dislocation line) are called mixed dislocations. A dislocation is characterized by the direction of the dislocation line, denoted by $\hat{\xi}$ and its Burgers vector **b**. For the two extreme cases, edge and screw dislocation, the following relations hold between

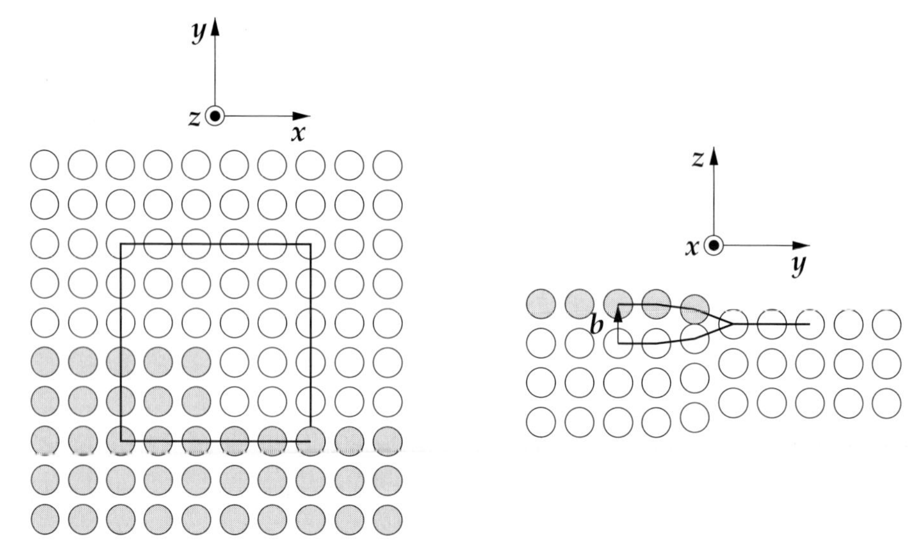

Figure 10.2. Illustration of a screw dislocation in a simple cubic crystal, in two views, top view on the left (along the dislocation line) and side view on the right. The path shows the Burgers vector construction: in the top view, starting at the lower right corner, we take five atomic-spacing steps in the $+y$ direction, five in the $-x$ direction, five in the $-y$ direction and five in the $+x$ direction; in this normally closed path, the end misses the beginning by the Burgers vector. The Burgers vector, shown in the side view, is indicated by the small arrow labeled **b** which lies along the z axis, parallel to the dislocation line. The shaded circles in the top view represent atoms that would normally lie on the same plane with white circles but are at higher positions on the z axis due to the presence of the dislocation.

the dislocation direction and Burgers vector:

$$\text{edge}: \quad \hat{\xi}_e \cdot \mathbf{b}_e = 0; \qquad \text{screw}: \quad \hat{\xi}_s \cdot \mathbf{b}_s = \pm\mathbf{b}_s$$

as is evident from Figs. 10.1 and 10.2. In general, dislocations can combine to form dislocations of a different type. For example, two edge dislocations of opposite Burgers vector can cancel each other; this so called "dislocation annihilation" can be easily rationalized if we consider one of them corresponding to an extra plane on the upper half of the crystal and the other to an extra plane on the lower half of the crystal: when the two extra planes are brought together, the defect in the crystal is annihilated. This can be generalized to the notion of reactions between dislocations, in which the resulting dislocation has a Burgers vector which is the vector sum of the two initial Burgers vectors. We return to this issue below.

Individual dislocations cannot begin or end within the solid without introducing additional defects. As a consequence, dislocations in a real solid must extend all the way to the surface, or form a closed loop, or form nodes at which they meet with other dislocations. Examples of a dislocation loop and dislocation nodes are shown in Fig. 10.3. Since a dislocation is characterized by a unique Burgers vector, in a

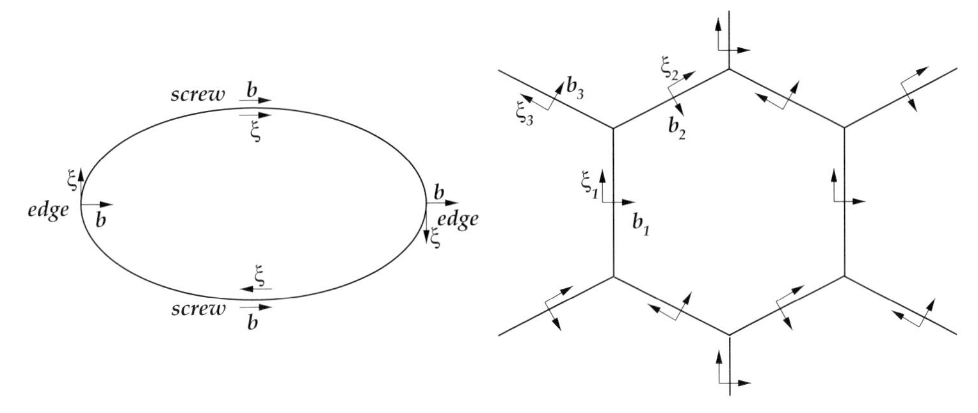

Figure 10.3. Illustration of a dislocation loop (**left**) and a network dislocations meeting at nodes (**right**), with $\mathbf{b}_i$ denoting Burgers vectors. In the dislocation loop there exist segments of edge character (**b** perpendicular to dislocation line), screw character (**b** parallel to dislocation line) and mixed character (**b** at some intermediate angle to dislocation line). The Burgers vectors at each node sum to zero.

dislocation loop the Burgers vector and the dislocation line will be parallel in certain segments, perpendicular in other segments and at some intermediate angle at other segments. Thus, the dislocation loop will consist of segments with screw, edge and mixed character, as indicated in Fig. 10.3. Dislocation nodes are defect structures at which finite segments of dislocations begin or end without creating any other extended defects in the crystal. Therefore, a path enclosing all the dislocations that meet at a node will not involve a Burgers vector displacement, or, equivalently, the sum of Burgers vectors of dislocations meeting at a node must be zero. Dislocations can form regular networks of lines and nodes, as illustrated in Fig. 10.3.

One of the most important features of dislocations is that they can move easily through the crystal. A closer look at the example of the edge dislocation demonstrates this point: a small displacement of the atomic columns near the dislocation core would move the dislocation from one position between adjacent lower-half vertical planes to the next equivalent position, as illustrated in Fig. 10.4. The energy cost for such a displacement, per unit length of the dislocation, is very small because the bonds of atoms in the dislocation core are already stretched and deformed. The total energy for moving the entire dislocation is of course infinite for an infinite dislocation line, but then the energy of the dislocation itself is infinite even at rest, due to the elastic distortion it induces to the crystal, without counting the energy cost of forming the dislocation core (see discussion in later sections of this chapter). These infinities are artifacts of the infinite crystal and the single, infinite dislocation line, which are both idealizations. In reality dislocations start and end at other defects, as mentioned already, while their motion involves the sequential displacement of small sections of the dislocation line. It is also easy to see

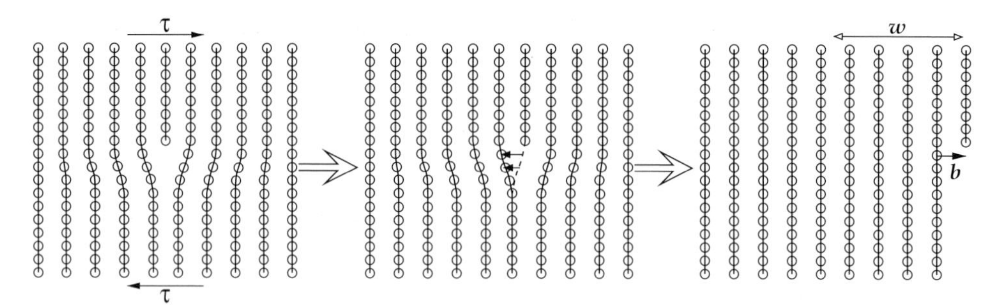

Figure 10.4. Illustration of how an edge dislocation can be moved by one lattice vector to the right by the slight displacement of few atomic columns near the core (left and middle panels), as indicated by the small arrows. Repeating this step eventually leads to a deformation of the solid, where the upper half and the lower half differ by one half plane (right panel), and the dislocation has been expelled from the crystal. τ is the external shear stress that forces the dislocation to move in the fashion indicated over a width w, and b is the Burgers vector of the dislocation; the length of the dislocation is l in the direction perpendicular to the plane of the figure.

how the displacement of an edge dislocation by steps similar to the one described above eventually leads to a permanent deformation of the solid: after a sufficient number of steps the dislocation will be expelled from the crystal and the upper half will differ from the lower half by a half plane, forming a ledge at the far end of the crystal, as shown in Fig. 10.4. This process is the main mechanism for plastic deformation of crystals.

Within the idealized situation of a single dislocation in an infinite crystal, it is possible to obtain the force per unit length of the dislocation due to the presence of external stress. If the width of the crystal is w and the length of the dislocation is l, then the work ΔW done by an external shear stress τ to deform the crystal by moving an edge dislocation through it, in the configuration of Fig. 10.4, is

$$\Delta W = (\tau w l) b$$

If we assume that this is accomplished by a constant force, called the "Peach–Koehler" force F_{PK}, acting uniformly along the dislocation line, then the work done by this force will be given by

$$\Delta W = F_{PK} w$$

Equating the two expressions for the work, we find that the force per unit length of the dislocation due to the external stress τ is given by

$$f_{PK} = \frac{1}{l} F_{PK} = \tau b$$

This expression can be generalized to an arbitrary dislocation line described by the vector $\hat{\xi}$, and external stress described by the tensor σ, to

$$\mathbf{f}_{PK} = (\sigma \cdot \mathbf{b}) \times \hat{\xi} \tag{10.1}$$

This force is evidently always perpendicular to the direction of the dislocation line $\hat{\xi}$, and it is non-zero if there is a component of the stress tensor σ parallel to the Burgers vector $\mathbf{b}$, as in the example of the edge dislocation under shear stress τ, shown in Fig. 10.4.

The ease with which dislocations move, and the distortions they induce throughout the crystal, make them very important defects for mediating the mechanical response of solids. Dislocations are also important in terms of the electronic properties of solids. In semiconductors, dislocations induce states in the gap which act like traps for electrons or holes. When the dislocation line lies in a direction that produces a short circuit, its effect can be disastrous to the operation of the device. Of the many important effects of dislocations we will only discuss briefly their mobility and its relation to mechanical behavior, such as brittle or ductile response to external loading. For more involved treatments we refer the reader to the specialized books mentioned at the end of the chapter.

Because dislocations induce long-range strain in crystals, and therefore respond to externally applied macroscopic stresses, they are typically described in the context of continuum elasticity theory. In this context, the atomic structure of the dislocation core does not enter directly. The basic concepts of elasticity theory are reviewed in Appendix E.

10.2 Elastic properties and motion of dislocations

Although many aspects of dislocation shape and motion depend on the crystal structure and the material in which the dislocation exists, some general features can be described by phenomenological models without specifying those details. The values that enter into these phenomenological models can then be fitted to reproduce, to the extent possible, the properties of dislocations in specific solids. A widely used phenomenological model is due to Peierls and Nabarro [129, 130]; this model actually provides some very powerful insights into dislocation properties, so we will examine its basic features in this section. It is interesting that although almost 60 years old, this model still serves as the basis of many contemporary quantitative attempts to describe dislocations in various solids. Before delving into the Peierls–Nabarro model we discuss some general results concerning the stress field and elastic energy of a dislocation.

10.2.1 Stress and strain fields

We examine first the stress fields for infinite straight edge and screw dislocations. To this end, we define the coordinate system to be such that the dislocation line is the z axis, the horizontal axis is the x axis and the vertical axis is the y axis (see Figs. 10.1 and 10.2). We also define the glide plane through its normal vector $\hat{\mathbf{n}}$ which is given by

$$\hat{\mathbf{n}} = \frac{\hat{\xi} \times \mathbf{b}}{|\mathbf{b}|}$$

For the edge dislocation shown in Fig. 10.1, the glide plane (also referred to as the slip plane) is the xz plane.

For the screw dislocation, the Burgers vector is $\mathbf{b}_s = b_s\hat{\mathbf{z}}$ while for the edge dislocation it is $\mathbf{b}_e = b_e\hat{\mathbf{x}}$. The magnitudes of the Burgers vectors, b_s, b_e, depend on the crystal. It is convenient to use the cylindrical coordinate system (r, θ, z) to express the stress fields of dislocations. The stress components for the screw and edge dislocations in these coordinates, as well as in the standard cartesian coordinates, are given in Table 10.1 for an isotropic solid. The derivation of these expressions is a straight-forward application of continuum elasticity theory, with certain assumptions for the symmetry of the problem and the long-range behavior of the stress fields (see Problems 1 and 2 for details). All components include the appropriate Burgers vectors b_s, b_e and the corresponding elastic constants, which are given by

$$K_s = \frac{\mu}{2\pi}, \qquad K_e = \frac{\mu}{2\pi(1-\nu)} \tag{10.2}$$

Table 10.1. *The stress fields for the screw and edge dislocations.* The components of the fields are given in polar (r, θ, z) and cartesian (x, y, z) coordinates.

Polar coordinates			Cartesian coordinates		
σ_{ij}	Screw	Edge	σ_{ij}	Screw	Edge
σ_{rr}	0	$-K_e b_e \frac{\sin\theta}{r}$	σ_{xx}	0	$-K_e b_e \frac{(3x^2+y^2)y}{(x^2+y^2)^2}$
$\sigma_{\theta\theta}$	0	$-K_e b_e \frac{\sin\theta}{r}$	σ_{yy}	0	$K_e b_e \frac{(x^2-y^2)y}{(x^2+y^2)^2}$
σ_{zz}	0	$-2\nu K_e b_e \frac{\sin\theta}{r}$	σ_{zz}	0	$-2\nu K_e b_e \frac{y}{(x^2+y^2)}$
$\sigma_{r\theta}$	0	$K_e b_e \frac{\cos\theta}{r}$	σ_{xy}	0	$K_e b_e \frac{x(x^2-y^2)}{(x^2+y^2)^2}$
$\sigma_{\theta z}$	$K_s b_s \frac{1}{r}$	0	σ_{yz}	$K_s b_s \frac{x}{(x^2+y^2)}$	0
σ_{zr}	0	0	σ_{zx}	$-K_s b_s \frac{y}{(x^2+y^2)}$	0

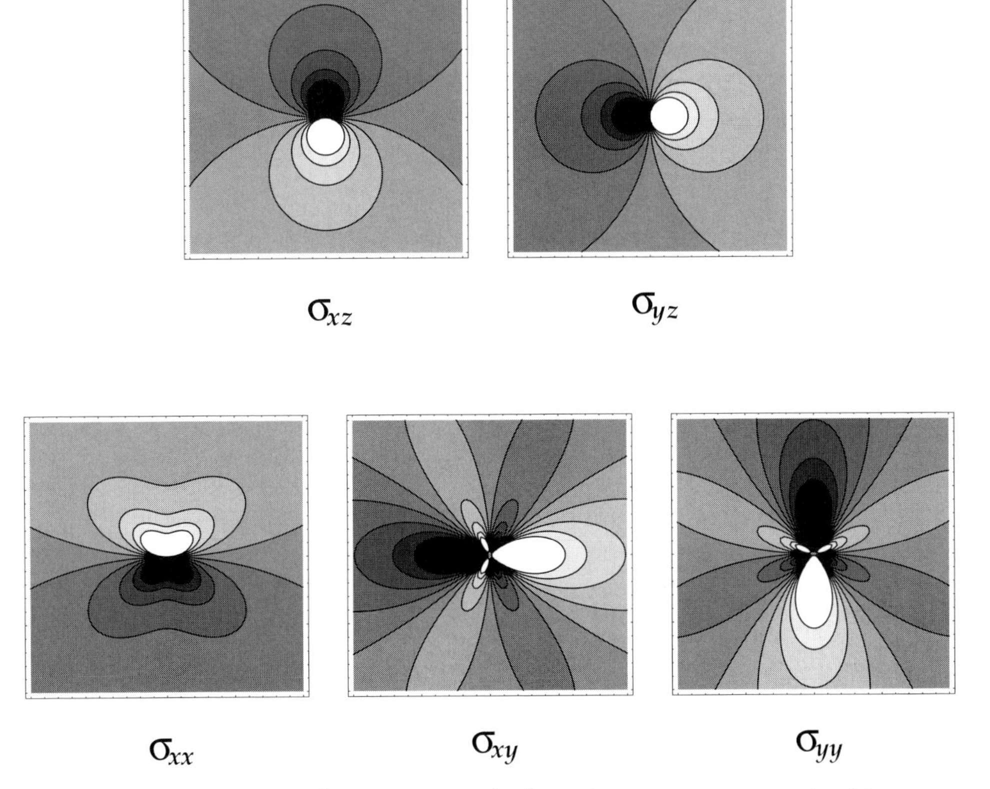

Figure 10.5. The contours of constant stress for the various stress components of the screw (top panel) and edge (bottom panel) dislocations, as given by the expressions of Table 10.1 in cartesian coordinates; white represents large positive values and black represents large negative values. The σ_{zz} component of the edge dislocation is identical in form to the σ_{xz} component of the screw dislocation.

for the screw and edge dislocations, respectively. These naturally involve the shear modulus μ and Poisson's ratio ν, defined for an isotropic solid (see Appendix E). Plots of constant stress contours for various components of the stress tensors are shown in Fig. 10.5.

An interesting consequence of these results is that the hydrostatic component of the stress, $\sigma_\Omega = (\sigma_{rr} + \sigma_{\theta\theta} + \sigma_{zz})/3$ is zero for the screw dislocation, but takes the value

$$\sigma_\Omega = -\frac{2}{3} K_e b_e (1 + \nu) \frac{\sin\theta}{r} \tag{10.3}$$

for the edge dislocation, that is, it is a compressive stress. For mixed dislocations with a screw and an edge component, the corresponding quantities are a combination of the results discussed so far. For instance, for a mixed dislocation, in which the angle between the dislocation line and the Burgers vector is θ, so that $b\cos\theta$ is the screw component and $b\sin\theta$ is the edge component, the corresponding elastic constant is given by

$$K_{mix} = \frac{\mu}{2\pi}\left[\frac{1}{1-\nu}\sin^2\theta + \cos^2\theta\right] \tag{10.4}$$

It is also interesting to analyze the displacement fields for the two types of dislocations. For the screw dislocation, the displacement field on the xy plane is given by

$$u_x(x, y) = u_y(x, y) = 0, \quad u_z(x, y) = \frac{b_s}{2\pi}\tan^{-1}\frac{y}{x} \tag{10.5}$$

a result derived from simple physical considerations, namely that far from the dislocation core u_z goes uniformly from zero to b_s as θ ranges from zero to 2π, while the other two components of the strain field vanish identically (see also Problem 1). For the edge dislocation, we can use for the stress field the stress–strain relations for an isotropic solid to obtain the strain field and from that, by integration, the displacement field. From the results given above we find for the diagonal components of the strain field of the edge dislocation

$$\epsilon_{xx} = \frac{-b_e}{4\pi(1-\nu)}\left[\frac{2x^2y}{(x^2+y^2)^2} + (1-2\nu)\frac{y}{(x^2+y^2)}\right] \tag{10.6}$$

$$\epsilon_{yy} = \frac{b_e}{4\pi(1-\nu)}\left[\frac{y(x^2-y^2)}{(x^2+y^2)^2} + 2\nu\frac{y}{(x^2+y^2)}\right] \tag{10.7}$$

and integrating the first with respect to x and the second with respect to y, we obtain

$$u_x(x, y) = \frac{b_e}{4\pi(1-\nu)}\left[2(1-\nu)\tan^{-1}\frac{y}{x} + \frac{xy}{(x^2+y^2)}\right] \tag{10.8}$$

$$u_y(x, y) = \frac{-b_e}{4\pi(1-\nu)}\left[\frac{1-2\nu}{2}\ln(x^2+y^2) + \frac{x^2-y^2}{2(x^2+y^2)}\right] \tag{10.9}$$

There are two constants of integration involved in obtaining these results: the one for u_x is chosen so that $u_x(x, 0) = 0$ and the one for u_y is chosen so that $u_y(x, y)$ is a symmetric expression in the variables x and y; these choices are of course not unique. Plots of $u_z(x, y)$ for the screw dislocation and of $u_x(x, y)$ for the edge dislocation are given in Fig. 10.6; for these examples we used a typical average value $\nu = 0.25$ (for most solids ν lies in the range 0.1–0.4; see Appendix E).

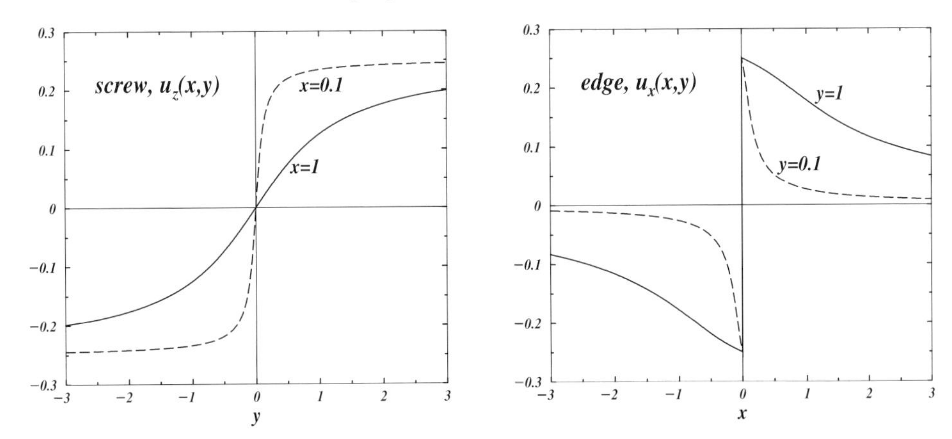

Figure 10.6. The u_z and u_x components of the displacement fields for the screw and edge dislocations, in units of the Burgers vectors. The values of the x and y variables are also scaled by the Burgers vectors. For negative values of x in the case of the screw dislocation, or of y in the case of the edge dislocation, the curves are reversed: they become their mirror images with respect to the vertical axis. For these examples we used a typical value of $\nu = 0.25$.

Certain features of these displacement fields are worth pointing out. Specifically, for the screw dislocation, we note first that there is a shift in u_z by $b_s/2$ when going from $-\infty$ to $+\infty$ along the y axis for a given value of $x > 0$; there is a similar shift of $b_s/2$ for the corresponding $x < 0$ value, thus completing a total shift by b_s along a Burgers circuit. The displacement field u_z is a smooth function of y and tends to a step function when $x \to 0$: this is sensible in the context of continuum elasticity, since for $x = 0$ there must be a jump in the displacement to accommodate the dislocation core, as is evident from the schematic representation of Fig. 10.2. For large values of x the displacement u_z is very gradual and the shift by $b_s/2$ takes place over a very wide range of y values.

For the edge dislocation, the displacement u_x is a discontinuous function of x. The discontinuity in this component occurs at $x = 0$ and is exactly $b_e/2$ for a given value of $y > 0$. The total shift by b_e is completed when the corresponding path at $y < 0$ is included in the Burgers circuit. For $y \to \pm\infty$, the displacement u_x becomes a one-step function, going abruptly from $\mp b_e/2$ to $\pm b_e/2$ at $x = 0$. For $y \to 0^+$ it becomes a three-step function, being zero all the way to $x \to 0^-$, then jumping to $-b_e/2$, next jumping to $+b_e/2$ and finally jumping back to zero for $x \to 0^+$ (the reverse jumps occur for $y \to 0^-$). The discontinuity is entirely due to the $\tan^{-1}(y/x)$ term in Eq. (10.8), the other term being equal to zero for $x = 0$. The other component of the displacement, u_y, is even more problematic at the origin, $(x, y) = (0, 0)$, due to the $\ln(x^2 + y^2)$ term which blows up. This pathological behavior is a reflection of the limitations of continuum elasticity: the u_y component must describe the presence of an extra plane of atoms in going from

$y < 0$ to $y > 0$ at $x = 0$, as indicated in the schematic representation of the edge dislocation in Fig. 10.1. But since there is no explicit information about atomic sites in continuum elasticity, this condition is reflected by the pathological behavior of the u_y component of the displacement field. In other words, the description of the physical system based on continuum elasticity fails when we approach the core region of the edge dislocation; in fact, the expressions for the stresses in this case were actually derived with the assumption that $r = \sqrt{x^2 + y^2}$ is very large on the scale of interatomic distances (see Problem 2), and thus are valid only far away from the dislocation core. A more realistic treatment, which takes into account the discrete nature of the atomic planes but retains some of the appealing features of the continuum approach, the Peierls–Nabarro model, is discussed below.

10.2.2 Elastic energy

We next examine the energy associated with the elastic distortion that an infinite straight dislocation induces in the crystal. We will consider first a simplified model to obtain the essential behavior of the elastic energy, that is, to motivate the presence of two terms: one arising from the strain far from the core and the other from the dislocation core. It turns out that the contribution of the first term is infinite. This has to do with the fact that both the strain and the stress induced by the dislocation fall off slowly, like $\sim 1/r$ (see Table 10.1 for the stress, while the strain is proportional to the stress in an isotropic solid, as shown in Appendix E). The contribution to the elastic energy is given by the product of the stress and strain which, when integrated over the plane perpendicular to the dislocation line, gives a logarithmically divergent term.

The idealized model we will consider is illustrated in Fig. 10.7: it consists of an edge dislocation in the same geometry as shown earlier in Figs. 10.1 and 10.4, but we

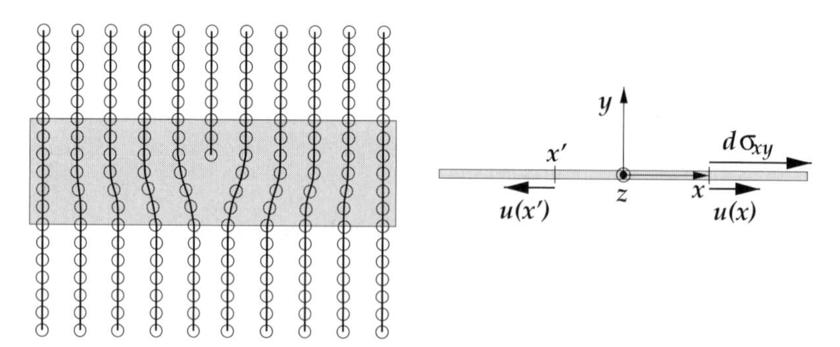

Figure 10.7. Idealized model of an edge dislocation for the calculation of the elastic energy: the displacement field, which spreads over many atomic sites in the x and y directions around the core (shaded area on left panel) is assumed to be confined on the glide plane (xz), as shown schematically on the right panel. The infinitesimal dislocation at x' gives rise to a shear stress $\sigma_{xy}(x, 0)$ at another point x, where the displacement is $u(x)$.

will assume that the displacement is confined to the glide plane, identified as the xz plane in the geometry of Fig. 10.1. With this assumption, symmetry considerations lead to the conclusion that the displacement is a scalar function, which we will denote as $u(x)$ and take to be a continuous function of the position x, with the dislocation core at $x = 0$. We will also find it useful to employ another function, the dislocation density $\rho(x)$, defined as

$$\rho(x) = -2\frac{\mathrm{d}u(x)}{\mathrm{d}x} \tag{10.10}$$

This quantity is useful because it describes the disregistry across the glide plane, which must integrate to the Burgers vector, b_e. The disregistry (also called the misfit) across the glide plane can be defined as

$$\overline{u}(x) = \lim_{y\to 0}\left[u_x(x, y) - u_x(x, -y)\right] = 2u_x(x, 0) \to 2u(x) \tag{10.11}$$

where we have used the expression for the x component of the displacement of an edge dislocation derived earlier, Eq. (10.8), which in the end we identified with the scalar displacement of the present idealized model. The definition of the disregistry leads to the equivalent definition of the dislocation density,

$$\rho(x) = -\frac{\mathrm{d}\overline{u}(x)}{\mathrm{d}x} \tag{10.12}$$

Integrating the dislocation density over all x we find

$$\begin{aligned}\int_{-\infty}^{\infty}\rho(x)\mathrm{d}x &= -2\lim_{\epsilon\to 0}\left[\int_{-\infty}^{-\epsilon}\frac{\mathrm{d}u(x)}{\mathrm{d}x}\mathrm{d}x + \int_{\epsilon}^{\infty}\frac{\mathrm{d}u(x)}{\mathrm{d}x}\mathrm{d}x\right]\\ &= -2\left[\int_{0}^{-b_e/4}\mathrm{d}u + \int_{b_e/4}^{0}\mathrm{d}u\right] = b_e\end{aligned} \tag{10.13}$$

where we have again used Eq. (10.8) to determine the displacement of an edge dislocation on the glide plane. The final result is what we expected from the definition of the dislocation density.

With the displacement a continuous function of x, we will treat the dislocation as if it were composed of a sequence of infinitesimal dislocations [131]: the infinitesimal dislocation between x' and $x' + dx'$ has a Burgers vector

$$\mathrm{d}b_e(x') = -2\left(\frac{\mathrm{d}u}{\mathrm{d}x}\right)_{x=x'}\mathrm{d}x' = \rho(x')\mathrm{d}x' \tag{10.14}$$

This infinitesimal dislocation produces a shear stress at some other point x which, from the expressions derived earlier (see Table 10.1), is given by

$$\sigma_{xy}(x, 0) = K_e\frac{\mathrm{d}b_e(x')}{x - x'}$$

where we think of the "core" of the infinitesimal dislocation as being located at x'. The shear stress at the point x is a force per unit area on the xz plane whose surface-normal unit vector is $\hat{y}$ (see definition of the stress tensor in Appendix E). The displacement $u(x)$ necessary to create the infinitesimal dislocation at x takes place in the presence of this force from the dislocation at x', giving the following contribution to the elastic energy from the latter infinitesimal dislocation:

$$\mathrm{d}U_e^{(el)} = K_e \frac{\mathrm{d}b_e(x')}{x - x'} u(x)$$

Integrating this expression over all values of x from $-L$ to L (with L large enough to accommodate the range of the displacement field), and over $\mathrm{d}b_e(x')$ to account for the contributions from all infinitesimal dislocations, we obtain for the elastic energy of the edge dislocation

$$U_e^{(el)}[u(x)] = \frac{K_e}{2} \int_L^{-L} \int_{-b_e/2}^{b_e/2} u(x) \frac{1}{x - x'} \mathrm{d}b_e(x') \mathrm{d}x$$

In the above expression, we have introduced a factor of 1/2 to account for the double-counting of the interactions between infinitesimal dislocations. We next employ the expression given in Eq. (10.14) for $\mathrm{d}b_e(x')$, we perform an integration by parts over the variable x, and we use the expression for the dislocation density from Eq. (10.10), to arrive at the following expression for the elastic energy:

$$U_e^{(el)}[u(x)] = \frac{K_e b_e^2}{2} \ln(L) - \frac{K_e}{2} \int_{-L}^{L} \int_{-L}^{L} \rho(x)\rho(x') \ln|x - x'| \mathrm{d}x \mathrm{d}x' \qquad (10.15)$$

This result clearly separates the contribution of the long-range elastic field of the dislocation, embodied in the first term, from the contribution of the large distortions at the dislocation core, embodied in the second term. The first term of the elastic energy in Eq. (10.15) is infinite for $L \to \infty$, which is an artifact of the assumption of a single, infinitely long straight dislocation in an infinite crystal. In practice, there are many dislocations in a crystal and they are not straight and infinitely long. A typical dislocation density is 10^5 cm of dislocation line per cm^3 of the crystal, expressed as 10^5 cm^{-2}. The dislocations tend to cancel the elastic fields of each other, providing a natural cutoff for the extent of the elastic field of any given segment of a dislocation. Thus, the first term in the elastic energy expression does not lead to an unphysical picture. Since the contribution of this term is essentially determined by the density of dislocations in the crystal, it is not an interesting term from the point of view of the atomic structure of the dislocation core. Accordingly, we will drop this first term, and concentrate on the second term, which includes the energy due to the dislocation core, as it depends exclusively on the distribution of the dislocation displacement $u(x)$ (or, equivalently, the disregistry $\overline{u}(x)$), through the dislocation density $\rho(x)$.

We outline yet another way to obtain an expression for the elastic energy associated with the dislocation, using the expressions for the stress fields provided in Table 10.1, and assuming that the dislocation interacts with its own average stress field as it is being created. This approach can be applied to both screw and edge dislocations. Imagine for example that we create a screw dislocation by cutting the crystal along the glide plane (xz plane) for $x > 0$ and then displacing the part above the cut relative to the part below, by an amount equal to the Burgers vector b_s along z. During this procedure the average stress at a distance r from the dislocation line will be half of the value $\sigma_{\theta z}$ for the screw dislocation, that is, $\frac{1}{2}K_s b_s/r$. In order to obtain the energy per unit length corresponding to this distortion, given by the stress $\times$ the corresponding strain, we must integrate over all values of the radius. We take as a measure of the strain the displacement b_s, by which the planes of atoms are misplaced. We will also need to introduce two limits for the integration over the radius, an inner limit r_c (called the core radius) and an outer limit L. The integration then gives

$$U_s^{(el)} = \frac{1}{2}K_s b_s \int_{r_c}^{L} \frac{b_s}{r} \mathrm{d}r = \frac{1}{2}K_s b_s^2 \ln(L) - \frac{1}{2}K_s b_s^2 \ln(r_c) \tag{10.16}$$

in which the first term is identical to the one in Eq. (10.15), and the second includes the core energy due to the disregistry $\overline{u}$. In the case of the edge dislocation, the cut is again on the glide plane (xz plane) for $x > 0$, and the part above this plane is displaced relative to the part below by an amount equal to the Burgers vector b_e along x. Since the misfit is along the x axis we only need to integrate the value of the $\sigma_{r\theta}$ component for $\theta = 0$ along x, from r_c to L. We obtain

$$U_e^{(el)} = \frac{1}{2}K_e b_e \int_{r_c}^{L} \frac{b_e}{r} \mathrm{d}r = \frac{1}{2}K_e b_e^2 \ln(L) - \frac{1}{2}K_e b_e^2 \ln(r_c) \tag{10.17}$$

a result identical in form to that for the screw dislocation.

If the dislocation is not pure edge or pure screw it can be thought of as having an edge component and a screw component. The Burgers vector is composed by the screw and edge components which are orthogonal, lying along the dislocation line and perpendicular to it, respectively. With the angle between the dislocation line and the Burgers vector defined as θ, the elastic energy of the mixed dislocation will be given by

$$U_{mix}^{(el)} = \frac{b^2}{2}K_{mix} \ln\left(\frac{L}{r_c}\right) = \frac{b^2}{2}\mu\left[\frac{1}{2\pi(1-\nu)}\cos^2\theta + \frac{1}{2\pi}\sin^2\theta\right]\ln\left(\frac{L}{r_c}\right) \tag{10.18}$$

where we have used the expression for the elastic constant of the mixed dislocation, Eq. (10.4). In all cases, edge, screw and mixed dislocations, the elastic energy is proportional to μb^2. This result justifies our earlier claim that the Burgers vector is the shortest lattice vector, since this corresponds to the lowest energy dislocation.

This result is also useful in interpreting the presence of so called "partial dislocations". These are dislocations which from very far away look like a single dislocation, but locally they look like two separate dislocations. Their Burgers vectors $\mathbf{b}_1, \mathbf{b}_2$ have magnitudes which are shorter than any lattice vector, but they add up to a total Burgers vector which, as usual, is a lattice vector, $\mathbf{b}_1 + \mathbf{b}_2 = \mathbf{b}$. The condition for the existence of partial dislocations is

$$b_1^2 + b_2^2 < b^2$$

because then they are energetically preferred over the full dislocation. This means that the angle between the vectors $\mathbf{b}_1$ and $\mathbf{b}_2$ must be greater than 90°, as illustrated in Fig. 10.8. When this condition is met, a full dislocation will be split into two partials, as long as the energy of the stacking fault which appears between the split dislocations, as shown in Fig. 10.8, does not overcompensate for the energy gain due to splitting. More specifically, if the stacking fault energy per unit area is γ_{sf}, the partial dislocations will be split by a distance d which must satisfy

$$K(b_1^2 + b_2^2) + \gamma_{sf} d \le K b^2$$

with K the relevant elastic constant. If the values of d in the above expression happen to be small on the interatomic scale because of a large γ_{sf}, then we cannot speak of splitting of partial dislocations; this situation occurs, for instance, in Al. An example of Burgers vectors for full and partial dislocations in an FCC crystal is shown in Fig. 10.8. The above argument assumes that the partial dislocations are

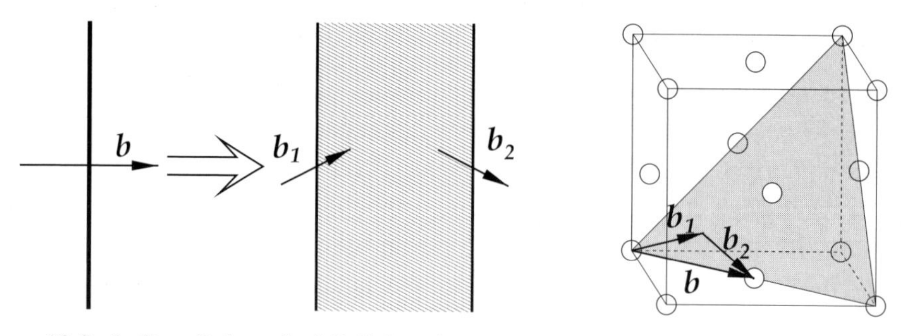

Figure 10.8. **Left:** splitting of a full dislocation with Burgers vector $\mathbf{b}$ into two partials with Burgers vectors $\mathbf{b}_1$ and $\mathbf{b}_2$, where $\mathbf{b}_1 + \mathbf{b}_2 = \mathbf{b}$; the partial dislocations are connected by a stacking fault (hatched area). **Right:** the conventional cubic unit cell of an FCC crystal with the slip plane shaded and the vectors $\mathbf{b}, \mathbf{b}_1, \mathbf{b}_2$ indicated for this example: note that $\mathbf{b}$ is the shortest lattice vector but $\mathbf{b}_1, \mathbf{b}_2$ are not lattice vectors.

not interacting and that the only contributions to the total energy of the system come from isolated dislocations. Since the interaction between the partial dislocations is repulsive, if the energetic conditions are met then splitting will occur.

10.2.3 Peierls–Nabarro model

The Peierls–Nabarro (PN) model relies on two important concepts.[1] The first concept is that the dislocation can be described in terms of a continuous displacement distribution $u(\mathbf{r})$, as was done above in deriving the expressions for the elastic energy. The second concept is that a misfit $\overline{u}$ between two planes of the crystal corresponds to an energy cost of $\gamma(\overline{u})$ per unit area on the plane of the misfit. This is an important quantity, called the generalized stacking fault energy or γ-surface, originally introduced by Vitek [132]. It has proven very useful in studying dislocation core properties from first-principles calculations [132–134]. In the PN theory, the $\gamma(\overline{u})$ energy cost is identified with the energy cost of displacing two semi-infinite halves of a crystal relative to each other uniformly by $\overline{u}$, across a crystal plane.

The crucial argument in the PN theory is that the elastic energy of the dislocation core is balanced by the energy cost of introducing the misfit in the lattice. In the following discussion we will drop, for the reasons we mentioned earlier, the infinite term $\frac{1}{2}Kb^2 \ln(L)$ which appears in all the expressions we derived above. We will also adopt the expression of Eq. (10.15) for the elastic energy of the dislocation core, which was derived above for an edge dislocation. Thus, our discussion of the PN model strictly speaking applies only to an edge dislocation, but the model can be generalized to other types of dislocations (see, for example, Eshelby's generalization to screw dislocations [131]). For simplicity of notation we will drop the subscript "e" denoting the edge character of the dislocation. Moreover, we can take advantage of the relation between the misfit $\overline{u}(x)$ and the displacement $u(x)$, Eq. (10.11), to express all quantities in terms of either the misfit or the displacement field. The energy cost due to the misfit will be given by an integral of $\gamma(\overline{u})$ over the range of the misfit. This leads to the following expression for the total energy of the dislocation:

$$U^{(tot)}[u(x)] = -\frac{K}{2}\int\int \rho(x)\rho(x')\ln|x-x'|\mathrm{d}x\mathrm{d}x' + \int \gamma(\overline{u}(x))\mathrm{d}x \quad (10.19)$$

A variational derivative of this expression with respect to the dislocation density

[1] The treatment presented here does not follow the traditional approach of guessing a sinusoidal expression for the shear stress (see for example the treatment by Hirth and Lothe, mentioned in the Further reading section). Instead, we adopt a more general point of view based on a variational argument for the total energy of the dislocation, and introduce the sinusoidal behavior as a possible simple choice for the displacement potential. The essence of the resulting equations is the same as in traditional approaches.

$\rho(x)$ leads to the PN integro-differential equation:

$$2K \int_{-b/2}^{b/2} \frac{1}{x - x'} \mathrm{d}u(x') + \frac{\mathrm{d}\gamma(\overline{u}(x))}{\mathrm{d}\overline{u}(x)} = 0 \tag{10.20}$$

The first term is the elastic stress at point x due to the infinitesimal dislocation $\rho(x')\mathrm{d}x'$ at point x'; the second term represents the restoring stress due to the non-linear misfit potential acting across the slip plane. This potential must be a periodic function of $\overline{u}(x)$ with a period equal to the Burgers vector of the dislocation. As the simplest possible model, we can assume a sinusoidal function for the misfit potential, which is referred to as the Frenkel model [136]. One choice is (see Problem 4 for a different possible choice)

$$\gamma(\overline{u}) = \frac{\gamma_{us}}{2}\left(1 - \cos\frac{2\pi\overline{u}}{b}\right) \tag{10.21}$$

where γ_{us} is the amplitude of the misfit energy variation (this is called the "unstable stacking energy" [137] and is a quantity important in determining the brittle or ductile character of the solid, as discussed in the following section). This form of the potential ensures that it vanishes when there is no misfit ($\overline{u} = 0$) or the misfit is an integral multiple of the Burgers vector (the latter case corresponds to having passed from one side of the dislocation core to the other). Using the sinusoidal potential of Eq. (10.21) in the PN integro-differential Eq. (10.20), we can obtain an analytic solution for the misfit, which is given by

$$\begin{aligned} \overline{u}(x) &= \frac{b}{\pi}\tan^{-1}\frac{x}{\zeta} + \frac{b}{2} \quad \left[\zeta = \frac{Kb^2}{2\pi\gamma_{us}}\right] \\ \rho(x) &= -\frac{\mathrm{d}\overline{u}}{\mathrm{d}x} = \frac{b\zeta}{\pi(\zeta^2 + x^2)} \end{aligned} \tag{10.22}$$

A typical dislocation profile is shown in Fig. 10.9. In this figure, we also present the dislocation profiles that each term in Eq. (10.20) would tend to produce, if acting alone. The elastic stress term would produce a very narrow dislocation to minimize the elastic energy, while the restoring stress would produce a very broad dislocation to minimize the misfit energy. The resulting dislocation profile is a compromise between these two tendencies.

One of the achievements of the PN theory is that it provides a reasonable estimate of the dislocation size. The optimal size of the dislocation core, characterized by the value of ζ, is a result of the competition between the two energy terms in Eq. (10.20), as shown schematically in Fig. 10.9: If the unstable stacking energy γ_{us} is high or the elastic moduli K are low, the misfit energy dominates and the dislocation becomes narrow (ζ is small) in order to minimize the misfit energy,

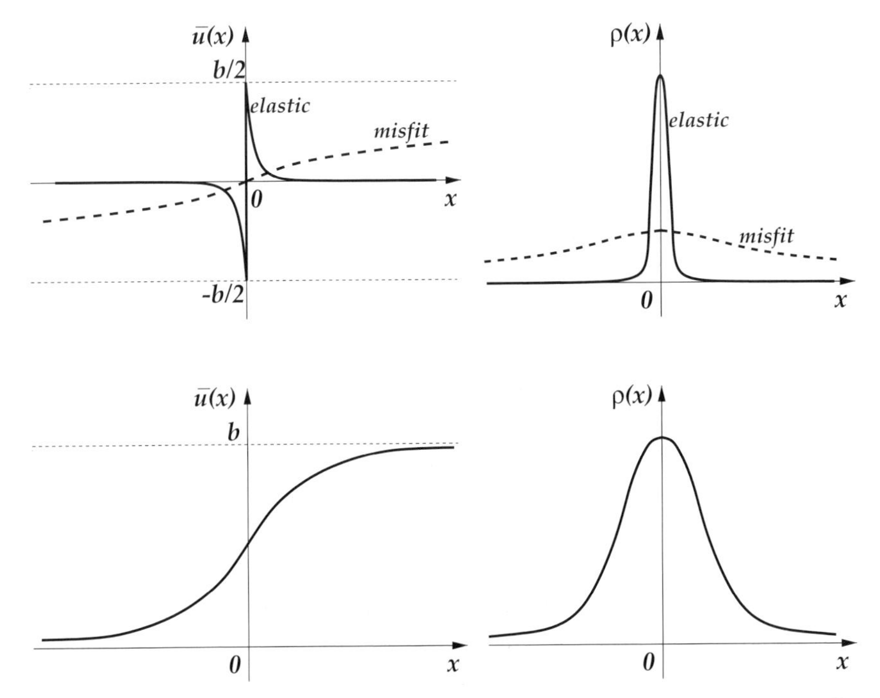

Figure 10.9. Profile of an edge dislocation: **Top:** the disregistry or misfit $\overline{u}(x)$ as dictated by the minimization of the elastic energy (solid line) or the misfit energy (dashed line) and the corresponding densities $\rho(x)$, given by Eq. (10.10). **Bottom:** the disregistry and density as obtained from the Peierls–Nabarro model, which represents a compromise between the two tendencies.

i.e. the second term in Eq. (10.20). In the opposite case, if γ_{us} is low or K is large, the dislocation spreads out in order to minimize the elastic energy, i.e. the first term in Eq. (10.20), which is dominant. In either case, the failures of the treatment based strictly on continuum elasticity theory discussed earlier are avoided, and there are no unphysical discontinuities in the displacement fields.

Yet another important achievement of PN theory is that it makes possible the calculation of the shear stress required to move a dislocation in a crystal. This, however, requires a modification of what we have discussed so far. The expression for the total energy of the dislocation, Eq. (10.19), is invariant with respect to arbitrary translation of the dislocation density $\rho(x) \rightarrow \rho(x + t)$. The dislocation described by the PN solution Eq. (10.22) does not experience any resistance as it moves through the lattice. This is clearly unrealistic, and is a consequence of neglecting the discrete nature of the lattice: the PN model views the solid as an isotropic continuous medium. The only effect of the lattice periodicity so far comes from the periodicity of the misfit potential with period b, which corresponds to a lattice vector of the crystal. In order to rectify this shortcoming and to introduce a resistance to the motion of dislocations through the lattice, the PN model was modified so that the misfit potential is not sampled continuously but only at the positions of the atomic planes. This amounts to the following modification of the second term in the total energy of the dislocation in Eq. (10.19):

$$\int \gamma(\overline{u}(x))\mathrm{d}x \rightarrow \sum_{n=-\infty}^{\infty} \gamma(\overline{u}(x_n))\Delta x \tag{10.23}$$

where x_n are the positions of atomic planes and Δx is the spacing of atomic rows in the lattice. With this modification, when the PN solution, Eq. (10.22), is translated through the lattice, the energy will have a periodic variation with period equal to the distance between two equivalent atomic rows in the crystal (this distance can be different from the Burgers vector b). The amplitude of this periodic variation in the energy is called the Peierls energy.

Having introduced an energy variation as a function of the dislocation position, which leads to an energy barrier to the motion of the dislocation, we can obtain the stress required to move the dislocation without any thermal activation. This stress can be defined as the maximum slope of the variation in the energy as a function of the translation.[2] Using this definition, Peierls and Nabarro showed that the shear stress for dislocation motion, the so called Peierls stress σ_P, is given by

$$\sigma_P = \frac{2\mu}{1-\nu} \exp\left(-\frac{2\pi a}{(1-\nu)b}\right) \tag{10.24}$$

with b the Burgers vector and a the lattice spacing across the glide plane. While this is a truly oversimplified model for dislocation motion, it does provide some interesting insight. Experimentally, the Peierls stress can be estimated by extrapolating to zero temperature the critical resolved yield stress, i.e. the stress beyond which plastic deformation (corresponding to dislocation motion) sets on. This gives Peierls stress values measured in terms of the shear modulus (σ_P/μ) of order 10^{-5} for close-packed FCC and HCP metals, 5×10^{-3} for BCC metals, 10^{-5}–10^{-4} for ionic crystals, and 10^{-2}–1 for compound and elemental semiconductors in the zincblende and diamond lattices. It therefore explains why in some crystals it is possible to have plastic deformation for shear stress values several orders of magnitude below the shear modulus: it is all due to dislocation motion! In particular, it is interesting to note that in covalently bonded crystals where dislocation activity is restricted by the strong directional bonds between atoms, the ratio σ_P/μ is of order unity, which implies that these crystals do not yield plastically, that is, they are brittle solids (see also the discussion in section 10.3).

[2] An alternative definition of this stress is the shear stress, which, when applied to the crystal, makes the energy barrier to dislocation motion vanish. Finding the stress through this definition relies on computational approaches and is not useful for obtaining an analytical expression.

As can be easily seen from Eq. (10.24), the value of the Peierls stress is extremely sensitive (exponential dependence) to the ratio (a/b), for fixed values of the elastic constants μ, ν. Therefore, in a given crystal the motion of dislocations corresponding to the largest value of (a/b) will dominate. Notice that there are two aspects to this criterion: the spacing between atomic planes across the glide plane a, and the Burgers vector of the dislocation b. Thus, according to this simple theory, the dislocations corresponding to the smallest Burgers vector and to the glide plane with the largest possible spacing between successive atomic planes will dominate. In close-packed metallic systems, the value of (a/b) is large, and these are the crystals exhibiting easy dislocation motion. In contrast to this, crystals with more complex unit cells have relatively large Burgers vectors and small spacing between atomic planes across the glide plane, giving large Peierls stress. In these solids, the shear stress for dislocation motion cannot be overcome before fracturing the solid.

The actual motion of dislocations is believed to take place through small segments of the dislocation moving over the Peierls barrier, and subsequent motion of the ensuing kink–antikink in the direction of the dislocation line. This is illustrated in Fig. 10.10: the dislocation line in the equilibrium configuration resides in the Peierls valley, where the energy is minimized. A section of the dislocation may overcome the Peierls energy barrier by creating a kink–antikink

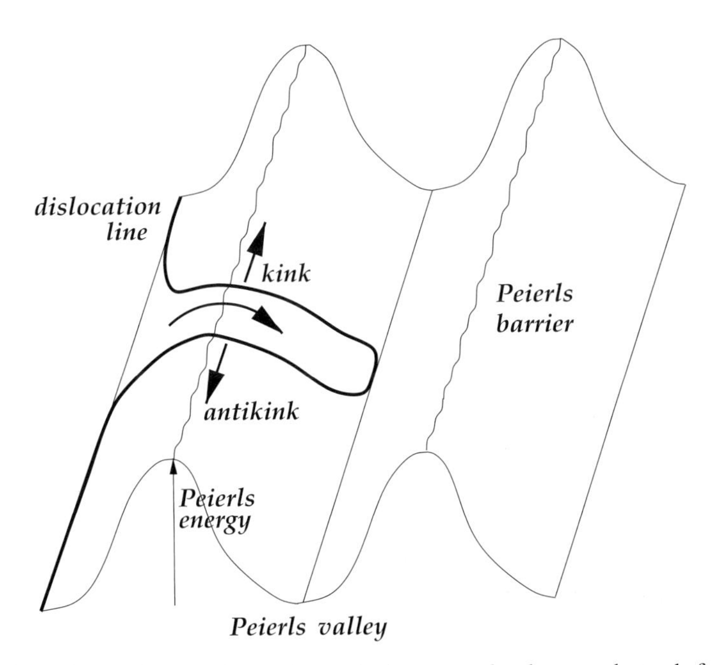

Figure 10.10. Dislocation motion in the Peierls energy landscape, through formation of kink–antikink pairs.

pair and moving into the next Peierls valley. The kinks can then move along the dislocation line, eventually displacing the entire dislocation over the Peierls energy barrier. Presumably it is much easier to move the kinks in the direction of the dislocation line rather than the entire line all at once in the direction perpendicular to it. The influence of the core structure, the effect of temperature, and the presence of impurities, all play an important role in the mobility of dislocations, which is central to the mechanical behavior of solids.

10.3 Brittle versus ductile behavior

The broadest classification of solids in terms of their mechanical properties divides them into two categories, brittle and ductile. Brittle solids fracture under the influence of external stresses. Ductile solids respond to external loads by deforming plastically. Typical stress–strain curves for brittle and ductile solids are shown in Fig. 10.11: a brittle solid is usually characterized by a large Young's modulus (see Appendix E), but can withstand only limited tensile strain, beyond which it fractures; it also remains in the elastic regime (linear stress-strain relation) up to the fracture point: if the external load is released the solid returns to its initial state. A ductile solid, on the other hand, has lower Young's modulus but does not break until much larger strain is applied. Beyond a certain amount of strain the solid starts deforming plastically, due to the introduction and motion of dislocations, as discussed in the previous section. The point beyond which the ductile solid is no longer elastic is called the yield point, characterized by the yield stress, σ_y, and yield strain, ϵ_y. If the external load is released after the yield point has been passed the solid does not return to its original state but has a permanent deformation. Often, the stress just above the yield point exhibits a dip as a function of strain, because dislocations at this point multiply fast, so that a smaller stress is needed to maintain a constant strain rate. This behavior is illustrated in Fig. 10.11.

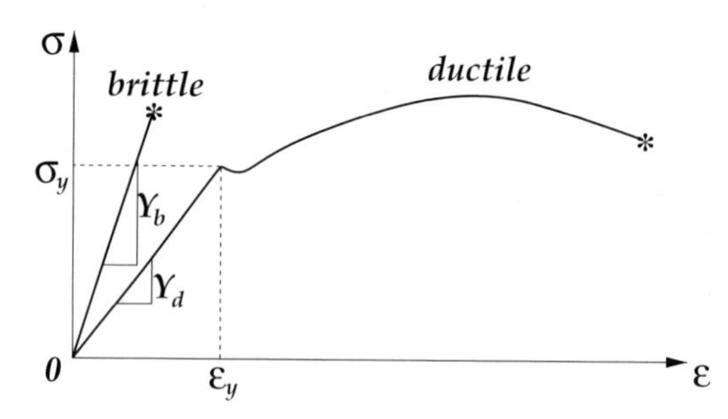

Figure 10.11. Stress σ versus strain ϵ relationships for typical brittle or ductile solids. The asterisks indicate the fracture points. The triangles in the elastic regime, of fixed length in the strain, indicate the corresponding Young's moduli, Y_b, Y_d. The yield point of the ductile solid is characterized by the yield stress σ_y and the yield strain ϵ_y.

Formulating a criterion to discriminate between brittle and ductile response based on atomistic features of the solid has been a long sought after goal in the mechanics of solids. At the phenomenological level, theories have been developed that characterize the two types of behavior in terms of cleavage of the crystal or the nucleation and motion of dislocations[138, 139]. We review here the basic elements of these notions.

10.3.1 Stress and strain under external load

We begin with some general considerations of how a solid responds to external loading. In all real solids there exist cracks of different sizes. The question of brittle or ductile response reduces to what happens to the cracks under external loading. The manner in which the external forces are applied to the crack geometry can lead to different types of loading, described as mode *I*, *II* and *III*; this is illustrated in Fig. 10.12. In mode *I* the applied stress is pure tension, while in modes *II* and *III* the applied stress is pure shear. The basic idea is that if the crack propagates into the solid under the influence of the external stress, the response is described as brittle, whereas if the crack blunts and absorbs the external load by deforming plastically, the response is described as ductile. The propagation of the crack involves the breaking of bonds between atoms at the very tip of the crack in a manner that leads to cleavage. The blunting of the crack requires the generation and motion of dislocations in the neighborhood of the crack tip; these are the defects that can lead to plastic deformation. Thus, the brittle or ductile nature of the solid is related to what happens at the atomistic level near the tip of pre-existing cracks under external loading.

Before we examine the phenomenological models that relate microscopic features to brittle or ductile behavior, we will present the continuum elasticity picture

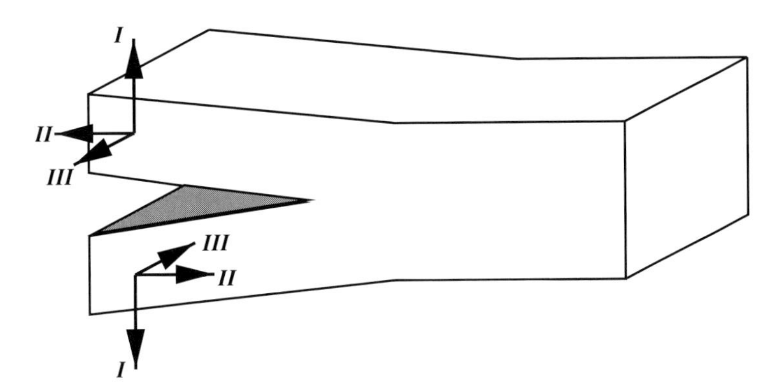

Figure 10.12. Definition of the three loading modes of a crack: mode *I* involves pure tension, and modes *II* and *III* involve pure shear in the two possible directions on the plane of the crack.

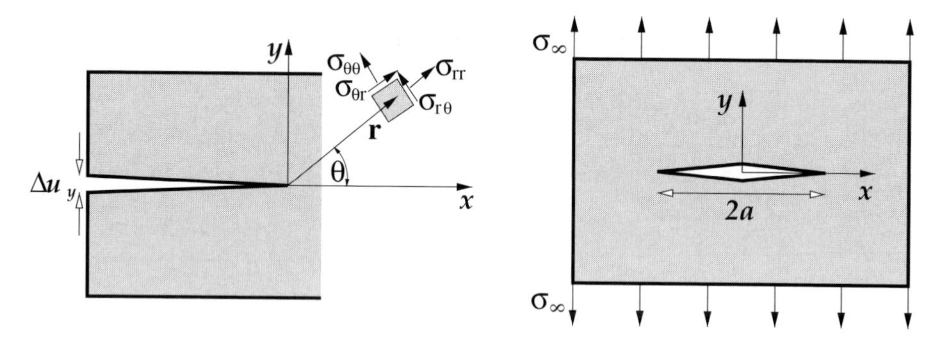

Figure 10.13. **Left:** definition of stress components at a distance **r** from the crack tip in polar coordinates, (r, θ). The crack runs along the z axis. **Right:** the penny crack geometry in a solid under uniform external loading σ_∞ very far from the crack.

of a loaded crack. The first solution of this problem was produced for an idealized 2D geometry consisting of a very narrow crack of length $2a$ (x direction), infinitesimal height (y direction) and infinite width (z direction), as illustrated in Fig. 10.13; this is usually called a "penny crack". The solid is loaded by a uniform stress σ_∞ very far from the penny crack, in what is essentially mode I loading for the crack. The solution to this problem gives a stress field in the direction of the loading and along the extension of the crack:

$$\left[\sigma_{yy}\right]_{y=0,x>a} = \frac{\sigma_\infty x}{\sqrt{x^2 - a^2}}$$

with the origin of the coordinate axes placed at the center of the crack. Letting $x = r + a$ and expanding the expression for σ_{yy} in powers of r, we find to lowest order

$$\left[\sigma_{yy}\right]_{y=0,r>0} = \frac{\sigma_\infty \sqrt{\pi a}}{\sqrt{2\pi r}}$$

This is an intriguing result, indicating that very near the crack tip, $r \to 0$, the stress diverges as $1/\sqrt{r}$. The general solution for the stress near the crack has the form

$$\sigma_{ij} = \frac{K}{\sqrt{2\pi r}} f_{ij}(\theta) \sum_{n=0}^{N} \alpha_{ij}^{(n)}(\theta) r^{n/2} \tag{10.25}$$

where $f_{ij}(\theta)$ is a universal function normalized so that $f_{ij}(0) = 1$. The constant K is called the "stress intensity factor". The higher order terms in $r^{n/2}$, involving the constants $\alpha_{ij}^{(n)}$, are bounded for $r \to 0$ and can be neglected in analyzing the behavior in the neighborhood of the crack tip.

For the geometry of Fig. 10.13, which is referred to as "plane strain", since by symmetry the strain is confined to the xy plane ($u_z = 0$), the above expression put in polar coordinates produces to lowest order in r

$$\sigma_{rr} = \frac{K_I}{\sqrt{2\pi r}} \cos\frac{\theta}{2}\left(2 - \cos^2\frac{\theta}{2}\right) \tag{10.26}$$

$$\sigma_{r\theta} = \frac{K_I}{\sqrt{2\pi r}} \cos^2\frac{\theta}{2} \sin\frac{\theta}{2} \tag{10.27}$$

$$\sigma_{\theta\theta} = \frac{K_I}{\sqrt{2\pi r}} \cos^3\frac{\theta}{2} \tag{10.28}$$

while the displacement field $\mathbf{u}$ to lowest order in r is

$$u_x = \frac{K_I(\kappa - \cos\theta)}{2\mu}\sqrt{\frac{r}{2\pi}} \cos\frac{\theta}{2} \tag{10.29}$$

$$u_y = \frac{K_I(\kappa - \cos\theta)}{2\mu}\sqrt{\frac{r}{2\pi}} \sin\frac{\theta}{2} \tag{10.30}$$

where $\kappa = 3 - 4\nu$ for plane strain (see Problem 5). The expression for the stress, evaluated at the plane which is an extension of the crack and ahead of the crack $((x > 0, y = 0)$ or $\theta = 0$, see Fig. 10.13), gives

$$\left[\sigma_{yy}\right]_{\theta=0} = \frac{K_I}{\sqrt{2\pi r}} \tag{10.31}$$

Comparing this result to that for the penny crack loaded in mode I, we find that in the latter case the stress intensity factor is given by $K_I = \sigma_\infty\sqrt{\pi a}$. The expression for the displacement field, evaluated on either side of the opening behind the crack $((x < 0, y = 0)$ or $\theta = \pm\pi$, see Fig. 10.13), gives

$$\Delta u_y = u_y|_{\theta=\pi} - u_y|_{\theta=-\pi} = \frac{4(1-\nu)}{\mu} K_I \sqrt{\frac{r}{2\pi}} \tag{10.32}$$

The generalization of these results to mode II and mode III loading gives

$$\text{mode } II: \ \sigma_{yx} = \frac{K_{II}}{\sqrt{2\pi r}}, \quad \Delta u_x = \frac{4(1-\nu)}{\mu} K_{II}\sqrt{\frac{r}{2\pi}}$$

$$\text{mode } III: \ \sigma_{yz} = \frac{K_{III}}{\sqrt{2\pi r}}, \quad \Delta u_z = \frac{4}{\mu} K_{III}\sqrt{\frac{r}{2\pi}}$$

where σ_{yx}, σ_{yz} are the dominant stress components on the plane which is an extension of the crack and ahead of the crack ($\theta = 0$), and $\Delta u_x, \Delta u_z$ refer to the displacement discontinuity behind the crack ($\theta = \pm\pi$).

These results have interesting physical implications. First, the divergence of the stress near the crack tip as $1/\sqrt{r}$ for $r \to 0$, Eq. (10.31), means that there are very large forces exerted on the atoms in the neighborhood of the crack tip. Of course in a real solid the forces on atoms cannot diverge, because beyond a certain point the bonds between atoms are broken and there is effectively no interaction between them. This bond breaking can lead to cleavage, that is, separation of the two surfaces on either side of the crack plane, or to the creation of dislocations, that is, plastic deformation of the solid. These two possibilities correspond to brittle or ductile response as already mentioned above. Which of the two possibilities will be preferred is dictated by the microscopic structure and bonding in the solid. Second, the displacement field is proportional to $r^{1/2}$, Eq. (10.32), indicating that the distortion of the solid can indeed be large far from the crack tip while right at the crack tip it is infinitesimal.

10.3.2 Brittle fracture – Griffith criterion

What remains to be established is a criterion that will differentiate between the tendency of the solid to respond to external loading by brittle fracture or ductile deformation. This is not an easy task, because it implies a connection between very complex processes at the atomistic level and the macroscopic response of the solid; in fact, this issue remains one of active research at present. Nevertheless, phenomenological theories do exist which capture the essence of this issue to a remarkable extent. In an early work, Griffith developed a criterion for the conditions under which brittle fracture will occur [140]. He showed that the critical rate[3] of energy per unit area G_b required to open an existing crack by an infinitesimal amount in mode I loading is given by

$$G_b = 2\gamma_s \tag{10.33}$$

where γ_s is the surface energy (energy per unit area on the exposed surface of each side of the crack). This result can be derived from a simple energy-balance argument. Consider a crack of length a and infinite width in the perpendicular direction, as illustrated in Fig. 10.14. The crack is loaded in mode I. If the load is P and the extension of the solid in the direction of the load is δy, then the change in internal energy U per unit width of the crack, for a fixed crack length will be given by

$$\delta U = P\delta y, \quad \delta a = 0$$

This can be generalized to the case of a small extension of the crack by δa, by introducing the energy release rate G_b related to this extension:

$$\delta U = P\delta y - G_b \delta a, \quad \delta a \neq 0 \tag{10.34}$$

[3] The word rate here is used not in a temporal sense but in a spatial sense, as in energy per unit crack area; this choice conforms with the terminology in the literature.

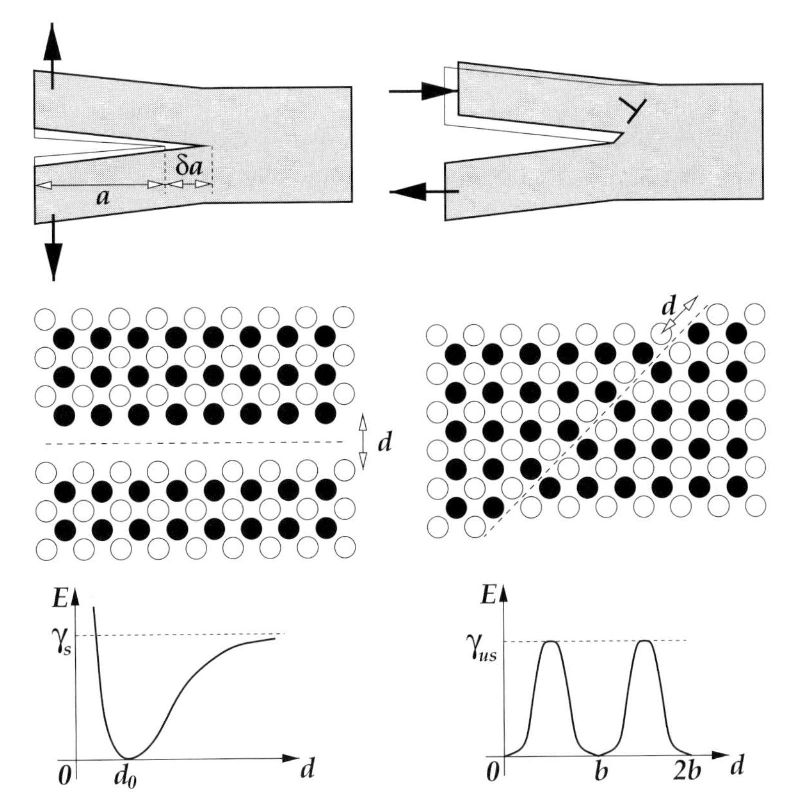

Figure 10.14. Schematic representation of key notions in brittle and ductile response. **Left:** the top panel illustrates how the crack opens in mode *I* loading (pure tension) by cleavage; the thin solid lines indicate the original position of the crack, the thicker solid lines indicate its final position after it has propagated by a small amount δa. The middle panel illustrates cleavage of the crystal along the cleavage plane. The bottom panel indicates the behavior of the energy during cleavage. **Right:** the top panel illustrates how the crack blunts in mode *II* loading (pure shear) by the emission of an edge dislocation (inverted T); the thin solid lines indicate the original position of the crack, the thicker solid lines indicate its final position. The middle panel illustrates sliding of two halves of the crystal on the glide plane (in general different than the cleavage plane). The bottom panel indicates the behavior of the energy during sliding.

The total energy of the solid E is given by the internal energy plus any additional energy cost introduced by the presence of surfaces on either side of the crack plane, which, per unit width of the crack, is

$$E = U + 2\gamma_s a$$

with γ_s the surface energy. From this last expression we obtain

$$\delta E = \delta U + 2\gamma_s \delta a = P\delta y - G_b \delta a + 2\gamma_s \delta a$$

where we have also used Eq. (10.34). At equilibrium, the total change in the energy must be equal to the total work by the external forces, $\delta E = P\delta y$, which leads directly to the Griffith criterion, Eq. (10.33).

Griffith's criterion for brittle fracture involves a remarkably simple expression. It is straightforward to relate the energy release rate to the stress intensity factors introduced above to describe the stresses and displacements in the neighborhood of the crack. In particular, for mode I loading in plane strain, the energy release rate G_I and the stress intensity factor K_I are related by

$$G_I = \frac{1-\nu}{2\mu}K_I^2$$

(see Problem 6). The Griffith criterion is obeyed well by extremely brittle solids, such as silica or bulk silicon. For most other solids, the energy required for fracture is considerably larger than $2\gamma_s$. The reason for the discrepancy is that, in addition to bond breaking at the crack tip, there is also plastic deformation of the solid ahead of the crack, which in this picture is not taken into account. The plastic deformation absorbs a large fraction of the externally imposed load and as a consequence a much larger load is required to actually break the solid.

10.3.3 Ductile response – Rice criterion

These considerations bring us to the next issue, that is, the formulation of a criterion for ductile response. As already mentioned, ductile response is related to dislocation activity. Nucleation of dislocations at a crack tip, and their subsequent motion away from it, is the mechanism that leads to blunting rather than opening of existing cracks, as shown schematically in Fig. 10.14. The blunting of the crack tip is the atomistic level process by which plastic deformation absorbs the external load, preventing the breaking of the solid. Formulating a criterion for the conditions under which nucleation of dislocations at a crack tip will occur is considerably more complicated than for brittle fracture. This issue has been the subject of much theoretical analysis. Early work by Rice and Thomson [141] attempted to put this process on a quantitative basis. The criteria they derived involved features of the dislocation such as the core radius r_c and the Burgers vector b; the core radius, however, is not a uniquely defined parameter. More recent work has been based on the Peierls framework for describing dislocation properties. This allowed the derivation of expressions that do not involve arbitrary parameters such as the dislocation core radius [137]. We briefly discuss the work of J.R. Rice and coworkers, which provides an appealingly simple and very powerful formulation of the problem, in the context of the Peierls framework [142].

The critical energy release rate G_d for dislocation nucleation at a crack tip, according to Rice's criterion [137], is given by

$$G_d = \alpha \gamma_{us} \tag{10.35}$$

γ_{us} is the unstable stacking energy, defined as the lowest energy barrier that must be overcome when one-half of an infinite crystal slides over the other half, while the crystal is brought from one equilibrium configuration to another equivalent one; α is a factor that depends on the geometry. For mode I and mode III loading $\alpha = 1$; for more general loading geometries α depends on two angles, the angle between the dislocation slip plane and the crack plane, and the angle between the Burgers vector and the crack line [142].

Rice's criterion can be rationalized in the special case of pure mode II loading, illustrated in Fig. 10.14: when the two halves of the crystal slide over each other on the slip plane, the energy goes through a maximum at a relative displacement $b/2$, where b is the Burgers vector for dislocations on this plane; this maximum value of the energy is γ_{us}. The variation in the energy is periodic with period b. Rice showed that the energy release rate in this case, which in general is given by

$$G_{II} = \frac{1-\nu}{2\mu} K_{II}^2$$

(see Problem 6), is also equal to the elastic energy associated with slip between the two halves of the crystal, $U(d_{tip})$, where d_{tip} is the position of the crack tip. When the crack tip reaches $d_{tip} = b/2$, the energy is at a local maximum in the direction of tip motion (this is actually a saddle point in the energy landscape associated with any relative displacement of the two halves of the crystal on the slip plane). Before this local maximum has been reached, if allowed to relax the solid will return to its original configuration. However, once this local maximum has been reached, the solid will relax to the next minimum of the energy situated at $d_{tip} = b$; this corresponds to the creation of an edge dislocation. Under the external shear stress, the dislocation will then move further into the solid through the types of processes we discussed in the previous section. In this manner, the work done by external forces on the solid is absorbed by the creation of dislocations at the tip of the crack and their motion away from it. In terms of structural changes in the solid, this process leads to a local change in the neighborhood of the crack tip but no breaking. In general, the dislocation will be created and will move on a plane which does not coincide with the crack plane, producing blunting of the crack as shown in Fig. 10.14. For a given crystal structure, a number of different possible glide planes and dislocation types (identified by their Burgers vectors) must be considered in order to determine the value of γ_{us}. For example, it is evident from the representation of Fig. 10.14 that in mode *II* loading the relevant dislocation is an edge one, but for mode *III* loading it is a screw dislocation.

The tendency for brittle versus ductile behavior can then be viewed as a competition between the G_b and G_d terms: their ratio will determine whether the crystal, when externally loaded, will undergo brittle fracture (high γ_{us}/γ_s), or whether it will absorb the load by creation and motion of dislocations (low γ_{us}/γ_s). There is, however, no guidance provided by these arguments as to what value of γ_{us}/γ_s differentiates between the tendency for brittle or ductile response. In fact, even when compared with atomistic simulations where the ratio γ_{us}/γ_s can be calculated directly and the response of the system is known, this ratio cannot be used as a predictive tool. The reason is that a number of more complicated issues come into play in realistic situations. Examples of such issues are the coupling of different modes of loading, the importance of thermal activation of dislocation processes, ledges effects, lattice trapping, dislocation loops, and other effects of the atomically discrete nature of real solids, which in the above picture of brittle or ductile response are not taken into account. All these issues are the subject of recent and on-going investigations (see, for example, Refs. [142–147]).

The real power of the, admittedly oversimplified, picture described above lies in its ability to give helpful hints and to establish trends of how the complex macroscopic phenomena we are considering can be related to atomistic level structural changes. In particular, both the surface energy γ_s for different cleavage planes of a crystal, as well as the unstable stacking energy γ_{us} for different glide planes, are intrinsic properties of the solid which can be calculated with high accuracy using modern computational methods of the type discussed in chapter 5. Changes in these quantities due to impurities, alloying with other elements, etc., can then provide an indication of how these structural alterations at the microscopic level can affect the large-scale mechanical behavior of the solid (see, for example, Ref. [149], where such an approach was successfully employed to predict changes in the brittleness of specific materials). At present, much remains to be resolved before these theories are able to capture all the complexities of the competition between crack blunting versus brittle fracture tendencies in real materials.

10.3.4 Dislocation–defect interactions

Up to this point we have been treating dislocations as essentially isolated line defects in solids. Obviously, dislocations in real solids coexist with other defects. The interaction of dislocations with defects is very important for the overall behavior of the solid and forms the basis for understanding several interesting phenomena. While a full discussion of these interactions is not possible in the context of the present treatment, some comments on the issue are warranted. We will consider selectively certain important aspects of these interactions, which we address in the order we adopted for the classification of defects: interaction between dislocations and zero-dimensional defects, interactions between dislocations themselves, and interactions between dislocations and two-dimensional defects (interfaces).

In the first category, we can distinguish between zero-dimensional defects which are of microscopic size, such as the point defects we encountered in chapter 9, and defects which have finite extent in all dimensions but are not necessarily of atomic-scale size. Atomic-scale point defects, such as vacancies, interstitials and impurities, typically experience long-range interactions with dislocations because of the strain field introduced by the dislocation to which the motion of point defects is sensitive. As a consequence, point defects can be either drawn toward the dislocation core or repelled away from it. If the point defects can diffuse easily in the bulk material, so that their equilibrium distribution can follow the dislocation as it moves in the crystal, they will alter significantly the behavior of the dislocation. A classic example is hydrogen impurities: H atoms, due to their small size, can indeed diffuse easily even in close-packed crystals and can therefore maintain the preferred equilibrium distribution in the neighborhood of the dislocation as it moves. This, in turn, can affect the overall response of the solid. It is known, for instance, that H impurities lead to embrittlement of many metals, including Al, the prototypical ductile metal. The actual mechanisms by which this effect occurs at the microscopic level remain open to investigation, but there is little doubt that the interaction of the H impurities with the dislocations is at the heart of this effect.

We turn next to the second type of zero-dimensional defects, those which are not of atomic-scale size. The interaction of a dislocation with such defects can be the source of dislocation multiplication. This process, known as the Frank–Read source, is illustrated in Fig. 10.15: a straight dislocation anchored at two such defects is made to bow out under the influence of an external stress which would normally make the dislocation move. At some point the bowing is so severe that two points of the dislocation meet and annihilate. This breaks the dislocation into two portions, the shorter of which shrinks to the original dislocation configuration between the defects while the larger moves away as a dislocation loop. The process can be continued indefinitely as long as the external stress is applied to the system, leading to multiple dislocations emanating from the original one.

As far as interactions of dislocations among themselves are concerned, these can take several forms. One example is an intersection, formed when two dislocations whose lines are at an angle meet. Depending on the type of dislocations and the angle at which they meet, this can lead to a junction (permanent lock between the two dislocations) or a jog (step on the dislocation line). The junctions make it difficult for dislocations to move past each other, hence they restrict the motion of the dislocation. We mentioned earlier how the motion of a dislocation through the crystal

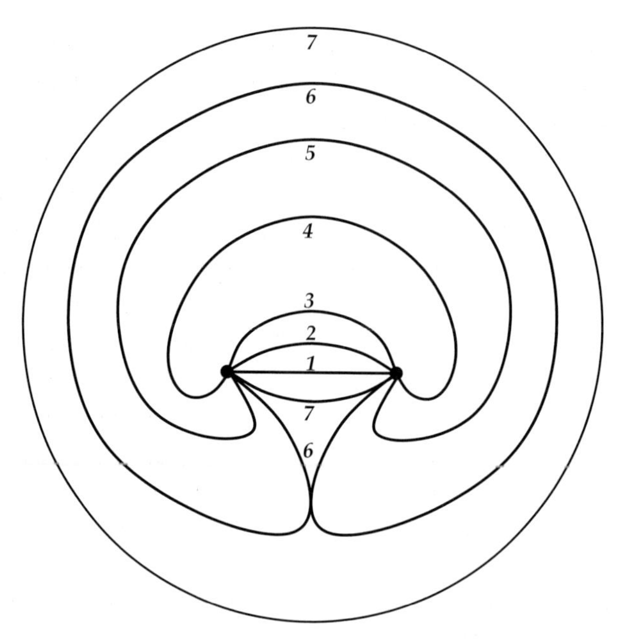

Figure 10.15. Illustration of Frank–Read source of dislocations: under external stress the original dislocation (labeled 1) between two zero-dimensional defects is made to bow out. After sufficient bowing, two parts of the dislocation meet and annihilate (configuration 6), at which point the shorter portion shrinks to the original configuration and the larger portion moves away as a dislocation loop (configuration 7, which has two components).

leads to plastic deformation (see Fig. 10.4 and accompanying discussion). When dislocation motion is inhibited, the plasticity of the solid is reduced, an effect known as hardening. The formation of junctions which correspond to attractive interactions between dislocations is one of the mechanisms for hardening. The junctions or jogs can be modeled as a new type of particle which has its own dynamics in the crystal environment. Another consequence of dislocation interactions is the formation of dislocation walls, that is, arrangements of many dislocations on a planar configuration. This type of interaction also restricts the motion of individual dislocations and produces changes in the mechanical behavior of the solid. This is one instance of a more general phenomenon, known as dislocation microstructure, in which interacting dislocations form well defined patterns. Detailed simulations of how dislocations behave in the presence of other dislocations are a subject of active research and reveal quite complex behavior at the atomistic level [150, 151].

Finally, we consider the interaction of dislocations with two-dimensional obstacles such as interfaces. Most materials are composed of small crystallites whose interfaces are called grain boundaries. Interestingly, grain boundaries themselves can be represented as arrays of dislocations (this topic is discussed in more detail in chapter 11). Here, we will examine what happens when dislocations which exist within the crystal meet a grain boundary. When this occurs, the dislocations become immobile and pile up at the boundary. As a consequence, the ability of the crystal to deform plastically is again diminished and the material becomes harder. It is easy to imagine that since the smaller the crystallite the higher the surface-to-volume ratio, materials composed of many small crystallites will be harder than materials composed of few large crystallites. This is actually observed experimentally, and is known as the Hall–Petch effect [152]. Interestingly, the trend continues down to a certain size of order 10 nm, below which the effect is reversed, that is, the material becomes softer as the grain size decreases. The reason for this reversal in behavior is that, for very small grains, sliding between adjacent grains at the grain boundaries becomes easy and this leads to a material that yields sooner to external stress [153].

Putting all the aspects of dislocation behavior together, from the simple motion of an isolated dislocation which underlies plastic deformation, to the mechanisms of dislocation nucleation at crack tips, to the interaction of dislocations with other zero-, one- and two-dimensional defects, is a daunting task. It is, nevertheless, an essential task in order to make the connection between the atomistic level structure and dynamics to the macroscopic behavior of materials as exemplified by fascinating phenomena like work hardening, fatigue, stress-induced corrosion, etc. To this end, recent computational approaches have set their sights at a more realistic connection between the atomistic and macroscopic regimes in what has become known as "multiscale modeling of materials" (for some representative examples see Refs. [154, 155]).

Further reading

1. *Theory of Dislocations*, J.P. Hirth and J. Lothe (Krieger, Malabar, 1992).
   This is the standard reference for the physics of dislocations, containing extensive and detailed discussions of every aspect of dislocations.
2. *The Theory of Crystal Dislocations*, F.R.N. Nabarro (Oxford University Press, Oxford, 1967).
   An older but classic account of dislocations by one of the pioneers in the field.
3. *Introduction to Dislocations*, D. Hull and D.J. Bacon (Pergamon Press, 1984).
   An accessible and thorough introduction to dislocations.
4. *Dislocations*, J. Friedel (Addison-Wesley, Reading, MA, 1964).
   A classic account of dislocations, with many insightful discussions.
5. *Elementary Dislocation Theory*, J. Weertman and J.R. Weertman (McMillan, New York, 1964).
6. *Theory of Crystal Dislocations*, A.H. Cottrell (Gordon and Breach, New York, 1964).
7. *Dislocations in Crystals*, W.T. Read (McGraw-Hill, New York, 1953).
8. *Dislocation Dynamics and Plasticity*, T. Suzuki, S. Takeuchi, H. Yoshinaga (Springer-Verlag, Berlin, 1991).
   This book offers an insightful connection between dislocations and plasticity.
9. "The dislocation core in crystalline materials", M.S. Duesbery and G.Y. Richardson, in *Solid State and Materials Science*, vol. 17, pp. 1–46 (CRC Press, 1991).
   This is a thorough, modern discussion of the properties of dislocation cores.

Problems

1. In order to obtain the stress field of a screw dislocation in an isotropic solid, we can define the displacement field as

$$u_x = u_y = 0, \quad u_z = \frac{b_s \theta}{2\pi} = \frac{b_s}{2\pi} \tan^{-1} \frac{y}{x} \tag{10.36}$$

with b_s the Burgers vector along the z axis. This is justified by the schematic representation of the screw dislocation in Fig. 10.2: sufficiently far from the dislocation core, u_z goes uniformly from zero to b_s as θ ranges from zero to 2π, while the other two components of the strain field vanish identically. Find the strain components of the screw dislocation in cartesian and polar coordinates. Then, using the the stress–strain relations for an isotropic solid, Eq. (E.32), find the stress components of the screw dislocation in cartesian and polar coordinates and compare the results to the expressions given in Table 10.1. (Hint: the shear stress components in cartesian and polar coordinates are related by:

$$\begin{aligned} \sigma_{rz} &= \sigma_{xz} \cos\theta + \sigma_{yz} \sin\theta \\ \sigma_{\theta z} &= \sigma_{yz} \cos\theta - \sigma_{xz} \sin\theta \end{aligned} \tag{10.37}$$

Similar relations hold for the shear strains.)

2. In order to obtain the stress field of an edge dislocation in an isotropic solid, we can use the equations of plane strain, discussed in detail in Appendix E. The geometry of Fig. 10.1 makes it clear that a single infinite edge dislocation in an isotropic solid satisfies the conditions of plane strain, with the strain ϵ_{zz} along the axis of the dislocation vanishing identically. The stress components for plane strain are given in terms of the Airy stress function, $A(r, \theta)$, by Eq. (E.49). We define the function

$$B = \sigma_{xx} + \sigma_{yy} = \nabla^2_{xy} A$$

where the laplacian with subscript xy indicates that only the in-plane components are used. The function B must obey Laplace's equation, since the function A obeys Eq. (E.50). Laplace's equation for B in polar coordinates in 2D reads

$$\left(\frac{\partial^2}{\partial r^2} + \frac{1}{r}\frac{\partial}{\partial r} + \frac{1}{r^2}\frac{\partial^2}{\partial \theta^2} \right) B(r, \theta) = 0$$

which is separable, that is, $B(r, \theta)$ can be written as a product of a function of r with a function of θ.

(a) Show that the four possible solutions to the above equation are

$$c, \quad \ln r, \quad r^{\pm n} \sin n\theta, \quad r^{\pm n} \cos n\theta$$

with c a real constant and n a positive integer. Of these, c and $\ln r$ must be rejected since they have no θ dependence, and the function B must surely depend on the variable θ from physical considerations. The solutions with positive powers of r must also be rejected since they blow up at large distances from the dislocation core, which is unphysical. Of the remaining solutions, argue that the dominant one for large r which makes physical sense from the geometry of the edge dislocation is

$$B(r,\theta) = \beta_1 \frac{1}{r} \sin\theta$$

(b) With this expression for $B(r, \theta)$, show that a solution for the Airy function is

$$A(r,\theta) = \alpha_1 r \sin\theta \ln r$$

Discuss why the solutions to the homogeneous equation $\nabla^2_{xy} A = 0$ can be neglected. From the above solution for $A(r, \theta)$, obtain the stresses as determined by Eq. (E.49) and use them to obtain the strains from the general strain–stress relations for an isotropic solid, Eq. (E.28). Then use the normalization of the integral of ϵ_{xx} to the Burgers vector b_e,

$$u_x(+\infty) - u_x(-\infty) = \int_{-\infty}^{\infty} [\epsilon_{xx}(x, -\delta y) - \epsilon_{xx}(x, +\delta y)]\, \mathrm{d}x = b_e$$

where $\delta y \to 0^+$, to determine the value of the constant α_1. This completes the derivation of the stress field of the edge dislocation.

(c) Express the stress components in both polar and cartesian coordinates and compare them to the expressions given in Table 10.1.

3. Derive the solution for the shape of the dislocation, Eq. (10.22), which satisfies the PN integro-differential equation, Eq. (10.20), with the assumption of a sinusoidal misfit potential, Eq. (10.21).
4. The original lattice restoring stress considered by Frenkel [136] was

$$\frac{\mathrm{d}\gamma(\overline{u})}{\mathrm{d}\overline{u}} = F_{max} \sin\left(\frac{2\pi \overline{u}(x)}{b}\right)$$

where b is the Burgers vector and F_{max} the maximum value of the stress. Does this choice satisfy the usual conditions that the restoring force should obey? Find the solution for $u(x)$ when this restoring force is substituted into the PN integro-differential equation, Eq. (10.20). From this solution obtain the dislocation density $\rho(x)$ and compare $u(x)$ and $\rho(x)$ to those of Eq. (10.22), obtained from the choice of potential in Eq. (10.21). What is the physical meaning of the two choices for the potential, and what are their differences?
5. We wish to derive the expressions for the stress, Eqs. (10.26)–(10.28), and the displacement field, Eqs. (10.29), (10.30), in the neighborhood of a crack loaded in mode I. We will use the results of the plane strain situation, discussed in detail in Appendix E. We are interested in solutions for the Airy function,

$$A(r,\theta) = r^2 f(r\theta) + g(r,\theta)$$

such that the resulting stress has the form $\sigma_{ij} \sim r^q$ near the crack tip: this implies $A \sim r^{q+2}$.

(a) Show that the proper choices for mode I symmetry are

$$f(r,\theta) = f_0 r^q \cos q\theta, \quad g(r,\theta) = g_0 r^{q+2} \cos(q+2)\theta$$

With these choices, obtain the stress components $\sigma_{\theta\theta}, \sigma_{r\theta}, \sigma_{rr}$.

(b) Determine the allowed values of q and the relations between the constants f_0, g_0 by requiring that $\sigma_{\theta\theta} = \sigma_{r\theta} = 0$ for $\theta = \pm\pi$.

(c) Show that by imposing the condition of bounded energy

$$\int_0^{2\pi} \int_0^R \sigma^2 r \mathrm{d}r \mathrm{d}\theta < \infty$$

and by discarding all terms r^q which give zero stress at the crack tip $r = 0$, we arrive at the solution given by Eqs. (10.26)–(10.28), as the only possibility.

(d) From the solution for the stress, obtain the displacement field given in Eqs. (10.29) and (10.30), using the standard stress–strain relations for an isotropic solid (see Appendix E).

6. Show that the energy release rate for the opening of a crack, per unit crack area, due to elastic forces is given by

$$G = \lim_{\delta a \to 0} \frac{1}{\delta a} \int_0^{\delta a} \frac{1}{2} \sigma_{yj} \Delta u_j \mathrm{d}r = \frac{1-\nu}{2\mu}(K_I^2 + K_{II}^2) + \frac{1}{2\mu} K_{III}^2$$

where σ_{yj}, $\Delta u_j (j = x, y, z)$ are the dominant stress components and displacement discontinuities behind the crack, in mode *I*, *II* and *III* loading and K_I, K_{II}, K_{III} are the corresponding stress intensity factors.

11

Defects III: surfaces and interfaces

Two-dimensional defects in crystals consist of planes of atomic sites where the solid terminates or meets a plane of another crystal. We refer to the first type of defects as surfaces, to the second as interfaces. Interfaces can occur between two entirely different solids or between two grains of the same crystal, in which case they are called grain boundaries.

Surfaces and interfaces of solids are extremely important from a fundamental as well as from a practical point of view. At the fundamental level, surfaces and interfaces are the primary systems where physics in two dimensions can be realized and investigated, opening a different view of the physical world. We have already seen that the confinement of electrons at the interface between a metal and a semiconductor or two semiconductors creates the conditions for the quantum Hall effects (see chapters 7 and 9). There exist several other phenomena particular to 2D: one interesting example is the nature of the melting transition, which in 2D is mediated by the unbinding of defects [156, 157]. Point defects in 2D are the equivalent of dislocations in 3D, and consequently have all the characteristics of dislocations discussed in chapter 10. In particular, dislocations in two dimensions are mobile and have long-range strain fields which lead to their binding in pairs of opposite Burgers vectors. Above a certain temperature (the melting temperature), the entropy term in the free energy wins and it becomes favorable to generate isolated dislocations; this produces enough disorder to cause melting of the 2D crystal. At a more practical level, there are several aspects of surfaces and interfaces that are extremely important for applications. For instance, grain boundaries, a type of interface very common in crystalline solids, are crucial to mechanical strength. Similarly, chemical reactions mediated by solid surfaces are the essence of catalysis, a process of huge practical significance.

Surfaces are the subject of a very broad and rich field of study called surface science, to which entire research journals are devoted (including *Surface Science* and *Surface Review and Letters*). It would not be possible to cover all the interesting phenomena related to crystal surfaces in a short chapter. Our aim here is to illustrate how the concepts and techniques we developed in earlier chapters, especially those dealing with the link between atomic and electronic structure, can be applied to study representative problems of surface and interface physics.

11.1 Experimental study of surfaces

We begin our discussion of surfaces with a brief review of experimental techniques for determining their atomic structure. The surfaces of solids under usual circumstances are covered by a large amount of foreign substances. This is due to the fact that a surface of a pure crystal is usually chemically reactive and easy to contaminate. For this reason, it has been very difficult to study the structure of surfaces quantitatively. The detailed study of crystalline surfaces became possible with the advent of ultra high vacuum (UHV) chambers, in which surfaces of solids could be cleaned and maintained in their pure form. Real surfaces of solids, even when they have no foreign contaminants, are not perfect two-dimensional planes. Rather, they contain many imperfections, such as steps, facets, islands, etc., as illustrated in Fig. 11.1. With proper care, flat regions on a solid surface can be prepared, called terraces, which are large on the atomic scale consisting of thousands to millions of interatomic distances on each side. These terraces are close approximations to the ideal 2D surface of an infinite 3D crystal. It is this latter type of surface that we discuss here, by analogy to the ideal, infinite, 3D crystal studied in earlier chapters.

Typically, scattering techniques, such as low-energy electron diffraction (referred to as LEED), reflection high-energy electron diffraction (RHEED), X-ray scattering, etc., have been used extensively to determine the structure of surfaces (see, for

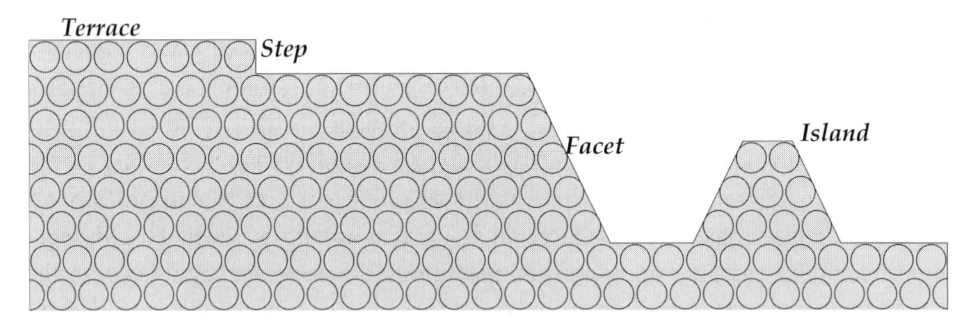

Figure 11.1. Various features of real surfaces shown in cross-section: reasonably flat regions are called terraces; changes in height of order a few atomic layers are called steps; terraces of a different orientation relative to the overall surface are referred to as facets; small features are referred to as islands. In the direction perpendicular to the plane of the figure, terraces, steps and facets can be extended over distances that are large on the atomic scale, but islands are typically small in all directions.

example, articles in the book by van Hove and Tong, mentioned in the Further reading section). Electron or X-ray scattering methods are based on the same principles used to determine crystal structure in 3D (see chapter 3). These methods, when combined with detailed analysis of the scattered signal as a function of incident-radiation energy, can be very powerful tools for determining surface structure. A different type of scattering measurement involves ions, which bounce off the sample atoms in trajectories whose nature depends on the incident energy; these methods are referred to as low-, medium-, or high-energy ion scattering (LEIS, MEIS and HEIS, respectively). Since surface atoms are often in positions different than those of a bulk-terminated plane (see section 11.2 for details), the pattern of scattered ions can be related to the surface structure. Yet a different type of measurement that reveals the structure of the surface on a local scale is referred to as "field ion microscope" (FIM). In this measurement a sample held at high voltage serves as an attraction center for ions, which bounce off it in a pattern that reflects certain aspects of the surface structure.

The methods mentioned so far concern measurements of the surface atomic structure, which can often be very different than the structure of a bulk plane of the same orientation, as discussed in section 11.2. Another interesting aspect of the surface is its chemical composition, which can also differ significantly from the composition of a bulk plane. This is usually established through a method called "Auger analysis", which consists of exciting core electrons of an atom and measuring the emitted X-ray spectrum when other electrons fall into the unoccupied core state: since core-state energies are characteristic of individual elements and the wavefunctions of core states are little affected by neighboring atoms, this spectrum can be used to identify the presence of specific atom types on the surface. The Auger signal of a particular atomic species is proportional to its concentration on the surface, and consequently it can be used to determine the chemical composition of the surface with great accuracy and sensitivity. The reason why this technique is particularly effective on surfaces is that the emitted X-rays are not subjected to any scattering when they are emitted by surface atoms.

Since the mid-1980s, a new technique called scanning tunneling microscopy (STM) has revolutionized the field of surface science by making it possible to determine the structure of surfaces by direct imaging of atomistic level details. This ingenious technique (its inventors, G. Binning and H. Rohrer were recognized with the 1986 Nobel prize for Physics), is based on a simple scheme (illustrated in Fig. 11.2). When an atomically sharp tip approaches a surface, and is held at a bias voltage relative to the sample, electrons can tunnel from the tip to the surface, or in the opposite direction, depending on the sign of the bias voltage. The tunneling current is extremely sensitive to the distance from the surface, with exponential dependence. Thus, in order to achieve constant tunneling current, a constant distance from the

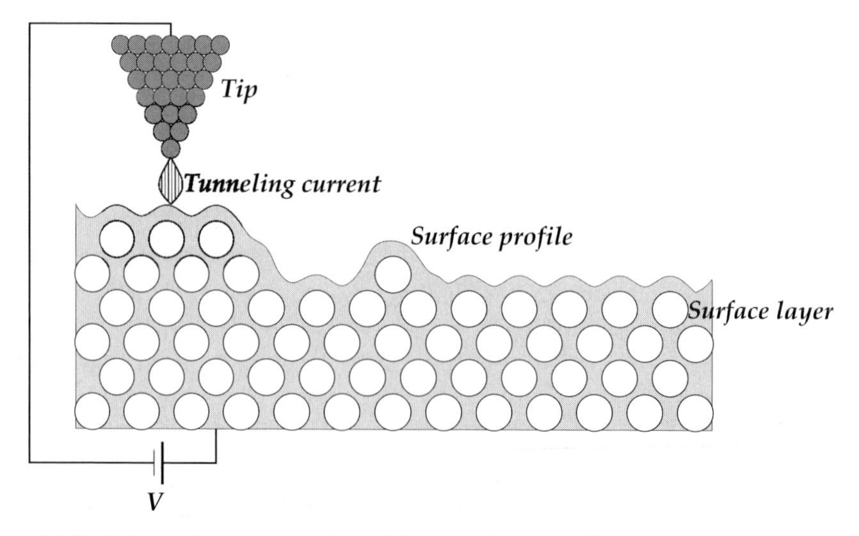

Figure 11.2. Schematic representation of the scanning tunneling microscope: a metal tip is held at a voltage bias V relative to the surface, and is moved up or down to maintain constant current of tunneling electrons. This produces a topographical profile of the electronic density associated with the surface.

surface must be maintained. A feedback loop can ensure this by moving the tip in the direction perpendicular to the surface while it is scanned over the surface. This leads to a scan of the surface at constant height. This height is actually determined by the electron density on the surface, since it is between electronic states of the sample and the tip that electrons can tunnel to and from.

We provide here an elementary discussion of how STM works, in order to illustrate its use in determining the surface structure. The theory of STM was developed by Tersoff and Hamann [158] and further extended by Chen [159]. The starting point is the general expression for the current I due to electrons tunneling between two sides, identified as left (L) and right (R):

$$I = \frac{2\pi e}{\hbar} \sum_{i,j} \left(n^{(L)}(\epsilon_i)[1 - n^{(R)}(\epsilon_j)] - n^{(R)}(\epsilon_j)[1 - n^{(L)}(\epsilon_i)] \right) |\mathcal{T}_{ij}|^2 \delta(\epsilon_i - \epsilon_j) \tag{11.1}$$

where $n^{(L)}(\epsilon)$ and $n^{(R)}(\epsilon)$ are the Fermi filling factors of the left and right sides, respectively, and $\mathcal{T}_{ij}$ is the tunneling matrix element between electronic states on the two sides identified by the indices i and j. The actual physical situation is illustrated in Fig. 11.3: for simplicity, we assume that both sides are metallic solids, with Fermi levels and work functions (the difference between the Fermi level and the vacuum level, which is common to both sides) defined as $\epsilon_F^{(L)}$, $\epsilon_F^{(R)}$ and $\phi^{(L)}$, $\phi^{(R)}$, respectively, when they are well separated and there is no tunneling current. When the two sides are brought close together and allowed to reach equilibrium with a

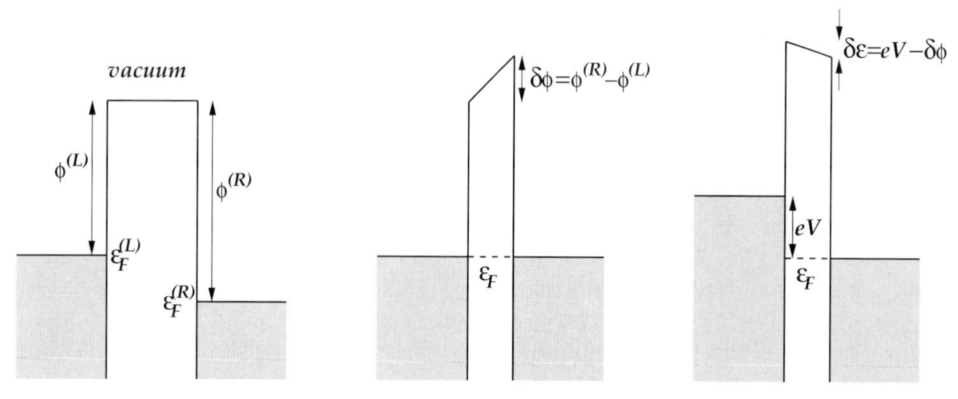

Figure 11.3. Energy level diagram for electron tunneling between two sides, identified as left (L) and right (R). **Left:** the situation corresponds to the two sides being far apart and having different Fermi levels denoted by $\epsilon_F^{(L)}$ and $\epsilon_F^{(R)}$; $\phi^{(L)}$ and $\phi^{(R)}$ are the work functions of the two sides. **Center:** the two sides are brought closer together so that equilibrium can be established by tunneling, which results in a common Fermi level $\epsilon_{\rm F}$ and an effective electric potential generated by the difference in work functions, $\delta\phi = \phi^{(R)} - \phi^{(L)}$. **Right:** one side is biased relative to the other by a potential difference V, resulting in a energy level shift and a new effective electric potential, generated by the energy difference $\delta\epsilon = eV - \delta\phi$.

common Fermi level, $\epsilon_{\rm F}$, there will be a barrier to tunneling due to the difference in work functions, given by

$$\delta\phi = \phi^{(R)} - \phi^{(L)}$$

This difference gives rise to an electric field, which in the case illustrated in Fig. 11.3 inhibits tunneling from the left to the right side. When a bias voltage V is applied, the inherent tunneling barrier can be changed. The energy shift introduced by the bias potential,

$$\delta\epsilon = eV - \delta\phi = eV - \phi^{(R)} + \phi^{(L)}$$

gives rise to an effective electric field, which in the case illustrated in Fig. 11.3 enhances tunneling relative to the zero-bias situation. Reversing the sign of the bias potential would have the opposite effect on tunneling.

At finite bias voltage V, one of the filling factor products appearing in Eq. (11.1) will give a non-vanishing contribution and the other one will give a vanishing contribution, as shown in Fig. 11.4. For the case illustrated in Fig. 11.3, the product that gives a non-vanishing contribution is $n^{(L)}(\epsilon_i)[1 - n^{(R)}(\epsilon_j)]$.

In order to fix ideas, in the following we will associate the left side with the sample, the surface of which is being studied, and the right side with the tip. In the limit of very small bias voltage and very low temperature, conditions which apply

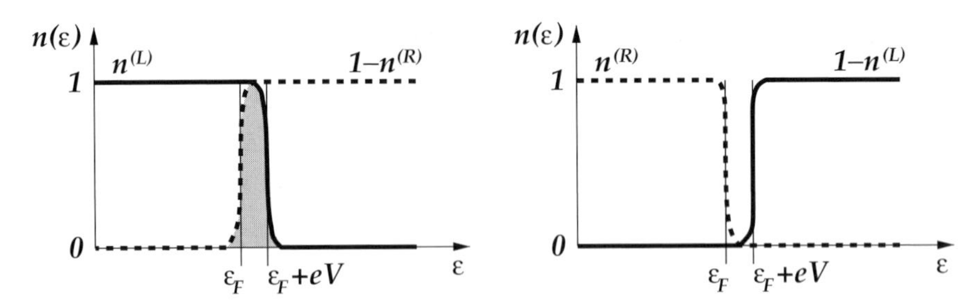

Figure 11.4. The filling factors entering in the expression for tunneling between the left $n^{(L)}(\epsilon)$, and right $n^{(R)}(\epsilon)$ sides, when there is a bias potential V on the left side: the product $n^{(L)}(\epsilon)[1 - n^{(R)}(\epsilon)]$ gives a non-vanishing contribution (shaded area), while $n^{(R)}(\epsilon)$ $[1 - n^{(L)}(\epsilon)]$ gives a vanishing contribution.

to most situations for STM experiments, the non-vanishing filling factor product divided by eV has the characteristic behavior of a δ-function (see Appendix G). Thus, in this limit we can write

$$I = \frac{2\pi e^2}{\hbar} V \sum_{i,j} |\mathcal{T}_{ij}|^2 \delta(\epsilon_i - \epsilon_F)\delta(\epsilon_j - \epsilon_F) \tag{11.2}$$

where we have used a symmetrized expression for the two δ-functions appearing under the summation over electronic states. The general form of the tunneling matrix element, as shown by Bardeen [160], is

$$\mathcal{T}_{ij} = \frac{\hbar^2}{2m_e} \int \left(\psi_i^*(\mathbf{r})\nabla_{\mathbf{r}}\psi_j(\mathbf{r}) - \psi_j(\mathbf{r})\nabla_{\mathbf{r}}\psi_i^*(\mathbf{r})\right) \cdot \mathbf{n}_s \mathrm{d}S \tag{11.3}$$

where $\psi_i(\mathbf{r})$ are the sample and $\psi_j(\mathbf{r})$ the tip electronic wavefunctions. The integral is evaluated on a surface S with surface-normal vector $\mathbf{n}_s$, which lies entirely between the tip and the sample. Tersoff and Hamann showed that, assuming a point source for the tip and a simple wavefunction of s character associated with it, the tunneling current takes the form

$$\lim_{V,T\to 0} I = I_0 \sum_i |\psi_i(\mathbf{r}_t)|^2 \delta(\epsilon_i - \epsilon_F) \tag{11.4}$$

where $\mathbf{r}_t$ is the position of the tip and I_0 is a constant which contains parameters describing the tip, such as the density of tip states at the Fermi level, $g_t(\epsilon_F)$, and its work function. The meaning of Eq. (11.4) is that the *spatial dependence* of the tunneling current is determined by the magnitude of electronic wavefunctions of the sample evaluated at the tip position. In other words, the tunneling current gives an exact topographical map of the sample electronic charge density evaluated at the position of a point-like tip. For a finite value of the bias voltage V, and in the limit of zero temperature, the corresponding expression for the tunneling current would involve a sum over sample states within eV of the Fermi level, as is evident from Fig. 11.3:

$$\lim_{T \to 0} I = I_0 \sum_i |\psi_i(\mathbf{r}_t)|^2 \theta(\epsilon_i - \epsilon_{\rm F})\theta(\epsilon_{\rm F} + eV - \epsilon_i) \tag{11.5}$$

where $\theta(x)$ is the Heavyside step function (see Appendix G). It is worthwhile mentioning that this expression is valid for either sign of the bias voltage, that is, whether one is probing occupied sample states (corresponding to $V > 0$, as indicated by Fig. 11.3), or unoccupied sample states (corresponding to $V < 0$); in the former case electrons are tunneling from the sample to the tip, in the latter from the tip to the sample.

In the preceding discussion we were careful to identify the electronic states on the left side as "sample" states without any reference to the surface. What remains to be established is that these sample states are, first, localized in space near the surface and, second, close in energy to the Fermi level. In order to argue that this is the case, we will employ simple one-dimensional models, in which the only relevant dimension is perpendicular to the surface plane. We call the free variable along this dimension z. In the following, since we are working in one dimension, we dispense with vector notation. First, we imagine that on the side of the sample, the electronic states are free-particle-like with wave-vector k, $\psi_k(z) \sim \exp(ikz)$, and energy $\epsilon_k = \hbar^2 k^2/2m_e$, and are separated by a potential barrier V_b (which we will take to be a constant) from the tip side; the solution to the single-particle Schrödinger equation in the barrier takes the form (see Appendix B)

$$\psi_k(z) \sim \mathrm{e}^{\pm\kappa z}, \quad \kappa = \sqrt{\frac{2m_e(V_b - \epsilon_k)}{\hbar^2}} \tag{11.6}$$

The tunneling current is proportional to the magnitude squared of this wavefunction evaluated at the tip position, assumed to be a distance d from the surface, which gives that

$$I \sim \mathrm{e}^{-2\kappa d}$$

where we have chosen the "−" sign for the solution along the positive z axis which points away from the surface, as the physically relevant solution that decays to zero far from the sample. This simple argument indicates that the tunneling current is exponentially sensitive to the sample–tip separation d. Thus, only states that have significant magnitude at the surface are relevant to tunneling.

We will next employ a slightly more elaborate model to argue that surface states are localized near the surface and have energies close to the Fermi level. We consider again a 1D model but this time with a weak periodic potential, that is, a nearly-free-electron model. The weak periodic potential in the sample will be taken to have only two non-vanishing components, the constant term V_0 and a term which

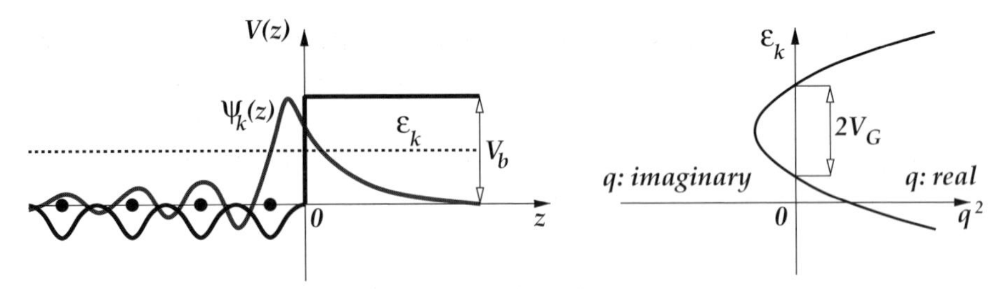

Figure 11.5. Illustration of the features of the nearly-free-electron model for surface states. **Left:** the potential $V(z)$ (thick black line), and wavefunction $\psi_k(z)$ of a surface state with energy ϵ_k, as functions of the variable z which is normal to the surface. The black dots represent the positions of ions in the model. **Right:** the energy ϵ_k as a function of q^2, where $q = k - G/2$. The energy gap is $2V_G$ at $q = 0$. Imaginary values of q give rise to the surface states.

involves the G reciprocal-lattice vector (see also the discussion of this model for the 3D case in chapter 3):

$$V(z) = V_0 + 2V_G \cos(Gz), \qquad z < 0 \tag{11.7}$$

Solving the 1D Schrödinger equation for this model gives for the energy ϵ_k and wavefunction $\psi_k(z < 0)$ (see Problem 1)

$$\epsilon_k = V_0 + \frac{\hbar^2}{2m_e}\left(\frac{G}{2}\right)^2 + \frac{\hbar^2 q^2}{2m_e} \pm \sqrt{V_G^2 + \left(\frac{\hbar^2}{2m_e} qG\right)^2} \tag{11.8}$$

$$\psi_k(z < 0) = c\mathrm{e}^{\mathrm{i}qz} \cos\left(\frac{G}{2}z + \phi\right), \qquad \mathrm{e}^{2\mathrm{i}\phi} = \frac{\epsilon_k - k^2}{V_G} \tag{11.9}$$

where we have introduced the variable $q = k - G/2$ which expresses the wave-vector relative to the Brillouin Zone edge corresponding to $G/2$; in the expression for the wavefunction, c is a constant that includes the normalization. For the potential outside the sample, we will assume a simple barrier of height V_b, which gives for the wavefunction

$$\psi_k(z > 0) = c'\mathrm{e}^{-\kappa z} \tag{11.10}$$

with c' another constant of normalization and κ defined in Eq. (11.6). The problem can then be solved completely by matching at $z = 0$ the values of the wavefunction and its derivative as given by the expressions for $z < 0$ and $z > 0$. The potential $V(z)$ and the wavefunction $\psi_k(z)$ as functions of z and the energy ϵ_k as a function of q^2 are shown in Fig. 11.5. The interesting feature of this solution is that it allows q to take imaginary values, $q = \pm\mathrm{i}|q|$. For such values, the wavefunction decays exponentially, both outside the sample ($z > 0$) as $\sim \exp(-\kappa z)$, as well as inside the sample ($z < 0$) as $\sim \exp[|q|z]$, that is, the state is *spatially confined to the surface*. Moreover, the imaginary values of q correspond to energies that lie within the forbidden energy gap for the 3D model (which is equal to $2V_G$ and occurs at $q = 0$, see Fig. 11.5), since those values of q would produce wavefunctions that grow exponentially for $z \rightarrow \pm\infty$. In other words, the surface states have energies within the energy gap. Assuming that all states below the energy gap are filled, when the surface states are occupied the Fermi level will intersect the surface energy band described by ϵ_k. Therefore, we have established both facts mentioned above, that is, the sample states probed by the STM tip are spatially localized near the surface and have energies close to the Fermi level. Although these facts were established within a simple 1D model, the results carry over to the 3D case in which the electronic wavefunctions have a dependence on the (x, y) coordinates of the surface plane as well, because their essential features in the direction perpendicular to the surface are captured by the preceding analysis.

Having established the basic picture underlying STM experiments, it is worthwhile considering the limitations of the theory as developed so far. This theory includes two important approximations. First, the use of Bardeen's expression for the tunneling current, Eq. (11.3), assumes that tunneling occurs between undistorted states of the tip and the sample. Put differently, the two sides between which electrons tunnel are considered to be far enough apart to not influence each other's electronic states. This is not necessarily the case for realistic situations encountered in STM experiments. In fact, the tip and sample surface often come to within a few angstroms of each other, in which case they certainly affect each other's electronic wavefunctions. The second important approximation is that the tip electronic states involved in the tunneling are simple s-like states. Actually, the tips commonly employed in STM experiments are made of transition (d-electron) metals, which are reasonably hard, making it possible to produce stable, very sharp tips (the STM signal is optimal when the tunneling takes place through a single atom at the very edge of the tip). Examples of metals typically used in STM tips are W, Pt and Ir. In all these cases the relevant electronic states of the tip at the Fermi level have d character. Chen [159] developed an extension of the Tersoff–Hamann theory which takes into account these features by employing Green's function techniques which are beyond the scope of the present treatment. The basic result is that the expression for the tunneling current, instead of being proportional to the surface wavefunction magnitude as given in Eqs. (11.4) and (11.5), it is proportional to the magnitude of derivatives of the wavefunctions, which depend on the nature of the tip electronic state. A summary of the relevant expressions is given in Table 11.1.

Thus, the Tersoff–Hamann–Chen theory establishes that the STM essentially produces an image of the electron density associated with the surface of the sample, constructed from sample electronic states within eV of the Fermi level. This was

Table 11.1. *Contribution of different types of orbitals associated with the STM tip to the tunneling current I.*

The ψ_i's are sample wavefunctions and $\partial_\alpha \equiv \partial/\partial\alpha(\alpha = x, y, z)$. All quantities are evaluated at the position of the tip orbital, and κ is given by Eq. (11.6).

Tip orbital	Contribution to I
s	$\lvert\psi_i\rvert^2$
$p_\alpha \ (\alpha = x, y, z)$	$3\kappa^{-2}\lvert\partial_\alpha\psi_i\rvert^2$
$d_{\alpha\beta} \ (\alpha, \beta = x, y, z)$	$15\kappa^{-4}\left\lvert\partial^2_{\alpha\beta}\psi_i\right\rvert^2$
$d_{x^2-y^2}$	$\frac{15}{4}\kappa^{-4}\lvert\partial^2_x\psi_i - \partial^2_y\psi_i\rvert^2$
$d_{3z^2-r^2}$	$\frac{5}{4}\kappa^{-4}\lvert 3\partial^2_z\psi_i - \psi_i\rvert^2$

a crucial step toward a proper interpretation of the images produced by STM experiments. To the extent that these states correspond to atomic positions on the surface, as they usually (but not always) do, the STM image can indeed be thought of as a "picture" of the atomic structure of the surface. The exceptions to this simple picture have to do with situations where the electronic states of the surface have a structure which is more complex than the underlying atomic structure. It is not hard to imagine how this might come about, since electrons are distributed so as to minimize the total energy of the system according to the rules discussed in chapter 2: while this usually involves shielding of the positive ionic cores, it can often produce more elaborate patterns of the electronic charge distribution which are significantly different from a simple envelope of the atomic positions. For this reason, it is important to compare STM images with theoretical electronic structure simulations, in order to establish the actual structure of the surface. This has become a standard approach in interpreting STM experiments, especially in situations which involve multi-component systems (for recent examples of such studies see, for example, Ref. [161]). In general, STM images provide an unprecedented amount of information about the structural and electronic properties of surfaces. STM techniques have even been used to manipulate atoms or molecules on a surface [161–163], to affect their chemical bonding [165], and to observe standing waves of electron density on the surface induced by microscopic structural features [166, 167].

11.2 Surface reconstruction

The presence of the surface produces an abrupt change in the external potential that the electrons feel: the potential is that of the bulk crystal below the surface and zero above the surface. The need of the system to minimize the energy given this change in the external potential leads to interesting effects both in the atomic and electronic structure near the surface. These changes in structure depend on the nature of the physical system (for instance, they are quite different in metals and semiconductors), and are discussed below in some detail for representative cases. Before embarking on this description, we consider some general features of the surface electronic structure.

At the simplest level, we can think of the external potential due to the ions as being constant within the crystal and zero outside; this is the generalization of the jellium model of the bulk crystal (see chapter 2). The presence of the surface is thus marked by an abrupt step in the ionic potential. The electronic charge distribution, which in general must follow the behavior of the ionic potential, will undergo a change from its constant value far into the bulk to zero far outside it. However, the electronic charge distribution does not change abruptly near the surface but goes smoothly from one limiting value to the other [168]. This smooth transition gives rise to a total charge imbalance, with a slightly positive net charge just below the surface and a slightly negative net charge just above the surface, as illustrated in Fig. 11.6. The charge imbalance leads to the formation of a dipole moment associated with the presence of the surface. This so called surface dipole is a common feature of all surfaces, but its nature in real systems is more complex than the simple picture we presented here based on the jellium model. In addition to the surface dipole, the change of the electronic charge density is not monotonic at the surface step, but involves oscillations which extend well into the bulk, as illustrated in Fig. 11.6. These are known as "Friedel oscillations" and have a characteristic wavelength of π/k_{F}, where k_{F} is the Fermi momentum, related to the average density $\overline{n}$ of the

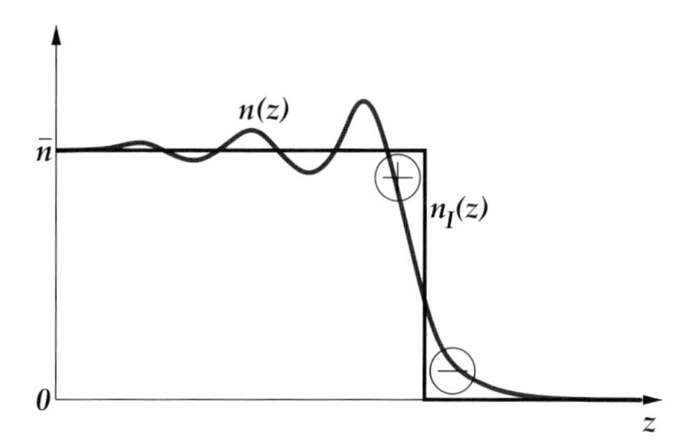

Figure 11.6. Illustration of surface features of the jellium model. The solid line is the ionic density $n_I(z)$, which is constant in the bulk and drops abruptly to zero outside the solid. The thicker wavy line is the electronic charge density $n(z)$, which changes smoothly near the surface and exhibits Friedel oscillations within the bulk. The $+$ and $-$ signs inside circles indicate the charge imbalance which gives rise to the surface dipole.

jellium model by $k_F = (3\pi^2\overline{n})^{1/3}$ (see Eq. (D.10) in Appendix D). The existence of the Friedel oscillations is a result of the plane wave nature of electronic states associated with the jellium model: the sharp feature in the ionic potential at the surface, which the electrons try to screen, has Fourier components of all wave-vectors, with the components corresponding to the Fermi momentum giving the largest contributions. For a detailed discussion of the surface physics of the jellium model see the review article by Lang, mentioned in the Further reading section.

We turn our attention next to specific examples of real crystal surfaces. An ideal crystal surface is characterized by two lattice vectors on the surface plane, $\mathbf{a}_1 = a_{1x}\hat{\mathbf{x}} + a_{1y}\hat{\mathbf{y}}$, and $\mathbf{a}_2 = a_{2x}\hat{\mathbf{x}} + a_{2y}\hat{\mathbf{y}}$. These vectors are multiples of lattice vectors of the three-dimensional crystal. The corresponding reciprocal space is also two dimensional, with vectors $\mathbf{b}_1, \mathbf{b}_2$ such that $\mathbf{b}_i \cdot \mathbf{a}_j = 2\pi\delta_{ij}$. Surfaces are identified by the bulk plane to which they correspond. The standard notation for this is the Miller indices of the conventional lattice. For example, the (001) surface of a simple cubic crystal corresponds to a plane perpendicular to the z axis of the cube. Since FCC and BCC crystals are part of the cubic system, surfaces of these lattices are denoted with respect to the conventional cubic cell, rather than the primitive unit cell which has shorter vectors but not along cubic directions (see chapter 3). Surfaces of lattices with more complex structure (such as the diamond or zincblende lattices which are FCC lattices with a two-atom basis), are also described by the Miller indices of the cubic lattice. For example, the (001) surface of the diamond lattice corresponds to a plane perpendicular to the z axis of the cube, which is a multiple of the PUC. The cube actually contains four PUCs of the diamond lattice and eight atoms. Similarly, the (111) surface of the diamond lattice corresponds to a plane perpendicular to the $\hat{\mathbf{x}} + \hat{\mathbf{y}} + \hat{\mathbf{z}}$ direction, that is, one of the main diagonals of the cube.

The characteristic feature of crystal surfaces is that the atoms on the surface assume positions different from those on a bulk-terminated plane. The differences can be small, which is referred to as "surface relaxation", or large, producing a structure that differs drastically from what is encountered in the bulk, which is referred to as "surface reconstruction". The changes in atomic positions can be such that the periodicity of the surface differs from the periodicity of atoms on a bulk-terminated plane of the same orientation. The standard way to describe the new periodicity of the surface is by multiples of the lattice vectors of the corresponding bulk-terminated plane. For instance, a $n_1 \times n_2$ reconstruction on the (klm) plane is one in which the lattice vectors on the plane are n_1 and n_2 times the primitive lattice vectors of the ideal, unreconstructed, bulk-terminated (klm) plane. Simple integer multiples of the primitive lattice vectors in the bulk-terminated plane often are not adequate to describe the reconstruction. It is possible, for example, to have reconstructions of the form $\sqrt{n_1} \times \sqrt{n_2}$, or $c(n_1 \times n_2)$, where c stands for "centered".

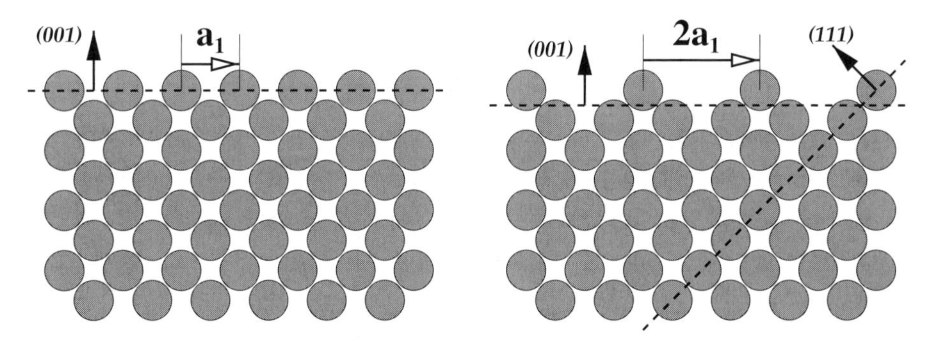

Figure 11.7. The missing row reconstruction in close-packed surfaces, illustrated in a 2D example. **Left:** the unreconstructed, bulk-terminated plane with surface atoms two-fold coordinated. The horizontal dashed line denotes the surface plane (average position of surface atoms) with surface unit cell vector $\mathbf{a}_1$. **Right:** the reconstructed surface with every second atom missing and the remaining atoms having either two-fold or three-fold coordination. The horizontal dashed line denotes the surface plane with surface unit cell vector $2\mathbf{a}_1$, while the inclined one indicates a plane of close-packed atoms. The labels of surface normal vectors denote the corresponding surfaces in the 3D FCC structure.

Surface reconstructions are common in both metal surfaces and semiconductor surfaces. The driving force behind the reconstruction is the need of the system to repair the damage done by the introduction of the surface, which severs the bonds of atoms on the exposed plane.

In metals with a close-packed bulk crystal structure, the surface atoms try to regain the number of neighbors they had in the bulk through the surface relaxation or reconstruction. Typically it is advantageous to undergo a surface reconstruction when the packing of atoms on the surface plane is not optimal. This is illustrated in Fig. 11.7. For simplicity and clarity we consider a two-dimensional example. We examine a particular cleaving of this 2D crystal which exposes a surface that does not correspond to close packing of the atoms on the surface plane. We define as surface atoms all those atoms which have fewer neighbors than in the bulk structure. The surface atoms in our example have two-fold coordination in the cleaved surface, rather than their usual four-fold coordination in the bulk. It is possible to increase the average coordination of surface atoms by introducing a reconstruction: removal of every other row of surface atoms (every other surface atom in our 2D example) leaves the rest of the atoms two-fold or three-fold coordinated; it also increases the size of the unit cell by a factor of 2, as shown in Fig. 11.7. The new surface unit cell contains one two-fold and two three-fold coordinated atoms, giving an *average* coordination of 8/3, a significant improvement over the unreconstructed bulk-terminated plane. What has happened due to the reconstruction is that *locally* the surface looks as if it were composed of smaller sections of a close-packed plane, on which every surface atom has three-fold coordination. This is actually a situation quite common on surfaces of FCC metals, in which case the close-packed plane is the (111) crystallographic plane while a plane with lower coordination of the surface atoms is the (001) plane. This is known as the "missing row" reconstruction.

When the 2D packing of surface atoms is already optimal, the effect of reconstruction cannot be very useful. In such cases, the surface layer simply recedes toward the bulk in an attempt to enhance the interactions of surface atoms with their remaining neighbors. This results in a shortening of the first-to-second layer spacing. To compensate for this distortion in bonding below the surface, the second-to-third layer spacing is expanded, but to a lesser extent. This oscillatory behavior continues for a few layers and eventually dies out. In general, the behavior of metal surfaces, as exhibited by their chemical and physical properties, is as much influenced by the presence of imperfections, such as steps and islands on the surface plane, as by the surface reconstruction.

In semiconductors, the surface atoms try to repair the broken covalent bonds to their missing neighbors by changing positions and creating new covalent bonds where possible. This may involve substantial rearrangement of the surface atoms, giving rise to interesting and characteristic patterns on the surface. Semiconductor surface reconstructions have a very pronounced effect on the chemical and physical properties of the surface. The general tendency is that the reconstruction restores the semiconducting character of the surface, which had been disturbed by the breaking of covalent bonds when the surface was created. There are a few simple and quite general structural patterns that allow semiconductor surfaces to regain their semiconducting character. We discuss next some examples of semiconductor surface reconstructions to illustrate these general themes.

11.2.1 Dimerization: the Si(001) surface

We begin with what is perhaps the most common and most extensively studied semiconductor surface, the Si(001) surface. The reason for its extensive study is that most electronic devices made out of silicon are built on crystalline substrates with the (001) surface. The (001) bulk-terminated plane consists of atoms that have two covalent bonds to the rest of the crystal, while the other two bonds on the surface side have been severed (see Fig. 11.8). The severed bonds are called dangling bonds. Each dangling bond is half-filled, containing one electron (since a proper covalent bond contains two electrons). If we consider a tight-binding approximation of the electronic structure, with an sp^3 basis of four orbitals associated with each atom, it follows that the dangling bond states have an energy in the middle of the band gap. This energy is also the Fermi level, since these are the highest occupied states.

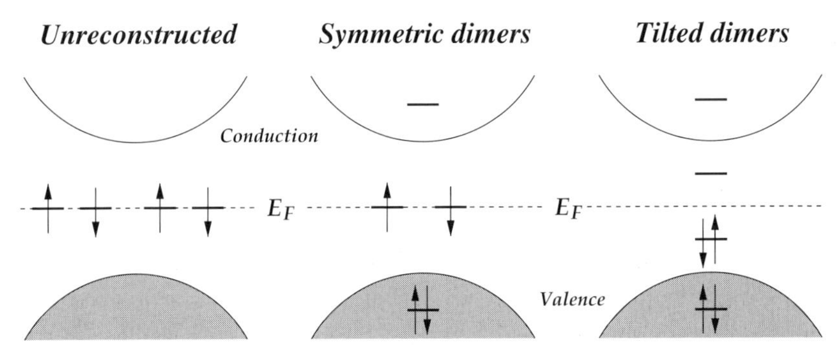

Figure 11.8. **Top panel:** reconstruction of the Si(001) surface. Left to right: the bulk-terminated (001) plane, with every surface atom having two broken (dangling) bonds; the dimerized surface, in which the surface atoms have come closer together in pairs to form the (2×1) dimer reconstruction, with symmetric dimers; the tilted dimer reconstruction in the (2×1) pattern. **Bottom panel:** schematic representation of the bands associated with the Si(001) dimer reconstruction. The shaded regions represent the projections of valence and conduction bands, while the up and down arrows represent single electrons in the two spin states. Left to right: the states associated with the bulk-terminated (001) plane, i.e. the degenerate, half-filled states of the dangling bonds in the mid-gap region which are coincident with the Fermi level, indicated by the dashed line; the states of the symmetric dimers, with the bonding (fully occupied) and antibonding (empty) combinations well within the valence and conduction bands, respectively, and the remaining half-filled dangling bond states in the mid-gap region; the states of the tilted dimer reconstruction, with the fully occupied and the empty surface state, separated by a small gap.

In reality, the unit cell of the surface reconstruction is (2×1) or multiples of that, which means that there are at least two surface atoms per surface unit cell (see Fig. 11.8). These atoms come together in pairs, hence the (2×1) periodicity, to form new bonds, called dimer bonds, with each bonded pair of atoms called a dimer. The formation of a dimer bond eliminates two of the dangling bonds in the unit cell, one per dimer atom. This leaves two dangling bonds, one per dimer atom, which for symmetric dimers are degenerate and half-filled each (they can accommodate two electrons of opposite spin but are occupied only by one). The energy of these states determines the position of the Fermi level, since they are the highest occupied states.

The dimers, however, do not have to be symmetric – there is no symmetry of the surface that requires this. Indeed, in the lowest energy configuration of the system, the dimers are tilted: one of the atoms is a little higher and the other a little lower relative to the average height of surface atoms, which is taken as the macroscopic definition of the surface plane. This tilting has an important effect on the electronic levels, which we analyze through the lens of the tight-binding approximation and is illustrated schematically in Fig. 11.8. The up-atom of the dimer has three bonds which form angles between them close to 90°. Therefore, it is in a bonding configuration close to p^3, that is, it forms covalent bonds through its three p orbitals, while its s orbital does not participate in bonding. At the same time, the down-atom of the dimer has three bonds which are almost on a plane. Therefore, it is in a bonding configuration close to sp^2, that is, it forms three bonding orbitals with its one s and two of its p states, while its third p state, the one perpendicular to the plane of the three bonds, does not participate in bonding. This situation is similar to graphite, discussed in chapters 1 and 4. Of the two orbitals that do not participate in bonding, the s orbital of the up-atom has lower energy than the p orbital of the down-atom. Consequently, the two remaining dangling bond electrons are accommodated by the up-atom s orbital, which becomes filled, while the down-atom p orbital is left empty. The net effect is that the surface has semiconducting character again, with a small band gap between the occupied up-atom s state and the unoccupied down-atom p state. The Fermi level is now situated in the middle of the surface band gap, which is smaller than the band gap of the bulk.

This example illustrates two important effects of surface reconstruction: a change in bonding character called rehybridization, and a transfer of electrons from one surface atom to another, in this case from the down-atom to the up-atom. The tilting of the dimer is another manifestation of the Jahn–Teller effect, which we discussed in connection to the Si vacancy in chapter 9. In this effect, a pair of degenerate, half-filled states are split to produce a filled and an empty state. All these changes in bonding geometry lower the total energy of the system, leading to a stable configuration. Highly elaborate first-principles calculations based on Density Functional Theory verify this simplified picture, as far as the atomic relaxation on the surface, the hybrids involved in the bonding of surface atoms, and the electronic levels associated with these hybrids, are concerned. We should also note that not all the dimers need to be tilted in the same way. In fact, alternating tilting of the dimers leads to more complex reconstruction patterns, which can become stable under certain conditions. The above picture is also verified experimentally. The reconstruction pattern and periodicity are established through scattering experiments, while the atomistic structure is established directly through STM images.

11.2.2 Relaxation: the GaAs(110) surface

We discuss next the surface structure of a compound semiconductor, GaAs, to illustrate the common features and the differences from elemental semiconductor surfaces. We examine the (110) surface of GaAs, which contains equal numbers of Ga and As atoms; other surface planes in this crystal, such as the (001) or (111) planes, contain only one of the two species of atoms present in the crystal. The former type of surface is called non-polar, the latter is called polar. The ratio of the two species of atoms is called stoichiometry; in non-polar planes the stoichiometry is the same as in the bulk, while in polar planes the stoichiometry deviates from its bulk value. Top and side views of the GaAs(110) surface are shown in Fig. 11.9.

The first interesting feature of this non-polar, compound semiconductor surface in its equilibrium state is that its unit cell remains the same as that of the bulk-terminated plane. Moreover, the number of atoms in the surface unit cell is the same as in the bulk-terminated plane. Accordingly, we speak of surface relaxation, rather than surface rcconstruction, for this case. The relaxation can be explained in simple chemical terms, as was done for the rehybridization of the Si(001) tilted-dimer surface. In the bulk-terminated plane, each atom has three bonds to other surface or subsurface atoms, and one of its bonds has been severed. The broken bonds of the Ga and As surface atoms are partially filled with electrons, in a situation similar to the broken bonds of the Si(001) surface. In the GaAs case, however, there is an important difference, namely that the electronic levels corresponding to the two broken bonds are not degenerate. This is because the levels, in a tight-binding sense, originate from sp^3 hybrids associated with the Ga and As atoms, and these are not equivalent. The hybrids associated with the As atoms lie lower in energy since As is more electronegative, and has a higher valence charge, than Ga. Consequently, we expect the electronic charge to move from the higher energy level, the Ga dangling bond, to the lower energy level, the As dangling bond.

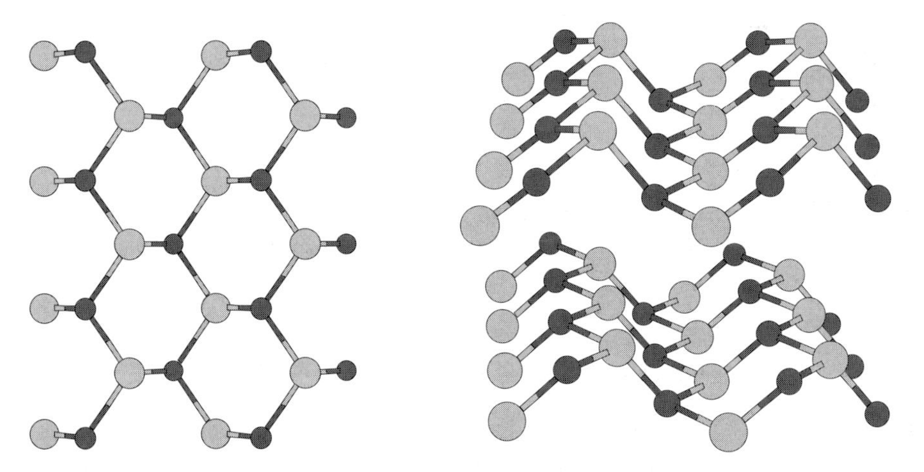

Figure 11.9. Structure of the GaAs(110) surface: **Left**: top view of the bulk-terminated (110) plane, containing equal numbers of Ga and As atoms. **Right**: side views of the surface before relaxation (below) and after relaxation (above). The surface unit cell remains unchanged after relaxation, the same as the unit cell in the bulk-terminated plane.

A relaxation of the two species of atoms on the surface enhances this difference. Specifically, the As atom is puckered upward from the mean height of surface atoms, while the Ga atom recedes toward the bulk. This places the As atom in a p^3-like bonding arrangement, and the Ga atom in an almost planar, sp^2-like bonding arrangement. As a result, the non-bonding electronic level of the As atom is essentially an s level, which lies even lower in energy than the As sp^3 hybrid corresponding to the unrelaxed As dangling bond. By the same token, the non-bonding electronic level of the Ga atom is essentially a p level, perpendicular to the plane of the three sp^2 bonding orbitals of Ga, which lies even higher in energy than the Ga sp^3 hybrid corresponding to the unrelaxed Ga dangling bond. These changes make the transfer of charge from the partially filled Ga dangling bond to the partially filled As dangling bond even more energetically favorable than in the unrelaxed surface. Moreover, this charge transfer is enough to induce semiconducting character to the surface, since it widens the gap between the occupied As level and the unoccupied Ga level. In fact, the two states, after relaxation, are separated enough to restore the full band gap of the bulk! In other words, after relaxation, the occupied As s level lies below the top of the bulk valence band, whereas the unoccupied Ga p level lies above the bottom of the bulk conduction band, as indicated in Fig. 11.10.

This picture needs to be complemented by a careful counting of the number of electrons in each state, to make sure that there are no unpaired electrons. The standard method for doing this is to associate a number of electrons per dangling

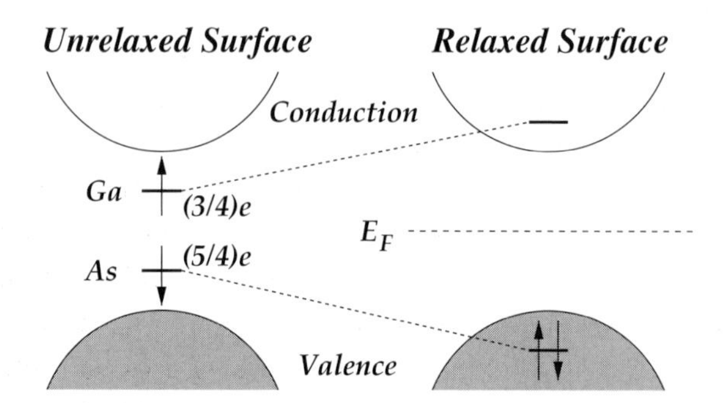

Figure 11.10. Schematic representation of the bands associated with the GaAs(110) surface relaxation. The shaded regions represent the projections of valence and conduction bands, while the up and down arrows represent partially occupied states in the two spin states. **Left**: the states associated with the bulk-terminated (110) plane; the Ga dangling bond state is higher in energy than the As dangling bond state, and both states lie inside the bulk band gap. **Right**: the states of the relaxed surface, with the fully occupied As s state below the top of the valence band, and the empty Ga p state above the top of the conduction band; the bulk band gap of the semiconductor has been fully restored by the relaxation.

bond equal to the valence of each atom divided by four, since each atom participates in the formation of four covalent bonds in the bulk structure of GaAs, the zincblende crystal (see chapter 1). With this scheme, each Ga dangling bond is assigned 3/4 of an electron since Ga has a valence of 3, whereas each As dangling bond is assigned 5/4 of an electron since As has a valence of 5. With these assignments, the above analysis of the energetics of individual electronic levels suggests that 3/4 of an electron leaves the surface Ga dangling bond and is transferred to the As surface dangling bond, which becomes fully occupied containing $3/4 + 5/4 = 2$ electrons. At the same time, rehybridization due to the relaxation drives the energy of these states beyond the limits of the bulk band gap. Indeed, this simple picture is verified by elaborate quantum mechanical calculations, which shows that there are no surface states in the bulk band gap, and that there is a relaxation of the surface atoms as described above. Confirmation comes also from experiment: STM images clearly indicate that the occupied states on the surface are associated with As atoms, while the unoccupied states are associated with Ga atoms [169].

11.2.3 Adatoms and passivation: the Si(111) surface

Finally, we discuss one last example of surface reconstruction which is qualitatively different than what we have seen so far, and will also help us introduce the idea of chemical passivation of surfaces. This example concerns the Si(111) surface. For this surface, the ideal bulk-terminated plane consists of atoms that have three neighbors on the side of the substrate and are missing one of their neighbors on the vacuum side (see Fig. 11.11). By analogy to what we discussed above for the Si(001) surface, the dangling bonds must contain one electron each in the unreconstructed surface. The energy of the dangling bond state lies in the middle of the band gap, and since this state is half filled, its energy is coincident with the Fermi level. This is a highly unstable situation, with high energy. There are two ways to remedy this and restore the semiconducting character of the surface: the first and simpler way is to introduce a layer of foreign atoms with just the right valence, without changing the basic surface geometry; the second and more complex way involves extra atoms called adatoms, which can be either intrinsic (Si) or extrinsic (foreign atoms), and which drastically change the surface geometry and the periodicity of the surface. We examine these two situations in turn.

If the atoms at the top layer on the Si(111) surface did not have four electrons but three, then there would not be a partially filled surface state due to the dangling bonds. This simple argument actually works in practice: under the proper conditions, it is possible to replace the surface layer of Si(111) with group-III atoms (such as Ga) which have only three valence electrons, and hence can form only three covalent

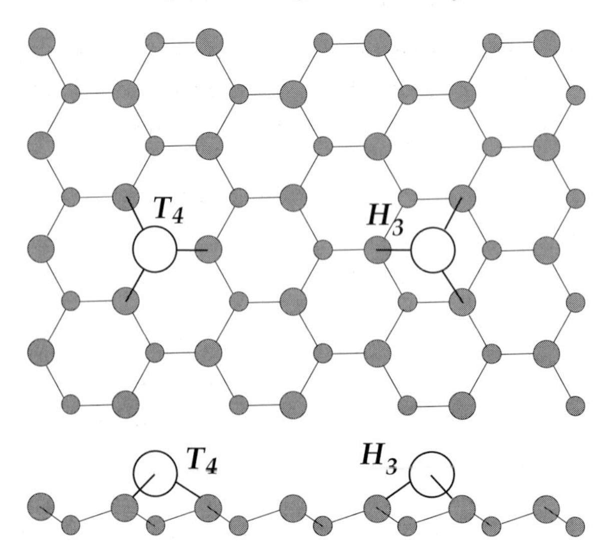

Figure 11.11. Top and side views of the surface bilayer of Si(111): the larger shaded circles are the surface atoms; the smaller shaded circles are the subsurface atoms which are bonded to the next layer below. The large open circles represent the adatoms in the two possible configurations, the H_3 and T_4 positions.

bonds, which is all that is required of the surface atoms. The resulting structure has the same periodicity as the bulk-terminated plane and has a surface-related electronic level which is empty. This level lies somewhat higher in energy than the Si dangling bond, since the Ga sp^3 hybrids have higher energy than the Si sp^3 hybrids. A different way of achieving the same effect is to replace the surface layer of Si atoms by atoms with one more valence electron (group-V atoms such as As). For these atoms the presence of the extra electron renders the dangling bond state full. The As-related electronic state lies somewhat lower in energy than the Si dangling bond, since the As sp^3 hybrids have lower energy than the Si sp^3 hybrids.

In both of these cases, slight relaxation of the foreign atoms on the surface helps move the surface-related states outside the gap region. The Ga atoms recede toward the bulk and become almost planar with their three neighbors in an sp^2 bonding arrangement, which makes the unoccupied level have p character and hence higher energy; this pushes the energy of the unoccupied level higher than the bottom of the conduction band, leaving the bulk band gap free. Similarly, the As atoms move away from the bulk and become almost pyramidal with their three neighbors in a p^3-like bonding arrangement, which makes the non-bonding occupied level have s character and hence lower energy; this pushes the energy of the occupied level lower than the top of the valence band, leaving the bulk band gap free. In both cases, the net result is a structure of low energy and much reduced chemical reactivity, that is, a passivated surface.

We can hypothesize that yet a different way of achieving chemical passivation of this surface is to saturate the surface Si dangling bonds through formation of covalent bonds to elements that prefer to have exactly one covalent bond. Such elements are the alkali metals (see chapter 1), because they have only one *s* valence electron. It turns out that this simple hypothesis works well for H: when each surface dangling bond is saturated by a H atom, a stable, chemically passive surface is obtained. Since it takes exactly one H atom to saturate one Si surface dangling bond, the resulting structure of the Si surface has the same periodicity, and in fact has the same atomic structure as the bulk-terminated plane. The Si–H bond is even stronger than the Si–Si bond in the bulk. Thus, by adding H, the structure of the bulk-terminated plane can be restored and maintained as a stable configuration. Similar effects can be obtained in other surfaces with the proper type of adsorbate atoms, which have been called "valence mending adsorbates" (see, for example, Ref. [170]). The simple hypothesis of saturating the Si(111) surface dangling bonds with monovalent elements does not work for the other alkalis, which have more complex interactions with the surface atoms, leading to more complicated surface reconstructions.

There is a way to passivate the Si(111) surface by adding extra atoms on the surface, at special sites. These atoms are called adatoms and can be either native Si atoms (intrinsic adatoms) or foreign atoms (extrinsic adatoms). There are two positions that adatoms can assume to form stable or metastable structures, as illustrated in Fig. 11.11. The first position involves placing the adatom directly above a second-layer Si atom, and bonding it to the three surface Si atoms which surround this second-layer atom; this position is called the T_4 site for being on *T*op of a second-layer atom and having four nearest neighbors, the three surface atoms to which it is bonded, and the second-layer atom directly below it. The second position involves placing the adatom above the center of a six-fold ring formed by three first-layer and three second-layer Si atoms, and bonding it to the three first-layer atoms of the ring; this position is called the H_3 site for being at the center of a *H*exagon of first- and second-layer atoms, and having three nearest neighbors. In both cases the adatom is three-fold bonded while the surface atoms to which it is bonded become four-fold coordinated. Thus, each adatom saturates three surface dangling bonds by forming covalent bonds with three surface Si atoms.

Now, if the adatom is of a chemical type that prefers to form exactly three covalent bonds like the group-III elements Al, Ga and In, then placing it at one of the two stable positions will result in a chemically passive and stable structure. This will be the case if the entire surface is covered by adatoms, which corresponds to a reconstruction with a unit cell containing one adatom and three surface atoms. The resulting periodicity is designated $(\sqrt{3} \times \sqrt{3})$, shown in Fig. 11.12, since the new surface lattice vectors are larger by a factor of $\sqrt{3}$ compared with the original lattice vectors of the bulk-terminated plane. The new lattice vectors are also rotated

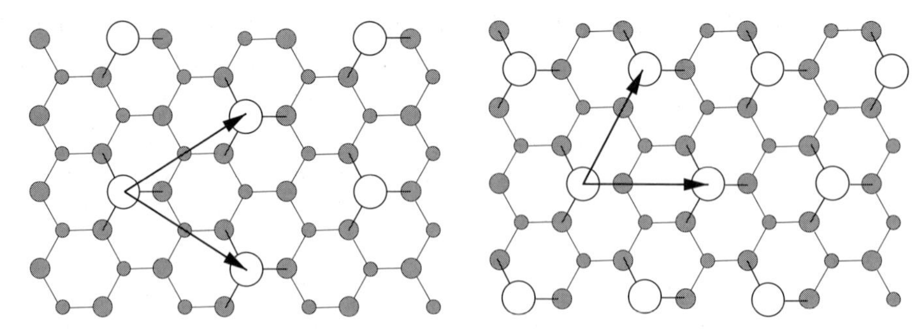

Figure 11.12. Adatom reconstructions of the Si(111) surface viewed from above. The small and medium sized shaded circles represent the atoms in the first bilayer; the large open circles represent the adatoms. The reconstructed unit cell vectors are indicated by arrows. **Left:** the (2×2) reconstruction with one adatom (at the T_4 site) and four surface atoms the three bonded to the adatom and a restatom; this version is appropriate for Si adatoms. **Right:** the $(\sqrt{3} \times \sqrt{3})$ reconstruction, with one adatom (at the T_4 site) and three surface atoms per unit cell; this version is appropriate for trivalent adatoms (eg. Al, Ga, In).

by 30° relative to the original lattice vectors, so this reconstruction is sometimes designated $(\sqrt{3} \times \sqrt{3})R30°$, but this is redundant, since there is only one way to form lattice vectors larger than the original ones by a factor of $\sqrt{3}$. Indeed, under certain deposition conditions, the Si(111) surface can be covered by group-III adatoms (Al, Ga, In) in a $(\sqrt{3} \times \sqrt{3})$ reconstruction involving one adatom per reconstructed surface unit cell. It turns out that the T_4 position has in all cases lower energy, so that it is the stable adatom position, whereas the H_3 position is metastable: it is higher in energy than the T_4 position, but still a local minimum of the energy.

We mentioned that native Si adatoms can also exist on this surface. These adatoms have four valence electrons, so even though they saturate three surface dangling bonds, they are left with a dangling bond of their own after forming three bonds to surface atoms. This dangling bond will contain an unpaired electron. Thus, we would expect that this situation is not chemically passive. If all the surface dangling bonds were saturated by Si adatoms, that is, if a $(\sqrt{3} \times \sqrt{3})$ reconstruction with Si adatoms were formed, the structure would be unstable for another reason. The Si adatoms prefer, just like group-III adatoms, the T_4 position; in order to form good covalent bonds to the surface atoms, i.e. bonds of the proper length and at proper angles among themselves, the adatoms pull the three surface atoms closer together, thereby inducing large compressive strain on the surface. This distortion is energetically costly.

Both the imbalance of electronic charge due to the unpaired electron in the Si adatom dangling bond, and the compressive strain due to the presence of the adatom, can be remedied by leaving one out of every four surface atoms not bonded to any adatom. This surface atom, called the restatom, will also have a dangling bond with one unpaired electron in it. The two unpaired electrons, one on the restatom and one on the adatom, can now be paired through charge transfer, ending up in the state with lower energy, and leaving the other state empty. As in the examples discussed above, this situation restores the semiconducting character of the surface by opening up a small gap between filled and empty surface states. Moreover, the presence of the restatom has beneficial effects on the surface strain. Specifically, the restatom relaxes by receding toward the bulk, which means that it pushes its three neighbors away and induces tensile strain to the surface. The amount of tensile strain introduced in this way is close to what is needed to compensate for the compressive strain due to the presence of the adatom, so that the net strain is very close to zero [171].

By creating a surface unit cell that contains one adatom and one restatom, we can then achieve essentially a perfect structure, in what concerns both the electronic features and the balance of strain on the reconstructed surface. This unit cell consists of four surface atoms plus the adatom. A natural choice for this unit cell is one with lattice vectors twice as large in each direction as the original lattice vectors of the bulk-terminated plane, that is, a (2×2) reconstruction, as shown in Fig. 11.12. It turns out that the actual reconstruction of the Si(111) surface is more complicated, and has in fact a (7×7) reconstruction, i.e. a unit cell 49 times larger than the original bulk-terminated plane! This reconstruction is quite remarkable, and consists of several interesting features, such as a stacking fault, dimers and a corner hole, as shown in Fig. 11.13: the stacking fault involves a 60° rotation of the surface layer relative to the subsurface layer, the dimers are formed by pairs of atoms in the subsurface layer which come together to make bonds, and at the corner holes three surface and one subsurface atoms of the first bilayer are missing. However, the main feature is the set of adatoms (a total of 12 in the unit cell), accompanied by restatoms (a total of six in the unit cell), which are *locally* arranged in a (2×2) pattern. Thus, the dominating features of the actual reconstruction of Si(111) are indeed adatom–restatom pairs, as discussed above. The reconstruction of the Ge(111) surface, another tetravalent element with the diamond bulk structure, is $c(2 \times 8)$, which is a simple variation of the (2×2) pattern, having as its dominant feature the adatom–restatom pair in each unit cell. Due to its large unit cell, it took a very long time to resolve the atomic structure of the Si(111) (7×7) reconstruction. In fact, the first STM images were crucial in establishing the atomic structure of this important reconstruction [172, 173]. The STM images of this surface basically pick out the adatom positions in both positive and negative bias [174]. Large-scale STM images of this surface give the impression of an incredibly intricate and beautiful lace pattern, as weaved by Nature, adatom by adatom (see Fig. 11.13).

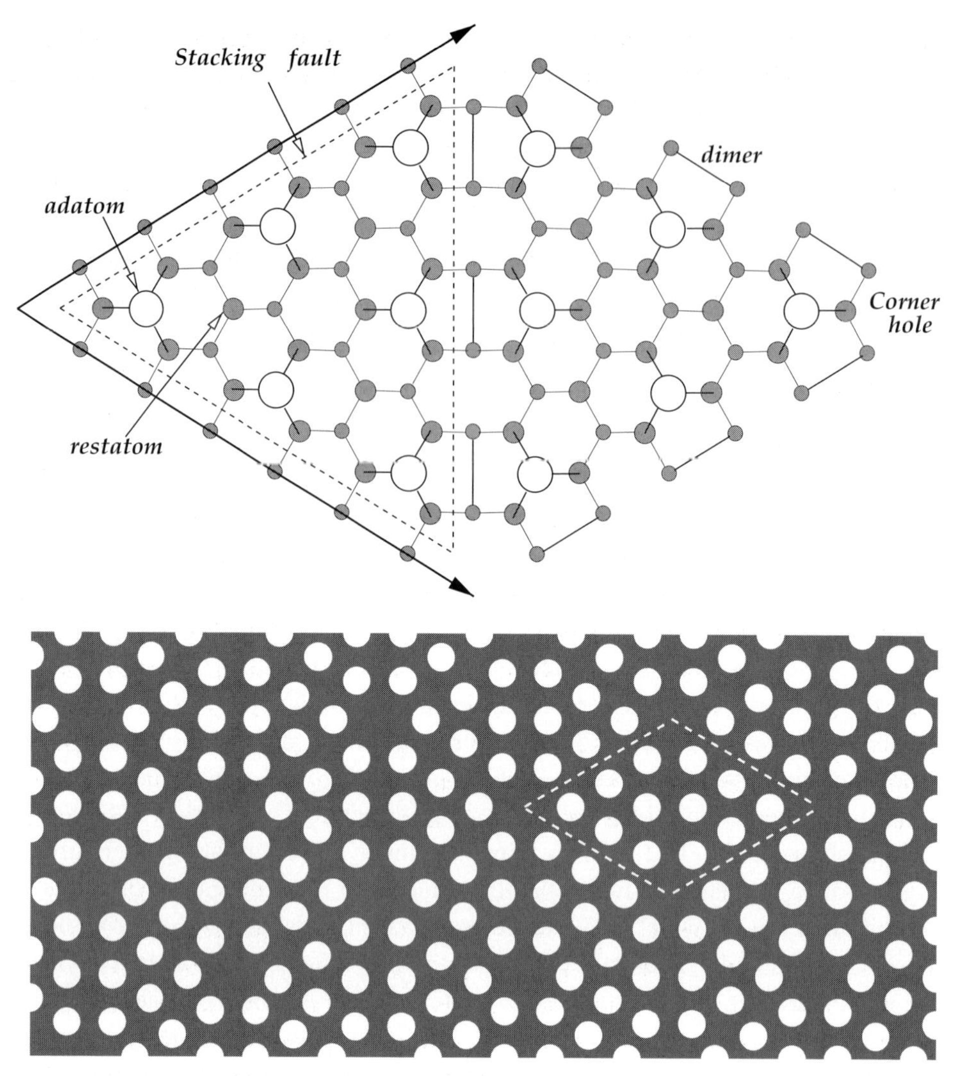

Figure 11.13. **Top:** the unit cell of the (7 × 7) reconstruction of the Si(111) surface viewed from above. The small and medium sized shaded circles represent the atoms in the first bilayer (the subsurface and surface atoms, respectively), and the large open circles represent the adatoms. The reconstructed unit cell vectors are indicated by arrows. The main features of the reconstruction are the corner holes, the 12 adatoms, the dimers, and the stacking fault on one-half of the unit cell (denoted by dashed lines). **Bottom:** a pattern of adatoms in an area containing several unit cells, as it would be observed in STM experiments; one of the unit cells is outlined in white dashed lines.

11.3 Growth phenomena

Great attention has also been paid to the dynamic evolution of crystal surfaces. This is in part sustained by the desire to control the growth of technologically important materials, which requires detailed knowledge of growth mechanisms starting at the atomistic level and reaching to the macroscopic scale. Equally important is the fundamental interest in dynamical phenomena on surfaces: these systems provide the opportunity to use elaborate theoretical tools for their study in an effort to gain deeper understanding of the physics. We describe briefly some of these aspects.

At the microscopic level, the relevant processes in growth are the deposition of atoms on the surface, usually described as an average flux F, and the subsequent motion of the atoms on the surface that allows them to be incorporated at lattice sites. This motion involves diffusion of the atoms on the surface, both on flat surface regions (terraces) as well as around steps between atomic planes. The precise way in which atoms move on the surface depends on the atomic structure of the terraces and steps, which is determined by the surface reconstruction. This justifies the attention paid to surface reconstructions, which can play a decisive role in determining the growth mechanisms. For instance, on the Si(001) surface, the dimer reconstruction produces long rows of dimers; motion along the dimer rows is much faster than motion across rows [174–176]). This in turn leads to highly anisotropic islands and anisotropic growth [178]. Studying the dynamics of atoms on surfaces and trying to infer the type of ensuing growth has become a cottage industry in surface physics, driven mostly by technological demands for growing ever smaller structures with specific features; such structures are expected to become the basis for next generation electronic and optical devices.

There is a different approach to the study of surface growth, in which the microscopic details are coarse-grained and attention is paid only to the macroscopic features of the evolving surface profile during growth. Such approaches are based on statistical mechanics theories, and aim at describing the surface evolution on a scale that can be directly compared to macroscopic experimental observations. The basic phenomenon of interest in such studies is the so called roughening transition, which leads to a surface with very irregular features on all length scales. To the extent that such surfaces are not useful in technological applications, the roughening transition is to be avoided during growth. Accordingly, understanding the physics of this transition is as important as understanding the microscopic mechanisms responsible for growth. A great deal of attention has been devoted to the statistical mechanical aspects of growth. Here we will give a very brief introduction to some basic concepts that are useful in describing growth phenomena of crystals (for more details see the books and monographs mentioned in the Further reading section).

In macroscopic treatments of growth, the main physical quantity is the height of the surface $h(\mathbf{r}, t)$, which depends on the position on the surface $\mathbf{r}$ and evolves with time t. The average height $\bar{h}(t)$ is defined as an integral over the surface S:

$$\bar{h}(t) = \frac{1}{A} \int_S h(\mathbf{r}, t) d\mathbf{r} \tag{11.11}$$

where A is the surface area. In terms of the average height $\bar{h}(t)$ and the local height $h(\mathbf{r}, t)$, the surface width is given by

$$w(L, t) = \left(\frac{1}{L^2} \int \left[h(\mathbf{r}, t) - \bar{h}(t) \right]^2 \right)^{1/2} d\mathbf{r} \tag{11.12}$$

where we have defined the length scale L of the surface as $L^2 = A$. In a macroscopic description of the surface height during growth, we assume that the surface width scales with the time t for some initial period, beyond which it scales with the system size L. This is based on empirical observations of how simple models of surface growth behave. The time at which the scaling switches from one to the other is called the crossover time, T_X. We therefore can write for the two regimes of growth

$$\begin{aligned} w(L, t) &\sim t^\beta, \quad t << T_X \\ w(L, t) &\sim L^\alpha, \quad t >> T_X \end{aligned} \tag{11.13}$$

where we have introduced the so called growth exponent β and roughness exponent α (both non-negative numbers), to describe the scaling in the two regimes. The meaning of the latter expression for large α is that the width of the surface becomes larger and larger with system size, which corresponds to wide variations in the height, comparable to the in-plane linear size of the system, that is, a very rough surface. For a system of fixed size, the width will saturate to the value L^α. The time it takes for the system to cross over from the time-scaling regime to the size-scaling regime depends on the system size, because the larger the system is, the longer it takes for the inhomogeneities in height to develop to the point where they scale with the linear system dimension. Thus, there is a relation between the crossover time T_X and the linear system size L, which we express by introducing another exponent, the so called dynamic exponent z,

$$T_X \sim L^z \tag{11.14}$$

From the above definitions, we can deduce that if the crossover point is approached from the time-scaling regime we will obtain

$$w(L, T_X) \sim T_X^\beta \tag{11.15}$$

whereas if it is approached from the size-scaling regime we will obtain

$$w(L, T_X) \sim L^\alpha \tag{11.16}$$

which, together with the definition of the dynamic exponent z, produce the relation

$$z = \frac{\alpha}{\beta} \tag{11.17}$$

From a physical point of view, the existence of a relationship between w and L implies that there must be some process (for instance, surface diffusion) which allows atoms to "explore" the entire size of the system, thereby linking its extent in the different directions to the surface width.

With these preliminary definitions we can now discuss some simple models of growth. The purpose of these models is to describe in a qualitative manner the evolution of real surfaces and to determine thc three exponents we introduced above; actually only two values are needed, since they are related by Eq. (11.17). The simplest model consists of a uniform flux of atoms being deposited on the surface. Each atom sticks wherever it happens to fall on the surface. For this, the so called random deposition model, the evolution of the surface height is given by

$$\frac{\partial h(\mathbf{r}, t)}{\partial t} = F(\mathbf{r}, t) \tag{11.18}$$

where $F(\mathbf{r}, t)$ is the uniform flux. While the flux is uniform on average, it will have fluctuations on some length scale, so we can write it as

$$F(\mathbf{r}, t) = F_0 + \eta(\mathbf{r}, t) \tag{11.19}$$

where F_0 represents the constant average flux and $\eta(\mathbf{r}, t)$ represents a noise term with zero average value and no correlation in space or time:

$$\langle \eta(\mathbf{r}, t) \rangle = 0, \quad \langle \eta(\mathbf{r}, t)\eta(\mathbf{r}', t') \rangle = F_n \delta^{(d)}(\mathbf{r} - \mathbf{r}')\delta(t - t') \tag{11.20}$$

where the brackets indicate averages over the entire surface and all times. In the above equation we have denoted the dimensionality d of the spatial δ-function explicitly as a superscript. This is done to allow for models with different spatial dimensionalities. Now we can integrate the random deposition model with respect to time to obtain

$$h(\mathbf{r}, t) = F_0 t + \int_0^t \eta(\mathbf{r}, t')\mathrm{d}t' \tag{11.21}$$

from which we find, using the properties of the noise term, Eq. (11.20),

$$\begin{aligned} \langle h(\mathbf{r}, t) \rangle &= F_0 t \\ w^2(t) = \langle h^2(\mathbf{r}, t) \rangle - \langle h(\mathbf{r}, t) \rangle^2 &= F_n t \end{aligned} \tag{11.22}$$

This result immediately gives the value of the growth exponent ($w \sim t^\beta$) for this model, $\beta = 0.5$. The other two exponents cannot be determined, because in this model there is no correlation in the noise term, which is the only term that can give

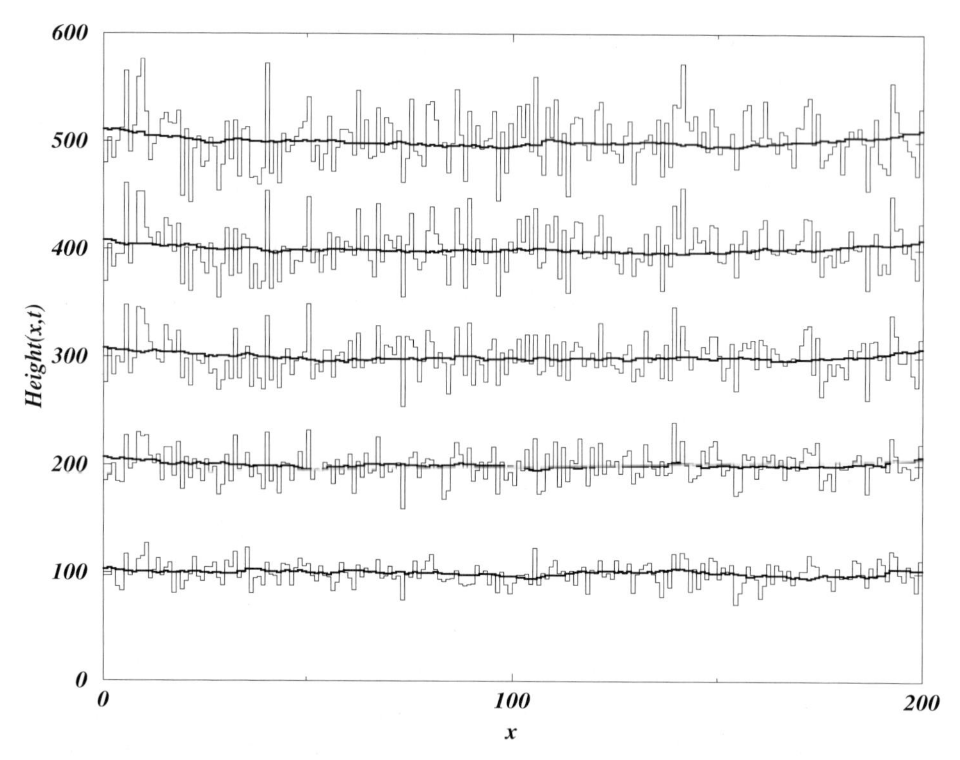

Figure 11.14. Simulation of one-dimensional growth models. The highly irregular lines correspond to the surface profile of the random deposition model. The smoother, thicker lines correspond to a model which includes random deposition plus diffusion to next neighbor sites, if this reduces the surface curvature locally. The two sets of data correspond to the same time instances in the two models, i.e. the same amount of deposited material, as indicated by the average height. It is evident how the diffusion step leads to a much smoother surface profile.

rise to roughening. Indeed, simulations of this model show that the width of the surface profile keeps growing indefinitely and does not saturate (see Fig. 11.14), as the definition of the roughness exponent, Eq. (11.13), would require. Consequently, for this case the roughness exponent α and the dynamic exponent $z = \alpha/\beta$ cannot be defined.

The next level of sophistication in growth models is the so called Edwards-Wilkinson (EW) model [179], defined by the equation:

$$\frac{\partial h(\mathbf{r}, t)}{\partial t} = \nu \nabla_{\mathbf{r}}^2 h(\mathbf{r}, t) + \eta(\mathbf{r}, t) \tag{11.23}$$

In this model, the first term on the right-hand side represents the surface tension, which tends to smooth out the surface: this term leads to a relative increase in the height locally where the curvature is positive and large, and to a relative decrease in the height locally where the curvature is negative. The net result is that points of positive curvature (valleys) will fill in, while points of negative curvature (hillocks) will decrease in relative height as the surface evolves, producing a smoother surface profile. The second term on the right-hand side in Eq. (11.23) is a noise term with the same properties as the noise term included in the random deposition model. Notice that for simplicity we have omitted the term corresponding to the uniform flux, because we can simply change variables $h \to h + F_0 t$, which eliminates the F_0 term from both sides of the equation. In other words, the EW model as defined in Eq. (11.23) deals with the variations in the moving average height.

This equation can be solved, at least as far as the values of the exponents are concerned, by a rescaling argument. Specifically, we assume that the length variable is rescaled by a factor b: $\mathbf{r} \to \mathbf{r}' = b\mathbf{r}$. By our scaling assumption, when the growth is in the length-scaling regime, we should have $h \to h' = b^{\alpha} h$. This is known as a "self-affine" shape; if $\alpha = 1$, so that the function is rescaled by the same factor as one of its variables, the shape is called "self-similar". We must also rescale the time variable in order to be able to write a similar equation to Eq. (11.23) for the rescaled function h'. Since space and time variables are related by the dynamic exponent z, Eq. (11.14), we conclude that the time variable must be rescaled as: $t \to t' = b^z t$. With these changes, and taking into account the general property of the δ-function in d dimensions $\delta^{(d)}(b\mathbf{r}) = b^{-d}\delta^{(d)}(\mathbf{r})$, we conclude that scaling of the noise term correlations will be given by

$$\langle \eta(b\mathbf{r}, b^z t)\eta(b\mathbf{r}', b^z t')\rangle = F_n b^{-(d+z)}\delta^{(d)}(\mathbf{r} - \mathbf{r}')\delta(t - t') \tag{11.24}$$

which implies that the noise term η should be rescaled by a factor $b^{-(d+z)/2}$. Putting all these together, we obtain that the equation for the height, after rescaling, will take the form

$$\begin{aligned} b^{\alpha - z}\partial_t h &= \nu b^{\alpha-2}\nabla^2 h + b^{-(d+z)/2}\eta \\ \Longrightarrow \partial_t h &= \nu b^{z-2}\nabla^2 h + b^{-d/2+z/2-\alpha}\eta \end{aligned}$$

where we have used the short-hand notation ∂_t for the partial derivative with respect to time. If we require this to be identical to the original equation (11.23), and assuming that the values of the constants involved (ν, F_n) are not changed by the rescaling, we conclude that the exponents of b on the right-hand side of the equation must vanish, which gives

$$\alpha = \frac{2-d}{2}, \quad \beta = \frac{2-d}{4}, \quad z = 2 \tag{11.25}$$

These relations fully determine the exponents for a given dimensionality d of the model. For example, in one dimension, $\alpha = 0.5$, $\beta = 0.25$, while in two dimensions $\alpha = \beta = 0$.

Let us consider briefly the implications of this model. We discuss the two-dimensional case ($d = 2$), which is closer to physical situations since it corresponds to a two-dimensional surface of a three-dimensional solid. In this case, the roughness exponent α is zero, meaning that the width of the surface profile does not grow as a power of the system size; this does not mean that the width does not increase, only that it increases slower than any power of the linear size of the system. The only possibility then is that the width increases logarithmically with system size, $w \sim \ln L$. This, however, is a very weak divergence of the surface profile width, implying that the surface is overall quite flat. The flatness of the surface in this model is a result of the surface tension term, which has the effect of smoothing out the surface, as we discussed earlier.

How can we better justify the surface tension term? We saw that it has the desired effect, namely it fills in the valleys and flattens out the hillocks, leading to a smoother surface. But where can such an effect arise from, on a microscopic scale? We can assign a chemical potential to characterize the condition of atoms at the microscopic level. We denote by $\mu(\mathbf{r}, t)$ the relative chemical potential between vapor and surface. It is natural to associate the local rate of change in concentration of atoms $C(\mathbf{r}, t)$ with minus the local chemical potential $C(\mathbf{r}, t) \sim -\mu(\mathbf{r}, t)$: an attractive chemical potential increases the concentration, and vice versa. It is also reasonable to set the local chemical potential proportional to minus the local curvature of the surface, as first suggested by Herring [180],

$$\mu \sim -\nabla_{\mathbf{r}}^2 h(\mathbf{r}, t) \tag{11.26}$$

because at the bottom of valleys where the curvature is positive, the surface atoms will tend to have more neighbors around them, and hence feel an attractive chemical potential; conversely, while at the top of hillocks where the curvature is negative, the atoms will tend to have fewer neighbors, and hence feel a repulsive chemical potential. A different way to express this is the following: adding material at the bottom of a valley reduces the surface area, which is energetically favorable, implying that the atoms will prefer to go toward the valley bottom, a situation described by a negative chemical potential; the opposite is true for the top of hillocks. This gives for the rate of change in concentration

$$C(\mathbf{r}, t) \sim -\mu(\mathbf{r}, t) \sim \nabla_{\mathbf{r}}^2 h(\mathbf{r}, t) \tag{11.27}$$

Notice that by $C(\mathbf{r}, t)$ we refer to variations of the rate of change in concentration relative to its average value, so that, depending on the curvature of $\mu(\mathbf{r}, t)$, this relative rate of change can be positive or negative. The rate of change in concentration can be related directly to changes in the surface height,

$$\frac{\partial h(\mathbf{r}, t)}{\partial t} = \nu C(\mathbf{r}, t) = \nu \nabla_{\mathbf{r}}^2 h(\mathbf{r}, t) \tag{11.28}$$

which produces the familiar surface tension term in the EW model, with the positive factor ν providing the proper units. In this sense, the mechanism that leads to a smoother surface in the EW model is the desorption of atoms from hillocks and the deposition of atoms in valleys. In this model, the atoms can come off the surface (they desorb) or attach to the surface (they are deposited) through the presence of a vapor above the surface, which acts as an atomic reservoir.

We can take this argument one step further by considering other possible atomistic level mechanisms that can have an effect on the surface profile. One obvious mechanism is surface diffusion. In this case, we associate a surface current with the negative gradient of the local chemical potential

$$\mathbf{j}(\mathbf{r}, t) \sim -\nabla_{\mathbf{r}} \mu(\mathbf{r}, t) \tag{11.29}$$

by an argument analogous to that discussed above. Namely, atoms move away from places with repulsive chemical potential (the hillocks, where they are weakly bound) and move toward places with attractive chemical potential (the valleys, where they are strongly bound). The change in the surface height will be proportional to the negative divergence of the current since the height decreases when there is positive change in the current, and increases when there is negative change in the current, giving

$$\frac{\partial h(\mathbf{r}, t)}{\partial t} \sim -\nabla_{\mathbf{r}} \cdot \mathbf{j}(\mathbf{r}, t) \tag{11.30}$$

Using again the relationship between chemical potential and surface curvature, given by Eq. (11.26), we obtain for the effect of diffusion

$$\frac{\partial h(\mathbf{r}, t)}{\partial t} = -q \nabla_{\mathbf{r}}^4 h(\mathbf{r}, t) \tag{11.31}$$

with q a positive factor that provides the proper units. Adding the usual noise term with zero average and no correlations in space or time, Eq. (11.20), leads to another statistical/stochastic model for surface growth, referred to as the Wolf–Villain (WV) model [181]. This model leads to a smooth surface under the proper conditions, but through a different mechanism than the EW model. The growth, roughening and dynamical exponents for the WV model can be derived through a simple rescaling argument, analogous to what we discussed for the EW model, which gives

$$\alpha = \frac{4-d}{2}, \quad \beta = \frac{4-d}{8}, \quad z = 4 \tag{11.32}$$

For a two-dimensional surface of a three-dimensional crystal ($d = 2$ in the above equation[1]), the roughness exponent is $\alpha = 1$, larger than in the EW model. Thus, the surface profile width in the WV model is more rough than in the EW model, that is, the surface diffusion is not as effective in reducing the roughness as the desorption/deposition mechanism. However, the surface profile will still be quite a bit smoother than in the random deposition model (see Fig. 11.14 for an example).

A model that takes this approach to a higher level of sophistication and, arguably, a higher level of realism, referred to as the Kardar–Parisi–Zhang (KPZ) model [182], assumes the following equation for the evolution of the surface height:

$$\frac{\partial h(\mathbf{r}, t)}{\partial t} = \nu \nabla_{\mathbf{r}}^2 h(\mathbf{r}, t) + \lambda \left[\nabla_{\mathbf{r}} h(\mathbf{r}, t)\right]^2 + \eta(\mathbf{r}, t) \tag{11.33}$$

This model has the familiar surface tension and noise terms, the first and third terms on the right-hand side, respectively, plus a new term which involves the square of the surface gradient. The presence of this term can be justified by a careful look at how the surface profile grows: if we assume that the surface grows in a direction which is perpendicular to the *local* surface normal, rather than along a constant growth direction, then the gradient squared term emerges as the lowest order term in a Taylor expansion of the surface height. This term has a significant effect on the behavior of the surface height. In fact, this extra term introduces a complex coupling with the other terms, so that it is no longer feasible to extract the values of the growth and roughness exponents through a simple rescaling argument as we did for the EW and WV models. This can be appreciated from the observation that the KPZ model is a non-linear model since it involves the first and second powers of h on the right-hand side of the equation, whereas the models we considered before are all linear models involving only linear powers of h on the right-hand side of the equation. Nevertheless, the KPZ model is believed to be quite realistic, and much theoretical work has gone into analyzing its behavior. Variations on this model have also been applied to other fields.

It turns out that the WV model is relevant to a type of growth that is used extensively in the growth of high-quality semiconductor crystals. This technique is called Molecular Beam Epitaxy (MBE);[2] it consists of sending a beam of atoms or molecules, under ultra high vacuum conditions and with very low kinetic energy,

[1] This is referred to in the literature as a "2+1" model, indicating that there is one spatial dimension in addition to the growth dimension.

[2] The word epitaxy comes from two Greek words, $\epsilon\pi\iota$ which means "on top" and $\tau\alpha\xi\iota\varsigma$, which means "order".

toward a surface. The experimental conditions are such that the atoms stick to the surface with essentially 100% probability (the low kinetic energy certainly enhances this tendency). Once on the surface, the atoms can diffuse around, hence the diffusion term ($\nabla^4 h$) is crucial, but cannot desorb, hence the deposition/desorption term ($\nabla^2 h$) is not relevant. The noise term is justified in terms of the random manner in which the atoms arrive at the surface both in position and in time. MBE has proven to be the technique of choice for growing high-quality crystals of various semiconductor materials, and especially for growing one type of crystal on top of a different type (the substrate); this is called "heteroepitaxy". Such combinations of materials joined by a smooth interface are extremely useful in optoelectronic devices, but are very difficult to achieve through any other means. Achieving a smooth interface between two high-quality crystals is not always possible: depending on the interactions between newly deposited atoms and the substrate, growth can proceed in a layer-by-layer mode, which favors a smooth interface, or in a 3D island mode which is detrimental to a smooth interface. In the first case, called the Frank–van der Merwe mode, the newly deposited atoms wet the substrate, that is, they prefer to cover as much of the substrate area as possible for a given amount of deposition, the interaction between deposited and substrate atoms being favorable. Moreover, this favorable interaction must not be adversely affected by the presence of large strain in the film, which implies that the film and substrate lattice constants must be very close to each other. In the second case, called the Volmer–Weber mode, the newly deposited atoms do no wet the substrate, that is, they prefer to form 3D islands among themselves and leave large portions of the substrate uncovered, even though enough material has been deposited to cover the substrate. When the 3D islands, which nucleate randomly on the substrate, eventually coalesce to form a film, their interfaces represent defects (see also section 11.4), which can destroy the desirable characteristics of the film. There is actually an intermediate case, in which growth begins in a layer-by-layer mode but quickly reverts to the 3D island mode. This case, called the Stranski–Krastanov mode, arises from favorable chemical interaction between deposited and substrate atoms at the initial stages of growth, which is later overcome by the energy cost due to strain in the growing film. This happens when the lattice constant of the substrate and the film are significantly different. The three epitaxial modes of growth are illustrated in Fig. 11.15. In these situations the stages of growth are identified quantitatively by the amount of deposited material, usually measured in units of the amount required to cover the entire substrate with one layer of deposited atoms, called a monolayer (ML). The problem with MBE is that it is rather slow and consequently very costly: crystals are grown at a rate of order a monolayer per minute, so this method of growth can be used only for the most demanding applications.

From the theoretical point of view, MBE is the simplest technique for studying crystal growth phenomena, and has been the subject of numerous theoretical

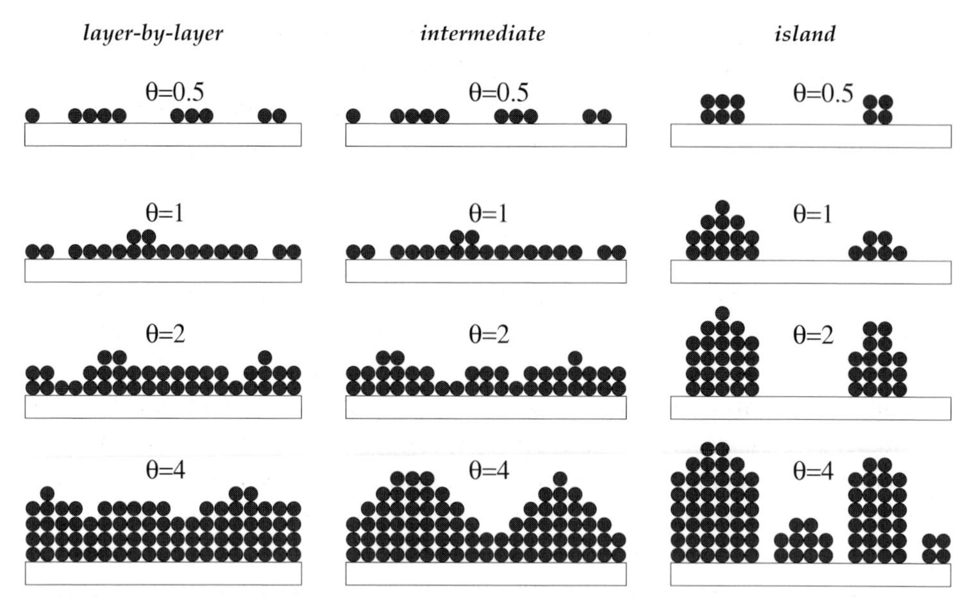

Figure 11.15. Illustration of the three different modes of growth in Molecular Beam Epitaxy. Growth is depicted at various stages, characterized by the total amount of deposited material which is measured in monolayer (ML) coverage θ. **Left:** layer-by-layer (Frank–van der Merwe) growth at $\theta = 0.5, 2, 1$ and 4 ML; in this case the deposited material wets the substrate and the film is not adversely affected by strain, continuing to grow in a layered fashion. **Center:** intermediate (Stranski–Krastanov) growth at $\theta = 0.5, 2, 1$ and 4 ML; in this case the deposited material wets the substrate but begins to grow in island mode after a critical thicknes h_c (in the example shown, $h_c = 2$). **Right:** 3D island (Volmer–Weber) growth at $\theta = 0.5, 2, 1$ and 4 ML; in this case the deposited material does not wet the substrate and grows in island mode from the very beginning, leaving a finite portion of the substrate uncovered, even though enough material has been deposited to cover the entire substrate.

investigations. The recent trend in such studies is to attempt to determine all the relevant atomistic level processes (such as atomic diffusion on terraces and steps, attachment–detachment of atoms from islands, etc.) using first-principles quantum mechanical calculations, and then to use this information to build stochastic models of growth, which can eventually be coarse-grained to produce continuum equations of the type discussed in this section.

It should be emphasized that in the continuum models discussed here, the surface height $h(\mathbf{r}, t)$ is assumed to vary slowly on the scale implied by the equation, so that asymptotic expansions can be meaningful. Therefore, this scale must be orders of magnitude larger than the atomic scale, since on this latter scale height variations can be very dramatic. It is in this sense that atomistic models must be coarse-grained in order to make contact with the statistical models implied by the continuum equations. The task of coarse-graining the atomistic behavior is not trivial, and the problem of how to approach it remains an open one. Nevertheless, the statistical models of the continuum equations can be useful in elucidating general features of growth. Even without the benefit of atomistic scale processes, much can be deduced about the terms that should enter into realistic continuum models, either through simple physical arguments as described above, or from symmetry principles.

11.4 Interfaces

An interface is a plane which joins two semi-infinite solids. Interfaces between two crystals exhibit some of the characteristic features of surfaces, such as the broken bonds suffered by the atoms on either side of the interface plane, and the tendency of atoms to rearrange themselves to restore their bonding environment. We can classify interfaces in two categories: those between two crystals of the same type, and those between two crystals of different types. The first type are referred to as grain boundaries, because they usually occur between two finite-size crystals (grains), in solids that are composed of many small crystallites. The second type are referred to as hetero-interfaces. We discuss some of the basic features of grain boundaries and hetero-interfaces next.

11.4.1 Grain boundaries

Depending on the orientation of equivalent planes in the two grains on either side of the boundary, it may be possible to match easily the atoms at the interface or not. This is illustrated in Fig. 11.16, for two simple cases involving cubic crystals. The plane of the boundary is defined as the (y, z) plane, with x the direction perpendicular to it. The cubic crystals we are considering, in projection on the (x, y) plane are represented by square lattices. In the first example, equivalent planes on either side of the boundary meet at a 45° angle, which makes it impossible to match atomic distances across more than one row at the interface. All atoms along the interface have missing or stretched bonds. This is referred to as an asymmetric tilt boundary, since the orientation of the two crystallites involves a relative tilt around the z axis. In the second example, the angle between equivalent planes on either side of the boundary is 28°. In this example we can distinguish four different types of sites on the boundary plane, labeled A, B, C, D in Fig. 11.16. An atom occupying site A will be under considerable compressive stress and will have five close neighbors; There is no atom at site B, because the neighboring atoms on either side of the interface are already too close to each other; each of these atoms has five nearest neighbors, counting the atom at site A as one, but their bonds are severely distorted. At site C there is plenty of space for an atom, which again will have five neighbors, but its bonds will be considerably stretched, so it is under tensile stress. The environment of the atom at site D seems less distorted, and this atom will have four neighbors at almost

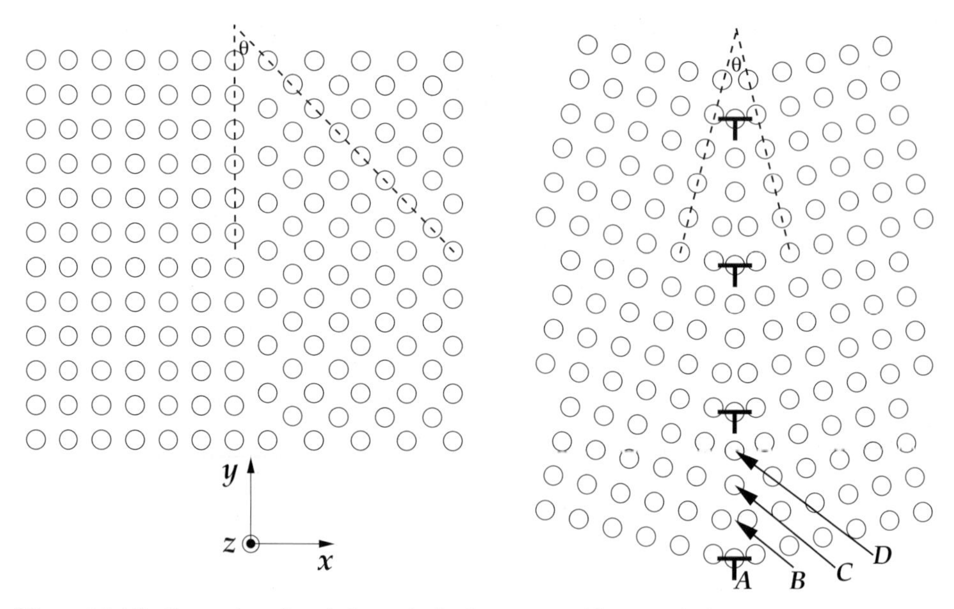

Figure 11.16. Examples of grain boundaries between cubic crystals for which the projection along one of the crystal axes, z, produces a 2D square lattice. **Left:** asymmetric tilt boundary at $\theta = 45°$ between two grains of a square lattice. **Right:** symmetric tilt boundary at $\theta = 28°$ between two grains of a square lattice. Four different sites are identified along the boundary, labeled A, B, C, D, which are repeated along the grain boundary periodically; the atoms at each of those sites have different coordination. This grain boundary is equivalent to a periodic array of edge dislocations, indicated by the T symbols.

regular distances, but will also have two more neighbors at slightly larger distance (the atoms closest to site A). This example is referred to as a symmetric tilt boundary.

The above examples illustrate some of the generic features of grain boundaries, namely: the presence of atoms with broken bonds, as in surfaces; the existence of sites with fewer or more neighbors than a regular site in the bulk, as in point defects; and the presence of local strain, as in dislocations. In fact, it is possible to model grain boundaries as a series of finite segments of a dislocation, as indicated in Fig. 11.16; this makes it possible to apply several of the concepts introduced in the theory of dislocations (see chapter 10). In particular, depending on how the two grains are oriented relative to each other at the boundary, we can distinguish between tilt boundaries, involving a relative rotation of the two crystallites around an axis parallel to the interface as in the examples of Fig. 11.16, or twist boundaries, involving a rotation of the two crystallites around an axis perpendicular to the interface. These two situations are reminiscent of the distinction between edge and screw dislocations. More generally, a grain boundary may be formed by rotating the two crystallites around axes both parallel and perpendicular to the interface, reminiscent of a mixed dislocation. Equally well, one can apply concepts from the theory of point defects, such as localized gap states (see chapter 9), or from the theory of surfaces, such as surface reconstruction (see the discussion in section 11.2). In general, due to the imperfections associated with them, grain boundaries provide a means for enhanced diffusion of atoms or give rise to electronic states in the band gap of doped semiconductors that act like traps for electrons or holes. As such, grain boundaries can have a major influence both on the mechanical properties and on the electronic behavior of real materials. We elaborate briefly on certain structural aspects of grain boundaries here and on the electronic aspects of interfaces in the next subsection. For more extensive discussions we refer the reader to the comprehensive treatment of interfaces by Sutton and Balluffi (see the Further reading section).

As indicated above, a tilt boundary can be viewed as a periodic array of edge dislocations, and a twist boundary can be viewed as a periodic array of screw dislocations. We discuss the first case in some detail; an example of a tilt boundary with angle $\theta = 28°$ between two cubic crystals is illustrated in Fig. 11.16. Taking the grain-boundary plane as the yz plane and the edge dislocation line along the z axis, and using the results derived in chapter 10 for the stress and strain fields of an isolated edge dislocation, we can obtain the grain-boundary stress field by adding the fields of an infinite set of ordered edge dislocations at a distance d from each other along the y axis (see Problem 7):

$$\sigma_{xx}^{tilt} = \frac{K_e b_e}{d} \sin(\bar{y}) \frac{[\cos(\bar{y}) - \cosh(\bar{x}) - \bar{x}\sinh(\bar{x})]}{[\cosh(\bar{x}) - \cos(\bar{y})]^2} \tag{11.34}$$

$$\sigma_{yy}^{tilt} = \frac{K_e b_e}{d} \sin(\bar{y}) \frac{[\cos(\bar{y}) - \cosh(\bar{y}) + \bar{x}\sinh(\bar{x})]}{[\cosh(\bar{x}) - \cos(\bar{y})]^2} \tag{11.35}$$

$$\sigma_{xy}^{tilt} = \frac{K_e b_e}{d} \frac{[\cosh(\bar{x})\cos(\bar{y}) - 1]}{[\cosh(\bar{x}) - \cos(\bar{y})]^2} \tag{11.36}$$

where the reduced variables are defined as $\bar{x} = 2\pi x/d$, $\bar{y} = 2\pi y/d$, and the Burgers vector b_e lies along the x axis. It is interesting to consider the asymptotic behavior of the stress components as given by these expressions. Far from the boundary in the direction perpendicular to it, $x \to \pm\infty$, all the stress components of the tilt grain boundary decay exponentially because of the presence of the $\cosh^2(\bar{x})$ term in the denominator.

11.4.2 Hetero-interfaces

As a final example of interfaces we consider a planar interface between two different crystals. We will assume that the two solids on either side of the interface have the same crystalline structure, but different lattice constants. This situation, called a hetero-interface, is illustrated in Fig. 11.17. The similarity between the two crystal structures makes it possible to match smoothly the two solids along a crystal plane.

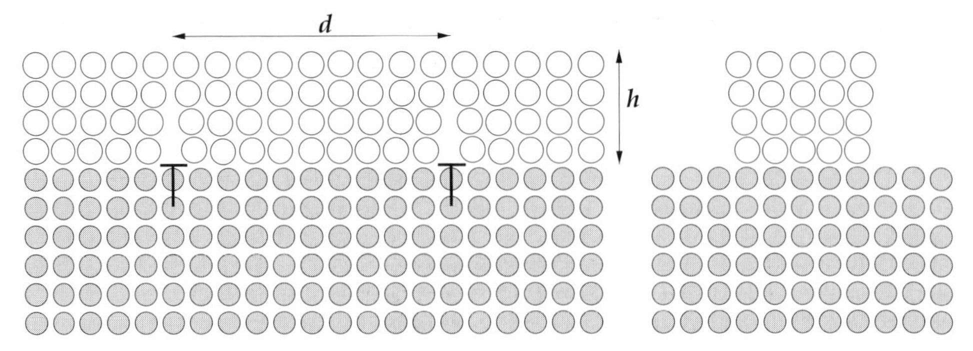

Figure 11.17. Schematic representation of hetero-interface. **Left:** the newly deposited film, which has height h, wets the substrate and the strain is relieved by the creation of misfit dislocations (indicated by a T), at a distance d from each other. **Right:** the newly deposited material does not wet the substrate, but instead forms islands which are coherently bonded to the substrate but relaxed to their preferred lattice constant at the top.

The difference in lattice constants, however, will produce strain along the interface. There are two ways to relieve the strain.

(i) Through the creation of what appears to be dislocations on one side of the interface, assuming that both crystals extend to infinity on the interface plane. These dislocations will be at regular intervals, dictated by the difference in the lattice constants; they are called "misfit dislocations".
(ii) Through the creation of finite-size structures on one side of the interface, which at their base are coherently bonded to the substrate (the crystal below the interface), but at their top are relaxed to their preferred lattice constant. These finite-size structures are called islands.

The introduction of misfit dislocations is actually related to the height h of the deposited film, and is only made possible when h exceeds a certain critical value, the critical height h_c. The reason is that without misfit dislocations the epilayer is strained to the lattice constant of the substrate, which costs elastic energy. This strain is relieved and the strain energy reduced by the introduction of the misfit dislocations, but the presence of these dislocations also costs elastic energy. The optimal situation is a balance between the two competing terms. To show this effect in a quantitative way, we consider a simple case of cubic crystals with an interface along one of the high-symmetry planes (the xy plane), perpendicular to one of the crystal axes (the z axis). In the absence of any misfit dislocations, the in-plane strain due to the difference in lattice constants of the film a_f and the substrate a_s, is given by

$$\epsilon_m \equiv \frac{a_s - a_f}{a_f} = \epsilon_{xx} = \epsilon_{yy} \tag{11.37}$$

and involves only diagonal components but no shear. For this situation, with cubic symmetry and only diagonal strain components, the stress–strain relations are

$$\begin{bmatrix} \sigma_{xx} \\ \sigma_{yy} \\ \sigma_{zz} \end{bmatrix} = \begin{bmatrix} c_{11} & c_{12} & c_{12} \\ c_{12} & c_{11} & c_{12} \\ c_{12} & c_{12} & c_{11} \end{bmatrix} \begin{bmatrix} \epsilon_{xx} \\ \epsilon_{yy} \\ \epsilon_{zz} \end{bmatrix} \tag{11.38}$$

However, since the film is free to relax in the direction perpendicular to the surface, $\sigma_{zz} = 0$, which, together with Eq. (11.37), leads to the following relation:

$$2c_{12}\epsilon_m + c_{11}\epsilon_{zz} = 0$$

From this, we obtain the strain in the z direction, ϵ_{zz}, and the misfit stress in the xy plane, $\sigma_m = \sigma_{xx} = \sigma_{yy}$, in terms of the misfit strain ϵ_m:

$$\epsilon_{zz} = -\frac{2c_{12}}{c_{11}}\epsilon_m \tag{11.39}$$

$$\sigma_m = \left[c_{11} + c_{12} - \frac{2c_{12}}{c_{11}}\right]\epsilon_m = \mu_m\epsilon_m \tag{11.40}$$

where we have defined the constant of proportionality between σ_m and ϵ_m as the effective elastic modulus for the misfit, μ_m. When misfit edge dislocations are introduced, as shown in Fig. 11.17, the strain is reduced by b_e/d, where b_e is the Burgers vector and d is the distance between the dislocations. The resulting strain energy per unit area of the interface is

$$\gamma_m = \mu_m\left(\epsilon_m - \frac{b_e}{d}\right)^2 h \tag{11.41}$$

The dislocation energy per unit length of dislocation is

$$U_d = \frac{\overline{K}_e b_e^2}{2} = \frac{b_e^2}{4\pi(1-\nu)}\overline{\mu}\ln\left(\frac{h}{b_e}\right) \tag{11.42}$$

with $\overline{K}_e$ the relevant elastic constant, $\overline{K}_e = \overline{\mu}/2\pi(1-\nu)$. In order to arrive at this result, we have used the expression derived in chapter 10 for the elastic energy of an edge dislocation, Eq. (10.17), with $\overline{\mu}$ the effective shear modulus at the interface where the misfit dislocations are created, which is given by

$$\overline{\mu} = \frac{2\mu_f\mu_s}{\mu_f + \mu_s},$$

where μ_f and μ_s are the shear moduli of the film and the substrate, respectively. For the relevant length, L, over which the dislocation field extends, we have taken $L \sim h$, while for the dislocation core radius, r_c, we have taken $r_c \sim b_e$, both reasonable approximations; the constants of proportionality in these approximations

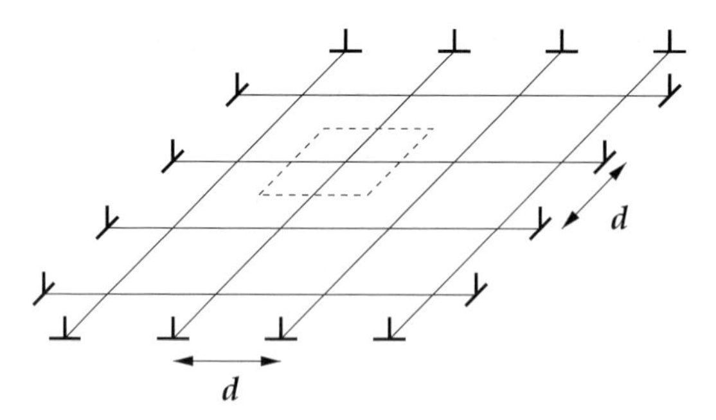

Figure 11.18. Illustration of a regular 2D array of misfit edge dislocations at the interface of two cubic crystals. The distance between dislocations in each direction is d. The total dislocation length in an area d^2, outlined by the dashed square, is $2d$.

are combined into a new constant which we will take to be unity (an assumption which is justified *a posteriori* by the final result, as explained below), giving $L/r_c = (h/b_e)$. What remains to be determined is the total misfit dislocation length per unit area of the interface, which is

$$l = \frac{2d}{d^2} = \frac{2}{d}$$

as illustrated in Fig. 11.18. With this, the dislocation energy per unit area takes the form

$$\gamma_d = lU_d = \frac{b_e^2}{2\pi(1-\nu)d}\overline{\mu}\ln\left(\frac{h}{b_e}\right) \tag{11.43}$$

which, combined with the misfit energy per unit area, γ_m, gives for the total energy per unit area of the interface

$$\gamma_{int}(\zeta) = \mu_m(\epsilon_m - \zeta)^2 h + \frac{b_e}{2\pi(1-\nu)}\overline{\mu}\ln\left(\frac{h}{b_e}\right)\zeta \tag{11.44}$$

where we have introduced the variable $\zeta = b_e/d$. The limit of no misfit dislocations at the interface corresponds to $\zeta \to 0$, because then their spacing $d \to \infty$ while the Burgers vector b_e is fixed. The condition for making misfit dislocations favorable is that the total interface energy decreases by their introduction, which can be expressed by

$$\left[\frac{\mathrm{d}\gamma_{int}(\zeta)}{\mathrm{d}\zeta}\right]_{\zeta=0} \leq 0$$

Consequently, the critical thickness of the film at which the introduction of misfit dislocations becomes energetically favorable is determined by the condition

$$\left[\frac{\mathrm{d}\gamma_{int}(\zeta)}{\mathrm{d}\zeta}\right]_{\zeta=0} = 0 \Longrightarrow \tilde{h}_c \frac{1}{\ln(\tilde{h}_c)} = \frac{\overline{\mu}}{4\pi(1-\nu)\mu_m\epsilon_m} \tag{11.45}$$

where in the last equation we have expressed the critical thickness in units of the Burgers vector $\tilde{h}_c = h_c/b_e$. All quantities on the right-hand side of the last equation are known for a given type of film deposited on a given substrate, thus the critical thickness $\tilde{h}_c$ can be uniquely determined. Here we comment on our choice of the constant of proportionality in the relation $L/r_c \sim (h/b_e)$, which we took to be unity: the logarithmic term in (h/b_e) makes the final value of the critical height insensitive to the precise value of this constant, as long as it is of order unity, which is the expected order of magnitude since $L \approx h$ and $r_c \approx b_e$ to within an order of magnitude. For film thickness $h < \tilde{h}_c b_e$ misfit dislocations are not energetically favorable, and thus are not stable; for $h > \tilde{h}_c b_e$ misfit dislocations are stable, but whether they form or not could also depend on kinetic factors.

The example of an interface between two cubic crystals with a regular array of misfit dislocations may appear somewhat contrived. As it turns out, it is very realistic and relevant to nanoscale structures for advanced electronic devices. As mentioned in section 11.3, crystals can be grown on a different substrate by MBE. When the substrate and the newly deposited material have the same crystal structure, it is possible to nucleate the new crystal on the substrate even if their lattice constants are different. As the newly deposited material grows into crystal layers, the strain energy due to the lattice constant difference also grows and eventually must be relieved. If the interactions between substrate and newly deposited atoms are favorable, so that the deposited atoms wet the substrate, then the first of the two situations described above arises where the strain is relieved by the creation of misfit dislocations beyond a critical film thickness determined by Eq. (11.45). If the chemical interactions do not allow wetting of the substrate, then finite-size islands are formed, which naturally relieve some of the strain through relaxation. These phenomena play a crucial role in the morphology and stability of the film that is formed by deposition on the substrate. In certain cases, the strain effects can lead to islands of fairly regular shapes and sizes. The resulting islands are called quantum dots and have sizes in the range of a few hundred to a few thousand angstroms. This phenomenon is called self-assembly of nanoscale quantum dots. Because of their finite size, the electronic properties of the dots are different than those of the corresponding bulk material. In particular, the confinement of electrons within the dots gives rise to levels whose energy depends sensitively on the size and shape of the dots. The hope that this type of structure will prove very useful in new electronic devices has generated great excitement recently (see, for example, the book by Barnham and Vvedensky, mentioned in the Further reading section).

We conclude this section on hetero-interfaces with a brief discussion of electronic states that are special to these systems. Of particular interest are interfaces between metals and semiconductors, structures that are of paramount importance in the operation of electronic devices. We will consider then an ideal metal surface and an ideal semiconductor surface in intimate contact, that is, as planes which are flat on the atomic scale and at a distance apart which is of order typical interatomic distances, without the presence of any contaminants such as oxide layers. It was first pointed out by Heine [183] that these conditions lead to the creation of Bloch states with energies in the gap of semiconductor and complex wave-vectors in the direction perpendicular to the interface. These are called "metal induced gap states" (MIGS). The MIGS decay exponentially into the semiconductor, a behavior similar to that of the surface states discussed earlier in relation to the 1D surface model, but they do not decay into the metal side, in contrast to the situation of the 1D surface model. A study of the nature of such states for a prototypical system, that between the (111) surfaces of Al and Si, in the context of DFT calculations, has been reported by Louie and Cohen [184]. The presence of these states will have important consequences for the electronic structure of the interface. To illustrate this, consider a junction between a metal and an n-type semiconductor, as discussed in chapter 9. Since the MIGS have energies within the semiconductor gap, they will be occupied when the junction is formed. The occupation of MIGS will draw electronic charge from the semiconductor side, but only in the immediate neighborhood of the interface since they decay exponentially on that side. This will give rise to a dipole moment at the interface, creating the conditions necessary to produce equilibrium between the Fermi levels on the two sides. However, since the dipole moment is due to MIGS, whose energy is fixed within the semiconductor gap, the induced barrier to electron transfer across the interface will not depend on the work function of the metal. This is in contrast to the Schottky model, where the barrier is directly proportional to the metal work function (see chapter 9). We had remarked there that the measured barrier is roughly proportional to the metal work function for semiconductors with large gap but essentially independent of the metal work function for semiconductors with small gap. This behavior can be interpreted in the context of MIGS: the wavefunctions of these states decay much more rapidly into the semiconductor side in a large-gap semiconductor, which makes them much less effective in creating the interface dipole. Thus, in the large-gap case the MIGS are not enough to produce the conditions necessary for electronic equilibrium, and the physics of the Schottky model are relevant. In the small-gap case, the MIGS are sufficient for creating equilibrium conditions. The picture outlined above is certainly an oversimplified view of real metal–semiconductor interfaces. Some aspects of the oversimplification are the presence of point defects at the interface and the fact that real surfaces are not atomically flat planes but contain features like steps and islands. All these issues further complicate the determination of the barrier to electron transfer across the interface, which is the physical quantity of interest from the practical point of view (it is relevant to the design of electronic devices). For more detailed discussion and extensive references the reader is directed to the book by Sutton and Balluffi (see the Further reading section).

Further reading

1. *Physics at Surfaces*, A. Zangwill, (Cambridge University Press, Cambridge, 1988).
   This is an excellent general book for the physics of surfaces, including many insightful discussions.
2. *Atomic and Electronic Structure of Surfaces*, M. Lannoo and P. Friedel, (Springer-Verlag, Berlin, 1991).
   An interesting book on surfaces in general, with extensive coverage of their electronic structure.
3. *The Structure of Surfaces*, M.A. Van Hove and S.Y. Tong, eds. (Springer-Verlag, Berlin, 1985).
   A collection of review articles on experimental and theoretical techniques for determining the atomic structure of surfaces, with emphasis on scattering methods such as LEED.
4. *Surface Science, The First Thirty Years*, C. Duke, ed. (Elsevier, Amsterdam, 1994).
   This is a collection of classic papers published in the journal *Surface Science* in its first 30 years.
5. "The density functional formalism and the electronic structure of metal surfaces", N.D. Lang, in *Solid State Physics*, vol. 28, pp. 225–300 (F. Seitz, D. Turnbull and H. Ehrenreich, eds., Academic Press, New York, 1973).
6. *Semiconductor Surfaces and Interfaces*, W.Mönch (Springer-Verlag, Berlin, 1995).
   This is a useful account of the structure and properties of semiconductor surfaces and interfaces.
7. "Basic mechanisms in the early stages of epitaxy", R. Kern, G. Le Lay and J.J. Metois, in *Current Topics in Materials Science*, vol. 3 (E. Kaldis, ed., North-Holland Publishing Co., Amsterdam, 1979).
   This is a very thorough treatment of early experimental work on epitaxial growth phenomena.
8. "Theory and simulation of crystal growth", A.C. Levi and M. Kotrla, *J. Phys.: Cond. Matt.*, **9**, pp. 299–344 (1997).
   This is a useful article describing the techniques and results of computer simulations of growth phenomena.
9. *Fractal Concepts in Crystal Growth*, A.L. Barabasi and H.E. Stanley (Cambridge University Press, Cambridge, 1995).
   A useful modern introduction to the statistical mechanics of growth phenomena on surfaces.
10. *Physics of Crystal Growth*, A. Pimpinelli and J. Villain (Cambridge University Press, Cambridge, 1998).
    A thorough account of modern theories of crystal growth phenomena.
11. *Low-dimensional Semiconductor Structures*, K. Barnham and D. Vvedensky, eds. (Cambridge University Press, Cambridge, 2000).

A collection of review articles highlighting problems in semiconductor devices based on low-dimensional structures, where surfaces and interfaces play a key role.

12. *Interfaces in Crystalline Materials*, A.P. Sutton and R.W. Balluffi (Clarendon Press, Oxford, 1995).

This is a thorough and detailed discussion of all aspects of interfaces in crystals.

Problems

1. We wish to solve the 1D surface model defined by Eq. (11.7). For this potential, the wavefunction will have components that involve the zero and G reciprocal-lattice vectors. A convenient choice for the wavefunction is

$$\psi_k(z < 0) = \mathrm{e}^{\mathrm{i}kz}\left[c_0 + c_1 \mathrm{e}^{-\mathrm{i}Gz}\right].$$

Write the Schrödinger equation for this model for $z < 0$ and produce a linear system of two equations in the unknown coefficients c_0, c_1. From this system show that the energy takes the values given in Eq. (11.8), while the wavefunction is given by the expression in Eq. (11.9).

2. For the simple cubic, body-centered cubic, face-centered cubic and diamond lattices, describe the structure of the bulk-terminated planes in the (001), (110) and (111) crystallographic directions: find the number of in-plane neighbors that each surface atom has, determine whether these neighbors are at the same distance as nearest neighbors in the bulk or not, show the surface lattice vectors, and determine how many "broken bonds" or "missing neighbors" each surface atom has compared to the bulk structure. Discuss which of these surfaces might be expected to undergo significant reconstruction and why.

3. Construct the schematic electronic structure of the Si(111) reconstructed surface, by analogy to what was done for Si(001) (see Fig. 11.8):

   (a) for the $(\sqrt{3} \times \sqrt{3})$ reconstruction with an adatom and three surface atoms per unit cell

   (b) for the (2×2) reconstruction with an adatom, three surface atoms, and one rest-atom per unit cell.

   Explain how the sp^3 states of the original bulk-terminated plane and those of the adatom and the restatom combine to form bonding and antibonding combinations, and how the occupation of these new states determines the metallic or semiconducting character of the surface.

4. Give an argument that describes the charge transfer in the (2×2) reconstruction of Si(111), with one adatom and one restatom per unit cell, using a simple tight-binding picture.

   (a) Assume that the adatom is at the lowest energy T_4 position, and is in a pyramidal bonding configuration with its three neighbors; what does that imply for its bonding?

   (b) Assume that the restatom is in a planar bonding configuration with its three neighbors; what does that imply for its bonding?

(c) Based on these assumptions, by analogy to what we discussed in the case of the Si(100) (2×1) tilted-dimer reconstruction, deduce what kind of electronic charge transfer ought to take place in the Si(111) (2×2) adatom–restatom reconstruction.

(d) It turns out that in a real surface the electronic charge transfer takes place from the adatom to the rest atom. This is also verified by detailed first-principles electronic structure calculations. Can you speculate on what went wrong in the simple tight-binding analysis?

5. Derive the roughness, growth and dynamical exponents given in Eq. (11.32), for the Wolf–Villain model through a rescaling argument, by analogy to what was done for the Edwards–Wilkinson model.

6. We will attempt to demonstrate the difficulty of determining, through the standard rescaling argument, the roughness and growth exponents in the KPZ model, Eq. (11.33).

(a) Suppose we use a rescaling argument as in the EW model; we would then obtain three equations for α, β, z which, together with the definition of the dynamical exponent Eq. (11.14), overdetermine the values of the exponents. Find these three equations and discuss the overdetermination problem.

(b) Suppose we neglect one of the terms in the equation to avoid the problem of overdetermination. Specifically, neglect the surface tension term, and derive the exponents for this case.

(c) Detailed computer simulations for this model give that, for $d = 1$, to a very good approximation $\alpha = 1/2$, $\beta = 1/3$. Is this result compatible with your values for α and β obtained from the rescaling argument? Discuss what may have gone wrong in this derivation.

7. Using the expression for the stress field of an isolated edge dislocation, given in chapter 10, Table 10.1, show that the stress field of an infinite array of edge dislocations along the y axis, separated by a distance d from each other, with the dislocation lines along the z axis, is given by Eqs. (11.34)–(11.36). Obtain the behavior of the various stress components far from the boundary in the direction perpendicular to it, that is, $x \to \pm\infty$. Also, using the expressions that relate stress and strain fields in an isotropic solid (see Appendix 1), calculate the strain field for this infinite array of edge dislocations. Comment on the physical meaning of the strain field far from the boundary.

12

Non-crystalline solids

While the crystalline state is convenient for describing many properties of real solids, there are a number of important cases where this picture cannot be used, even as a starting point for more elaborate descriptions, as was done for defects in chapters 9–11. As an example, we look at solids in which certain symmetries such as rotations, reflections, or an underlying regular pattern are present, but these symmetries are not compatible with three-dimensional periodicity. Such solids are called *quasicrystals*. Another example involves solids where the local arrangement of atoms, embodied in the number of neighbors and the preferred bonding distances, has a certain degree of regularity, but there is no long-range order of any type. Such solids are called *amorphous*. Amorphous solids are very common in nature; glass, based on SiO_2, is a familiar example.

In a different class of non-crystalline solids, the presence of local order in bonding leads to large units which underlie the overall structure and determine its properties. In such cases, the local structure is determined by strong covalent interactions, while the variations in large-scale structure are due to other types of interactions (ionic, hydrogen bonding, van der Waals) among the larger units. These types of structures are usually based on carbon, hydrogen and a few other elements, mostly from the first and second rows of the Periodic Table (such as N, O, P, S). This is no accident: carbon is the most versatile element in terms of forming bonds to other elements, including itself. For instance, carbon atoms can form a range of bonds among themselves, from single covalent bonds, to multiple bonds to van der Waals interactions. A class of widely used and quite familiar solids based on such structures are plastics; in plastics, the underlying large units are *polymer chains*.

12.1 Quasicrystals

As was mentioned in chapter 1, the discovery of five-fold and ten-fold rotational symmetry in certain metal alloys [5] was a shocking surprise. This is because perfect periodic order in three (or, for that matter, two) dimensions, which these unusual solids *appeared* to possess, is not compatible with five-fold or ten-fold rotational symmetry (see Problem 5 in chapter 3). Accordingly, these solids were named "quasicrystals". Theoretical studies have revealed that it is indeed possible to create, for example, structures that have five-fold rotational symmetry and can tile the two-dimensional plane, by combining certain simple geometric shapes according to specific local rules. This means that such structures can extend to infinity without any defects, maintaining the perfect rotational local order while having no translational long-range order. These structures are referred to as Penrose tilings (in honor of R. Penrose who first studied them). We will discuss a one-dimensional version of this type of structure, in order to illustrate how they give rise to a diffraction pattern with features reminiscent of a solid with crystalline order.

Our quasicrystal example in one dimension is called a Fibonacci sequence, and is based on a very simple construction: consider two line segments, one long and one short, denoted by L and S, respectively. We use these to form an infinite, perfectly periodic solid in one dimension, as follows:

$$LSLSLSLSLS \cdots$$

This sequence has a unit cell consisting of two parts, L and S; we refer to it as a two-part unit cell. Suppose now that we change this sequence using the following rule: we replace L by LS and S by L. The new sequence is

$$LSLLSLLSLLSLLSL \cdots$$

This sequence has a unit cell consisting of three parts, L, S and L; we refer to it as a three-part unit cell. If we keep applying the replacement rule to each new sequence, in the next iteration we obtain

$$LSLLSLSLLSLSLLSLSLLSLSLLS \cdots$$

i.e., a sequence with a five-part unit cell ($LSLLS$), which generates the following sequence:

$$LSLLSLSLLSLLSLSLLSLLSLSLLSLLSLSLLSLLSLSL \cdots$$

i.e., a sequence with an eight-part unit cell ($LSLLSLSL$), and so on *ad infinitum*.

It is easy to show that in the sequence with an n-part unit cell, with $n \to \infty$, the ratio of L to S segments tends to the golden mean value, $\tau = (1 + \sqrt{5})/2$. Using this formula, we can determine the position of the nth point on the infinite line (a point determines the beginning of a new segment, either long or short), as

$$x_n = n + \frac{1}{\tau} NINT \left[\frac{n}{\tau} \right] \qquad (12.1)$$

where we use the notation $NINT[x]$ to denote the largest integer less than or equal to x. We can generalize this expression to

$$x_n = \lambda \left(n + a + \frac{1}{\tau} NINT \left[\frac{n}{\tau} + b \right] \right) \tag{12.2}$$

where a, b, λ are arbitrary numbers. This change simply shifts the origin to a, alters the ordering of the segments through b, and introduces an overall scale factor λ. The shifting and scaling does not alter the essential feature of the sequence, namely that the larger the number of L and S segments in the unit cell, the more it appears that any *finite* section of the sequence is a random succession of the two components. We know, however, that by construction the sequence is not random at all, but was produced by a very systematic augmentation of the original perfectly periodic sequence. In this sense, the Fibonacci sequence captures the characteristics of quasicrystals in one dimension.

The question which arises now is, how can the quasicrystalline structure be distinguished from a perfectly ordered or a completely disordered one? In particular, what is the experimental signature of the quasicrystalline structure? We recall that the experimental hallmark of crystalline structure was the Bragg diffraction planes which scatter the incident waves coherently. Its signature is the set of spots in reciprocal space at multiples of the reciprocal-lattice vectors, which are the vectors that satisfy the coherent diffraction condition. Specifically, the Fourier Transform (FT) of a Bravais lattice is a set of δ-functions in reciprocal space with arguments $(\mathbf{k} - \mathbf{G})$, where $\mathbf{G} = m_1\mathbf{b}_1 + m_2\mathbf{b}_2 + m_3\mathbf{b}_3$, m_1, m_2, m_3 are integers and $\mathbf{b}_1, \mathbf{b}_2, \mathbf{b}_3$ are the reciprocal-lattice vectors. Our aim is to explore what the FT of the Fibonacci sequence is.

In order to calculate the FT of the Fibonacci sequence we will use a trick: we will generate it through a construction in a two-dimensional square lattice on the xy plane, as illustrated in Fig. 12.1. Starting at some lattice point (x_0, y_0) (which we take to be the origin of the coordinate system), we draw a line at an angle θ to the x axis, where $\tan\theta = 1/\tau$ (this also implies $\sin\theta = 1/\sqrt{1+\tau^2}$, and $\cos\theta = \tau/\sqrt{1+\tau^2}$). We define this new line as the x' axis, and the line perpendicular to it as the y' axis. Next, we define the distance w between (x_0, y_0) and a line parallel to the x' axis passing through the point diametrically opposite (x_0, y_0) in the lattice, on the negative x' and positive y' sides. We draw two lines parallel to the x' axis at distances $\pm w/2$ along the y' axis. We consider all the points of the original square lattice that lie between these last two lines (a strip of width w centered at the x' axis), and project them onto the x' axis, along the y' direction. The projected points form a Fibonacci sequence on the x' axis, as can be easily seen from the fact that the

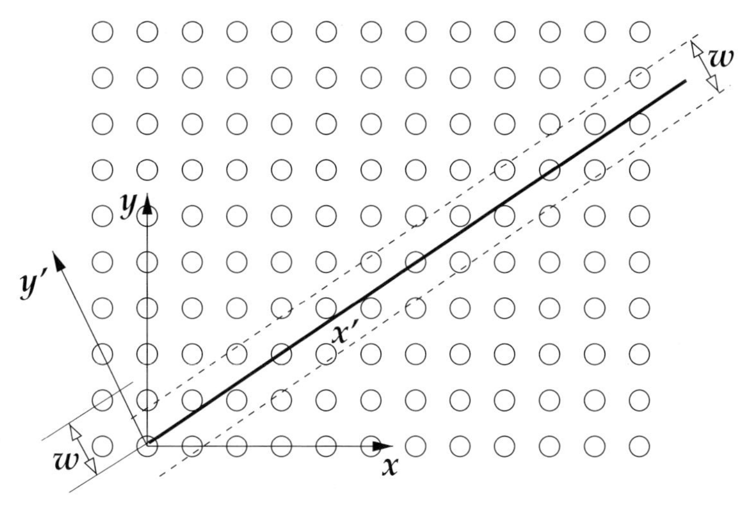

Figure 12.1. Construction of the Fibonacci sequence by using a square lattice in two dimensions: the thick line and corresponding rotated axes x', y' are at angle θ ($\tan\theta = 1/\tau$) with respect to the original x, y axes. The square lattice points within the strip of width w, outlined by the two dashed lines, are projected onto the thick line to produce the Fibonacci sequence.

positions of the projections on the x' axis are given by

$$x'_n = \sin\theta\left(n + \frac{1}{\tau}NINT\left[\frac{n}{\tau} + \frac{1}{2}\right]\right) \tag{12.3}$$

which is identical to Eq. (12.2), with $a = 0$, $b = 1/2$, $\lambda = \sin\theta$. This way of creating the Fibonacci sequence is very useful for deriving its FT, which we do next.

We begin by defining the following functions:

$$\bar{g}(x, y) = \sum_{nm} \delta(x - m)\delta(y - n) \tag{12.4}$$

$$h(y') = \begin{cases} 0 & \text{for } |y'| < w/2 \\ 1 & \text{for } |y'| \geq w/2 \end{cases} \tag{12.5}$$

where m, n are integers. The first function produces all the points of the original square lattice on the xy plane. The second function identifies all the points that lie within the strip of width w around the rotated x' axis. In terms of these two functions, the points on the x' axis that constitute the Fibonacci sequence are given by

$$\begin{aligned} f(x') &= \int g(x', y')h(y')\mathrm{d}y' \\ g(x', y') &= \bar{g}(x(x', y'), y(x', y')) \end{aligned} \tag{12.6}$$

where x and y have become functions of x', y' through

$$x = x' \cos\theta - y' \sin\theta, \quad y = x' \sin\theta + y' \cos\theta$$

The FT, $\tilde{f}(p)$, of $f(x')$ is given in terms of the FTs, $\tilde{g}(p, q)$, and $\tilde{h}(q)$, of $g(x', y')$ and $h(y')$, respectively, as

$$\tilde{f}(p) = \frac{1}{2\pi} \int \tilde{g}(p, q)\tilde{h}(-q)\mathrm{d}q \tag{12.7}$$

so all we need to do is calculate the FTs $\tilde{g}(p, q)$ and $\tilde{h}(p)$ of the functions $g(x', y')$ and $h(y')$. These are obtained from the standard definition of the FT and the functions defined in Eqs. (12.4) and (12.5):

$$\tilde{g}(p, q) = \frac{1}{2\pi} \sum_{nm} \delta(p - 2\pi(m \cos\theta + n \sin\theta))\delta(q - 2\pi(-m \sin\theta + n \cos\theta))$$

$$\tilde{h}(p) = w \frac{\sin(wp/2)}{wp/2} \tag{12.8}$$

which, when inserted into Eq. (12.7) give

$$\begin{aligned}\tilde{f}(p) &= \frac{1}{2\pi} \sum_{nm} \left[w \frac{\sin(\pi w(m \sin\theta - n \cos\theta))}{\pi w(m \sin\theta - n \cos\theta)} \right] \delta(p - 2\pi(m \cos\theta + n \sin\theta)) \\ &= \frac{1}{2\pi} \sum_{nm} \left[w \frac{\sin\left(\frac{\pi w(m-n\tau)}{\sqrt{1+\tau^2}}\right)}{\frac{\pi w(m-n\tau)}{\sqrt{1+\tau^2}}} \right] \delta\left(p - \frac{2\pi(m\tau + n)}{\sqrt{1+\tau^2}}\right)\end{aligned} \tag{12.9}$$

where in the last expression we have substituted the values of $\sin\theta$ and $\cos\theta$ in terms of the slope of the x' axis, as mentioned earlier (for a proof of the above relations see Problem 1).

The final result is quite interesting: the FT of the Fibonacci sequence contains two independent indices, n and m, as opposed to what we would expect for the FT of a periodic lattice in one dimension; the latter is the sum of δ-functions over the single index m, with arguments $(p - b_m)$, where $b_m = m(2\pi/a)$, with a the lattice vector in real space. Moreover, at the values of p for which the argument of the δ-functions in $\tilde{f}(p)$ vanishes, the magnitude of the corresponding term is not constant, but is given by the factor in square brackets in front of the δ-function in Eq. (12.9). This means that some of the spots in the diffraction pattern will be prominent (those corresponding to large values of the pre-δ-function factor), while others will be vanishingly small (those corresponding to vanishing values of the pre-δ-function factor). Finally, because of the two independent integer indices that enter into Eq. (12.9), there will be a dense sequence of spots, but, since not all of them are prominent, a few values of p will stand out in a diffraction pattern giving the

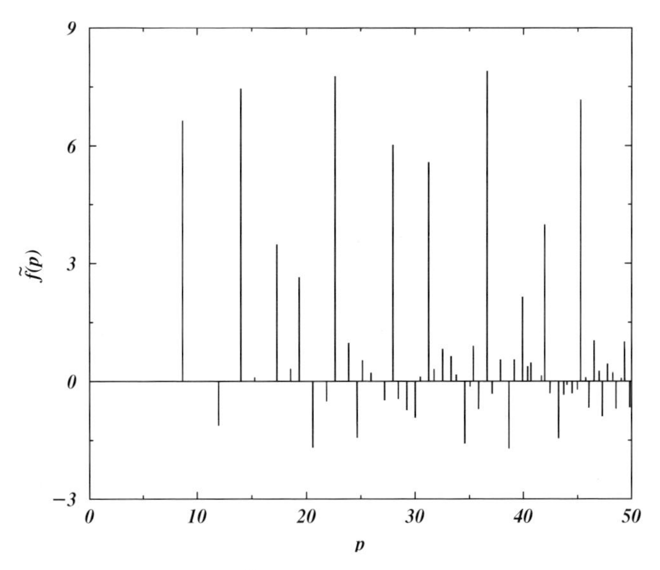

Figure 12.2. The Fourier Transform $\tilde{f}(p)$ of the Fibonacci sequence as given by Eq. (12.9), with $w = 1$ and the δ-function represented by a window of width 0.02 (cf. Eq. (G.53)).

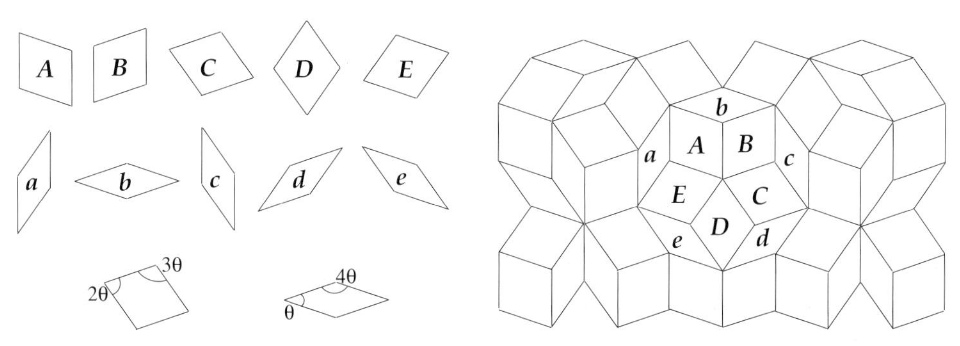

Figure 12.3. Construction of 2-dimensional quasicrystal pattern with "fat" (A, B, C, D, E) and "skinny" (a, b, c, d, e) rhombuses. A seed consisting of the five fat and five skinny rhombuses is shown labeled on the right-hand side. The pattern can be extended to cover the entire two-dimensional plane without defects (overlaps or gaps); this maintains the five-fold rotational symmetry of the seed but has no long-range translational periodicity.

appearance of order in the structure, which justifies the name "quasicrystal". This is illustrated in Fig. 12.2.

Similar issues arise for quasicrystals in two and three dimensions, the only difference being that the construction of the corresponding sequences requires more complicated building blocks and more elaborate rules for creating the patterns. An example in two dimensions is shown in Fig. 12.3: the elemental building blocks here are rhombuses (so called "fat" and "skinny") with equal sides and angles which are multiples of $\theta = 36°$. This is one scheme proposed by R. Ammann and

R. Penrose; other schemes have also been proposed by Penrose for this type of two-dimensional pattern. Note how the elemental building blocks can be arranged in a structure that has perfect five-fold symmetry, and this structure can be extended to tile the infinite two-dimensional plane without any overlaps or gaps in the tiling. This, however, is only achieved when strict rules are followed on how the elemental building blocks should be arranged at each level of extension, otherwise mistakes are generated which make it impossible to continue the extension of the pattern. The properly constructed pattern maintains the five-fold rotational symmetry but has no long-range translational periodicity.

In real solids the elementary building blocks are three-dimensional structural units with five-fold or higher symmetry (like the icosahedron, see chapter 1). An important question that arose early on was: can the perfect patterns be built based on local rules for matching the building blocks, or are global rules necessary? The answer to this question has crucial implications for the structure of real solids: if global rules were necessary, it would be difficult to accept that the elementary building blocks in real solids are able to communicate their relative orientation across distances of thousands or millions of angstroms. This would cast a doubt on the validity of using the quasicrystal picture for real solids. Fortunately, it was demonstrated through simulations that essentially infinite structures can be built using local rules only. It is far easier to argue that local rules can be obeyed by the real building blocks: such rules correspond, for instance, to a preferred local relative orientation for bonding, which a small unit of atoms (the building block) can readily find. Another interesting aspect of the construction of quasicrystals is that a projection of a regular lattice from a higher dimensional space, as was done above to produce the one-dimensional Fibonacci sequence from the two-dimensional square lattice, can also be devised in two and three dimensions.

12.2 Amorphous solids

Many real solids lack any type of long-range order in their structure, even the type of order we discussed for quasicrystals. We refer to these solids as amorphous. We can distinguish two general types of amorphous solids.

(i) Solids composed of one type of atom (or a very small number of atom types) which *locally* see the same environment throughout the solid, just like in simple crystals. The local environment cannot be identical for all atoms (this would produce a crystal), so it exhibits small deviations from site to site, which are enough to destroy the long-range order. Amorphous semiconductors and insulators (such as Si, Ge, SiO_2), chalcogenide glasses (such as As_2S_3, As_2Se_3), and amorphous metals are some examples.

(ii) Solids composed of many different types of atoms which have complicated patterns of bonding. The class of structures based on long polymeric chains, that we refer to as plastics, is a familiar example.

We will examine in some detail representative examples of both types of amorphous solids. We begin with a discussion of amorphous solids of the first category, those in which the local bonding environment of all atoms is essentially the same, with small deviations. This makes it possible to define a regular type of site as well as defects in structure, akin to the definitions familiar from the crystal environment. Recalling the discussion of chapter 1, we can further distinguish the solids of this category into two classes, those with close-packed structures (like the crystalline metals), and those with open, covalently bonded structures (like crystalline semiconductors and insulators). A model that successfully represents the structure of close-packed amorphous solids is the so-called random close packing (RCP) model. The idea behind this model is that atoms behave essentially like hard spheres and try to optimize the energy by having as many close neighbors as possible. We know from earlier discussion that for crystalline arrangements this leads to the FCC (with 12 nearest neighbors) or to the BCC (with eight nearest neighbors) lattices. In amorphous close-packed structures, the atoms are not given the opportunity (due, for instance, to rapid quenching from the liquid state) to find the perfect crystalline close-packed arrangement, so they are stuck in some arbitrary configuration, as close to the perfect structure as is feasible within the constraints of how the structure was created. Analogous considerations hold for the open structures that resemble the crystal structure of covalently bonded semiconductors and insulators; in this case, the corresponding model is the so called continuous random network (CRN). The random packing of hard spheres, under conditions that do not permit the formation of the close-packed crystals, is an intuitive idea which will not be examined further. The formation of the CRN from covalently bonded atoms is more subtle, and will be discussed in some detail next.

12.2.1 Continuous random network

Amorphous Si usually serves as the prototypical covalently bonded amorphous solid. In this case, the bonding between atoms locally resembles that of the ideal crystalline Si, i.e. the diamond lattice, with tetrahedral coordination. The resemblance extends to the number of nearest neighbors (four), the bond legnth (2.35 Å) and the angles between the bonds (109°); of these, the most important is the number of nearest neighbors, while the values of the bond lengths and bond angles can deviate somewhat from the ideal values (a few percent for the bond length, significantly more for the bond angle). The resemblance to crystalline Si ends there, however, since there is no long-range order. Thus, each atom sees an environment very similar to that in the ideal crystalline structure, but the small distortions allow a random arrangement of the atomic units so that the crystalline order is destroyed. The reason for this close resemblance between the ideal crystalline structure and the amorphous structure is the strong preference of Si to form exactly four covalent bonds at tetrahedral directions with its neighbors, which arises from its electronic structure (the four sp^3 orbitals associated with each atom; see the discussion in chapter 1).

The idea that an amorphous solid has a structure *locally* similar to the corresponding crystalline structure seems reasonable. It took, however, a long time to establish the validity of this idea with quantitative arguments. The first convincing evidence that a local bonding arrangement resembling that of the crystal could be extended to large numbers of atoms, without imposing the regularity of the crystal, came from hand-built models by Polk [185]. What could have gone wrong in forming such an extended structure? It was thought that the strain in the bonds as the structure was extended (due to deviations from the ideal value) might make it impossible for it to continue growing. Polk's hand-built model and subsequent computer refinement showed that this concern was not justified, at least for reasonably large, free-standing structures consisting of several hundred atoms.

Modern simulations attempt to create models of the tetrahedral continuous random network by quenching a liquid structure very rapidly, in supercells with periodic boundary conditions. Such simulations, based on empirical interatomic potentials that capture accurately the physics of covalent bonding, can create models consisting of several thousand atoms (see, for example, Ref. [186], and Fig. 12.4). A more realistic picture of bonding (but not necessarily of the structure itself) can be obtained by using a smaller number of atoms in the periodic supercell (fewer than 100) and employing first-principles electronic structure methods for the quenching simulation [187]. Alternatively, models have been created by neighbor-switching in the crystal and relaxing the distorted structure with empirical potentials [188]. These structures are quite realistic, in the sense that they are good approximations of the infinite tetrahedral CRN. The difference from the original idea for the CRN is that the simulated structures contain some defects. The defects consist of mis-coordinated atoms, which can be either three-fold bonded (they are missing one neighbor, and are called "dangling bond" sites), or five-fold bonded (they have an extra bond, and are called "floating bond" sites). This is actually in agreement with experiment, which shows of order $\sim 1\%$ mis-coordinated atoms in pure amorphous Si. Very often amorphous Si is built in an atmosphere of hydrogen, which has the ability to saturate the single missing bonds, and reduces dramatically the number of defects (to the level of 0.01%). There is on-going debate as to what exactly the defects in amorphous Si are: there are strong experimental indications that the three-fold coordinated defects are dominant, but there are also compelling reasons to expect that five-fold coordinated defects are equally important [189]. It is intriguing that all modern simulations of pure amorphous Si suggest a predominance of five-fold coordinated defects.

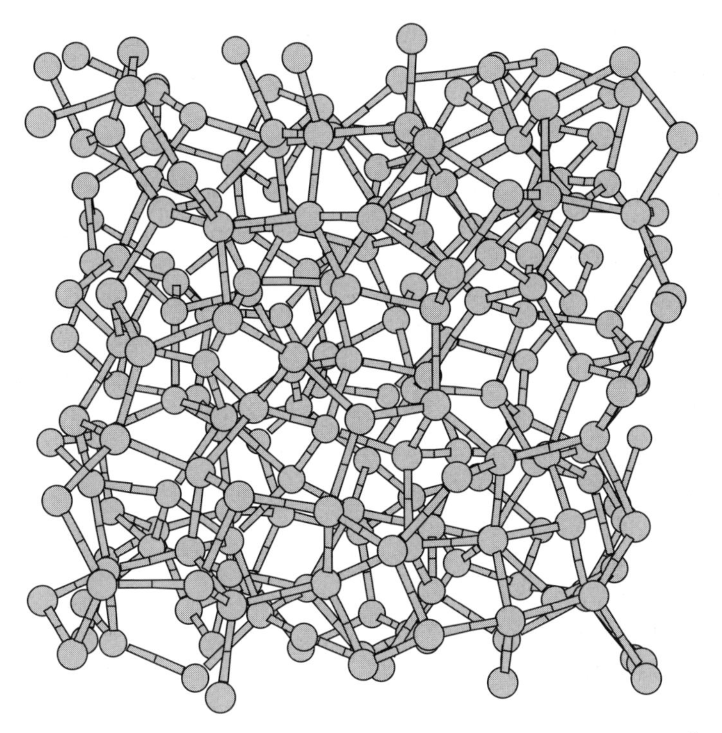

Figure 12.4. Example of a continuous random network of tetrahedrally coordinated atoms. This particular structure is a model for amorphous silicon, it contains 216 atoms and has only six mis-coordinated atoms (five-fold bonded), which represent fewer than 3% defects.

So far we have been discussing the tetrahedral version of the continuous random network (we will refer to it as t-CRN). This can be generalized to model amorphous solids with related structures. In particular, if we consider putting an oxygen atom at the center of each Si–Si bond in a t-CRN and allowing the bond lengths to relax to their optimal length by uniform expansion of the structure, then we would have a rudimentary model for the amorphous SiO_2 structure. In this structure each Si atom is at the center of a tetrahedron of O atoms and is therefore tetrahedrally coordinated, assuming that the original t-CRN contains no defects. At the same time, each O atom has two Si neighbors, so that all atoms have exactly the coordination and bonding they prefer. This structure is actually much more flexible than the original t-CRN, because the Si–O–Si angles can be easily distorted with very small energy cost, which allows the Si-centered tetrahedra to move considerable distances relative to each other. This flexibility gives rise to a very stable structure, which can exhibit much wider diversity in the arrangement of the tetrahedral units than what would be allowed in the oxygen-decorated t-CRN. The corresponding amorphous SiO_2 structures are good models of common glass. Analogous extensions of the t-CRN can be made to model certain chalcogenide glasses, which contain covalently bonded group-V and group-VI atoms (for more details see the book by Zallen, mentioned in the Further reading section).

12.2.2 Radial distribution function

How could the amorphous structure be characterized in a way that would allow comparison to experimentally measurable quantities? In the case of crystals and quasicrystals we considered scattering experiments in order to determine the signature of the structure. We will do the same here. Suppose that radiation is incident on the amorphous solid in plane wave form with a wave-vector $\mathbf{q}$. Since there is no periodic or regular structure of any form in the solid, we now have to treat each atom as a point from which the incident radiation is scattered. We consider that the detector is situated at a position $\mathbf{R}$ well outside the solid, and that the scattered radiation arrives at the detector with a wave-vector $\mathbf{q}'$; in this case the directions of $\mathbf{R}$ and $\mathbf{q}'$ must be the same. Due to the lack of order, we have to assume that the incident wave, which has an amplitude $\exp(\mathrm{i}\mathbf{q}\cdot\mathbf{r}_n)$ at the position $\mathbf{r}_n$ of an atom, is scattered into a spherical wave $\exp(\mathrm{i}|\mathbf{q}'||\mathbf{R}-\mathbf{r}_n|)/|\mathbf{R}-\mathbf{r}_n|$. We then have to sum the contributions of all these waves at the detector to find the total amplitude $A(\mathbf{q},\mathbf{q}';\mathbf{R})$. This procedure gives

$$A(\mathbf{q},\mathbf{q}';\mathbf{R}) = \sum_n \mathrm{e}^{\mathrm{i}\mathbf{q}\cdot\mathbf{r}_n} \frac{\mathrm{e}^{\mathrm{i}|\mathbf{q}'||\mathbf{R}-\mathbf{r}_n|}}{|\mathbf{R}-\mathbf{r}_n|} \tag{12.10}$$

The directions of vectors $\mathbf{R}$ and $\mathbf{q}'$ are the same, therefore we can write

$$\begin{aligned}|\mathbf{q}'||\mathbf{R}-\mathbf{r}_n| &= |\mathbf{q}'||\mathbf{R}|\left(1-\frac{2}{|\mathbf{R}|^2}\mathbf{R}\cdot\mathbf{r}_n+\frac{1}{|\mathbf{R}|^2}|\mathbf{r}_n|^2\right)^{1/2}\\ &\approx |\mathbf{q}'||\mathbf{R}| - |\mathbf{q}'|\hat{\mathbf{R}}\cdot\mathbf{r}_n = |\mathbf{q}'||\mathbf{R}| - \mathbf{q}'\cdot\mathbf{r}_n\end{aligned} \tag{12.11}$$

where we have neglected the term $(|\mathbf{r}_n|/|\mathbf{R}|)^2$ since the vector $\mathbf{r}_n$, which lies within the solid, has a much smaller magnitude than the vector $\mathbf{R}$, the detector being far away from the solid. Similarly, in the denominator of the spherical wave we can neglect the vector $\mathbf{r}_n$ relative to $\mathbf{R}$, to obtain for the amplitude in the detector:

$$A(\mathbf{q},\mathbf{q}';\mathbf{R}) \approx \frac{\mathrm{e}^{\mathrm{i}|\mathbf{q}'||\mathbf{R}|}}{|\mathbf{R}|}\sum_n \mathrm{e}^{\mathrm{i}(\mathbf{q}-\mathbf{q}')\cdot\mathbf{r}_n} = \frac{\mathrm{e}^{\mathrm{i}|\mathbf{q}'||\mathbf{R}|}}{|\mathbf{R}|}\sum_n \mathrm{e}^{-\mathrm{i}\mathbf{k}\cdot\mathbf{r}_n} \tag{12.12}$$

where we have defined the scattering vector $\mathbf{k} = \mathbf{q}' - \mathbf{q}$. The signal $f(\mathbf{k},\mathbf{R})$ at the detector will be proportional to $|A(\mathbf{q},\mathbf{q}';\mathbf{R})|^2$, which gives

$$f(\mathbf{k},\mathbf{R}) = \frac{A_0}{|\mathbf{R}|^2}\sum_{nm} \mathrm{e}^{-\mathrm{i}\mathbf{k}\cdot(\mathbf{r}_n-\mathbf{r}_m)} \tag{12.13}$$

with the two independent indices n and m running over all the atoms in the solid. We can put this into our familiar form of a FT:

$$f(\mathbf{k}, \mathbf{R}) = \frac{A_0}{|\mathbf{R}|^2} \int e^{-i\mathbf{k}\cdot\mathbf{r}} d\mathbf{r} \sum_{nm} \delta(\mathbf{r} - (\mathbf{r}_n - \mathbf{r}_m)) \tag{12.14}$$

so that, except for the factor $A_0/|\mathbf{R}|^2$, the signal is the FT of the quantity

$$\sum_{nm} \delta(\mathbf{r} - (\mathbf{r}_n - \mathbf{r}_m)) \tag{12.15}$$

which contains all the information about the structure of the solid, since it depends on all the interatomic distances $(\mathbf{r}_n - \mathbf{r}_m)$. We notice first that the diagonal part of this sum, that is, the terms corresponding to $n = m$, is

$$\sum_{n=m} \delta(\mathbf{r} - (\mathbf{r}_n - \mathbf{r}_m)) = N\delta(\mathbf{r}) \tag{12.16}$$

where N is the total number of atoms in the solid. For the off-diagonal part ($n \neq m$), usually we are not interested in the specific orientation of the interatomic distances in space, which also depends on the orientation of the solid relative to the detector; a more interesting feature is the magnitude of the interatomic distances, so that we can define the spherically averaged version of the off-diagonal part in the sum Eq. (12.15):

$$\frac{1}{4\pi} \int \sum_{n \neq m} \delta(\mathbf{r} - (\mathbf{r}_n - \mathbf{r}_m)) d\hat{\mathbf{r}} = \sum_{n \neq m} \delta(r - |\mathbf{r}_n - \mathbf{r}_m|) \tag{12.17}$$

with the obvious definition $r = |\mathbf{r}|$. Putting together the two parts we obtain

$$\frac{1}{4\pi} \int \sum_{nm} \delta(\mathbf{r} - (\mathbf{r}_n - \mathbf{r}_m)) d\hat{\mathbf{r}} = N \left[\delta(r) + \frac{1}{N} \sum_{n \neq m} \delta(r - |\mathbf{r}_n - \mathbf{r}_m|) \right] \tag{12.18}$$

Next we introduce a function to represent the sum of δ-functions:

$$g(r) = \frac{1}{\Omega} \sum_{n \neq m} \delta(r - |\mathbf{r}_n - \mathbf{r}_m|) \tag{12.19}$$

where Ω is the volume of the solid. The function $g(r)$ is called the *radial distribution function*. In terms of this function, the signal at the detector takes the form

$$f(\mathbf{k}, \mathbf{R}) = \frac{A_0 N}{|\mathbf{R}|^2} \left[1 + \rho \int e^{-i\mathbf{k}\cdot\mathbf{r}} g(r) d\mathbf{r} \right] \tag{12.20}$$

where $\rho = \Omega/N$ is the density of the solid. This expression provides the desired link between the microscopic structure of the amorphous solid (implicit in $g(r)$) and the measured signal in the scattering experiment. An example of $g(r)$ is given in Fig. 12.5.

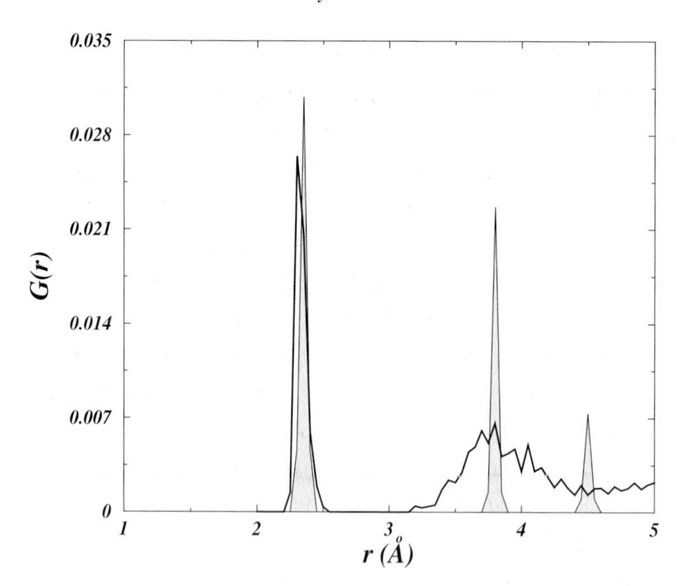

Figure 12.5. The radial distribution function for crystalline and amorphous Si. The curves show the quantity $G(r) = g(r)/(4\pi r^2 \mathrm{d}r)$, that is, the radial distribution function defined in Eq.(12.19) divided by the volume of the elementary spherical shell $(4\pi r^2 \mathrm{d}r)$ at each value of r. The thick solid line corresponds to a model of the amorphous solid, the thin shaded lines to the crystalline solid. The atomic positions in the crystalline solid are randomized with an amplitude of 0.02 Å, so that $G(r)$ has finite peaks rather than δ-functions at the various neighbor distances: the peaks corresponding to the first, second and third neighbor distances are evident, centered at $r = 2.35$, 3.84 and 4.50 Å, respectively . The values of $G(r)$ for the crystal have been divided by a factor of 10 to bring them on the same scale as the values for the amorphous model. In the results for the amorphous model, the first neighbor peak is clear (centered also at 2.35 Å), but the second neighbor peak has been considerably broadened and there is no discernible third neighbor peak.

The experimental signal is typically expressed in units of $A_0 N/|\mathbf{R}|^2$, which conveniently eliminates the factor in front of the integral in Eq. (12.20). The resulting quantity, called the *structure factor* $S(k)$, does not depend on the direction of the scattering vector $\mathbf{k}$, but only on its magnitude $k = |\mathbf{k}|$, because the function $g(r)$ has already been averaged over the directions of interatomic distances:

$$S(k) = \frac{f(\mathbf{k}, \mathbf{R})}{A_0 N/|\mathbf{R}|^2} = 1 + \rho \int \mathrm{e}^{-\mathrm{i}\mathbf{k}\cdot\mathbf{r}} g(r) \mathrm{d}\mathbf{r} \tag{12.21}$$

The structure factor is the quantity obtained directly by scattering experiments, which can then be used to deduce information about the structure of the amorphous solid, through $g(r)$. This is done by *assuming* a structure for the amorphous solid, calculating the $g(r)$ for that structure, and comparing it with the experimentally determined $g(r)$ extracted from $S(k)$. This procedure illustrates the importance of good structural models for the amorphous structure.

12.2.3 Electron localization due to disorder

We discuss next the effect of disorder on the density of electronic states (DOS), denoted as $g(\epsilon)$, and on the electronic wavefunctions. The density of states of disordered structures is characterized by the lack of any sharp features like those we encountered in the study of crystalline structures (the van Hove singularities, see chapter 5). In analyzing those sharp features, we discussed how they are related to the vanishing of $\nabla_{\mathbf{k}}\epsilon_{\mathbf{k}}$, where $\mathbf{k}$ is the reciprocal-space vector that labels the electronic levels. The existence of the reciprocal-space vector $\mathbf{k}$ as an index for the electron wavefunctions is a direct consequence of the crystalline long-range order (see chapter 3). We expect that in disordered structures the lack of any long-range order implies that we cannot use the concept of reciprocal-space vectors, and consequently we cannot expect any sharp features in $g(\epsilon)$. Indeed, the DOS of amorphous solids whose *local* structure resembles a crystalline solid, such as amorphous Si, has similar overall behavior to the crystalline DOS but with the sharp features smoothed out. This statement applies also to the band edges, i.e. the top of the valence band (corresponding to the bonding states) and the bottom of the conduction band (corresponding to the antibonding states). The fact that there exist regions in the amorphous DOS which can be identified with the valence and conduction bands of the crystalline DOS, is in itself quite remarkable. It is a consequence of the close similarity between the amorphous and crystalline structures at the local level (the nature of local bonding is very similar), which gives rise to similar manifolds of bonding and antibonding states in the two structures.

The lack of sharp features in the amorphous DOS has important consequences in the particular example of Si: the band gap of the crystalline DOS can no longer exist as a well-defined range of energy with no electronic states in it, because this implies the existence of sharp minimum and maximum features in the DOS. Instead, in the amorphous system there is a range of energies where the DOS is very small, almost negligible compared with the DOS in the bonding and antibonding regions. Moreover, the DOS on either side of this region decays smoothly, as shown in Fig. 12.6. Interestingly, the states within this range of very small DOS tend to be localized, while states well within the bonding and antibonding manifolds are extended. Accordingly, the values of the energy that separate the extended states in the bonding and antibonding regions from the localized states are referred to as "mobility edges"; the energy range between these values is referred to as the "mobility gap".

In amorphous Si, many of the localized states in the mobility gap (especially those far from the mobility edges, near the middle of the gap) are related to defects, such as the dangling bond and floating bond sites mentioned above. In this case, it is easy to understand the origin of localization, since these defect-related states

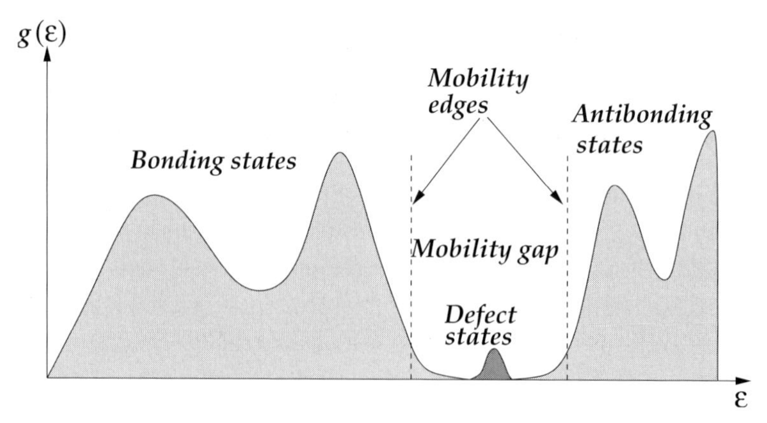

Figure 12.6. Schematic representation of the density of states $g(\epsilon)$ as a function of the energy ϵ, of an amorphous semiconductor (such as Si) near the region that corresponds to the band gap of the crystalline structure.

do not have large overlap with states in neighboring sites, and therefore cannot couple to wavefunctions extending throughout the system. There is, however, a different type of disorder-related localization of electronic wavefunctions, called *Anderson localization*, which does not depend on the presence of defects. This type of localization applies to states in the mobility gap of amorphous Si which are close to the mobility edges, as well as to many other systems. The theoretical explanation of this type of localization, first proposed by P. W. Anderson [190], was one of the major achievements of modern condensed matter theory (it was recognized by the 1977 Nobel prize for Physics, together with N.F. Mott and J. H. van Vleck). Here we will provide only a crude argument to illustrate some key concepts of the theory. One of its important features is that it is based on the single-particle picture, which places it well within the context of the theory of solids presented so far.

The basic model consists of a set of electronic states with energies ϵ_i which take values within an interval W. Since we are considering a model for a disordered structure, the values of the electronic states correspond to a random distribution in the interval W. If the system had long-range order and all the individual states had the same energy, the corresponding band structure would produce a band width B. Anderson argued that if $W > \kappa B$, then the electronic wavefunctions in the disordered system will be localized (where κ is an unspecified numerical constant of order unity). The meaning of localization here is that the wavefunction corresponding to a particular site decays exponentially beyond a certain range, as illustrated in Fig. 12.7. In the ordered case, the crystalline wavefunctions are delocalized since they have equal weight in any unit cell of the system due to the periodicity (see also chapters 1 and 3).

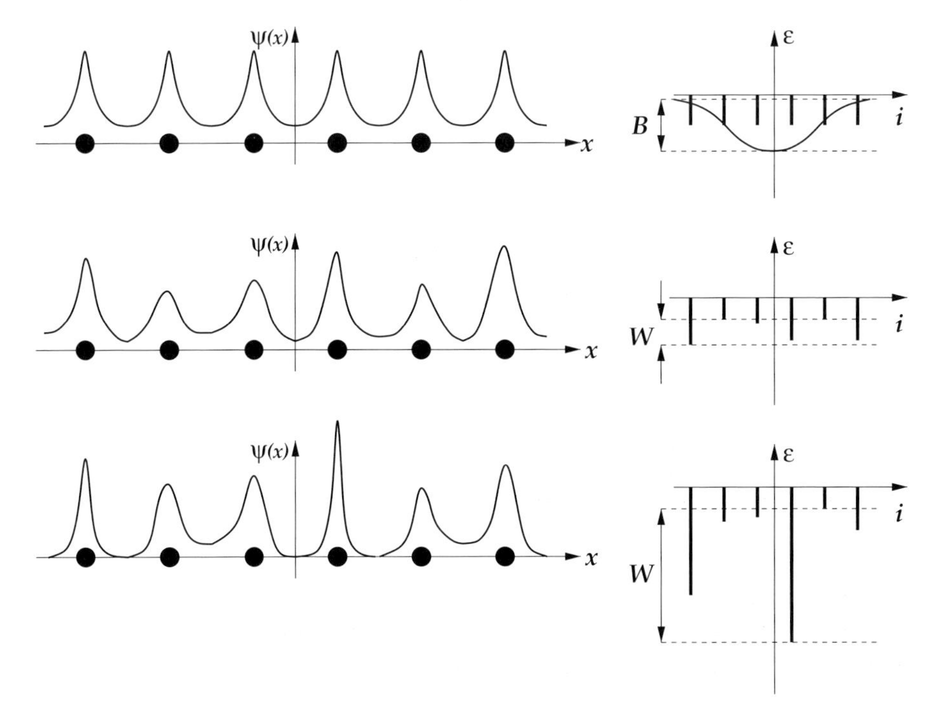

Figure 12.7. Schematic representation of Anderson localization in a one-dimensional periodic system of atoms. The left panels show the atomic positions (indicated by the filled circles) and electronic wavefunctions $\psi(x)$. The right panels show the distribution of electronic energies ϵ associated with the different sites labeled by the index i. **Top:** *delocalized* wavefunctions, which arise from identical energies at each atomic site and a band structure of width B. **Center:** *delocalized* wavefunctions, which arise from random on-site energies distributed over a range $W \leq B$. **Bottom:** *localized* wavefunctions, which arise from random on-site energies distributed over a range $W \gg B$.

We can view the transition from the localized to the delocalized state in the context of perturbation theory within the tight-binding approximation (the following discussion follows closely the arguments given in the book by Zallen, pp. 239–242, mentioned in the Further reading section). First, we note that for the case of long-range order, assuming a simple periodic structure with z nearest neighbors and one electronic level per site, the band structure will be given by

$$\epsilon_{\mathbf{k}} = \epsilon_0 + 2V_{nn} \sum_i \cos(\mathbf{k} \cdot \mathbf{a}_i) \tag{12.22}$$

where ϵ_0 is the on-site energy (the same for all sites), V_{nn} is the nearest neighbor hamiltonian matrix element, and $\mathbf{a}_i$ are the lattice vectors. The band width from this model is easily seen to be $B = 2zV_{nn}$. This result was derived explicitly for the linear chain, the square lattice and the cubic lattice in one, two and three dimensions in chapter 4. For the *disordered structure*, we can think of the sites as being on a lattice, while the disorder is provided by the random variation of the on-site energies ϵ_i, which are no longer equal to ϵ_0 but are randomly distributed in the interval W. In the limit where V_{nn} is vanishingly small, there is no interaction between electronic states and each one will be localized at the corresponding site. Treating V_{nn} as a perturbation, we see that the terms in the perturbation expansion will have the form

$$\frac{V_{nn}}{(\epsilon_i - \epsilon_j)} \tag{12.23}$$

where ϵ_j is the energy of a nearest neighbor site of i. On average, these energy differences will be equal to $W/2z$, since for a given value of ϵ_i the values of ϵ_j will be distributed randomly in over the range W, in z equal intervals. Thus, the perturbation expansion will involve factors of typical value

$$\frac{V_{nn}}{W/2z} = \frac{2zV_{nn}}{W} = \frac{B}{W} \tag{12.24}$$

In order for the perturbation expansion to converge, we must have

$$\frac{B}{W} < 1 \Rightarrow B < W \tag{12.25}$$

in which case the states remain localized since the resulting solution is a small perturbation of the original state of the system (taken to correspond to well localized states). If the perturbation expansion fails to converge, we can interpret the state of the system as corresponding to a qualitatively different physical situation, that is, to delocalized states. This oversimplified picture provides some crude argument for the origin of the relation between the band width B and the range of on-site energies in the disordered system W.

Anderson localization is only one possible way of inducing a transition from a localized to a delocalized state. Since the localized state is associated with insulating behavior and the delocalized state is associated with metallic behavior, this transition is also referred to as the metal–insulator transition. In the case of Anderson localization this is purely a consequence of disorder. The metal–insulator transition can be observed in other physical situations as well, but is due to more subtle many-body effects, such as correlations between electrons; depending on the precise nature of the transition, these situations are referred to as the Mott transition or the Wigner crystallization.

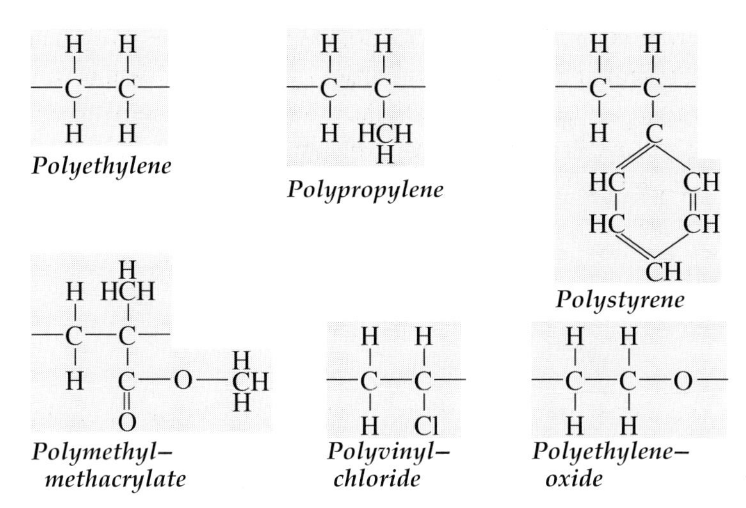

Figure 12.8. Examples of constitutional repeating units that form important polymers. **Top row:** units containing only C and H atoms, polyethylene (PE, used in film, bottles, cups, electrical insulation), polypropylene (PP) and polystyrene (PS, used in coffee cups, packaging). **Bottom row:** units containing other atoms as well, polymethylmethacrylate (PMMA, used in transparent sheets, aircraft windows), polyvinylchloride (PVC) and polyethyleneoxide (PEO).

12.3 Polymers

Polymers[1] are extremely interesting and important structures. They consist of very long chains produced by linking together many small units of organic molecules (the "monomers" or constitutional repeating units). These units are mostly hydrocarbons with a few other atoms (typically O, N, Cl, F, S, Si). Examples of repeating units that give rise to commonly used polymers are shown in Fig. 12.8. Two examples of linear chains consisting of the simplest repeating units are shown in Fig. 12.9. A single polymer chain may consist of 10^3 to 10^6 repeating units. The study of polymers is mostly the domain of organic chemistry (as far as their synthesis is concerned), statistical physics (as far as their physical properties are concerned) and chemical engineering (as far as their production and practical applications are concerned). Polymers form solids and liquids which have very intriguing properties due to the unusual nature of the basic units of which they are composed. The fields of polymer physics and chemistry are enormous and cannot be adequately covered here. We will attempt to give a brief overview of polymer physics, with emphasis on connections to the material we have discussed in earlier parts of the book. A detailed discussion of polymer science is given in the book by Gedde (see the Further reading section).

[1] The term polymer comes from the Greek words $\pi o\lambda\upsilon$ = many and $\mu\epsilon\rho\eta$ = parts; two related terms are oligomer, $o\lambda\iota\gamma o$ = few, and monomer, $\mu o\nu o$ = single.

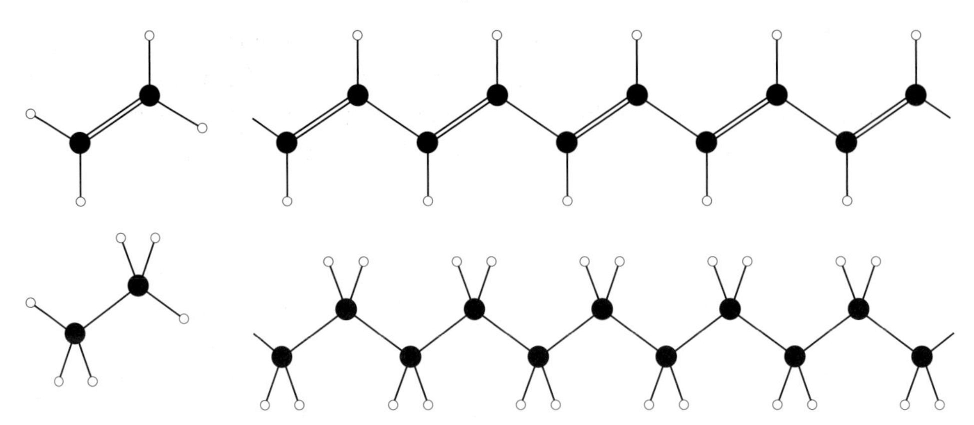

Figure 12.9. Schematic representation of single straight chains of polyacetylene (**top**) and polyethylene (**bottom**). Black circles represent the C atoms; white spheres denote H atoms; and single or double lines represent the single or double covalent bonds between atoms. The corresponding stable molecules, ethylene and ethane, are shown on the left. In polyacetylene all atoms lie on the same plane. In polyethylene the H atoms lie in front and behind the plane of the C-atom chain. Note the different angles between C–C atoms in the two structures: $\sim 120°$ for the polyacetylene, $\sim 109°$ for polyethylene.

12.3.1 Structure of polymer chains and solids

We begin with a discussion of the structure of polymer solids with the two prototypical infinite chains, polyacetylene and polyethylene, which are shown in Fig. 12.9. The first of these is the 1D equivalent of the graphite structure while the second is the 1D equivalent of the diamond structure. These infinite chains are formed by joining stable molecules with multiple C–C bonds to form an extended structure. For example, acetylene H$-$C$\equiv$C$-$H, with a triple bond between the C atoms, when polymerized forms polyacetylene, in which the bond between the C atoms is a double bond. The structure of the resulting polymer is closely related to that of the ethylene molecule, $H_2-C=C-H_2$, which is a planar structure with sp^2 bonds between the C atoms and a π-bond from the unpaired p_z orbitals. The bond angles, both in ethylene and polyacetylene, are equal to 120°, reflecting the sp^2 nature of bonds (the actual value of the bond angle may be slightly off 120° because the C–C and C–H bonds are not equivalent). The electronic structure of this chain is similar to that of graphite (see chapter 1 and Problem 2). As Fig. 12.9 illustrates, there are single and double bonds between the C atoms along the chain, but of course there are two degenerate states depending on where the double bond is placed. This is a representative case of resonant bonding (for more details see Pauling's book, mentioned in chapter 1). It is possible to create a topological defect in this structure which at some point switches the order of double–single–double–single bonds to single–double–single–double bonds. The two half chains on either side of the defect have the same energy, making the energy cost for the introduction of such defects very low. This type of defect is called a "soliton" and has attracted considerable attention with regard to electron transport phenomena [191, 192] in the context of many-body physics which lies beyond the scope of the present treatment.

The other simple 1D hydrocarbon chain, polyethylene, has a structure which involves only single covalent bonds between the C atoms, because the remaining orbitals are saturated by H atoms (two per C atom). Each C atom in polyethylene is four-fold coordinated, with two C and two H neighbors in an sp^3 bonding configuration. Consequently, the bond angles are close to the tetrahedral angle of 109.47° (the actual value of the bond angles along the chain may be slightly off the ideal tetrahedral angle because the C–C and C–H bonds are not equivalent). As a result of its atomic structure, polyethylene has an electronic structure which involves a large gap between occupied and unoccupied electronic states, much like in the diamond crystal (see Problem 3). This makes it chemically stable and less reactive than polyacetylene and as such a more appropriate building block for solids.

There is another property of this macromolecule which makes it well suited for building interesting solids. This property is its remarkable ability to bend with very small energy cost. We will demonstrate this property for polyethylene because of its particularly simple structure, but we emphasize that it is a property common to most polymers. The main idea is illustrated in Fig. 12.10: for a given chain atom, rotations of one of its C neighbors and the two H neighbors around the axis defined by the bond to its other C neighbor, cost very little energy because they do not disturb the lengths of the covalent bonds. The energy cost due to these rotations is of order few kcal/mol. The distortions from the ideal configuration are described in terms of two angles, θ and ϕ. The two covalent bonds that remain fixed during the distortion define a plane, identified as the xy plane; ϕ is the angle of rotation around the x-axis, called the torsion angle. Rotations of the third bond by $\Delta\phi = \pm 120°$

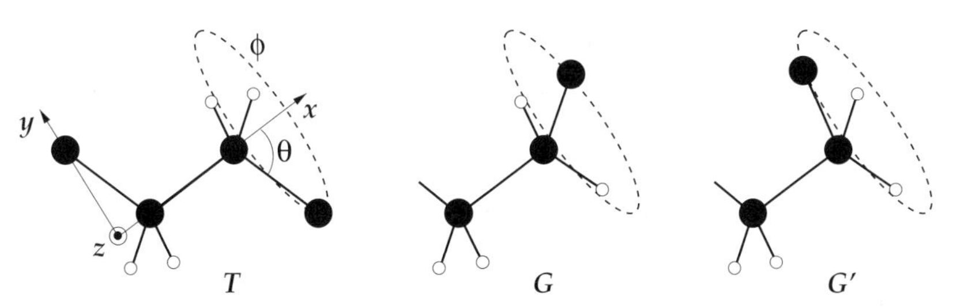

Figure 12.10. Illustration of the trans (T) and the two gauche (G, G') conformations of a short section of a polymer with four-fold coordinated C atoms. The xy plane is defined by the two bonds that remain fixed and the torsion angle ϕ is the angle of rotation around the x axis. The angle θ is not affected if the bond angles are not altered in the distortion, as shown in the gauche conformations.

relative to the most stable configuration do not disturb the bond angles between covalent bonds. These distortions are called "gauche" (denoted by G, G') and the undistorted structure is called "trans" (denoted by T). As long as these distortions do not bring other units further down the chain into close contact, which would generate strong repulsion called "steric hindrance" in chemical parlance, they are higher in energy than the trans configuration by only ~ 1 kcal/mol, and should therefore be quite common. The structures that are generated by allowing such distortions between adjacent repeating units are called "conformations". The huge number of possible conformations available to a polymer chain with 10^6 repeating units gives it a fantastic flexibility, which is manifested in the unique properties of the solids and fluids made out of it.

It is worthwhile emphasizing the difference in flexibility between polyacetylene and polyethylene. In polyacetylene, the planarity of the ideal structure makes it possible to have strong π-bonds between the p_z orbitals in adjacent C atoms. These π-bonds derive their strength from the parallel orientation of the contributing p_z orbitals. If the repeat units were to be rotated relative to each other around an axis defined by one of the in-plane bonds, the π-bonds between nearest neighbors would be severely disturbed because the p orbitals which do not participate in σ-bonds would no longer be parallel. There is one such p orbital per C atom, which we do not designate with the index z to emphasize the fact that it is no longer perpendicular to the xy plane once the C atom has been rotated. As a consequence of the disturbance of π-bonds, which is associated with a high energy cost, polyacetylene is much less prone to rotations of repeat units relative to each other, and hence much more rigid. This makes it also less useful as a building block of solids.

Polymer solids are formed when the macromolecules condense and are stabilized by secondary interactions between the chains. These involve bonding of the van der Waals type or hydrogen-bond type (see chapter 1). This has several consequences. First, the crystals formed in this manner are highly anisotropic, because the strength of bonds along the polymer chains is vastly different than the strength of bonds across chains. A typical example of a crystal formed by polyethylene chains is shown in Fig. 12.11; Young's modulus for such crystals along the chain direction is approximately 300 GPa, whereas perpendicular to the chain direction it is approximately 3 GPa. An interesting feature of polymer crystals is that the degree of crystallinity is never 100%: for polyethylene molecules with relatively short length it can reach 90%, and for polyvinylchloride it can be as low as 5%. This is indicative of the issues that come into play when polymer chains condense to form solids. Of paramount importance is the chemical composition and chain structure, which makes it very difficult for a polymer with complex chain structure to form a crystal. Such polymers typically form fully amorphous solids. Even polymers with the simplest chains, like polyethylene, have a difficult time forming perfect

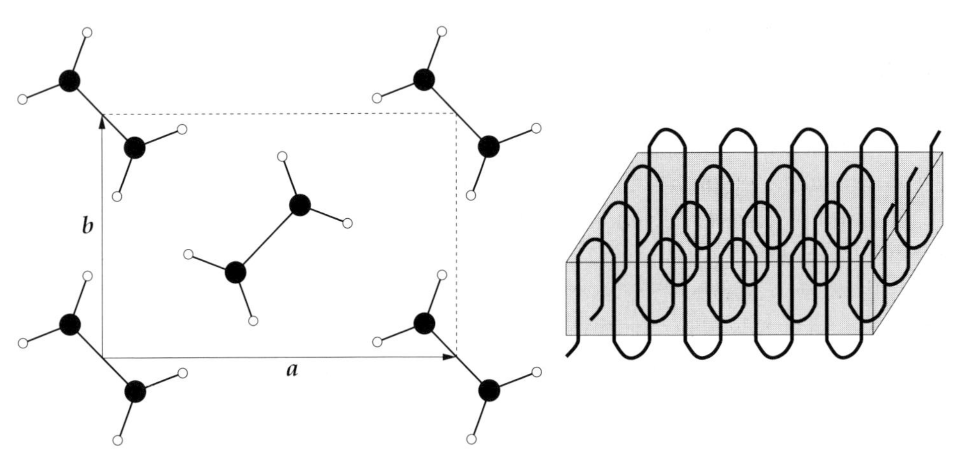

Figure 12.11. Examples of polymer crystal structures. **Left:** planar view of crystalline polyethylene on a plane perpendicular to the molecular axis, in which the repeat distances are $a = 7.40$ Å and $b = 4.95$ Å, while the repeat distance along the molecular axis is $c = 2.53$ Å. **Right:** illustration of a platelet (lamella) of a polymer crystal with folded chains.

crystals because of the flexibility of the chains which refuse to remain straight to facilitate crystalline order. In fact, the crystalline portions of polymers typically involve extensive folding of the chains, resulting in flat platelet-like crystallites called lamellae (see Fig. 12.11). Accordingly, the best situation as far as crystalline order is concerned corresponds to some part of the solid consisting of lamellae and the rest comprising amorphous material. Thus, polymer solids with some degree of crystallinity are described as being in a "semicrystalline" state. In certain situations, crosslinks of covalent bonds can develop between the chains in the solid, which enhance its strength and render it more like an isotropic 3D material.

12.3.2 The glass and rubber states

The solids formed by polymers have certain peculiar properties which owe their origin to the unusual structure of the basic building blocks. There are two solid states typical of polymers, referred to as the "glass" and the "rubber" states, by analogy to the two common, naturally occurring solids. Before we describe the nature of these two states we mention briefly some aspects of polymer structure in the molten state which are useful in understanding the properties of the solid state. To begin, we consider a polymer chain in a solvent which does not strongly affect its properties but merely provides a medium for its existence. In this state, the polymer can be modeled as a long chain consisting of non-interacting units which can have almost arbitrary orientation relative to each other. The different relative orientations of adjacent units along the polymer chain reflects its flexibility due to the ease of re-orienting C atoms by rotations around the C–C bonds, as described above for polyethylene. If the units were truly non-interacting, the structure would be modeled by a random walk, starting at one end of the polymer chain and taking steps of a given length a (related to the monomer size) in random directions. This would lead to an average end-to-end distance, called the radius of gyration R_G, which is given by

$$R_G \sim N^{1/2} a \tag{12.26}$$

with N the number of steps (identified with the number of repeat units in the chain). In reality, the repeating units have interactions, the most important of which is that they cannot fold back onto themselves. This can be taken into account by a self-avoiding random walk, in which case the radius of gyration scales as

$$R_G \sim N^{3/5} a$$

The larger value of the exponent in this case, as compared to the simple random walk, reflects the swelling of the chain due to the condition of self-avoidance. The exponent of N in the self-avoiding random walk is very close to the experimentally measured value of 0.59. In any case, the polymer chain is highly coiled, as illustrated in Fig. 12.12. This description of the polymer applies equally well to the molten state, as was first postulated by Flory (see the Further reading section). The reasons behind this behavior can be attributed to the interactions between polymer chains in the molten state, which are described in the book by de Gennes (see the Further reading section).

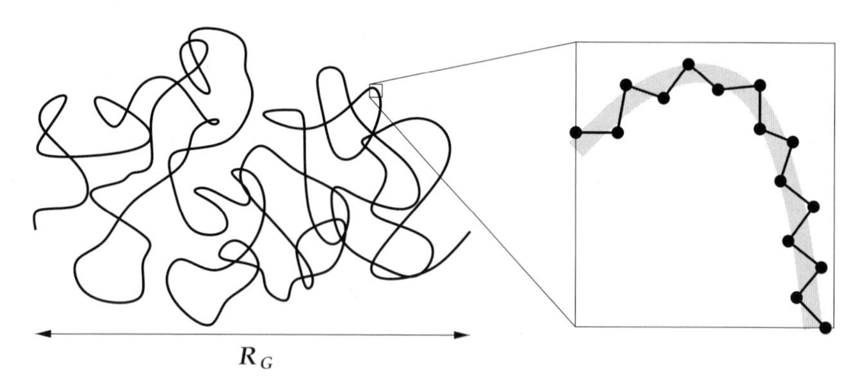

Figure 12.12. Illustration of a coiled polymer with a radius of gyration R_G. The diagram on the right indicates how a small part of the chain is composed of repeating units bonded by covalent bonds (straight-line segments) between C atoms (black dots), which can form the tight turns as a result of the flexibility of the polymer.

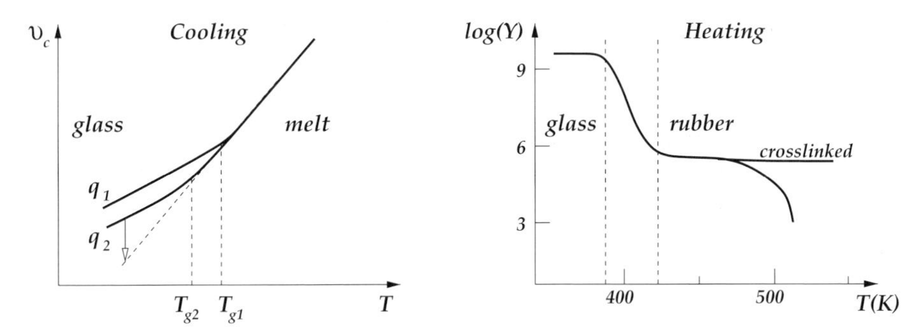

Figure 12.13. **Left:** the glass transition in polymers upon cooling from the melt, as described in terms of changes in the specific volume v_c as a function of temperature. Different cooling rates q produce different glass transition temperatures T_g. In the case illustrated, $|q_1| > |q_2|$ and $T_{g1} > T_{g2}$. The vertical arrow indicates the relaxation over long time scales. **Right:** the transition from the glass state to the rubber state upon heating, as described by changes in the Young's modulus Y on a logarithmic scale. Three regions are identified, the glass region, the transition region and the rubber region, the extent of the latter depending on the nature of the polymer (crosslinked or not, length of polymer chains), before melting begins (indicated by the sudden drop in Y). The numbers on the axes are indicative and correspond roughly to the measured values for amorphous polystyrene.

We consider next the solid state of polymers. Starting from the melt, when a polymer is cooled it undergoes a transition to the glass state, which is an amorphous structure. This transition is most easily identified by changes in the specific volume, v_c: the slope of the decrease in v_c as the temperature falls changes abruptly and significantly at a certain temperature called the glass transition temperature T_g. This is illustrated in Fig. 12.13. The abrupt change in the slope of the specific volume has some features of a second order phase transition. For example, the volume and enthalpy are continuous functions across the transition, but their temperature derivatives, that is, the thermal expansion coefficient and the specific heat, are discontinuous. However, the transition also has strong features of kinetic effects and in this sense it is not a true thermodynamic phase transition (see Appendix C). One very obvious effect is that the glass transition temperature depends strongly on the cooling rate, as shown in Fig. 12.13. The essence of the transition from the molten to the glass state is that atomic motion freezes during this transition. The polymer chains that are highly mobile in the molten state are essentially fixed in the glass state. There is still significant local motion of the atoms in the glass state, but this leads to changes in the structure over much larger time scales. In fact, allowing for a long enough relaxation time, the specific volume of the glass state can decrease further to reach the point where the extrapolation of v_c for the molten state would be (see Fig. 12.13). This makes it evident that the glass transition involves the quenching of free volume within the solid, into which atoms can eventually move but at a much slower pace than in the molten state. The behavior described so far is rather typical of polymer solids, but in some cases a slow enough cooling rate can lead to the semicrystalline state. Even in that state, however, a significant portion of the solid is in an amorphous state, as already mentioned above.

The opposite transition, which takes place upon heating the solid starting from the glass state, is also quite interesting. This transition is typically described in terms of the elastic modulus Y, as shown in Fig. 12.13, which changes by several orders of magnitude. The transition is rather broad in temperature and does not lead directly to the molten state, but to an intermediate state called the rubber state. The name for this state comes from the fact that its peculiar properties are exemplified by natural rubber, a polymer-based solid (the polymer involved in natural rubber is called *cis*-1,4-polyisopropene). This state has a Young's modulus roughly three orders of magnitude lower than that of the glass state and exhibits rather remarkable properties. The solid in the rubber state can be strained by a large amount, with extensions by a factor of 10, that is, strains of 1000%, being quite common. The deformation is elastic, with the solid returning to its original length when the external stress is removed. This is in stark contrast to common solids whose building blocks are individual atoms; such solids behave elastically for strains up to $\sim 0.2-0.3\%$ and undergo plastic deformation beyond that point (see chapter 10). Moreover, a solid in the rubber state gives out heat reversibly when it is stretched, or, conversely, it contracts reversibly when it is heated while in the stretched configuration. The reason for this peculiar behavior is that the elastic response of the rubber state is entropy-driven, while in normal solids it is energy-driven. The implication of this statement is that the internal energy of the rubber state changes very little when it is stretched, therefore the work $\mathrm{d}W$ done on the solid goes into production of heat, $\mathrm{d}Q = -\mathrm{d}W$, which changes the temperature.

To demonstrate the entropic origin of the elastic response of the rubber state, we consider the thermodynamics of this state, in which changes in the internal energy E are given by

$$\mathrm{d}E = T\mathrm{d}S - P\mathrm{d}\Omega + f\mathrm{d}L \tag{12.27}$$

with S, Ω, L being the entropy, volume and length of the solid and T, P, f the temperature, pressure and external force. Using the definitions of the enthalpy, $\Theta = E + P\Omega$, and the Gibbs free energy, $G = \Theta - TS$, and standard thermodynamic relations (see Appendix C and Problem 5), we find

$$\left(\frac{\partial G}{\partial L}\right)_{P,T} = f, \quad \left(\frac{\partial G}{\partial T}\right)_{L,P} = -S \tag{12.28}$$

From the first of these, we can relate the force to thermodynamic derivatives with respect to L as

$$f = \left(\frac{\partial E}{\partial L}\right)_{P,T} + P\left(\frac{\partial \Omega}{\partial L}\right)_{P,T} - T\left(\frac{\partial S}{\partial L}\right)_{P,T} \tag{12.29}$$

The second term on the right-hand side is usually negligible because the change in volume with extension of the solid is very small, $(\partial\Omega/\partial L)_{P,T} \approx 0$. This leaves two important contributions to the force, one being energy-related, the other entropy-related. It turns out that in the rubber state

$$\left(\frac{\partial S}{\partial L}\right)_{P,T} < 0$$

because an increase in the length is accompanied by a decrease in the entropy. This is a direct consequence of the structure of polymers, as described earlier. In the rubber state, the polymer chains are highly coiled in order to maximize the entropy and thus reduce the free energy. The coiled structure corresponds to many more conformations with the same overall size (characterized by the radius of gyration) than the fully extended configuration, which is a unique conformation. Therefore, there is a large amount of entropy associated with the coiled configuration, and this entropy decreases as the chain is stretched, which is what happens at the microscopic scale when the solid is stressed. The internal energy, on the other hand, can either increase or decrease, depending on the nature of the polymer chain. When the chain is composed of units which have attractive interactions in the coiled configuration, then extending its length reduces the number of attractive interactions leading to $(\partial E/\partial L)_{P,T} > 0$. However, there are also examples of chains which assume their lowest energy in the fully extended configuration, as in polyethylene discussed above; for these cases, $(\partial E/\partial L)_{P,T} < 0$. If we define the first term in Eq. (12.29) as the energy term, f_e, then, using standard thermodynamic arguments to manipulate the expressions in Eq. (12.29), we can write the ratio of this part to the total force as

$$\frac{f_e}{f} = 1 - T\left(\frac{\partial f}{\partial T}\right)_{P,L} - \frac{\alpha T}{1 - (L/L_0)^3} \tag{12.30}$$

where α is the thermal expansion coefficient and L_0 is the natural length at temperature T. Experimental measurements give a value of $f_e/f = -0.42$ for polyethylene in agreement with the description of its conformational energy, and $f_e/f = 0.17$ for natural rubber. In both cases, the contribution of the two other terms, and especially the entropy term, to the total force is very large ($\sim 0.80\%$ in natural rubber where the last term is not significant).

The actual microscopic motions that are behind these effects are quite complicated. The basic ideas were already mentioned, namely that in the glass state the motion of the atoms is very restricted, while in the rubber state the molecules are stretched from their coiled configuration in response to the external stress. The ability of the solid to remain in the rubber state depends on the length and the degree of crosslinking of the polymers. As indicated in Fig. 12.13, a highly crosslinked solid exhibits a much wider rubber plateau than one in which there is no crosslinking. In the semicrystalline case, the rubber plateau is also very extended, the presence of crystallites acting as the equivalent of strong links between the parts of chains that are in the amorphous regions. Finally, the length of the chains themselves is important in maintaining the rubber state as the temperature increases. Solids consisting of very long polymers remain in the rubber state much longer, because the chains are highly entangled, which delays the onset of the molten state in which the chains move freely.

What is truly remarkable about the rubber state is that the response of the solid to a step stress is instantaneous, which implies an extremely fast rate of conformational changes. Moreover, these changes are reversible so the response is elastic. This can only happen if large numbers of atoms can move easily while the polymer is in a coiled configuration, allowing it to adopt the length dictated by the external stress conditions. Taking into account that all the chains in the solid are coiled and that they are interdispersed, hence highly entangled, their ability to respond so readily to external forces is nothing short of spectacular. One idea providing insight into this motion is that individual polymers undergo a "reptation", moving in a snake-like fashion in the space they occupy between all the other polymers. The notion of reptation was proposed by P.-G. de Gennes, who was awarded the 1991 Nobel prize for Physics for elucidating the physics of polymers and liquid crystals. These issues are still the subject of intense experimental investigation. On the theoretical side, significant insights are being provided by elaborate simulations which try to span the very wide range of length scales and time scales involved in polymer physics [193].

Further reading

1. *Quasicrystals: The State of the Art*, D.P. DiVincenzo and P.J. Steinhardt, eds. (2nd edn, World Scientific, Singapore, 1999).
   This book contains an extensive collection of articles that cover most aspects of the physics of quasicrystals.
2. *Quasicrystals: A Primer*, C. Janot, ed. (2nd edn, Oxford University Press, Oxford, 1994).
3. "Polyhedral order in condensed matter", D.R. Nelson and F. Spaepen, in vol. 42 of *Solid State Physics* (H. Ehrenreich and D. Turnbull, eds., Academic Press, 1989).

This review article contains a comprehensive and insightful discussion of issues related to the packing of basic units that fill space, including various polyhedra (such as the icosahedron) and their relation to quasicrystals.

4. *Beyond the Crystalline State*, G. Venkataraman, D. Sahoo and V. Balakrishnan (Springer-Verlag, Berlin, 1989).
   This book contains a discussion of several topics on the physics of non-crystalline solids, including quasicrystals.
5. *The Physics of Amorphous Solids*, R. Zallen (J. Wiley, New York, 1983).
   This is a detailed and extensive account of the structure and properties of amorphous solids.
6. *Models of Disorder*, J.M. Ziman (Cambridge University Press, Cambridge, 1979).
   This book contains interesting insights on the effects of disorder on the properties of solids.
7. *Polymer Physics*, U.W. Gedde (Chapman & Hall, London, 1995).
   This book is a very detailed account of all aspects of polymers, with extensive technical discussions of both theory and experiment, as well as interesting historical perspectives.
8. *Statistical Mechanics of Chain Molecules*, P.J. Flory (Hanser, New York, 1989).
9. *Scaling Concepts in Polymer Physics*, P.-G. de Gennes (Cornell University Press, Ithaca, NY, 1979).
10. *Condensed Matter Physics*, A. Ishihara (Oxford University Press, New York, 1991).
    This book contains useful discussions on 1D systems and polymers from the statistical physics point of view.

Problems

1. Derive the relations in Eqs. (12.7) (12.8) and (12.9), for the Fourier Transform of the Fibonacci series, using the definitions of the functions $\bar{g}(x, y)$, $h(y')$ and $f(x')$ given in Eqs. (12.4), (12.5) and (12.7).
2. Calculate the band structure of the infinite polyacetylene chain and compare it to the band structure of graphite. Use the nearest neighbor tight-binding model employed for the band structure of graphite, Problem 3 in chapter 4, with a basis consisting of s, p_x, p_y, p_z orbitals for the C atoms and an s orbital for the H atoms; for the matrix elements between the H and C atoms choose a value that will make the corresponding electronic states lie well below the rest of the bands. Consider a small structural change in polyacetylene which makes the two distances between C atoms in the repeat unit inequivalent. Discuss the implications of this structural change on the band structure. This type of structural change is called a Peierls distortion and has important implications for electron–phonon coupling (see the book by Ishihara mentioned in the Further reading section).
3. Using the same tight-binding model as in the previous problem, calculate the band structure of polyethylene and of the diamond crystal and compare the two. Obtain the density of states for polyethylene with sufficient accuracy to resolve the van Hove singularities as discussed in chapter 5.
4. Show that the average end-to-end distance for a random walk consisting of N steps, each of length a, is given by Eq. (12.26). For simplicity, you can assume that the walk takes place on a 1D lattice, with possible steps $\pm a$. To prove the desired result, relate $\langle x^2(N)\rangle$ to $\langle x^2(N-1)\rangle$.

5. Starting with the expression for changes in the internal energy of a polymer, Eq. (12.27), and the definitions of the enthalpy and the Gibbs free energy, derive the expressions for f and S given in Eq. (12.28). Show that these expressions imply

$$\left(\frac{\partial f}{\partial T}\right)_{P,L} = -\left(\frac{\partial S}{\partial L}\right)_{P,T}$$

Then derive the expressions for f as given in Eq. (12.29) and the ratio f_e/f, as given in Eq. (12.30).

13

Finite structures

In this final chapter we deal with certain structures which, while not of macroscopic size in 3D, have certain common characteristics with solids. One such example is what has become known as "clusters". These are relatively small structures, consisting of a few tens to a few thousands of atoms. The common feature between clusters and solids is that in both cases the change in size by addition or subtraction of a few atoms does not change the basic character of the structure. Obviously, such a change in size is negligible for the properties of a macroscopic solid, but affects the properties of a cluster significantly. Nevertheless, the change in the properties of the cluster is quantitative, not qualitative. In this sense clusters are distinct from molecules, where a change by even one atom can drastically alter all the physical and chemical properties. A good way to view clusters is as embryonic solids, in which the evolution from the atom to the macroscopic solid was arrested at a very early stage.

Clusters composed of either metallic or covalent elements have been studied extensively since the 1980s and are even being considered as possible building blocks for new types of solids (see the collection of articles edited by Sattler, mentioned in the Further reading section). In certain cases, crystals made of these units have already been synthesized and they exhibit intriguing properties. One example is crystals of C_{60} clusters, which when doped with certain metallic elements become high-temperature superconductors. There are also interesting examples of elongated structures of carbon, called carbon nanotubes, which can reach a size of several micrometers in one dimension. These share many common structural features with carbon clusters, and they too show intriguing behavior, acting like 1D wires whose metallic or insulating character is very sensitive to the structure. The study of carbon clusters and nanotubes has dominated the field of clusters because of their interesting properties and their many possible applications (see the book by Dresselhaus, Dresselhaus and Eklund in the Further reading section).

A different class of very important finite-size structures are large biological molecules, such as the nucleic acids (DNA and RNA) and proteins. Although these are not typically viewed as solids, being regular sequences of simpler structural units (for example, the four bases in nucleic acids or the 20 aminoacids in proteins), they maintain a certain characteristic identity which is not destroyed by changing a few structural units. Such changes can of course have dramatic effects in their biological function, but do not alter the essential nature of the structure, such as the DNA double helix or the folding patterns in a protein (α-helices and β-sheets). In this sense, these finite structures are also extensions of the notion of solids. Moreover, many of the techniques applied to the study of solids are employed widely in characterizing these structures, a prominent example being X-ray crystallography. These systems are also attracting attention as possible components of new materials which will have tailored and desirable properties. For instance, DNA molecules are being considered as 1D wires for use in future devices, although their electronic transport properties are still the subject of debate and investigation [194, 195].

13.1 Clusters

We begin the discussion of clusters by separating them into three broad categories: metallic clusters, clusters of carbon atoms and clusters of other elements that tend to form covalent bonds. The reason for this separation is that each category represents different types of structures and different physics.

13.1.1 Metallic clusters

Metallic clusters were first produced and studied by Knight and coworkers [196, 197]. These clusters were formed by supersonic expansion of a gas of sodium atoms, or by sputtering the surface of a solid, and then equilibrated in an inert carrier gas, such as Ar. The striking result of these types of studies is a shell structure which is also revealed by the relative abundance of the clusters. Specifically, when the relative number of clusters is measured as a function of their size, there are very pronounced peaks at numbers that correspond to the filling of electronic shells in a simple external confining potential [196, 197]. For example, the experimentally measured relative abundance of Na clusters shows very pronounced peaks for the sizes $n = 8, 20, 40, 58, 92$. A calculation of the single-particle electronic levels in a spherical potential well [196, 197] shows that these sizes correspond to the following electronic shells being filled: $n = 8 \rightarrow [1s1p]$, $n = 20 \rightarrow [1s1p1d2s]$, $n = 40 \rightarrow [1s1p1d2s1f2p]$, $n = 58 \rightarrow [1s1p1d2s1f2p1g]$, $n = 92 \rightarrow [1s1p1d2s1f2p1g2d3s1h]$, where the standard notation for atomic shells is used (see Appendix B). The exceptional stability of the closed electronic shell configurations can then be used to interpret and justify the higher abundance of clusters of the corresponding size. Because of this behavior, these sizes are called "magic numbers". The sequence of magic numbers continues to higher values, but becomes more difficult to determine experimentally as the size grows because variations in the properties of large clusters are less dramatic with size changes. It is quite remarkable that these same magic number sizes are encountered for several metallic elements, including Ag, Au and Cs, in addition to Na. The nominal valence of all these elements is unity.

More elaborate calculations, using mean-field potentials or self-consistent calculations (such as the jellium model within Density Functional Theory and the Local Density Approximation, described in chapter 2), also find the same sequence of magic numbers for closed electronic shells [198, 199]. These results have prompted interesting comparisons with closed shells in nuclear matter, although not much can be learned by such analogies due to the vast differences in the nature of bonding in atomic and nuclear systems. In all these calculations, a spherical potential is assumed. The fact that the calculations reproduce well the experimental measurements with a simple spherical potential is significant. It implies that the *atomic structure* of these clusters is not very important for their stability. In fact, a detailed study of the energetics of Na clusters, examining both electronic and geometric contributions to their stability, found that the electronic contribution is always dominant [200]. In particular, sizes that might be geometrically preferred are the shells of nearest neighbors in close-packed crystal structures, such as the FCC or BCC crystals. The sequence of these sizes with $n < 100$ is for FCC: 13, 19, 43, 55, 79, 87, and for BCC: 9, 15, 27, 51, 59, 65, 89. Neither of these sequences contains any of the magic numbers observed in experiments. It is also worth mentioning that when the energy of the closed-geometric-shell structures is optimized as a function of atomic positions, these structures do not remain in the form of crystalline fragments but adopt forms without any apparent order [200].

There is another possibility for constructing compact geometric structures of successive neighbor shells, based on icosahedral order. This order is not compatible with 3D periodicity and cannot occur in crystals, but does occur in quasicrystalline solids (see chapter 12). Of course, the finite size of clusters makes the icosahedral order perfectly legitimate as a candidate for low-energy structures. It turns out that cluster structures of alkali atoms with icosahedral order have lower energy than the corresponding crystal-fragment structures. However, these structures do not provide any solution to the problem of magic numbers observed in experiment.

The question of icosahedral versus crystalline order in clusters has also been studied for clusters of Al. There are two sizes, $n = 13$ and $n = 55$, corresponding to closed neighbor structural shells of both the icosahedral and the cuboctahedral

(FCC-like) order. The question of which structure is preferred is interesting from the point of view of crystal growth, in the sense that a cluster represents an early stage of a solid. Early calculations [200–202], not taking into account complete relaxation of the atomic positions, suggested that the icosahedral order is preferred by the smaller size and the cuboctahedral order is preferred for the larger size, suggesting a transition between icosahedral order to crystalline order somewhere between these two sizes. More extensive investigation, including full relaxation of atomic coordinates, showed that the icosahedral-like structures are energetically preferred for both sizes [204]. Perhaps more significant is the fact that full optimization of the total energy by relaxation of the atomic positions tends to destroy the perfect order of model structures for large sizes: the structure of the Al_{13} cluster is not affected by relaxation, but that of Al_{55} is affected significantly. This is a general result in the study of single-element clusters (except for C), which we will encounter again in the discussion of clusters consisting of covalently bonded elements. It is interesting that the icosahedral Al_{13} cluster has a number of electrons (39) which almost corresponds to the closing of an electronic shell. This has inspired studies of derivative clusters, in which one of the Al atoms is replaced by a tetravalent element such as C or Si, bringing the number of electrons in the cluster to 40 and thus producing a closed electronic shell and a potentially more stable cluster [205]. Theoretical investigations indeed show that the $Al_{12}C$ cluster, with a C atom at the center of the icosahedron, is chemically inert compared with the Al_{13} cluster [206].

13.1.2 Carbon clusters

By far the most important cluster structures are those composed of C atoms. The field of carbon clusters was born with the discovery of a perfectly symmetric and exquisitely beautiful structure consisting of 60 C atoms [207]. The discoverers of this structure, R.F. Curl, H.W. Kroto and R.E. Smalley, were awarded the 1996 Nobel prize for Chemistry. The subsequent production of this cluster in macroscopic quantities [208], made possible both its detailed study as well as its use in practical applications such as the formation of C_{60} crystals. Following the discovery of C_{60}, several other clusters composed of C atoms were found, as well as interesting variations of these structures. We will mention briefly these other clusters and structures, but will concentrate the discussion mostly on C_{60}, which remains the center of attention in this field.

Structure of C_{60} and other fullerenes

The geometric features of this structure are very simple. The 60 C atoms form 12 pentagons and 20 hexagons, with each pentagon surrounded entirely by hexagons, as

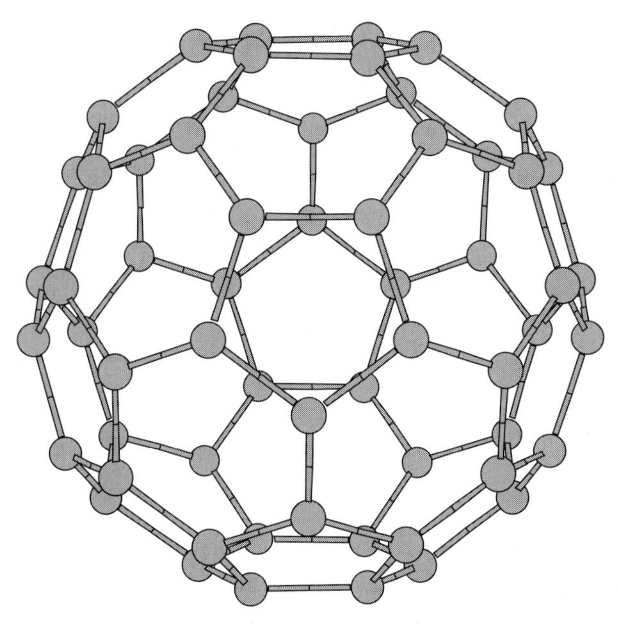

Figure 13.1. The C_{60} cluster. The C atoms form 12 pentagons and 20 hexagons. Two of the pentagons, on the front and the back sides, are immediately obvious at the center of the structure (they are rotated by 180° relative to each other), while the other 10 pentagons are distributed in the perimeter of the perspective view presented here.

shown in Fig. 13.1. The use of pentagons and hexagons to form large geodesic domes was introduced in architecture by Buckminster Fuller, the American inventor of the late 19th/early 20th century. The C_{60} structure was nicknamed "buckminsterfullerene" by its discoverers, and has become known in the literature as "fullerene" (more casually it is also referred to as a "bucky-ball"). This name is actually misleading, since this structure is one of the archimedean solids which are obtained from the platonic solids by symmetric truncation of their corners. The platonic solids are those formed by perfect 2D polygons, which are shapes with all their sides equal; some examples of platonic solids are the tetrahedron, composed of four equilateral triangles, the cube, composed of six squares, the dodecahedron, composed of 12 pentagons, the icosahedron, composed of 20 equilateral triangles, etc. A dodecahedron composed of C atoms is shown in Fig. 13.2; the icosahedron was discussed in chapter 1 and shown in Fig. 1.6. The structure of C_{60} corresponds to the truncated icosahedron. As is explicitly stated in ancient Greek literature,[1] this structure was among the many cases of truncated platonic solids originally studied by Archimedes. The truncated icosahedron had also been discussed by Johannes Kepler and had even been drawn in an elegant perspective view by Leonardo da

[1] See, for instance, the writings of Pappus of Alexandria, a mathematician who lived in the third century and discussed in detail the works of Archimedes.

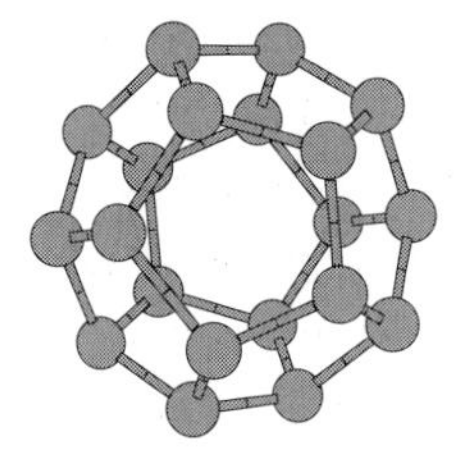
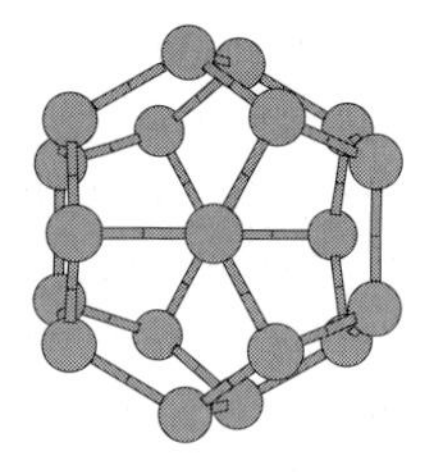
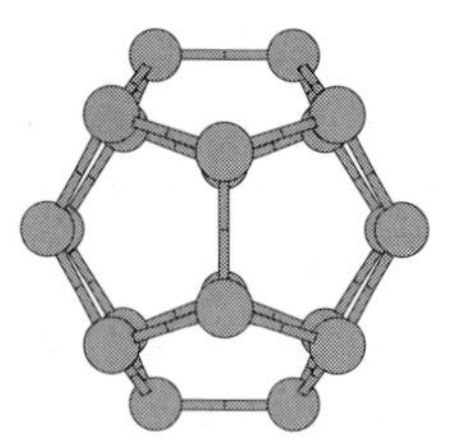
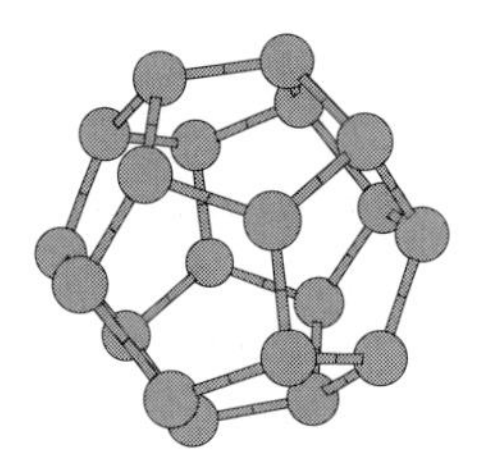

Figure 13.2. Four views of the C_{20} cluster, which is a perfect dodecahedron, with the C atoms forming 12 pentagons. **From left to right:** view along a five-fold rotational axis, along a three-fold axis, along a two-fold axis (equivalent to a reflection plane) and a perspective view along a random direction.

Vinci [209]. Perhaps, instead of the misnomer that has prevailed in the literature, a more appropriate nickname for the C_{60} cluster might have been "archimedene".

The combination of pentagons and hexagons found in the fullerene is also identical to that of a soccer ball, suggesting that the cluster is as close to a spherical shape as one can get for this size of physical object (the diameter of C_{60} is 1.034 nm). The fullerene possesses a very high degree of symmetry. In fact, its point group has the highest possible number of symmetry operations, a total of 120. These are the symmetries of the icosahedral group. In terms of operations that leave the fullerene invariant, grouped by class, they are as listed below.

1. Six sets of five-fold axes of rotation, each axis going through the centers of two diametrically opposite pentagons; there are 24 independent such operations.
2. Ten sets of three-fold axes of rotation, each axis going through the centers of two diametrically opposite hexagons; there are 20 independent such operations.
3. Fifteen sets of two-fold rotation axes, each axis going through the centers of two diametrically opposite bonds between neighboring pentagons; there are 15 independent such operations (these are also equivalent to reflection planes which pass through the pairs of the diametrically opposite bonds between neighboring pentagons).
4. The identity operation.

These symmetries are evident in the dodecahedron, Fig. 13.2, which has the full symmetry of the icosahedral group. The structure also has inversion symmetry about its geometric center, which, when combined with all the above operations, generates a total of 120 different symmetry operations. This high degree of symmetry makes all the atoms in the C_{60} cluster geometrically equivalent.

There is a whole sequence of other C clusters which bear resemblance to the fullerene, but consist of higher numbers of atoms. Some examples are C_{70} and C_{80}, whose structure is an elongated version of the fullerene. The C_{70} cluster can be constructed by cutting the C_{60} cluster in half along one of its equatorial planes so that there are six pentagons in each half, and then rotating one of the two halves by 180° around their common five-fold rotational axis and adding an extra five bonded pairs of atoms in the equatorial plane. To form the C_{80} cluster, ten bonded pairs of atoms are added to a C_{60} cut in half, five pairs on each side of the equatorial plane. In this case, the two halves are not rotated by 180°, maintaining the same relative orientation as in C_{60}. There is also a different isomer of the C_{80} cluster with icosahedral symmetry. In all these cases the structure consists of 12 pentagons and an increasing number of hexagons (20 in C_{60}, 25 in C_{70}, 30 in C_{80}). The occurrence of exactly 12 pentagons is not accidental, but a geometric necessity for structures composed exclusively of hexagons and pentagons. Euler's theorem for polyhedra states that the following relation is obeyed between the number of faces, n_f, the number of vertices, n_v and the number of edges n_e:

$$n_f + n_v - n_e = 2.$$

If the structure consists of n_h hexagons and n_p pentagons, then the number of faces is given by $n_f = n_h + n_p$, the number of vertices by $n_v = (5n_p + 6n_h)/3$ (each vertex is shared by three faces), and the number of edges by $n_e = (5n_p + 6n_h)/2$ (each edge is shared by two faces). Substituting these expressions into Euler's relation, we find that $n_p = 12$. The smallest structure that can be constructed consistent with this rule is C_{20}, in which the C atoms form 12 pentagons in a dodecahedron (see Fig. 13.2). This structure, however, does not produce a stable cluster. The reason is simple: all bond angles in the dodecahedron are exactly 108°, giving to the bonds a character very close to tetrahedral sp^3 (the tetrahedral bond angle is 109.47°), but each atom only forms three covalent bonds, leaving a dangling bond per atom. Since these dangling bonds are pointing radially outward from the cluster center, they are not able to form bonding combinations. This leads to a highly reactive cluster which will quickly interact with other atoms or clusters to alter its structure. The only possibility for stabilizing this structure is by saturating simultaneously all the dangling bonds with H atoms, producing $C_{20}H_{20}$, which is a stable molecule with icosahedral symmetry. In order to allow π-bonding between the dangling bonds around a pentagon, the pentagons must be separated from each other by hexagons. This requirement for producing stable structures out of pentagons and hexagons is known as the "isolated pentagon rule". C_{60} corresponds to the smallest structure in which all pentagons are separated from each other, sharing no edges or corners. The theme of 12 pentagons and n_h hexagons can be continued *ad infinitum*, producing ever larger structures. Some of these structures with full icosahedral symmetry and fewer than 1000 C atoms are: C_{140}, C_{180}, C_{240}, C_{260}, C_{320}, C_{380}, C_{420}, C_{500}, C_{540}, C_{560}, C_{620}, C_{720}, C_{740}, C_{780}, C_{860}, C_{960}, C_{980}. All carbon clusters with this general structure, whether they have icosahedral symmetry or not, are collectively referred to as fullerenes.

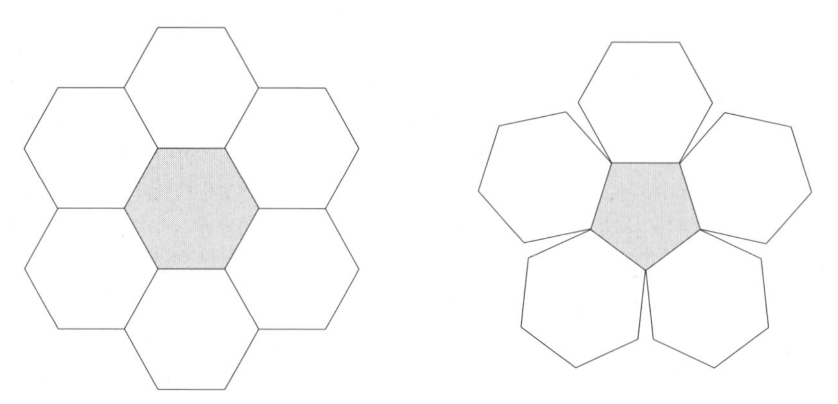

Figure 13.3. Illustration of the curvature induced in the graphite structure by replacing hexagons with pentagons. **Left:** part of the ideal graphite lattice where a hexagon at the center which is surrounded by six other hexagons. **Right:** the effect of replacing the central hexagon by a pentagon: the mismatch between the surrounding hexagons can be mended by curving each of those hexagons upward, thereby distorting neither their internal bonds nor their bond angles.

What is special about the fullerene that makes it so stable and versatile? From the structural point of view, the fullerene resembles a section of a graphite sheet, which wraps upon itself to produce a closed shell. The function of the pentagons is to introduce curvature in the graphite sheet, which allows it to close. The graphite sheet in its ideal configuration is composed exclusively of hexagonal rings of three-fold coordinated C atoms. When a hexagon is replaced by a pentagon in the graphite sheet, the structure must become curved to maintain the proper bond lengths and bond angles of the surrounding hexagons, as illustrated in Fig. 13.3. The combination of 12 pentagons and 20 hexagons turns out to be precisely what is needed to form a perfectly closed structure.

Electronic structure of C_{60}

The geometric perfection of the fullerene is accompanied by a fortuitous closing of electronic shells, which is essential to its stability. A hint of the importance of closed electronic shells was given above, in the discussion of dangling bonds arising from pentagonal rings which are not surrounded by hexagons. More specifically, the three-fold coordinated C atoms in C_{60} form strong σ-bonds between themselves from the sp^2 hybrids that are pointing toward each other, just like in graphite (see chapter 1). These bonds are not all equivalent by symmetry: there are two different types of bonds, one which is part of a pentagon and another which links two neighboring pentagons. Every hexagon on the fullerene is formed by three bonds of each type. The C–C bond on the pentagon is 1.46 Å long, whereas that between two neighboring pentagons is 1.40 Å long; for comparison, the bond length of C–C bonds in graphite is 1.42 Å, whereas that in diamond is 1.53 Å. This strongly suggests that the bonds between C atoms in C_{60} also have a significant π-component from the bonding combinations of p_z orbitals. In fact, the lengths of the two inequivalent bonds suggest that the bond between two neighboring pentagons has stronger π-character and is therefore closer to a double bond (the bond in graphite), which is shorter than a single covalent bond (the bond in diamond).

From these considerations, we can construct the following simple picture of bonding in C_{60}. Of the 240 valence electrons (four from each C atom), 180 are accommodated in the 90 σ-bonding states while the remaining 60 are accommodated in bonding combinations of the p_z orbitals. The preceding analysis suggests that there are 30 bonding combinations of p_z orbitals, associated with the bonds between neighboring pentagons (there are exactly 30 such bonds in C_{60}). This happens to be the right number of π-bonds to accommodate those valence electrons not in σ-bonds. This, however, is overly simplistic, because it would lead to purely single covalent bonds within each pentagon and purely double bonds between pentagons. Surely, there must be some π-bonding between p_z orbitals on the same pentagon, even if it is considerably lower than that between orbitals in neighboring pentagons. In order to produce a more realistic description of bonding, we must take into account the symmetry of the structure, which usually plays a crucial role in determining the electronic structure.

Before we do this, we address a simpler question which provides some insight into the nature of bonds and the chemical reactivity of C_{60}. This question, which arises from the above analysis, is why are the double bonds between neighboring pentagons and not within the pentagons? This can be answered in two ways. First, the bond angles in a pentagon are 108°, which is much closer to the tetrahedral angle of 109.47° rather than the 120° angle which corresponds to pure sp^2 bonding. This observation argues that the orbitals that form bonds around a pentagon are closer to sp^3 character (the bonding configuration in a tetrahedral arrangement, see chapter 1), which makes the pentagon bonds close to single covalent bonds. Second, the formation of π-bonds usually involves resonance between single and double bonds. This notion is not crucial for understanding the physics of graphite, where we can think of the π-bonding states as extending through the entire structure and being shared equally by all the atoms. It becomes more important when dealing with 1D or finite structures, where it is useful to know where the single or double bond character lies in order to gain insight on the behavior of electrons. We have already encountered an aspect of this behavior in the discussion of electron transport in polyacetylene (see chapter 12). The importance of this notion in chemistry is exemplified by the fact that L.C. Pauling was awarded the 1954 Nobel prize for Chemistry for the elucidation of resonance and its role in chemical bonding. The prototypical structure for describing resonance is the benzene molecule, a planar

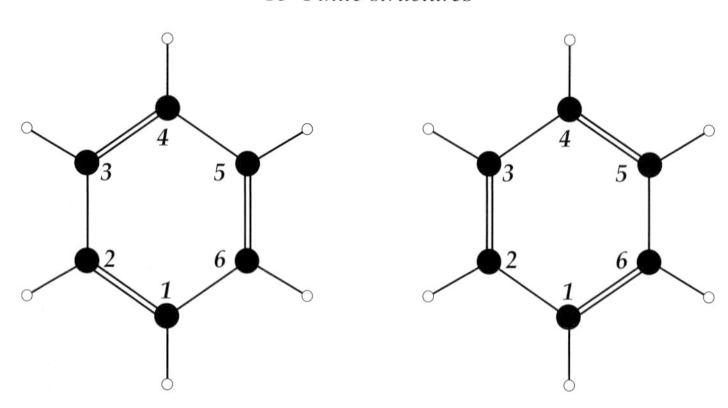

Figure 13.4. Single–double bond resonance in the benzene molecule. The black dots are C atoms; the white dots are H atoms; and the lines represent the single and double bonds, which alternate between pairs of C atoms.

structure consisting of six C and six H atoms, shown in Fig. 13.4. In this structure the orbitals of the C atoms are in purely sp^2 and p_z configurations, the first set contributing to the σ-bonds between C atoms as well as to the C–H bonds, while the second set contributes to π-bonding. The latter can be viewed as involving three pairs of atoms, with stronger (double) intra-pair bonds as opposed to inter-pair bonds. The double bonds are between the pairs of atoms 1–2, 3–4, 5–6 half of the time, and between the pairs 2–3, 4–5, 6–1, the other half; this makes all the bonds in the structure equivalent. Getting back to the fullerene, it is obvious that this type of arrangement cannot be applied to a pentagon, where the odd number of bonds prohibits alternating double and single bonds. As far as the hexagons in the fullerene are concerned, they have two types of inequivalent bonds, one which belongs also to a pentagon and one which connects two neighboring pentagons. Since we have argued that the pentagon bonds are not good candidates for resonance, the three double bonds of the hexagons must be those connecting neighboring pentagons. As already mentioned, this is only a crude picture which must be refined by taking into account the symmetries of the structure.

The most interesting aspect of the electronic structure of C_{60} is what happens to the electrons in the p_z orbitals, since the electrons in sp^2 orbitals form the σ-bonds that correspond to low-energy states, well separated from the corresponding antibonding states. We will concentrate then on what happens to these 60 p_z states, one for each C atom. The simplest approximation is to imagine that these states are subject to a spherical effective potential, as implied by the overall shape of the cluster. This effective potential would be the result of the ionic cores and the valence electrons in the σ-manifold. A spherical potential would produce eigenstates of definite angular momentum (see Appendix B), with the sequence of states shown in Fig. 13.5. However, there is also icosahedral symmetry introduced by the atomic

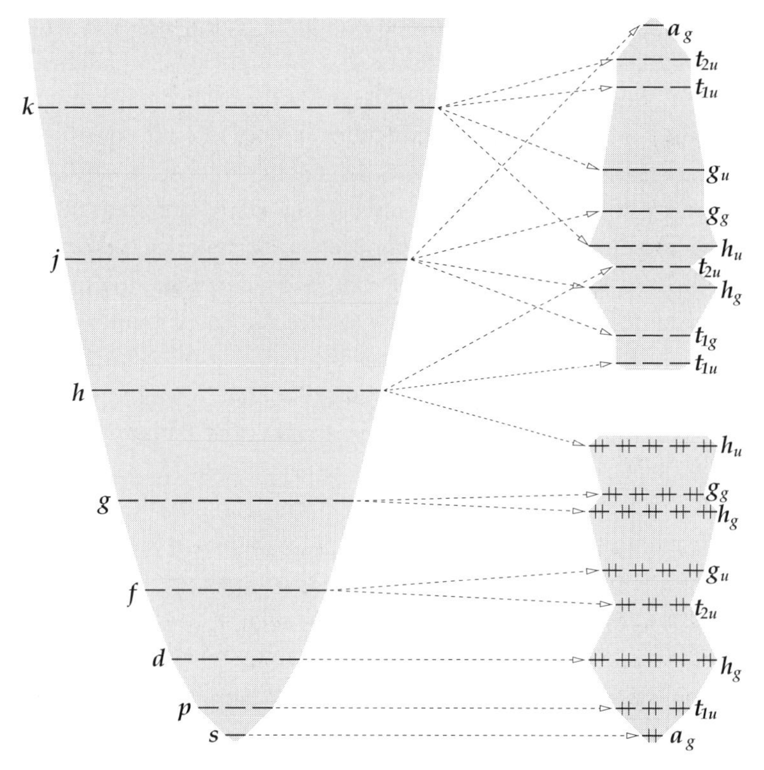

Figure 13.5. Schematic representation of the electronic structure of π states in C_{60}. **Left:** eigenstates of the angular momentum in a potential with spherical symmetry; states with angular momentum $l = 0(s), 1(p), 2(d), 3(f), 4(g), 5(h), 6(j), 7(k)$ are shown, each with degeneracy $(2l + 1)$. **Right:** sequence of states in C_{60} with the appropriate labels from the irreducible representations of the icosahedral group. The filled states are marked by two short vertical lines, indicating occupancy by two electrons of opposite spin. The sequence of levels reflects that obtained from detailed calculations [210, 211], but their relative spacing is not faithful and was chosen mainly to illustrate the correspondence between the spherical and icosahedral sets.

positions in the fullerene structure, which will break the full rotational symmetry of the spherical potential. Consequently, the angular momentum states will be split into levels which are compatible with the symmetry of the icosahedral group. The icosahedral group has five irreducible representations of dimensions 1, 3, 3, 4 and 5, which are the allowed degeneracies of the electronic levels. These are denoted in the literature as a, t_1, t_2, g and h, respectively. The levels are also characterized by an additional index, denoted by the subscripts g and u, depending on whether they have even (g) or odd (u) parity[2] with respect to inversion. The splitting of the spherical potential levels to icosahedral group levels is shown in Fig. 13.5. From this diagram

[2] From the German words "gerade" = symmetric and "ungerade" = antisymmetric.

it is evident that one of Nature's wonderful accidents occurs in C_{60}, in which there is gap between filled and empty electronic levels: the highest occupied level is a five-fold degenerate state of h_u character in terms of icosahedral symmetry, which derives from the $l = 5(h)$ angular momentum state, while the lowest unoccupied level is a three-fold degenerate state of t_{1u} character in terms of icosahedral symmetry deriving from the same angular momentum state. Note that this separation between filled and empty states does not occur in the spherical potential levels; in other words, the icosahedral symmetry is a crucial aspect of the structure. Moreover, symmetry considerations alone are enough to determine that there is a gap between the filled and empty states, assuming that the icosahedral symmetry is a small perturbation on top of the spherical potential. Elaborate quantum mechanical calculations [210, 211], support this picture and show that the gap is about 1.5 eV. These calculations also show that the icosahedral symmetry is actually a large perturbation of the spherical potential, leading to significant reordering of the angular momentum levels for the higher l values ($l \geq 5$), as shown in Fig. 13.5; fortunately, it does not alter the picture for levels up to the Fermi energy.

The closed electronic shell structure and relatively large gap between filled and empty levels in C_{60} makes it chemically inert: forming bonds with other molecules would involve the excitation of electrons across the gap, which would disrupt the perfect bonding arrangement, an energetically costly process. Its inert chemical nature makes C_{60} a good candidate for a building block of more complex structures. Two major categories of such structures have emerged. The first, known as "fullerides", involves *crystals* composed of C_{60} clusters and other atoms, the second concerns *clusters* with additional atoms inside or outside the hollow C_{60} sphere, which are known as "endohedral" or "exohedral" fullerenes. We discuss these two possibilities next.

Solid forms of fullerenes

It is not surprising that the solid form of the C_{60} fullerenes is a close-packed structure, specifically an FCC crystal, given that the basic unit has essentially a spherical shape and a closed electronic shell (see the general discussion of the solids formed by closed-shell atoms, in chapter 1). This solid form is known as "fullerite". The fact that the basic unit in the fullerite has a rich internal structure beyond that of a sphere makes it quite interesting: the various possible orientations of the C_{60} units relative to the cubic axes produce complex ordering patterns which are sensitive to temperature. At high enough temperature we would expect the C_{60} units to rotate freely, giving rise to a true FCC structure. This is indeed the case, for temperatures above the orientational order temperature $T_o = 261$ K. Below this critical value, the interactions between neighboring C_{60} units become sufficiently strong to induce orientational order in addition to the translational order of the cubic lattice. To

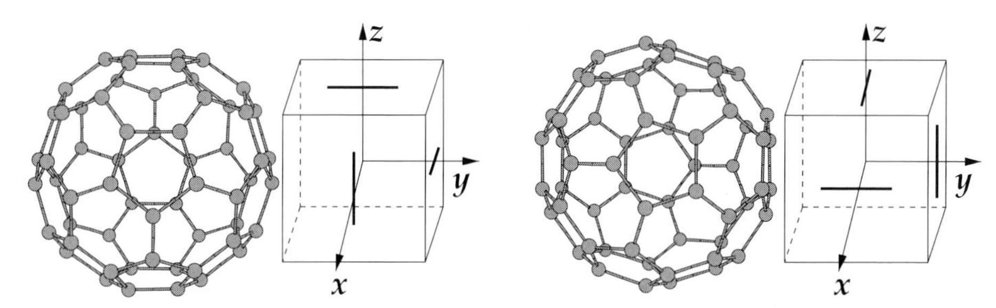

Figure 13.6. The two possible relative orientations of the C_{60} cluster with respect to the cartesian axes which do not disrupt the cubic symmetry. The bonds between neighboring pentagons which are bisected by the two-fold rotation axes are indicated by the thick short lines on the faces of the cube.

explain the order in this structure, we first note that a single C_{60} unit can be oriented so that the three pairs of bonds between neighboring hexagons are bisected by the cartesian axes, which thus become identified with the two-fold rotational symmetry axes of the cluster. There are two such possible orientations, shown in Fig. 13.6, which do not disrupt the cubic symmetry. In these orientations, the three-fold rotational symmetry axes of the cluster become identified with the $\{111\}$ axes of the cube (the major diagonals), which are three-fold rotational symmetry axes of the cubic system (see chapter 3). Since there are four such major axes in a cube, and there are four units in the conventional cubic cell of an FCC crystal, we might expect that at low temperature the four C_{60} units would be oriented in exactly this way, producing a simple cubic structure.

This picture is essentially correct, with two relatively minor modifications. First, the clusters are actually rotated by 22°–26° around their $\{111\}$ three-fold symmetry axes. The reason for this change in orientation is that it produces an optimal interaction between neighboring units, with the electron-rich double bond between neighboring hexagons on one cluster facing the electron-deficient region of a hexagon center in the next cluster. Second, the two possible orientations of a single unit relative to the cubic axes mentioned above produce a certain amount of disorder in the perfectly ordered structure, called "merohedral disorder". It should be pointed out that below T_o the clusters are actually jumping between equivalent orientations consistent with the constrains described, that is, their interactions are quite strong so that they produce correlation between their relative orientations but not strong enough to freeze all rotational motion. Complete freezing of rotational motion occurs at a still lower temperature, but this change does not have the characteristics of a phase transition as the onset of orientational order below T_o does.

The solid formed by close-packed C_{60} units has a large amount of free volume, because the basic units are large by atomic dimensions (recall that the diameter of

Table 13.1. *M_3C_{60} compounds that exhibit superconductivity.*

a is the lattice constant of the FCC lattice. The second and third columns denote the occupation of the octahedral and tetrahedral sites of the FCC lattice (see Fig. 13.7).

Compound	Octahedral	Tetrahedral	a (Å)	T_c (K)
C_{60}	–	–	14.161	
Na_2RbC_{60}	Rb	Na	14.028	2.5
Li_2CsC_{60}	Cs	Li	14.120	12.0
Na_2CsC_{60}	Cs	Na	14.134	12.5
K_3C_{60}	K	K	14.240	19.3
K_2RbC_{60}	Rb	K	14.243	23.0
K_2CsC_{60}	Cs	K	14.292	24.0
Rb_2KC_{60}	Rb,K	Rb,K	14.323	27.0
Rb_3C_{60}	Rb	Rb	14.384	29.0
Rb_2CsC_{60}	Cs	Rb	14.431	31.3
$RbCs_2C_{60}$	Cs	Rb,Cs	14.555	33.0

Source: Ref. [210].

C_{60} is $d = 1.034$ nm $= 10.34$ Å). The lattice constant of the FCC crystal composed of C_{60} units is 14.16 Å, just 3% smaller than the value $d\sqrt{2} = 14.62$ Å, which is what we would expect for incompressible spheres in contact with their nearest neighbors. The free space between such spheres is enough to accommodate other atoms. In fact, structures in which this space is occupied by alkali atoms (Li, Na, K, Rb and Cs) have proven quite remarkable: they exhibit superconductivity at relatively high transition temperatures (see Table 13.1). The positions occupied by the alkali atoms are the eight tetrahedral and four octahedral sites of the cubic cell, shown in Fig. 13.7. When all these sites are occupied by alkali atoms the structure has the composition M_3C_{60}. Solids with one, two, four and six alkali atoms per C_{60} have also been observed but they correspond to structures with lower symmetry or to close-packed structures of a different type (like the BCC crystal). All these crystals are collectively referred to as "doped fullerides".

By far the most widely studied of the doped fullerides are the FCC-based M_3C_{60} crystals with M being one of the alkali elements, so we elaborate briefly on their electronic properties. The C_{60} cluster is significantly more electronegative than the alkali atoms, so it acts as an acceptor of the alkali's valence electrons. Since there are three alkali atoms per C_{60} cluster, contributing one valence electron each, we expect that three units of electronic charge will be added to each cluster. We have seen earlier that the first unoccupied state of an isolated C_{60} cluster is a three-fold degenerate state of t_{1u} character, deriving from the $l = 5$ angular momentum state (see Fig. 13.5). Placing three electrons in this state makes it exactly half full, so the system

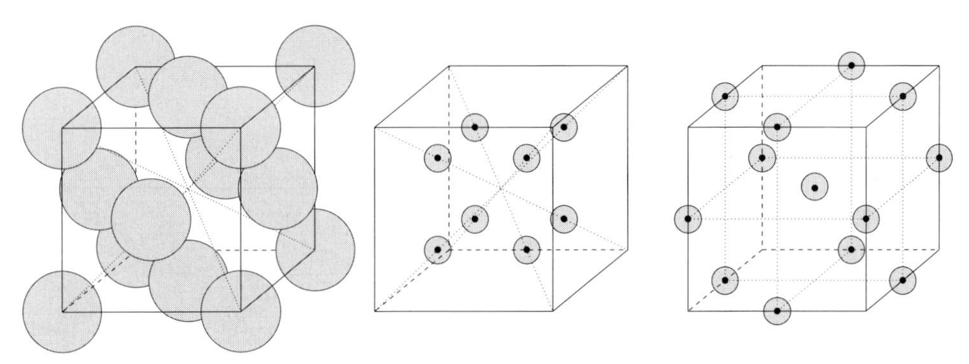

Figure 13.7. The arrangement of C_{60} clusters and dopant alkali atoms in the cubic lattice. **Left:** the C_{60} clusters with their centers at the FCC sites; there are four distinct such sites in the cubic unit cell. **Center:** the tetrahedral sites which are at the centers of tetrahedra formed by the FCC sites; there are eight distinct such sites in the cubic cell. **Right:** the octahedral sites which are at the centers of octahedra formed by the FCC sites; there are four distinct such sites in the cubic cell.

will have metallic character. These states will of course acquire band dispersion in the solid, but since the interaction between C_{60} units is relatively weak the dispersion will be small. The Fermi energy will then intersect the four bands arising from the four C_{60} units in the cubic cell, with each band being on average half filled. This produces a metallic solid with high density of states at the Fermi level, a situation conducive to superconductivity. There is a striking correlation between the lattice constant of the doped fulleride crystals and their T_c: the larger the lattice constant, the larger the T_c. This can be interpreted as due to band-structure effects, since the larger lattice constant implies weaker interaction between the clusters and hence narrower band widths, which leads to higher density of states at the Fermi level, $g_F = g(\epsilon_F)$. Additional evidence of the importance of g_F to the superconducting state is that the T_c of doped fullerides is very sensitive to external pressure: when the pressure is increased, T_c drops fast. The interpretation of the pressure effect is similar to that of the lattice constant effect, namely that higher pressure brings the clusters closer together, thus increasing their interaction and the band width of electronic states, which reduces the value of g_F. It is worthwhile noting that the M_6C_{60} solids are insulators, since the number of valence electrons of the alkali atoms is exactly what is needed to fill the first three unoccupied levels of each cluster, which are separated by a gap from the next set of unoccupied levels (see Fig. 13.5).

The mechanism of superconductivity in doped fullerides is not perfectly clear, because the presence of extra electrons on each C_{60} unit introduces complications to electron correlations. This has motivated theoretical studies based on the Hubbard model, in addition to models based on electron–phonon interactions. In

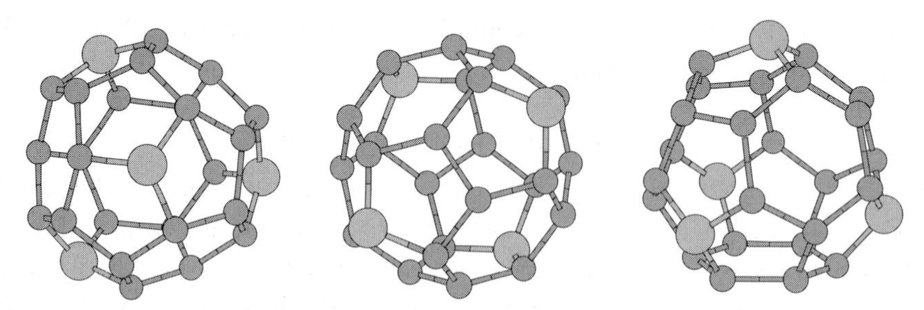

Figure 13.8. Three views of the C_{28} cluster which has tetrahedral symmetry and consists of 12 pentagons and 4 hexagons. **From left to right:** view along a three-fold rotational axis, view along a two-fold axis (equivalent to a reflection plane) and a perspective view. The four apex atoms, where triplets of pentagonal rings meet, are shown larger and in lighter shading.

the context of the latter, it appears that the relevant phonon modes are related to the C_{60} units exclusively, and that they involve mostly intramolecular modes. It is clear from experimental measurements that the role of the alkali atoms is simply to donate their valence electrons to the clusters, because there is no isotope effect associated with the dopant atoms (when they are replaced by radioactive isotopes there is no change in T_c). There is, however, a pronounced isotope effect related to substitution of radioactive ^{13}C for the normal ^{12}C isotope in C_{60} [212]. In the superconducting compounds, certain intramolecular modes are significantly different from the corresponding modes of the pure C_{60} solid or the insulating compound M_6C_{60}, suggesting that these modes are involved in the superconducting state.

An altogether different type of solid based on the C_{28} fullerene has also been proposed and studied theoretically [213, 214]. This cluster is considerably smaller in size than C_{60} and its structure cannot fulfil the isolated pentagon rule. The pentagonal rings are arranged in triplets in which all three pentagons share one corner and the three pairs share one edge. There are four such triplets of pentagons, with the common corner atoms forming the appexes of a tetrahedron. The structure is completed by four hexagonal rings which lie diametrically opposite from the apex atoms, as shown in Fig. 13.8. This cluster represents an interesting coincidence of electronic and structural features. The four apex atoms, due to the three 108° bond angles that form as part of three pentagons, have a hybridization very close to sp^3 and consequently have a dangling bond which is unable to combine with any other electronic state. Moreover, these four dangling bonds are pointing outwardly in perfect tetrahedral directions, much like the sp^3 hybrids that an isolated C or Si atom would form. We can then consider the C_{28} cluster as a large version of a tetravalent element in sp^3 hybridization. This immediately suggests the possibility of forming diamond-like solids out of this cluster, by creating covalent bonds between the dangling tetrahedral orbitals of neighboring clusters. Such solids, if they could be formed, would have interesting electronic and optical properties [213, 214], but their realization remains a dream.

Endohedral and exohedral fullerenes

Finally, we mention briefly structures derived from the fullerenes by adding other atoms, either inside the cage of C atoms or outside it. Some of these structures involve chemical reactions which break certain C–C bonds on the fullerene cage, while others involve weaker interactions between the added atoms and the cluster which do not affect the cage structure. We will concentrate on the latter type of structures, since in those the character of the fullerene remains essentially intact. Some of the most intriguing structures of this type contain one metal atom inside the cage, called endohedral fullerenes or metallofullerenes and represented by $M@C_{60}$ [214–216]. A case that has been studied considerably is $La@C_{60}$. The result of including this trivalent metal in the cage is that its valence electrons hybridize with the lowest unoccupied level of the cluster, producing occupied electronic states of novel character. Another interesting finding is that endohedral atoms can stabilize clusters which by themselves are not stable because they violate the isolated pentagon rule. A case in point is $La@C_{44}$ [218, 219]. Other endohedral fullerenes which have been observed include higher fullerenes, such as $M@C_{76}$, $M@C_{82}$ and $M@C_{84}$, or structures with more than one endohedral atom, such as $M_2@C_{82}$ and $M_3@C_{82}$, with M being one of the lanthanide elements such as La, Y, Sc.

As far as fullerenes decorated by a shell of metal atoms are concerned, which we refer to as exohedral fullerenes, we can identify several possibilities which maintain the full symmetry of the bare structure. For C_{60}, there are $n_1 = 12$ sites above the centers of the pentagons, $n_2 = 20$ sites above the centers of the hexagons, $n_3 = 30$ sites above the centers of the electron-rich bonds between neighboring pentagons, $n_4 = 60$ sites above the C atoms, and $n_5 = 60$ sites above the electron-poor bonds of the pentagons. We can place metal atoms at any of these sets of sites to form a shell of

$$N = \sum_{i=1}^{5} f_i n_i$$

where the occupation numbers f_i can take the values zero or unity. From packing considerations, it becomes evident that not all of these shells can be occupied at the same time (this would bring the metal atoms too close to each other). Some possible shell sizes are then: $N = 32$ ($f_1 = f_2 = 1$), $N = 50$ ($f_2 = f_3 = 1$), $N = 62$ ($f_1 = f_2 = f_3 = 1$), $N = 72$ ($f_1 = f_4 = 1$), $N = 80$ ($f_2 = f_4 = 1$). It is intriguing that several of these possibilities have been observed experimentally by T.P. Martin and coworkers, for instance $N = 32$ for alkali atoms, $N = 62$ for Ti, V, Zr, and $N = 50, 72, 80$ for V [220, 221]. Theoretical investigations of some of these exohedral fullerenes indicate that the metal shell inhibits π-bonding, leading to a slightly enlarged fullerene cage, while the metal atoms are bonded both to the fullerene cage by mixed ionic and covalent bonding as well as among themselves by covalent bonds [222].

13.1.3 Carbon nanotubes

In the early 1990s two new classes of structures closely related to the fullerenes were discovered experimentally: the first, found by Iijima [223], consists of long tubular structures referred to as carbon nanotubes; the second, found by Ugarte [224], consists of concentric shells of fullerenes. Of these, the first type has proven quite interesting, being highly stable, versatile, and exhibiting intriguing 1D physics phenomena, possibly even high-temperature superconductivity. We will concentrate on the structure and properties of these nanotubes. In experimental observations the nanotubes are often nested within each other as coaxial cylinders and they have closed ends. Their ends are often half cages formed from a half fullerene whose diameter and structure is compatible with the diameter of the tube. Given that the inter-tube interactions are weak, similar to the interaction between graphite sheets, and that their length is of order ~1 μm, it is reasonable to consider them as single infinite tubes in order to gain a basic understanding of their properties. This is the picture we adopt below.

The simplest way to visualize the structure of C nanotubes is by considering a sheet of graphite, referred to as graphene, and studying how it can be rolled into a cylindrical shape. This is illustrated in Fig. 13.9. The types of cylindrical shapes that can be formed in this way are described in terms of the multiples n and m of the in-plane primitive lattice vectors, $\mathbf{a}_1, \mathbf{a}_2$, that form the vector connecting two atoms which become identical under the operation of rolling the graphene into a cylinder. We define this vector, which is perpendicular to the tube axis, as $\mathbf{a}_\perp^{(n,m)}$. The vector perpendicular to it, which is along the tube axis, is defined as $\mathbf{a}_\parallel^{(n,m)}$. In terms of cartesian components on the graphene plane, in the orientation shown in Fig. 13.9, these two vectors are expressed as

$$\mathbf{a}_\perp^{(n,m)} = n\mathbf{a}_1 + m\mathbf{a}_2 = na\left[\frac{(1+\kappa)\sqrt{3}}{2}\hat{\mathbf{x}} + \frac{(1-\kappa)}{2}\hat{\mathbf{y}}\right] \tag{13.1}$$

$$\mathbf{a}_\parallel^{(n,m)} = a\left[\frac{(\kappa-1)}{2}\hat{\mathbf{x}} + \frac{(\kappa+1)\sqrt{3}}{2}\hat{\mathbf{y}}\right]\frac{\sqrt{3}\lambda}{(1+\kappa+\kappa^2)} \tag{13.2}$$

where we have defined two variables, $\kappa = n/m$ and λ, to produce a more compact notation. The second variable is defined as the smallest rational number that produces a vector $\mathbf{a}_\parallel^{(n,m)}$ which is a graphene lattice vector. Such a rational number can always be found; for example the choice $\lambda = (1+\kappa+\kappa^2)$, which is rational

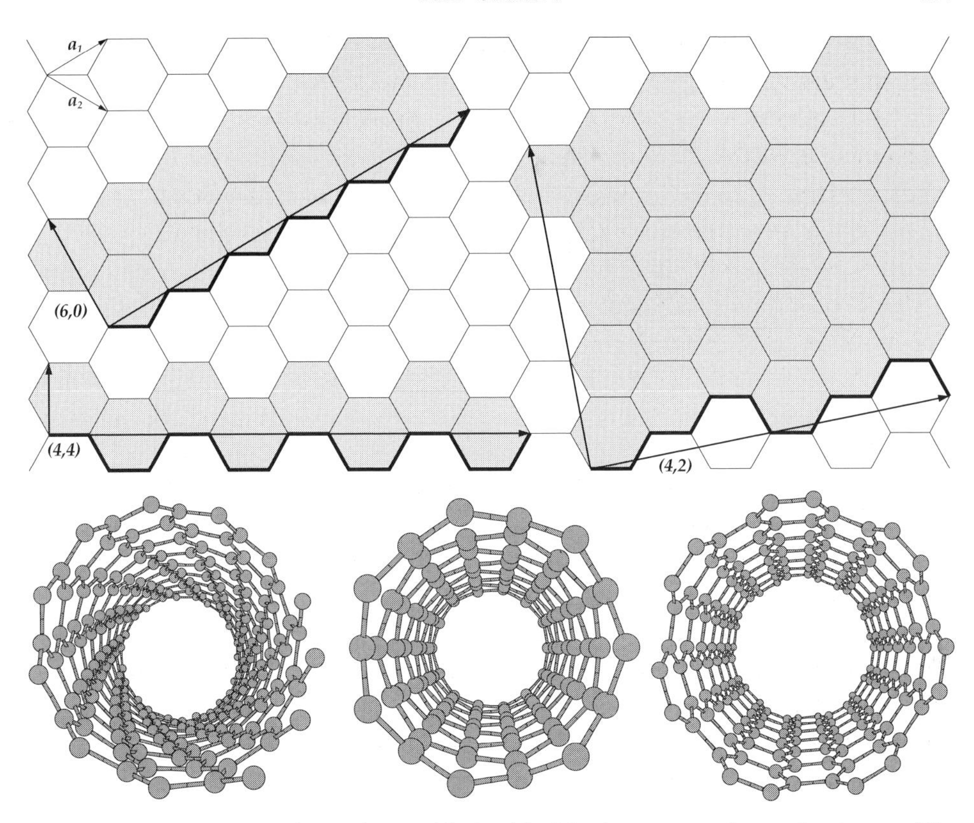

Figure 13.9. **Top:** a graphene sheet with the ideal lattice vectors denoted as $\mathbf{a}_1, \mathbf{a}_2$. The thicker lines show the edge profile of a (6, 0) (zig-zag), a (4, 4) (armchair), and a (4, 2) (chiral) tube. The tubes are formed by matching the end-points of these profiles. The hexagons that form the basic repeat unit of each tube are shaded, and the thicker arrows indicate the repeat vectors along the axis of the tube and perpendicular to it, when the tube is unfolded. **Bottom:** perspective views of the (8, 4) chiral tube, the (7, 0) zig-zag tube and the (7, 7) armchair tube along their axes.

from the definition of κ, always produces a graphene lattice vector. The reason for introducing λ as an additional variable is to allow for the possibility that a smaller number than $(1 + \kappa + \kappa^2)$ can be found which makes $\mathbf{a}_{\parallel}^{(n,m)}$ a graphene lattice vector, thus reducing the size of the basic repeat unit that produces the tube. The length of the $\mathbf{a}_{\perp}^{(n,m)}$ vector cannot be reduced and corresponds to the perimeter of the tube. The diameter of the tube can be inferred from this length divided by 2π. We can also define the corresponding vectors in reciprocal space:

$$\mathbf{b}_{\perp}^{(n,m)} = \frac{2\pi}{na}\left[\frac{(1+\kappa)\sqrt{3}}{2}\hat{\mathbf{x}} + \frac{(1-\kappa)}{2}\hat{\mathbf{y}}\right]\frac{1}{(1+\kappa+\kappa^2)} \tag{13.3}$$

$$\mathbf{b}_{\parallel}^{(n,m)} = \frac{2\pi}{a}\left[\frac{(\kappa-1)}{2}\hat{\mathbf{x}} + \frac{(\kappa+1)\sqrt{3}}{2}\hat{\mathbf{y}}\right]\frac{1}{\sqrt{3}\lambda} \tag{13.4}$$

With these definitions, we can then visualize both the atomic structure and the electronic structure of C nanotubes.

There are three types of tubular structures: the first corresponds to $m = 0$ or $(n, 0)$, which are referred to as "zig-zag" tubes; the second corresponds to $m = n$ or (n, n), which are referred to as "armchair" tubes; and the third corresponds to $m \neq n$ or (n, m), which are referred to as "chiral" tubes. Since there are several ways to define the same chiral tube with different sets of indices, we will adopt the convention that $m \leq n$, which produces a unique identification for every tube. Examples of the three types of tubes and the corresponding vectors along the tube axis and perpendicular to it are shown in Fig. 13.9. The first two types of tubes are quite simple and correspond to regular cylindrical shapes with small basic repeat units. The third type is more elaborate because the hexagons on the surface of the cylinder form a helical structure. This is the reason why the basic repeat units are larger for these tubes.

The fact that the tubes can be described in terms of the two new vectors that are parallel and perpendicular to the tube axis, and are both multiples of the primitive lattice vectors of graphene, also helps to determine the electronic structure of the tubes. To first approximation, this will be the same as the electronic structure of a graphite plane folded into the Brillouin Zone determined by the reciprocal lattice vectors $\mathbf{b}_{\parallel}^{(n,m)}$ and $\mathbf{b}_{\perp}^{(n,m)}$. Since these vectors are uniquely defined for a pair of indices (n, m) with $m \leq n$, it is in principle straightforward to take the band structure of the graphite plane and fold it into the appropriate part of the original BZ to produce the desired band structure of the tube (n, m). This becomes somewhat complicated for the general case of chiral tubes, but it is quite simple for zig-zag and armchair tubes. We discuss these cases in more detail to illustrate the general ideas.

It is actually convenient to define two new vectors, in terms of which both zig-zag and armchair tubes can be easily described. These vectors and their corresponding reciprocal lattice vectors are

$$\mathbf{a}_3 = \mathbf{a}_1 + \mathbf{a}_2 = \sqrt{3}a\hat{\mathbf{x}}, \quad \mathbf{b}_3 = \frac{1}{2}(\mathbf{b}_1 + \mathbf{b}_2) = \frac{2\pi}{\sqrt{3}a}\hat{\mathbf{x}} \tag{13.5}$$

$$\mathbf{a}_4 = \mathbf{a}_1 - \mathbf{a}_2 = a\hat{\mathbf{y}}, \quad \mathbf{b}_4 = \frac{1}{2}(\mathbf{b}_1 - \mathbf{b}_2) = \frac{2\pi}{a}\hat{\mathbf{y}} \tag{13.6}$$

In terms of these vectors, the zig-zag and armchair tubes are described as

$$\text{zig-zag}: \quad \mathbf{a}_{\parallel}^{(n,0)} = \mathbf{a}_3, \ \mathbf{b}_{\parallel}^{(n,0)} = \mathbf{b}_3, \quad \mathbf{a}_{\perp}^{(n,0)} = n\mathbf{a}_4, \ \mathbf{b}_{\perp}^{(n,0)} = \frac{1}{n}\mathbf{b}_4$$

$$\text{armchair}: \quad \mathbf{a}_{\parallel}^{(n,n)} = \mathbf{a}_4, \ \mathbf{b}_{\parallel}^{(n,n)} = \mathbf{b}_4, \quad \mathbf{a}_{\perp}^{(n,n)} = n\mathbf{a}_3, \ \mathbf{b}_{\perp}^{(n,n)} = \frac{1}{n}\mathbf{b}_3$$

Now it becomes simple to describe the folding of the graphene BZ into the smaller area implied by the tube structure, as shown in the examples of Fig. 13.10. In

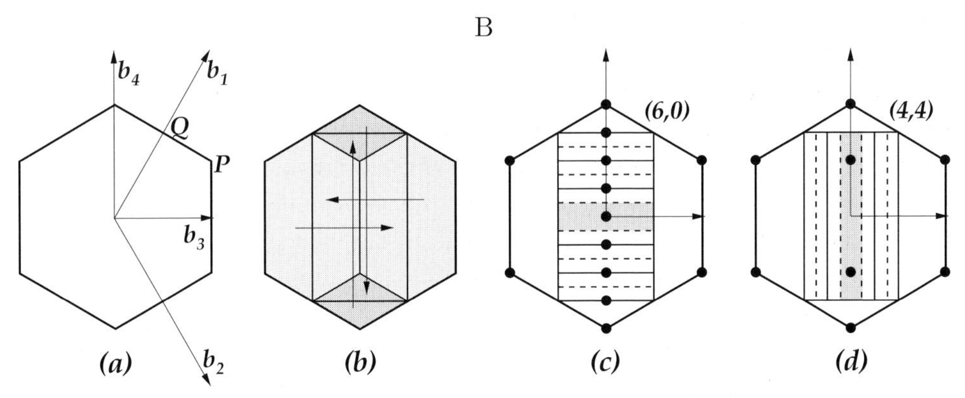

Figure 13.10. (a) The graphene Brillouin Zone, with the reciprocal lattice vectors $\mathbf{b}_1, \mathbf{b}_2$ and the tube-related vectors $\mathbf{b}_3, \mathbf{b}_4$. (b) The folding of the full zone into the reduced zone, determined by the vectors $\mathbf{b}_3, \mathbf{b}_4$. (c) The Brillouin Zone for $(n, 0)$ tubes, and the example of the (6,0) tube: solid lines indicate sets of points equivalent to the $k_y = 0$ line, and dashed lines indicate zone boundaries. (d) The Brillouin Zone for (n, n) tubes, and the example of the (4,4) tube: solid lines indicate sets of points equivalent to the $k_x = 0$ line, and dashed lines indicate zone boundaries. The black dots in (c) and (d) are the images of the point P of the graphene BZ under the folding introduced by the tube structure.

both cases, the smallest possible folding (which does *not* correspond to physically observable tubes because of the extremely small diameter), that is, $(1, 0)$ and $(1, 1)$, produce a BZ which is half the size of the graphene BZ. The larger foldings, $(n, 0)$ and (n, n), further reduce the size of the tube BZ by creating stripes parallel to the $\mathbf{b}_3$ vector for the zig-zag tubes or to the $\mathbf{b}_4$ vector for the armchair tubes.

Combining the results of the above analysis with the actual band structure of graphene as discussed in chapter 4, we can draw several conclusions about the electronic structure of nanotubes. Graphene is a semimetal, with the occupied and unoccupied bands of π-character meeting at a single point of the BZ, labeled P (see Fig. 4.6). From Fig. 13.10, it is evident that this point is mapped onto the point $(k_x, k_y) = \mathbf{b}_4/3$, which is always within the first BZ of the (n, n) armchair tubes. Therefore, all the armchair tubes are metallic, with two bands crossing the Fermi level at $k_y = 2\pi/3a$, that is, two-thirds of the way from the center to the edge of their BZ in the direction parallel to the tube axis. It is also evident from Fig. 13.10 that in the $(n, 0)$ zig-zag tubes, if n is a multiple of 3, P is mapped onto the center of the BZ, which makes these tubes metallic, whereas if n is not a multiple of 3 these tubes can have semiconducting character. Analogous considerations applied to the chiral tubes of small diameter (10–20Å) lead to the conclusion that about one-third of them have metallic character while the other two-thirds are semiconducting. The chiral tubes of metallic character are those in which the indices n and m satisfy the relation $2n + m = 3l$, with l an integer.

What we have described above is a simplified but essentially correct picture of C nanotube electronic states. The true band structure is also affected by the curvature of the tube and variations in the bond lengths which are not all equivalent. The effects are more pronounced in the tubes of small diameter, but they do not alter significantly the simple picture based on the electronic structure of graphene. Examples of band structures of various tubes are shown in Fig. 13.11. These band

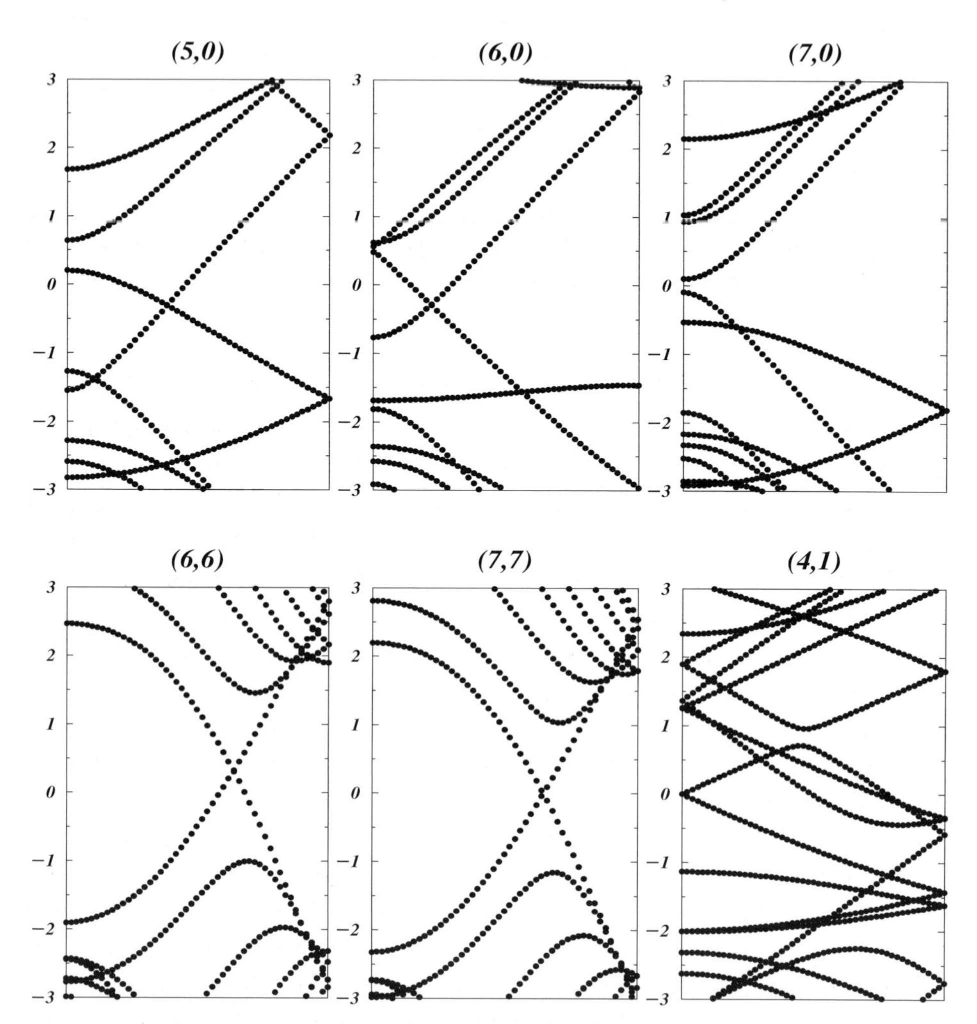

Figure 13.11. Examples of C nanotube band structures obtained with the tight-binding approximation. In all cases the horizontal axis runs from the center to the edge of the BZ in the direction parallel to the tube axis and the energy levels are shifted so that the Fermi level is at zero; the energy scale is in eV. **Top:** three $(n, 0)$ tubes, of which $(5, 0)$ is metallic by accident, (6,0) is metallic by necessity and (7,0) is semiconducting. **Bottom:** two (n, n) tubes, showing the characteristic crossing of two levels which occurs at two-thirds of the way from the center to the edge of the BZ, and one (n, m) tube for which the rule $2n + m = 3l$ holds, giving it metallic character.

structures, plotted from the center to the edge of the BZ along the direction parallel to the tube axis, were obtained with a sophisticated version of the tight-binding approximation [40], which reproduces well the band structures of graphite and diamond; for these calculations the bond length was kept fixed at 1.42 Å (the bond length of graphite). The band structures contain both σ- and π-states bonding and antibonding states. Of these examples, the (5,0) tube is metallic by accident (by this we mean that the metallic character is not dictated by symmetry), the (6,0) tube is metallic by necessity, as mentioned above, and the (7,0) tube is semiconducting, with a small band gap of about 0.2 eV. The two (n, n) tubes exhibit the characteristic crossing of two bands which occurs at two-thirds of the way from the center to the edge of the BZ, a feature which renders them metallic. Finally, the (4,1) tube, for which the rule $2n + m = 3l$ holds, is metallic in character with two bands meeting at the Fermi level at the center of the BZ, as predicted by the simple analysis of zone folding in graphene (see also Problem 3).

13.1.4 Other covalent and mixed clusters

Clusters of several elements or compounds have also been produced and studied in some detail. Two cases that have attracted considerable attention are clusters of Si and the so called metallo-carbohedrenes or "met-cars". These two classes of clusters exhibit certain similarities to the fullerenes and have been considered as possible building blocks of novel materials, so we discuss them briefly here.

Clusters of Si atoms were first shown to exhibit magic number behavior by Smalley and coworkers [225] and Jarrold and coworkers [226], who found that the reactivity of clusters can change by two to three orders of magnitude when their size changes by one atom. The sizes 33, 39 and 45, corresponding to deep minima in the reactivity as a function of size, are magic numbers in the 20–50 size range. The changes in the reactivity do not depend on the reactant (NH_3, C_2H_4, O_2 and H_2O have been used) suggesting that the low reactivity of the magic number clusters is a property inherent to the cluster. This extraordinary behavior has been attributed to the formation of closed-shell structures, in which the interior consists of tetrahedrally bonded Si atoms resembling bulk Si, while the exterior consists of four-fold or three-fold coordinated Si atoms in configurations closely resembling surface reconstructions of bulk Si [227]. Quantum mechanical calculations show that these models for the magic number Si clusters, referred to as Surface-Reconstruction Induced Geometries (SRIGs), have relatively low energy while their low chemical reactivity is explained by the elimination of surface dangling bonds induced by surface reconstruction (see chapter 11 for a detailed discussion of this effect). In Fig. 13.12 we present some of these models for the sizes 33, 39 and 45, which illustrates the concept of surface reconstruction applied to clusters. For example,

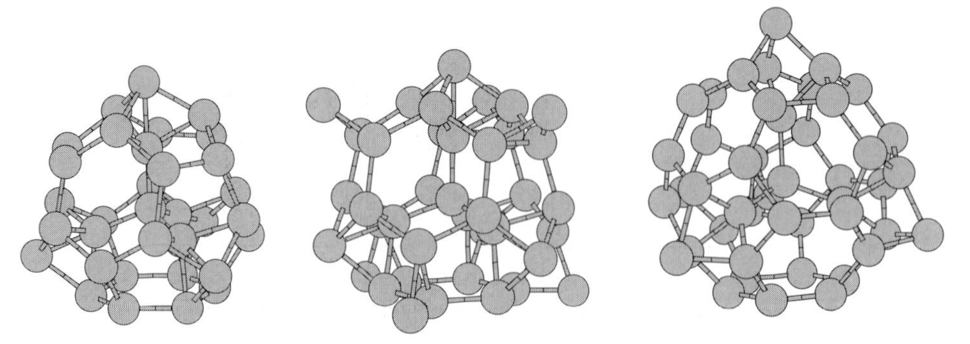

Figure 13.12. The Surface-Reconstruction Induced Geometry models for the Si clusters of sizes 33, 39 and 45. In all clusters there is a core of five atoms which are tetrahedrally coordinated as in bulk Si and are surrounded by atoms bonded in geometries reminiscent of surface reconstructions of Si, like adatoms (found on the Si(111) (7×7) surface reconstruction) or dimers (found on the Si(100) (2×1) surface reconstruction).

the 33-atom cluster has a core of five atoms which are tetrahedrally coordinated as in bulk Si, with one atom at the center of the cluster. Of the core atoms, all but the central one are each capped by a group of four atoms in a geometry resembling the adatom reconstruction of the Si(111) (7×7) surface. The structure is completed by six pairs of atoms bonded in a geometry akin to the dimer reconstruction of the Si(001) (2×1) surface. This theme is extended to the sizes 39 and 45, where the larger size actually allows for additional possibilities of SRIGs [227].

We should emphasize that the SRIG idea is not a generally accepted view of the structure of Si clusters, which remains a topic of debate. Unlike the case of the fullerenes, the Si clusters have not been produced in large enough amounts or in condensed forms to make feasible detailed experimental studies of their structural features. It is extremely difficult to find the correct structure for such large clusters from simulations alone [227, 228]. The problem is that the energy landscape generated when the atomic positions are varied can be very complicated, with many local minima separated by energy barriers that are typically large, since they involve significant rearrangement of bonds. Even if the energy barriers were not large, the optimal structure may correspond to a very narrow and deep well in the energy landscape that is very difficult to locate by simulations, which by computational necessity take large steps in configurational space. It is worth mentioning that unrestricted optimization of the energy of any given size cluster tends to produce a more compact structure than the SRIG models, with the Si atoms having coordination higher than four [229]. These compact structures, however, have no discernible geometric features which could justify the magic number behavior (for a review of the issue see articles in the book edited by Sattler, in the Further reading section).

Lastly, we discuss briefly the case of met-car clusters which were discovered by Castleman and coworkers [230]. There are two interesting aspects to these clusters.

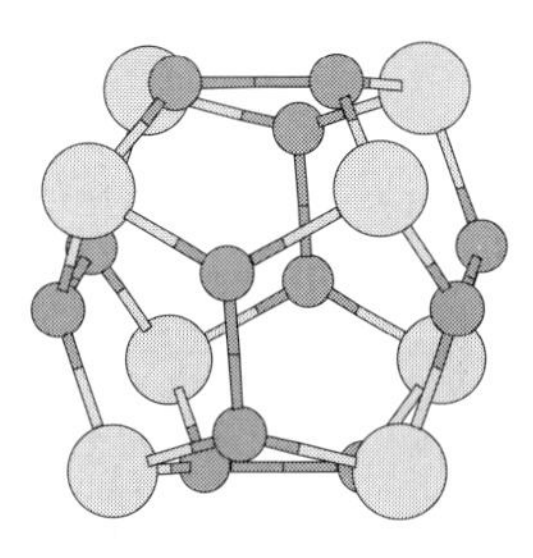
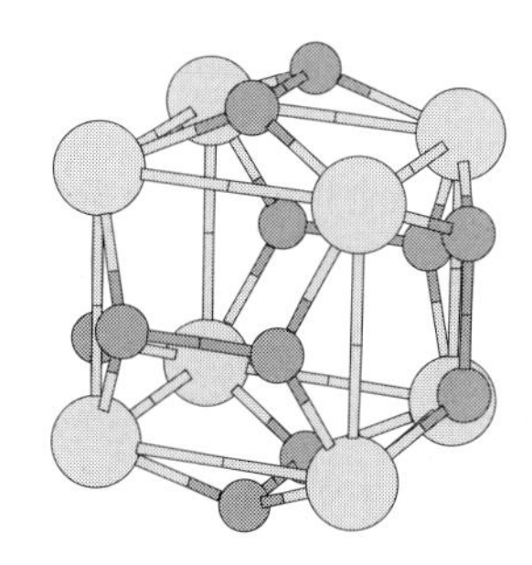

Figure 13.13. The structure of the metallo-carbohedrene (met-car) M_8C_{12} clusters: the larger spheres represent the metal atoms (M = Ti, Zr, Hf, V) and the smaller spheres the C atoms. **Left:** this emphasizes the relation to the dodecahedron, with only nearest neighbor bonds drawn; several pentagonal rings, consisting of two metal and three carbon atoms each, are evident. **Right:** this emphasizes the relation to a cube of metal atoms with its faces decorated by carbon dimers; additional bonds between the metal atoms are drawn.

The first is the structure of the basic unit, which consists of eight transition metal atoms and 12 carbon atoms: examples that have been observed experimentally include Ti_8C_{12}, Zr_8C_{12}, Hf_8C_{12} and V_8C_{12}. The atoms in these units are arranged in a distorted dodecahedron structure, as shown in Fig. 13.13. The eight metal atoms form a cube while the C atoms form dimers situated above the six faces of the cube. Each C atom has three nearest neighbors, another C atom and two metal atoms, while each metal atom has three C nearest neighbors. The structure maintains many but not all of the symmetries of the icosahedron. In particular, the two-fold and three-fold axes of rotation are still present, but the five-fold axes of rotation are absent since there are no perfect pentagons: the pentagonal rings are composed of three C and two metal atoms, breaking their full symmetry. The second interesting aspect of these clusters is that they can be joined at their pentagonal faces to produce stable larger structures. For example, joining two such units on a pentagonal face has the effect of making the C atoms four-fold coordinated with bond angles close to the tetrahedral angle of 109.47°. This pattern can be extended to larger sizes, possibly leading to interesting solid forms.

13.2 Biological molecules and structures

We turn our attention next to a class of structures that form the basis for many biological macromolecules. We will concentrate here on DNA, RNA and proteins, the central molecules of life. Our goal is to discuss their structural features which in many ways resemble those of solids and clusters, and to bring out the common themes in the structure of these different systems. There is no attempt made to discuss the biological functions of these macromolecules, save for some very basic notions of genetic coding and some general comments. The biological function of these structures is much too complex an issue and in most cases is the subject of ongoing research; as such, it lies well beyond the scope of the present book. We refer the reader to standard texts of molecular biology and biochemistry for more extensive discussions, examples of which are mentioned in the Further reading section.

As far as the structure of the biological molecules is concerned, one common feature with solids is that it can be deduced by X-ray scattering from crystals formed by these molecules. In fact, crystallizing biological molecules in order to deduce their structure represents a big step toward understanding their function. The application of X-ray scattering techniques to the study of such complex systems requires sophisticated analytical methods for inverting the scattering pattern to obtain the atomic positions; the development of such analytical methods represented a major breakthrough in the study of complex crystals, and was recognized by the 1985 Nobel prize for Chemistry, awarded to H.A. Hauptman and J. Karle. More recently, nuclear magnetic resonance experiments have also been applied to the study of the structure of biological macromolecules. For many complex molecules the structure has been completely determined; extensive data on the structure of proteins, as well as a discussion of experimental methods used in such studies, can be found in the Protein Data Bank (PDB)[3], and for nucleic acids like DNA and RNA such information can be obtained from the Nucleic acid Data Bank (NDB).[4]

Dealing with biological or chemical structures can be intimidating because of the complicated nomenclature: even though the basic constituents of the structure are just a few elements (typically C, H, O, N, P and S), there are many standard subunits formed by these atoms which have characteristic structures and properties. It is much more convenient to think in terms of these subunits rather than in terms of the constituent atoms, but this requires inventing a name for each useful subunit. We will use bold face letters for the name of various subunits the first time they are introduced in the text, hoping that this will facilitate the acquaintance of the reader with these structures.

13.2.1 The structure of DNA and RNA

We first consider the structure of the macromolecules DNA and RNA that carry genetic information. We will view these macromolecules as another type of polymer; accordingly, we will refer to the subunits from which they are composed as

[3] `http://www.rcsb.org/pdb/`

[4] `http://ndbserver.rutgers.edu/`

The publisher has used its best endeavors to ensure that the above URLs are correct and active at the time of going to press. However, the publisher has no responsibility for the websites, and can make no guarantee that a site will remain live, or that the content is or will remain appropriate.

monomers. The monomers in the case of DNA and RNA have two components: the first component is referred to as the **base**, the second as the **sugar-phosphate** group. The sugar-phosphate groups form the backbone of the DNA and RNA polymers, while the bases stack on top of one another due to attractive interactions of the van der Waals type. In the case of DNA, the bases form hydrogen bonds in pairs, linking the two strands of the macromolecule; the resulting structure is a long and interwined double helix with remarkable properties.

The bases

The bases are relatively simple molecules composed of H, C, N and O. The C and N atoms form closed rings with six or five sides (hexagons and pentagons), while the O and H atoms are attached to one of the atoms of the rings. There are two classes of bases, the **pyrimidines**, which contain only one hexagon, and the **purines**, which contain a hexagon and a pentagon sharing one edge.

Pyrimidine is the simplest molecule in the first class; it contains only C, N and H atoms and consists of a hexagonal ring of four C and two N atoms which closely resembles the benzene molecule (see Fig. 13.14). To facilitate identification of more complex structures formed by bonding to other molecules, the atoms in the hexagonal ring are numbered and referred to as N1, C2, N3, etc. Each C atom has a H atom bonded to it, while there are no H atoms bonded to the N atoms. The presence of the N atoms breaks the six-fold symmetry of the hexagonal ring. There is, however, a reflection symmetry about a plane perpendicular to the plane of the hexagon, going through atoms C2 and C5, or equivalently a two-fold rotation axis through these two atoms. This symmetry implies that resonance between the double bonds and single bonds, analogous to the case of benzene, will be present in pyrimidine as well. This symmetry is broken in the derivatives of the pyrimidine molecule discussed below, so the resonance is lost in those other molecules.

There are three interesting derivatives of the pyrimidine molecule.

Uracil (U), in which two of the H atoms of pyrimidine which were bonded to C atoms are replaced by O atoms. There is a double bond between the C and O atoms, which eliminates two of the double bonds in the hexagonal ring.
Cytosine (C), in which one of the H atoms of pyrimidine which was bonded to a C atom is replaced by an O atom and another one is replaced by a NH_2 unit.
Thymine (T), in which two of the H atoms of pyrimidine which were bonded to C atoms are replaced by O atoms and a third one is replaced by a CH_3 unit.

By convention we refer to these molecules by their first letter (in italicized capitals to distinguish them from the symbols of chemical elements). A remarkable feature of these molecules is that all atoms lie on the same plane, the plane defined by the hexagonal ring, except for the two H atoms in the NH_2 unit of C and the three H

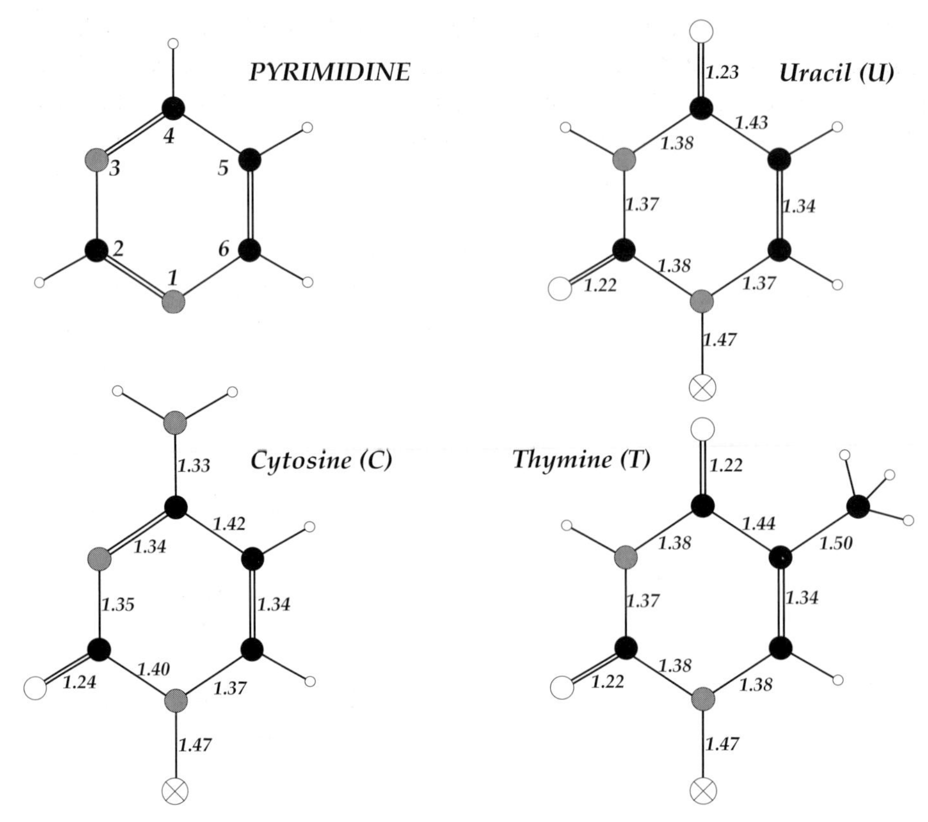

Figure 13.14. The structure of the neutral bases in the pyrimidine class. The black circles represent C atoms, the gray circles are N atoms and the white circles are O atoms. The smaller circles are the H atoms. Single and double bonds are indicated by lines joining the circles. The interatomic distances for the bases uracil (U), cytosine (C) and thymine (T) which take part in DNA and RNA formation are shown in angstroms. All atoms lie on the same plane, except for the two H atoms in the NH_2 unit bonded to C4 in cytosine and the three H atoms in the CH_3 unit bonded to C5 in thymine. All distances are indicative of typical bond lengths which vary somewhat depending on the environment; the C–H bond distance is 1.10 Å and the N–H bond distance is 1.00 Å. In the case of U, C and T the H atom normally attached to N1 is replaced by the symbol ⊗, denoting the C atom that connects the base to the sugar-phosphate backbone; the distance indicated in these cases corresponds to the N1–C bond.

atoms in the CH_3 unit of T. This makes it possible to stack these molecules on top of each other so that the van der Waals type attraction between them, similar to the interaction between graphitic planes, produces low-energy configurations. The attractive stacking interaction between these bases is one of the factors that stabilize the DNA polymer.

It is also worth noting here that the bases are typically bonded to other molecules, the sugars, forming units that are called **nucleosides**. For the bases in the pyrimidine class, the sugar molecule is always bonded to the N1 atom in the hexagonal ring. A nucleoside combined with one or more phosphate molecules is called a **nucleotide**; nucleotides are linked together through covalent bonds to form the extremely long chains of the DNA and RNA macromolecules, as will be explained below. Thus, what is more interesting is the structure of the bases in the nucleoside configuration rather than in their pure form: this structure is given in Fig. 13.14 for the bases *U*, *C* and *T*.

Purine is the simplest molecule in the second class of bases; it also contains only C, N and H atoms, but has slightly more complex structure than pyrimidine: it consists of a hexagon and a pentagon which share one side, as shown in Fig. 13.15. In the purine molecule there are four H atoms, two attached to C atoms in the

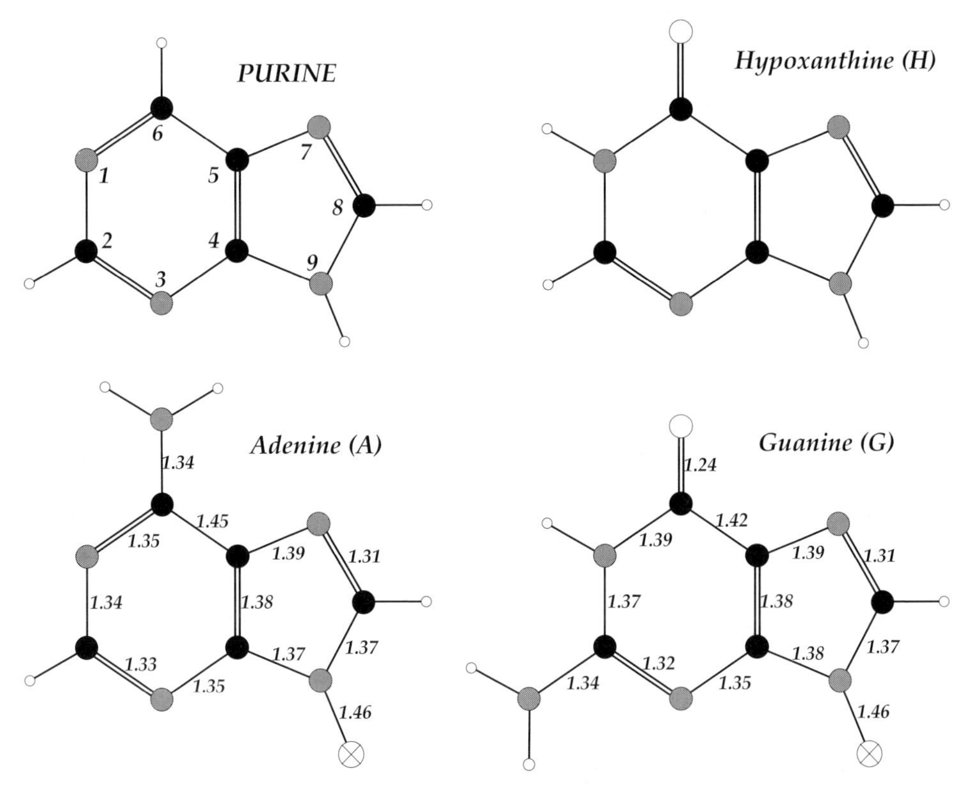

Figure 13.15. The structure of the neutral bases in the purine class. The notation and symbols are the same as in the pyrimidine class. The interatomic distances for the bases adenine (*A*) and guanine (*G*) are shown in angstroms. All atoms lie on the same plane, except for the two H atoms of the NH_2 units attached to C6 in adenine and to C2 in guanine. All distances are indicative of typical bond lengths which vary somewhat depending on the environment; the C–H bond distance is 1.10 Å and the N–H bond distance is 1.00 Å. In the case of *A* and *G* the H atom normally attached to N9 is replaced by the symbol ⊗, denoting the C atom that connects to the sugar-phosphate backbone; the distance indicated in these cases corresponds to the N9–C bond.

hexagon and two in the pentagon, one attached to a N atom and one to a C atom. There are four double bonds in total, three between C–N pairs and one between the two C atoms that form the common side of the pentagon and the hexagon. The atoms in the hexagon and the pentagon are numbered 1–9, as indicated in Fig. 13.15.

There are three interesting derivatives of purine.

Hypoxanthine (*H*), in which one of the H atoms of purine which was bonded to a C atom in the hexagon is replaced by an O atom; this eliminates one of the double bonds in the hexagon.
Adenine (*A*), in which one of the H atoms of purine which was bonded to a C atom in the hexagon is replaced by a NH_2 unit.
Guanine (*G*), in which one of the H atoms of purine which was bonded to a C atom in the hexagon is replaced by an O atom, eliminating one of the double bonds in the hexagon, and another is replaced by a NH_2 unit.

The four molecules in the purine class are also planar (except for the two H atoms of the NH_2 units in *A* and *G*), leading to efficient stacking interactions. These molecules form nucleosides by bonding to a sugar molecule at the N9 position. The structure of *A* and *G* in nucleosides is shown in Fig. 13.15.

In addition to the attractive stacking interactions, the bases can form hydrogen bonds between a N$-$H unit on one base and a C$=$O unit in another or between a N$-$H unit in one base and a two-fold coordinated N atom in another. In particular, the TA pair can form two hydrogen bonds, one between a N$-$H and a C$=$O unit at a distance of 2.82 Å, and one between a N$-$H unit and a two-fold bonded N atom at a distance of 2.91 Å. These distances are not very different from the O$-$O distance between water molecules which are hydrogen bonded in ice: in that case the O$-$O distance was 2.75 Å as discussed in chapter 1. As a result of hydrogen bonding, the sites at which the *T* and *A* bases are attached to the sugar molecules happen to be 10.85 Å apart (see Fig. 13.16). In the CG pair, there are three hydrogen bonds, two of the N$-$H-to-C$=$O type and one of the N$-$H-to-N type, with bond lengths of 2.84 Å and 2.92 Å, respectively. For the CG pair of bases, the distance between the sites at which they are bonded to the sugar molecules is 10.85 Å (see Fig. 13.16), exactly equal to the corresponding distance for the TA pair. This remarkable coincidence makes it possible to stack the CG and TA pairs exactly on top of each other with the sugar molecules forming the backbone of the structure, as discussed in more detail below. Such stacked pairs form DNA, while in RNA *T* is replaced by *U* in the base pair, and the sugar involved in the backbone of the molecule is different.

The strength of hydrogen bonds in the CG and TA pairs is approximately 0.15 eV (or 3 kcal/mol, the more commonly used unit for energy comparisons in biological systems; 1 eV = 23.06 kcal/mol). For comparison, the stacking energy of various

Table 13.2. *Comparison of stacking energies of base pairs and single covalent bond energies between pairs of atoms.*

	Stacking energy			Bond energy	
Base pair	(eV)	(kcal/mol)	Atom pair	(eV)	(kcal/mol)
$GC \cdot GC$	0.633	14.59	H–H	4.52	104.2
$AC \cdot GT$	0.456	10.51	C–H	4.28	98.8
$TC \cdot GA$	0.425	9.81	O–H	4.80	110.6
$CG \cdot CG$	0.420	9.69	N–H	4.05	93.4
$GG \cdot CC$	0.358	8.26	C–C	3.60	83.1
$AT \cdot AT$	0.285	6.57	C–N	3.02	69.7
$TG \cdot CA$	0.285	6.57	C–O	3.64	84.0
$AG \cdot CT$	0.294	6.78	C–S	2.69	62.0
$AA \cdot TT$	0.233	5.37	S–S	2.21	50.9
$TA \cdot TA$	0.166	3.82	O–O	1.44	30.2

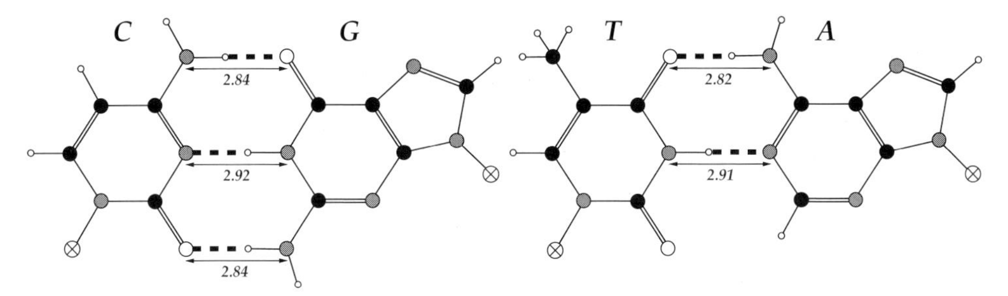

Figure 13.16. Illustration of hydrogen bonding between the CG and TA base pairs: the distances between the N–O atoms among which the hydrogen bonds are formed are given in angstroms. The symbols and notation are the same as in Figs. 13.14 and 13.15. There are three hydrogen bonds in the CG pair and two hydrogen bonds in the TA pair. The symbols ⊗ attached to the lower left corner of C and T and to the lower right corner of the pentagon of G and A represent the C atoms of the sugars through which the bases connect to the sugar-phosphate backbone. The distance between the sites denoted by ⊗ is 10.85 Å in both the CG and TA base pairs. U and A form a hydrogen-bonded complex similar to TA.

base pairs ranges from 4 to 15 kcal/mol, while the strength of single covalent bonds between pairs of atoms that appear in biological structures ranges from 30 to 110 kcal/mol. In Table 13.2 we give the values of the stacking energy for various base pairs and those of typical single covalent bonds.

One important aspect of the bases is that they also exist in forms which involve a change in the position of a H atom relative to those described above. These forms are called **tautomers** and have slightly higher energy than the stable forms described already, that is, the tautomers are metastable. The stable forms are referred to as **keto**, when the H atom in question is part of an NH group that has two more

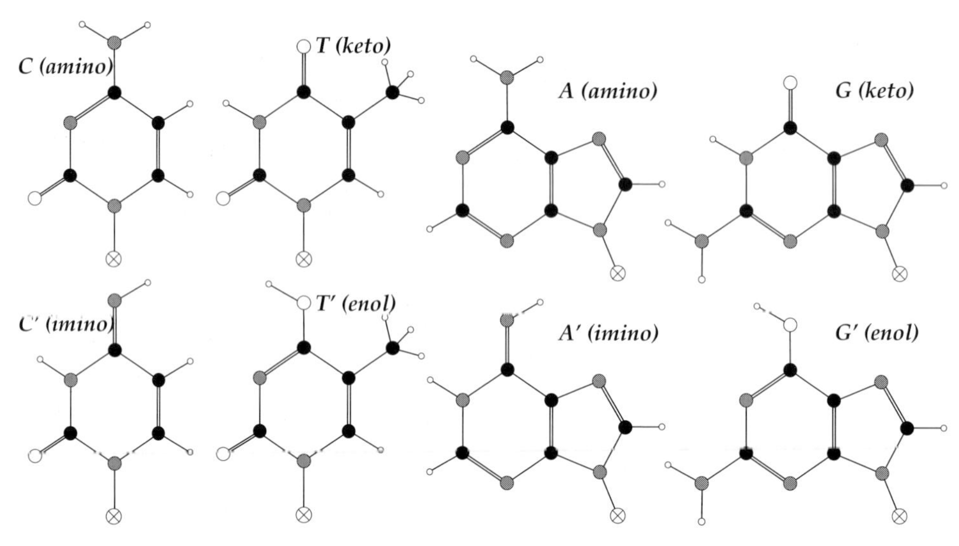

Figure 13.17. The four bases of DNA in their normal form (top row) and tautomeric form (bottom row). The tautomers involve a change in the position of a H atom from an NH group (keto form) to an OH group (enol form) or from an NH_2 group (amine form) to an NH group (imine form). The change in position of the H atom is accompanied by changes in bonding of the atoms related to it.

covalent bonds to other atoms in the ring of the base, or **amino** when the H atom is part of an NH_2 group that has one more covalent bond to another atom in the ring. The metastable forms are referred to as **enol** when the H atom that has left the NH group is attached to an O atom to form an OH group, or **imino** when the H atom that has left the NH_2 group is attached to a N atom to form an NH group. In both cases, the change in position of the H atom is accompanied by changes in the bonding of the atoms related to it in the stable or the metastable form. All this is illustrated in Fig. 13.17 for the amino, imino, keto and enol forms of the four bases that participate in the formation of DNA.

The tautomers can also form hydrogen-bonded pairs with the regular bases, but because of their slightly different structure they form "wrong" pairs: specifically, instead of the regular pairs CG and TA, the tautomers lead to pairs $C'A$ or CA' and $T'G$ or TG' where we have used primed symbols to denote the tautomeric forms. These pairs look structurally very similar to the regular pairs in terms of the number and arrangement of the hydrogen bonds between the two bases involved, as illustrated in Fig. 13.18. The significance of the wrong pairs due to the presence of tautomers is that when DNA is transcribed the wrong pairs can lead to the wrong message (see section below on the relation between DNA, RNA and proteins). It has been suggested that such errors in DNA transcription are related to mutations, although it appears that the simple errors introduced by the tautomeric bases are

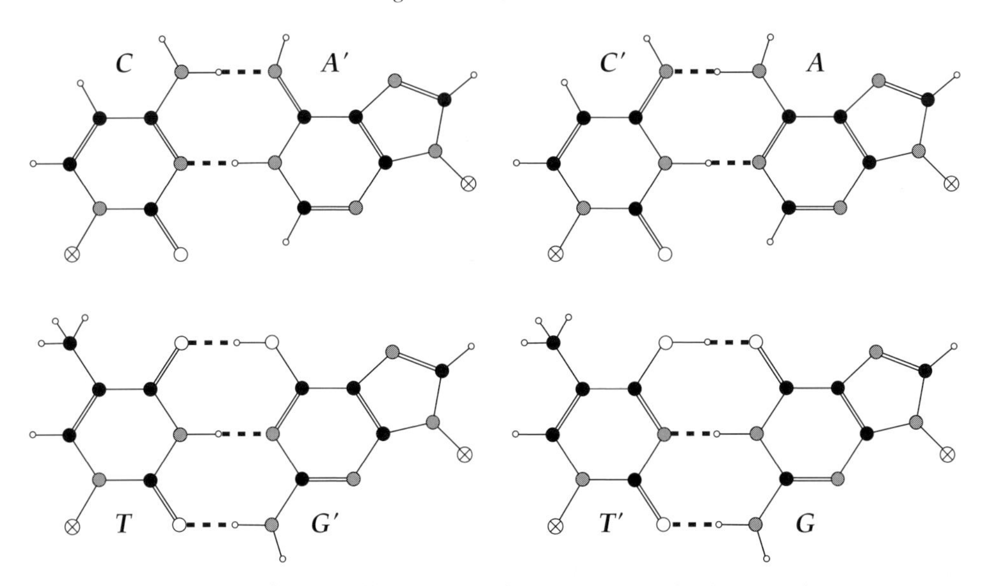

Figure 13.18. The four possible wrong pairs between regular bases and tautomers.

readily corrected by certain enzymes and it is only more severe changes in the base structure that can lead to mutations.

It is worth mentioning that the side of the bases which does not participate in formation of base pairs by hydrogen bonding is also very important. The atoms on that side can also form hydrogen bonds to other molecules on the outer side of the DNA double helix. The number and direction of these outward hydrogen bonds can serve to identify the type of base within the helix. This, in turn, can be very useful in identifying the sequence of bases along a portion of the double helix, a feature that can promote specific interactions between DNA and proteins or enzymes which result in important biological functions.

The backbone

To complete the discussion of the structure of the DNA and RNA macromolecules, we consider next the sugar and phosphate molecules which form the backbone. The sugar molecules in DNA and RNA are shown in Fig. 13.19. These are also simple structures composed of C, O and H. The basic unit is a pentagonal ring formed by four C and one O atom. To facilitate the description of the structure of DNA and RNA, the C atoms in the sugar molecules are numbered in a counterclockwise sense, starting with the C atom in the pentagonal ring to the right of the O atom; the numbers in this case are primed to distinguish them from the labels of the atoms in the bases themselves. There exist right-handed or left-handed isomers of the sugar molecules which are labeled D and L. The **ribose** sugar has OH units attached to

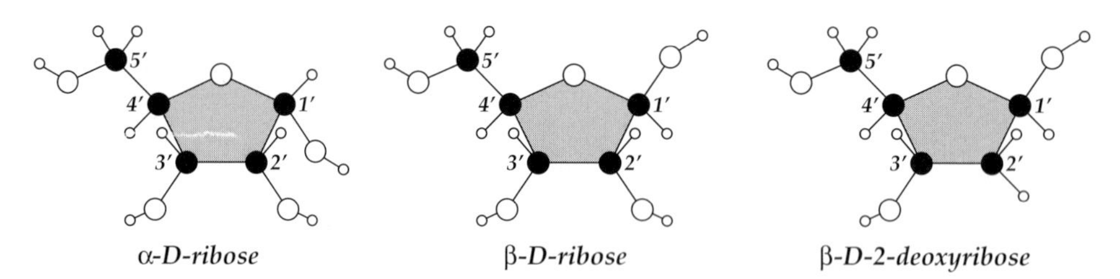

Figure 13.19. The structure of the α-D-ribose, β-D-ribose, and β-D-2-deoxyribose molecules: black circles represent C atoms, open circles O atoms and smaller open circles H atoms. The planar pentagonal unit consisting of one O and four C atoms is shaded. The conventional labels of the C atoms are also indicated, in a counterclockwise direction starting with the C bonded to the right of the O atom in the pentagonal ring. β-D-2-deoxyribose is missing one O atom in comparison to β-D-ribose, which was bonded to the C2′ atom.

the C1′, C2′ and C3′ atoms and a CH_2OH unit attached to the C4′ atom (the carbon atom in that unit is labeled C5′). There are two versions of this molecule, depending on whether the OH units attached to the C1′, C2′ and C3′ atoms lie on the same side of the pentagonal ring (called α isomer) or not (called β isomer). The structures of the α-D-ribose and β-D-ribose molecules are shown in Fig. 13.19. Another sugar that enters the structure of the biological macromolecules is β-D-2-deoxyribose. This has a very simple relation to β-D-ribose, namely they are the same except for one O atom which is missing from position C2′ in deoxyribose (hence the name of this molecule).

When a base is bonded to a sugar molecule, the resulting nucleoside is referred to by the name of the base with the ending changed to "sine" or "dine": thus, the bases adenine, guanine, cytosine, uracil and thymine produce the nucleosides **adenosine, guanosine, cytidine, uridine** and **thymidine**.

The sugar molecules form links to PO_4^{3-} units that correspond to ionized phosphoric acid molecules. It is worthwhile exploring this structure a bit further, since it has certain analogies to the structure of group-V solids discussed in chapter 1. The phosphoric acid molecule H_3PO_4 consists of a P atom bonded to three OH groups by single covalent bonds and to another O atom by a double bond, as shown in Fig. 13.20. This is consistent with the valence of P, which possesses five valence electrons. Of these, it shares three with the O molecules of the OH units in the formation of the single covalent bonds. The other two electrons of P are in the so called "lone-pair" orbital, which is used to form the double, partially ionic, bond to the fourth O atom. This arrangement places the P atom at the center of a tetrahedron of O atoms; since these atoms are not chemically equivalent, the tetrahedron is slightly distorted from its perfect geometrical shape. The O atoms in the OH units have two covalent bonds each, one to P and one to H, thus producing a stable structure.

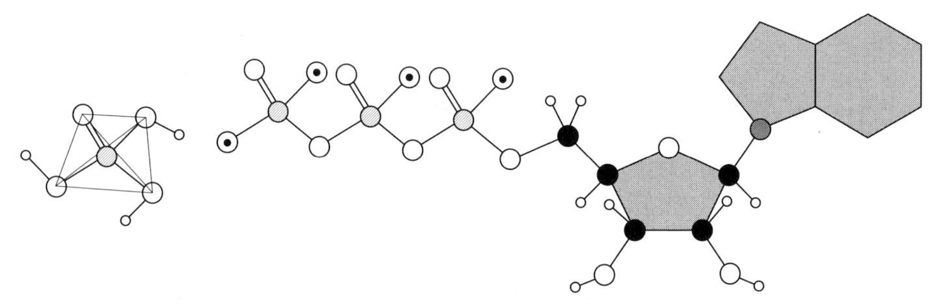

Figure 13.20. **Left:** perspective view of a phosphoric acid, H_3PO_4, molecule. The thin lines outline the distorted tetrahedron whose center is occupied by the P atom, shown as a shaded circle. The other symbols are the same as in Fig. 13.19. **Right:** the structure of adenosine tri-phosphate (ATP). The O atoms which have only one covalent bond and an extra electron are denoted by smaller filled circles at their center. The adenine base is shown schematically as a gray unit consisting of a hexagon and pentagon joined along one side; it is attached through its N9 site to the C1′ site of the β-D-ribose sugar.

In water solution, the protons of the H atoms in the OH units can be easily removed, producing a charged structure in which three of the O atoms are negatively charged, having kept the electrons of the H atoms. This structure can react with the OH unit attached to the C5′ atom of the ribose sugar, producing a structure with one or more phosphate units, as shown in Fig. 13.20. The O atoms in the phosphate units which do not have a double bond to the P atom and do not have another P neighbor are negatively charged, that is, they have one extra electron. The combinations of phosphate units plus sugar plus base, i.e. the nucleotides, are denoted by the first letter of the nucleoside and the number of phosphate units: MP for one (mono-phosphate), DP for two (di-phosphate) and TP for three (tri-phosphate). For instance, ATP, adenosine tri-phosphate, is a molecule with a very important biological function: it is the molecule whose breakdown into simpler units provides energy to the cells.

Two nucleotides can be linked together with one phosphate unit bonded to the C5′ atom of the first β-D-2-deoxyribose and the C3′ atom of the second. This is illustrated in Fig. 13.21. The bonding produces a phosphate unit shared between the two nucleotides which has only one ionized O atom. The remaining phosphate unit attached to the C5′ atom of the second nucleotide has two ionized O atoms. This process can be repeated, with nucleotides bonding always by sharing phosphate units between the C3′ and C5′ sites. This leads to a long chain with one C3′ and one C5′ end which can be extended indefinitely. A single strand of the DNA macromolecule is exactly such a long chain of nucleotides with the β-D-2-deoxyribose sugar and the T, A, C, G bases; the acronym DNA stands for **deoxyribose nucleic acid**. RNA is a similar polynucleotide chain, with the β-D-ribose sugar and the U, A, C, G bases; the acronym RNA stands for **ribose nucleic acid**. Both molecules are

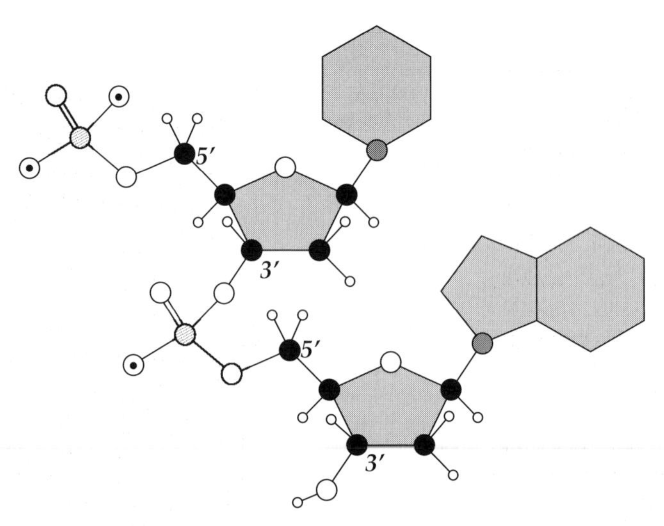

Figure 13.21. Formation of a chain by linking β-D-2-deoxyribose molecules with PO_4^{3-} units. The symbols are the same as in Fig. 13.20. The linkage takes place between the O atoms attached to the C3′ atom of one sugar molecule and the C5′ atom of the other sugar molecule, leaving one C3′ and one C5′ atom at either end free for further extension of the chain; this is called phosphodiester linkage. The β-D-2-deoxyribose molecules are shown with bases (one of the pyrimidine and one of the purine class) attached to them, as they would exist in nucleosides.

referred to as acids because of the negatively charged O atoms in the sugar-phosphate backbone.

The double helix

Since the bases can form hydrogen-bonded pairs in the combinations TA (or UA in RNA) and CG, a given sequence of nucleotides will have its complementary sequence to which it can be hydrogen bonded. Two such complementary sequences form the DNA right-handed double helix, in which both hydrogen bonding and the attractive stacking interactions between the bases are optimized. This structure is illustrated in Fig. 13.22.

There are several types of DNA double helices, distinguished by structural differences in the specific features of the helix, such as the diameter, pitch (i.e. the period along the axis of the helix) and the number of base pairs within one period of the helix. For example, in the common form, called B-DNA, there are 10.5 base pairs per period along the axis with a distance between them of 3.4 Å which results in a diameter of 20 Å, a pitch of 35.7 Å and a rotation angle of 34.3° between subsequent base pairs. This form is referred to as the Watson–Crick model after the scientists who first elucidated its structure. The work of F.H.C. Crick, J.D. Watson and M.H.F. Wilkins was recognized by the 1962 Nobel prize for Medicine as a

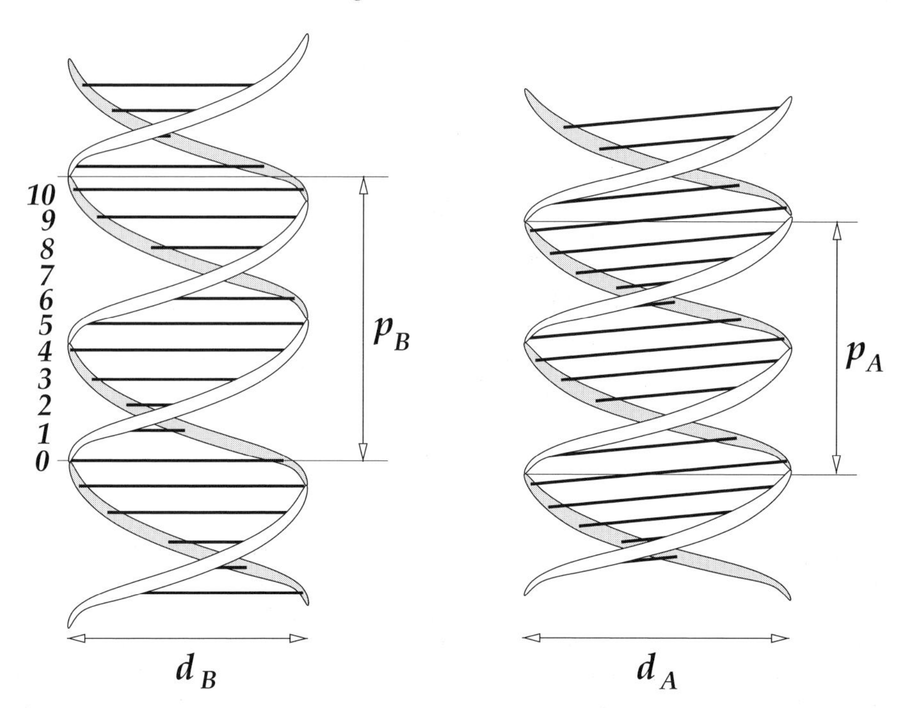

Figure 13.22. Schematic representation of the main features of the two most common forms of DNA double helices: B-DNA (**left**) and A-DNA (**right**). B-DNA has a diameter $d_B = 20$ Å, a pitch $p_B = 35.7$ Å, and there are 10.5 base pairs per full turn of the helix, corresponding to a rotation angle of 34.3° and an axial rise (i.e. distance along the axis between adjacent pairs) of 3.4 Å (the base pairs are indicated by horizontal lines and numbered from 0 to 10 within one full helical turn). A-DNA, which is less common, has a diameter $d_A = 23$ Å, a pitch $p_A = 28$ Å, and there are 11 base pairs per full turn of the helix, corresponding to a rotation angle of 33° and an axial rise of 2.55 Å. In B-DNA the planes of the bases are tilted by a small angle (6°) relative to the axis of the helix, which is not shown; in A-DNA the tilt of the base planes relative to the axis is much larger (20°), as indicated. The overall shape of B-DNA has two pronounced grooves of different size (the major and minor groove); in A-DNA the grooves are not very different in size but one is much deeper than the other.

major breakthrough in understanding the molecular design of life.[5] A less common form, called A-DNA, with somewhat different features of the helix is also shown in Fig. 13.22. In B-DNA there are pronounced grooves of two different sizes, called the major and minor groove. The grooves are also present in A-DNA but their size is almost the same, while their depth is very different: one of the two grooves penetrates even deeper than the central axis of the helix. Another difference

[5] It is of historical interest that the scientist who did the pioneering X-ray studies of DNA which led to the understanding of its structure was Rosalind Franklin, who died at age 37 before the Nobel prize for the DNA discovery was awarded; a personal and very accessible account of the research that led to the discovery is given by J.D. Watson in his short book *The Double Helix: A Personal Account of the Discovery of the Structure of DNA* [231].

between these two forms is that the planes of the bases are tilted by a small angle (6°) relative to the axis of the helix in B-DNA but by a much larger angle (20°) in A-DNA. All these differences have important consequences in how DNA reacts with its environment. There are several other less common forms of DNA. One interesting form is called Z-DNA and is a left-handed helix. In contrast to DNA, pairs of single-strand RNA molecules typically do not form a double helix; however, a single-strand RNA molecule can fold back onto itself and form hydrogen bonds between complementary bases (UA or CG) which lie far from each other along the linear sequence in the polynucleotide chain.

Although we have described its structure as if it were an infinite 1D structure, DNA is usually coiled and twisted in various ways. A finite section of DNA often forms a closed loop. In such a loop, one can define the "linking number" L, which is the number of times one strand crosses the other if the DNA molecule were made to lie flat on a plane. Assuming that we are dealing with B-DNA in its relaxed form, the equilibrium linking number is given by

$$L_0 = \frac{N_{bp}}{10.5}$$

where N_{bp} is the number of base pairs in the molecule. When the linking number L is different than L_0 the DNA is called "supercoiled". If $L < L_0$ it is underwound, that is, it has fewer helical turns than normal per unit length along its axis, and it is referred to as "negatively supercoiled". If $L > L_0$ it is overwound, with more helical turns than normal per unit length along its axis, and it is called "positively supercoiled". The supercoiling introduces torsional strain in the double helix which can be relieved by writhing. This feature is characterized by another variable, the "writhing number" W. If we define the total number of helical turns as T, then the three variables are connected by the simple equation

$$L = T + W$$

Typically, W is such that it restores the total number of turns T in the supercoiled DNA to what it should be in the relaxed form. Thus, negative supercoiling is characterized by a negative value of W and positive supercoiling by a positive value. For example, assume that we are dealing with a DNA section containing $N_{bp} = 840$ base pairs, therefore $L_0 = 80$. A negatively supercoiled molecule with one fewer turn per ten repeat units (105 bp) would give $L = 72$; the total number of turns can be restored by introducing eight additional crossings of the double helix in the same right-handed sense, producing a structure with $T = 80$ and $W = -8$. Alternatively, the same DNA section could be positively supercoiled with one extra turn per 20 repeat units (210 bp), giving $L = 84$; the total number of turns can be restored by introducing four additional crossings in the opposite (left-handed) sense,

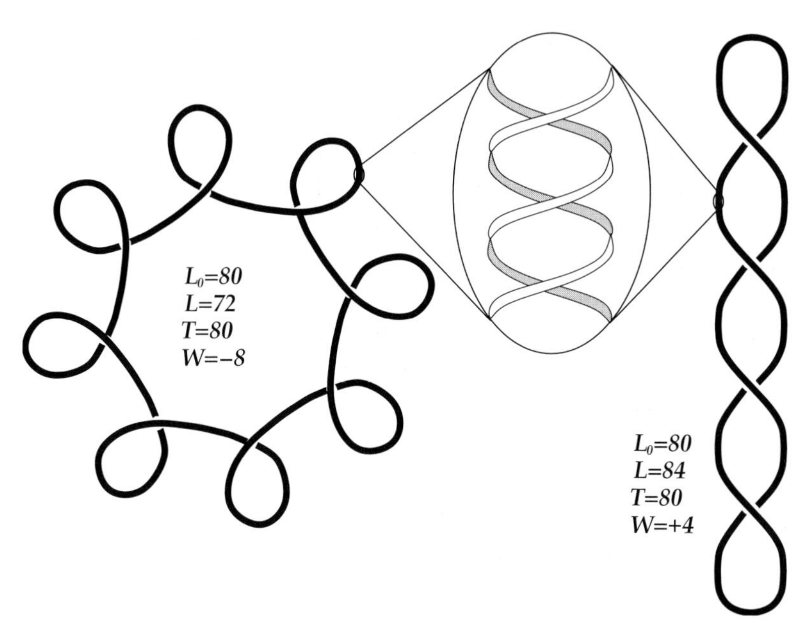

Figure 13.23. Supercoiled DNA in toroidal (**left**) and interwound (**right**) forms. The example on the left is negatively supercoiled (underwound, $L < L_0$), while the one on the right is positively supercoiled (overwound, $L > L_0$). The sense of winding is determined by the orientation of the underlying double helix, as indicated in the magnified section.

producing a structure with $T = 80$ and $W = +4$. The writhing can be introduced in toroidal or interwound forms, as illustrated in Fig. 13.23. The interwound form has lower energy because it introduces smaller curvature along the molecule except at the ends (called apical loops), and is therefore the more common form of supercoiled DNA.

DNA can also be bent and coiled by interacting with proteins. A particular set of proteins, called **histones**, produces a very compact, tightly bound unit, which is very common when DNA is not transcriptionally active (this is explained in the following section). The organization of the DNA molecule into these compact units makes it possible for it to fit inside the nucleus of eucaryotic cells. In the most common units there are eight histones that form the **nucleosome** around which a section of DNA about 146 base pairs long is wrapped. The nucleosome has the shape of a cylinder with a diameter of 110 Å and a length of 57 Å. One more histone provides the means for wrapping an additional DNA section of 21 base pairs; this tightly wrapped total of 167 DNA base pairs and the nine histones compose the **chromatosome**. The DNA section in the chromatosome is almost 16 full periods long and forms two complete helical turns around the nucleosome. This is illustrated in Fig. 13.24.

The electronic states associated with DNA structures have become the subject of intense investigation recently, since these macromolecules are being considered as

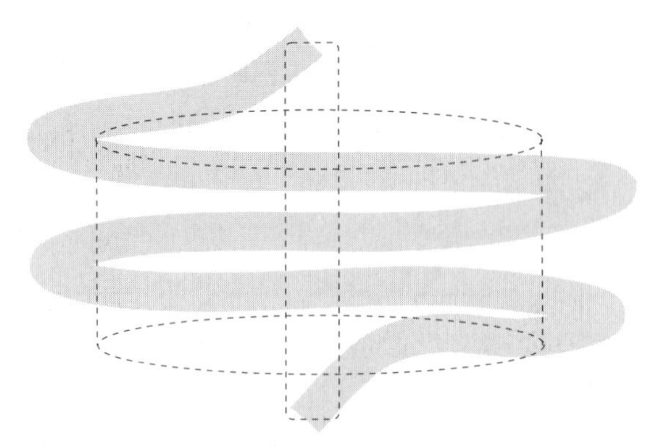

Figure 13.24. Illustration of DNA coiling in the chromatosome. The cylinder represents the nucleosome, which has a diameter of 110 Å and a length of 57 Å and consists of eight histones. A section of DNA (the thick grey line) consisting of 146 base pairs forms 1.75 helical turns arounds this cylinder. An additional 21 base pairs, bound to the ninth histone, which is denoted by the vertical box, complete a total of two helical turns. The length of the coiled DNA section is 167 base pairs or almost 16 repeat periods (16×10.5 bp $= 168$ bp).

building blocks of future electronic devices. At present it is not even clear whether the macromolecules have metallic or semiconducting character. Electronic structure calculations of the type discussed in detail in chapter 4 reveal that simple periodic DNA structures in a dry, isolated state have a large band gap of about 2 eV [232]. These calculations also show that electronic states near the top of the occupied manifold are associated with the G and A bases, whereas those near the bottom of the unoccupied manifold are associated with the C and T bases. It is, however, unclear to what extent the electronic behavior of DNA molecules is influenced by the environment, including the water molecules and other ions commonly found in the solution where DNA exists, or by the actual configuration of the helical chain, which may be stretched or folded in various ways with little cost in energy, or even by their interaction with the contact leads employed in the electrical measurements.

13.2.2 The structure of proteins

Proteins are the macromolecules responsible for essentially all biological processes.[6] Proteins are composed of aminoacids, which are relatively simple molecules; the aminoacids are linked together by peptide bonds. The exact structure of proteins has several levels of complexity, which we will discuss briefly at the end of the present subsection. We consider first the structure of aminoacids and the way in which they are linked to form proteins.

The general structure of aminoacids is shown in Fig. 13.25: there is a central C atom with four bonds to: (1) an **amino** (NH_2) group; (2) a **carboxyl** (COOH)

[6] The name protein derives from the Greek word πρωτειος (proteios), which means "of the first rank".

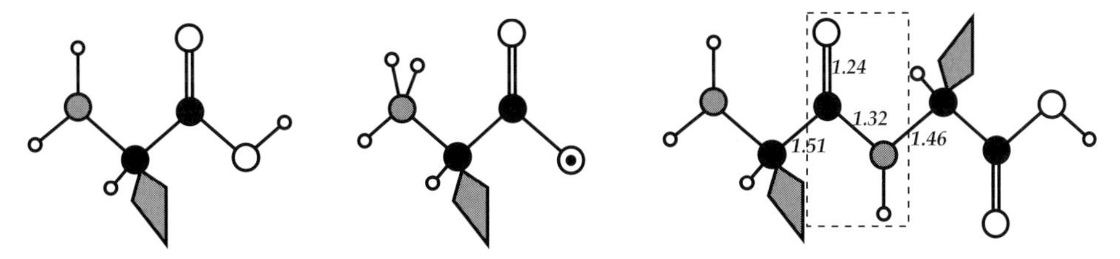

Figure 13.25. The peptide bond in aminoacids: **Left:** the general structure of the aminoacid, with black circles representing C atoms, open circles O atoms, gray circles N atoms and small open circles H atoms; the gray rhombus represents the side chain of the aminoacid. **Center:** the aminoacid is shown in its ionized state at neutral pH: the amino group (NH_2) has gained a proton and is positively charged, while the carboxyl group (COOH) has lost a proton leaving one of the O atoms negatively charged, indicated by a smaller filled circle at its center. **Right:** two aminoacids are bonded through a peptide bond, which involves a C=O and an N−H group in a planar configuration, identified by the rectangle in dashed lines. The distances between the C and N atoms involved in the peptide bond and their neighbors are shown in angstroms.

group; (3) a specific side chain of atoms that uniquely identifies the aminoacid; and (4) a H atom. We discuss the features of these subunits (other than the H atom) in some detail as follows.

(1) The N atom in the amino group is three-fold bonded with covalent bonds to the central C atom and to two H atoms. In this arrangement, the N atom uses three of its five valence electrons to form covalent bonds and retains the two other electrons in a lone-pair orbital, in the usual bonding pattern of group-V elements; in the case of N, this bonding arrangement is close to planar, with the lone-pair electrons occupying predominantly a p orbital which is perpendicular to the plane of the covalent bonds.
(2) In the carboxyl group, the C atom forms a double bond with one of the oxygen atoms and a single bond to the other oxygen atom; taking into account the bond to the central C atom, we conclude that the C atom in the carboxyl group is three-fold coordinated in a planar configuration, which leaves a p orbital perpendicular to the plane of the covalent bonds. The O atom which has a single bond to the C atom of the carboxyl unit is also covalently bonded to a H atom, which thus satisfies its valence.
(3) The side chain is bonded to the central C atom by a single covalent bond. The simplest chain consists of a single H atom, while some of the more complicated chains contain C, H, O, N and S atoms; the chains are referred to as **residues**.

There are a total of 20 different residues, giving rise to 20 different aminoacids, which are shown in Fig. 13.26. The aminoacids are usually referred to by a single-letter abbreviation; a different labeling scheme uses the first three letters of their names. In the following, we will use the latter scheme to avoid any confusion with the single-letter labels introduced for the bases; the three-letter symbols and the full names of the 20 aminoacids are given in Table 13.3.

Under normal conditions in an aqueous solution the aminoacids are ionized, with one proton removed from the OH of the carboxyl group and transferred to

Table 13.3. *Names of the 20 aminoacids and their three-letter symbols.*

Ala: alanine	**Arg:** arginine	**Asn:** asparagine	**Asp:** aspartic acid
Gly: glycine	**Cys:** cysteine	**Gln:** glutamine	**Glu:** glutamic acid
Lys: lysine	**His:** histidine	**Leu:** leucine	**Ile:** isoleucine
Ser: serine	**Pro:** proline	**Met:** methionine	**Phe:** phenylalanine
Val: valine	**Tyr:** tyrosine	**Thr:** threonine	**Trp:** tryptophan

the amino group; thus, the carboxyl group is negatively charged while the amino group is positively charged. In certain aminoacids, such as aspartic acid and glutamic acid, there are protons missing from the carboxyls in the residue which have an additional negative charge. In certain other aminoacids, namely histidine, arginine and lysine, there are amino groups in the residue which can gain a proton to become positively charged and thus act as bases. These structures are shown in the last row of Fig. 13.26.

Two aminoacids can be linked together by forming a **peptide bond**: the amino group of one aminoacid reacts with the carboxyl group of a different aminoacid to form a covalent bond between the N and the C atoms. The amino group loses a H atom and the carboxyl group loses an OH unit: the net result is the formation of a water molecule during the reaction. The formation of peptide bonds can be repeated, leading to very long chains, called **polypeptides**; proteins are such polypeptide sequences of aminoacids.

The peptide bond has the property that the C=O and N−H units involved in it form a planar configuration, illustrated in Fig. 13.25. This is a result of π-bonding between the p orbitals of the N and C atoms, which are both perpendicular to the plane of the covalent bonds formed by these two atoms; any deviation from the planar configuration disrupts the π-bond, which costs significant energy. Thus the planar section of the peptide bond is very rigid. In contrast to this, there is little hindrance to rotation around the covalent bonds between the central C atom and its N and C neighbors. This allows for considerable freedom in the structure of proteins: the aminoacids can rotate almost freely around the two covalent bonds of the central C atom to which the chain is linked, while maintaining the planar character of the peptide bonds. This flexible structural pattern leads to a wonderful diversity of large-scale structures, which in turn determines the amazing diversity of protein functions. Some basic features of this range of structures are described next.

We notice first that the H atom of the N−H unit and the O atom of the C=O unit in two different polypeptide chains can form hydrogen bonds, in close analogy to the hydrogen bonds formed by the bases in the double helix of DNA. The simplest arrangement which allows for hydrogen bonding between all the C=O and N−H units in neighboring polypeptides is shown in Fig. 13.27. This arrangement can

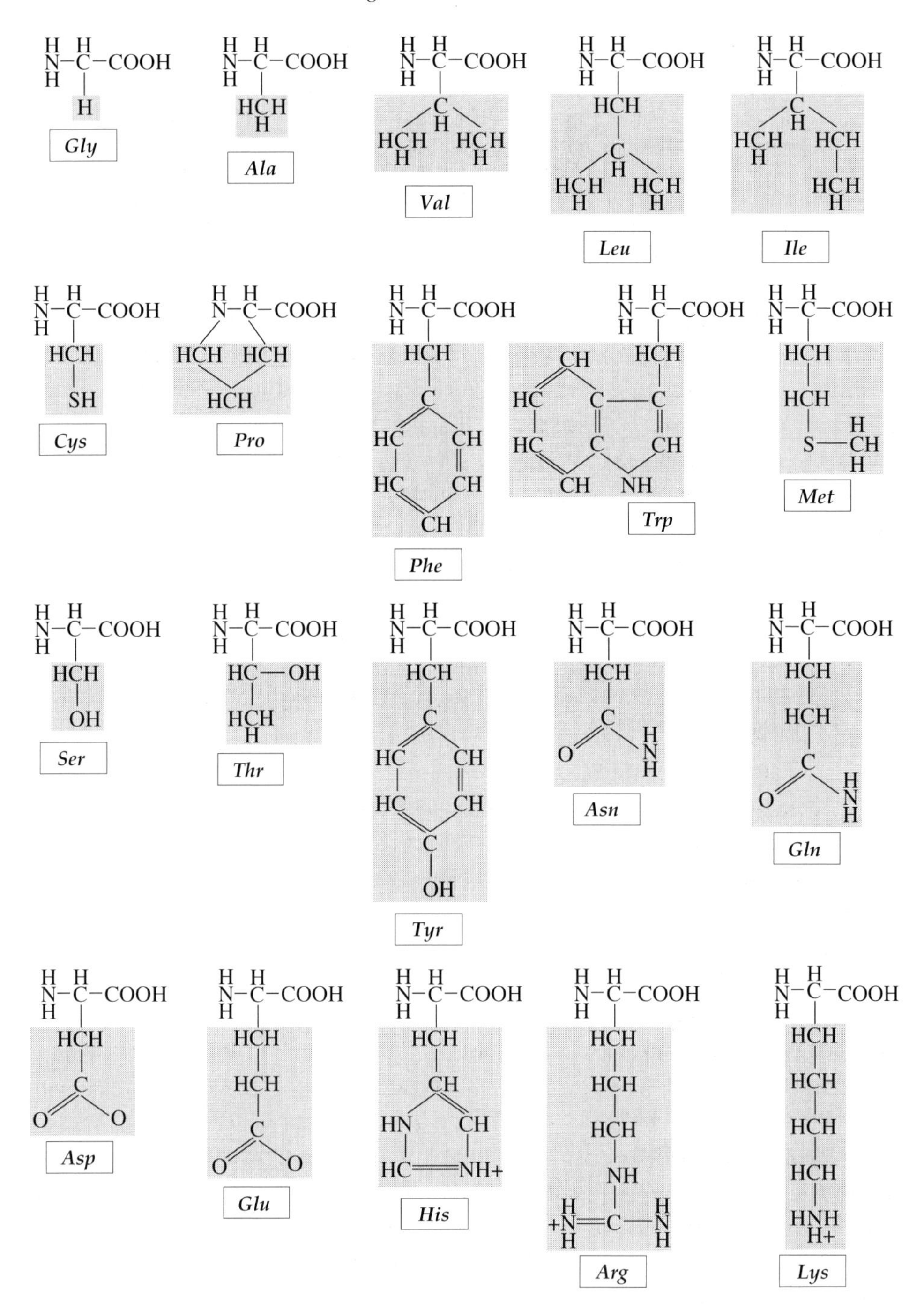

Figure 13.26. Schematic representation of the structure of the 20 aminoacids. The side chains are highlighted. Covalent bonds are shown as lines, except for the bonds between H and other atoms.

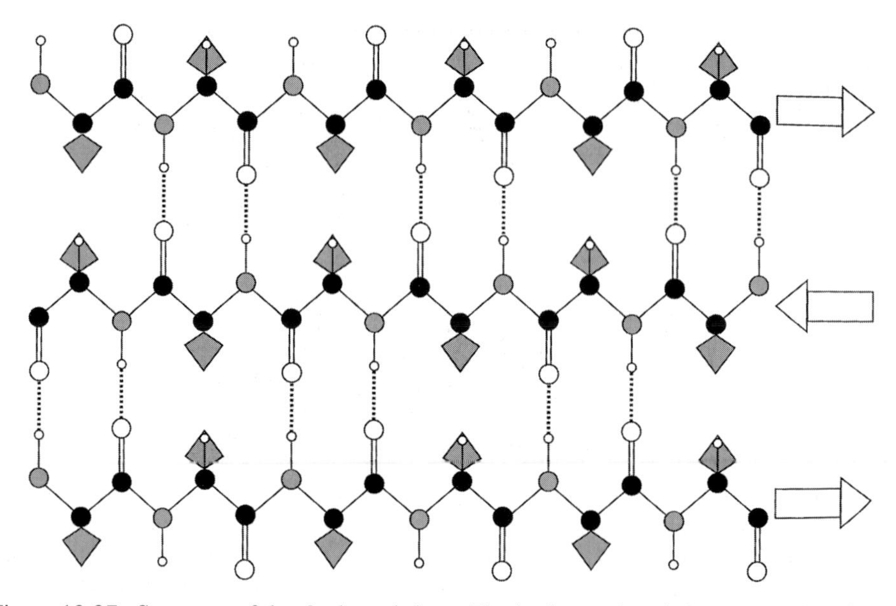

Figure 13.27. Structure of the β-pleated sheet. The hydrogen bonds between pairs of C=O and N−H units are indicated by thick dashed lines. Consecutive chains run in opposite directions, as indicated by the arrows on the right. The aminoacid residues are indicated by the shaded rhombuses which are above or below the plane of the sheet. In this perspective the H atom bonded to the central C atom is hidden by the aminoacid residue in half of the units, when the residue is in front of the C−H bond.

be repeated to form hydrogen bonds across several polypeptide chains, which then form a sheet, referred to as the **β-pleated sheet**. Notice that if we assign a direction to a polypeptide chain, for instance going from the C=O to the N−H unit, then the next polypeptide chain in the β-pleated sheet must run in the opposite direction to allow for hydrogen bonding between all the peptide units, as shown in Fig. 13.27.

A different arrangement of the peptide units allows for hydrogen bonding *within* the same polypeptide chain. This is accomplished by forming a helical structure from the polypeptide chain, so that the N−H unit of a peptide bond next to a given residue can hydrogen bond to the C=O unit of the peptide bond three residues down the chain. This structure, which was predicted on theoretical grounds by L.C. Pauling, is called the **α-helix**. All the residues lie on the outside of the helix, which is very tightly wound to optimize the hydrogen-bonding interactions between the peptide units, as illustrated in Fig. 13.28.

The structure of a protein determines its overall shape and the profile it presents to the outside world. This in turn determines its biological function, through the types of chemical reactions it can perform. Thus, the structure of proteins is crucial in understanding their function. This structure is very complex: even though proteins are very long polypeptide molecules, in their natural stable form they have shapes

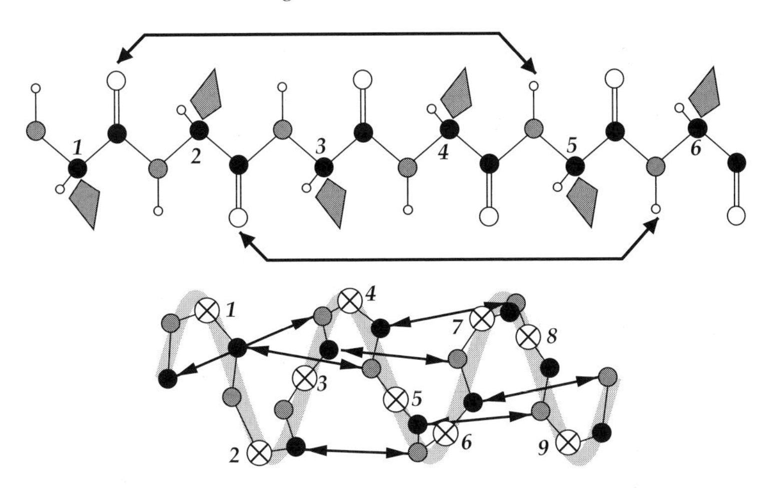

Figure 13.28. The structure of the α-helix: **Top:** a polypeptide chain stretched out with arrows indicating which pairs of C=O and N−H units are hydrogen bonded. **Bottom:** the chain coiled into a helix, with double-headed arrows indicating the hydrogen-bonded pairs of C=O and N−H units. In this figure the central C atoms to which the aminoacid chains are attached are indicated by the symbols $\otimes$; all the aminoacid residues lie on the outside of the helix.

that are very compact, with all the aminoacids neatly folded into densely packed structures; this is referred to as **protein folding**. Additional hydrogen bonds can be formed in folded proteins between the amino group of a peptide unit and the hydroxyl (OH) group of a residue, or between the hydroxyl group of one residue and the carboxyl group of another one. Yet a different type of bond that can be formed between aminoacids in a protein is the sulfide bond; this is a bond between two S atoms in different residues. In particular, the aminoacids Cys and Met contain S−H units in their residues, which can form a covalent bond once they have lost one of their H atoms. All these possibilities give rise to a great variety of structural patterns in proteins. For illustration, we show in Fig. 13.29 how several units, such as α-helices or sections of β-pleated sheets, can be joined by hairpin junctions along the protein to form compact substructures.

Here we describe briefly the four levels of structure in folded proteins.

- The **primary** structure is simply the aminoacid sequence, that is, the specific arrangement of the aminoacid residues along the polypeptide chain.
- The **secondary** structure is the more complicated spatial arrangements that aminoacids close to each other can form. The α-helix and β-pleated sheet are examples of secondary structure. There are usually many helices and sheets in a single protein.
- The **tertiary** structure refers to the spatial arrangement of aminoacids that are relatively far from each other so that they can not be considered parts of the same α-helix or β-pleated sheet. In essence, the tertiary structure refers to the 3D arrangement of α-helices, β-sheets

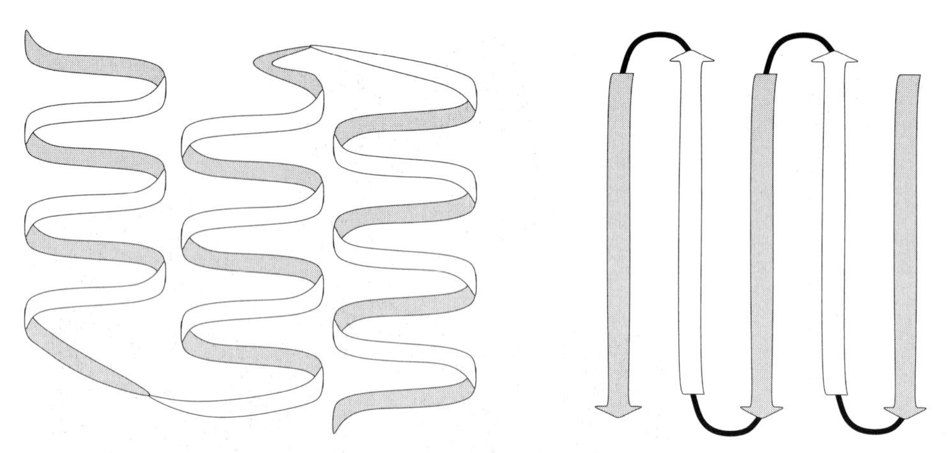

Figure 13.29. Examples of combinations of α-helices and β-pleated sheets joined by hairpin turns, as they might appear in folded proteins. The arrows in the β-pleated sheets indicate the directional character of this structure (see Fig. 13.27).

and the intervening portions which are not part of sheets or helices, among themselves to form the large-scale structure of the folded protein.

- The **quaternary** structure refers to proteins that have more than one large polypeptide subunit, each subunit representing a well-defined structural entity; the subunits are then organized into a single complex structure.

The four levels of structure are illustrated in Fig. 13.30 for the potassium channel KcsA protein [233]. It is quite remarkable that no matter how complicated the molecular structure of proteins is, they can quickly find their natural folded state once they have been synthesized. The way in which proteins fold into their stable structure when they are in their natural environment (aqueous solution of specific pH) remains one of the most intriguing and difficult unsolved problems in molecular biology and has attracted much attention in recent years.

13.2.3 Relationship between DNA, RNA and proteins

Even though we stated at the beginning of this section that it is not our purpose to discuss the biological function of the macromolecules described, it is hard to resist saying a few words about some of their most important interactions. First, it is easy to visualize how the structure of DNA leads to its replication and hence the transfer of genetic information which is coded in the sequence of bases along the chain. Specifically, when the DNA double helix opens up it exposes the sequence of bases in each strand of the molecule. In a solution where bases, sugars and phosphates are available or can be synthesized, it is possible to form the complementary strand of each of the two strands of the parent molecule. This results in

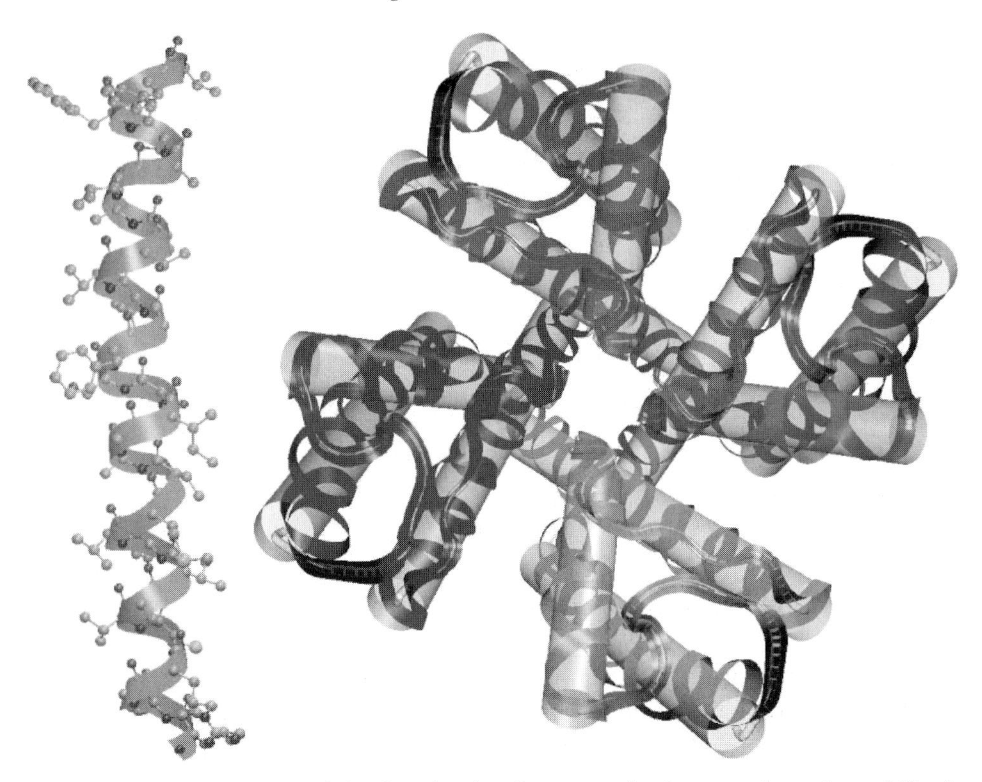

Figure 13.30. Illustration of the four levels of structure in the potassium channel KcsA protein. **Left:** a small section of the protein showing the primary structure, that is, the sequence of aminoacids identified by the side chains, and the secondary structure, that is, the formation of an α-helix. **Right:** the tertiary structure, that is, the formation of larger subunits conisting of several α helices (in this case each subunit consists of three helices), and the quaternary structure, that is, the arrangement of the three-helix subunits in a pattern with four-fold rotational symmetry around a central axis. The α-helix on the left is the longest helix in each of the subunits, which are each shown in slightly different shades. The ion channel that this protein forms is at the center of structure. (Figure provided by P.L. Maragakis based on data from Ref. [232].)

two identical copies of the parent molecule which have exactly the same sequence of base pairs. Further duplication can be achieved in the same way. Biologically important chemical reactions are typically catalyzed by certain proteins called **enzymes**. The formation of the complementary strand is accomplished by the enzyme called **DNA polymerase**, a truly impressive molecular factory that puts the right bases together with the corresponding sugar and phosphate units while "reading" the base sequence of the parent molecule. Even more astonishing is the fact that DNA polymerase has the ability to back-track and correct mistakes in the sequence when such mistakes are introduced! The mechanisms for these processes are under intense investigation by molecular biologists.

In a process similar to DNA duplication by complementary copies of each strand, a molecule of RNA can be produced from a single strand of DNA. This is accomplished by a different enzyme, called **RNA polymerase**. Recall that a strand of RNA is similar to a strand of DNA, the only differences being that the base U is involved in RNA in place of the base T, and the sugar in the backbone is a ribose rather than a deoxyribose. Thus, genetic information is transferred from DNA to RNA. A DNA single strand does not map onto a single RNA molecule but rather on many short RNA molecules. A DNA double strand can be quite long, depending on the amount of genetic information it carries. For example, the DNA of viruses can contain anywhere from a few thousand to a few hundred thousand base pairs (the actual lengths range from roughly 1 to 100 μm); the DNA of bacteria like *E. coli* contains 4 million base pairs ($\sim$1.36 mm long); the DNA of the common fly contains 165 million base pairs ($\sim$56 mm long); the DNA of humans contains about 3×10^9 base pairs, reaching in length almost one meter! Of course the molecule is highly coiled so that it can fit inside the nucleus of each cell. In contrast, RNA molecules vary widely in size. The reason for this huge difference is that DNA contains all the genetic code of the organism it belongs to while RNA is only used to transfer this information outside the nucleus of the cell in order to build proteins. The RNA molecules are as long as they need to be in order to express a certain protein or group of proteins. The sections of DNA that give rise to these RNA molecules are referred to as "genes".

There are actually three types of RNA molecules needed for the translation, **messenger RNA** (mRNA), **transfer RNA** (tRNA) and **ribosomal RNA** (rRNA). The mRNA molecule is determined from the DNA template and carries the genetic information. The other two RNA molecules are also determined by the DNA template but do not carry genetic information by themselves. The size of all three types of RNA molecules is much smaller than the parent DNA strand. For example, in *E. coli* bacteria, which have a DNA of 4 million base pairs, the average size of mRNA is 1200 bases, tRNA consists of about 75 bases and rRNA can range from about 100 to a few thousand bases. The three types of RNA molecules together with the **ribosome** work in a complex way to produce the proteins. The ribosome is another impressive molecular factory composed of many proteins itself, whose function is to translate the genetic information of the RNA molecules into proteins, as shown symbolically in the following scheme:

$$\text{DNA} \stackrel{\text{polymerase}}{\longrightarrow} [\text{mRNA, tRNA, rRNA}] \stackrel{\text{ribosome}}{\longrightarrow} \text{proteins}$$

The first part of this process is referred to as **transcription**, and the second part is referred to as **translation**. While the process by which the whole task is accomplished is very complex, the rules for producing specific proteins from the sequence of bases in DNA are quite simple. The discovery of the rules for protein synthesis from the DNA code was another milestone in our understanding of the molecular basis of life; R.W. Holley, H.G. Khorana and M.W. Nirenberg were awarded the 1968 Nobel prize for Medicine for their work leading to this discovery. We discuss these rules next.

We consider first how many bases are needed so that a specific sequence of bases can be translated into an aminoacid. A group of n bases can produce 4^n distinct ordered combinations since there are four different bases in DNA and RNA. $n = 2$ would not work because a two-base group can produce only $4^2 = 16$ distinct ordered base combinations but there are 20 different aminoacids. Therefore n must be at least 3; this turns out to be the size of the base group chosen by Nature to translate the genetic code into aminoacid sequences. There are $4^3 = 64$ distinct ordered three-base combinations that can be formed by the four bases of RNA, a much larger number than the 20 aminoacids. Thus the genetic code is highly degenerate.

A three-base group that corresponds to a specific aminoacid is called a **codon**. The correspondence between the 64 codons and the 20 aminoacids is given in Table 13.4. Notice that certain codons, specifically UAA, UGA, UAG, do not form aminoacids but correspond to the `STOP` signal: when these codons are encountered, the aminoacid sequence is terminated and the protein is complete. Only one codon corresponds to the aminoacid **Met**, namely AUG; this codon signifies the start of the protein when it is at the beginning of the sequence, as well as the aminoacid **Met** when it is in the middle. The following is an example of a DNA section that has been transcribed into an mRNA section, which is then translated into an aminoacid sequence which forms part of a protein:

$$\underbrace{GGG}_{\text{Gly}}|\underbrace{UUC}_{\text{Phe}}|\underbrace{UUG}_{\text{Leu}}|\underbrace{GGA}_{\text{Gly}}|\underbrace{GCA}_{\text{Ala}}|\underbrace{GCA}_{\text{Ala}}|\underbrace{GGA}_{\text{Gly}}|\underbrace{AGC}_{\text{Ser}}|\underbrace{ACA}_{\text{Thr}}|\underbrace{AUG}_{\text{Met}}|\underbrace{GGG}_{\text{Gly}}|\underbrace{GCA}_{\text{Ala}}$$

This is not a random sequence but turns out to be part of the RNA of the AIDS virus. Thus, although the process of protein formation involves three types of RNA molecules and a remarkably complex molecular factory of proteins, the ribosome, the transcription code is beautifully simple. This genetic code, including the bases, aminoacids and codons, is essentially universal for all living organisms!

We need to clarify one important point in the transcription of genetic information discussed above. This has to do with the precise sequence of bases in DNA that is transcribed into mRNA and then translated into a protein or group of proteins. Organisms are classified into two broad categories, those whose cells contain a nucleus, called **eucaryotes**, and those whose cells do not have a nucleus, called

Table 13.4. *The rules for translation of RNA to proteins.*

The correspondence is given between the 20 aminoacids and the 64 three-base combinations (codons) that can be formed by the four bases U, C, A, G. The aminoacids are given by their three-letter symbols.

	Second base				
First base	U	C	A	G	Third base
U	**Phe**	**Ser**	**Tyr**	**Cys**	U
U	**Phe**	**Ser**	**Tyr**	**Cys**	C
U	**Leu**	**Ser**	STOP	STOP	A
U	**Leu**	**Ser**	STOP	**Trp**	G
C	**Leu**	**Pro**	**His**	**Arg**	U
C	**Leu**	**Pro**	**His**	**Arg**	C
C	**Leu**	**Pro**	**Gln**	**Arg**	A
C	**Leu**	**Pro**	**Gln**	**Arg**	G
A	**Ile**	**Thr**	**Asn**	**Ser**	U
A	**Ile**	**Thr**	**Asn**	**Ser**	C
A	**Ile**	**Thr**	**Lys**	**Arg**	A
A	**Met**	**Thr**	**Lys**	**Arg**	G
G	**Val**	**Ala**	**Asp**	**Gly**	U
G	**Val**	**Ala**	**Asp**	**Gly**	C
G	**Val**	**Ala**	**Glu**	**Gly**	A
G	**Val**	**Ala**	**Glu**	**Gly**	G

procaryotes. In procaryotic cells the sequence of bases in DNA is directly transcribed into the sequences of bases in mRNA. In eucaryotic cells the process is somewhat more complex: the set of bases which correspond to a protein are not necessarily consecutive bases in the RNA sequence as it is read from the DNA. Instead, there are certain sections in this sequence which contain the genetic information necessary to produce the protein; these sections are called **exons**. But there are also intervening sections in the sequence which contain no genetic information; these are called **introns**, in distinction to the exons. An intron is recognized as a sequence of bases starting with GU and ending with AG immediately preceded by a pyrimidine-rich tract. Before RNA can be used by the ribosome to express a protein, it must be spliced and rejoined at the proper points so that only the exon sequences remain in the mRNA. This task is accomplished by the **spliceosomes** in combination with small units of RNA called small nuclear RNA (snRNA). Particular exons of the RNA sequence often correspond to entire subdomains of a protein. This suggests that including both exons and introns in the gene is not just a nuisance but can add flexibility to how proteins are built by useful subunits. The impressive complexity of the translation process introduced by the presence of exons and introns must have been a key feature in the ability of living organisms to evolve.

13.2.4 Protein structure and function

In closing this chapter we mention briefly the interplay between structure and function of proteins. We emphasized already that this topic is very broad and represents an active field of research involving many scientific disciplines, from biology to chemistry, physics and even applied mathematics. It would be difficult to cover this topic adequately, even if the entire book had been devoted to it. The discussion of the following paragraph should be considered only as a minuscule sample of the structure–function relationship in proteins, which will hopefully inspire the reader to further exploration. An interesting trend in such studies is the increasingly important contributions of computer simulations, which are able to shed light onto complex processes by taking into account all the relevant degrees of freedom (for details see *Computational Biochemistry and Biophysics*, in the Further reading section).

A fascinating example of the structure–function interplay in proteins is the case of the so called KcsA potassium channel, whose structure was solved using X-ray crystallography by MacKinnon and coworkers [233]. This protein, shown in Fig. 13.30, forms a channel which makes it possible for ions to pass through the membrane of cells. The ion movement is crucial for a number of biological processes, an example being the transmission of electrical signals between neurons, the cells of the nervous system. The exchange of electrical signals between neurons underlies all of the cognitive processes. What is truly impressive about the KcsA channel is its superb selectivity: its permeability for the large alkali ions, K^+ and Rb^+, is very high, but for the small alkali ions, Na^+ and Li^+, is *smaller* by at least four orders of magnitude! This selectivity, which at first sight appears counterintuitive and is essential to the function of the channel in a biological context, is the result of the structure of the protein. Briefly, the size of the K^+ ions is such that they neatly fit in the channel due to favorable electrostatic interactions, whereas the smaller Na^+ ions experience unfavorable interactions with the channel walls and, even though they take less space, they are actually repelled from the channel. The K^+ ions venture into the channel more than one at a time, and the repulsive interaction between themselves is sufficient to overcome the attractive interactions with the protein, leading to rapid conduction across the membrane. The ability of Nature to create such exquisitely tuned and efficient filters at the microscopic level, which are hard at work at the very moment that the reader is contemplating these words, is nothing short of miraculous. This level of elegant sophistication and sheer beauty is manifested in all microscopic aspects of the wondrous phenomenon we call life.

Further reading

1. *Physics and Chemistry of Finite Systems: From Clusters to Crystals*, P. Jena, S.N. Khanna and B.K. Rao, vols. 1 and 2 (Kluwer Academic, Amsterdam, 1992).
2. *Cluster-Assembled Materials*, K. Sattler, ed., Materials Science Forum, vol. 232 (1996).
   This book is a collection of review articles discussing the properties of different classes of clusters and the prospects of assembling new solids from them.
3. *Science of Fullerenes and Carbon Nanotubes*, M.S. Dresselhaus, G. Dresselhaus and P.C. Eklund (Academic Press, San Diego, 1995).
   This book is the most comprehensive compilation of experimental and theoretical studies on carbon clusters and nanotubes, with extensive references to the research literature.
4. *Molecular Biology of the Cell*, J. Lewis M. Raff K. Roberts B. Alberts, D. Bray and J.D. Watson (3rd edn, Garland Publishing, New York, 1994).
   This is a standard introductory text on molecular biology.
5. *DNA Structure and Function*, R.R. Sinden. (Academic Press, San Diego, 1994).
6. *Nucleic Acids in Chemistry and Biology* (G.M. Blackburn and M.J. Gait, eds., Oxford University Press, New York, 1996).
7. *Nucleic Acids: Structures, Properties and Functions*, V. A. Bloomfield Jr, D. M. Crothers and I. Tinoco (University Science Books, Sausalito, 1999).
8. *Proteins: Structure and Molecular Properties*, T.E. Creighton (W.H. Freeman, New York, 1993).
9. *Introduction to Protein Structure*, C. Branden and J. Tooze (Garland Publishing, New York, 1998).
10. *Computational Biochemistry and Biophysics* (O.M. Becker, A.D. MacKerell, Jr, B. Roux and M. Watanabe, eds., M. Dekker, New York, 2001).

Problems

1. Using the conventions of Problem 3 in chapter 4 for the band structure of graphene, show that the π-bands of the carbon nanotubes can be approximated by the following expressions: For the $(n, 0)$ zig-zag tubes,

$$\epsilon_l^{(n,0)}(k) = \epsilon_0 \pm t \left[1 \pm 4\cos\left(\frac{\sqrt{3}a}{2}k\right)\cos\left(\frac{l\pi}{n}\right) + 4\cos^2\left(\frac{l\pi}{n}\right)\right]^{1/2} \tag{13.7}$$

where $l = 1, \ldots, n$ and $-\pi < ka\sqrt{3} < \pi$. For the (n, n) armchair tubes,

$$\epsilon_l^{(n,n)}(k) = \epsilon_0 \pm t \left[1 \pm 4\cos\left(\frac{a}{2}k\right)\cos\left(\frac{l\pi}{n}\right) + 4\cos^2\left(\frac{a}{2}k\right)\right]^{1/2} \tag{13.8}$$

where $l = 1, \ldots, n$ and $-\pi < ka < \pi$. For the (n, m) chiral tubes,

$$\epsilon_l^{(n,m)}(k) = \epsilon_0 \pm t \left[1 \pm 4\cos\left(\frac{l\pi}{n} - \frac{mka}{2n}\right)\cos\left(\frac{a}{2}k\right) + 4\cos^2\left(\frac{a}{2}k\right)\right]^{1/2} \tag{13.9}$$

where l is an integer determined by the condition

$$\sqrt{3}nk_x a + mk_y a = 2\pi l$$

and we have set $k = k_y$ with $-\pi < ka < \pi$. Compare these approximate bands to the band structure obtained from the tight-binding calculation for the tubes (6,0), (6,6).

2. Construct an algorithm for obtaining the structure of an arbitrary carbon nanotube based on folding of the graphene sheet. Implement this algorithm in a computer code and obtain the atomic positions of the repeat unit for nanotubes of various types and sizes.
3. Describe the basic repeat unit on the graphene plane of the chiral tubes with $(n, m) = (4, 1), (4, 3)$. Determine the BZ of the chiral $(n, m) = (4, 1), (4, 2), (4, 3)$ tubes using the zone-folding scheme described in relation to Fig. 13.10. Comment on why the tube for which the relation $2n + m = 3l$ (l: integer) holds, must exhibit metallic character. Calculate the band structure of the (4,1) tube using the results of the previous problem and compare it to the tight-binding band structure shown in Fig. 13.11 (note that the σ-bands will be missing from the approximate description of Eq. (13.9)).
4. Discuss whether the assignment of single and double bonds around the hexagonal rings in the pyrimidines and the purines shown in Figs. 13.14 and 13.15, respectively, is unique, or whether there is another equivalent assignment. In other words, is it possible to have bond resonance in the pyrimidines and the purines analogous to that of the benzene molecule, as discussed in relation to Fig. 13.4? Based on simple tight-binding type arguments, can you determine the nature of electronic states at the top of the occupied manifold (Highest Occupied Molecular Orbital or HOMO) and the bottom of the unoccupied manifold (Lowest Unoccupied Molecular Orbital or LUMO) of a DNA double strand consisting of only one type of base pairs (all $C - G$ or all $A - T$)?
5. Consider the two base pairs in DNA, CG and AT, each pair hydrogen bonded within the chain, and determine the possible hydrogen bonds that the pair can form on the *outside* of the chain, in the minor and major grooves. For a sequence $GACT$ along one of the two strands, determine the hydrogen-bond pattern in the two grooves on the outside of the double helix.

Part III

Appendices

Appendix A

Elements of classical electrodynamics

Electrodynamics is the theory of fields and forces associated with stationary or moving electric charges. The classical theory is fully described by Maxwell's equations, the crowning achievement of 19th century physics. There is also a quantum version of the theory which reconciles quantum mechanics with special relativity, but the scales of phenomena associated with electromagnetic fields in solids, that is, the energy, length and time scale, are such that it is not necessary to invoke quantum electrodynamics. For instance, the scale of electron velocities in solids, set by the Fermi velocity $v_F = \hbar k_F / m_e$, is well below the speed of light, so electrons behave as non-relativistic point particles. We certainly have to take into account the quantized nature of electrons in a solid, embodied in the wavefunctions and energy eigenvalues that characterize the electronic states, but we can treat the electromagnetic fields as classical variables. It is often convenient to incorporate the effects of electromagnetic fields on solids using perturbation theory; this is explicitly treated in Appendix B. Accordingly, we provide here a brief account of the basic concepts and equations of classical electrodynamics. For detailed discussions, proofs and applications, we refer the reader to standard textbooks on the subject, a couple of which are mentioned in the Further reading section.

A.1 Electrostatics and magnetostatics

The force on a charge q at $\mathbf{r}$ due to the presence of a point charge q' at $\mathbf{r}'$ is given by

$$\mathbf{F} = \frac{qq'}{|\mathbf{r} - \mathbf{r}'|^3}(\mathbf{r} - \mathbf{r}') \tag{A.1}$$

which is known as Coulomb's force law. The corresponding electric field is defined as the force on a unit charge at $\mathbf{r}$, or, taking q' to be at the origin,

$$\mathbf{E}(\mathbf{r}) = \frac{q'}{|\mathbf{r}|^2}\hat{\mathbf{r}} \tag{A.2}$$

Forces in electrostatics are additive, so that the total electric field at $\mathbf{r}$ due to a continuous charge distribution is given by

$$\mathbf{E}(\mathbf{r}) = \int \frac{\rho(\mathbf{r}')}{|\mathbf{r} - \mathbf{r}'|^3}(\mathbf{r} - \mathbf{r}')\mathrm{d}\mathbf{r}' \tag{A.3}$$

with $\rho(\mathbf{r})$ the charge density (which has dimensions electric charge per unit volume). Taking the divergence of both sides of this equation and using Gauss's theorem (see Eq. (G.22) in Appendix G), we find

$$\nabla_{\mathbf{r}} \cdot \mathbf{E}(\mathbf{r}) = 4\pi\rho(\mathbf{r}) \tag{A.4}$$

where we have used the fact that the divergence of $(\mathbf{r} - \mathbf{r}')/|\mathbf{r} - \mathbf{r}'|^3$ is equal to $4\pi\delta(\mathbf{r} - \mathbf{r}')$ (see Appendix G). Integrating both sides of the above equation over the volume enclosed by a surface S, we obtain

$$\int \nabla_{\mathbf{r}} \cdot \mathbf{E}(\mathbf{r})\mathrm{d}\mathbf{r} = \oint_S \mathbf{E}(\mathbf{r}) \cdot \hat{\mathbf{n}}_s \, \mathrm{d}S = 4\pi Q \tag{A.5}$$

where Q is the total electric charge enclosed by the surface S and $\hat{\mathbf{n}}_s$ is the unit vector normal to the surface element $\mathrm{d}S$; this expression is known as Gauss's law. The electrostatic potential $\Phi(\mathbf{r})$ is defined through

$$\mathbf{E}(\mathbf{r}) = -\nabla_{\mathbf{r}}\Phi(\mathbf{r}) \tag{A.6}$$

The potential is defined up to a constant, since the gradient of a constant is always zero; we can choose this constant to be zero, which is referred to as the Coulomb gauge. In terms of the potential, Eq. (A.4) becomes

$$\nabla_{\mathbf{r}}^2\Phi(\mathbf{r}) = -4\pi\rho(\mathbf{r}) \tag{A.7}$$

a relation known as Poisson's equation. We note that the definition of the electrostatic potential, Eq. (A.6), implies that the curl of $\mathbf{E}$ must be zero, since

$$\nabla_{\mathbf{r}} \times \mathbf{E} = -\nabla_{\mathbf{r}} \times \nabla_{\mathbf{r}}\Phi$$

and the curl of a gradient is identically zero (see Appendix G). This is indeed true for the electric field defined in Eq. (A.2), as can be proven straightforwardly by calculating the line integral of $\mathbf{E}(\mathbf{r})$ around any closed loop and invoking Stokes's theorem (see Eq. (G.23) in Appendix G) to relate this integral to $\nabla_{\mathbf{r}} \times \mathbf{E}$. Because of the additive nature of electrostatic fields, this result applies to any field deriving from an arbitrary distribution of charges, hence it is always possible to express such a field in terms of the potential as indicated by Eq. (A.6). In particular, the potential of a continuous charge distribution $\rho(\mathbf{r})$ turns out to be

$$\Phi(\mathbf{r}) = \int \frac{\rho(\mathbf{r}')}{|\mathbf{r} - \mathbf{r}'|}\mathrm{d}\mathbf{r}' \tag{A.8}$$

which immediately leads to Eq. (A.3) for the corresponding electrostatic field.

From the above definitions it is simple to show that the energy W required to assemble a set of point charges q_i at positions $\mathbf{r}_i$ is given by

$$W = \frac{1}{2}\sum_i q_i\Phi(\mathbf{r}_i)$$

where $\Phi(\mathbf{r})$ is the total potential due to the charges. If we now generalize this expression to a continuous distribution of charge $\rho(\mathbf{r})$, use Eq. (A.4) to relate $\rho(\mathbf{r})$ to $\mathbf{E}(\mathbf{r})$, integrate by parts and assume that the field dies at infinity, we find that the electrostatic energy W associated with an electric field $\mathbf{E}$ is given by

$$W_e = \frac{1}{8\pi}\int |\mathbf{E}(\mathbf{r})|^2\mathrm{d}\mathbf{r} \tag{A.9}$$

The force on a charge q moving with velocity $\mathbf{v}$ in a magnetic field $\mathbf{B}$ is given by

$$\mathbf{F} = q\left(\frac{\mathbf{v}}{c} \times \mathbf{B}\right) \tag{A.10}$$

where c is the speed of light; this is known as the Lorentz force law. The motion of a charge is associated with a current, whose density per unit area perpendicular to its flow is defined as $\mathbf{J}(\mathbf{r})$ (the dimensions of $\mathbf{J}$ are electric charge per unit area per unit time). The current density and the charge density $\rho(\mathbf{r})$ are related by

$$\nabla_{\mathbf{r}} \cdot \mathbf{J}(\mathbf{r}) = -\frac{\partial \rho(\mathbf{r})}{\partial t} \tag{A.11}$$

which is known as the continuity relation. This relation is a consequence of the conservation of electric charge (a simple application of the divergence theorem produces the above expression). The magnetic field due to a current density $\mathbf{J}(\mathbf{r})$ is given by

$$\mathbf{B}(\mathbf{r}) = \frac{1}{c}\int \frac{1}{|\mathbf{r}-\mathbf{r}'|^3}\mathbf{J}(\mathbf{r}') \times (\mathbf{r}-\mathbf{r}')\mathrm{d}\mathbf{r}' \tag{A.12}$$

which is known as the Biot–Savart law. From this equation, it is a straightforward exercise in differential calculus to show that the divergence of $\mathbf{B}$ vanishes identically,

$$\nabla_{\mathbf{r}} \cdot \mathbf{B}(\mathbf{r}) = 0$$

while its curl is related to the current density:

$$\nabla_{\mathbf{r}} \times \mathbf{B}(\mathbf{r}) = \frac{4\pi}{c}\mathbf{J}(\mathbf{r}) \tag{A.13}$$

Integrating the second equation over an area S which is bounded by a contour C and using Stokes's theorem leads to

$$\int (\nabla_{\mathbf{r}} \times \mathbf{B}(\mathbf{r})) \cdot \hat{\mathbf{n}}_s \, \mathrm{d}S = \frac{4\pi}{c}\int \mathbf{J}(\mathbf{r}) \cdot \hat{\mathbf{n}}_s \, \mathrm{d}S \Rightarrow \oint_C \mathbf{B}(\mathbf{r}) \cdot \mathrm{d}\mathbf{l} = \frac{4\pi}{c} I \tag{A.14}$$

where I is the total current passing through the area enclosed by the loop C; the last expression is known as Ampère's law.

By analogy to the electrostatic case, we define a vector potential $\mathbf{A}(\mathbf{r})$ through which we obtain the magnetic field as

$$\mathbf{B}(\mathbf{r}) = \nabla_{\mathbf{r}} \times \mathbf{A}(\mathbf{r}) \tag{A.15}$$

The vector form of the magnetic potential is dictated by the fact that $\nabla_{\mathbf{r}} \times \mathbf{B}$ does not vanish, which is in contrast to what we had found for the electrostatic potential $\mathbf{E}$. Similar to that situation, however, the magnetic potential $\mathbf{A}$ is defined up to a function whose curl vanishes; we exploit this ambiguity by choosing the vector field so that

$$\nabla_{\mathbf{r}} \cdot \mathbf{A}(\mathbf{r}) = 0 \tag{A.16}$$

which is known as the Coulomb gauge. With this choice, and using the relations between $\mathbf{B}$ and $\mathbf{J}$ that we derived above, we find that the laplacian of the vector potential is given by

$$\nabla_{\mathbf{r}}^2 \mathbf{A}(\mathbf{r}) = -\frac{4\pi}{c}\mathbf{J}(\mathbf{r}) \tag{A.17}$$

which is formally the same as the Poisson equation, Eq. (A.7), only applied to vector rather than scalar quantities. From this relation, the vector potential itself can be expressed in terms of the current density as

$$\mathbf{A}(\mathbf{r}) = \frac{1}{c}\int \frac{\mathbf{J}(\mathbf{r}')}{|\mathbf{r}-\mathbf{r}'|}\mathrm{d}\mathbf{r}'$$

Finally, by analogy to the derivation of the electrostatic energy associated with an electric field $\mathbf{E}$, it can be shown that the magnetostatic energy associated with a magnetic field $\mathbf{B}$ is

$$W_m = \frac{1}{8\pi}\int |\mathbf{B}(\mathbf{r})|^2 \mathrm{d}\mathbf{r} \tag{A.18}$$

Having obtained the expressions that relate the electric and magnetic potentials to the distributions of charges and currents, we can then calculate all the physically relevant quantities (such as the fields and from those the forces), which completely determine the behavior of the system. The only things missing are the boundary conditions that are required to identify uniquely the solution to the differential equations involved; these are determined by the nature and the geometry of the physical system under consideration. For example, for conductors we have the following boundary conditions.

(a) The electric field $\mathbf{E}(\mathbf{r})$ must vanish inside the conductor.
(b) The electrostatic potential $\Phi(\mathbf{r})$ must be a constant inside the conductor.
(c) The charge density $\rho(\mathbf{r})$ must vanish inside the conductor and any net charge must reside on the surface.
(d) The only non-vanishing electric field $\mathbf{E}(\mathbf{r})$ must be perpendicular to the surface just outside the conductor.

These conditions are a consequence of the presence of free electric charges which can move within the conductor to shield any non-vanishing electric fields.

A.2 Fields in polarizable matter

In many situations the application of an external electric or magnetic field on a substance can instigate a response which is referred to as "polarization". Usually, the response is proportional to the applied field, which is called linear response. We refer to the polarization of electric quantities as simply the polarization and to the polarization of magnetic quantities as the magnetization. We will also refer to the polarization or magnetization of the unit volume (or elementary unit, such as the unit cell of a crystal), as the induced dipole moment $\mathbf{p}(\mathbf{r})$ or the induced magnetic moment $\mathbf{m}(\mathbf{r})$, respectively.

To conform to historical conventions (which are actually motivated by physical considerations), we define the total, induced and net electric fields as:

$$\mathbf{E}(\mathbf{r}) : \text{total electric field}$$

$$-4\pi\mathbf{p}(\mathbf{r}) : \text{induced electric field}$$

$$\mathbf{D}(\mathbf{r}) : \text{net electric field}$$

where the net field is defined as the total minus the induced field. ($\mathbf{D}$ is also called the electric displacement). These conventions lead to the following expressions:

$$\mathbf{D}(\mathbf{r}) = \mathbf{E}(\mathbf{r}) + 4\pi\mathbf{p}(\mathbf{r}) \Rightarrow \mathbf{E}(\mathbf{r}) = \mathbf{D}(\mathbf{r}) - 4\pi\mathbf{p}(\mathbf{r})$$

For linear response, we define the electric susceptibility χ_e through the relation[1]

$$-4\pi \mathbf{p}(\mathbf{r}) = -\chi_e \mathbf{E}(\mathbf{r}) \Rightarrow \mathbf{p}(\mathbf{r}) = \frac{\chi_e}{4\pi} \mathbf{E}(\mathbf{r})$$

It is also customary to define the dielectric constant ε as the factor that relates the total field to the net field through

$$\mathbf{D} = \varepsilon \mathbf{E} \Rightarrow \varepsilon = 1 + \chi_e$$

where the last expression between the dielectric constant and the electric susceptibility holds for linear response. The historical and physical reason for this set of definitions is that, typically, an external field produces an electric dipole moment that tends to shield the applied field (the dipole moment produces an opposite field), hence it is natural to define the induced field with a negative sign.

The corresponding definitions for the magnetic field are as follows:

$$\mathbf{B}(\mathbf{r}) : \text{total magnetic field}$$

$$4\pi \mathbf{m}(\mathbf{r}) : \text{induced magnetic field}$$

$$\mathbf{H}(\mathbf{r}) : \text{net magnetic field}$$

where, as before, the net field is defined as the total minus the induced field. These conventions lead to the following expressions:

$$\mathbf{H}(\mathbf{r}) = \mathbf{B}(\mathbf{r}) - 4\pi \mathbf{m}(\mathbf{r}) \Rightarrow \mathbf{B}(\mathbf{r}) = \mathbf{H}(\mathbf{r}) + 4\pi \mathbf{m}(\mathbf{r})$$

For linear response, we define the magnetic susceptibility χ_m through the relation

$$4\pi \mathbf{m}(\mathbf{r}) = \chi_m \mathbf{H}(\mathbf{r}) \Rightarrow \mathbf{m}(\mathbf{r}) = \frac{\chi_m}{4\pi} \mathbf{H}(\mathbf{r})$$

It is also customary to define the magnetic permeability μ as the factor that relates the total field to the net field through

$$\mathbf{B} = \mu \mathbf{H} \Rightarrow \mu = 1 + \chi_m$$

where the last expression between the magnetic permeability and the magnetic susceptibility holds for linear response. This set of definitions is not exactly analogous to the definitions concerning electric polarization; the reason is that in a substance which exhibits magnetic polarization the induced field, typically, tends to be aligned with the applied field (it enhances it), hence the natural definition of the induced field has the same sign as the applied field.

The net fields are associated with the presence of charges and currents which are called "free", while the electric and magnetic polarization are associated with induced charges and currents that are called "bound". Thus, in the case of electric polarization which produces an electric dipole moment $\mathbf{p}(\mathbf{r})$, the bound charges are given by

$$\sigma_b(\mathbf{r}) = 4\pi \mathbf{p}(\mathbf{r}) \cdot \hat{\mathbf{n}}_s(\mathbf{r}), \qquad \rho_b(\mathbf{r}) = -4\pi \nabla_{\mathbf{r}} \cdot \mathbf{p}(\mathbf{r})$$

where σ_b is a surface charge density, $\hat{\mathbf{n}}_s$ is the surface-normal unit vector and ρ_b is a bulk charge density. The potential due to the induced dipole moment can then be

[1] Here we treat the susceptibility as a scalar quantity, which is appropriate for isotropic solids. In anisotropic solids, such as crystals, the susceptibility must be generalized to a second rank tensor, $\chi_{\alpha\beta}$, with α, β taking three independent values each in 3D.

expressed in terms of these charges as

$$\Phi^{ind}(\mathbf{r}) = 4\pi \int \frac{\mathbf{p}(\mathbf{r}') \cdot (\mathbf{r} - \mathbf{r}')}{|\mathbf{r} - \mathbf{r}'|^3} \mathrm{d}\mathbf{r}' = \oint_{S'} \frac{\sigma_b(\mathbf{r}')}{|\mathbf{r} - \mathbf{r}'|} \mathrm{d}S' + \int_{\Omega'} \frac{\rho_b(\mathbf{r}')}{|\mathbf{r} - \mathbf{r}'|} \mathrm{d}\mathbf{r}'$$

where Ω' is the volume of the polarized substance and S' is the surface that encloses it. Similarly, in the case of magnetic polarization which produces a magnetic dipole moment $\mathbf{m}(\mathbf{r})$, the bound currents are given by

$$\mathbf{K}_b(\mathbf{r}) = 4\pi \mathbf{m}(\mathbf{r}) \times \hat{\mathbf{n}}_s(\mathbf{r}), \quad \mathbf{J}_b(\mathbf{r}) = 4\pi \nabla_{\mathbf{r}} \times \mathbf{m}(\mathbf{r})$$

where $\mathbf{K}_b$ is a surface current density, $\hat{\mathbf{n}}_s$ is the surface-normal unit vector and $\mathbf{J}_b$ is a bulk current density. The potential due to the induced dipole moment can then be expressed in terms of these currents as

$$\mathbf{A}^{ind}(\mathbf{r}) - 4\pi \int \frac{\mathbf{m}(\mathbf{r}') \times (\mathbf{r} - \mathbf{r}')}{|\mathbf{r} - \mathbf{r}'|^3} \mathrm{d}\mathbf{r}' = \oint_{S'} \frac{\mathbf{K}_b(\mathbf{r}')}{|\mathbf{r} - \mathbf{r}'|} \mathrm{d}S' \mid \int_{\Omega'} \frac{\mathbf{J}_b(\mathbf{r}')}{|\mathbf{r} - \mathbf{r}'|} \mathrm{d}\mathbf{r}'$$

with the meaning of Ω', S' the same as in the electric case. The values of the bound charges and currents are determined by the nature of the physical system and its geometrical features. Having identified the induced fields with the bound charges or currents, we can now obtain the net fields in terms of the free charges described by the charge density $\rho_f(\mathbf{r})$ or free currents described by the current density $\mathbf{J}_f(\mathbf{r})$:

$$\nabla_{\mathbf{r}} \cdot \mathbf{D}(\mathbf{r}) = 4\pi \rho_f(\mathbf{r}), \quad \nabla_{\mathbf{r}} \times \mathbf{H}(\mathbf{r}) = \frac{4\pi}{c} \mathbf{J}_f(\mathbf{r})$$

These are identical expressions to those relating the total fields $\mathbf{E}$, $\mathbf{B}$ to the charge or current densities in free space, Eqs. (A.4) and (A.13), respectively.

A.3 Electrodynamics

The combined presence of electric charges and currents gives rise to electric and magnetic fields which are related to each other. Our discussion of electrostatics and magnetostatics leads to the following expression for the total force on a charge q moving with velocity $\mathbf{v}$ in the presence of external electric and magnetic fields $\mathbf{E}$ and $\mathbf{B}$, respectively:

$$\mathbf{F} = q \left(\mathbf{E} + \frac{\mathbf{v}}{c} \times \mathbf{B} \right) \tag{A.19}$$

Since this charge is moving, it corresponds to a current density $\mathbf{J}$. If the current density is proportional to the force per unit charge,

$$\mathbf{J} = \sigma \frac{\mathbf{F}}{q} = \sigma \left(\mathbf{E} + \frac{\mathbf{v}}{c} \times \mathbf{B} \right)$$

we call the behavior ohmic, with the constant of proportionality σ called the conductivity (not to be confused with the surface charge density). The inverse of the conductivity is called the resistivity $\rho = 1/\sigma$ (not to be confused with the bulk charge density). When the velocity is much smaller than the speed of light, or the magnetic field is much weaker than the electric field, the above relation reduces to

$$\mathbf{J} = \sigma \mathbf{E} \tag{A.20}$$

which is known as Ohm's law. The conductivity is a characteristic material property and can be calculated if a detailed knowledge of the electronic states (their eigenfunctions and corresponding energies) are known (see chapter 5).

The expressions we derived relating the fields to static charge and current density distributions are not adequate to describe the physics if we allow the densities and the fields to have a time dependence. These have to be augmented by two additional relations, known as Faraday's law and Maxwell's extension of Ampère's law. To motivate Faraday's law we introduce first the notion of the magnetic flux ϕ: it is the projection of the magnetic field $\mathbf{B}$ onto a surface element $\hat{\mathbf{n}}_s \, dS$, integrated over some finite surface area:

$$\phi = \int \mathbf{B}(\mathbf{r}) \cdot \hat{\mathbf{n}}_s \, dS$$

where $\hat{\mathbf{n}}_s$ is the surface-normal unit vector corresponding to the surface element dS; the magnetic flux is a measure of the amount of magnetic field passing through the surface. Now let us consider the change with respect to time of the magnetic flux through a fixed surface, when the magnetic field is a time-dependent quantity:

$$\frac{d\phi}{dt} = \int \frac{\partial \mathbf{B}(\mathbf{r}, t)}{\partial t} \cdot \hat{\mathbf{n}}_s \, dS$$

The electromagnetic fields propagate with the speed of light, so if we divide both sides of the above equation by c we will obtain the change with respect to time of the total magnetic flux passing through the surface, normalized by the speed of propagation of the field:

$$\frac{1}{c}\frac{d\phi}{dt} = \frac{1}{c}\int \frac{\partial \mathbf{B}(\mathbf{r}, t)}{\partial t} \cdot \hat{\mathbf{n}}_s \, dS$$

From the definition of the force due to a magnetic field, Eq. (A.10), which shows that the magnetic field has the dimensions of a force per unit charge, we conclude that the expression on the right-hand side of the last equation has the dimensions of energy per unit charge. Therefore, the time derivative of the magnetic flux divided by c is a measure of the energy per unit charge passing through the surface due to changes in the magnetic field with respect to time. To counter this expense of energy when the magnetic field changes, a current can be set up along the boundary C of the surface S. In order to induce such a current, an electric field $\mathbf{E}$ must be introduced which will move charges along C. The definition of the electric field as the force per unit charge implies that $\mathbf{E} \cdot \mathbf{dl}$ is the elementary energy per unit charge moving along the contour C, with $\mathbf{dl}$ denoting the length element along C. Requiring that the energy needed to move a unit charge around C under the influence of $\mathbf{E}$ exactly counters the energy per unit charge passing through the surface S due to changes in the magnetic field, leads to

$$\oint_C \mathbf{E}(\mathbf{r}, t) \cdot \mathbf{dl} + \frac{1}{c}\frac{d\phi}{dt} = 0 \Rightarrow \oint_C \mathbf{E}(\mathbf{r}, t) \cdot \mathbf{dl} = -\frac{1}{c}\int \frac{\partial \mathbf{B}(\mathbf{r}, t)}{\partial t} \cdot \hat{\mathbf{n}}_s \, dS$$

Turning this last relation into differential form by using Stokes's theorem, we find

$$\nabla_{\mathbf{r}} \times \mathbf{E}(\mathbf{r}, t) = -\frac{1}{c}\frac{\partial \mathbf{B}(\mathbf{r}, t)}{\partial t}$$

which is Faraday's law.

To motivate Maxwell's extension of Ampère's law, we start from the continuity relation, Eq. (A.11), and allow for both spatial and temporal variations of the charge and current densities. Relating the charge density $\rho(\mathbf{r}, t)$ to the electric field $\mathbf{E}(\mathbf{r}, t)$ through Eq. (A.4), we find

$$\nabla_{\mathbf{r}} \cdot \mathbf{J}(\mathbf{r}, t) = -\frac{\partial \rho(\mathbf{r}, t)}{\partial t} = -\frac{\partial}{\partial t}\left(\frac{1}{4\pi}\nabla_{\mathbf{r}} \cdot \mathbf{E}(\mathbf{r}, t)\right) = -\nabla_{\mathbf{r}} \cdot \left(\frac{1}{4\pi}\frac{\partial \mathbf{E}(\mathbf{r}, t)}{\partial t}\right)$$

A comparison of the first and last expressions in this set of equalities shows that the quantity in parentheses on the right-hand side is equivalent to a current density and as such it should be included in Eq. (A.13). In particular, this current density is generated by the temporal variation of the fields induced by the presence of external current densities, so its role will be to counteract those currents and therefore must be subtracted from the external current density, which leads to

$$\nabla_{\mathbf{r}} \times \mathbf{B}(\mathbf{r}, t) = \frac{4\pi}{c}\mathbf{J}(\mathbf{r}) + \frac{1}{c}\frac{\partial \mathbf{E}(\mathbf{r}, t)}{\partial t}$$

This is Ampère's law augmented by the second term, as argued by Maxwell.

The set of equations which relates the spatially and temporally varying charge and current densities to the corresponding electric and magnetic fields are known as Maxwell's equations:

$$\nabla_{\mathbf{r}} \cdot \mathbf{E}(\mathbf{r}, t) = 4\pi\rho(\mathbf{r}, t) \tag{A.21}$$

$$\nabla_{\mathbf{r}} \cdot \mathbf{B}(\mathbf{r}, t) = 0 \tag{A.22}$$

$$\nabla_{\mathbf{r}} \times \mathbf{E}(\mathbf{r}, t) = -\frac{1}{c}\frac{\partial \mathbf{B}(\mathbf{r}, t)}{\partial t} \tag{A.23}$$

$$\nabla_{\mathbf{r}} \times \mathbf{B}(\mathbf{r}, t) = \frac{4\pi}{c}\mathbf{J}(\mathbf{r}) + \frac{1}{c}\frac{\partial \mathbf{E}(\mathbf{r}, t)}{\partial t} \tag{A.24}$$

The second and third of these equations are not changed in polarizable matter, but the first and fourth can be expressed in terms of the net fields **D**, **H** and the corresponding free charge and current densities in an exactly analogous manner:

$$\nabla_{\mathbf{r}} \cdot \mathbf{D}(\mathbf{r}, t) = 4\pi\rho_f(\mathbf{r}, t) \tag{A.25}$$

$$\nabla_{\mathbf{r}} \times \mathbf{H}(\mathbf{r}, t) = \frac{4\pi}{c}\mathbf{J}_f(\mathbf{r}) + \frac{1}{c}\frac{\partial \mathbf{D}(\mathbf{r}, t)}{\partial t} \tag{A.26}$$

Maxwell's equations can also be put in integral form. Specifically, integrating both sides of each equation, over a volume enclosed by a surface S for the first two equations and over a surface enclosed by a curve C for the last two equations, and using Gauss's theorem or Stokes's theorem (see Appendix G) as appropriate, we find that the equations in polarizable matter take the form

$$\oint_S \mathbf{D}(\mathbf{r}, t) \cdot \hat{\mathbf{n}}_s \, \mathrm{d}S = 4\pi Q_f \tag{A.27}$$

$$\oint_S \mathbf{B}(\mathbf{r}, t) \cdot \hat{\mathbf{n}}_s \, \mathrm{d}S = 0 \tag{A.28}$$

$$\oint_C \mathbf{E}(\mathbf{r}, t) \cdot \mathrm{d}\mathbf{l} = -\frac{1}{c}\frac{\partial}{\partial t}\int \mathbf{B}(\mathbf{r}, t) \cdot \hat{\mathbf{n}}_s \, \mathrm{d}S \tag{A.29}$$

$$\oint_C \mathbf{H}(\mathbf{r}, t) \cdot \mathrm{d}\mathbf{l} = \frac{4\pi}{c}I_f + \frac{1}{c}\frac{\partial}{\partial t}\int \mathbf{D}(\mathbf{r}, t) \cdot \hat{\mathbf{n}}_s \, \mathrm{d}S \tag{A.30}$$

where $\hat{\mathbf{n}}_s$ is the surface-normal unit vector associated with the surface element of S and $\mathbf{dl}$ is the length element along the curve C; Q_f is the total free charge in the volume of integration and I_f is the total free current passing through the surface of integration.

A direct consequence of Maxwell's equations is that the electric and magnetic fields can be expressed in terms of the scalar and vector potentials Φ, $\mathbf{A}$, which now include both spatial and temporal dependence. From the second Maxwell equation, Eq. (A.22), we conclude that we can express the magnetic field as the curl of the vector potential $\mathbf{A}(\mathbf{r}, t)$, since the divergence of the curl of a vector potential vanishes identically (see Appendix G):

$$\mathbf{B}(\mathbf{r}, t) = \nabla_{\mathbf{r}} \times \mathbf{A}(\mathbf{r}, t) \tag{A.31}$$

Substituting this into the third Maxwell equation, Eq. (A.23), we find

$$\nabla_{\mathbf{r}} \times \mathbf{E}(\mathbf{r}, t) = -\frac{1}{c}\frac{\partial}{\partial t}(\nabla_{\mathbf{r}} \times \mathbf{A}(\mathbf{r}, t)) \Rightarrow \nabla_{\mathbf{r}} \times \left(\mathbf{E}(\mathbf{r}, t) + \frac{1}{c}\frac{\partial \mathbf{A}(\mathbf{r}, t)}{\partial t}\right) = 0$$

which allows us to define the expression in the last parenthesis as the gradient of a scalar field, since its curl vanishes identically (see Appendix G):

$$\mathbf{E}(\mathbf{r}, t) + \frac{1}{c}\frac{\partial \mathbf{A}(\mathbf{r}, t)}{\partial t} = -\nabla_{\mathbf{r}}\Phi(\mathbf{r}, t) \Rightarrow \mathbf{E}(\mathbf{r}, t) = -\nabla_{\mathbf{r}}\Phi(\mathbf{r}, t) - \frac{1}{c}\frac{\partial \mathbf{A}(\mathbf{r}, t)}{\partial t} \tag{A.32}$$

Inserting this expression into the first Maxwell equation, Eq. (A.21), we obtain

$$\nabla_{\mathbf{r}}^2\Phi(\mathbf{r}, t) + \frac{1}{c}\frac{\partial}{\partial t}(\nabla_{\mathbf{r}} \cdot \mathbf{A}(\mathbf{r}, t)) = -4\pi\rho(\mathbf{r}, t) \tag{A.33}$$

which is the generalization of Poisson's equation to spatially and temporally varying potentials and charge densities. Inserting the expressions for $\mathbf{B}(\mathbf{r}, t)$ and $\mathbf{E}(\mathbf{r}, t)$ in terms of the vector and scalar potentials, Eqs. (A.31), (A.32), in the last Maxwell equation, Eq. (A.24), and using the identity which relates the curl of the curl of a vector potential to its divergence and its laplacian, Eq. (G.21), we obtain

$$\left(\nabla_{\mathbf{r}}^2\mathbf{A}(\mathbf{r}, t) - \frac{1}{c^2}\frac{\partial^2\mathbf{A}(\mathbf{r}, t)}{\partial t^2}\right) - \nabla_{\mathbf{r}}\left(\nabla_{\mathbf{r}} \cdot \mathbf{A}(\mathbf{r}, t) + \frac{1}{c}\frac{\partial\Phi(\mathbf{r}, t)}{\partial t}\right) = -\frac{4\pi}{c}\mathbf{J}(\mathbf{r}, t) \tag{A.34}$$

Equations (A.33) and (A.34), can be used to obtain the vector and scalar potentials, $\mathbf{A}(\mathbf{r}, t)$, $\Phi(\mathbf{r}, t)$ when the charge and current density distributions $\rho(\mathbf{r}, t)$, $\mathbf{J}(\mathbf{r}, t)$ are known. In the Coulomb gauge, Eq. (A.16), the first of these equations reduces to the familiar Poisson equation, which, as in the electrostatic case, leads to

$$\Phi(\mathbf{r}, t) = \int \frac{\rho(\mathbf{r}', t)}{|\mathbf{r} - \mathbf{r}'|}\mathrm{d}\mathbf{r}' \tag{A.35}$$

and the second equation then takes the form

$$\nabla_{\mathbf{r}}^2\mathbf{A}(\mathbf{r}, t) - \frac{1}{c^2}\frac{\partial^2\mathbf{A}(\mathbf{r}, t)}{\partial t^2} = \frac{1}{c}\nabla_{\mathbf{r}}\left(\frac{\partial\Phi(\mathbf{r}, t)}{\partial t}\right) - \frac{4\pi}{c}\mathbf{J}(\mathbf{r}, t) \tag{A.36}$$

As a final point, we note that the total energy of the electromagnetic field including spatial and temporal variations is given by

$$E^{em} = \frac{1}{8\pi}\int \left(|\mathbf{E}(\mathbf{r}, t)|^2 + |\mathbf{B}(\mathbf{r}, t)|^2\right)\mathrm{d}\mathbf{r} \tag{A.37}$$

which is a simple generalization of the expressions we gave for the static electric and magnetic fields.

A.4 Electromagnetic radiation

The propagation of electromagnetic fields in space and time is referred to as electromagnetic radiation. In vacuum, Maxwell's equations become

$$\nabla_{\mathbf{r}} \cdot \mathbf{E}(\mathbf{r}, t) = 0, \quad \nabla_{\mathbf{r}} \cdot \mathbf{B}(\mathbf{r}, t) = 0$$

$$\nabla_{\mathbf{r}} \times \mathbf{E}(\mathbf{r}, t) = -\frac{1}{c}\frac{\partial \mathbf{B}(\mathbf{r}, t)}{\partial t}, \quad \nabla_{\mathbf{r}} \times \mathbf{B}(\mathbf{r}, t) = \frac{1}{c}\frac{\partial \mathbf{E}(\mathbf{r}, t)}{\partial t}$$

Taking the curl of both sides of the third and fourth Maxwell equations and using the identity that relates the curl of the curl of a vector field to its divergence and its laplacian, Eq. (G.21), we find

$$\nabla^2_{\mathbf{r}}\mathbf{E}(\mathbf{r}, t) = \frac{1}{c^2}\frac{\partial^2 \mathbf{E}(\mathbf{r}, t)}{\partial t^2}, \quad \nabla^2_{\mathbf{r}}\mathbf{B}(\mathbf{r}, t) = \frac{1}{c^2}\frac{\partial^2 \mathbf{B}(\mathbf{r}, t)}{\partial t^2}$$

where we have also used the first and second Maxwell equations to eliminate the divergence of the fields. Thus, both the electric and the magnetic field obey the wave equation with speed c. The plane wave solution to these equations is

$$\mathbf{E}(\mathbf{r}, t) = \mathbf{E}_0 \mathrm{e}^{\mathrm{i}(\mathbf{k}\cdot\mathbf{r}-\omega t)}, \quad \mathbf{B}(\mathbf{r}, t) = \mathbf{B}_0 \mathrm{e}^{\mathrm{i}(\mathbf{k}\cdot\mathbf{r}-\omega t)}$$

with $\mathbf{k}$ the wave-vector and ω the frequency of the radiation, which are related by

$$|\mathbf{k}| = \frac{\omega}{c} \quad \text{(free space)}$$

With these expressions for the fields, from the first and second Maxwell equations we deduce that

$$\mathbf{k} \cdot \mathbf{E}_0 = \mathbf{k} \cdot \mathbf{B}_0 = 0 \quad \text{(free space)} \tag{A.38}$$

that is, the vectors $\mathbf{E}_0$, $\mathbf{B}_0$ are perpendicular to the direction of propagation of radiation which is determined by $\mathbf{k}$; in other words, the fields have only transverse and no longitudinal components. Moreover, from the third Maxwell equation, we obtain the relation

$$\mathbf{k} \times \mathbf{E}_0 = \frac{\omega}{c}\mathbf{B}_0 \quad \text{(free space)} \tag{A.39}$$

which implies that the vectors $\mathbf{E}_0$ and $\mathbf{B}_0$ are also perpendicular to each other and have the same magnitude, since $|\mathbf{k}| = \omega/c$. The fourth Maxwell equation leads to the same result.

Inside a material with dielectric constant ε and magnetic permeability μ, in the absence of any free charges or currents, Maxwell's equations become

$$\nabla_{\mathbf{r}} \cdot \mathbf{D}(\mathbf{r}, t) = 0, \quad \nabla_{\mathbf{r}} \cdot \mathbf{B}(\mathbf{r}, t) = 0$$

$$\nabla_{\mathbf{r}} \times \mathbf{E}(\mathbf{r}, t) = -\frac{1}{c}\frac{\partial \mathbf{B}(\mathbf{r}, t)}{\partial t}, \quad \nabla_{\mathbf{r}} \times \mathbf{H}(\mathbf{r}, t) = \frac{1}{c}\frac{\partial \mathbf{D}(\mathbf{r}, t)}{\partial t}$$

Using the relations $\mathbf{D} = \varepsilon\mathbf{E}$, $\mathbf{B} = \mu\mathbf{H}$, the above equations lead to the same wave equations for $\mathbf{E}$ and $\mathbf{B}$ as in free space, except for a factor $\varepsilon\mu$:

$$\nabla^2_{\mathbf{r}}\mathbf{E}(\mathbf{r}, t) = \frac{\varepsilon\mu}{c^2}\frac{\partial^2 \mathbf{E}(\mathbf{r}, t)}{\partial t^2}, \quad \nabla^2_{\mathbf{r}}\mathbf{B}(\mathbf{r}, t) = \frac{\varepsilon\mu}{c^2}\frac{\partial^2 \mathbf{B}(\mathbf{r}, t)}{\partial t^2}$$

which implies that the speed of the electromagnetic radiation in the solid is reduced by a factor $\sqrt{\varepsilon\mu}$. This has important consequences. In particular, assuming as before plane wave solutions for $\mathbf{E}$ and $\mathbf{B}$ and using the equations which relate $\mathbf{E}$ to $\mathbf{B}$ and $\mathbf{H}$ to $\mathbf{D}$, we arrive at the following relations between the electric and magnetic field vectors and the wave-vector $\mathbf{k}$:

$$\mathbf{k} \times \mathbf{E}_0 = \frac{\omega}{c}\mathbf{B}_0, \quad \mathbf{k} \times \mathbf{B}_0 = -\varepsilon\mu\frac{\omega}{c}\mathbf{E}_0$$

which, in order to be compatible, require

$$|\mathbf{k}| = \frac{\omega}{c}\sqrt{\varepsilon\mu}, \quad |\mathbf{B}_0| = \sqrt{\varepsilon\mu}|\mathbf{E}_0| \tag{A.40}$$

As an application relevant to the optical properties of solids we consider a situation where electromagnetic radiation is incident on a solid from the vacuum, with the wave-vector of the radiation at a 90° angle to the surface plane (this is called normal incidence). First, we review the relevant boundary conditions. We denote the vacuum side by the index 1 and the solid side by the index 2. For the first two Maxwell equations, (A.27) and (A.28), we take the volume of integration to consist of an infinitesimal volume element with two surfaces parallel to the interface and negligible extent in the perpendicular direction, which gives

$$D_\perp^{(1)} - D_\perp^{(2)} = 4\pi\sigma_f, \quad B_\perp^{(1)} - B_\perp^{(2)} = 0$$

where σ_f is the free charge per unit area at the interface. Similarly, for the last two Maxwell equations, Eqs. (A.29) and (A.30), we take the surface of integration to consist of an infinitesimal surface element with two sides parallel to the interface and negligible extent in the perpendicular direction, which gives

$$E_\parallel^{(1)} - E_\parallel^{(2)} = 0, \quad H_\parallel^{(1)} - H_\parallel^{(2)} = \frac{4\pi}{c}\mathbf{K}_f \times \hat{\mathbf{n}}_s$$

where $\mathbf{K}_f$ is the free current per unit area at the interface and $\hat{\mathbf{n}}_s$ is the unit vector perpendicular to the surface element. We can also express the net fields $\mathbf{D}$ and $\mathbf{H}$ in terms of the total fields $\mathbf{E}$ and $\mathbf{B}$, using the dielectric constant and the magnetic permeability; only the first and last equations change, giving

$$\varepsilon^{(1)}E_\perp^{(1)} - \varepsilon^{(2)}E_\perp^{(2)} = 4\pi\sigma_f$$

$$\frac{1}{\mu^{(1)}}B_\parallel^{(1)} - \frac{1}{\mu^{(2)}}B_\parallel^{(2)} = \frac{4\pi}{c}\mathbf{K}_f \times \hat{\mathbf{n}}_s$$

We will next specify the physical situation to side 1 of the interface being the vacuum region, with $\varepsilon^{(1)} = 1$ and $\mu^{(1)} = 1$, and side 2 of the interface being the solid, with $\varepsilon^{(2)} = \varepsilon$ and $\mu^{(2)} \approx 1$ (most solids show negligible magnetic response). The direction of propagation of the radiation $\mathbf{k}$ will be taken perpendicular to the interface for normal incidence. We also assume that there are no free charges or free currents at the interface. This assumption is reasonable for metals where the presence of free carriers eliminates any charge accumulation. It also makes sense for semiconductors whose passivated, reconstructed surfaces correspond to filled bands which cannot carry current, hence there can be only bound charges. It is convenient to define the direction of propagation of the radiation as the z axis and the interface as the xy plane. Moreover, since the electric and magnetic field vectors are perpendicular to the direction of propagation and perpendicular to each other, we can choose them to define the x axis and y axis. The incident radiation will then be described by the fields

$$\mathbf{E}^{(I)}(\mathbf{r}, t) = E_0^{(I)}\mathrm{e}^{\mathrm{i}(kz-\omega t)}\hat{\mathbf{x}}, \quad \mathbf{B}^{(I)}(\mathbf{r}, t) = E_0^{(I)}\mathrm{e}^{\mathrm{i}(kz-\omega t)}\hat{\mathbf{y}}$$

The reflected radiation will propagate in the opposite direction with the same wave-vector and frequency:

$$\mathbf{E}^{(R)}(\mathbf{r}, t) = E_0^{(R)} \mathrm{e}^{\mathrm{i}(-kz-\omega t)} \hat{\mathbf{x}}, \quad \mathbf{B}^{(R)}(\mathbf{r}, t) = -E_0^{(R)} \mathrm{e}^{\mathrm{i}(-kz-\omega t)} \hat{\mathbf{y}}$$

where the negative sign in the expression for $\mathbf{B}^{(R)}$ is dictated by Eq. (A.39). Finally, the transmitted radiation will propagate in the same direction as the incident radiation and will have the same frequency but a different wave-vector given by $k' = k\sqrt{\varepsilon}$ because the speed of propagation has been reduced by a factor $\sqrt{\varepsilon}$, as we argued above. Therefore, the transmitted fields will be given by

$$\mathbf{E}^{(T)}(\mathbf{r}, t) = E_0^{(T)} \mathrm{e}^{\mathrm{i}(k'z-\omega t)} \hat{\mathbf{x}}, \quad \mathbf{B}^{(T)}(\mathbf{r}, t) = \sqrt{\varepsilon} E_0^{(T)} \mathrm{e}^{\mathrm{i}(k'z-\omega t)} \hat{\mathbf{y}}$$

where we have taken advantage of the result we derived above, Eq. (A.40), to express the magnitude of the magnetic field as $\sqrt{\varepsilon}$ times the magnitude of the electric field in the solid. The general boundary conditions we derived above applied to the situation at hand give

$$E_0^{(I)} + E_0^{(R)} = E_0^{(T)}, \quad E_0^{(I)} - E_0^{(R)} = \sqrt{\varepsilon} E_0^{(T)}$$

where we have used only the equations for the components parallel to the interface since there are no components perpendicular to the interface because those would correspond to longitudinal components in the electromagnetic waves. The equations we obtained can be easily solved for the amplitude of the transmitted and the reflected radiation in terms of the amplitude of the incident radiation, leading to

$$\left| \frac{E_0^{(T)}}{E_0^{(I)}} \right| = \left| \frac{2\sqrt{\varepsilon}}{\sqrt{\varepsilon}+1} \right|, \quad \left| \frac{E_0^{(R)}}{E_0^{(I)}} \right| = \left| \frac{\sqrt{\varepsilon}-1}{\sqrt{\varepsilon}+1} \right| \tag{A.41}$$

These ratios of the amplitudes are referred to as the transmission and reflection coefficients; their squares give the relative power of transmitted and reflected radiation.

As a final exercise, we consider the electromagnetic fields inside a solid in the presence of free charges and currents. We will assume again that the fields, as well as the free charge and current densities, can be described by plane waves:

$$\begin{aligned} \rho_f(\mathbf{r}, t) &= \rho_0 \mathrm{e}^{\mathrm{i}(\mathbf{k}\cdot\mathbf{r}-\omega t)}, \quad \mathbf{J}_f(\mathbf{r}, t) = \mathbf{J}_0 \mathrm{e}^{\mathrm{i}(\mathbf{k}\cdot\mathbf{r}-\omega t)} \\ \mathbf{E}(\mathbf{r}, t) &= \mathbf{E}_0 \mathrm{e}^{\mathrm{i}(\mathbf{k}\cdot\mathbf{r}-\omega t)}, \quad \mathbf{B}(\mathbf{r}, t) = \mathbf{B}_0 \mathrm{e}^{\mathrm{i}(\mathbf{k}\cdot\mathbf{r}-\omega t)} \end{aligned} \tag{A.42}$$

where all the quantities with subscript zero are functions of $\mathbf{k}$ and ω. We will also separate the fields in longitudinal (parallel to the wave-vector $\mathbf{k}$) and transverse (perpendicular to the wave-vector $\mathbf{k}$) components: $\mathbf{E}_0 = \mathbf{E}_{0,\parallel} + \mathbf{E}_{0,\perp}$, $\mathbf{B}_0 = \mathbf{B}_{0,\parallel} + \mathbf{B}_{0,\perp}$. From the general relations we derived earlier, we expect the transverse components of the fields to be perpendicular to each other. Accordingly, it is convenient to choose the direction of $\mathbf{k}$ as the z axis, the direction of $\mathbf{E}_{0,\perp}$ as the x axis and the direction of $\mathbf{B}_{0,\perp}$ as the y axis. We will also separate the current density into longitudinal and transverse components, $\mathbf{J}_0 = \mathbf{J}_{0,\parallel} + \mathbf{J}_{0,\perp}$, the direction of the latter component to be determined by Maxwell's equations. In the following, all quantities that are not in boldface represent the magnitude of the corresponding vectors, for instance, $k = |\mathbf{k}|$.

From the first Maxwell equation we obtain

$$\nabla_{\mathbf{r}} \cdot \mathbf{D}(\mathbf{r}, t) = 4\pi\rho_f(\mathbf{r}, t) \Rightarrow \mathrm{i}\mathbf{k} \cdot \varepsilon(\mathbf{E}_{0,\parallel} + \mathbf{E}_{0,\perp}) = 4\pi\rho_0 \Rightarrow E_{0,\parallel} = -\mathrm{i}\frac{4\pi}{\varepsilon}\frac{1}{k}\rho_0 \tag{A.43}$$

From the second Maxwell equation we obtain

$$\nabla_{\mathbf{r}} \cdot \mathbf{B}(\mathbf{r}, t) = 0 \Rightarrow \mathrm{i}\mathbf{k} \cdot (\mathbf{B}_{0,\parallel} + \mathbf{B}_{0,\perp}) = 0 \Rightarrow B_{0,\parallel} = 0 \tag{A.44}$$

From the third Maxwell equation we obtain

$$\nabla_{\mathbf{r}} \times \mathbf{E}(\mathbf{r}, t) = -\frac{1}{c}\frac{\partial \mathbf{B}(\mathbf{r}, t)}{\partial t} \Rightarrow \mathrm{i}\mathbf{k} \times (\mathbf{E}_{0,\parallel} + \mathbf{E}_{0,\perp}) = \mathrm{i}\frac{\omega}{c}\mathbf{B}_{0,\perp} \Rightarrow B_{0,\perp} = \frac{ck}{\omega}E_{0,\perp} \tag{A.45}$$

Finally, from the fourth Maxwell equation we obtain

$$\nabla_{\mathbf{r}} \times \mathbf{H}(\mathbf{r}, t) = \frac{4\pi}{c}\mathbf{J}_f(\mathbf{r}, t) + \frac{1}{c}\frac{\partial \mathbf{D}(\mathbf{r}, t)}{\partial t}$$

$$\Rightarrow \mathrm{i}\mathbf{k} \times \frac{1}{\mu}\mathbf{B}_{0,\perp} = \frac{4\pi}{c}(\mathbf{J}_{0,\parallel} + \mathbf{J}_{0,\perp}) - \mathrm{i}\frac{\omega}{c}\varepsilon(\mathbf{E}_{0,\parallel} + \mathbf{E}_{0,\perp}) \tag{A.46}$$

Separating components in the last equation, we find that $\mathbf{J}_{0,\perp}$ is only in the $\hat{\mathbf{x}}$ direction and that

$$E_{0,\parallel} = -\mathrm{i}\frac{4\pi}{\varepsilon}\frac{1}{\omega}J_{0,\parallel} \tag{A.47}$$

$$B_{0,\perp} = \mathrm{i}\frac{4\pi\mu}{kc}J_{0,\perp} + \frac{\omega}{kc}\varepsilon\mu E_{0,\perp} \tag{A.48}$$

In the last expression we use the result of Eq. (A.45) to obtain

$$E_{0,\perp} = -\mathrm{i}\frac{4\pi\omega}{c^2}\frac{\mu}{\omega^2\varepsilon\mu/c^2 - k^2}J_{0,\perp} \tag{A.49}$$

With this last result we have managed to determine all the field components in terms of the charge and current densities.

It is instructive to explore the consequences of this solution. First, we note that we have obtained $E_{0,\parallel}$ as two different expressions, Eq. (A.43) and (A.47), which must be compatible, requiring that

$$\frac{1}{k}\rho_0 = \frac{1}{\omega}J_{0,\parallel} \Rightarrow J_{0,\parallel}k = \omega\rho_0$$

which is of course true due to the charge conservation equation:

$$\nabla_{\mathbf{r}}\mathbf{J}_f(\mathbf{r}, t) = -\frac{\partial \rho_f(\mathbf{r}, t)}{\partial t} \Rightarrow \mathrm{i}\mathbf{k} \cdot (\mathbf{J}_{0,\parallel} + \mathbf{J}_{0,\perp}) = \mathrm{i}\omega\rho_0$$

the last equality implying a relation between $J_{0,\parallel}$ and ρ_0 identical to the previous one. Another interesting aspect of the solution we obtained is that the denominator appearing in the expression for $E_{0,\perp}$ in Eq. (A.49) cannot vanish for a physically meaningful solution. Thus, for these fields $k \neq (\omega/c)\sqrt{\varepsilon\mu}$, in contrast to what we found for the case of zero charge and current densities. We will define this denominator to be equal to $-\kappa^2$:

$$-\kappa^2 \equiv \frac{\omega^2}{c^2}\varepsilon\mu - k^2 \tag{A.50}$$

We also want to relate the electric field to the current density through Ohm's law, Eq. (A.20). We will use this requirement, as it applies to the transverse components, to relate the conductivity σ to κ through Eq. (A.49):

$$\sigma^{-1} = \mathrm{i}\frac{4\pi\omega}{c^2}\frac{\mu}{\kappa^2} \Rightarrow \kappa^2 = \mathrm{i}\sigma\frac{4\pi}{c^2}\mu\omega$$

The last expression for κ when substituted into its definition, Eq. (A.50), yields

$$k^2 = \frac{\omega^2}{c^2}\mu\varepsilon + \mathrm{i}\mu\omega\frac{4\pi}{c^2}\sigma$$

which is an interesting result, revealing that the wave-vector has now acquired an imaginary component. Indeed, expressing the wave-vector in terms of its real and imaginary parts, $k = k_R + \mathrm{i}k_I$, and using the above equation, we find

$$k_R = \frac{\omega}{c}\sqrt{\varepsilon\mu}\left[\frac{1}{2}\left(1 + \left(\frac{4\pi\sigma}{\varepsilon\omega}\right)^2\right)^{1/2} + \frac{1}{2}\right]^{1/2}$$

$$k_I = \frac{\omega}{c}\sqrt{\varepsilon\mu}\left[\frac{1}{2}\left(1 + \left(\frac{4\pi\sigma}{\varepsilon\omega}\right)^2\right)^{1/2} - \frac{1}{2}\right]^{1/2}$$

These expressions show that, for finite conductivity σ, when $\omega \to \infty$ the wave-vector reverts to a real quantity only ($k_I = 0$). The presence of an imaginary component in the wave-vector has important physical implications: it means that the fields decay exponentially inside the solid as $\sim \exp(-k_I z)$. For large enough frequency the imaginary component is negligible, that is, the solid is transparent to such radiation. We note incidentally that with the definition of κ given above, the longitudinal components of the electric field and the current density obey Ohm's law but with an extra factor multiplying the conductivity σ:

$$-\mathrm{i}\frac{4\pi}{\varepsilon}\frac{1}{\omega} = -\frac{1}{\sigma}\frac{\kappa^2 c^2}{\omega^2\mu\varepsilon} = \frac{1}{\sigma}\left(1 - \frac{k^2c^2}{\omega^2\mu\varepsilon}\right) \Rightarrow J_{0,l} = \sigma\left(1 - \frac{k^2c^2}{\omega^2\mu\varepsilon}\right)^{-1} E_{0,\parallel}$$

Finally, we will determine the vector and scalar potentials which can describe the electric and magnetic fields. We define the potentials in plane wave form as

$$\mathbf{A}(\mathbf{r}, t) = \mathbf{A}_0 \mathrm{e}^{\mathrm{i}(\mathbf{k}\cdot\mathbf{r}-\omega t)}, \quad \Phi(\mathbf{r}, t) = \Phi_0 \mathrm{e}^{\mathrm{i}(\mathbf{k}\cdot\mathbf{r}-\omega t)} \tag{A.51}$$

From the standard definition of the magnetic field in terms of the vector potential, Eq. (A.31) and using the Coulomb gauge, Eq. (A.16), we obtain that the vector potential must have only a transverse component in the same direction as the transverse magnetic field:

$$\mathbf{A}_{0,\perp} = -\mathrm{i}\frac{1}{k}\mathbf{B}_{0,\perp} \tag{A.52}$$

From the standard definition of the electric field in terms of the scalar and vector potentials, Eq. (A.32), we then deduce that the transverse components must obey

$$\mathbf{E}_{0,\perp} = \frac{\mathrm{i}\omega}{c}\mathbf{A}_{0,\perp} = \frac{\omega}{kc}\mathbf{B}_{0,\perp}$$

which is automatically satisfied because of Eq. (A.45), while the longitudinal components must obey

$$\mathbf{E}_{0,\parallel} = -\mathrm{i}\mathbf{k}\Phi_0 \Rightarrow \Phi_0 = \frac{4\pi}{\varepsilon}\frac{1}{k^2}\rho_0 \tag{A.53}$$

where for the last step we have used Eq. (A.43). It is evident that the last expression for the magnitude of the scalar potential is also compatible with Poisson's equation:

$$\nabla_{\mathbf{r}}^2 \Phi(\mathbf{r}, t) = -\frac{4\pi}{\varepsilon} \rho_f(\mathbf{r}, t) \Rightarrow \Phi_0 = \frac{4\pi}{\varepsilon} \frac{1}{|\mathbf{k}|^2} \rho_0$$

These results are useful in making the connection between the dielectric function and the conductivity from microscopic considerations, as discussed in chapter 5.

Further reading

1. *Introduction to Electrodynamics*, D.J. Griffiths (3rd edn, Prentice-Hall, New Jersey, 1999).
2. *Classical Electrodynamics*, J.D. Jackson (3rd edn, J. Wiley, New York, 1999).

Appendix B

Elements of quantum mechanics

Quantum mechanics is the theory that captures the particle–wave duality of matter. Quantum mechanics applies in the microscopic realm, that is, at length scales and at time scales relevant to subatomic particles like electrons and nuclei. It is the most successful physical theory: it has been verified by every experiment performed to check its validity. It is also the most counter-intuitive physical theory, since its premises are at variance with our everyday experience, which is based on macroscopic observations that obey the laws of classical physics. When the properties of physical objects (such as solids, clusters and molecules) are studied at a resolution at which the atomic degrees of freedom are explicitly involved, the use of quantum mechanics becomes necessary.

In this Appendix we attempt to give the basic concepts of quantum mechanics relevant to the study of solids, clusters and molecules, in a reasonably self-contained form but avoiding detailed discussions. We refer the reader to standard texts of quantum mechanics for more extensive discussion and proper justification of the statements that we present here, a couple of which are mentioned in the Further reading section.

B.1 The Schrödinger equation

There are different ways to formulate the theory of quantum mechanics. In the following we will discuss the Schrödinger wave mechanics picture. The starting point is the form of a free traveling wave

$$\psi(\mathbf{r}, t) = e^{i(\mathbf{k}\cdot\mathbf{r}-\omega t)} \tag{B.1}$$

of wave-vector $\mathbf{k}$ and frequency ω. The free traveling wave satisfies the equation

$$i\hbar\frac{\partial\psi(\mathbf{r}, t)}{\partial t} = -\frac{\hbar^2}{2m}\nabla_{\mathbf{r}}^2\psi(\mathbf{r}, t) \tag{B.2}$$

if the wave-vector and the frequency are related by

$$\hbar\omega = \frac{\hbar^2\mathbf{k}^2}{2m}$$

where $\hbar$, m are constants (for the definition of the operator $\nabla_{\mathbf{r}}^2$ see Appendix G). Schrödinger postulated that $\psi(\mathbf{r}, t)$ can also be considered to describe the motion of a free particle with mass m and momentum $\mathbf{p} = \hbar\mathbf{k}$, with $\hbar\omega = \mathbf{p}^2/2m$ the energy of the free particle. Thus, identifying

$$\mathrm{i}\hbar\frac{\partial}{\partial t} \rightarrow \epsilon : \text{energy}, \quad -\mathrm{i}\hbar\nabla_{\mathbf{r}} \rightarrow \mathbf{p} : \text{momentum}$$

introduces the quantum mechanical operators for the energy and the momentum for a free particle of mass m; $\hbar$ is related to Planck's constant h by

$$\hbar = \frac{h}{2\pi}$$

$\psi(\mathbf{r}, t)$ is the *wavefunction* whose absolute value squared, $|\psi(\mathbf{r}, t)|^2$, is interpreted as the probability of finding the particle at position $\mathbf{r}$ and time t.

When the particle is not free, we add to the wave equation the potential energy term $V(\mathbf{r}, t)\psi(\mathbf{r}, t)$, so that the equation obeyed by the wavefunction $\psi(\mathbf{r}, t)$ reads

$$\left[-\frac{\hbar^2\nabla_{\mathbf{r}}^2}{2m} + V(\mathbf{r}, t)\right]\psi(\mathbf{r}, t) = \mathrm{i}\hbar\frac{\partial\psi(\mathbf{r}, t)}{\partial t} \tag{B.3}$$

which is known as the time-dependent Schrödinger equation. If the absolute value squared of the wavefunction $\psi(\mathbf{r}, t)$ is to represent a probability, it must be properly normalized, that is,

$$\int |\psi(\mathbf{r}, t)|^2 \mathrm{d}\mathbf{r} = 1$$

so that the probability of finding the particle anywhere in space is unity. The wavefunction and its gradient must also be continuous and finite everywhere for this interpretation to have physical meaning. One other requirement on the wavefunction is that it decays to zero at infinity. If the external potential is independent of time, we can write the wavefunction as

$$\psi(\mathbf{r}, t) = \mathrm{e}^{-\mathrm{i}\epsilon t/\hbar}\phi(\mathbf{r}) \tag{B.4}$$

which, when substituted into the time-dependent Schrödinger equation gives

$$\left[-\frac{\hbar^2\nabla_{\mathbf{r}}^2}{2m} + V(\mathbf{r})\right]\phi(\mathbf{r}) = \epsilon\phi(\mathbf{r}) \tag{B.5}$$

This is known as the time-independent Schrödinger equation (TISE). The quantity inside the square brackets is called the hamiltonian, $\mathcal{H}$. In the TISE the hamiltonian corresponds to the energy operator. Notice that the energy ϵ has now become the eigenvalue of the wavefunction $\phi(\mathbf{r})$ in the second-order differential equation represented by the TISE.

In most situations of interest we are faced with the problem of solving the TISE once the potential $V(\mathbf{r})$ has been specified. The solution gives the eigenvalues ϵ and eigenfunctions $\phi(\mathbf{r})$, which together provide a complete description of the physical system. There are usually many (often infinite) solutions to the TISE, which are identified by their eigenvalues labeled by some index or set of indices, denoted here collectively by the subscript i:

$$\mathcal{H}\phi_i(\mathbf{r}) = \epsilon_i\phi_i(\mathbf{r})$$

It is convenient to choose the wavefunctions that correspond to different eigenvalues of the energy to be orthonormal:

$$\int \phi_i^*(\mathbf{r})\phi_j(\mathbf{r})\mathrm{d}\mathbf{r} = \delta_{ij} \tag{B.6}$$

Such a set of eigenfunctions is referred to as a complete basis set, spanning the Hilbert space of the hamiltonian $\mathcal{H}$. We can then use this set to express a general state of the system $\chi(\mathbf{r})$ as

$$\chi(\mathbf{r}) = \sum_i c_i \phi_i(\mathbf{r})$$

where the coefficients c_i, due to the orthonormality of the wavefunctions Eq. (B.6), are given by

$$c_i = \int \phi_i^*(\mathbf{r})\chi(\mathbf{r})\mathrm{d}\mathbf{r}$$

The notion of the Hilbert space of the hamiltonian is a very useful one: we imagine the eigenfunctions of the hamiltonian as the axes in a multi-dimensional space (the Hilbert space), in which the state of the system is a point. The position of this point is given by its projection on the axes, just like the position of a point in 3D space is given by its cartesian coordinates, which are the projections of the point on the x, y, z axes. In this sense, the coefficients c_i defined above are the projections of the state of the system on the basis set comprising the eigenfunctions of the hamiltonian.

A general feature of the wavefunction is that it has oscillating wave-like character when the potential energy is lower than the total energy (in which case the kinetic energy is positive) and decaying exponential behavior when the potential energy is higher than the total energy (in which case the kinetic energy is negative). This is most easily seen in a one-dimensional example, where the TISE can be written as

$$-\frac{\hbar^2}{2m}\frac{\mathrm{d}^2\phi(x)}{\mathrm{d}x^2} = [\epsilon - V(x)]\phi(x) \Longrightarrow \frac{\mathrm{d}^2\phi(x)}{\mathrm{d}x^2} = -[k(x)]^2\phi(x)$$

where we have defined the function

$$k(x) = \sqrt{\frac{2m}{\hbar^2}[\epsilon - V(x)]}$$

The above expression then shows that, if we treat k as constant,

$$\text{for } k^2 > 0 \longrightarrow \phi(x) \sim \mathrm{e}^{\pm \mathrm{i}|k|x}$$
$$\text{for } k^2 < 0 \longrightarrow \phi(x) \sim \mathrm{e}^{\pm |k|x}$$

and in the last expression we choose the sign that makes the wavefunction vanish for $x \to \pm\infty$ as the only physically plausible choice. This is illustrated in Fig. B.1 for a square barrier and a square well, so that in both cases the function $[k(x)]^2$ is a positive or

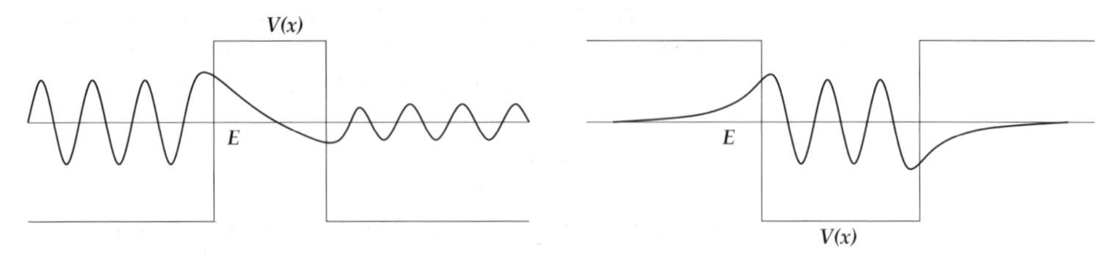

Figure B.1. Illustration of the oscillatory and decaying exponential nature of the wavefunction in regions where the potential energy $V(x)$ is lower than or higher than the total energy E. **Left:** a square barrier; **right:** a square well.

negative constant everywhere. Notice that in the square barrier, the wavefunction before and after the barrier has the same wave-vector (k takes the same value before and after the barrier), but the amplitude of the oscillation has decreased, because only part of the wave is transmitted while another part is reflected from the barrier. The points at which $[k(x)]^2$ changes sign are called the "turning points", because they correspond to the positions where a classical particle would be reflected at the walls of the barrier or the well. The quantum mechanical nature of the particle allows it to penetrate the walls of the barrier or leak out of the walls of the well, as a wave would. In terms of the wavefunction, a turning point corresponds to a value of x at which the curvature changes sign, that is, it is an inflection point.

B.2 Bras, kets and operators

Once the wavefunction of a state has been determined, the value of any physical observable can be calculated by taking the expectation value of the corresponding operator between the wavefunction and its complex conjugate. The expectation value of an operator $\mathcal{O}(\mathbf{r})$ in a state described by wavefunction $\phi(\mathbf{r})$ is defined as

$$\int \phi^*(\mathbf{r})\mathcal{O}(\mathbf{r})\phi(\mathbf{r})\mathrm{d}\mathbf{r}$$

An example of an operator is $-\mathrm{i}\hbar\nabla_{\mathbf{r}}$ for the momentum $\mathbf{p}$, as we saw earlier.

As an application of these concepts we consider the following operators in one dimension:

$$X = x - \bar{x}, \quad P = p - \bar{p}, \quad p = -\mathrm{i}\hbar\frac{\partial}{\partial x}$$

with $\bar{x}$ and $\bar{p}$ denoting the expectation values of x and p, respectively:

$$\bar{x} \equiv \int \phi^*(x)x\phi(x)\mathrm{d}x, \quad \bar{p} \equiv \int \phi^*(x)p\phi(x)\mathrm{d}x$$

First, notice that the expectation values of X and P, vanish identically. With the assumption that the wavefunction vanishes at infinity, it is possible to show that the expectation values of the squares of the operators X and P obey the following relation:

$$\int \phi^*(x)X^2\phi(x)\mathrm{d}x \int \phi^*(x)P^2\phi(x)\mathrm{d}x \geq \frac{\hbar^2}{4} \tag{B.7}$$

as shown in Problem 1. We can interpret the expectation value as the average, in which case the expectation value of X^2 is the standard deviation of x, also denoted as $(\Delta x)^2$, and similarly for P^2, with $(\Delta p)^2$ the standard deviation of p. Then Eq. (B.7) becomes

$$(\Delta x)(\Delta p) \geq \frac{\hbar}{2}$$

which is known as the **Heisenberg uncertainty relation**. This is seemingly abstract relation between the position and momentum variables can be used to yield very practical results. For instance, electrons associated with an atom are confined by the Coulomb potential of the nucleus to a region of ~ 1 Å, that is, $\Delta x \sim 1$ Å, which means that their typical momentum will be of order $p = \hbar/2\Delta x$. This gives a direct estimate of the energy scale for these electronic states:

$$\frac{p^2}{2m_e} = \frac{\hbar^2}{2m_e(2\Delta x)^2} \sim 1 \text{ eV}$$

which is very close to the binding energy scale for valence electrons (the core electrons are more tightly bound to the nucleus and therefore their binding energy is higher). The two variables linked by the Heisenberg uncertainty relation are referred to as conjugate variables. There exist other pairs of conjugate variables linked by the same relation, such as the energy ϵ and the time t:

$$(\Delta\epsilon)(\Delta t) \geq \frac{\hbar}{2}$$

In the calculation of expectation values it is convenient to introduce the so called "bra" $\langle\phi|$ and "ket" $|\phi\rangle$ notation with the first representing the wavefunction and the second its complex conjugate. In the bra and ket expressions the spatial coordinate **r** is left deliberately unspecified, so that they can be considered as wavefunctions independent of the representation; when the coordinate **r** is specified, the wavefunctions are considered to be expressed in the "position representation". Thus, the expectation value of an operator $\mathcal{O}$ in state ϕ is

$$\langle\phi|\mathcal{O}|\phi\rangle \equiv \int \phi^*(\mathbf{r})\mathcal{O}(\mathbf{r})\phi(\mathbf{r})\mathrm{d}\mathbf{r}$$

where the left-hand-side expression is independent of representation and the right-hand-side expression is in the position representation. In terms of the bra and ket notation, the orthonormality of the energy eigenfunctions can be expressed as

$$\langle\phi_i|\phi_j\rangle = \delta_{ij}$$

and the general state of the system χ can be expressed as

$$|\chi\rangle = \sum_i \langle\phi_i|\chi\rangle|\phi_i\rangle$$

from which we can deduce that the expression

$$\sum_i |\phi_i\rangle\langle\phi_i| = 1$$

is the identity operator. The usefulness of the above expressions is their general form, which is independent of representation.

This representation-independent notation can be extended to the time-dependent wavefunction $\psi(\mathbf{r}, t)$, leading to an elegant expression. We take advantage of the series expansion of the exponential to define the following operator:

$$\mathrm{e}^{-\mathrm{i}\mathcal{H}t/\hbar} = \sum_{n=0}^{\infty} \frac{(-\mathrm{i}\mathcal{H}t/\hbar)^n}{n!} \tag{B.8}$$

where $\mathcal{H}$ is the hamiltonian, which we assume to contain a time-independent potential. When this operator is applied to the time-independent part of the wavefunction, it gives

$$\mathrm{e}^{-\mathrm{i}\mathcal{H}t/\hbar}\phi(\mathbf{r}) = \sum_{n=0}^{\infty} \frac{(-\mathrm{i}\mathcal{H}t/\hbar)^n}{n!}\phi(\mathbf{r}) = \sum_{n=0}^{\infty} \frac{(-\mathrm{i}\epsilon t/\hbar)^n}{n!}\phi(\mathbf{r})$$
$$= \mathrm{e}^{-\mathrm{i}\epsilon t/\hbar}\phi(\mathbf{r}) = \psi(\mathbf{r}, t) \tag{B.9}$$

where we have used Eq. (B.5) and the definition of the time-dependent wavefunction Eq. (B.4). This shows that in general we can write the time-dependent wavefunction as the operator $\exp(-\mathrm{i}\mathcal{H}t/\hbar)$ applied to the wavefunction at $t = 0$. In a representation-independent expression, this statement gives

$$|\psi(t)\rangle = \mathrm{e}^{-\mathrm{i}\mathcal{H}t/\hbar}|\psi(0)\rangle \rightarrow \langle\psi(t)| = \langle\psi(0)|\mathrm{e}^{\mathrm{i}\mathcal{H}t/\hbar} \tag{B.10}$$

with the convention that for a bra, the operator next to it acts to the left.

Now consider a general operator $\mathcal{O}$ corresponding to a physical observable; we assume that $\mathcal{O}$ itself is a time-independent operator, but the value of the observable changes with time because of changes in the wavefunction:

$$\langle\mathcal{O}\rangle(t) \equiv \langle\psi(t)|\mathcal{O}|\psi(t)\rangle = \langle\psi(0)|\mathrm{e}^{\mathrm{i}\mathcal{H}t/\hbar}\mathcal{O}\mathrm{e}^{-\mathrm{i}\mathcal{H}t/\hbar}|\psi(0)\rangle \tag{B.11}$$

We now define a new operator

$$\mathcal{O}(t) \equiv \mathrm{e}^{\mathrm{i}\mathcal{H}t/\hbar}\mathcal{O}\mathrm{e}^{-\mathrm{i}\mathcal{H}t/\hbar}$$

which includes explicitly the time dependence, and whose expectation value in the state $|\psi(0)\rangle$ is exactly the same as the expectation value of the original time-independent operator in the state $|\psi(t)\rangle$. Working with the operator $\mathcal{O}(t)$ and the state $|\psi(0)\rangle$ is called the "Heisenberg picture" while working with the operator $\mathcal{O}$ and the state $|\psi(t)\rangle$ is called the "Schrödinger picture". The two pictures give identical results as far as the values of physical observables are concerned, as Eq. (B.11) shows, so the choice of one over the other is a matter of convenience. In the Heisenberg picture the basis is fixed and the operator evolves in time, whereas in the Schrödinger picture the basis evolves and the operator is independent of time. We can also determine the evolution of the time-dependent operator from its definition, as follows:

$$\frac{\mathrm{d}}{\mathrm{d}t}\mathcal{O}(t) = \frac{\mathrm{i}\mathcal{H}}{\hbar}\mathrm{e}^{\mathrm{i}\mathcal{H}t/\hbar}\mathcal{O}\mathrm{e}^{-\mathrm{i}\mathcal{H}t/\hbar} - \frac{\mathrm{i}}{\hbar}\mathrm{e}^{\mathrm{i}\mathcal{H}t/\hbar}\mathcal{O}\mathcal{H}\mathrm{e}^{-\mathrm{i}\mathcal{H}t/\hbar} = \frac{\mathrm{i}}{\hbar}\mathcal{H}\mathcal{O}(t) - \frac{\mathrm{i}}{\hbar}\mathcal{O}(t)\mathcal{H}$$

$$\Longrightarrow \frac{\mathrm{d}}{\mathrm{d}t}\mathcal{O}(t) = \frac{\mathrm{i}}{\hbar}[\mathcal{H}, \mathcal{O}(t)] \tag{B.12}$$

The last expression is defined as "the commutator" of the hamiltonian with the time-dependent operator $\mathcal{O}(t)$. The commutator is a general concept that applies to any pair of operators $\mathcal{O}_1$, $\mathcal{O}_2$:

$$\text{commutator} : [\mathcal{O}_1, \mathcal{O}_2] \equiv \mathcal{O}_1\mathcal{O}_2 - \mathcal{O}_2\mathcal{O}_1$$

The bra and ket notation can be extended to situations that involve more than one particle, as in the many-body wavefunction relevant to electrons in a solid. For example, such a many-body wavefunction may be denoted by $|\Psi\rangle$, and when expressed in the position representation it takes the form

$$\langle\mathbf{r}_1, \ldots, \mathbf{r}_N|\Psi\rangle \equiv \Psi(\mathbf{r}_1, \ldots, \mathbf{r}_N)$$

where N is the total number of particles in the system. When such a many-body wavefunction refers to a system of indistinguishable particles it must have certain symmetries: in the case of fermions it is antisymmetric (it changes sign upon interchange of all the coordinates of any two of the particles) while for bosons it is symmetric (it is the same upon interchange of all the coordinates of any two of the particles). We can define operators relevant to many-body wavefunctions in the usual way. A useful example is the density operator: it represents the probability of finding any of the particles involved in the wavefunction at a certain position in space. For one particle by itself, in state $\phi(\mathbf{r})$, the meaning we assigned to the wavefunction already gives this probability as

$$|\phi(\mathbf{r})|^2 = n(\mathbf{r})$$

which can also be thought of as the density at $\mathbf{r}$, since the integral over all space gives the total number of particles (in this case, unity):

$$\int n(\mathbf{r})\mathrm{d}\mathbf{r} = \int \phi^*(\mathbf{r})\phi(\mathbf{r})\mathrm{d}\mathbf{r} = 1$$

The corresponding operator must be defined as $\delta(\mathbf{r} - \mathbf{r}')$, with the second variable an arbitrary position in space. This choice of the density operator, when we take its matrix elements in the state $|\phi\rangle$ by inserting a complete set of states in the position representation, gives

$$\begin{aligned}\langle\phi|\delta(\mathbf{r} - \mathbf{r}')|\phi\rangle &= \int \langle\phi|\mathbf{r}'\rangle\delta(\mathbf{r} - \mathbf{r}')\langle\mathbf{r}'|\phi\rangle\mathrm{d}\mathbf{r}' \\ &= \int \phi^*(\mathbf{r}')\delta(\mathbf{r} - \mathbf{r}')\phi(\mathbf{r}')\mathrm{d}\mathbf{r}' = |\phi(\mathbf{r})|^2 = n(\mathbf{r})\end{aligned}$$

as desired. Generalizing this result to the N-particle system, we define the density operator as

$$\mathcal{N}(\mathbf{r}) = \sum_{i=1}^{N} \delta(\mathbf{r} - \mathbf{r}_i)$$

with $\mathbf{r}_i$, $i = 1, \ldots, N$ the variables describing the positions of the particles. The expectation value of this operator in the many-body wavefunction gives the particle density at $\mathbf{r}$:

$$n(\mathbf{r}) = \langle\Psi|\mathcal{N}(\mathbf{r})|\Psi\rangle$$

In the position representation this takes the form

$$\begin{aligned}n(\mathbf{r}) &= \int \langle\Psi|\mathbf{r}_1, \ldots, \mathbf{r}_N\rangle\mathcal{N}(\mathbf{r})\langle\mathbf{r}_1, \ldots, \mathbf{r}_N|\Psi\rangle\mathrm{d}\mathbf{r}_1 \cdots \mathrm{d}\mathbf{r}_N \\ &= \int \Psi^*(\mathbf{r}_1, \ldots, \mathbf{r}_N)\sum_{i=1}^{N}\delta(\mathbf{r} - \mathbf{r}_i)\Psi(\mathbf{r}_1, \ldots, \mathbf{r}_N)\mathrm{d}\mathbf{r}_1 \cdots \mathrm{d}\mathbf{r}_N \\ &= N\int \Psi^*(\mathbf{r}, \mathbf{r}_2, \ldots, \mathbf{r}_N)\Psi(\mathbf{r}, \mathbf{r}_2, \ldots, \mathbf{r}_N)\mathrm{d}\mathbf{r}_2 \cdots \mathrm{d}\mathbf{r}_N \qquad \text{(B.13)}\end{aligned}$$

where the last equation applies to a system of N indistinguishable particles. By analogy to the expression for the density of a system of indistinguishable particles, we can define a function of two independent variables $\mathbf{r}$ and $\mathbf{r}'$, the so called one-particle density matrix $\gamma(\mathbf{r}, \mathbf{r}')$:

$$\gamma(\mathbf{r}, \mathbf{r}') \equiv N\int \Psi^*(\mathbf{r}, \mathbf{r}_2, \ldots, \mathbf{r}_N)\Psi(\mathbf{r}', \mathbf{r}_2, \ldots, \mathbf{r}_N)\mathrm{d}\mathbf{r}_2 \cdots \mathrm{d}\mathbf{r}_N \qquad \text{(B.14)}$$

whose diagonal components are equal to the density: $\gamma(\mathbf{r}, \mathbf{r}) = n(\mathbf{r})$. An extension of these concepts is the pair correlation function, which describes the probability of finding two particles simultaneously at positions $\mathbf{r}$ and $\mathbf{r}'$. The operator for the pair correlation function is

$$\mathcal{G}(\mathbf{r}, \mathbf{r}') = \frac{1}{2} \sum_{i \neq j=1}^{N} \delta(\mathbf{r} - \mathbf{r}_i)\delta(\mathbf{r}' - \mathbf{r}_j)$$

and its expectation value in the many-body wavefunction in the position representation gives $g(\mathbf{r}, \mathbf{r}')$:

$$\begin{aligned} g(\mathbf{r}, \mathbf{r}') &= \int \langle \Psi | \mathbf{r}_1, \ldots, \mathbf{r}_N \rangle \mathcal{G}(\mathbf{r}, \mathbf{r}') \langle \mathbf{r}_1, \ldots, \mathbf{r}_N | \Psi \rangle \mathrm{d}\mathbf{r}_1 \cdots \mathrm{d}\mathbf{r}_N \\ &= \int \Psi^*(\mathbf{r}_1, \ldots, \mathbf{r}_N) \frac{1}{2} \sum_{i \neq j=1}^{N} \delta(\mathbf{r} - \mathbf{r}_i)\delta(\mathbf{r}' - \mathbf{r}_j) \Psi(\mathbf{r}_1, \ldots, \mathbf{r}_N) \mathrm{d}\mathbf{r}_1 \cdots \mathrm{d}\mathbf{r}_N \\ &= \frac{N(N-1)}{2} \int \Psi^*(\mathbf{r}, \mathbf{r}', \mathbf{r}_3, \ldots, \mathbf{r}_N) \Psi(\mathbf{r}, \mathbf{r}', \mathbf{r}_3, \ldots, \mathbf{r}_N) \mathrm{d}\mathbf{r}_3 \cdots \mathrm{d}\mathbf{r}_N \end{aligned}$$

where the last equation applies to a system of N indistinguishable particles. By analogy to the expression for the pair-correlation function of a system of indistinguishable particles, we can define a function of four independent variables $\mathbf{r}_1, \mathbf{r}_2, \mathbf{r}_1', \mathbf{r}_2'$, the so called two-particle density matrix $\Gamma(\mathbf{r}_1, \mathbf{r}_2 | \mathbf{r}_1', \mathbf{r}_2')$:

$$\Gamma(\mathbf{r}_1, \mathbf{r}_2 | \mathbf{r}_1', \mathbf{r}_2') \equiv \frac{N(N-1)}{2} \int \Psi^*(\mathbf{r}_1, \mathbf{r}_2, \mathbf{r}_3, \ldots, \mathbf{r}_N) \Psi(\mathbf{r}_1', \mathbf{r}_2', \mathbf{r}_3, \ldots, \mathbf{r}_N) \mathrm{d}\mathbf{r}_3 \cdots \mathrm{d}\mathbf{r}_N \quad \text{(B.15)}$$

whose diagonal components are equal to the pair-correlation function: $\Gamma(\mathbf{r}, \mathbf{r}' | \mathbf{r}, \mathbf{r}') = g(\mathbf{r}, \mathbf{r}')$. These functions are useful when dealing with one-body and two-body operators in the hamiltonian of the many-body system. An example of this use is given below, after we have defined the many-body wavefunction in terms of single-particle states.

The density matrix concept can be generalized to n particles with $n \leq N$ (see Problem 3):

$$\begin{aligned} \Gamma(\mathbf{r}_1, \mathbf{r}_2, \ldots, \mathbf{r}_n | \mathbf{r}_1', \mathbf{r}_2', \ldots, \mathbf{r}_n') &= \frac{N!}{n!(N-n)!} \int \int \cdots \int \Psi^*(\mathbf{r}_1, \mathbf{r}_2, \ldots, \mathbf{r}_n, \mathbf{r}_{n+1}, \ldots, \mathbf{r}_N) \\ &\quad \times \Psi(\mathbf{r}_1', \mathbf{r}_2', \ldots, \mathbf{r}_n', \mathbf{r}_{n+1}, \ldots, \mathbf{r}_N) d\mathbf{r}_{n+1} \cdots d\mathbf{r}_N \end{aligned}$$

The many-body wavefunction is often expressed as a product of single-particle wavefunctions, giving rise to expressions that include many single-particle states in the bra and the ket, as in the Hartree and Hartree–Fock theories discussed in chapter 2:

$$|\Psi\rangle = |\phi_1 \cdots \phi_N\rangle$$

In such cases we adopt the convention that the order of the single-particle states in the bra or the ket is meaningful, that is, when expressed in a certain representation the nth independent variable of the representation is associated with the nth single-particle state in the order it appears in the many-body wavefunction; for example, in the position representation we will have

$$\langle \mathbf{r}_1, \ldots, \mathbf{r}_N | \Psi \rangle = \langle \mathbf{r}_1, \ldots, \mathbf{r}_N | \phi_1 \cdots \phi_N \rangle \equiv \phi_1(\mathbf{r}_1) \cdots \phi_N(\mathbf{r}_N)$$

Thus, when expressing matrix elements of the many-body wavefunction in the position representation, the set of variables appearing as arguments in the single-particle states of the bra and the ket must be in exactly the same order; for example, in the Hartree theory, Eq. (2.10), the Coulomb repulsion term is represented by

$$\langle \phi_i \phi_j \mid \frac{1}{\mid \mathbf{r} - \mathbf{r}' \mid} \mid \phi_i \phi_j \rangle \equiv \int \phi_i^*(\mathbf{r}) \phi_j^*(\mathbf{r}') \frac{1}{\mid \mathbf{r} - \mathbf{r}' \mid} \phi_i(\mathbf{r}) \phi_j(\mathbf{r}') d\mathbf{r} d\mathbf{r}'$$

Similarly, in the Hartree–Fock theory, the exchange term is given by

$$\langle \phi_i \phi_j \mid \frac{1}{\mid \mathbf{r} - \mathbf{r}' \mid} \mid \phi_j \phi_i \rangle \equiv \int \phi_i^*(\mathbf{r}) \phi_j^*(\mathbf{r}') \frac{1}{\mid \mathbf{r} - \mathbf{r}' \mid} \phi_j(\mathbf{r}) \phi_i(\mathbf{r}') d\mathbf{r} d\mathbf{r}'$$

When only one single-particle state is involved in the bra and the ket but more than one variables appear in the bracketed operator, the variable of integration is evident from the implied remaining free variable; for example, in Eq. (2.11) all terms in the square brackets are functions of $\mathbf{r}$, therefore the term involving the operator $1/|\mathbf{r} - \mathbf{r}'|$, the so called Hartree potential $V^H(\mathbf{r})$, must be

$$V^H(\mathbf{r}) = e^2 \sum_{j \neq i} \langle \phi_j \mid \frac{1}{\mid \mathbf{r} - \mathbf{r}' \mid} \mid \phi_j \rangle = e^2 \sum_{j \neq i} \int \phi_j^*(\mathbf{r}') \frac{1}{\mid \mathbf{r} - \mathbf{r}' \mid} \phi_j(\mathbf{r}') d\mathbf{r}'$$

An expression for the many-body wavefunction in terms of products of single-particle states which is by construction totally antisymmetric, that is, it changes sign upon interchange of the coordinates of any two particles, is the so called Slater determinant:

$$\Psi(\{\mathbf{r}_i\}) = \frac{1}{\sqrt{N!}} \begin{vmatrix} \phi_1(\mathbf{r}_1) & \phi_1(\mathbf{r}_2) & \cdots & \phi_1(\mathbf{r}_N) \\ \phi_2(\mathbf{r}_1) & \phi_2(\mathbf{r}_2) & \cdots & \phi_2(\mathbf{r}_N) \\ \cdot & \cdot & & \cdot \\ \cdot & \cdot & & \cdot \\ \cdot & \cdot & & \cdot \\ \phi_N(\mathbf{r}_1) & \phi_N(\mathbf{r}_2) & \cdots & \phi_N(\mathbf{r}_N) \end{vmatrix} \tag{B.16}$$

The antisymmetric nature of this expression comes from the fact that if two rows or two columns of a determinant are interchanged, which corresponds to interchanging the coordinates of two particles, then the determinant changes sign. This expression is particularly useful when dealing with systems of fermions.

As an example of the various concepts introduced above we describe how the expectation value of an operator $\mathcal{O}$ of a many-body system, that can be expressed as a sum of single-particle operators

$$\mathcal{O}(\{\mathbf{r}_i\}) = \sum_{i=1}^{N} o(\mathbf{r}_i)$$

can be obtained from the matrix elements of the single-particle operators in the single-particle states used to express the many-body wavefunction. We will assume that we are dealing with a system of fermions described by a many-body wavefunction which has the form of a Slater determinant. In this case, the many-body wavefunction can be expanded as

$$\Psi^N(\mathbf{r}_1, \ldots, \mathbf{r}_N) = \frac{1}{\sqrt{N}} \left[\phi_1(\mathbf{r}_i) \Psi_{1,i}^{N-1} - \phi_2(\mathbf{r}_i) \Psi_{2,i}^{N-1} + \cdots \right] \tag{B.17}$$

where the $\Psi_{n,i}^{N-1}$ are determinants of size $N - 1$ from which the row and column corresponding to states $\phi_n(\mathbf{r}_i)$ are missing. With this, the expectation value of $\mathcal{O}$ takes the form

$$\langle \mathcal{O} \rangle \equiv \langle \Psi^N | \mathcal{O} | \Psi^N \rangle$$
$$= \frac{1}{N} \sum_i \int \left[\phi_1^*(\mathbf{r}_i) o(\mathbf{r}_i) \phi_1(\mathbf{r}_i) + \phi_2^*(\mathbf{r}_i) o(\mathbf{r}_i) \phi_2(\mathbf{r}_i) + \cdots \right] \mathrm{d}\mathbf{r}_i$$

where the integration over all variables other than $\mathbf{r}_i$, which is involved in $o(\mathbf{r}_i)$, gives unity for properly normalized single-particle states. The one-particle density matrix in the single-particle basis is expressed as

$$\gamma(\mathbf{r}, \mathbf{r}') = \sum_{n=1}^{N} \phi_n(\mathbf{r}) \phi_n^*(\mathbf{r}') = \sum_{n=1}^{N} \langle \mathbf{r} | \phi_n \rangle \langle \phi_n | \mathbf{r}' \rangle \tag{B.18}$$

which gives for the expectation value of $\mathcal{O}$

$$\langle \mathcal{O} \rangle = \int \left[o(\mathbf{r}) \gamma(\mathbf{r}, \mathbf{r}') \right]_{\mathbf{r}'=\mathbf{r}} \mathrm{d}\mathbf{r} \tag{B.19}$$

With the following definitions of $\gamma_{n,n'}$, $o_{n,n'}$:

$$\gamma_{n',n} = \langle \phi_{n'} | \gamma(\mathbf{r}', \mathbf{r}) | \phi_n \rangle, \qquad o_{n',n} = \langle \phi_{n'} | o(\mathbf{r}) | \phi_n \rangle \tag{B.20}$$

where the brackets imply integration over all real-space variables that appear in the operators, we obtain the general expression for the expectation value of $\mathcal{O}$:

$$\langle \mathcal{O} \rangle = \sum_{n,n'} o_{n,n'} \gamma_{n',n} \tag{B.21}$$

This expression involves exclusively matrix elements of the single-particle operators $o(\mathbf{r})$ and the single-particle density matrix $\gamma(\mathbf{r}, \mathbf{r}')$ in the single-particle states $\phi_n(\mathbf{r})$, which is very convenient for actual calculations of physical properties (see, for example, the discussion of the dielectric function in chapter 6).

B.3 Solution of the TISE

We discuss here some representative examples of how the TISE is solved to determine the eigenvalues and eigenfunctions of some potentials that appear frequently in relation to the physics of solids. These include free particles and particles in a harmonic oscillator potential or a Coulomb potential.

B.3.1 Free particles

For free particles the external potential is zero everywhere. We have already seen that in this case the time-independent part of the wavefunction is

$$\phi(\mathbf{r}) = C \mathrm{e}^{\mathrm{i}\mathbf{k}\cdot\mathbf{r}}$$

which describes the spatial variation of a plane wave; the constant C is the normalization. The energy eigenvalue corresponding to such a wavefunction is simply

$$\epsilon_{\mathbf{k}} = \frac{\hbar^2 \mathbf{k}^2}{2m}$$

which is obtained directly by substituting $\phi(\mathbf{r})$ in the TISE. All that remains is to determine the constant of normalization. To this end we assume that the particle is inside a box of dimensions $(2L_x, 2L_y, 2L_z)$ in cartesian coordinates, that is, the values of x range between $-L_x$, L_x and similarly for the other two coordinates. The wavefunction must vanish at the boundaries of the box, or equivalently it must have the same value at the two edges in each direction, which implies that

$$\mathbf{k} = k_x\hat{\mathbf{x}} + k_y\hat{\mathbf{y}} + k_z\hat{\mathbf{z}}$$

with

$$k_x = \frac{\pi n_x}{L_x},\ k_y = \frac{\pi n_y}{L_y},\ k_z = \frac{\pi n_z}{L_z}$$

where n_x, n_y, n_z are integers. From the form of the wavefunction we find that

$$\int |\phi(\mathbf{r})|^2 d\mathbf{r} = \Omega|C|^2$$

where $\Omega = (2L_x)(2L_y)(2L_z)$ is the volume of the box. This shows that we can choose

$$C = \frac{1}{\sqrt{\Omega}}$$

for the normalization, which completes the description of wavefunctions for free particles in a box. For $L_x, L_y, L_z \to \infty$ the spacing of values of k_x, k_y, k_z becomes infinitesimal, that is, $\mathbf{k}$ becomes a continuous variable. Since the value of $\mathbf{k}$ specifies the wavefunction, we can use it as the only index to identify the wavefunctions of free particles:

$$\phi_{\mathbf{k}}(\mathbf{r}) = \frac{1}{\sqrt{\Omega}} e^{i\mathbf{k}\cdot\mathbf{r}}$$

These results are also related to the Fourier and inverse Fourier transforms, which are discussed in Appendix G. We also notice that

$$\langle\phi_{\mathbf{k}'}|\phi_{\mathbf{k}}\rangle = \frac{1}{\Omega}\int e^{-i\mathbf{k}'\cdot\mathbf{r}} e^{i\mathbf{k}\cdot\mathbf{r}} d\mathbf{r} = 0, \quad \text{unless } \mathbf{k} = \mathbf{k}'$$

which we express by the statement that wavefunctions corresponding to different wave-vectors are *orthogonal* (see also Appendix G, the discussion of the δ-function and its Fourier representation). The wavefunctions we found above are also eigenfunctions of the momentum operator with momentum eigenvalues $\mathbf{p} = \hbar\mathbf{k}$:

$$-i\hbar\nabla_{\mathbf{r}}\phi_{\mathbf{k}}(\mathbf{r}) = -i\hbar\nabla_{\mathbf{r}}\frac{1}{\sqrt{\Omega}} e^{i\mathbf{k}\cdot\mathbf{r}} = (\hbar\mathbf{k})\phi_{\mathbf{k}}(\mathbf{r})$$

Thus, the free-particle eigenfunctions are an example where the energy eigenfunctions are also eigenfunctions of some other operator; in such cases, the hamiltonian and this other operator commute, that is,

$$[\mathcal{H}, \mathcal{O}] = \mathcal{H}\mathcal{O} - \mathcal{O}\mathcal{H} = 0$$

B.3.2 Harmonic oscillator potential

We consider a particle of mass m in a harmonic oscillator potential in one dimension:

$$\left[-\frac{\hbar^2}{2m}\frac{d^2}{dx^2} + \frac{1}{2}\kappa x^2\right]\phi(x) = \epsilon\phi(x) \tag{B.22}$$

where κ is the spring constant. This is a potential that arises frequently in realistic applications because near the minimum of the potential energy at $\mathbf{r} = \mathbf{r}_0$ the behavior is typically quadratic for small deviations, from a Taylor expansion:

$$V(\mathbf{r}) = V(\mathbf{r}_0) + \frac{1}{2}\left[\nabla_{\mathbf{r}}^2 V(\mathbf{r})\right]_{\mathbf{r}=\mathbf{r}_0} (\mathbf{r} - \mathbf{r}_0)^2$$

with the first derivative of the potential vanishing by definition at the minimum. We can take the position of the minimum $\mathbf{r}_0$ to define the origin of the coordinate system, and use a separable form of the wavefunction $\phi(\mathbf{r}) = \phi_1(x)\phi_2(y)\phi_3(z)$ to arrive at Eq. (B.22) for each spatial coordinate separately with $\left[\nabla_{\mathbf{r}}^2 V(\mathbf{r})\right]_0 = \kappa$. We will find it convenient to introduce the frequency of the oscillator,

$$\omega = \sqrt{\frac{\kappa}{m}}$$

The following change of variables:

$$\alpha = \sqrt{\frac{\omega m}{\hbar}}, \quad \gamma = \frac{2\epsilon}{\hbar\omega}, \quad u = \alpha x, \quad \phi(x) = C H(u) \mathrm{e}^{-u^2/2}$$

produces a differential equation of the form

$$\frac{\mathrm{d}^2 H(u)}{\mathrm{d}u^2} - 2u\frac{\mathrm{d}H(u)}{\mathrm{d}u} + (\gamma - 1)H(u) = 0$$

which, with $\gamma = 2n + 1$ and n an integer, is solved by the so called Hermite polynomials, defined recursively as

$$H_{n+1}(u) = 2u H_n(u) - 2n H_{n-1}(u)$$
$$H_0(u) = 1, \quad H_1(u) = 2u$$

With these polynomials, the wavefunction of the harmonic oscillator becomes

$$\phi_n(x) = C_n H_n(\alpha x)\mathrm{e}^{-\alpha^2 x^2/2} \tag{B.23}$$

The Hermite polynomials satisfy the relation

$$\int_{-\infty}^{\infty} H_n(u) H_m(u) \mathrm{e}^{-u^2} \mathrm{d}u = \delta_{nm}\sqrt{\pi}2^n n! \tag{B.24}$$

that is, they are orthogonal with a weight function $\exp(-u^2)$, which makes the wavefunctions $\phi_n(x)$ orthogonal. The above relation also allows us to determine the normalization C_n:

$$C_n = \sqrt{\frac{\alpha}{\sqrt{\pi}2^n n!}} \tag{B.25}$$

which completely specifies the wavefunction with index n. The lowest six wavefunctions ($n = 0 - 5$) are given explicitly in Table B.1 and shown in Fig. B.2. We note that the original equation then takes the form

$$\left[-\frac{\hbar^2}{2m}\frac{\mathrm{d}^2}{\mathrm{d}x^2} + \frac{1}{2}\kappa x^2\right]\phi_n(x) = \left(n + \frac{1}{2}\right)\hbar\omega\phi_n(x) \tag{B.26}$$

Table B.1. *Solutions of the one-dimensional harmonic oscillator potential.*

The lowest six states ($n = 0 - 5$) are given; ϵ_n is the energy, $H_n(u)$ is the Hermite polynomial and $\phi_n(x)$ is the full wavefunction, including the normalization.

n	ϵ_n	$H_n(u)$	$\phi_n(x)$
0	$\frac{1}{2}\hbar\omega$	1	$\sqrt{\frac{\alpha}{\sqrt{\pi}}}\mathrm{e}^{-\alpha^2x^2/2}$
1	$\frac{3}{2}\hbar\omega$	$2u$	$\sqrt{\frac{2\alpha}{\sqrt{\pi}}}\alpha x\mathrm{e}^{-\alpha^2x^2/2}$
2	$\frac{5}{2}\hbar\omega$	$4(u^2 - \frac{1}{2})$	$\sqrt{\frac{2\alpha}{\sqrt{\pi}}}(\alpha^2x^2 - \frac{1}{2})\mathrm{e}^{-\alpha^2x^2/2}$
3	$\frac{7}{2}\hbar\omega$	$8(u^3 - \frac{3}{2}u)$	$\sqrt{\frac{4\alpha}{3\sqrt{\pi}}}(\alpha^3x^3 - \frac{3}{2}\alpha x)\mathrm{e}^{-\alpha^2x^2/2}$
4	$\frac{9}{2}\hbar\omega$	$16(u^4 - 3u^2 + \frac{3}{4})$	$\sqrt{\frac{2\alpha}{3\sqrt{\pi}}}(\alpha^4x^4 - 3\alpha^2x^2 + \frac{3}{4})\mathrm{e}^{-\alpha^2x^2/2}$
5	$\frac{11}{2}\hbar\omega$	$32(u^5 - 5u^3 + \frac{15}{4}u)$	$\sqrt{\frac{4\alpha}{15\sqrt{\pi}}}(\alpha^5x^5 - 5\alpha^3x^3 + \frac{15}{4}\alpha x)\mathrm{e}^{-\alpha^2x^2/2}$

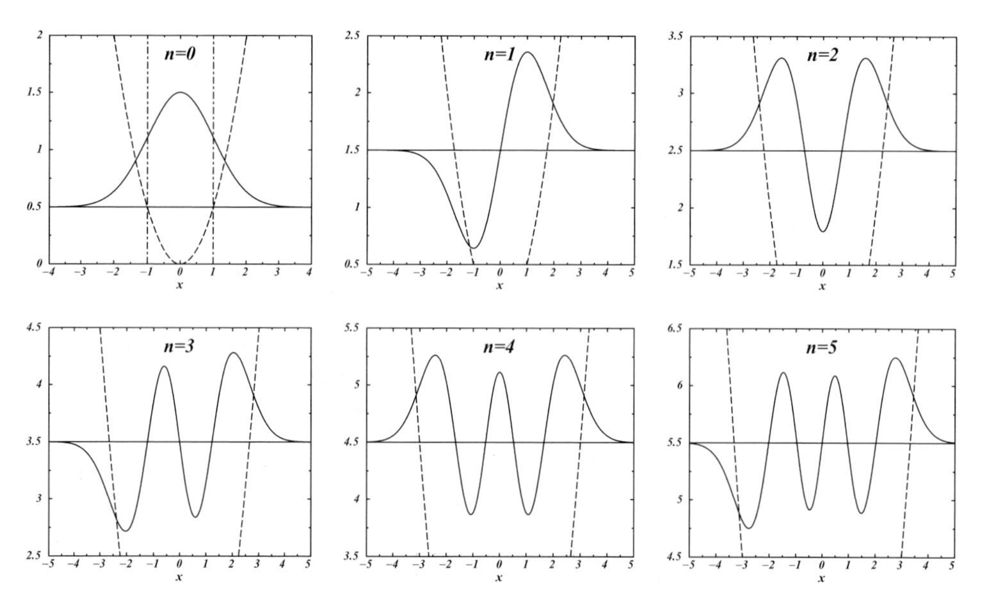

Figure B.2. The lowest six eigenfunctions ($n = 0 - 5$) of the one-dimensional harmonic oscillator potential. The units used are such that all the constants appearing in the wavefunctions and the potential are equal to unity. The wavefunctions have been shifted up by their energy in each case, which is shown as a horizontal thin line. The harmonic oscillator potential is also shown in dashed line. Notice the inflection points in the wavefunctions at the values of x where the energy eigenvalue becomes equal to the value of the harmonic potential (shown explicitly as vertical lines for the lowest eigenfunction).

that is, the eigenvalues that correspond to the wavefunctions we have calculated are quantized,

$$\epsilon_n = \left(n + \frac{1}{2}\right) \hbar\omega \tag{B.27}$$

B.3.3 Coulomb potential

Another very common case is a potential that behaves as $1/|\mathbf{r}|$, known as the Coulomb potential, from the electrostatic interaction between particles with electrical charges at distance $|\mathbf{r}|$. We will discuss this case for the simplest physical system where it applies, the hydrogen atom. For simplicity, we take the proton fixed at the origin of the coordinate system and the electron at $\mathbf{r}$. The hamiltonian for this system takes the form

$$\mathcal{H} = -\frac{\hbar^2}{2m_e}\nabla^2_{\mathbf{r}} - \frac{e^2}{|\mathbf{r}|} \tag{B.28}$$

with m_e the mass of the electron and $\pm e$ the proton and electron charges. For this problem, we change coordinates from the cartesian x, y, z to the spherical r, θ, ϕ, which are related by

$$x = r\sin\theta\cos\phi, \quad y = r\sin\theta\sin\phi, \quad z = r\cos\theta$$

and write the wavefunction as a product of two terms, one that depends only on r, called $R(r)$, and one that depends only on θ and ϕ, called $Y(\theta, \phi)$. Then the hamiltonian becomes

$$-\frac{\hbar^2}{2m_e}\left[\frac{1}{r^2}\frac{\partial}{\partial r}\left(r^2\frac{\partial}{\partial r}\right) + \frac{1}{r^2\sin\theta}\frac{\partial}{\partial\theta}\left(\sin\theta\frac{\partial}{\partial\theta}\right) + \frac{1}{r^2\sin^2\theta}\frac{\partial^2}{\partial\phi^2}\right] - \frac{e^2}{r}$$

and the TISE, with ϵ the energy eigenvalue, takes the form

$$\frac{1}{R}\frac{\mathrm{d}}{\mathrm{d}r}\left(r^2\frac{\mathrm{d}R}{\mathrm{d}r}\right) + \frac{2m_e r^2}{\hbar^2}\left[\epsilon + \frac{e^2}{r}\right] = -\frac{1}{Y}\left[\frac{1}{\sin\theta}\frac{\partial}{\partial\theta}\left(\sin\theta\frac{\partial Y}{\partial\theta}\right) + \frac{1}{\sin^2\theta}\frac{\partial^2 Y}{\partial\phi^2}\right]$$

Since in the above equation the left-hand side is exclusively a function of r while the right-hand side is exclusively a function of θ, ϕ, they each must be equal to a constant, which we denote by λ, giving rise to the following two differential equations:

$$\frac{1}{r^2}\frac{\mathrm{d}}{\mathrm{d}r}\left(r^2\frac{\mathrm{d}R}{\mathrm{d}r}\right) + \frac{2m_e}{\hbar^2}\left[\epsilon + \frac{e^2}{r}\right]R = \frac{\lambda}{r^2}R \tag{B.29}$$

$$\left[\frac{1}{\sin\theta}\frac{\partial}{\partial\theta}\left(\sin\theta\frac{\partial Y}{\partial\theta}\right) + \frac{1}{\sin^2\theta}\frac{\partial^2 Y}{\partial\phi^2}\right] = -\lambda Y \tag{B.30}$$

We consider the equation for $Y(\theta, \phi)$ first. This equation is solved by the functions

$$Y(\theta, \phi) = (-1)^{(|m|+m)/2}\left[\frac{(2l+1)(l-|m|)!}{4\pi(l+|m|)!}\right]^{1/2} P_l^m(\cos\theta)\mathrm{e}^{\mathrm{i}m\phi} \tag{B.31}$$

where l and m are integers with the following range:

$$l \geq 0, \quad m = -l, -l+1, \ldots, 0, \ldots, l-1, l$$

and $P_l^m(w)$ are the functions

$$P_l^m(w) = (1-w^2)^{|m|/2}\frac{\mathrm{d}^{|m|}}{\mathrm{d}w^{|m|}}P_l(w) \tag{B.32}$$

with $P_l(w)$ the Legendre polynomials defined recursively as

$$P_{l+1}(w) = \frac{2l+1}{l+1}w\,P_l(w) - \frac{l}{l+1}P_{l-1}(w)$$
$$P_0(w) = 1, \quad P_1(w) = w$$

The $Y_{lm}(\theta, \phi)$ functions are called "spherical harmonics".

The spherical harmonics are the functions that give the anisotropic character of the eigenfunctions of the Coulomb potential, since the remaining part $R(r)$ is spherically symmetric. Taking into account the correspondence between the (x, y, z) cartesian coordinates and the (r, θ, ϕ) spherical coordinates, we can relate the spherical harmonics to functions of x/r, y/r and z/r. Y_{00} is a constant which has spherical symmetry and is referred to as an s state; it represents a state of zero angular momentum. Linear combinations of higher spherical harmonics that correspond to $l = 1$ are referred to as p states, those for $l = 2$ as d states and those for $l = 3$ as f states. Still higher angular momentum states are labeled $g, h, j, k, \ldots$, for $l = 4, 5, 6, 7, \ldots$. The linear combinations

Table B.2. *The spherical harmonics $Y_{lm}(\theta, \phi)$ for $l = 0, 1, 2, 3$.*

The x, y, z representation of the linear combinations for given l and $|m|$ and the identification of those representations as s, p, d, f orbitals are also given.

l	m	$Y_{lm}(\theta,\phi)$	x, y, z representation	s, p, d, f orbitals
0	0	$\left(\frac{1}{4\pi}\right)^{1/2}$	1	s
1	0	$\left(\frac{3}{4\pi}\right)^{1/2}\cos\theta$	z/r	p_z
1	±1	$\mp\left(\frac{3}{8\pi}\right)^{1/2}\sin\theta e^{\pm i\phi}$	$x/r \sim Y_{1-1} - Y_{1+1}$	p_x
			$y/r \sim Y_{1+1} + Y_{1-1}$	p_y
2	0	$\left(\frac{5}{16\pi}\right)^{1/2}(3\cos^2\theta - 1)$	$(3z^2 - r^2)/r^2$	$d_{3z^2-r^2}$
2	±1	$\mp\left(\frac{15}{8\pi}\right)^{1/2}\sin\theta\cos\theta e^{\pm i\phi}$	$(xz)/r^2$	d_{xz}
			$(yz)/r^2$	d_{yz}
2	±2	$\left(\frac{15}{32\pi}\right)^{1/2}\sin^2\theta e^{\pm 2i\phi}$	$(xy)/r^2$	d_{xy}
			$(x^2 - y^2)/r^2$	$d_{x^2-y^2}$
3	0	$\left(\frac{7}{16\pi}\right)^{1/2}(5\cos^3\theta - 3\cos\theta)$	$(5z^3 - zr^2)/r^3$	$f_{5z^3-zr^2}$
3	±1	$\mp\left(\frac{21}{64\pi}\right)^{1/2}(5\cos^2\theta - 1)\sin\theta e^{\pm i\phi}$	$(5xz^2 - xr^2)/r^3$	$f_{5xz^2-xr^2}$
			$(5yz^2 - yr^2)/r^3$	$f_{5yz^2-yr^2}$
3	±2	$\left(\frac{105}{32\pi}\right)^{1/2}\sin^2\theta\cos\theta e^{\pm i2\phi}$	$(zx^2 - zy^2)/r^3$	$f_{zx^2-zy^2}$
			$(xyz)/r^3$	f_{xyz}
3	±3	$\mp\left(\frac{35}{64\pi}\right)^{1/2}\sin^3\theta e^{\pm i3\phi}$	$(3yx^2 - y^3)/r^3$	$f_{3yx^2-y^3}$
			$(3xy^2 - x^3)/r^3$	$f_{3xy^2-x^3}$

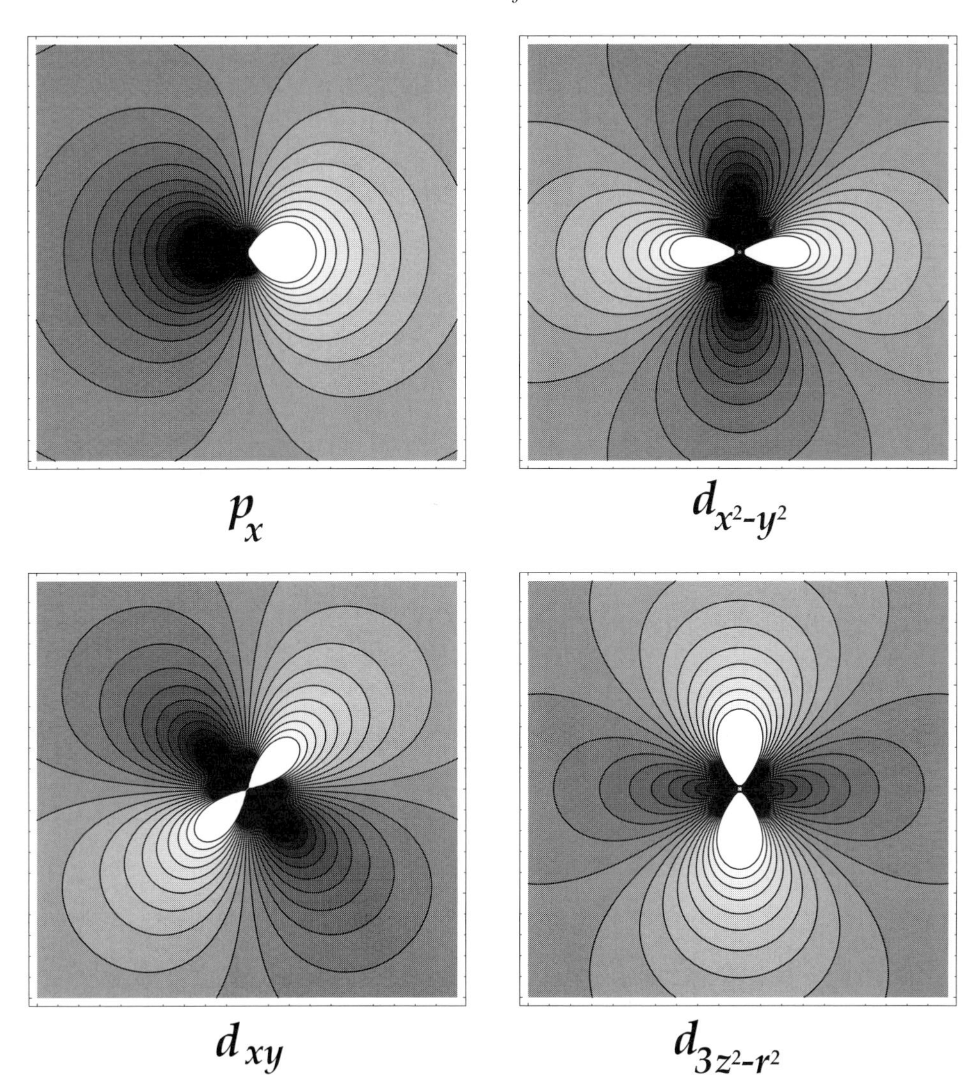

Figure B.3. Contours of constant value of the p_x (on the xy plane), $d_{x^2-y^2}$ (on the xy plane), d_{xy} (on the xy plane) and $d_{3z^2-r^2}$ (on the xz plane) orbitals. The positive values are shown in white, the negative values in black. The lobes used in the simplified representations in the text are evident.

of spherical harmonics for $l = 1$ that correspond to the usual p states as expressed in terms of x/r, y/r, z/r (up to constant factors that ensure proper normalization) are given in Table B.2. Similar combinations of spherical harmonics for higher values of l correspond to the usual d, f states defined as functions of x/r, y/r, z/r. The character of selected p, d and f states is shown in Fig. B.3 and Fig. B.4. The connection between the spherical harmonics and angular momentum is explained next.

The eigenvalue of the spherical harmonic Y_{lm} in Eq. (B.30) is $\lambda = -l(l+1)$. When this is substituted into Eq. (B.29) and $R(r)$ is expressed as $R(r) = Q(r)/r$, this equation takes

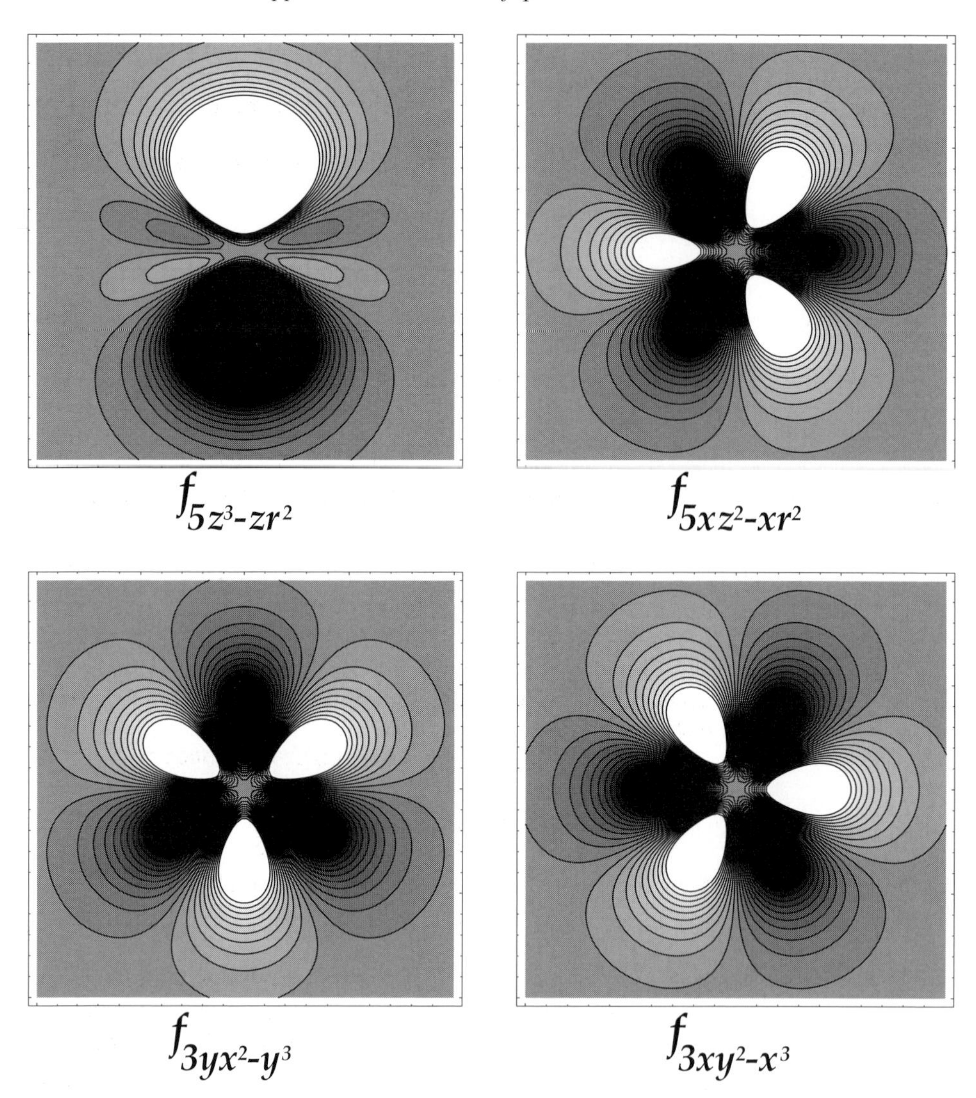

Figure B.4. Contours of constant value of the $f_{5z^3-zr^2}$ (on the xz plane), $f_{5xz^2-xr^2}$ (on the xz plane), $f_{3yx^2-y^3}$ (on the xy plane) and $f_{3xy^2-x^3}$ (on the xy plane) orbitals. The positive values are shown in white, the negative values in black.

the form

$$\left[-\frac{h^2}{2m_e} \frac{\mathrm{d}^2}{\mathrm{d}r^2} + \frac{\hbar^2 l(l+1)}{2m_e r^2} - \frac{e^2}{r} \right] Q(r) = \epsilon Q(r) \tag{B.33}$$

which has the form of a one-dimensional TISE. We notice that the original potential between the particles has been changed by the factor $\hbar^2 l(l+1)/2m_e r^2$, which corresponds to a centrifugal term if we take $\hbar^2 l(l+1)$ to be the square of the angular momentum. By analogy to classical mechanics, we define the quantum mechanical angular momentum operator as $\mathbf{L} = \mathbf{r} \times \mathbf{p}$, which, using the quantum mechanical operator for the momentum $\mathbf{p} = -\mathrm{i}\hbar\nabla_{\mathbf{r}}$, has the following components (expressed both in cartesian and spherical coordinates):

$$L_x = yp_z - zp_y = -\mathrm{i}\hbar\left(y\frac{\partial}{\partial z} - z\frac{\partial}{\partial y}\right) = \mathrm{i}\hbar\left(\sin\phi\frac{\partial}{\partial\theta} + \cot\theta\cos\phi\frac{\partial}{\partial\phi}\right)$$

$$L_y = zp_x - xp_z = -\mathrm{i}\hbar\left(z\frac{\partial}{\partial x} - x\frac{\partial}{\partial z}\right) = -\mathrm{i}\hbar\left(\cos\phi\frac{\partial}{\partial\theta} - \cot\theta\sin\phi\frac{\partial}{\partial\phi}\right)$$

$$L_z = xp_y - yp_x = -\mathrm{i}\hbar\left(x\frac{\partial}{\partial z} - z\frac{\partial}{\partial x}\right) = -\mathrm{i}\hbar\frac{\partial}{\partial\phi} \tag{B.34}$$

It is a straightforward exercise to show from these expressions that

$$\mathbf{L}^2 = L_x^2 + L_y^2 + L_z^2 = -\hbar^2\left[\frac{1}{\sin\theta}\frac{\partial}{\partial\theta}\left(\sin\theta\frac{\partial}{\partial\theta}\right) + \frac{1}{\sin^2\theta}\frac{\partial^2}{\partial\phi^2}\right] \tag{B.35}$$

and that the spherical harmonics are eigenfunctions of the operators $\mathbf{L}^2$ and L_z, with eigenvalues

$$\mathbf{L}^2 Y_{lm}(\theta,\phi) = l(l+1)\hbar^2 Y_{lm}(\theta,\phi), \quad L_z Y_{lm}(\theta,\phi) = m\hbar Y_{lm}(\theta,\phi) \tag{B.36}$$

as might have been expected from our earlier identification of the quantity $\hbar^2 l(l+1)$ with the square of the angular momentum. This is another example of simultaneous eigenfunctions of two operators, which according to our earlier discussion must commute: $[\mathbf{L}^2, L_z] = 0$. Thus, the spherical harmonics determine the angular momentum l of a state and its z component, which is equal to $m\hbar$.

In addition to L_x, L_y, L_z, there are two more interesting operators defined as

$$L_\pm = L_x \pm \mathrm{i}L_y = \hbar e^{\pm\mathrm{i}\phi}\left[\pm\frac{\partial}{\partial\theta} + \mathrm{i}\cot\theta\frac{\partial}{\partial\phi}\right] \tag{B.37}$$

which when applied to the spherical harmonics give the following result:

$$L_\pm Y_{lm}(\theta,\phi) = C_{lm}^{(\pm)} Y_{lm\pm1}(\theta,\phi) \tag{B.38}$$

that is, they raise or lower the value of the z component by one unit. For this reason these are called the raising and lowering operators. The value of the constants $C_{lm}^{(\pm)}$ can be obtained from the following considerations. First we notice from the definition of L_+, L_- that

$$L_+L_- = (L_x + \mathrm{i}L_y)(L_x - \mathrm{i}L_y) = L_x^2 + L_y^2 - \mathrm{i}L_xL_y + \mathrm{i}L_yL_x = L_x^2 + L_y^2 + \mathrm{i}[L_x, L_y]$$

where in the last expression we have introduced the commutator of L_x, L_y. Next we can use the definition of L_x, L_y to show that their commutator is equal to $\mathrm{i}\hbar L_z$:

$$[L_x, L_y] = L_xL_y - L_yL_x = \mathrm{i}\hbar L_z$$

which gives the following relation:

$$\mathbf{L}^2 = L_z^2 + L_+L_- - \hbar L_z$$

Using this last relation, we can take the expectation value of $L_\pm Y_{lm}$ with itself to obtain

$$\begin{aligned}
\langle L_\pm Y_{lm}|L_\pm Y_{lm}\rangle &= |C_{lm}^{(\pm)}|^2\langle Y_{lm\pm1}|Y_{lm\pm1}\rangle = \langle Y_{lm}|L_\mp L_\pm Y_{lm}\rangle \\
&= \langle Y_{lm}|(\mathbf{L}^2 - L_z^2 \mp \hbar L_z)Y_{lm}\rangle = \hbar^2\left[l(l+1) - m(m\pm1)\right] \\
&\Longrightarrow C_{lm}^{(\pm)} = \hbar\left[l(l+1) - m(m\pm1)\right]^{1/2}
\end{aligned} \tag{B.39}$$

This is convenient because it provides an explicit expression for the result of applying the raising or lowering operators to spherical harmonics: it can be used to generate all spherical harmonics of a given l starting with one of them.

Finally, we consider the solution to the radial equation, Eq. (B.29). This equation is solved by the functions

$$R_{nl}(r) = -\left[\left(\frac{2}{n(n+l)!a_0}\right)^3 \frac{(n-l-1)!}{2n}\right]^{1/2} e^{-\rho/2}\rho^l L_{n+l}^{2l+1}(\rho) \tag{B.40}$$

where we have defined two new variables

$$a_0 \equiv \frac{\hbar^2}{e^2 m_e}, \quad \rho \equiv \frac{2r}{na_0}$$

and the functions $L_n^l(r)$ are given by

$$L_n^l(r) - \frac{d^l}{dr^l} L_n(r) \tag{B.41}$$

with $L_n(r)$ the Laguerre polynomials defined recursively as

$$L_{n+1}(r) = (2n+1-r)L_n(r) - n^2 L_{n-1}(r)$$
$$L_0(r) = 1, \quad L_1(r) = 1 - r \tag{B.42}$$

The index n is an integer that takes the values $n = 1, 2, \ldots$, while for a given n the index l is allowed to take the values $l = 0, \ldots, n-1$, and the energy eigenvalues are given by:

$$\epsilon_n = -\frac{e^2}{2a_0 n^2}$$

The first few radial wavefunctions are given in Table B.3. The nature of these wavefunctions is illustrated in Fig. B.5. It is trivial to extend this description to a nucleus of charge Ze with a single electron around it: the factor a_0 is replaced everywhere by a_0/Z, and there is an extra factor of Z in the energy to account for the charge of the

Table B.3. *Radial wavefunctions of the Coulomb potential.*

The radial wavefunctions for $n = 1, 2, 3$, are given, together with the associated Laguerre polynomials $L_{n+l}^{2l+1}(r)$ used in their definition.

n	l	$L_{n+l}^{2l+1}(r)$	$a_0^{3/2} R_{nl}(r)$
1	0	-1	$2e^{-r/a_0}$
2	0	$-2(2-r)$	$\frac{1}{\sqrt{2}}e^{-r/2a_0}\left(1 - \frac{r}{2a_0}\right)$
2	1	$-3!$	$\frac{1}{\sqrt{6}}e^{-r/2a_0}\left(\frac{r}{2a_0}\right)$
3	0	$-3(6-6r+r^2)$	$\frac{2}{9\sqrt{3}}e^{-r/3a_0}\left(3 - 6\left(\frac{r}{3a_0}\right) + 2\left(\frac{r}{3a_0}\right)^2\right)$
3	1	$-24(4-r)$	$\frac{2\sqrt{2}}{9\sqrt{3}}e^{-r/3a_0}\left(\frac{r}{3a_0}\right)\left(2 - \frac{r}{3a_0}\right)$
3	2	$-5!$	$\frac{2\sqrt{2}}{9\sqrt{15}}e^{-r/3a_0}\left(\frac{r}{3a_0}\right)^2$

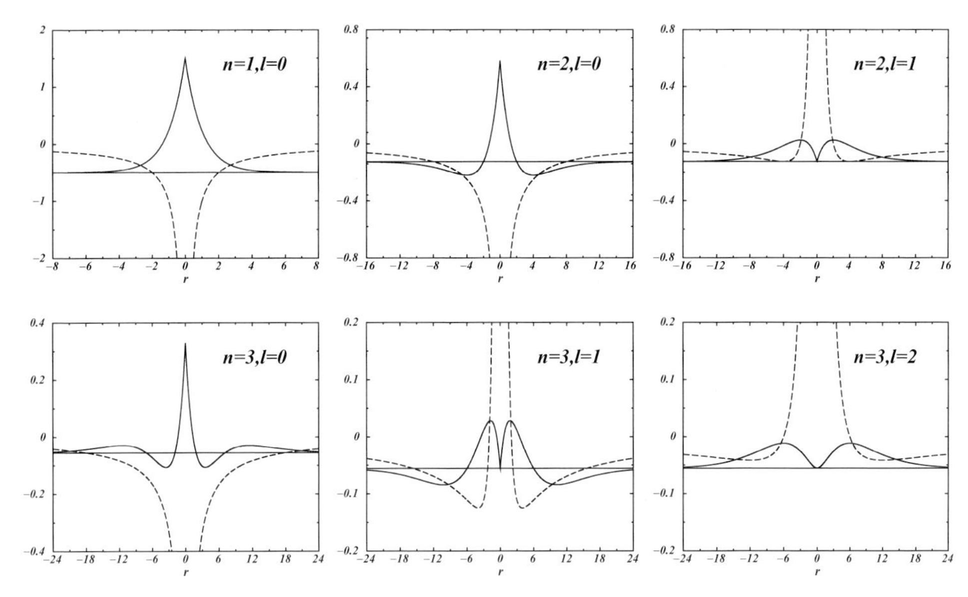

Figure B.5. The lowest six radial eigenfunctions of the Coulomb potential for the hydrogen atom $R_{nl}(r)$ $[(n, l) = (1,0), (2,0), (2,1), (3,0), (3,1), (3,2)]$. The horizontal axis is extended to negative values to indicate the spherically symmetric nature of the potential and wavefunctions; this range corresponds to $\theta = \pi$. The units used are such that all the constants appearing in the wavefunctions and the potential are equal to unity. The wavefunctions have been shifted by their energy in each case, which is shown as a horizontal thin line. The total radial potential, including the Coulomb part and the angular momentum part, is also shown as a dashed line: notice in particular the large repulsive potential near the origin for $l = 1$ and $l = 2$, arising from the angular momentum part. The scale for each case has been adjusted to make the features of the wavefunctions visible.

nucleus. In our treatment we have also considered the nucleus to be infinitely heavy and fixed at the origin of the coordinate system. If we wish to include the finite mass m_n of the nucleus then the mass of the electron in the above equations is replaced by the reduced mass of nucleus and electron, $\mu = m_n m_e/(m_n + m_e)$. Since nuclei are much heavier than electrons, $\mu \approx m_e$ is a good approximation.

B.4 Spin angular momentum

In addition to the usual terms of kinetic energy and potential energy, quantum mechanical particles possess another property called spin, which has the dimensions of angular momentum. The values of spin are quantized to half integer or integer multiples of $\hbar$. In the following we omit the factor of $\hbar$ when we discuss spin values, for brevity. If the total spin of a particle is s then there are $2s + 1$ states associated with it, because the projection of the spin onto a particular axis can have that many possible values, ranging from $+s$ to $-s$ in increments of 1. The axis of spin projection is usually labeled the z axis, so a spin of $s = 1/2$ can have projections on the z axis $s_z = +1/2$ and $s_z = -1/2$; a spin of $s = 1$ can have $s_z = -1, 0, +1$, and so on.

The spin of quantum particles determines their statistics. Particles with half-integer spin are called fermions and obey the **Pauli exclusion principle**, that is, no two of them can be in exactly the same quantum mechanical state as it is defined by all the relevant quantum numbers. A typical example is the electron which is a fermion with spin $s = 1/2$. In the Coulomb potential there are three quantum numbers n, l, m associated with the spatial degrees of freedom as determined by the radial and angular parts of the wavefunction $R_{nl}(r)$ and $Y_{lm}(\theta, \phi)$. A particle with spin $s = 1/2$ can only have two states associated with a particular set of n, l, m values, corresponding to $s_z = \pm 1/2$. Thus, each state of the Coulomb potential characterized by a set of values n, l, m can accommodate two electrons with spin $\pm 1/2$, to which we refer as the "spin-up" and "spin-down" states. This rule basically explains the sequence of elements in the Periodic Table, as follows.

$n = 1$ This state can only have the angular momentum state with $l = 0, m = 0$ (the s state), which can be occupied by one or two electrons when spin is taken into consideration; the first corresponds to H, the second to He, with atomic numbers 1 and 2.

$n = 2$ This state can have angular momentum states with $l = 0, m = 0$ (an s state) or $l = 1, m = \pm 1, 0$ (a p state), which account for a total of eight possible states when the spin is taken into consideration, corresponding to the elements Li, Be, B, C, N, O, F and Ne with atomic numbers 3–10.

$n = 3$ This state can have angular momentum states with $l = 0, m = 0$ (an s state), $l = 1, m = \pm 1, 0$ (a p state), or $l = 2, m = \pm 2, \pm 1, 0$ (a d state), which account for a total of 18 possible states when the spin is taken into consideration, corresponding to the elements Na, Mg, Al, Si, P, S, Cl and Ar with atomic numbers 11–18, (in which the s and p states are gradually filled) and elements Sc, Ti, V, Cr, Mn, Fe, Co, Ni, Cu and Zn with atomic numbers 21–30 (in which the d state is gradually filled).

There is a jump in the sequential occupation of states of the Coulomb potential, namely we pass from atomic number 18 with the $n = 3, l = 0$ and $l = 1$ states filled to atomic number 21, in which the $n = 3, l = 2$ states begin to be filled. The reason for this jump has to do with the fact that each electron in an atom does not experience only the pure Coulomb potential of the nucleus, but a more complex potential which is also due to the presence of all the other electrons. For this reason the true states in the atoms are somewhat different than the states of the pure Coulomb potential, which is what makes the states with $n = 4$ and $l = 0$ start filling before the states with $n = 3$ and $l = 2$; the $n = 4, l = 0, m = 0$ states correspond to the elements K and Ca, with atomic numbers 19 and 20. Overall, however, the sequence of states in real atoms is remarkably close to what would be expected from the pure Coulomb potential. The same pattern is followed for states with higher n and l values.

The other type of particles, with integer spin values, do not obey the Pauli exclusion principle. They are called bosons and obey different statistics than fermions. For instance, under the proper conditions, all bosons can collapse into a single quantum mechanical state, a phenomenon referred to as Bose–Einstein condensation, that has been observed experimentally.

Since spin is a feature of the quantum mechanical nature of particles we need to introduce a way to represent it and include it in the hamiltonian as appropriate. Spins are represented by "spinors", which are one-dimensional vectors of zeros and ones. The spinors identify the exact state of the spin, including its magnitude s and its projection on the z axis, s_z: the magnitude is included in the length of the spinor, which is $2s + 1$, while the s_z value is given by the non-zero entry of the spinor in a sequential manner going through the values s to $-s$. The spinor of spin 0 is a vector of length 1, with a single entry [1]. The two spinors that identify the states corresponding to total spin $s = 1/2$

have length 2 and are

$$\begin{bmatrix} 1 \\ 0 \end{bmatrix} \rightarrow \quad s_z = +\frac{1}{2} \ (\uparrow), \qquad \begin{bmatrix} 0 \\ 1 \end{bmatrix} \rightarrow \quad s_z = -\frac{1}{2} \ (\downarrow)$$

where we have also included the usual notation for spins 1/2, as up and down arrows. The three spinors that identify the states corresponding to total spin $s = 1$, which have length 3, are

$$\begin{bmatrix} 1 \\ 0 \\ 0 \end{bmatrix} \rightarrow \quad s_z = +1 \qquad \begin{bmatrix} 0 \\ 1 \\ 0 \end{bmatrix} \rightarrow \quad s_z = 0, \qquad \begin{bmatrix} 0 \\ 0 \\ 1 \end{bmatrix} \rightarrow \quad s_z = -1$$

The same pattern applies to higher spin values.

When spin needs to be explicitly included in the hamiltonian, the appropriate operators must be used that can act on spinors. These are square matrices of size $(2s + 1) \times (2s + 1)$ which multiply the spinors to produce other spinors or combinations of spinors. Since there are three components of angular momentum in 3D space, there must exist three matrices corresponding to each component of the spin. For spin $s = 1/2$ these are the following 2×2 matrices:

$$J_x = \frac{\hbar}{2} \begin{bmatrix} 0 & 1 \\ 1 & 0 \end{bmatrix}, \quad J_y = \frac{\hbar}{2} \begin{bmatrix} 0 & -\mathrm{i} \\ \mathrm{i} & 0 \end{bmatrix}, \quad J_z = \frac{\hbar}{2} \begin{bmatrix} 1 & 0 \\ 0 & -1 \end{bmatrix}$$

The matrices in square brackets without the constants in front are called the Pauli matrices and are denoted by $\sigma_x, \sigma_y, \sigma_z$. For spin $s = 1$ the spin operators are the following 3×3 matrices:

$$J_x = \frac{\hbar}{\sqrt{2}} \begin{bmatrix} 0 & 1 & 0 \\ 1 & 0 & 1 \\ 0 & 1 & 0 \end{bmatrix}, \quad J_y = \frac{\hbar}{\sqrt{2}} \begin{bmatrix} 0 & -\mathrm{i} & 0 \\ \mathrm{i} & 0 & -\mathrm{i} \\ 0 & \mathrm{i} & 0 \end{bmatrix}, \quad J_z = \hbar \begin{bmatrix} 1 & 0 & 0 \\ 0 & 0 & 0 \\ 0 & 0 & -1 \end{bmatrix}$$

It is easy to show that the matrix defined by $J^2 = J_x^2 + J_y^2 + J_z^2$ in each case is a diagonal matrix with all diagonal elements equal to one multiplied by the factor $s(s + 1)\hbar 2$. Thus, the spinors are eigenvectors of the matrix J^2 with eigenvalue $s(s + 1)\hbar^2$; for this reason, it is customary to attribute the value $s(s + 1)$ to the square of the spin magnitude. It is also evident from the definitions given above that each spinor is an eigenvector of the matrix J_z, with an eigenvalue equal to the s_z value to which the spinor corresponds. It is easy to show that the linear combination of matrices $J_+ = (J_x + \mathrm{i}J_y)$ is a matrix which when multiplied with a spinor corresponding to s_z gives the spinor corresponding to $s_z + 1$, while the linear combination $J_- = (J_x - \mathrm{i}J_y)$ is a matrix which when multiplied with a spinor corresponding to s_z gives the spinor corresponding to $s_z - 1$; for example, for the $s = 1/2$ case:

$$J_+ = \hbar \begin{bmatrix} 0 & 1 \\ 0 & 0 \end{bmatrix}, \quad J_- = \hbar \begin{bmatrix} 0 & 0 \\ 1 & 0 \end{bmatrix}$$

$$J_+ \begin{bmatrix} 1 \\ 0 \end{bmatrix} = 0, \quad J_+ \begin{bmatrix} 0 \\ 1 \end{bmatrix} = \hbar \begin{bmatrix} 1 \\ 0 \end{bmatrix}, \quad J_- \begin{bmatrix} 1 \\ 0 \end{bmatrix} = \hbar \begin{bmatrix} 0 \\ 1 \end{bmatrix}, \quad J_- \begin{bmatrix} 0 \\ 1 \end{bmatrix} = 0$$

where the operator J_+ applied to the up-spin gives zero because there is no state with higher s_z, and similarly the operator J_- applied to the down-spin gives zero because there is no state with lower s_z. These two matrices are therefore the raising and lowering operators, as might be expected by the close analogy to the definition of the corresponding operators for angular momentum, given in Eqs. (B.37) and (B.38).

Finally, we consider the addition of spin angular momenta. We start with the case of two spin-1/2 particles, as the most common situation encountered in the study of solids;

the generalization to arbitrary values is straightforward and will also be discussed briefly. We denote the spin of the combined state and its z component by capital letters (S, S_z), in distinction to the spins of the constituent particles, denoted by $(s^{(1)}, s_z^{(1)})$ and $(s^{(2)}, s_z^{(2)})$. We can start with the state in which both spinors have $s_z = 1/2$, that is, they are the up-spins. The combined spin will obviously have projection on the z axis $S_z = 1$. We will show that this is one of the states of the total spin $S = 1$ manifold. We can apply the lowering operator, denoted by $S_- = s_-^{(1)} + s_-^{(2)}$, which is of course composed of the two individual lowering operators, to the state with $S_z = 1$ to obtain a state of lower S_z value. Notice that $s_-^{(i)}$ applies only to particle i:

$$S_- [\uparrow\uparrow] = \left[(s_-^{(1)} \uparrow) \uparrow + \uparrow (s_-^{(2)} \uparrow)\right] = \hbar\sqrt{2} \times \frac{1}{\sqrt{2}} [\downarrow\uparrow + \uparrow\downarrow]$$

where we have used up and down arrows as shorthand notation for spinors with $s_z = \pm 1/2$ and the convention that the first arrow corresponds to the spin of the first particle and the second arrow corresponds to the spin of the second particle. Now notice that the state in square brackets with the factor $1/\sqrt{2}$ in front is a properly normalized spinor state with $S_z = 0$, and its coefficient is $\hbar\sqrt{2}$, precisely what we would expect to get by applying the lowering operator on a state with angular momentum 1 and z projection 1, according to Eq. (B.39). If we apply the lowering operator once again to the new state with $S_z = 0$, we obtain

$$S_- \frac{1}{\sqrt{2}} [\downarrow\uparrow + \uparrow\downarrow] = \frac{1}{\sqrt{2}} \left[\downarrow (s_-^{(2)} \uparrow) + (s_-^{(1)} \uparrow) \downarrow\right] = \hbar\sqrt{2} [\downarrow\downarrow]$$

which is a state with $S_z = -1$ and coefficient $\hbar\sqrt{2}$, again consistent with what we expected from applying the lowering operator on a state with angular momentum 1 and z projection 0. Thus, we have generated the three states in the $S = 1$ manifold with $S_z = +1, 0, -1$. We can construct one more state, which is orthogonal to the $S_z = 0$ state found above and also has $S_z = 0$, by taking the linear combination of up–down spins with opposite relative sign:

$$\frac{1}{\sqrt{2}} [\uparrow\downarrow - \downarrow\uparrow]$$

Moreover, we can determine that the total spin value of this new state is $S = 0$, because applying the lowering or the raising operator on it gives 0. This completes the analysis of the possible spin states obtained by combining two spin-1/2 particles which are the following:

$$S = 1: \quad S_z = +1 \to [\uparrow\uparrow], \quad S_z = 0 \to \frac{1}{\sqrt{2}} [\uparrow\downarrow + \downarrow\uparrow], \quad S_z = -1 \to [\downarrow\downarrow]$$

$$S = 0: \quad S_z = 0 \to \frac{1}{\sqrt{2}} [\uparrow\downarrow - \downarrow\uparrow]$$

The generalization of these results to addition of two angular momenta with arbitrary values is as follows. Consider the two angular momenta to be S and L; this notation usually refers to the situation where S is the spin and L is the orbital angular momentum, a case to which we specify the following discussion as the most relevant to the physics of solids. The z projection of the first component will range from $S_z = +S$ to $-S$, and of the second component from $L_z = +L$ to $-L$. The resulting total angular momentum J will take values ranging from $L + S$ to zero, because the possible z components will range from a maximum of $S_z + L_z = S + L$ to a minimum of $-(S + L)$, with all the intermediate values, each differing by a unit:

$$J = L + S: \quad J_z = L + S, L + S - 1, \ldots, -L - S;$$

$$J = L + S - 1: \quad J_z = L + S - 1, L + S - 2, \ldots, -L - S + 1; \text{ etc.}$$

An important application of these rules is the calculation of expectation values of operators in the basis of states with definite J and J_z. Since these states are obtained by combinations of states with spin S and orbital angular momentum L, we denote them as $|JLSJ_z\rangle$, which form a complete set of $(2J + 1)$ states for each value of J, and a complete set for all possible states resulting from the addition of L and S when all allowed values of J are included:

$$\sum_{JJ_z} |JLSJ_z\rangle\langle JLSJ_z| = 1 \tag{B.43}$$

A general theorem, known as the Wigner–Eckart theorem (see, for example, Schiff, p. 222; see the Further reading section), states that the expectation value of any vector operator in the space of the $|JLSJ_z\rangle$ states is proportional to the expectation value of $\hat{\mathbf{J}}$ itself and does not depend on J_z (in the following we use bold symbols with a hat, $\hat{\mathbf{O}}$, to denote vector operators and bold symbols, $\mathbf{O}$, to denote their expectation values). Let us consider as an example a vector operator which is a linear combination of $\hat{\mathbf{L}}$ and $\hat{\mathbf{S}}$, $(\hat{\mathbf{L}} + \lambda\hat{\mathbf{S}})$, with λ an arbitrary constant. From the Wigner–Eckart theorem we will have

$$\langle JLSJ'_z|(\hat{\mathbf{L}} + \lambda\hat{\mathbf{S}})|JLSJ_z\rangle = g_\lambda(JLS)\langle JLSJ'_z|\hat{\mathbf{J}}|JLSJ_z\rangle$$

where we have written the constants of proportionality as $g_\lambda(JLS)$. In order to obtain the values of these constants, we note that the expectation value of the dot product $(\hat{\mathbf{L}} + \lambda\hat{\mathbf{S}}) \cdot \hat{\mathbf{J}}$ will be given by

$$\begin{aligned}
&\langle JLSJ'_z|(\hat{\mathbf{L}} + \lambda\hat{\mathbf{S}}) \cdot \hat{\mathbf{J}}|JLSJ_z\rangle \\
&\quad = \langle JLSJ'_z|(\hat{\mathbf{L}} + \lambda\hat{\mathbf{S}}) \cdot \sum_{J''J''_z} |J''LSJ''_z\rangle\langle J''LSJ''_z|\hat{\mathbf{J}}|JLSJ_z\rangle \\
&\quad = \langle JLSJ'_z|(\hat{\mathbf{L}} + \lambda\hat{\mathbf{S}}) \cdot \sum_{J''J''_z} |J''LSJ''_z\rangle\mathbf{J}\delta_{JJ''}\delta_{J_zJ''_z} \\
&\quad = g_\lambda(JLS)\langle JLSJ'_z|\hat{\mathbf{J}}^2|JLSJ_z\rangle = g_\lambda(JLS)J(J+1)
\end{aligned} \tag{B.44}$$

where in the first step we have inserted the unity operator of Eq. (B.43) between the two vectors of the dot product; in the second step we used the fact that the states $|JLSJ_z\rangle$ are eigenstates of the operator $\hat{\mathbf{J}}$ and therefore $\hat{\mathbf{J}}|JLSJ_z\rangle = |JLSJ_z\rangle\mathbf{J}$; in the third step we used the Wigner–Eckart theorem; and in the last step we used the fact that the expectation value of $\hat{\mathbf{J}}^2$ in the basis $|JLSJ_z\rangle$ is simply $J(J+1)$. Thus, for the expectation value of the original operator we obtain

$$\langle JLSJ'_z|\hat{\mathbf{L}} \cdot \hat{\mathbf{J}}|JLSJ_z\rangle + \lambda\langle JLSJ'_z|\hat{\mathbf{S}} \cdot \hat{\mathbf{J}}|JLSJ_z\rangle = g_\lambda(JLS)J(J+1).$$

From the vector addition of $\hat{\mathbf{L}}$ and $\hat{\mathbf{S}}$, namely $\hat{\mathbf{J}} = \hat{\mathbf{L}} + \hat{\mathbf{S}}$, we find

$$\hat{\mathbf{S}} = \hat{\mathbf{J}} - \hat{\mathbf{L}} \Rightarrow \hat{\mathbf{L}} \cdot \hat{\mathbf{J}} = \frac{1}{2}\left[\hat{\mathbf{J}}^2 + \hat{\mathbf{L}}^2 - \hat{\mathbf{S}}^2\right]$$

$$\hat{\mathbf{L}} = \hat{\mathbf{J}} - \hat{\mathbf{S}} \Rightarrow \hat{\mathbf{S}} \cdot \hat{\mathbf{J}} = \frac{1}{2}\left[\hat{\mathbf{J}}^2 + \hat{\mathbf{S}}^2 - \hat{\mathbf{L}}^2\right]$$

while the expectation values of $\hat{\mathbf{L}}^2$ and $\hat{\mathbf{S}}^2$ in the basis $|JLSJ_z\rangle$ are $L(L+1)$ and $S(S+1)$. These results, employed in the above equation for $g_\lambda(JLS)$, lead to

$$g_\lambda(JLS) = \frac{1}{2}(\lambda + 1) + \frac{1}{2}(\lambda - 1)\left[\frac{S(S+1) - L(L+1)}{J(J+1)}\right] \tag{B.45}$$

These expressions are known as the "Landé g-factors" in the context of the total angular momentum of atoms or ions with partially filled electronic shells, which are relevant to the magnetic behavior of insulators.

B.5 Stationary perturbation theory

Perturbation theory in general is a very useful method in quantum mechanics: it allows us to find approximate solutions to problems that do not have simple analytic solutions. In stationary perturbation theory (SPT), we assume that we can view the problem at hand as a slight change from another problem, called the unperturbed case, which we can solve exactly. The basic idea is that we wish to find the eigenvalues and eigenfunctions of a hamiltonian $\mathcal{H}$ which can be written as two parts:

$$\mathcal{H} = \mathcal{H}_0 + \mathcal{H}_1$$

where $\mathcal{H}_0$ is a hamiltonian whose solutions we know analytically (as in one of the examples we discussed above) and $\mathcal{H}_1$ is a small perturbation of $\mathcal{H}_0$. We can use the solutions of $\mathcal{H}_0$ to express the unknown solutions of $\mathcal{H}$:

$$\mathcal{H}|\phi_i\rangle = \epsilon_i|\phi_i\rangle \tag{B.46}$$

$$|\phi_i\rangle = |\phi_i^{(0)}\rangle + |\delta\phi_i\rangle \tag{B.47}$$

$$\epsilon_i = \epsilon_i^{(0)} + \delta\epsilon_i \tag{B.48}$$

where quantities with superscript (0) identify the solutions of the hamiltonian $\mathcal{H}_0$, which form a complete orthonormal set,

$$\mathcal{H}_0|\phi_i^{(0)}\rangle = \epsilon_i|\phi_i^{(0)}\rangle, \quad \langle\phi_i^{(0)}|\phi_j^{(0)}\rangle = \delta_{ij} \tag{B.49}$$

and quantities without superscripts correspond to the solutions of the hamiltonian $\mathcal{H}$. The problem reduces to finding the quantities $|\delta\phi_i\rangle$ and $\delta\epsilon_i$, assuming that they are small compared with the corresponding wavefunctions and eigenvalues. Notice in particular that the wavefunctions $|\phi_i\rangle$ are not normalized in the way we wrote them above, but their normalization is a trivial matter once we have calculated $|\delta\phi_i\rangle$. Moreover, each $|\delta\phi_i\rangle$ will only include the part which is orthogonal to $|\phi_i^{(0)}\rangle$, that is, it can include any wavefunction $|\phi_j^{(0)}\rangle$ with $j \neq i$. Our goal then is to express $|\delta\phi_i\rangle$ and $\delta\epsilon_i$ in terms of the known wavefunctions and eigenvalues of $\mathcal{H}_0$, which we will do as a power series expansion:

$$|\delta\phi_i\rangle = |\delta\phi_i^{(1)}\rangle + |\delta\phi_i^{(2)}\rangle + \cdots$$

$$\delta\epsilon_i = \delta\epsilon_i^{(1)} + \delta\epsilon_i^{(2)} + \cdots$$

where the superscripts indicate the order of the approximation, that is, superscript (1) is the first order approximation which includes the perturbation $\mathcal{H}_1$ only to first power, etc.

B.5.1 Non-degenerate perturbation theory

We start with the simplest case, assuming that the eigenfunctions of the unperturbed hamiltonian are non-degenerate, that is, $\epsilon_i^{(0)} \neq \epsilon_j^{(0)}$ for $i \neq j$. When we substitute

Eqs. (B.47) and (B.48) into Eq. (B.46), we find

$$(\mathcal{H}_0 + \mathcal{H}_1)\left(|\phi_i^{(0)}\rangle + |\delta\phi_i\rangle\right) = \left(\epsilon_i^{(0)} + \delta\epsilon_i\right)\left(|\phi_i^{(0)}\rangle + |\delta\phi_i\rangle\right)$$

Next, we multiply from the left both sides of the above equation by $\langle\phi_j^{(0)}|$ and use Eq. (B.49) to find

$$\begin{aligned}\epsilon_i^{(0)}\delta_{ij} + \langle\phi_j^{(0)}|\mathcal{H}_1|\phi_i^{(0)}\rangle + \epsilon_j^{(0)}\langle\phi_j^{(0)}|\delta\phi_i\rangle + \langle\phi_j^{(0)}|\mathcal{H}_1|\delta\phi_i\rangle \\ = \epsilon_i^{(0)}\delta_{ij} + \delta\epsilon_i\delta_{ij} + \epsilon_i^{(0)}\langle\phi_j^{(0)}|\delta\phi_i\rangle + \delta\epsilon_i\langle\phi_j^{(0)}|\delta\phi_i\rangle\end{aligned} \tag{B.50}$$

If in Eq. (B.50) we take $j = i$ and keep only first order terms, that is, terms that involve only one of the small quantities $\mathcal{H}_1$, $|\delta\phi_i\rangle$, $\delta\epsilon_i$, we find

$$\delta\epsilon_i^{(1)} = \langle\phi_i^{(0)}|\mathcal{H}_1|\phi_i^{(0)}\rangle \tag{B.51}$$

where we have introduced the superscript (1) in the energy correction to indicate the order of the approximation. This is a simple and important result: the change in the energy of state i to first order in the perturbation $\mathcal{H}_1$ is simply the expectation value of $\mathcal{H}_1$ in the unperturbed wavefunction of that state $|\phi_i^{(0)}\rangle$. If in Eq. (B.50) we take $j \neq i$ and keep only first order terms, we obtain

$$\langle\phi_j^{(0)}|\delta\phi_i\rangle = -\frac{\langle\phi_j^{(0)}|\mathcal{H}_1|\phi_i^{(0)}\rangle}{\epsilon_j^{(0)} - \epsilon_i^{(0)}} \tag{B.52}$$

which is simply the projection of $|\delta\phi_i\rangle$ on the unperturbed state $\langle\phi_j^{(0)}|$. Therefore, $|\delta\phi_i\rangle$ will involve a summation over all such terms, each multiplied by the corresponding state $|\phi_j^{(0)}\rangle$, which gives for the first order correction to the wavefunction of state i:

$$|\delta\phi_i^{(1)}\rangle = \sum_{j\neq i} \frac{\langle\phi_j^{(0)}|\mathcal{H}_1|\phi_i^{(0)}\rangle}{\epsilon_i^{(0)} - \epsilon_j^{(0)}} |\phi_j^{(0)}\rangle \tag{B.53}$$

This is another important result, showing that the change in the wavefunction of state i to first order in the perturbation $\mathcal{H}_1$ involves the matrix elements of $\mathcal{H}_1$ between the unperturbed state i and all the other unperturbed states j, divided by the unperturbed energy difference between state i and j.

If in Eq. (B.50) we take $j = i$ and keep first and second order terms, we obtain

$$\delta\epsilon_i^{(2)} = \langle\phi_i^{(0)}|\mathcal{H}_1|\phi_i^{(0)}\rangle + \langle\phi_i^{(0)}|\mathcal{H}_1|\delta\phi_i\rangle$$

where we have used the fact that by construction $\langle\phi_i^{(0)}|\delta\phi_i\rangle = 0$, from the orthogonality of the $|\phi_i^{(0)}\rangle$'s. Substituting into the above equation the expression that we found for $|\delta\phi_i\rangle$ to first order, Eq. (B.53), we obtain for the change in the energy of state i to second order in $\mathcal{H}_1$:

$$\delta\epsilon_i^{(2)} = \sum_{j\neq i} \frac{\langle\phi_j^{(0)}|\mathcal{H}_1|\phi_i^{(0)}\rangle\langle\phi_i^{(0)}|\mathcal{H}_1|\phi_j^{(0)}\rangle}{\epsilon_i^{(0)} - \epsilon_j^{(0)}} \tag{B.54}$$

It is a simple extension of these arguments to obtain $|\delta\phi_i\rangle$ to second order, which turns out to be

$$|\delta\phi_i^{(2)}\rangle = -\sum_{j\neq i} \frac{\langle\phi_j^{(0)}|\mathcal{H}_1|\phi_i^{(0)}\rangle\langle\phi_i^{(0)}|\mathcal{H}_1|\phi_i^{(0)}\rangle}{(\epsilon_i^{(0)} - \epsilon_j^{(0)})^2}|\phi_j^{(0)}\rangle$$

$$+ \sum_{j,k\neq i} \frac{\langle\phi_j^{(0)}|\mathcal{H}_1|\phi_k^{(0)}\rangle\langle\phi_k^{(0)}|\mathcal{H}_1|\phi_i^{(0)}\rangle}{(\epsilon_i^{(0)} - \epsilon_j^{(0)})(\epsilon_i^{(0)} - \epsilon_k^{(0)})}|\phi_j^{(0)}\rangle \quad \text{(B.55)}$$

It is also possible to go to even higher orders in both $\delta\epsilon_i$ and $|\delta\phi_i\rangle$. Usually the first order approximation in the wavefunction and the second order approximation in the energy are adequate.

B.5.2 Degenerate perturbation theory

The approach we developed above will not work if the original set of states involves degeneracies, that is, states whose unperturbed energies are the same, because then the factors that appear in the denominators in Eqs. (B.53) and (B.54) will vanish. In such cases we need to apply degenerate perturbation theory. We outline here the simplest case of degenerate perturbation theory, which involves two degenerate states, labeled i and j. By assumption the two unperturbed states $|\phi^{(0)}{}_i\rangle$ and $|\phi^{(0)}{}_j\rangle$ have the same unperturbed energy $\epsilon^{(0)}{}_i = \epsilon^{(0)}{}_j$. We consider a linear combination of the two degenerate states and try to find the effect of the perturbation on the energy and the wavefunction of this state, denoted by $\delta\epsilon$ and $|\delta\phi\rangle$. We will then have for the unperturbed case

$$\mathcal{H}_0\left(a_i|\phi_i^{(0)}\rangle + a_j|\phi_j^{(0)}\rangle\right) = \epsilon_i^{(0)}\left(a_i|\phi_i^{(0)}\rangle + a_j|\phi_j^{(0)}\rangle\right)$$

and for the perturbed case

$$(\mathcal{H}_0 + \mathcal{H}_1)\left(a_i|\phi_i^{(0)}\rangle + a_j|\phi_j^{(0)}\rangle + |\delta\phi\rangle\right) = (\epsilon_i^{(0)} + \delta\epsilon)\left(a_i|\phi_i^{(0)}\rangle + a_j|\phi_j^{(0)}\rangle + |\delta\phi\rangle\right)$$

Closing both sides of the last equation first with $\langle\phi_i^{(0)}|$ and then with $\langle\phi_j^{(0)}|$, keeping first order terms only in the small quantities $\mathcal{H}_1$, $\delta\epsilon$, $|\delta\phi\rangle$, and taking into account that the original unperturbed states are orthonormal, we obtain the following two equations:

$$a_i\langle\phi_i^{(0)}|\mathcal{H}_1|\phi_i^{(0)}\rangle + a_j\langle\phi_i^{(0)}|\mathcal{H}_1|\phi_j^{(0)}\rangle = a_i\delta\epsilon$$

$$a_i\langle\phi_j^{(0)}|\mathcal{H}_1|\phi_i^{(0)}\rangle + a_j\langle\phi_j^{(0)}|\mathcal{H}_1|\phi_j^{(0)}\rangle = a_j\delta\epsilon$$

In order for this system of linear equations to have a non-trivial solution, the determinant of the coefficients a_i, a_j must vanish, which gives

$$\delta\epsilon = \frac{1}{2}\left[\langle\phi_i^{(0)}|\mathcal{H}_1|\phi_i^{(0)}\rangle + \langle\phi_j^{(0)}|\mathcal{H}_1|\phi_j^{(0)}\rangle\right]$$

$$\pm\frac{1}{2}\left[\left(\langle\phi_i^{(0)}|\mathcal{H}_1|\phi_i^{(0)}\rangle + \langle\phi_j^{(0)}|\mathcal{H}_1|\phi_j^{(0)}\rangle\right)^2 - 4\left|\langle\phi_j^{(0)}|\mathcal{H}_1|\phi_i^{(0)}\rangle\right|^2\right]^{\frac{1}{2}} \quad \text{(B.56)}$$

This last equation gives two possible values for the first order correction to the energy of the two degenerate states. In general these two values are different, and we associate them with the change in energy of the two degenerate states; this is referred to as "splitting of the degeneracy" by the perturbation term $\mathcal{H}_1$. If the two possible values for $\delta\epsilon$ are not different, which implies that in the above equation the expression under the square root vanishes, then we need to go to higher orders of perturbation theory to find how the degeneracy is split by the perturbation. A similar approach can also yield the splitting in the energy of states with higher degeneracy, in which case the subspace of states which must be included involves all the unperturbed degenerate states.

B.6 Time-dependent perturbation theory

In time-dependent perturbation theory we begin with a hamiltonian $\mathcal{H}_0$ whose eigenfunctions $|\phi_k^{(0)}\rangle$ and eigenvalues $\epsilon_k^{(0)}$ are known and form a complete orthonormal set. We then turn on a time-dependent perturbation $\mathcal{H}_1(t)$ and express the new time-dependent wavefunction in terms of the $|\phi_k^{(0)}\rangle$'s:

$$|\psi(t)\rangle = \sum_k c_k(t) \mathrm{e}^{-\mathrm{i}\epsilon_k^{(0)} t/\hbar} |\phi_k^{(0)}\rangle \tag{B.57}$$

where we have explicitly included a factor $\exp(-\mathrm{i}\epsilon_k^{(0)} t/\hbar)$ in the time-dependent coefficient that accompanies the unperturbed wavefunction $|\phi_k^{(0)}\rangle$ in the sum. The wavefunction $|\psi(t)\rangle$ satisfies the time-dependent Schrödinger equation for the full hamiltonian:

$$[\mathcal{H}_0 + \mathcal{H}_1(t)]\,|\psi(t)\rangle = \mathrm{i}\hbar \frac{\partial}{\partial t} |\psi(t)\rangle$$

Introducing the expression of Eq. (B.57) on the left-hand side of this equation we obtain

$$[\mathcal{H}_0 + \mathcal{H}_1(t)] \sum_k c_k(t) \mathrm{e}^{-\mathrm{i}\epsilon_k^{(0)} t/\hbar} |\phi_k^{(0)}\rangle$$
$$= \sum_k c_k(t) \mathrm{e}^{-\mathrm{i}\epsilon_k^{(0)} t/\hbar} \epsilon_k^{(0)} |\phi_k^{(0)}\rangle + \sum_k c_k(t) \mathrm{e}^{-\mathrm{i}\epsilon_k^{(0)} t/\hbar} \mathcal{H}_1(t) |\phi_k^{(0)}\rangle$$

while the same expression introduced on the right-hand side of the above equation gives

$$\mathrm{i}\hbar \frac{\partial}{\partial t} \sum_k c_k(t) \mathrm{e}^{-\mathrm{i}\epsilon_k^{(0)} t/\hbar} |\phi_k^{(0)}\rangle$$
$$= \sum_k c_k(t) \mathrm{e}^{-\mathrm{i}\epsilon_k^{(0)} t/\hbar} \epsilon_k^{(0)} |\phi_k^{(0)}\rangle + \mathrm{i}\hbar \sum_k \frac{\mathrm{d}c_k(t)}{\mathrm{d}t} \mathrm{e}^{-\mathrm{i}\epsilon_k^{(0)} t/\hbar} |\phi_k^{(0)}\rangle$$

By comparing the two results, we arrive at

$$\mathrm{i}\hbar \sum_k \frac{\mathrm{d}c_k(t)}{\mathrm{d}t} \mathrm{e}^{-\mathrm{i}\epsilon_k^{(0)} t/\hbar} |\phi_k^{(0)}\rangle = \sum_k c_k(t) \mathrm{e}^{-\mathrm{i}\epsilon_k^{(0)} t/\hbar} \mathcal{H}_1(t) |\phi_k^{(0)}\rangle \tag{B.58}$$

Now, multiplying both sides of this last equation from the left by $\langle\phi_j^{(0}|$ and using the completeness and orthonormality properties of the unperturbed wavefunctions, we obtain

$$\mathrm{i}\hbar \frac{\mathrm{d}c_j(t)}{\mathrm{d}t} = \sum_k \langle\phi_j^{(0)}|\mathcal{H}_1(t)|\phi_k^{(0)}\rangle \mathrm{e}^{\mathrm{i}(\epsilon_j^{(0)} - \epsilon_k^{(0)})t/\hbar} c_k(t) \tag{B.59}$$

This is very useful for finding the transition probability between states of the system induced by the perturbation: Suppose that at $t = 0$ the system is in the eigenstate i of $\mathcal{H}_0$, in which case $c_i(0) = 1$ and $c_j(0) = 0$ for $j \neq i$. Then, writing the above expression in differential form at time $\mathrm{d}t$ we obtain

$$\mathrm{d}c_j(t) = -\frac{\mathrm{i}}{\hbar}\langle\phi_j^{(0)}|\mathcal{H}_1(t)|\phi_i^{(0)}\rangle \mathrm{e}^{\mathrm{i}(\epsilon_j^{(0)}-\epsilon_i^{(0)})t/\hbar}\mathrm{d}t$$

which, with the definitions

$$\hbar\omega_{ji} \equiv \epsilon_j^{(0)} - \epsilon_i^{(0)}, \quad V_{ji}(t) \equiv \langle\phi_j^{(0)}|\mathcal{H}_1(t)|\phi_i^{(0)}\rangle$$

can be integrated over time to produce

$$c_j(t) = -\frac{\mathrm{i}}{\hbar}\int_0^t V_{ji}(t')\mathrm{e}^{\mathrm{i}\omega_{ji}t'}\mathrm{d}t' \tag{B.60}$$

If we assume that the perturbation $\mathcal{H}_1(t)$ has the simple time dependence

$$\mathcal{H}_1(t) = V_1\mathrm{e}^{-\mathrm{i}\omega t}$$

then the integration over t' can be performed easily to give

$$c_j(t) = \frac{\langle\phi_j^{(0)}|V_1|\phi_i^{(0)}\rangle}{\epsilon_j^{(0)} - \epsilon_i^{(0)}}\left(1 - \mathrm{e}^{\mathrm{i}(\omega_{ji}-\omega)t}\right)$$

Typically, the quantity of interest is the probability for the transition from the initial state of the system at time 0, in the present example identified with the state $|\phi_i^{(0)}\rangle$, to another state such as $|\phi_j^{(0)}\rangle$ at time t; this probability is precisely equal to $|c_j(t)|^2$, which from the above expression takes the form

$$|c_j(t)|^2 = 2\frac{\left|\langle\phi_j^{(0)}|V_1|\phi_i^{(0)}\rangle\right|^2}{\left[\epsilon_j^{(0)} - \epsilon_i^{(0)}\right]^2}\left[1 - \cos(\omega_{ji} - \omega)t\right] \tag{B.61}$$

In this discussion we have made the implicit assumption that the states $|\phi_j^{(0)}\rangle$ have a discrete spectrum. Generally, the spectrum of the unperturbed hamiltonian $\mathcal{H}_0$ may be a continuum with density of states $g(\epsilon)$, in which case we need to include in the expression for the transition probability all possible final states with energy $\epsilon_j^{(0)}$; the number of such states in an interval $\mathrm{d}\epsilon_j$ is $g(\epsilon_j)\mathrm{d}\epsilon_j$. These considerations lead to the following expression for the *rate* of the transition between an initial state $|\phi_i^{(0)}\rangle$ with energy $\epsilon_i^{(0)}$ and a continuum of final states $|\phi_f^{(0)}\rangle$ with energy $\epsilon_f^{(0)}$ in an interval $\mathrm{d}\epsilon_f$:

$$\mathrm{d}P_{i\to f} = \frac{\mathrm{d}}{\mathrm{d}t}|c_f(t)|^2 g(\epsilon_f)\mathrm{d}\epsilon_f = \frac{2}{\hbar^2}\left|\langle\phi_f^{(0)}|V_1|\phi_i^{(0)}\rangle\right|^2\frac{\sin(\omega_{fi} - \omega)t}{(\omega_{fi} - \omega)}g(\epsilon_f)\mathrm{d}\epsilon_f$$

If we now let the time of the transition be very long, $t \to \infty$, the function $\sin(\omega_{fi} - \omega)t/(\omega_{fi} - \omega)$ that appears in the above expression becomes a δ-function (see Appendix G, Eq. (G.56)),

$$\lim_{t\to\infty}\frac{\sin(\omega_{fi} - \omega)t}{(\omega_{fi} - \omega)} = \pi\delta(\omega_{fi} - \omega)$$

which leads to

$$\frac{\mathrm{d}P_{i\to f}}{\mathrm{d}\epsilon_f} = \frac{2\pi}{\hbar}\left|\langle\phi_f^{(0)}|V_1|\phi_i^{(0)}\rangle\right|^2\delta(\epsilon_i^{(0)} - \epsilon_f^{(0)} - \hbar\omega)g(\epsilon_f) \tag{B.62}$$

This last expression is known as **Fermi's golden rule**. For transitions from one single-particle state $|\phi_i^{(0)}\rangle$ to another single-particle state $|\phi_f^{(0)}\rangle$, in which case neither the density of states $g(\epsilon_f)$ nor the dependence of the transition probability on ϵ_f enter, the transition rate takes the form

$$P_{i\to f}(\omega) = \frac{2\pi}{\hbar}\left|\langle\phi_f^{(0)}|V_1|\phi_i^{(0)}\rangle\right|^2 \delta(\epsilon_i^{(0)} - \epsilon_f^{(0)} - \hbar\omega) \quad \text{(B.63)}$$

B.7 The electromagnetic field term

An example of the application of perturbation theory is the motion of a particle of mass m and charge q in an external electromagnetic field. The electric field, $\mathbf{E}$, is given in terms of the scalar and vector potentials Φ, $\mathbf{A}$ as

$$\mathbf{E}(\mathbf{r}, t) = -\frac{1}{c}\frac{\partial \mathbf{A}(\mathbf{r}, t)}{\partial t} - \nabla_{\mathbf{r}}\Phi(\mathbf{r}, t)$$

where c is the speed of light. In terms of these potentials, the classical hamiltonian which describes the motion of the particle is given by

$$\mathcal{H} = \frac{1}{2m}\left[\mathbf{p} - \frac{q}{c}\mathbf{A}\right]^2 + q\Phi.$$

We will adopt this expression as the quantum mechanical hamiltonian of the particle, by analogy to that for the free particle. Moreover, we can choose the vector and scalar potentials of the electromagnetic fields such that $\nabla_{\mathbf{r}} \cdot \mathbf{A} = 0$, $\Phi = 0$, a choice called the Coulomb gauge (see Appendix A).

Our goal now is to determine the interaction part of the hamiltonian, that is, the part that describes the interaction of the charged particle with the external electromagnetic field, assuming that the latter is a small perturbation in the motion of the free particle. We will also specialize the discussion to electrons, in which case $m = m_e$, $q = -e$. In the Coulomb gauge, we express the vector potential as

$$\mathbf{A}(\mathbf{r}, t) = \mathbf{A}_0[\mathrm{e}^{\mathrm{i}(\mathbf{k}\cdot\mathbf{r}-\omega t)} + \mathrm{e}^{-\mathrm{i}(\mathbf{k}\cdot\mathbf{r}-\omega t)}] \text{ and } \mathbf{k}\cdot\mathbf{A}_0 = 0 \quad \text{(B.64)}$$

where $\mathbf{A}_0$ is a real, constant vector. Expanding $[\mathbf{p} + (e/c)\mathbf{A}]^2$ and keeping only up to first order terms in $\mathbf{A}$, we find

$$\begin{aligned}
\frac{1}{2m_e}\left[\mathbf{p} + \frac{e}{c}\mathbf{A}\right]^2\psi &= \frac{1}{2m_e}\left[\mathbf{p}^2 + \frac{e}{c}\mathbf{p}\cdot\mathbf{A} + \frac{e}{c}\mathbf{A}\cdot\mathbf{p}\right]\psi \\
&= \frac{1}{2m_e}\mathbf{p}^2\psi + \frac{e\hbar}{2\mathrm{i}m_e c}\left[\nabla_{\mathbf{r}}\cdot(\mathbf{A}\psi) + \mathbf{A}\cdot(\nabla_{\mathbf{r}}\psi)\right] \\
&= \frac{1}{2m_e}\mathbf{p}^2\psi + \frac{e\hbar}{2\mathrm{i}m_e c}\left[2\mathbf{A}\cdot(\nabla_{\mathbf{r}}\psi) + (\nabla_{\mathbf{r}}\cdot\mathbf{A})\psi\right] \quad = \left[\hat{T} + \mathcal{H}^{int}\right]\psi
\end{aligned}$$

where ψ is the wavefunction on which the hamiltonian operates and $\hat{T} = \mathbf{p}^2/2m_e$ is the kinetic energy operator. Now we take into account that, due to our choice of the Coulomb gauge Eq. (B.64),

$$\nabla_{\mathbf{r}}\cdot\mathbf{A} = \mathrm{i}\mathbf{k}\cdot\mathbf{A}_0[\mathrm{e}^{\mathrm{i}(\mathbf{k}\cdot\mathbf{r}-\omega t)} + \mathrm{e}^{-\mathrm{i}(\mathbf{k}\cdot\mathbf{r}-\omega t)}] = 0$$

which gives for the interaction hamiltonian the expression

$$\mathcal{H}^{int}(t) = \frac{e}{m_e c} \mathrm{e}^{\mathrm{i}(\mathbf{k}\cdot\mathbf{r}-\omega t)} \mathbf{A}_0 \cdot \mathbf{p} + \text{c.c.} \tag{B.65}$$

since $(\hbar/\mathrm{i})\nabla_{\mathbf{r}} = \mathbf{p}$, and c.c. denotes complex conjugate.

Further reading

1. *Quantum Mechanics*, L.I. Schiff (McGraw-Hill, New York, 1968).
   This is one of the standard references, containing an advanced and comprehensive treatment of quantum mechanics.
2. *Quantum Mechanics*, L.D. Landau and L. Lifshitz (3rd edn, Pergamon Press, Oxford, 1977).
   This is another standard reference with an advanced but somewhat terse treatment of quantum mechanics.

Problems

1. We wish to prove Eq. (B.7) from which the Heisenberg uncertainty relation follows.[1] Show that we can write the product of the standard deviations as follows:

$$\begin{aligned}(\Delta x)^2(\Delta p)^2 &= \int \phi^* X^2 \phi \mathrm{d}x \int \phi^* P^2 \phi \mathrm{d}x \\ &= \int (X^*\phi^*)(X\phi)\mathrm{d}x \int (P^*\phi^*)(P\phi)\mathrm{d}x \\ &= \int |X\phi|^2 \mathrm{d}x \int |P\phi|^2 \mathrm{d}x\end{aligned}$$

which involves doing an integration by parts and using the fact that the wavefunction vanishes at infinity. Next show that

$$\begin{aligned}&\int \left| X\phi - P\phi \frac{\int X\phi P^*\phi^* \mathrm{d}x}{\int |P\phi|^2 \mathrm{d}x} \right|^2 \mathrm{d}x \geq 0 \\ &\Longrightarrow \int |X\phi|^2 \mathrm{d}x \int |P\phi|^2 \mathrm{d}x \geq \left| \int X^*\phi^* P\phi \mathrm{d}x \right|^2 = \left| \int \phi^* X P \phi \mathrm{d}x \right|^2\end{aligned}$$

Next prove that the last term in the above equation can be rewritten as

$$\begin{aligned}&\left| \int \phi^* \left[\frac{1}{2}(XP - PX) + \frac{1}{2}(XP + PX) \right] \phi \mathrm{d}x \right|^2 \\ &= \frac{1}{4} \left| \int \phi^*(XP - PX)\phi \mathrm{d}x \right|^2 + \frac{1}{4} \left| \int \phi^*(XP + PX)\phi \mathrm{d}x \right|^2\end{aligned}$$

by doing integration by parts as above. Finally, using the definition of the momentum operator, show that

$$(XP - PX)\phi = \mathrm{i}\hbar\phi$$

[1] This problem is discussed in more detail in Schiff (see the Further reading section, above), where also references to the original work of Heisenberg are given.

and taking into account the earlier relations derive obtain the desired result:

$$(\Delta x)^2(\Delta p)^2 \geq \frac{1}{4}\hbar^2$$

2. Consider a one-dimensional square barrier with height $V_0 > 0$ in the range $0 < x < L$, and an incident plane wave of energy ϵ. Write the wavefunction as a plane wave with wave-vector q in the region where the potential is zero, and an exponential with decay constant κ for $\epsilon < V_0$ or a plane wave with wave-vector k for $\epsilon > V_0$. Use $\exp(iqx)$ to represent the incident wave, $R\exp(-iqx)$ for the reflected wave, both in the range $x < 0$, and $T\exp(iqx)$ for the transmitted wave in the range $x > L$; use $A\exp(\kappa x) + B\exp(-\kappa x)$ for $\epsilon < V_0$ and $C\exp(ikx) + D\exp(-ikx)$ for $\epsilon > V_0$ to represent the wavefunction in the range $0 < x < L$. Employ the conditions on the continuity of the wavefunction and its derivative to show that the transmission coefficient $|T|^2$ is given by

$$|T|^2 = \frac{1}{1 + \frac{1}{4\lambda(1-\lambda)}\sinh^2(\kappa L)}, \quad \text{for} \quad \epsilon < V_0$$

$$|T|^2 = \frac{1}{1 + \frac{1}{4\lambda(\lambda-1)}\sin^2(kL)}, \quad \text{for} \quad \epsilon > V_0$$

with $\lambda = \epsilon/V_0$. Discuss the behavior of $|T|^2$ as a function of λ for a given value of $V_0 L^2$. For what values of λ is the transmission coefficient unity? Interpret this result in simple physical terms.

3. (a) For a system of $N = 3$ particles, prove the relationship between the one-particle and two-particle density matrices, as given in Eq. (2.81), for the case when the four variables in the two-particle density matrix are related by $\mathbf{r}_1 = \mathbf{r}_2 = \mathbf{r}$, $\mathbf{r}'_1 = \mathbf{r}'_2 = \mathbf{r}'$. Write this relationship as a determinant in terms of the density matrices $\gamma(\mathbf{r}, \mathbf{r})$ and $\gamma(\mathbf{r}, \mathbf{r}')$, and generalize it to the case where the four variables in the two-particle density matrix are independent.
   (b) Write the expression for the n-particle density matrix in terms of the many-body wavefunction $\Psi(\mathbf{r}_1, \mathbf{r}_2, \ldots, \mathbf{r}_N)$, and express it as a determinant in terms of the density matrix $\gamma(\mathbf{r}, \mathbf{r}')$; this relation shows the isomorphism between the many-body wavefunction and the density matrix representations; see for example Ref. [234].
4. Show that the commutator of the hamiltonian with the potential energy $[\mathcal{H}, V(\mathbf{r})]$ is not zero, except when the potential is a constant $V(\mathbf{r}) = V_0$. Based on this result, provide an argument to the effect that the energy eigenfunctions can be simultaneous eigenfunctions of the momentum operator only for particles in a constant potential.
5. Find all the spin states, including the total spin and its projection on the z axis, obtained by combining four spin-1/2 particles; use arguments analogous to what was discussed in the text for the case of two spin with 1/2 particles. Compare the results what you would get by combining two particles with spin 1.
6. We wish to determine the eigenvalues of the Morse potential, discussed in chapter 1 (see Eq. (1.6). One method is to consider an expansion in powers of $(r - r_0)$ near the minimum and relate it to the harmonic oscillator potential with higher order terms. Specifically, the potential

$$V(r) = \frac{1}{2}m\omega^2(r - r_0)^2 - \alpha(r - r_0)^3 + \beta(r - r_0)^4 \tag{B.66}$$

has eigenvalues

$$\epsilon_n = \left(n + \frac{1}{2}\right)\hbar\omega\left[1 - \lambda\left(n + \frac{1}{2}\right)\right]$$

where

$$\lambda = \frac{3}{2\hbar\omega}\left(\frac{\hbar}{m\omega}\right)^2\left[\frac{5}{2}\frac{\alpha^2}{m\omega^2} - \beta\right]$$

as discussed in Landau and Lifshitz, p. 318 (see the Further reading section). First, check to what extent the expansion (B.66) with up to fourth order terms in $(r - r_0)$ is a good representation of the Morse potential; what are the values of α and β in terms of the parameters of the Morse potential? Use this approach to show that the eigenvalues of the Morse potential are given by Eq. (1.9).

7. Calculate the splitting of the degeneracy of the $2p$ states of the hydrogen atom ($n = 2, l = 1$) due to a weak external electromagnetic field.
8. An important simple model that demonstrates some of the properties of electron states in periodic solids is the so called Kronig–Penney model. In this model, a particle of mass m experiences a one-dimensional periodic potential with period a:

$$\begin{aligned} V(x) &= 0, \quad 0 < x < (a - l) \\ &= V_0, \quad (a - l) < x < a \\ V(x + a) &= V(x) \end{aligned}$$

where we will take $V_0 > 0$. The wavefunction $\psi(x)$ obeys the Schrödinger equation

$$\left[-\frac{\hbar^2}{2m}\frac{\mathrm{d}^2}{\mathrm{d}x^2} + V(x)\right]\psi(x) = \epsilon\psi(x)$$

(a) Choose the following expression for the particle wavefunction

$$\psi(x) = \mathrm{e}^{\mathrm{i}kx}u(x)$$

and show that the function $u(x)$ must obey the equation

$$\frac{\mathrm{d}^2u(x)}{\mathrm{d}x^2} + 2\mathrm{i}k\frac{\mathrm{d}u(x)}{\mathrm{d}x} - \left[k^2 - \frac{2m\epsilon}{\hbar^2} + \frac{2mV(x)}{\hbar^2}\right]u(x) = 0$$

Assuming that $u(x)$ is finite for $x \to \pm\infty$, the variable k must be real so that the wavefunction $\psi(x)$ is finite for all x.

(b) We first examine the case $\epsilon > V_0 > 0$. Consider two solutions, $u_0(x)$, $u_1(x)$ for the ranges $0 < x < (a - l)$ and $(a - l) < x < a$, respectively, which obey the equations

$$\frac{\mathrm{d}^2u_0(x)}{\mathrm{d}x^2} + 2\mathrm{i}k\frac{\mathrm{d}u_0(x)}{\mathrm{d}x} - [k^2 - \kappa^2]u_0(x) = 0, \quad 0 < x < (a - l)$$

$$\frac{\mathrm{d}^2u_1(x)}{\mathrm{d}x^2} + 2\mathrm{i}k\frac{\mathrm{d}u_1(x)}{\mathrm{d}x} - [k^2 - \lambda^2]u_1(x) = 0, \quad (a - l) < x < a$$

where we have defined the quantities

$$\kappa = \sqrt{\frac{2m\epsilon}{\hbar^2}}, \qquad \lambda = \sqrt{\frac{2m(\epsilon - V_0)}{\hbar^2}}$$

which are both real for $\epsilon > 0$ and $(\epsilon - V_0) > 0$. Show that the solutions to these equations can be written as

$$u_0(x) = c_0\mathrm{e}^{\mathrm{i}(\kappa - k)x} + d_0\mathrm{e}^{-\mathrm{i}(\kappa + k)x}, \quad 0 < x < (a - l)$$

$$u_1(x) = c_1\mathrm{e}^{\mathrm{i}(\lambda - k)x} + d_1\mathrm{e}^{-\mathrm{i}(\lambda + k)x}, \quad (a - l) < x < a$$

By matching the values of these solutions and of their first derivatives at $x = 0$ and $x = a - l$, find a system of four equations for the four unknowns, c_0, d_0, c_1, d_1. Show that requiring this system to have a non-trivial solution leads to the following condition:

$$-\frac{\kappa^2 + \lambda^2}{2\kappa\lambda} \sin(\kappa(a - l)) \sin(\lambda l) + \cos(\kappa(a - l)) \cos(\lambda l) = \cos(ka)$$

Next, show that with the definition

$$\tan(\theta) = -\frac{\kappa^2 + \lambda^2}{2\kappa\lambda}$$

the above condition can be written as

$$\left[1 + \frac{(\kappa^2 - \lambda^2)^2}{4\kappa^2\lambda^2} \sin^2(\lambda l)\right]^{1/2} \cos(\kappa(a - l) - \theta) = \cos(ka)$$

Show that this last equation admits real solutions for k only in certain intervals of ϵ; determine these intervals of ϵ and the corresponding values of k. Plot the values of ϵ as a function of k and interpret the physical meaning of these solutions. How does the solution depend on the ratio ϵ / V_0? How does it depend on the ratio l/a?

(c) Repeat the above problem for the case $V_0 > \epsilon > 0$. Discuss the differences between the two cases.

Appendix C

Elements of thermodynamics

C.1 The laws of thermodynamics

Thermodynamics is the empirical science that describes the state of macroscopic systems without reference to their microscopic structure. The laws of thermodynamics are based on experimental observations. The physical systems described by thermodynamics are considered to be composed of a very large number of microscopic particles (atoms or molecules). In the context of thermodynamics, a macroscopic system is described in terms of the external conditions that are imposed on it, determined by scalar or vector fields, and the values of the corresponding variables that specify the state of the system for given external conditions. The usual fields and corresponding thermodynamic variables are

$$\begin{aligned}
\text{temperature} : T &\longleftrightarrow S : \text{entropy} \\
\text{pressure} : P &\longleftrightarrow \Omega : \text{volume} \\
\text{chemical potential} : \mu &\longleftrightarrow N : \text{number of particles} \\
\text{magnetic field} : \mathbf{H} &\longleftrightarrow \mathbf{M} : \text{magnetization} \\
\text{electric field} : \mathbf{E} &\longleftrightarrow \mathbf{P} : \text{polarization}
\end{aligned}$$

with the last two variables referring to systems that possess internal magnetic or electric dipole moments, so that they can respond to the application of external magnetic or electric fields. The temperature, pressure[1] and chemical potential fields are determined by putting the system in contact with an appropriate reservoir; the values of these fields are set by the values they have in the reservoir. The fields are *intensive* quantities (they do not depend on the amount of substance) while the variables are *extensive* quantities (they are proportional to the amount of substance) of the system. Finally, each system is characterized by its internal energy E, which is a state function, that is, it depends on the state of the system and not on the path that the system takes to arrive at a certain state. Consequently, the quantity $\mathrm{d}E$ is an exact differential.

Thermodynamics is based on three laws.

(1) **The first law of thermodynamics** states that:

Heat is a form of energy

This can be put into a quantitative expression as follows: the sum of the change in the internal energy ΔE of a system and the work done by the system ΔW is equal to the heat

[1] If the pressure field is not homogeneous, we can introduce a tensor to describe it, called the stress; in this case the corresponding variable is the strain tensor field. These notions are discussed in Appendix E.

ΔQ absorbed by the system,

$$\Delta Q = \Delta E + \Delta W$$

Heat is not a concept that can be related to the microscopic structure of the system, and therefore Q cannot be considered a state function, and $\mathrm{d}Q$ is not an exact differential. From the first law, heat is equal to the increase in the internal energy of the system when its temperature goes up without the system having done any work.

(2) **The second law of thermodynamics** states that:

> *There can be no thermodynamic process whose sole effect is to transform heat entirely to work*

A direct consequence of the second law is that, for a cyclic process which brings the system back to its initial state,

$$\oint \frac{\mathrm{d}Q}{T} \leq 0 \tag{C.1}$$

where the equality holds for a *reversible* process. This must be true because otherwise the released heat could be used to produce work, this being the sole effect of the cyclic process since the system returns to its initial state, which would contradict the second law. For a reversible process, the heat absorbed during some infinitesimal part of the process must be released at some other infinitesimal part, the net heat exchange summing to zero, or the process would not be reversible. From this we draw the conclusion that for a reversible process we can define a state function S through

$$\int_{\text{initial}}^{\text{final}} \frac{\mathrm{d}Q}{T} = S_{\text{final}} - S_{\text{initial}}: \quad \text{reversible process} \tag{C.2}$$

which is called the entropy; its differential is given by

$$\mathrm{d}S = \frac{\mathrm{d}Q}{T} \tag{C.3}$$

Thus, even though heat is not a state function, and therefore $\mathrm{d}Q$ is not an exact differential, the entropy S defined through Eq. (C.2) *is* a state function and $\mathrm{d}S$ *is* an exact differential. Since the entropy is defined with reference to a reversible process, for an arbitrary process we will have

$$\int_{\text{initial}}^{\text{final}} \frac{\mathrm{d}Q}{T} \leq S_{\text{final}} - S_{\text{initial}}: \quad \text{arbitrary process} \tag{C.4}$$

This inequality is justified by the following argument. We can imagine the arbitrary process between initial and final states to be combined with a reversible process between the final and initial states, which together form a cyclic process for which the inequality (C.1) holds, and $S_{\text{final}} - S_{\text{initial}}$ is defined by the reversible part. Hence, for the combined cyclic process

$$\begin{aligned}\oint \frac{\mathrm{d}Q}{T} &= \int_{\text{initial}}^{\text{final}} \left(\frac{\mathrm{d}Q}{T}\right)_{\text{arbitrary}} + \int_{\text{final}}^{\text{initial}} \left(\frac{\mathrm{d}Q}{T}\right)_{\text{reversible}} \\ &= \int_{\text{initial}}^{\text{final}} \left(\frac{\mathrm{d}Q}{T}\right)_{\text{arbitrary}} - (S_{\text{final}} - S_{\text{initial}}) \leq 0\end{aligned}$$

which leads to Eq. (C.4). Having defined the inequality (C.4) for an arbitrary process, the second law can now be cast in a more convenient form: in any thermodynamic process which takes an isolated system (for which $\mathrm{d}Q = 0$) from an initial to a final state the following inequality holds for the difference in the entropy between the two states:

$$\Delta S = S_{\text{final}} - S_{\text{initial}} \geq 0 \tag{C.5}$$

where the equality holds for reversible processes. Therefore, *the equilibrium state of an isolated system is a state of maximum entropy.*

(3) **The third law of thermodynamics** states that:

The entropy at the absolute zero of the temperature is a universal constant.

This statement holds for all substances. We can choose this universal constant to be zero:

$$S(T = 0) = S_0 = 0 : \quad \text{universal constant}$$

The three laws of thermodynamics must be supplemented by the equation of state (EOS) of a system, which together provide a complete description of all the possible states in which the system can exist, as well the possible transformations between such states. Since these states are characterized by macroscopic experimental measurements, it is often convenient to introduce and use quantities such as:

$$C_\Omega \equiv \left(\frac{\mathrm{d}Q}{\mathrm{d}T}\right)_\Omega \quad : \quad \text{constant-volume specific heat}$$

$$C_P \equiv \left(\frac{\mathrm{d}Q}{\mathrm{d}T}\right)_P \quad : \quad \text{constant-pressure specific heat}$$

$$\alpha \equiv \frac{1}{\Omega}\left(\frac{\partial \Omega}{\partial T}\right)_P \quad : \quad \text{thermal expansion coefficient}$$

$$\kappa_T \equiv -\frac{1}{\Omega}\left(\frac{\partial \Omega}{\partial P}\right)_T \quad : \quad \text{isothermal compressibility}$$

$$\kappa_S \equiv -\frac{1}{\Omega}\left(\frac{\partial \Omega}{\partial P}\right)_S \quad : \quad \text{adiabatic compressibility}$$

A familiar EOS is that of an ideal gas:

$$P\Omega = Nk_{\mathrm{B}}T \tag{C.6}$$

where k_{B} is Boltzmann's constant. For such a system it is possible to show, simply by manipulating the above expressions and using the laws of thermodynamics, that

$$E = E(T) \tag{C.7}$$

$$C_P - C_\Omega = \frac{T\Omega\alpha^2}{\kappa_T} \tag{C.8}$$

$$C_\Omega = \frac{T\Omega\alpha^2\kappa_S}{(\kappa_T - \kappa_S)\kappa_T} \tag{C.9}$$

$$C_P = \frac{T\Omega\alpha^2}{(\kappa_T - \kappa_S)} \tag{C.10}$$

$$\tag{C.11}$$

Example We prove the first of the above relations, Eq. (C.7), that is, for the ideal gas the internal energy E is a function of the temperature only. The first law of thermodynamics, for the case when the work done on the system is mechanical, becomes

$$dQ = dE + dW = dE + Pd\Omega$$

Using as independent variables Ω and T, we obtain

$$dQ = \left(\frac{\partial E}{\partial T}\right)_\Omega dT + \left(\frac{\partial E}{\partial \Omega}\right)_T d\Omega + Pd\Omega$$

The definition of the entropy Eq. (C.2) allows us to write dQ as TdS, so that the above equation takes the form

$$dS = \frac{1}{T}\left(\frac{\partial E}{\partial T}\right)_\Omega dT + \left[\frac{1}{T}\left(\frac{\partial E}{\partial \Omega}\right)_T + \frac{P}{T}\right] d\Omega$$

and the second law of thermodynamics tells us that the left-hand side is an exact differential, so that taking the cross-derivatives of the the two terms on the right-hand side we should have

$$\left(\frac{\partial}{\partial \Omega}\right)_T \left[\frac{1}{T}\left(\frac{\partial E}{\partial T}\right)_\Omega\right] = \left(\frac{\partial}{\partial T}\right)_\Omega \left[\frac{1}{T}\left(\frac{\partial E}{\partial \Omega}\right)_T + \frac{P}{T}\right]$$

Carrying out the differentiations and using the EOS of the ideal gas, Eq. (C.6),

$$P = \frac{Nk_B T}{\Omega} \Rightarrow \left(\frac{\partial P}{\partial T}\right)_\Omega = \frac{Nk_B}{\Omega}$$

we obtain as the final result

$$\left(\frac{\partial E}{\partial \Omega}\right)_T = 0 \Rightarrow E = E(T)$$

QED

C.2 Thermodynamic potentials

Another very useful concept in thermodynamics is the definition of "thermodynamic potentials" or "free energies", appropriate for different situations. For example, the definition of the

$$\textbf{enthalpy}: \ \Theta = E + P\Omega \tag{C.12}$$

is useful because in terms of it the specific heat at constant pressure takes the form

$$C_P = \left(\frac{dQ}{dT}\right)_P = \left(\frac{\partial E}{\partial T}\right)_P + P\left(\frac{\partial \Omega}{\partial T}\right)_P = \left(\frac{\partial \Theta}{\partial T}\right)_P \tag{C.13}$$

that is, the enthalpy is the appropriate free energy which we need to differentiate with temperature in order to obtain the specific heat at constant pressure. In the case of the ideal gas the enthalpy takes the form

$$\Theta = (C_\Omega + Nk_B)T \tag{C.14}$$

where we have used the result derived above, namely $E = E(T)$, to express the specific heat at constant volume as

$$C_\Omega = \left(\frac{dQ}{dT}\right)_\Omega = \left(\frac{\partial E}{\partial T}\right)_\Omega = \frac{dE}{dT} \Rightarrow E(T) = C_\Omega T \tag{C.15}$$

Now we can take advantage of these expressions to obtain the following relation between specific heats of the the ideal gas at constant pressure and constant volume:

$$C_P - C_\Omega = Nk_B$$

The statement of the second law in terms of the entropy motivates the definition of another thermodynamic potential, the

$$\textbf{Helmholtz free energy}:\ F = E - TS \tag{C.16}$$

Specifically, according to the second law as expressed in Eq. (C.4), for an arbitrary process at constant temperature we will have

$$\int \frac{\mathrm{d}Q}{T} \leq \Delta S \Rightarrow \frac{\Delta Q}{T} = \frac{\Delta E + \Delta W}{T} \leq \Delta S \tag{C.17}$$

For a system that is mechanically isolated, $\Delta W = 0$, and consequently

$$0 \leq -\Delta E + T\Delta S = -\Delta F \Rightarrow \Delta F \leq 0 \tag{C.18}$$

which proves that *the equilibrium state of a mechanically isolated system at constant temperature is one of minimum Helmholtz free energy*. More explicitly, the above relation tells us that changes in F due to changes in the state of the system (which of course must be consistent with the conditions of constant temperature and no mechanical work done on or by the system), can only decrease the value of F; therefore at equilibrium, when F does not change any longer, its value must be at a minimum.

Finally, if in addition to the temperature the pressure is also constant, then by similar arguments we will have the following relation:

$$\int \frac{\mathrm{d}Q}{T} \leq \Delta S \Rightarrow \frac{\Delta Q}{T} = \frac{\Delta E + \Delta W}{T} \leq \Delta S \Rightarrow P\Delta\Omega \leq -\Delta E + T\Delta S = -\Delta F \tag{C.19}$$

We can then define a new thermodynamic potential, called the

$$\textbf{Gibbs free energy}:\ G = F + P\Omega \tag{C.20}$$

for which the above relation implies that

$$\Delta G = \Delta F + P\Delta\Omega \leq 0 \tag{C.21}$$

which proves that *the equilibrium state of a system at constant temperature and pressure is one of minimum Gibbs free energy*. The logical argument that leads to this conclusion is identical to the one invoked for the Helmholtz free energy.

The thermodynamic potentials are also useful because the various thermodynamic fields (or variables) can be expressed as partial derivatives of the potentials with respect to the corresponding variable (or field) under proper conditions. Specifically, it can be shown directly from the definition of the thermodynamic potentials that the following relations hold:

$$P = -\left(\frac{\partial F}{\partial \Omega}\right)_T, \quad P = -\left(\frac{\partial E}{\partial \Omega}\right)_S \tag{C.22}$$

$$\Omega = +\left(\frac{\partial G}{\partial P}\right)_T, \quad \Omega = +\left(\frac{\partial \Theta}{\partial P}\right)_S \tag{C.23}$$

$$S = -\left(\frac{\partial F}{\partial T}\right)_\Omega, \quad S = -\left(\frac{\partial G}{\partial T}\right)_P \tag{C.24}$$

Table C.1. *Thermodynamic potentials and their derivatives.*

In each case the potential, the variable(s) or field(s) associated with it, and the relations connecting fields and variables through partial derivatives of the potential are given.

Internal energy	E	S, Ω	$P = -\left(\frac{\partial E}{\partial \Omega}\right)_S$	$T = +\left(\frac{\partial E}{\partial S}\right)_\Omega$
Helmholtz free energy	$F = E - TS$	T, Ω	$P = -\left(\frac{\partial F}{\partial \Omega}\right)_T$	$S = -\left(\frac{\partial F}{\partial T}\right)_\Omega$
Gibbs free energy	$G = F + P\Omega$	T, P	$\Omega = +\left(\frac{\partial G}{\partial P}\right)_T$	$S = -\left(\frac{\partial G}{\partial T}\right)_P$
Enthalpy	$\Theta = E + P\Omega$	S, P	$\Omega = +\left(\frac{\partial \Theta}{\partial P}\right)_S$	$T = +\left(\frac{\partial \Theta}{\partial S}\right)_P$

$$T = +\left(\frac{\partial \Theta}{\partial S}\right)_P, \quad T = +\left(\frac{\partial E}{\partial S}\right)_\Omega \tag{C.25}$$

For example, from the first and second laws of thermodynamics we have

$$\mathrm{d}E = \mathrm{d}Q - P\mathrm{d}\Omega = T\mathrm{d}S - P\mathrm{d}\Omega$$

and E is a state function, so its differential using as free variables S and Ω must be expressed as

$$\mathrm{d}E = \left(\frac{\partial E}{\partial S}\right)_\Omega \mathrm{d}S + \left(\frac{\partial E}{\partial \Omega}\right)_S \mathrm{d}\Omega$$

Comparing the last two equations we obtain the second of Eq. (C.22) and the second of Eq. (C.25). Notice that in all cases a variable and its corresponding field are connected by Eqs. (C.22)–(C.25), that is, Ω is given as a partial derivative of a potential with respect to P and vice versa, and S is given as a partial derivative of a potential with respect to T and vice versa. The thermodynamic potentials and their derivatives which relate thermodynamic variables and fields are collected together in Table C.1. In cases when the work is of magnetic nature the relevant thermodynamic field is the magnetic field $\mathbf{H}$ and the relevant thermodynamic variable is the magnetization $\mathbf{M}$; similarly, for electrical work the relevant thermodynamic field is the electric field $\mathbf{E}$ and the relevant thermodynamic variable is the polarization $\mathbf{P}$. Note, however, that the identification of the thermodynamic field and variable is not always straightforward: in certain cases the proper analogy is to identify the magnetization as the relevant thermodynamic *field* and the external magnetic field is the relevant thermodynamic *variable* (for instance, in a system of non-interacting spins on a lattice under the influence of an external magnetic field, discussed in Appendix D).

As an example, we consider the case of a magnetic system. For simplicity, we assume that the fields and the magnetic moments are homogeneous, so we can work with scalar rather than vector quantities. The magnetic moment is defined as $m(\mathbf{r})$, in terms of which the magnetization M is given by

$$M = \int m(\mathbf{r})\mathrm{d}\mathbf{r} = \Omega m \tag{C.26}$$

where the last expression applies to a system of volume Ω in which the magnetic moment is constant $m(\mathbf{r}) = m$. The differential of the internal energy $\mathrm{d}E$ for such a system in an external field H is given by

$$\mathrm{d}E = T\mathrm{d}S + H\mathrm{d}M$$

where now H plays the role of the pressure in a mechanical system, and $-M$ plays the role of the volume. The sign of the HM term in the magnetic system is opposite from that of the $P\Omega$ term in the mechanical system because an increase in the magnetic field H increases the magnetization M, whereas an increase of the pressure P decreases the volume Ω. By analogy to the mechanical system, we define the thermodynamic potentials Helmholtz free energy F, Gibbs free energy G and enthalpy Θ as

$$F = E - TS \Longrightarrow \mathrm{d}F = \mathrm{d}E - T\mathrm{d}S - S\mathrm{d}T = -S\mathrm{d}T + H\mathrm{d}M$$
$$G = F - HM \Longrightarrow \mathrm{d}G = \mathrm{d}F - H\mathrm{d}M - M\mathrm{d}H = -S\mathrm{d}T - M\mathrm{d}H$$
$$\Theta = E - HM \Longrightarrow \mathrm{d}\Theta = \mathrm{d}E - H\mathrm{d}M - M\mathrm{d}H = T\mathrm{d}S - M\mathrm{d}H$$

which give the following relations between the thermodynamic potentials, variables and fields:

$$T = +\left(\frac{\partial E}{\partial S}\right)_M, \quad H = +\left(\frac{\partial E}{\partial M}\right)_S \Longrightarrow \left(\frac{\partial T}{\partial M}\right)_S = \left(\frac{\partial H}{\partial S}\right)_M$$
$$S = -\left(\frac{\partial F}{\partial T}\right)_M, \quad H = +\left(\frac{\partial F}{\partial M}\right)_T \Longrightarrow \left(\frac{\partial S}{\partial M}\right)_T = -\left(\frac{\partial H}{\partial T}\right)_M$$
$$S = -\left(\frac{\partial G}{\partial T}\right)_H, \quad M = -\left(\frac{\partial G}{\partial H}\right)_T \Longrightarrow \left(\frac{\partial S}{\partial H}\right)_T = \left(\frac{\partial M}{\partial T}\right)_H$$
$$T = +\left(\frac{\partial \Theta}{\partial S}\right)_H, \quad M = -\left(\frac{\partial \Theta}{\partial H}\right)_S \Longrightarrow \left(\frac{\partial T}{\partial H}\right)_S = -\left(\frac{\partial M}{\partial S}\right)_H$$

When there is significant interaction at the microscopic level among the elementary moments induced by the external field (electric or magnetic), then the thermodynamic field should be taken as the total field, including the external and induced contributions. Specifically, the total magnetic field is given by

$$\mathbf{B}(\mathbf{r}) = \mathbf{H} + 4\pi\mathbf{m}(\mathbf{r})$$

and the total electric field is given by

$$\mathbf{D}(\mathbf{r}) = \mathbf{E} + 4\pi\mathbf{p}(\mathbf{r})$$

with $\mathbf{m}(\mathbf{r})$, $\mathbf{p}(\mathbf{r})$ the local magnetic or electric moment and $\mathbf{H}$, $\mathbf{E}$ the external magnetic or electric fields, usually considered to be constant.

C.3 Application: phase transitions

As an application of the concepts of thermodynamics we consider the case of the van der Waals gas and show how the minimization of the free energy leads to the idea of a first order phase transition. In the van der Waals gas the particles interact with an attractive potential at close range. The effect of the attractive interaction is to reduce the pressure that the kinetic energy P_{kin} of the particles would produce. The interaction between particles changes from attractive to repulsive if the particles get too close. Consistent with the macroscopic view of thermodynamics, we attempt to describe these effects not from a detailed microscopic description but through empirical parameters. Thus, we write for the actual pressure P exerted by the gas on the walls of the container,

$$P = P_{kin} - \frac{\eta}{\Omega^2}$$

where η is a positive constant with dimensions energy $\times$ volume and Ω is the volume of the box. This equation describes the reduction in pressure due to the attractive interaction between particles; the term $1/\Omega^2$ comes from the fact that collisions between *pairs* of particles are responsible for the reduction in pressure and the probability of finding two particles within the range of the attractive potential is proportional to the square of the probability of finding one particle at a certain point within an infinitesimal volume, the latter probability being proportional to $1/\Omega$ for a homogeneous gas. The effective volume of the gas will be equal to

$$\Omega_{eff} = \Omega - \Omega_0$$

where Ω_0 is a constant equal to the excluded volume due to the repulsive interaction between particles at very small distances. The van der Waals equation of state asserts that the same relation exists between the effective volume and the pressure due to kinetic energy, as in the ideal gas between volume and pressure, that is

$$P_{kin}\Omega_{eff} = Nk_{\mathrm{B}}T$$

In this equation we substitute the values of P and Ω which are the measurable thermodynamic field and variable, using the expressions we discussed above for P_{kin} and Ω_{eff}, to obtain

$$\left(P + \frac{\eta}{\Omega^2}\right)(\Omega - \Omega_0) = Nk_{\mathrm{B}}T$$

This equation can be written as a third order polynomial in Ω:

$$\Omega^3 - \left(\Omega_0 + \frac{Nk_{\mathrm{B}}T}{P}\right)\Omega^2 + \frac{\eta}{P}\Omega - \frac{\eta\Omega_0}{P} = 0 \qquad \text{(C.27)}$$

For a fixed temperature, this is an equation that relates pressure and volume, called an isotherm. In general it has three roots, as indicated graphically in Fig. C.1: there is a range of pressures in which a horizontal line corresponding to a given value of the pressure intersects the isotherm at three points.

From our general thermodynamic relations we have

$$\mathrm{d}F = \mathrm{d}E - T\mathrm{d}S = \mathrm{d}E - \mathrm{d}Q = -\mathrm{d}W = -P\mathrm{d}\Omega$$

This shows that we can calculate the free energy as

$$F = -\int P\mathrm{d}\Omega$$

which produces the second plot in Fig. C.1. Notice that since the pressure P and volume Ω are always positive the free energy F is always negative and monotonically decreasing. The free energy must be minimized for a mechanically isolated system. As the free energy plot of Fig. C.1 shows, it is possible to reduce the free energy of the system between the states 1 and 2 (with volumes Ω_1 and Ω_2) by taking a mixture of them rather than having a pure state at any volume between those two states. The mixture of the two states will have a free energy which lies along the common tangent between the two states. When the system follows the common tangent in the free energy plot, the pressure will be constant,

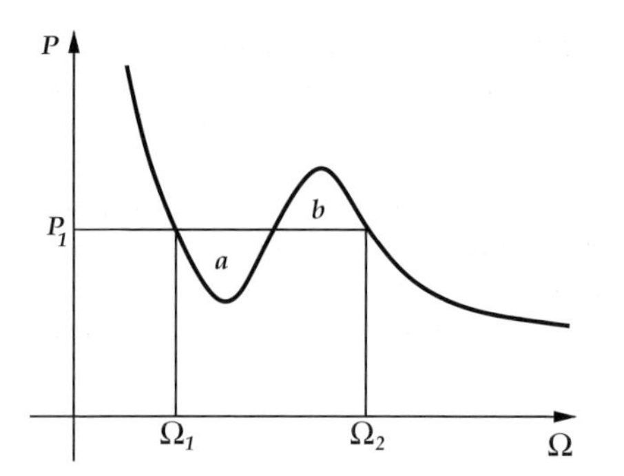

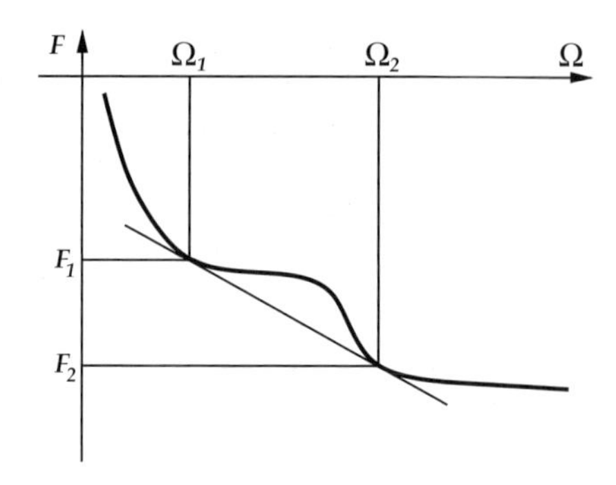

Figure C.1. The Maxwell construction argument. **Left:** pressure versus volume curve for the van der Waals gas at fixed temperature. **Right:** corresponding free energy versus volume curve. The common tangent construction on the free energy curve determines the mixture of two phases corresponding to states 1 and 2 that will give lower free energy than the single-phase condition. The common tangent of the free energy corresponds to a horizontal line in the pressure–volume curve, which is determined by the requirement that the areas in regions a and b are equal.

since

$$P = -\left(\frac{\partial F}{\partial \Omega}\right)_T$$

The constant value of the pressure is determined by the common tangent and equal pressure conditions, which together imply

$$\frac{F_2 - F_1}{\Omega_2 - \Omega_1} = \left(\frac{\partial F}{\partial \Omega}\right)_{\Omega=\Omega_1} = \left(\frac{\partial F}{\partial \Omega}\right)_{\Omega=\Omega_2} \Longrightarrow -\left(\frac{\partial F}{\partial \Omega}\right)_{\Omega=\Omega_1} (\Omega_2 - \Omega_1) = -(F_2 - F_1)$$

which in turn produces the equation

$$P_1(\Omega_2 - \Omega_1) = \int_{\Omega_1}^{\Omega_2} P \mathrm{d}\Omega$$

The graphical interpretation of this equation is that the areas a and b between the pressure–volume curve and the constant pressure line must be equal. This is known as the Maxwell construction.

The meaning of this derivation is that the system can lower its free energy along an isotherm by forming a mixture at constant pressure between two phases along the isotherm, rather than being in a homogeneous single phase. This corresponds to the transition between the liquid phase on the left (the smaller volume, higher pressure phase) and the gas phase on the right (with higher volume and lower pressure). This particular transition in which pressure and temperature remain constant while the volume changes significantly during the transition, because of the different volumes of the two phases, is referred to as a *first order phase transition*. The difference in volume between the two phases leads to a discontinuity in the first derivative of the Gibbs free energy as a function of pressure for fixed temperature. Related to this behavior is a discontinuity in the first derivative of the Gibbs free energy as a function of temperature for fixed pressure, which is due to a difference in the entropy of the two phases. If the first derivatives of the free energy are continuous, then the transition is referred to as *second order*; second order phase transitions usually have discontinuities in higher derivatives of the relevant free energy. A characteristic feature of a first order phase transition is the existence of a *latent heat*, that is, an amount of heat which is released when going from one phase to the other at constant temperature and pressure.

The above statements can be proven by considering the Gibbs free energy of the system consisting of two phases and recalling that, as discussed above, it must be at a minimum for constant temperature and pressure. We denote the Gibbs free energy per unit mass as g_1 for the liquid and g_2 for the gas, and the corresponding mass in each phase as m_1 and m_2. At the phase transition the total Gibbs free energy G will be equal to the sum of the two parts,

$$G = g_1 m_1 + g_2 m_2$$

and it will be a minimum, therefore variations of it with respect to changes other than in temperature or pressure must vanish, $\delta G = 0$. The only relevant variation at the phase transition is transfer of mass from one phase to the other, but because of conservation of mass we must have $\delta m_1 = -\delta m_2$; therefore

$$\delta G = g_1 \delta m_1 + g_2 \delta m_2 = (g_1 - g_2)\delta m_1 = 0 \Longrightarrow g_1 = g_2$$

since the mass changes are arbitrary. But using the relations between thermodynamic potentials and variables from Table C.1, we can write

$$\left(\frac{\partial(g_2 - g_1)}{\partial T}\right)_P = -(s_2 - s_1), \qquad \left(\frac{\partial(g_2 - g_1)}{\partial P}\right)_T = (\omega_2 - \omega_1)$$

where s_1, s_2 are the entropies per unit mass of the two phases and ω_1, ω_2 are the volumes per unit mass; these are the discontinuities in the first derivatives of the Gibbs free energy as a function of temperature for fixed pressure, or as a function of pressure for fixed temperature, that we mentioned above.

Denoting the differences in Gibbs free energy, entropy and volume per unit mass between the two phases as $\Delta g, \Delta s, \Delta\omega$, we can write the above relationships as

$$\left(\frac{\partial \Delta g}{\partial T}\right)_P \left(\frac{\partial \Delta g}{\partial P}\right)_T^{-1} = -\frac{\Delta s}{\Delta \omega}$$

Now the Gibbs free energy difference Δg between the two phases is a function of T and P, which can be inverted to produce expressions for T as a function of Δg and P or for P as a function of Δg and T; from these relations it is easy to show that the following relation holds between the partial derivatives:

$$\left(\frac{\partial \Delta g}{\partial T}\right)_P \left(\frac{\partial P}{\partial \Delta g}\right)_T = -\left(\frac{\partial P}{\partial T}\right)_{\Delta g} = \left(\frac{\partial \Delta g}{\partial T}\right)_P \left(\frac{\partial \Delta g}{\partial P}\right)_T^{-1}$$

which, when combined with the earlier equation, gives

$$\left(\frac{\partial P}{\partial T}\right)_{\Delta g} = \frac{\Delta s}{\Delta \omega} \Longrightarrow \frac{\mathrm{d}P(T)}{\mathrm{d}T} = \frac{\Delta s}{\Delta \omega}$$

where in the last relation we have used the fact that at the phase transition $\Delta g = g_2 - g_1 = 0$ is constant and therefore the pressure P is simply a function of T. We

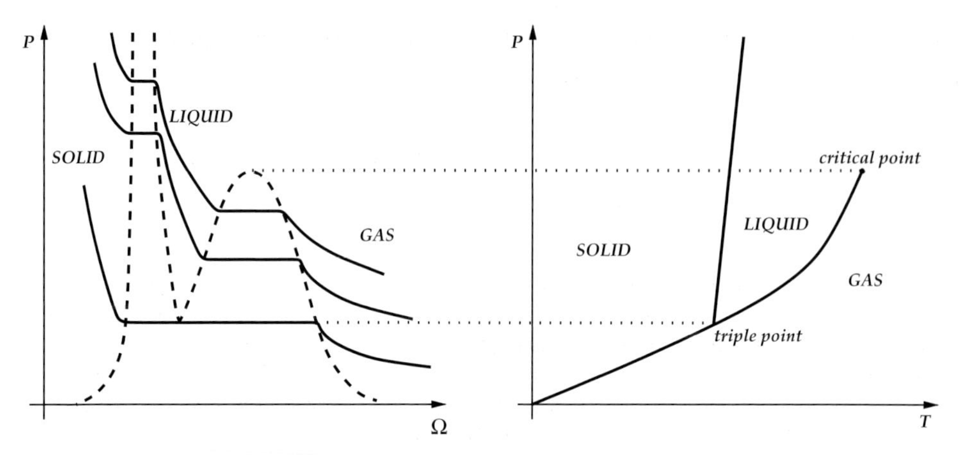

Figure C.2. Isotherms on the $P-\Omega$ plane (**left plot**) and phase boundaries between the solid, liquid and gas phases on the $P-T$ plane (**right plot**). The dashed lines in the $P-\Omega$ plane identify the regions that correspond to the transitions between the phases which occur at constant pressure (horizontal lines). The triple point and critical point are identified on both diagrams.

can define the latent heat L through the difference in entropy of the two phases to obtain an expression for the derivative of the pressure,

$$L = T\Delta s \Longrightarrow \frac{\mathrm{d}P}{\mathrm{d}T} = \frac{L}{T\Delta\omega} \tag{C.28}$$

known as the Clausius–Clapeyron equation:[2] a similar analysis applies to the solid–liquid phase transition boundary. The change in volume per unit mass in going from the liquid to the gas phase is positive and large, while that going from the solid to the liquid phase is much smaller and can even be negative (see discussion below).

A general diagram of typical isotherms on the $P-\Omega$ plane and the corresponding curves on the $P-T$ plane that separate different phases is shown in Fig. C.2. Some interesting features in these plots are the so called *triple point* and *critical point*. At the triple point, there is coexistence between all three phases and there is no liquid phase for lower temperatures. At the critical point the distinction between liquid and gas phases has disappeared. This last situation corresponds to the case when all three roots of Eq. (C.27) have collapsed to one. At this point, the equation takes the form

$$(\Omega - \Omega_c)^3 = \Omega^3 - 3\Omega_c\Omega + 3\Omega_c^2\Omega - \Omega_c^3 = 0 \tag{C.29}$$

where Ω_c is the volume at the critical point. We can express the volume Ω_c, pressure P_c and temperature T_c at the critical point in terms of the constants that appear in the van der Waals equation of state:

$$\Omega_c = 3\Omega_0, \quad P_c = \frac{\eta}{27\Omega_0^2}, \quad T_c = \frac{8\eta}{27Nk_B\Omega_0}$$

[2] The latent heat and the change of volume are defined here per unit mass, so they can be both scaled by the same factor without changing the Clausius–Clapeyron equation.

Moreover, if we use the reduced variables for volume, pressure and temperature,

$$\omega = \frac{\Omega}{\Omega_c}, \quad p = \frac{P}{P_c}, \quad t = \frac{T}{T_c}$$

we obtain the following expression for the van der Waals equation of state:

$$\left(p + \frac{3}{\omega^3}\right)\left(\omega - \frac{1}{3}\right) = \frac{8}{3}t$$

which is the same for all substances since it does not involve any substance-specific constants. This equation is called the "law of corresponding states" and is obeyed rather accurately by a number of substances which behave like a van der Waals gas.

In Fig. C.2 we have shown the boundary between the solid and liquid phases with a large positive slope, as one might expect for a substance that expands when it melts: a higher melting temperature is needed to compensate for the additional constraint imposed on the volume by the increased pressure. The slope of curves that separate two phases is equal to the latent heat divided by the temperature and the change in volume, as determined by the Clausius–Clapeyron relation, Eq. (C.28). For solids that expand when they melt, the change in volume is positive, and of course so are the latent heat and temperature, leading to a curve with large positive slope, since the change in volume going from solid to liquid is usually small. Interestingly, there are several solids that contract upon melting, in which case the corresponding phase boundary on the $P-T$ plane has *negative* slope. Examples of such solids are ice and silicon. The reason for this unusual behavior is that the bonds in the solid are of a nature that actually makes the structure of the solid rather open, in contrast to most common solids where the atoms are closely packed. In the latter case, when the solid melts the atoms have larger average distance to their neighbors in the liquid than in the solid due to the increased kinetic energy, leading to an increase in the volume. In the case of solids with open structures, melting actually reduces the average distance between atoms when the special bonds that keep them at fixed distances from each other in the solid disintegrate due to the increased kinetic energy in the liquid, and the volume decreases.

Finally, we discuss the phenomena of *supersaturation* and *supercooling*. Under certain circumstances, it is possible to follow the homogeneous phase isotherm in Fig. C.1 beyond the point where the phase transition should have occurred. This can be done in the gas phase, in which case the substance does not liquefy even though the volume has been reduced below the critical value, or in the liquid phase, in which case the substance does not solidify even though the volume has been reduced below the critical value;[3] the first is referred to as supersaturation, the second as supercooling. We analyze the phenomenon of supersaturation only, which involves an imbalance between the gas and liquid phases, supercooling being similar in nature but involving the liquid and solid phases instead. In supersaturation it is necessary to invoke surface effects for the condensed phase (liquid): in addition to the usual bulk term we have to include a surface term for the internal energy, which is proportional to the surface tension σ and the area of the body A. Accordingly, in the condensed phase the internal energy of the substance will be given by

$$E_l = \frac{4\pi}{3}R^3\epsilon + 4\pi\sigma R^2$$

[3] We are concerned here with substances that behave in the usual manner, that is, the volume expands upon melting.

and consequently the Gibbs free energy will be given by

$$G_1 = \frac{4\pi}{3} R^3 \zeta + 4\pi\sigma R^2$$

where ϵ, ζ are the internal energy and Gibbs free energy per unit volume of the infinite bulk phase and R is the radius, assuming a simple spherical shape for the liquid droplet. Using the same notation as above, with g_1 and g_2 the free energy per unit mass of the liquid and gas phases at equilibrium, we will have

$$\delta(G_1 + G_2) = 0 \Longrightarrow \delta(g_1 m_1 + 4\pi\sigma R^2 + g_2 m_2) = 0 \Longrightarrow g_1 + 8\pi\sigma R \frac{\partial R}{\partial m_1} - g_2 = 0$$

where we have used again the fact that at equilibrium mass is transferred between the two phases, so that $\delta m_2 = -\delta m_1$. From the definition of the mass density ρ_1 of the condensed phase we obtain

$$\rho_1 = \frac{m_1}{\frac{4\pi}{3} R^3} \Longrightarrow \frac{\partial R}{\partial m_1} = \frac{1}{4\pi\rho_1 R^2}$$

which, when substituted into the above equation, gives

$$g_2 - g_1 = \frac{2\sigma}{\rho_1 R}$$

Using the relations between thermodynamic variables and potentials from Table C.1, we can rewrite the mass density of each phase in terms of a partial derivative of the Gibbs free energy per unit mass:

$$\left(\frac{\partial G_i}{\partial P}\right)_T = \Omega_i \Longrightarrow \left(\frac{\partial g_i}{\partial P}\right)_T = \frac{\Omega_i}{m_i} = \frac{1}{\rho_i}, \quad i = 1, 2$$

with the help of which the previous equation is rewritten as

$$\left(\frac{\partial g_2}{\partial P}\right)_T - \left(\frac{\partial g_1}{\partial P}\right)_T = -\frac{2\sigma}{\rho_1 R^2}\left(\frac{\partial R}{\partial P}\right)_T - \frac{2\sigma}{\rho_1^2 R}\left(\frac{\partial \rho_1}{\partial P}\right)_T = \frac{1}{\rho_2} - \frac{1}{\rho_1}$$

We will make the following approximations, which are justified by physical considerations:

$$\frac{1}{\rho_1} << \frac{1}{\rho_2}, \quad \frac{\partial \rho_1}{\partial P} \approx 0$$

The first is simply the statement that the gas density ρ_2 is much smaller than the liquid density ρ_1, and the second implies a negligible compressibility for the liquid, which is true by comparison to the compressibility of the gas. With these approximations, and using the equation of state for the ideal gas $P = (k_B T/m_2)\rho_2$, the previous equation becomes

$$\left(\frac{\partial R}{\partial P}\right)_T = -\frac{k_B T}{m_2}\frac{\rho_1 R^2}{2\sigma P}$$

This can be easily solved to produce the following equation for the equilibrium pressure $P(R, T)$ for a liquid droplet of a given radius R at temperature T, or equivalently the

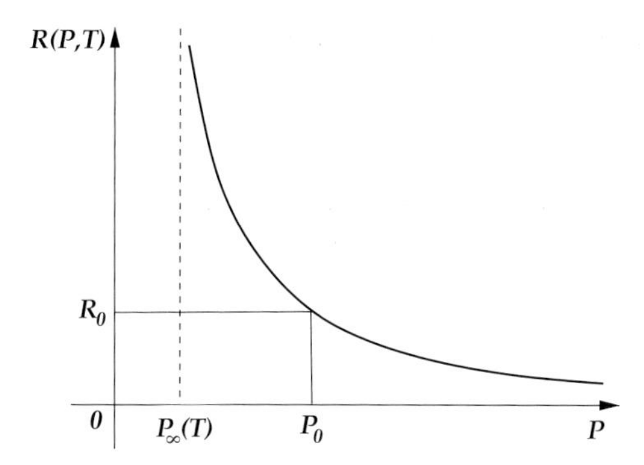

Figure C.3. The equilibrium radius $R(P, T)$ of a droplet as a function of pressure P at fixed temperature T; $P_\infty(T)$ is the pressure for the bulk gas to liquid phase transition.

equilibrium radius $R(P, T)$ for a droplet at given external pressure P and temperature T, shown in Fig. C.3:

$$P(R, T) = P_\infty(T) \exp\left[\frac{2\sigma m_2}{\rho_1 k_B T R}\right] \Longrightarrow R(P, T) = \left(\frac{2\sigma m_2}{\rho_1 k_B T}\right) \frac{1}{\ln[P/P_\infty(T)]} \quad \text{(C.30)}$$

with $P_\infty(T)$ a constant as far as the radius is concerned, which corresponds to the equilibrium pressure for the phase transition between the gas and the infinite liquid.

An analysis of this result reveals the following. For a given external pressure $P_0 > P_\infty(T)$, droplets of radius smaller than R_0 would require a larger pressure to be in equilibrium; the only mechanism available to such droplets to increase the external pressure they feel, is to evaporate some of their mass into the vapor around them, since for the ideal gas $P = (k_B T/m_2)\rho_2$, so the evaporation of mass from the droplet into the vapor increases the vapor density and therefore the pressure that the vapor exerts. However, by evaporating some of their mass, these droplets become even smaller, therefore requiring even larger pressure to be in equilibrium, and the process continues until the droplets disappear by evaporation. Conversely, droplets of radius larger than R_0 would require a smaller pressure to be in equilibrium; the only mechanism available to such droplets to reduce the external pressure they feel, is to absorb some of the vapor around them, since condensation of the vapor reduces its density and therefore the pressure it exerts. However, by absorbing some vapor these droplets become even larger, therefore requiring even smaller pressure to be in equilibrium and the process continues until all the vapor has been consumed and turned into liquid. This argument shows that the existence of a single droplet of size larger than the equilibrium size dictated by the external pressure is sufficient to induce condensation of the vapor to the liquid phase. If all droplets are smaller than this equilibrium size, they evaporate. The creation of a droplet of size slightly larger than equilibrium size is a random event, brought about by the incorporation of a small amount of additional mass into the droplet, by random collisions of molecules in the vapor phase with it. Until such droplets have formed, the substance is predominantly in the gas phase, since droplets smaller than the equilibrium size quickly evaporate. This allows the system to be in an unstable equilibrium phase until a large enough droplet has formed; this is called supersaturation. The transition to the condensed phase is very abrupt once a large enough droplet has formed.

Problems

1. Using arguments similar to the proof of Eq. (C.7), and the definitions of the thermal expansion coefficient α, isothermal compressibility κ_T and adiabatic compressibility κ_S, prove the relations given in Eqs. (C.8)–(C.10).
2. Define the thermodynamic potentials for a system with electric moment $\mathbf{p}(\mathbf{r})$ in an external electric field $\mathbf{E}$, and derive the relations between the thermodynamic potentials, variables and fields.

Appendix D

Elements of statistical mechanics

Statistical mechanics is the theory that describes the behavior of macroscopic systems in terms of thermodynamic variables (such as the entropy, volume, average number of particles, etc.), using as a starting point the microscopic structure of the physical system of interest. The difference between thermodynamics and statistical mechanics is that the first theory is based on empirical observations, whereas the second theory is based on knowledge (true or assumed) of the microscopic constituents and their interactions. The similarity between the two theories is that they both address the macroscopic behavior of the system: thermodynamics does it by dealing exclusively with macroscopic quantities and using empirical laws, and statistical mechanics does it by constructing averages over all states consistent with the external conditions imposed on the system (such as temperature, pressure, chemical potential, etc.). Thus, the central theme in statistical mechanics is to identify all the possible states of the system in terms of their microscopic structure, and take an average of the physical quantities of interest over those states that are consistent with the external conditions. The average must involve the proper weight for each state, which is related to the likelihood of this state to occur, given the external conditions.

As in thermodynamics, statistical mechanics assumes that we are dealing with systems composed of a very large number of microscopic particles (typically atoms or molecules). The variables that determine the state of the system are the position, momentum, electric charge and magnetic moment of the particles. All these are microscopic variables, and their values determine the state of individual particles. A natural thing to consider within statistical mechanics is the average occupation number of microscopic states by particles. The space that corresponds to all the allowed values of these microscopic variables is called the *phase space* of the system. A central postulate of statistical mechanics is that all relevant portions of phase must be sampled properly in the average. This is formalized through the notion of *ergodicity*: a sampling procedure is called ergotic if it does not exclude in principle any state of the system that is consistent with the imposed external conditions. There exists a theorem, called Poincaré's theorem, which says that given enough time a system will come arbitrarily close to any of the states consistent with the external conditions. Sampling the phase space of a system composed of many particles is exceedingly difficult; while Poincaré's theorem assures us that a system will visit all the states that are consistent with the imposed external conditions, it will take a huge amount of time to sample all the relevant states by evolving the system in a causal manner between states. To circumvent this difficulty, the idea of *ensembles* was developed, which makes calculations feasible in the context of statistical mechanics. In this chapter we develop the notions of average occupation numbers and of different types of ensembles, and give some elementary examples of how these notions are applied to simple systems.

D.1 Average occupation numbers

The average occupation numbers can be obtained in the simplest manner by making the assumption that the system exists in its most probable state, that is, the state with the highest probability to occur given the external conditions. We define our physical system as consisting of particles which can exist in a (possibly infinite) number of microscopic states labeled by their energy, ϵ_i. The energy of the microscopic states is bounded from below but not necessarily bounded from above. If there are n_i particles in the microscopic state i, then the total number N of particles in the system and its total energy E will be given by

$$N = \sum_i n_i, \quad E = \sum_i n_i \epsilon_i \tag{D.1}$$

We will denote by $\{n_i\}$ the distribution of particles that consists of a particular set of values n_i that determine the occupation of the levels ϵ_i. The number of states of the entire system corresponding to a particular distribution $\{n_i\}$ will be denoted as $W(\{n_i\})$. This number is proportional to the volume in phase space occupied by this particular distribution $\{n_i\}$. The most probable distribution must correspond to the largest volume in phase space. That is, if we denote the most probable distribution of particles in microscopic states by $\{\bar{f}_i\}$, then $W(\{\bar{f}_i\})$ is the maximum value of W. Let us suppose that the degeneracy or multiplicity of the microscopic state labeled i is g_i, that is, there are g_i individual microscopic states with the same energy ϵ_i. With these definitions, and with the restriction of constant N and E, we can now derive the average occupation of level i, which will be the same as $\bar{f}_i$. There are three possibilities, which we examine separately.

D.1.1 Classical Maxwell–Boltzmann statistics

In the case of classical distinguishable particles, there are $g_i^{n_i}$ ways to put n_i particles in the same level i, and there are

$$W(\{n_i\}) = \frac{N!}{n_1!n_2!\cdots n_i!\cdots} n_1^{n_1} n_2^{n_2} \cdots g_i^{n_i} \cdots$$

ways of arranging the particles in the levels with energy ϵ_i. The ratio of factorials gives the number of permutations for putting n_i particles in level i, since the particles are distinguishable (it does not matter which n_i of the N particles were put in level i). Since we are dealing with large numbers of particles, we will use the following approximation:

$$\textbf{Stirling's formula}: \quad \ln(N!) = N \ln N - N \tag{D.2}$$

which is very accurate for large N. With the help of this, we can now try to find the maximum of $W(\{n_i\})$ by considering variations in n_i, under the constraints of Eq. (D.1). It is actually more convenient to find the maximum of $\ln W$, which will give the maximum of W, since W is a positive quantity ≥ 1. With Stirling's formula, $\ln W$ takes the form

$$\ln W = N \ln N - N + \sum_i n_i \ln g_i - \sum_i (n_i \ln n_i - n_i) = N \ln N + \sum_i n_i (\ln g_i - \ln n_i)$$

We include the constraints through the Lagrange multipliers α and β, and perform the variation of W with respect to n_i to obtain

$$\begin{aligned} 0 &= \delta \ln W - \alpha \delta N - \beta \delta E \\ \Longrightarrow 0 &= \delta \left[N \ln N + \sum_i n_i (\ln g_i - \ln n_i) \right] - \alpha \delta \sum_i n_i - \beta \delta \sum_i n_i \epsilon_i \\ \Longrightarrow 0 &= \sum_i \delta n_i \left[\ln g_i - \ln n_i - \alpha - \beta \epsilon_i + \ln N \right] \end{aligned} \tag{D.3}$$

where we have used $N = \sum_i n_i \Rightarrow \delta N = \sum_i \delta n_i$. Since Eq. (D.3) must hold for arbitrary variations δn_i, and it applies to the maximum value of W which is obtained for the distribution $\{\bar{f}_i\}$, we conclude that

$$0 = \left[\ln g_i - \ln \bar{f}_i - \gamma - \beta \epsilon_i \right] \Longrightarrow \bar{f}_i^{MB} = g_i \mathrm{e}^{-\gamma} \mathrm{e}^{-\beta \epsilon_i} \tag{D.4}$$

where we have defined $\gamma = \alpha - \ln N$. Thus, we have derived the Maxwell–Boltzmann distribution, in which the average occupation number of level i with energy ϵ_i is proportional to $\exp(-\beta \epsilon_i)$. All that remains to do in order to have the exact distribution is to determine the values of the constants that appear in Eq. (D.4). Recall that these constants were introduced as the Lagrange multipliers that take care of the constraints of constant number of particles and constant energy. The constant β must have the dimensions of inverse energy. The only other energy scale in the system is the temperature, so we conclude that $\beta = 1/k_B T$. It is actually possible to show that this must be the value of β through a much more elaborate argument based on the Boltzmann transport equation (see for example the book by Huang, in the Further reading section). The other constant, γ, is obtained by normalization, that is, by requiring that when we sum $\{\bar{f}_i\}$ over all the values of the index i we obtain the total number of particles in the system. This can only be done explicitly for specific systems where we can evaluate g_i and ϵ_i.

Example We consider the case of the classical ideal gas which consists of particles of mass m, with the only interaction between the particles being binary hard-sphere collisions. The particles are contained in a volume Ω, and the gas has density $n = N/\Omega$. In this case, the energy ϵ_i of a particle, with momentum $\mathbf{p}$ and position $\mathbf{r}$, and its multiplicity g_i are given by

$$\epsilon_i = \frac{\mathbf{p}^2}{2m}, \quad g_i = \mathbf{drdp}$$

Then summation over all the values of the index i gives

$$\begin{aligned} N &= \sum_i n_i = \int \mathrm{d}\mathbf{r} \int \mathrm{e}^{-\gamma} \mathrm{e}^{-\beta \mathbf{p}^2/2m} \mathrm{d}\mathbf{p} = \Omega \mathrm{e}^{-\gamma} (2\pi m/\beta)^{\frac{3}{2}} \\ \Rightarrow \mathrm{e}^{-\gamma} &= \frac{n}{(2\pi m/\beta)^{\frac{3}{2}}} \end{aligned} \tag{D.5}$$

which completely specifies the average occupation number for the classical ideal gas. With this we can calculate the total energy of this system as

$$E = \sum_i n_i \epsilon_i = \frac{n}{(2\pi m/\beta)^{\frac{3}{2}}} \int d\mathbf{r} \int \frac{\mathbf{p}^2}{2m} e^{-\beta \mathbf{p}^2/2m} d\mathbf{p}$$

$$= \Omega \frac{n}{(2\pi m/\beta)^{\frac{3}{2}}} \left[-\frac{\partial}{\partial \beta} \int e^{-\beta \mathbf{p}^2/2m} d\mathbf{p} \right] = \frac{3}{2}\frac{N}{\beta} \quad \text{(D.6)}$$

D.1.2 Quantum Fermi–Dirac statistics

In the case of quantum mechanical particles that obey Fermi–Dirac statistics, each level i has occupation 0 or 1, so that the number of particles that can be accommodated in a level of energy ϵ_i is $n_i \leq g_i$, where g_i is the degeneracy of the level due to the existence of additional good quantum numbers. Since the particles are indistinguishable, there are

$$\frac{g_i!}{n_i!(g_i - n_i)!}$$

ways of distributing n_i particles in the g_i states of level i. The total number of ways of distributing N particles in all the levels is given by

$$W(\{n_i\}) = \prod_i \frac{g_i!}{n_i!(g_i - n_i)!} \Longrightarrow$$

$$\ln W(\{n_i\}) = \sum_i [g_i(\ln g_i - 1) - n_i(\ln n_i - 1) - (g_i - n_i)(\ln(g_i - n_i) - 1)]$$

where we have used Stirling's formula for $\ln(g_i!)$, $\ln(n_i!)$, $\ln(g_i - n_i!)$. Using the same variational argument as before, in terms of δn_i, we obtain for the most probable distribution $\{\bar{f}_i\}$ which corresponds to the maximum of W:

$$\bar{f}_i^{FD} = g_i \frac{1}{e^{\beta(\epsilon_i - \mu_{FD})} + 1} \quad \text{(D.7)}$$

where $\mu_{FD} = -\alpha/\beta$. The constants are fixed by normalization, that is, summation of n_i over all values of i gives N and summation of $n_i \epsilon_i$ gives E. This is the Fermi–Dirac distribution, with μ_{FD} the highest energy level which is occupied by particles at zero temperature, called the Fermi energy; μ_{FD} is also referred to as the chemical potential, since its value is related to the total number of particles in the system.

Example We calculate the value of μ_{FD} for the case of a gas of non-interacting particles, with $g_i = 2$. This situation corresponds, for instance, to a gas of electrons if we ignore their Coulomb repulsion and any other exchange or correlation effects associated with their quantum mechanical nature and interactions. The multiplicity of 2 for each level comes from their spin, which allows each electron to exist in two possible states, spin-up or -down, for a given energy; we specialize the discussion to the electron gas next, and take the mass of the particles to be m_e. We identify the state of such an electron by its wave-vector $\mathbf{k}$, which takes values from zero up to a maximum magnitude k_F, which we call the Fermi momentum. The average occupation $n_{\mathbf{k}}$ of the state with energy $\epsilon_{\mathbf{k}}$ at zero temperature ($\beta \to +\infty$) is:

$$n_{\mathbf{k}} = \frac{2}{e^{\beta(\epsilon_{\mathbf{k}} - \mu_{FD})} + 1} \to 2 \;\; [\text{for } \epsilon_{\mathbf{k}} < \mu_{FD}], \quad 0 \;\; [\text{for } \epsilon_{\mathbf{k}} > \mu_{FD}]$$

In three-dimensional cartesian coordinates the wave-vector is $\mathbf{k} = k_x\hat{\mathbf{x}} + k_y\hat{\mathbf{y}} + k_z\hat{\mathbf{z}}$ and the energy of the state with wave-vector $\mathbf{k}$ is

$$\epsilon_{\mathbf{k}} = \frac{\hbar^2\mathbf{k}^2}{2m_e}$$

We consider the particles enclosed in a box with sides $2L_x, 2L_y, 2L_z$, in which case a relation similar to Eq. (G.46) holds for $\mathrm{d}k_x, \mathrm{d}k_y, \mathrm{d}k_z$:

$$\mathrm{d}k_x\mathrm{d}k_y\mathrm{d}k_z = \frac{(2\pi)^3}{(2L_x)(2L_y)(2L_z)} \Longrightarrow \frac{\mathrm{d}\mathbf{k}}{(2\pi)^3} = \frac{1}{\Omega} \Longrightarrow \lim_{L_x,L_y,L_z\to\infty} \sum_{\mathbf{k}} = \Omega\int\frac{\mathrm{d}\mathbf{k}}{(2\pi)^3} \tag{D.8}$$

with $\Omega = (2L_x)(2L_y)(2L_z)$ the total volume of the box. Therefore, the total number of states is

$$2\sum_{|\mathbf{k}|\leq k_{\mathrm{F}}} = 2\Omega\int_0^{k_{\mathrm{F}}}\frac{\mathrm{d}\mathbf{k}}{(2\pi)^3} = \frac{k_{\mathrm{F}}^3}{3\pi^2}\Omega \tag{D.9}$$

which of course must be equal to the total number of electrons N in the box, giving the following relation between the density $n = N/\Omega$ and the Fermi momentum k_{F}:

$$n = \frac{k_{\mathrm{F}}^3}{3\pi^2} \tag{D.10}$$

which in turn gives for the the energy of the highest occupied state

$$\epsilon_{\mathrm{F}} = \frac{\hbar^2k_{\mathrm{F}}^2}{2m_e} = \frac{\hbar^2(3\pi^2n)^{3/2}}{2m_e} = \mu_{FD} \tag{D.11}$$

If the electrons have only kinetic energy, the total energy of the system will be

$$E^{kin} = 2\sum_{|\mathbf{k}|\leq k_{\mathrm{F}}}\epsilon_{\mathbf{k}} = 2\Omega\int_0^{k_{\mathrm{F}}}\frac{\mathrm{d}\mathbf{k}}{(2\pi)^3}\frac{\hbar^2k^2}{2m_e} = \frac{\Omega}{\pi^2}\frac{\hbar^2k_{\mathrm{F}}^5}{10m_e} = \frac{3}{5}N\epsilon_{\mathrm{F}} \tag{D.12}$$

D.1.3 Quantum Bose–Einstein statistics

In the case of quantum mechanical particles that obey Bose statistics, any number of particles can be at each of the g_i states associated with level i. We can think of the g_i states as identical boxes in which we place a total of n_i particles, in which case the system is equivalent to one consisting of $n_i + g_i$ objects which can be arranged in a total of

$$\frac{(n_i + g_i)!}{n_i!g_i!}$$

ways, since both boxes and particles can be interchanged among themselves arbitrarily. This gives for the total number of states $W(\{n_i\})$ associated with a particular distribution $\{n_i\}$ in the levels with energies ϵ_i

$$W(\{n_i\}) = \prod_i \frac{(n_i + g_i)!}{n_i!(g_i)!} \Longrightarrow$$

$$\ln W(\{n_i\}) = \sum_i [(n_i + g_i)(\ln(n_i + g_i) - 1) - n_i(\ln n_i - 1) - g_i(\ln g_i - 1)]$$

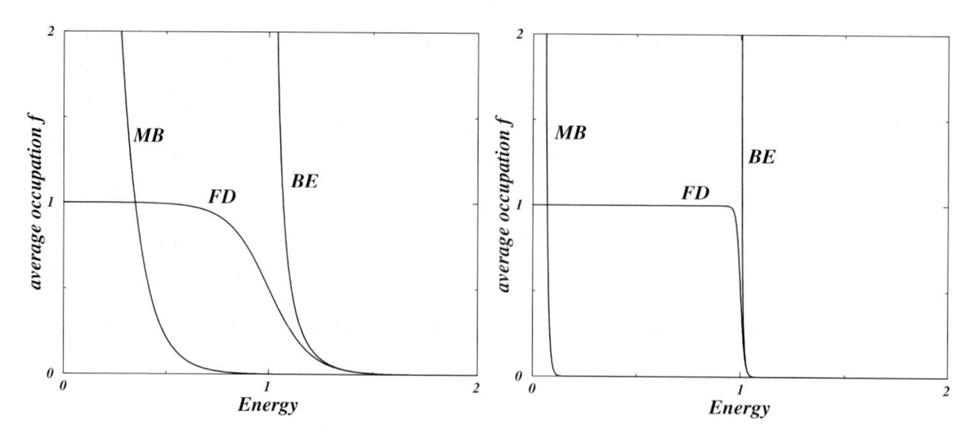

Figure D.1. Average occupation numbers in the Maxwell–Boltzmann (MB), Fermi–Dirac (FD) and Bose–Einstein (BE) distributions, for $\beta = 10$ (**left**) and $\beta = 100$ (**right**), on an energy scale in which $\mu_{FD} = \mu_{BE} = 1$ for the FD and BE distributions. The factor $n/(2\pi m)^{3/2}$ in the Maxwell–Boltzmann distribution of the ideal gas is equal to unity.

where again we have used Stirling's formula for $\ln(g_i!)$, $\ln(n_i!)$, $\ln((n_i + g_i)!)$. Through a similar variational argument as in the previous two cases, we obtain for the most probable distribution $\{\bar{f}_i\}$ which corresponds to the maximum of W:

$$\bar{f}_i^{BE} = g_i \frac{1}{e^{\beta(\epsilon_i - \mu_{BE})} - 1} \tag{D.13}$$

where $\mu_{BE} = -\alpha/\beta$. The constants are once more fixed by the usual normalization conditions, as in the previous two cases. In this case, μ_{BE} is the value of energy just below the lowest occupied level: the values of ϵ_i of occupied states must be above μ_{BE} at finite temperature.

The three distributions are compared in Fig.D.1, for two different values of the temperature ($\beta = 10$ and 100), on an energy scale in which $\mu_{FD} = \mu_{BE} = 1$. The multiplicities g_i for the FD and BE distributions are taken to be unity. For the MB distribution we use the expression derived for the ideal gas, with the factor $n/(2\pi m)^{3/2} = 1$. As these plots show, the FD distribution approaches a step function, that is, a function equal to unity for $\epsilon < \mu_{FD}$ and 0 for $\epsilon > \mu_{FD}$, when the temperature is very low (β very large). For finite temperature T, the FD distribution has a width of order $1/\beta = k_B T$ around μ_{FD} over which it decays from unity to zero. The BE distribution also shows interesting behavior close to μ_{BE}: as the temperature decreases, f changes very sharply as ϵ approaches μ_{BE} from above, becoming infinite near μ_{BE}. This is suggestive of a phase transition, in which all particles occupy the lowest allowed energy level $\epsilon = \mu_{BE}$ for sufficiently low temperature. Indeed, in systems composed of particles obeying Bose statistics, there is a phase transition at low temperature in which all particles collapse to the lowest energy level. This is known as the Bose–Einstein condensation and has received considerable attention recently because it has been demonstrated in systems consisting of trapped atoms.

D.2 Ensemble theory

As mentioned earlier, constructing an average of a physical quantity (such as the energy of the system) is very difficult if the system has many particles and we insist on evolving the system between states until we have included enough states to obtain an accurate estimate.

For this reason, the idea of an ensemble of states has been developed: an ensemble is a set of states of the system which are consistent with the imposed external conditions. These states do not have to be causally connected but can be at very different regions of phase space; as long as they satisfy the external conditions they are admissible as members of the ensemble. Taking an average of a physical quantity over the members of the ensemble is a more efficient way of obtaining accurate averages: with a sample consisting of $\mathcal{N}$ images in an ensemble it is possible to sample disparate regions of phase space, while a sample based on the same number $\mathcal{N}$ of states obtained by causal evolution from a given state is likely to be restricted to a much smaller region of the relevant phase space. A key notion of ensemble theory is that the selection of the images of the system included in the ensemble is not biased in any way, known formally as the *postulate of equal a priori probabilities*.

D.2.1 Definition of ensembles

We consider systems which consist of a large number N of particles and their states are described in terms of the microscopic variables $(\{s\})$ and the total volume Ω that the system occupies. These variables can be, for example, the $3N$ positions $(\{q\})$ and $3N$ momenta $(\{p\})$ of the particles in 3D space, or the N values of the spin or the dipole moment of a system of particles in an external magnetic field. The energy E of a state of the system is the value that the hamiltonian $\mathcal{H}_N(\{s\})$ of the N-particle system takes for the values of $(\{s\})$ that correspond to that particular state.

An ensemble is defined by the density of states ρ included in it. In principle all states that satisfy the imposed external conditions are considered to be members of the ensemble. There are three types of ensembles: the microcanonical, canonical and grand canonical. The precise definition of the density ρ for each ensemble depends on the classical or quantum mechanical nature of the system.

Classical case

Microcanonical ensemble In the microcanonical ensemble, the density of states is defined as

$$\begin{aligned}\rho(\{s\}) &= \frac{1}{Q} \quad \text{for} \quad E \leq \mathcal{H}_N(\{s\}) \leq E + \delta E \\ &= 0 \quad \text{otherwise}\end{aligned} \tag{D.14}$$

with δE an infinitesimal quantity on the scale of E. Thus, in the microcanonical ensemble the energy of the system is fixed within an infinitesimal range. We assume that in this ensemble the total number of particles N and total volume Ω of the system are also fixed. The value of Q is chosen so that summation over all the values of $\rho(\{s\})$ for all the allowed states $\{s\}$ gives unity. For example, if the system consists of N indistinguishable particles there are $N!$ equivalent ways of arranging them in a given configuration $\{s\}$, so the weight of a particular configuration must be $1/Q = 1/N!$ This counting of states has its origin in quantum mechanical considerations: the wavefunction of the system contains a factor of $1/\sqrt{N!}$, and any average involves expectation values of an operator within the system wavefunction, which produces a factor $1/N!$ for every configuration. Without this factor we would be led to inconsistencies. Similarly, if the system is composed of n types of indistinguishable particles, then the relevant factor would be $1/(N_1!N_2! \cdots N_n!)$, where N_i is the number of indistinguishable particles of type i $(i = 1, 2, \ldots, n)$. If the particles are described by their positions and momenta in 3D space, that is, the state of the system is specified by $3N$ position values $(\{q\})$ and $3N$ momentum values $(\{p\})$, an additional factor is needed to cancel the dimensions of the integrals over positions and momenta when averages are taken. This factor is taken to be h^{3N}, where h is Planck's constant. This choice is a consequence of the fact that the elementary volume in position–momentum space is h, by Heisenberg's uncertainty relation (see Appendix B). Again, the proper normalization comes from quantum mechanical considerations.

Canonical ensemble In the canonical ensemble the total volume Ω and total number of particles N in the system are fixed, but the energy E is allowed to vary. The system is considered to be in contact with an external reservoir with temperature T. The density of states in the canonical ensemble is

$$\rho(\{s\}) = \frac{1}{Q} \exp\left[-\beta \mathcal{H}_N(\{s\})\right] \tag{D.15}$$

where $\beta = 1/k_B T$ with T the externally imposed temperature. The factor $1/Q$ serves the purpose of normalization, as in the case of the microcanonical ensemble. For a system of N identical particles $Q = N!$, as in the microcanonical ensemble.

Grand canonical ensemble In the grand canonical ensemble we allow for fluctuations in the volume and in the total number of particles in the system, as well as in the energy. The system is considered to be in contact with an external reservoir with temperature T and chemical potential μ, and under external pressure P. The density of states in the grand canonical ensemble is

$$\rho(\{s\}, N, \Omega) = \frac{1}{Q} \exp\left[-\beta(\mathcal{H}_N(\{s\}) - \mu N + P\Omega)\right] \tag{D.16}$$

with Q the same factor as in the canonical ensemble.

It is a straightforward exercise to show that even though the energy varies in the canonical ensemble, its variations around the average are extremely small for large enough systems. Specifically, if $\langle \mathcal{H} \rangle$ is the average energy and $\langle \mathcal{H}^2 \rangle$ is the average of the energy squared, then

$$\frac{\langle \mathcal{H} \rangle^2 - \langle \mathcal{H}^2 \rangle}{\langle \mathcal{H} \rangle^2} \sim \frac{1}{N} \tag{D.17}$$

which shows that for large systems the deviation of the energy from its average value is negligible. Similarly, in the grand canonical ensemble, the deviation of the number of particles from its average is negligible. Thus, in the limit $N \to \infty$ the three types of ensembles are equivalent.

Quantum mechanical case

Microcanonical ensemble We consider a system with N particles and volume Ω, both fixed. For this system, the quantum mechanical microcanonical ensemble is defined through the wavefunctions of the states of the system whose energies lie in a narrow range δE which is much smaller than the energy E, by analogy to the classical case:

$$\rho(\{s\}) = \sum_{K=1}^{\mathcal{N}} |\Psi_K(\{s\})\rangle\langle\Psi_K(\{s\})| \quad \text{for} \quad E \leq \langle\Psi_K|\mathcal{H}|\Psi_K\rangle \leq E + \delta E \tag{D.18}$$

where the wavefunctions $|\Psi_K(\{s\})\rangle$ represent the quantum mechanical states of the system with energy $\langle\psi_K(\{s\})|\mathcal{H}(\{s\})|\psi_K(\{s\})\rangle$ in the range $[E, E + \delta E]$; the set of variables $\{s\}$ includes all the relevant degrees of freedom in the system which enter in the description of the hamiltonian $\mathcal{H}$ and the wavefunctions (to simplify the notation we will not include these variables in association with the hamiltonian and the wavefunctions in most of what follows).

The thermodynamic average of an operator $\mathcal{O}$ is given by the average of its expectation value over all the states in the ensemble, properly normalized by the sum of the norms of these states:

$$\overline{\mathcal{O}} = \sum_{K=1}^{\mathcal{N}} \langle \Psi_K | \mathcal{O} | \Psi_K \rangle \left[\sum_{K=1}^{\mathcal{N}} \langle \Psi_K | \Psi_K \rangle \right]^{-1} \tag{D.19}$$

If we assume that the wavefunctions $|\Psi_K\rangle$ are properly normalized, $\langle \Psi_K | \Psi_K \rangle = 1$, then the normalization factor that enters in the average becomes

$$\left[\sum_{K=1}^{\mathcal{N}} \langle \Psi_K | \Psi_K \rangle \right]^{-1} = \frac{1}{\mathcal{N}}$$

The form of the normalization factor used in Eq. (D.19) is more general. In the definition of the quantum ensembles we will not need to introduce factors of $1/Q$ as we did for the classical ensembles, because these factors come out naturally when we include in the averages the normalization factor mentioned above.

Inserting complete sets of states $\sum_I |\Phi_I\rangle\langle\Phi_I|$ between the operator and the wavefunctions $\langle \Psi_K|$ and $|\Psi_K\rangle$, we obtain

$$\overline{\mathcal{O}} = \sum_{K=1}^{\mathcal{N}} \sum_{IJ} \langle \Psi_K | \Phi_I \rangle \langle \Phi_I | \mathcal{O} | \Phi_J \rangle \langle \Phi_J | \Psi_K \rangle \left[\sum_{K=1}^{\mathcal{N}} \langle \Psi_K | \Psi_K \rangle \right]^{-1}$$

which can be rewritten in terms of matrix notation, with the following definitions of matrix elements:

$$\begin{aligned} \mathcal{O}_{IJ} &= \langle \Phi_I | \mathcal{O} | \Phi_J \rangle \\ \rho_{JI} &= \langle \Phi_J | \left[\sum_{K=1}^{\mathcal{N}} | \Psi_K \rangle \langle \Psi_K | \right] | \Phi_I \rangle \end{aligned} \tag{D.20}$$

In terms of these matrix elements, the sum of the norms of the states $|\Psi_K\rangle$ becomes

$$\begin{aligned} \sum_{K=1}^{\mathcal{N}} \langle \Psi_K | \Psi_K \rangle &= \sum_{K=1}^{\mathcal{N}} \langle \Psi_K | \left[\sum_I | \Phi_I \rangle \langle \Phi_I | \right] | \Psi_K \rangle \\ &= \sum_I \langle \Phi_I | \left[\sum_{K=1}^{\mathcal{N}} | \Psi_K \rangle \langle \Psi_K | \right] | \Phi_I \rangle = \sum_I \rho_{II} = \mathrm{Tr}[\rho] \end{aligned} \tag{D.21}$$

where Tr denotes the trace, that is, the sum of diagonal matrix elements. With this expression the thermodynamic average of the operator $\mathcal{O}$ takes the form

$$\begin{aligned} \overline{\mathcal{O}} &= \frac{1}{\mathrm{Tr}[\rho]} \sum_{K=1}^{\mathcal{N}} \sum_{IJ} \langle \Psi_K | \Phi_I \rangle \mathcal{O}_{IJ} \langle \Phi_J | \Psi_K \rangle \\ &= \frac{1}{\mathrm{Tr}[\rho]} \sum_{IJ} \mathcal{O}_{IJ} \langle \Phi_J | \left[\sum_{K=1}^{\mathcal{N}} | \Psi_K \rangle \langle \Psi_K | \right] | \Phi_I \rangle \\ &= \frac{1}{\mathrm{Tr}[\rho]} \sum_J \left[\sum_I \rho_{JI} \mathcal{O}_{IJ} \right] = \frac{1}{\mathrm{Tr}[\rho]} \sum_J [\rho \mathcal{O}]_{JJ} = \frac{\mathrm{Tr}[\rho \mathcal{O}]}{\mathrm{Tr}[\rho]} \end{aligned} \tag{D.22}$$

Notice that this last expression is general, and depends on the states included in the ensemble only through the definition of the density matrix, Eq. (D.20). The new element that was introduced in the derivation of this general expression was the complete state of states $|\Phi_I\rangle$, which, in principle, can be found for a given quantum mechanical system. In most situations it is convenient to choose these states to be the energy eigenstates of the system. Moreover, assuming that the energy interval δE is smaller than the spacing between energy eigenvalues, the states $|\Psi_K\rangle$ are themselves energy eigenstates. In this case, we can identify the set of states $|\Phi_I\rangle$ with the set of states $|\Psi_K\rangle$ since the two sets span the same Hilbert space and are therefore related to each other by at most a rotation. Having made this identification, we see that the matrix elements of the density operator become diagonal,

$$\rho_{JI} = \langle \Psi_J | \left[\sum_{K=1}^{\mathcal{N}} |\Psi_K\rangle\langle\Psi_K| \right] |\Psi_I\rangle = \sum_{K=1}^{\mathcal{N}} \delta_{JK}\delta_{KI} = \delta_{JI} \tag{D.23}$$

if the set of states $|\Psi_K\rangle$ is orthonormal. This choice of a complete set of states allows also straightforward definitions of the canonical and grand canonical ensembles, by analogy to the classical case.

Canonical ensemble We consider a system with N particles and volume Ω, both fixed, at a temperature T. The quantum canonical ensemble for this system in the energy representation, that is, using as basis a complete set of states $|\Phi_I\rangle$ which are eigenfunctions of the hamiltonian $\mathcal{H}$ is defined through the density:

$$\rho(\{s\}) = \sum_I |\Phi_I\rangle e^{-\beta E_I(\{s\})} \langle\Phi_I| = \sum_I |\Phi_I\rangle e^{-\beta\mathcal{H}(\{s\})} \langle\Phi_I| = e^{-\beta\mathcal{H}(\{s\})} \tag{D.24}$$

where we have taken advantage of the relations

$$\mathcal{H}|\Phi_I\rangle = E_I|\Phi_I\rangle, \quad \sum_I |\Phi_I\rangle\langle\Phi_I| = 1$$

that hold for this complete set of energy eigenstates. Then the average of a quantum mechanical operator $\mathcal{O}$ in this ensemble is calculated as

$$\overline{\mathcal{O}} = \frac{\mathrm{Tr}[\rho\mathcal{O}]}{\mathrm{Tr}[\rho]} = \frac{\mathrm{Tr}[e^{-\beta\mathcal{H}}\mathcal{O}]}{\mathrm{Tr}[e^{-\beta\mathcal{H}}]} \tag{D.25}$$

Grand canonical ensemble Finally, for the quantum grand canonical ensemble we define the density by analogy to the classical case as

$$\rho(\{s\}, N, \Omega) = e^{-\beta(\mathcal{H}(\{s\}) - \mu N + P\Omega)} \tag{D.26}$$

where μ is the chemical potential and P is the pressure, both determined by the reservoir with which the system under consideration is in contact. In this case the number of particles in the system N and its volume Ω are allowed to fluctuate, and their average values are determined by the chemical potential and pressure imposed by the reservoir; for simplicity we consider only fluctuations in the number of particles, the extension to volume fluctuations being straightforward. In principle we should define an operator that counts the number of particles in each quantum mechanical state and use it in the expression for the density $\rho(\{s\}, N, \Omega)$. However, matrix elements of either the hamiltonian or the particle number operators between states with different numbers of particles vanish identically, so in effect we can take N in the above expression to be the number of particles in the system and then sum over all values of N when we take averages. Therefore, the proper definition of the average of the operator $\mathcal{O}$ in this ensemble is given by

$$\overline{\mathcal{O}} = \frac{\mathrm{Tr}[\rho\mathcal{O}]}{\mathrm{Tr}[\rho]} = \frac{\sum_{N=0}^{\infty} \mathrm{Tr}[\mathrm{e}^{-\beta\mathcal{H}}\mathcal{O}]_N \mathrm{e}^{\beta\mu N}}{\sum_{N=0}^{\infty} \mathrm{Tr}[\mathrm{e}^{-\beta\mathcal{H}}]_N \mathrm{e}^{\beta\mu N}} \tag{D.27}$$

where the traces inside the summations over N are taken for fixed value of N, as indicated by the subscripts.

D.2.2 Derivation of thermodynamics

The average values of physical quantities calculated through ensemble theory should obey the laws of thermodynamics. In order to derive thermodynamics from ensemble theory we have to make the proper identification between the usual thermodynamic variables and quantities that can be obtained directly from the ensemble. In the microcanonical ensemble the basic equation that relates thermodynamic variables to ensemble quantities is

$$S(E, \Omega) = k_{\mathrm{B}} \ln \sigma(E, \Omega) \tag{D.28}$$

where S is the entropy and $\sigma(E, \Omega)$ is the number of microscopic states with energy E and volume Ω. This definition of the entropy is the only possible one that satisfies all the requirements, such as the extensive nature of S with E and Ω and the second and third laws of thermodynamics. Having defined the entropy, we can further define the temperature T and the pressure P as

$$\frac{1}{T} = \left(\frac{\partial S(E, \Omega)}{\partial E}\right)_{\Omega}, \quad P = T\left(\frac{\partial S(E, \Omega)}{\partial \Omega}\right)_{E} \tag{D.29}$$

and through these we can calculate all the other quantities of interest. Although it is simple, the microcanonical ensemble is not very useful because it is rather difficult to calculate $\sigma(E, \Omega)$, as Eq. (D.28) requires. The other two ensembles are much more convenient for calculations, so we concentrate our attention on those.

For calculations within the canonical ensemble, it is convenient to introduce the partition function,

$$Z_N(\Omega, T) = \sum_{\{s\}} \rho(\{s\}) = \sum_{\{s\}} \frac{1}{Q} \mathrm{e}^{-\beta\mathcal{H}_N(\{s\})} \tag{D.30}$$

which is simply a sum over all values of the density $\rho(\{s\})$. We have used the subscript N, the total number of particles in the system (which is fixed), to denote that we are in the canonical ensemble. The partition function in the case of a system of N indistinguishable particles with momenta $\{p\}$ and coordinates $\{q\}$ takes the form

$$Z_N(\Omega, T) = \frac{1}{N! h^{3N}} \int \mathrm{d}\{q\} \int \mathrm{d}\{p\} \mathrm{e}^{-\beta\mathcal{H}_N(\{p\},\{q\})} \tag{D.31}$$

Using the partition function, which embodies the density of states relevant to the canonical ensemble, we define the free energy as

$$F_N(\Omega, T) = -k_{\mathrm{B}} T \ln Z_N(\Omega, T) \tag{D.32}$$

and through it the entropy and pressure:

$$S_N = -\left(\frac{\partial F_N}{\partial T}\right)_\Omega, \quad P_N = -\left(\frac{\partial F_N}{\partial \Omega}\right)_T \tag{D.33}$$

from which we can obtain all other thermodynamic quantities of interest.

For the grand canonical ensemble we define the grand partition function as

$$Z(\mu, \Omega, T) = \sum_{\{s\},N} \frac{1}{Q}\rho(\{s\}, N, \Omega)\mathrm{e}^{\beta P\Omega} = \sum_{N=0}^{\infty} \mathrm{e}^{\beta\mu N}\frac{1}{Q}\sum_{\{s\}} \mathrm{e}^{-\beta\mathcal{H}_N(\{s\})} \tag{D.34}$$

A more convenient representation of the grand partition function is based on the introduction of the variable z, called the fugacity:

$$z = \mathrm{e}^{\beta\mu}$$

With this, the grand partition becomes

$$Z(z, \Omega, T) = \sum_{N=0}^{\infty} z^N Z_N(\Omega, T) \tag{D.35}$$

where we have also used the partition function in the canonical ensemble $Z_N(\Omega, T)$ that we defined earlier. In the case of a system consisting of N indistinguishable particles with positions $\{q\}$ and momenta $\{p\}$, the grand partition function takes the form

$$Z(\mu, \Omega, T) = \sum_{N=0}^{\infty} \mathrm{e}^{\beta\mu N}\frac{1}{N!h^{3N}}\int \mathrm{d}\{q\}\int \mathrm{d}\{p\}\mathrm{e}^{-\beta\mathcal{H}_N(\{p\},\{q\})} \tag{D.36}$$

In terms of the grand partition function we can calculate the average energy $\bar{E}$ and the average number of particles in the system $\bar{N}$:

$$\bar{E}(z, \Omega, T) = -\left(\frac{\partial \ln Z(z, \Omega, T)}{\partial \beta}\right)_{z,\Omega}, \quad \bar{N}(z, \Omega, T) = z\left(\frac{\partial \ln Z(z, \Omega, T)}{\partial z}\right)_{\Omega,T} \tag{D.37}$$

We can then define the free energy as

$$F(z, \Omega, T) = -k_{\mathrm{B}}T \ln\left[\frac{Z(z, \Omega, T)}{z^{\bar{N}}}\right] \tag{D.38}$$

which can be considered as either a function of z or a function of $\bar{N}$ since these two variables are related by Eq. (D.37). From the free energy we can obtain all other thermodynamic quantities. For example, the pressure P, entropy S and chemical potential μ are given as

$$P = -\left(\frac{\partial F}{\partial \Omega}\right)_{\bar{N},T}, \quad S = -\left(\frac{\partial F}{\partial T}\right)_{\bar{N},\Omega}, \quad \mu = \left(\frac{\partial F}{\partial \bar{N}}\right)_{\Omega,T} \tag{D.39}$$

Finally, we find from the definition of the grand partition function and the fact that $\rho(\{s\}, N, \Omega)$ is normalized to unity, that

$$\frac{P\Omega}{k_{\mathrm{B}}T} = \ln[Z(z, \Omega, T)] \tag{D.40}$$

which determines the equation of state of the system.

The above expressions hold also for the quantum mechanical ensembles. The only difference is that the partition function is in this case defined as the trace of the density matrix; for example, in the quantum canonical ensemble, the partition function becomes

$$Z_N(\Omega, T) = \mathrm{Tr}[\rho(\{s\})] = \sum_I \rho_{II} = \sum_I \langle \Phi_I | \mathrm{e}^{-\beta \mathcal{H}(\{s\})} | \Phi_I \rangle = \sum_I \mathrm{e}^{-\beta E_I(\{s\})}$$

where we have assumed a complete set of energy eigenstates $|\Phi_I\rangle$ as the basis for calculating the density matrix elements. This form of the partition function is essentially the same as in the classical case, except for the normalization factors which will enter automatically in the averages from the proper definition of the normalized wavefunctions $|\Phi_I\rangle$. Similarly, the partition function for the quantum grand canonical ensemble will have the same form as in Eq. (D.35), with $Z_N(\Omega, T)$ the partition function of the quantum canonical system with N particles as defined above, and $z = \exp(\beta\mu)$ the fugacity.

D.3 Applications of ensemble theory

D.3.1 Equipartition and the Virial

A general result that is a direct consequence of ensemble theory is the equipartition theorem, which we state here without proof (for a proof see the book by Huang, in the Further reading section). For a system consisting of particles with coordinates $q_i, i = 1, 2, \ldots, 3N$ and momenta $p_i, i = 1, 2, \ldots, 3N$, the equipartition theorem is

$$\langle p_i \frac{\partial \mathcal{H}}{\partial p_i} \rangle = \langle q_i \frac{\partial \mathcal{H}}{\partial q_i} \rangle = k_{\mathrm{B}} T \tag{D.41}$$

A direct consequence of the equipartition theorem is the following: using the hamiltonian equations of motion for such a system

$$\frac{\partial p_i}{\partial t} = -\frac{\partial \mathcal{H}}{\partial q_i}$$

we obtain the expression

$$\mathcal{V} = \langle \sum_{i=1}^{3N} q_i \frac{\partial p_i}{\partial t} \rangle = -\langle \sum_{i=1}^{3N} q_i \frac{\partial \mathcal{H}}{\partial q_i} \rangle = -3N k_{\mathrm{B}} T \tag{D.42}$$

where $\mathcal{V}$ is known as the "Virial". These general relations are useful in checking the behavior of complex systems in a simple way. For example, for a hamiltonian which is quadratic in the degrees of freedom, like that of harmonic oscillators with spring constants κ,

$$\mathcal{H}_{HO} = \sum_{i=1}^{3N} \frac{p_i^2}{2m} + \sum_{i=1}^{3N} \frac{1}{2} \kappa q_i^2$$

there is a thermal energy equal to $k_{\mathrm{B}} T/2$ associated with each harmonic degree of freedom, because from the equipartition theorem we have

$$\langle q_i \frac{\partial \mathcal{H}}{\partial q_i} \rangle = \langle \kappa q_i^2 \rangle = k_{\mathrm{B}} T \Longrightarrow \langle \frac{1}{2} \kappa q_i^2 \rangle = \frac{1}{2} k_{\mathrm{B}} T$$

and similarly for the momentum variables.

D.3.2 Ideal gases

The ideal gas is defined as consisting of N particles confined in a volume Ω, with the only interaction between particles being binary elastic collisions, that is, collisions which preserve energy and momentum and take place at a single point in space when two particles meet. This model is a reasonable approximation for dilute systems of atoms or molecules which interact very weakly. Depending on the nature of the particles that compose the ideal gas, we distinguish between the classical and quantum mechanical cases. Of course, all atoms or molecules should ultimately be treated quantum mechanically, but at sufficiently high temperatures their quantum mechanical nature is not apparent. Therefore, the classical ideal gas is really the limit of quantum mechanical ideal gases at high temperature.

Classical ideal gas

We begin with the classical ideal gas, in which any quantum mechanical features of the constituent particles are neglected. This is actually one of the few systems that can be treated in the microcanonical ensemble. In order to obtain the thermodynamics of the classical ideal gas in the microcanonical ensemble we need to calculate the entropy from the total number of states $\sigma(E, \Omega)$ with energy E and volume Ω, according to Eq. (D.28). The quantity $\sigma(E, \Omega)$ is given by

$$\sigma(E, \Omega) = \frac{1}{N!h^{3N}} \int_{E \leq \mathcal{H}(\{p\},\{q\}) \leq E+\delta E} \mathrm{d}\{p\}\mathrm{d}\{q\}$$

However, it is more convenient to use the quantity

$$\Sigma(E, \Omega) = \frac{1}{N!h^{3N}} \int_{0 \leq \mathcal{H}(\{p\},\{q\}) \leq E} \mathrm{d}\{p\}\mathrm{d}\{q\}$$

instead of $\sigma(E, \Omega)$ in the calculation of the entropy: it can be shown that this gives a difference in the value of the entropy of order $\ln N$, a quantity negligible relative to N, to which the entropy is proportional as an extensive variable. In the above expression we have taken the value of zero to be the lower bound of the energy spectrum.

$\Sigma(E, \Omega)$ is much easier to calculate: each integral over a position variable $\mathbf{q}$ gives a factor of Ω, while each variable $\mathbf{p}$ ranges in magnitude from a minimum value of zero to a maximum value of $\sqrt{2mE}$, with E the total energy and m the mass of the particles. Thus, the integration over $\{q\}$ gives a factor Ω^N, while the integration over $\{p\}$ gives the volume of a sphere in $3N$ dimensions with radius $\sqrt{2mE}$. This volume in momentum space is given by

$$\Pi_{3N} = \frac{\pi^{3N/2}}{(\frac{3N}{2}+1)!}(2mE)^{3N/2}$$

so that the total number of states with energy up to E becomes

$$\Sigma(E, \Omega) = \frac{1}{N!h^{3N}} \Pi_{3N} \Omega^N$$

and using Stirling's formula we obtain for the entropy

$$S(E, \Omega) = N k_{\mathrm{B}} \ln\left[\frac{\Omega}{N}\left(\frac{E}{N}\right)^{3/2}\right] + \frac{3}{2} N k_{\mathrm{B}} \left[\frac{5}{3} + \ln\left(\frac{4\pi m}{3h^2}\right)\right] \quad \text{(D.43)}$$

From this expression we can calculate the temperature and pressure as discussed earlier, obtaining

$$\frac{1}{T} = \frac{\partial S(E, \Omega)}{\partial E} \Longrightarrow T = \frac{2}{3}\frac{E}{Nk_B} \tag{D.44}$$

$$P = T\frac{\partial S(E, \Omega)}{\partial \Omega} \Longrightarrow P = \frac{Nk_B T}{\Omega} \tag{D.45}$$

with the last expression being the familiar equation of state of the ideal gas.

An interesting application of the above results is the calculation of the entropy of mixing of two ideal gases of the same temperature and density. We first rewrite the entropy of the ideal gas in terms of the density n, the energy per particle u and the constant entropy per particle s_0, defined through

$$n = \frac{N}{\Omega}, \quad u = \frac{E}{N}, \quad s_0 = \frac{3}{2}k_B \left[\frac{5}{3} + \ln\left(\frac{4\pi m}{3h^2}\right)\right]$$

With these definitions the entropy becomes

$$S = \frac{3}{2}Nk_B \ln(u) - Nk_B \ln(n) + Ns_0$$

Now consider two gases at the same density $n_1 = n_2 = n$ and temperature, which implies $u_1 = u_2 = u$, from the result we obtained above, Eq. (D.44). The first gas occupies volume Ω_1 and the second gas occupies volume Ω_2. We assume that the particles in the two gases have the same mass m. If the gases are different[1] and we allow them to mix and occupy the total volume $\Omega = \Omega_1 + \Omega_2$, the new density of each gas will be $n_1' = N_1/\Omega \neq n$, $n_2' = N_2/\Omega \neq n$, while their energy will not have changed after mixing, since the temperature remains constant (no work is done on or by either gas). Then the entropy of mixing ΔS, defined as the entropy of the system after mixing minus the entropy before, will be

$$\Delta S = N_1 k_B \ln\left(1 + \frac{\Omega_2}{\Omega_1}\right) + N_2 k_B \ln\left(1 + \frac{\Omega_1}{\Omega_2}\right) > 0$$

If the gases were the same, then the density of the gas after mixing will be the same because

$$n = \frac{N_1}{\Omega_1} = \frac{N_2}{\Omega_2} = \frac{N_1 + N_2}{\Omega_1 + \Omega_2}$$

and the temperature will remain the same as before, which together imply that $\Delta S = 0$, that is, there is no entropy of mixing. This is the expected result since the state of the gas before and after mixing is exactly the same. We note that if we had not included the $1/N!$ factor in the definition of the ensemble density, this last result would be different, that is, the entropy of mixing of two parts of the same gas at the same temperature and density would be positive, an unphysical result. This is known as the "Gibbs paradox", which had led Gibbs to hypothesize the existence of the $1/N!$ factor in the ensemble density long before its quantum mechanical origin was understood.

[1] This can be achieved experimentally, for instance, by the right-hand and left-hand isomers of the same molecule.

Finally, we calculate for future reference the canonical partition function of the N-particle classical ideal gas:

$$
\begin{aligned}
Z_N(\beta, \Omega) &= \frac{1}{N!h^{3N}} \int \mathrm{d}\mathbf{r}_1 \cdots \mathrm{d}\mathbf{r}_N \int \mathrm{d}\mathbf{p}_1 \cdots \mathrm{d}\mathbf{p}_N \exp\left[-\beta \sum_i \frac{\mathbf{p}_i^2}{2m}\right] \\
&= \frac{1}{N!h^{3N}} \Omega^N \prod_{i=1}^{N} \int \mathrm{d}\mathbf{p}_i \exp\left[-\beta \frac{\mathbf{p}_i^2}{2m}\right] \\
&= \frac{1}{N!} \Omega^N \left(\frac{mk_{\mathrm{B}}T}{2\pi\hbar^2}\right)^{3N/2}
\end{aligned}
\tag{D.46}
$$

where we have used the relations $\mathbf{p} = \hbar\mathbf{k}$, with $\hbar = h/2\pi$, for the momentum and $\beta = 1/k_{\mathrm{B}}T$ for the temperature.

Quantum ideal gases

For the quantum mechanical ideal gases, we begin with the construction of a properly normalized wavefunction for N indistinguishable free particles of mass m. In this case the hamiltonian of the system consist of only the kinetic energy term,

$$
\mathcal{H}_N = -\sum_{i=1}^{N} \frac{\hbar^2 \nabla_{\mathbf{r}_i}^2}{2m}
$$

The individual free particles are in plane wave states with momenta $\mathbf{k}_1, \mathbf{k}_2, \ldots, \mathbf{k}_N$:

$$
|\mathbf{k}_i\rangle = \frac{1}{\sqrt{\Omega}} \mathrm{e}^{\mathrm{i}\mathbf{k}_i \cdot \mathbf{r}_i}
$$

with Ω the volume of the system. A product over all such states is an eigenfunction of the hamiltonian

$$
\langle \mathbf{r}_1, \ldots, \mathbf{r}_N | \mathbf{k}_1, \ldots, \mathbf{k}_N \rangle = \frac{1}{\sqrt{\Omega^N}} \mathrm{e}^{\mathrm{i}(\mathbf{k}_1 \cdot \mathbf{r}_1 + \cdots + \mathbf{k}_N \cdot \mathbf{r}_N)}
$$

with total energy given by

$$
\sum_{i=1}^{N} \frac{\hbar^2 \mathbf{k}_i^2}{2m} = \frac{\hbar^2 K^2}{2m} = E_K
$$

where K is taken to be a positive number. However, a single such product is not a proper wavefunction for the system of indistinguishable particles, because it does not embody the requisite symmetries. The N-body wavefunction $|\Psi_K^{(N)}\rangle$ must include all possible permutations of the individual momenta, which are $N!$ in number:

$$
|\Psi_K^{(N)}\rangle = \frac{1}{\sqrt{N!}} \sum_P s_P |P[\mathbf{k}_1, \ldots, \mathbf{k}_N]\rangle = \frac{1}{\sqrt{N!}} \sum_P s_P |P\mathbf{k}_1, \ldots, P\mathbf{k}_N\rangle \tag{D.47}
$$

where P denotes a particular permutation of the individual momenta and s_P the sign associated with it; $P\mathbf{k}_i$ is the value of the momentum of the particle i under permutation P, which must be equal to one of the other values of the individual momenta. For bosons, all signs $s_P = 1$, so that the wavefunction is symmetric with respect to an interchange of any two particles. For fermions, the signs of odd permutations are -1, while those for even permutations are $+1$, so that the exchange of any two particles will produce an overall minus sign for the entire wavefunction.

As a first step in the analysis of the thermodynamic behavior of this system, we calculate the partition function in the canonical ensemble and in the position representation. A matrix element of the canonical density in the position representation is given by

$$\langle \mathbf{r}_1, \ldots, \mathbf{r}_N | \rho | \mathbf{r}'_1, \ldots, \mathbf{r}'_N \rangle = \langle \mathbf{r}_1, \ldots, \mathbf{r}_N | \sum_K |\Psi_K^{(N)}\rangle e^{-\beta E_K} \langle \Psi_K^{(N)} | \mathbf{r}'_1, \ldots, \mathbf{r}'_N \rangle \quad \text{(D.48)}$$

We will manipulate this matrix element to obtain a useful expression by performing the following steps, which we explain below:

$$[1]\ \langle \{\mathbf{r}_i\} | \sum_K |\Psi_K^{(N)}\rangle e^{-\beta E_K} \langle \Psi_K^{(N)} | \{\mathbf{r}'_i\}\rangle =$$

$$[2]\ \frac{1}{N!} \sum_K \sum_{PP'} s_P s_{P'} \langle \{\mathbf{r}_i\} | \{P\mathbf{k}_i\}\rangle e^{-\beta E_K} \langle \{P'\mathbf{k}_i\} | \{\mathbf{r}'_i\}\rangle =$$

$$[3]\ \frac{1}{N!} \sum_P s_P \langle \{P\mathbf{r}_i\} | \sum_K \sum_{P'} s_{P'} | \{\mathbf{k}_i\}\rangle e^{-\beta E_K} \langle \{P'\mathbf{k}_i\} | \{\mathbf{r}'_i\}\rangle =$$

$$[4]\ \frac{1}{(N!)^2} \sum_P s_P \langle \{P\mathbf{r}_i\} | \sum_{\mathbf{k}_1,\ldots,\mathbf{k}_N} \sum_{P'} s_{P'} | \{\mathbf{k}_i\}\rangle e^{-\beta E_K} \langle \{P'\mathbf{k}_i\} | \{\mathbf{r}'_i\}\rangle =$$

$$[5]\ \frac{1}{N!} \sum_P s_P \langle \{P\mathbf{r}_i\} | \sum_{\mathbf{k}_1,\ldots,\mathbf{k}_N} | \{\mathbf{k}_i\}\rangle e^{-\beta E_K} \langle \{\mathbf{k}_i\} | \{\mathbf{r}'_i\}\rangle =$$

$$[6]\ \frac{1}{N!} \sum_P s_P \sum_{\mathbf{k}_1,\ldots,\mathbf{k}_N} \langle \{\mathbf{r}_i\} | \{P\mathbf{k}_i\}\rangle e^{-\beta E_K} \langle \{\mathbf{k}_i\} | \{\mathbf{r}'_i\}\rangle =$$

$$[7]\ \frac{1}{N!} \sum_P s_P \sum_{\mathbf{k}_1,\ldots,\mathbf{k}_N} e^{-\beta E_K} \prod_{n=1}^{N} \langle \mathbf{r}_n | P\mathbf{k}_n\rangle \langle \mathbf{k}_n | \mathbf{r}'_n\rangle =$$

$$[8]\ \frac{1}{N!} \sum_P s_P \sum_{\mathbf{k}_1,\ldots,\mathbf{k}_N} \frac{1}{\Omega^N} \prod_{n=1}^{N} \exp\left[i(P\mathbf{k}_n \cdot \mathbf{r}_n - \mathbf{k}_n \cdot \mathbf{r}'_n) - \frac{\beta\hbar^2}{2m}\mathbf{k}_n^2 \right]$$

In step [1] we rewrite Eq. (D.48) with slightly different notation, denoting as $\{\mathbf{r}_i\}$, $\{\mathbf{r}'_i\}$ the unprimed and primed coordinate sets. In step [2] we use the expression of the wavefunction from Eq. (D.47), which introduces the set of wave-vectors $\{\mathbf{k}_i\}$ for the bra and ket, and the associated sums over permutations P and P' with signs s_P and s'_P, respectively. In step [3] we make use of the fact that in the position representation the individual single-particle states are plane waves $\exp(\mathrm{i}P\mathbf{k}_i \cdot \mathbf{r}_i)$ which involve the dot products between the *permuted* $\mathbf{k}_i$ vectors and the $\mathbf{r}_i$ vectors, but the sum over all these dot products is equal to the sum over dot products between the $\mathbf{k}_i$ vectors and the *permuted* $\mathbf{r}_i$ vectors, with the same signs for each permutation; we can therefore transfer the permutation from the $\{\mathbf{k}_i\}$ vectors to the $\{\mathbf{r}_i\}$ vectors. In step [4] we use the fact that we can obtain the value of K^2 as the sum over any of the $N!$ permutations of the $\{\mathbf{k}_i\}$ vectors, therefore if we replace the sum over values of K by the N sums over values of the individual $\mathbf{k}_i$ vectors we are overcounting by a factor of $N!$; accordingly, we introduce an extra factor of $N!$ in the denominator to compensate for the overcounting. In step [5] we use the fact that all permutations P' of the vectors $\mathbf{k}_i$ give the same result, because now the the $\mathbf{k}_i$'s have become dummy summation variables since the sums are taken over all of them independently; therefore, we can replace the $N!$ terms from the permutations P' by a single term multiplied by $N!$, and since this term can be any of the possible permutations we pick the first permutation (with unchanged order of $\mathbf{k}_j$'s) for convenience. In this way we get rid of the summation over the permutations P' and restore the factor $1/N!$ in front of the entire sum, rather than the $1/(N!)^2$ we had in step [4]. In step [6] we move the permutations P from the $\mathbf{r}_i$'s to the $\mathbf{k}_i$'s, by the same arguments that we had used above for the reverse move. In step [7] we write explicitly the bra and ket of the many-body free-particle wavefunction as products of single-particle states, and in step [8] we use the plane wave expression for these single-particle states.

Let us consider how this general result applies to $N = 1$ and $N = 2$ particles for illustration. For $N = 1$ it takes the form

$$\langle \mathbf{r}|\mathrm{e}^{-\beta\mathcal{H}_1}|\mathbf{r}'\rangle = \int \frac{\mathrm{d}\mathbf{k}}{(2\pi)^3} \exp\left[-\frac{\beta\hbar^2}{2m}k^2 + \mathrm{i}\mathbf{k}\cdot(\mathbf{r}-\mathbf{r}')\right]$$
$$= \left(\frac{m}{2\pi\beta\hbar^2}\right)^{3/2} \mathrm{e}^{-m(\mathbf{r}-\mathbf{r}')^2/2\beta\hbar^2} \qquad \text{(D.49)}$$

where we have used the usual replacement of $(1/\Omega)\sum_{\mathbf{k}} \to \int \mathrm{d}\mathbf{k}/(2\pi)^3$, and obtained the last expression by completing the square in the integral. We define the thermal wavelength λ as

$$\lambda = \sqrt{\frac{2\pi\hbar^2}{mk_\mathrm{B}T}} \qquad \text{(D.50)}$$

in terms of which the matrix element of $\exp(-\beta\mathcal{H}_1)$ in the position representation takes the form

$$\langle \mathbf{r}|\mathrm{e}^{-\beta\mathcal{H}_1}|\mathbf{r}'\rangle = \frac{1}{\lambda^3}\mathrm{e}^{-\pi(\mathbf{r}-\mathbf{r}')^2/\lambda^2} \qquad \text{(D.51)}$$

The physical interpretation of this expression is that it represents the probability of finding the particle at positions $\mathbf{r}$ and $\mathbf{r}'$, or the spread of the particle's wavefunction in space; the thermal wavelength gives the scale over which this probability is non-vanishing. When $\lambda \to 0$ this expression tends to a δ-function, which is the expected limit since it corresponds to $T \to \infty$, at which point the behavior becomes classical and the particle is localized in space. Thus, λ gives the extent of the particle's wavefunction.

The normalization factor required for thermodynamic averages becomes

$$\mathrm{Tr}[\mathrm{e}^{-\beta\mathcal{H}_1}] = \int \langle \mathbf{r}|\mathrm{e}^{-\beta\mathcal{H}_1}|\mathbf{r}\rangle \mathrm{d}\mathbf{r} = \Omega\left(\frac{m}{2\pi\beta\hbar^2}\right)^{3/2} = \frac{\Omega}{\lambda^3}$$

With these expressions we can calculate thermodynamic averages, for instance the energy of the quantum mechanical free particle, which involves matrix elements of the hamiltonian in the position representation given by

$$\langle \mathbf{r}|\mathcal{H}_1|\mathbf{r}'\rangle = \left[-\frac{\hbar^2}{2m}\nabla^2_{\mathbf{r}'}\right]_{\mathbf{r}'=\mathbf{r}}$$

giving for the average energy

$$E_1 = \overline{\mathcal{H}_1} = \frac{\mathrm{Tr}[\mathcal{H}_1\mathrm{e}^{-\beta\mathcal{H}_1}]}{\mathrm{Tr}[\mathrm{e}^{-\beta\mathcal{H}_1}]} = \frac{3}{2}k_\mathrm{B}T \qquad \text{(D.52)}$$

as expected for a particle with only kinetic energy at temperature T.

For the case of two quantum mechanical particles, $N = 2$, the density matrix becomes

$$\langle \mathbf{r}_1, \mathbf{r}_2 | e^{-\beta \mathcal{H}_2} | \mathbf{r}'_1, \mathbf{r}'_2 \rangle = \int \frac{d\mathbf{k}_1 d\mathbf{k}_2}{2(2\pi)^6} \exp\left[-\frac{\beta \hbar^2}{2m} (\mathbf{k}_1^2 + \mathbf{k}_2^2) \right]$$

$$\times \left[e^{i(\mathbf{k}_1 \cdot \mathbf{r}_1 - \mathbf{k}_1 \cdot \mathbf{r}'_1)} e^{i(\mathbf{k}_2 \cdot \mathbf{r}_2 - \mathbf{k}_2 \cdot \mathbf{r}'_2)} \pm e^{i(\mathbf{k}_2 \cdot \mathbf{r}_1 - \mathbf{k}_1 \cdot \mathbf{r}'_1)} e^{i(\mathbf{k}_1 \cdot \mathbf{r}_2 - \mathbf{k}_2 \cdot \mathbf{r}'_2)} \right]$$

where the $+$ sign corresponds to bosons and the $-$ sign to fermions. With this expression, the normalization factor becomes

$$\mathrm{Tr}[e^{-\beta \mathcal{H}_2}] = \int \langle \mathbf{r}_1, \mathbf{r}_2 | e^{-\beta \mathcal{H}_2} | \mathbf{r}_1, \mathbf{r}_2 \rangle d\mathbf{r}_1 d\mathbf{r}_2 = \frac{1}{2} \left(\frac{\Omega}{\lambda^3} \right)^2 \left[1 \pm \frac{\lambda^3}{2^{3/2} \Omega} \right]$$

so that the normalized diagonal matrix element of the density takes the form

$$\frac{\langle \mathbf{r}_1, \mathbf{r}_2 | e^{-\beta \mathcal{H}_2} | \mathbf{r}_1, \mathbf{r}_2 \rangle}{\mathrm{Tr}[e^{-\beta \mathcal{H}_2}]} = \frac{1}{\Omega^2} \frac{1 \pm \exp\left[-2\pi (\mathbf{r}_1 - \mathbf{r}_2)^2 / \lambda^2 \right]}{1 \pm \lambda^3 / 2^{3/2} \Omega} \tag{D.53}$$

with the $+$ signs for bosons and the $-$ signs for fermions. This matrix element represents the probability of finding the two particles situated at positions $\mathbf{r}_1$ and $\mathbf{r}_2$ at temperature T, which is equivalent to the finite-temperature pair correlation function. The expression we obtained for this probability is interesting because it reveals that, even though we have assumed the particles to be non-interacting, there is an effective interaction between them: if there were no interaction at all, this probability should be simply $1/\Omega^2$, the product of the independent probabilities of finding each particle at a position $\mathbf{r}$ in space. Therefore, the remaining term represents the effective interaction which is due to the quantum mechanical nature of the particles. To explore this feature further, we first note that we can neglect the denominator of the remaining term, since $\lambda^3 << \Omega$: this is justified on the grounds that λ^3 is the volume over which the single-particle wavefunction is non-vanishing, as we found above, which we assume to be much smaller than the volume of the system Ω. We can then define an effective interaction potential $V(r)$ through the numerator of the remaining term in Eq. (D.53),

$$e^{-\beta V(r)} = 1 \pm \exp\left(-\frac{2\pi r^2}{\lambda^2} \right) \Longrightarrow V(r) = -k_B T \ln \left[1 \pm \exp\left(-\frac{2\pi r^2}{\lambda^2} \right) \right]$$

where we have expressed the potential in terms of the relative distance between the particles $r = |\mathbf{r}_1 - \mathbf{r}_2|$. A plot of $V(r)/k_B T$ as a function of r (in units of λ), is shown in Fig. D.2. From this plot it is evident that the effective interaction is *repulsive* for fermions and *attractive* for bosons. The effective interaction is negligible for distances $r > \lambda$, which indicates that quantum mechanical effects become noticeable only for inter-particle distances within the thermal wavelength. As the temperature increases, λ decreases, making the range over which quantum effects are important vanishingly small in the large T limit, thus reaching the classical regime as expected.

Generalizing the above results, it is easy to show that for an N-particle system the diagonal matrix element of the density takes the form

$$\langle \mathbf{r}_1, \ldots, \mathbf{r}_N | e^{-\beta \mathcal{H}_N} | \mathbf{r}_1, \ldots, \mathbf{r}_N \rangle = \frac{1}{N! \lambda^{3N}} \left[1 \pm \sum_{i<j} f_{ij} f_{ji} + \sum_{i<j<k} f_{ij} f_{jk} f_{ki} \pm \cdots \right] \tag{D.54}$$

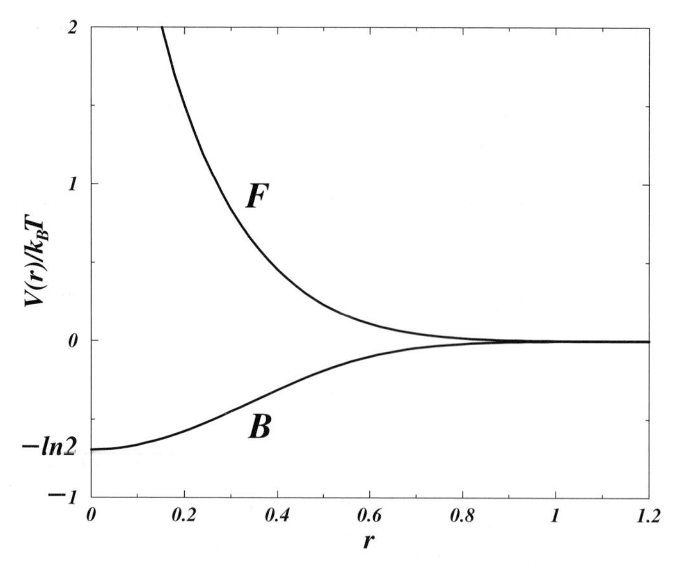

Figure D.2. Effective interaction potential for a pair of bosons or fermions: the inter-particle distance r is measured in units of the thermal wavelength λ.

with + signs for bosons and − signs for fermions, and the factors f_{ij} defined as

$$f_{ij} = \exp\left[-\frac{\pi}{\lambda^2}(\mathbf{r}_i - \mathbf{r}_j)^2\right]$$

The average inter-particle distance in this system is $|\mathbf{r}_i - \mathbf{r}_j| \sim (\Omega/N)^{1/3}$ while the extent of single-particle wavefunctions is λ; for

$$\lambda << (\Omega/N)^{1/3} \sim |\mathbf{r}_i - \mathbf{r}_j|$$

which corresponds to the classical limit, the exponentials f_{ij} are vanishingly small and the normalizing factor becomes

$$\mathrm{Tr}[\mathrm{e}^{-\beta\mathcal{H}_N}] = \int \langle \mathbf{r}_1, \ldots, \mathbf{r}_N|\mathrm{e}^{-\beta\mathcal{H}_N}|\mathbf{r}_1, \ldots, \mathbf{r}_N\rangle \mathrm{d}\mathbf{r}_1 \cdots \mathrm{d}\mathbf{r}_N = \frac{1}{N!}\left(\frac{\Omega}{\lambda^3}\right)^N \quad \text{(D.55)}$$

which is exactly the same as the partition function $Z_N(\beta, \Omega)$ that we calculated for the classical free-particle gas, Eq. (D.46), only now the factors $N!$ and h^{3N} needed in the denominator have appeared naturally from the underlying quantum mechanical formulation of the problem.

As a last application of ensemble theory to the quantum mechanical ideal gas, we obtain the equation of state for fermions and bosons. To this end, the most convenient approach is to use the grand canonical partition function and the momentum representation, in which the matrix elements of the density are diagonal. This gives for the canonical partition function

$$\begin{aligned} Z_N(\beta, \Omega) &= \sum_{\mathbf{k}_1,\ldots,\mathbf{k}_N} \langle \mathbf{k}_1, \ldots, \mathbf{k}_N|\mathrm{e}^{-\beta\mathcal{H}_N}|\mathbf{k}_1, \ldots, \mathbf{k}_N\rangle \\ &= \sum_{\mathbf{k}_1,\ldots,\mathbf{k}_N} \exp\left[-\beta E(n_{\mathbf{k}_1}, \ldots, n_{\mathbf{k}_N})\right] \\ &= \sum_{\{n_\mathbf{k}\}} \exp\left[-\beta E(\{n_\mathbf{k}\})\right] \end{aligned} \quad \text{(D.56)}$$

where we have used the occupation numbers $n_{\mathbf{k}}$ of states $\mathbf{k}$ to express the energy of the system as

$$E(\{n_{\mathbf{k}}\}) = \sum_{i=1}^{N} n_{\mathbf{k}_i} \epsilon_{\mathbf{k}_i}, \qquad \epsilon_{\mathbf{k}_i} = \frac{\hbar^2 \mathbf{k}_i^2}{2m}, \qquad \sum_{i=1}^{N} n_{\mathbf{k}_i} = N$$

With this result for the canonical partition function, the grand canonical partition function becomes:

$$\begin{aligned} Z(z, \beta, \Omega) &= \sum_{N=0}^{\infty} z^N Z_N(\beta, \Omega) = \sum_{N=0}^{\infty} z^N \sum_{\{n_{\mathbf{k}}\}}' \exp\left[-\beta \sum_{\mathbf{k}} n_{\mathbf{k}} \epsilon_{\mathbf{k}}\right] \\ &= \sum_{N=0}^{\infty} \sum_{\{n_{\mathbf{k}}\}}' \prod_{\mathbf{k}} \left(z\mathrm{e}^{-\beta\epsilon_{\mathbf{k}}}\right)^{n_{\mathbf{k}}} \end{aligned} \tag{D.57}$$

where the symbol $\sum'_{\{n_{\mathbf{k}}\}}$ is used to denote that the summation over the occupation numbers $\{n_{\mathbf{k}}\}$ must be done with the restriction $\sum_{\mathbf{k}} n_{\mathbf{k}} = N$. However, since we also have an additional summation over all the values of N, we can simply substitute the restricted summation over $\{n_{\mathbf{k}}\}$ with an unrestricted summation and omit the summation over the values of N, thus obtaining

$$\begin{aligned} Z(z, \beta, \Omega) &= \sum_{\{n_{\mathbf{k}}\}} \prod_{\mathbf{k}} (z\mathrm{e}^{-\beta\epsilon_{\mathbf{k}}})^{n_{\mathbf{k}}} = \sum_{n_{\mathbf{k}_1}} \sum_{n_{\mathbf{k}_2}} \cdots (z\mathrm{e}^{-\beta\epsilon_{\mathbf{k}_1}})^{n_{\mathbf{k}_1}} (z\mathrm{e}^{-\beta\epsilon_{\mathbf{k}_2}})^{n_{\mathbf{k}_2}} \cdots \\ &= \prod_{\mathbf{k}} \sum_{n_{\mathbf{k}}} (z\mathrm{e}^{-\beta\epsilon_{\mathbf{k}}})^{n_{\mathbf{k}}} \end{aligned} \tag{D.58}$$

The individual occupation numbers of indistinguishable particles can be

$$\begin{aligned} n_{\mathbf{k}} &= 0, 1, 2, \ldots \quad \text{for bosons} \\ n_{\mathbf{k}} &= 0, 1, \ldots, (2s+1) \ \text{for fermions with spin } s \end{aligned}$$

For simplicity we only discuss below the case of spin $s = 0$ particles, the extension to the general case being straightforward. In this case, for spinless fermions $n_{\mathbf{k}} = 0, 1$ and the grand partition function becomes

$$Z^{(F)}(z, \beta, \Omega) = \prod_{\mathbf{k}} (1 + z\mathrm{e}^{-\beta\epsilon_{\mathbf{k}}}) \tag{D.59}$$

and the logarithm of this expression gives the equation of state,

$$\frac{P\Omega}{k_{\mathrm{B}}T} = \ln Z^{(F)}(z, \beta, \Omega) = \sum_{\mathbf{k}} \ln(1 + z\mathrm{e}^{-\beta\epsilon_{\mathbf{k}}}) \tag{D.60}$$

while the average occupation number of state $\mathbf{k}$ becomes

$$\bar{n}_{\mathbf{k}}^{(F)} = -\frac{1}{\beta} \frac{\partial}{\partial \epsilon_{\mathbf{k}}} \ln Z^{(F)}(z, \beta, \Omega) = \frac{1}{z^{-1}\mathrm{e}^{\beta\epsilon_{\mathbf{k}}} + 1} \tag{D.61}$$

which, with $z = \exp(\beta\mu_{FD})$, is identical to the expression derived earlier, Eq. (D.7), through the argument based on the most probable distribution. The average total number of particles in the system, since we are dealing with the grand canonical ensemble which allows fluctuations in this number, is given by

$$\bar{N}^{(F)} = z\frac{\partial}{\partial z} \ln Z^{(F)}(z, \beta, \Omega) = \sum_{\mathbf{k}} \frac{z\mathrm{e}^{-\beta\epsilon_{\mathbf{k}}}}{1 + z\mathrm{e}^{-\beta\epsilon_{\mathbf{k}}}} = \sum_{\mathbf{k}} \bar{n}_{\mathbf{k}}^{(F)} \tag{D.62}$$

In terms of these average occupation numbers, the entropy of a gas of spinless fermions is found to be (see Problem 1)

$$S = k_B \sum_{\mathbf{k}} [\bar{n}_{\mathbf{k}} \ln \bar{n}_{\mathbf{k}} + (1 - \bar{n}_{\mathbf{k}}) \ln(1 - \bar{n}_{\mathbf{k}})] \tag{D.63}$$

The grand partition function for bosons contains a geometric sum which can be summed to give

$$Z^{(B)}(z, \beta, \Omega) = \prod_{\mathbf{k}} \frac{1}{1 - ze^{-\beta\epsilon_{\mathbf{k}}}} \tag{D.64}$$

and the logarithm of this expression gives the equation of state,

$$\frac{P\Omega}{k_B T} = \ln Z^{(B)}(z, \beta, \Omega) = -\sum_{\mathbf{k}} \ln(1 - ze^{-\beta\epsilon_{\mathbf{k}}}) \tag{D.65}$$

while the average occupation number of state $\mathbf{k}$ becomes

$$\bar{n}_{\mathbf{k}}^{(B)} = -\frac{1}{\beta}\frac{\partial}{\partial \epsilon_{\mathbf{k}}} \ln Z^{(B)}(z, \beta, \Omega) = \frac{1}{z^{-1}e^{\beta\epsilon_{\mathbf{k}}} - 1} \tag{D.66}$$

which, with $z = \exp(\beta\mu_{BE})$, is identical to the expression derived earlier, Eq. (D.13). The average total number of particles in the system is

$$\bar{N}^{(B)} = z\frac{\partial}{\partial z} \ln Z^{(B)}(z, \beta, \Omega) = \sum_{\mathbf{k}} \frac{ze^{-\beta\epsilon_{\mathbf{k}}}}{1 - ze^{-\beta\epsilon_{\mathbf{k}}}} = \sum_{\mathbf{k}} \bar{n}_{\mathbf{k}}^{(B)} \tag{D.67}$$

We elaborate on the nature of the equation of state for the two quantum gases. We use the usual substitution $(1/\Omega)\sum_{\mathbf{k}} \to \int d\mathbf{k}/(2\pi)^3$ and define the volume per particle $\omega = \Omega/\bar{N}$ to obtain for the fermion gas from Eqs. (D.60) and (D.62):

$$\frac{P}{k_B T} = \int \ln\left(z \exp\left[-\beta\frac{\hbar^2\mathbf{k}^2}{2m}\right] + 1\right)\frac{d\mathbf{k}}{(2\pi)^3} = \frac{1}{\lambda^3}h_5(z) \tag{D.68}$$

$$h_5(z) \equiv \frac{4}{\sqrt{\pi}}\int_0^\infty u^2 \ln(1 + ze^{-u^2})du$$

$$\frac{1}{\omega} = \int \left(z^{-1}\exp\left[\beta\frac{\hbar^2\mathbf{k}^2}{2m}\right] + 1\right)^{-1}\frac{d\mathbf{k}}{(2\pi)^3} = \frac{1}{\lambda^3}h_3(z) \tag{D.69}$$

$$h_3(z) \equiv \frac{4}{\sqrt{\pi}}\int_0^\infty u^2 \frac{ze^{-u^2}}{1 + ze^{-u^2}}du$$

where we have introduced the variable

$$u = \left(\frac{\beta\hbar^2}{2m}\right)^{1/2}|\mathbf{k}| = \frac{1}{\sqrt{4\pi}}\lambda|\mathbf{k}|$$

with λ the thermal wavelength. We have also defined the integrals over $\mathbf{k}$ in the previous equations as the functions h_5 and h_3 of z, which is the only remaining parameter; the subscripts of these two functions come from their series expansions in z:

$$h_n(z) = \sum_{m=1}^{\infty} \frac{(-1)^{m+1} z^m}{m^{n/2}}, \quad n = 3, 5 \tag{D.70}$$

From these expressions it is possible to extract the behavior of this system in the classical limit and in the extreme quantum limit (see Problem 2).

For the free-particle Bose gas, by similar calculations we obtain from Eqs. (D.65) and (D.67):

$$\frac{P}{k_B T} = \int \ln\left(z \exp\left[-\beta\frac{\hbar^2\mathbf{k}^2}{2m}\right] - 1\right)\frac{d\mathbf{k}}{(2\pi)^3}$$

$$= \frac{1}{\lambda^3}g_5(z) - \frac{\ln(1-z)}{\Omega} \tag{D.71}$$

$$g_5(z) \equiv -\frac{4}{\sqrt{\pi}}\int_0^\infty u^2 \ln(1 - z e^{-u^2}) du$$

$$\frac{1}{\omega} = \int \left(z^{-1}\exp\left[\beta\frac{\hbar^2\mathbf{k}^2}{2m}\right] - 1\right)^{-1}\frac{d\mathbf{k}}{(2\pi)^3}$$

$$= \frac{1}{\lambda^3}g_3(z) + \frac{z}{\Omega(1-z)} \tag{D.72}$$

$$g_3(z) \equiv -\frac{4}{\sqrt{\pi}}\int_0^\infty u^2 \frac{z e^{-u^2}}{1 - z e^{-u^2}} du$$

where the functions g_5 and g_3 of z are given as series expansions by

$$g_n(z) = \sum_{m=1}^{\infty}\frac{z^m}{m^{n/2}}, \quad n = 3, 5 \tag{D.73}$$

Notice that there is an important difference between these expressions and the ones for the free-particle fermion gas: due to the nature of the Bose average occupation number, the argument of the logarithm in the integrand of Eq. (D.71) or the denominator in the integrand of Eq. (D.72) can become zero for $\mathbf{k} = 0$, so we explicitly included these terms in the final result; the factors $1/\Omega$ that accompany these terms come from the original summation over all values of $\mathbf{k}$, out of which we have separated the $\mathbf{k} = 0$ terms. These factors are actually crucial in describing the behavior of the ideal Bose gas at low temperature, as they give rise to a very interesting phenomenon called Bose–Einstein condensation. To see this, first notice that the average occupation numbers must be non-negative quantities:

$$\bar{n}_{\mathbf{k}} = \frac{z\exp[-\beta\epsilon_{\mathbf{k}}]}{1 - z\exp[-\beta\epsilon_{\mathbf{k}}]} \geq 0 \Longrightarrow 1 \geq z\exp[-\beta\epsilon_{\mathbf{k}}]$$

and this must be true for *every* value of $\mathbf{k}$ for a given z. But since $\epsilon_{\mathbf{k}} = \hbar^2\mathbf{k}^2/2m$ for any finite temperature, the factors $\exp(-\beta\epsilon_{\mathbf{k}}) \leq 1$ with the equality applying to the $\mathbf{k} = 0$ state which shows that z must lie within the limits

$$0 \leq z \leq 1$$

Next consider the average occupation number of the state with $\mathbf{k} = 0$:

$$\bar{n}_0 = \frac{z}{1-z}$$

With this expression, Eq. (D.72) takes the form

$$\frac{1}{\omega} = \frac{1}{\lambda^3}g_3(z) + \frac{1}{\Omega}\bar{n}_0 \Longrightarrow \frac{\bar{n}_0}{\Omega} = \frac{1}{\omega} - \frac{1}{\lambda^3}g_3(z)$$

but from the series expansion of $g_3(z)$ we have

$$z \leq 1 \Longrightarrow g_3(z) \leq g_3(1)$$

which implies that

$$\frac{\bar{n}_0}{\Omega} \geq \frac{1}{\omega} - \frac{1}{\lambda^3} g_3(1)$$

The value of the function $g_3(z)$ for $z = 1$ turns out to be 2.612... so that if the right-hand side of the above inequality happens to be finite and positive it implies that $\bar{n}_0/\Omega$ is a finite and positive quantity. But Ω is a macroscopic quantity, while $\bar{n}_0$ refers to the average occupation of a microscopic state (that with $\mathbf{k} = 0$) and is usually of order unity. In the thermodynamic limit all macroscopic quantities such as Ω and N are taken to be infinite but their ratios, like the density $n = N/\Omega$, may be finite; accordingly, we would expect that

$$\frac{\bar{n}_0}{\Omega} \to 0 \quad \text{for} \quad \Omega \to \infty$$

The fact that for temperature and density conditions such that

$$\frac{1}{\omega} - \frac{1}{\lambda^3} g_3(1) \geq 0 \tag{D.74}$$

the quantity $\bar{n}_0/\Omega$ is finite and positive means that $\bar{n}_0$ is itself a macroscopic quantity, that is, a finite fraction of the total number of particles N. Thus, when these conditions are satisfied a macroscopic number of particles has condensed to the $\mathbf{k} = 0$ state, hence the term Bose–Einstein condensation (BEC). The volume per particle (inverse density) $\omega_c = 1/n_c$ and temperature T_c at which this begins to occur are called the critical values. The critical value of the volume is a function of the temperature, and the critical temperature is a function of the volume, as Eq. (D.74) implies:

$$\omega_c = \frac{1}{g_3(1)} \lambda^3 = \left[\left(\frac{2\pi\hbar^2}{mk_\mathrm{B}} \right)^{3/2} g_3(1) \right] \frac{1}{T^{3/2}} \Longrightarrow T_c = \left(\frac{2\pi\hbar^2}{mk_\mathrm{B}\,[g_3(1)]^{2/3}} \right) \frac{1}{\omega^{2/3}} \tag{D.75}$$

BEC occurs for $T \leq T_c$ at fixed volume per particle or for $\omega \leq \omega_c$ at fixed temperature. We can use these values to express $\bar{n}_0/N$, the condensate fraction, as a function of temperature or inverse density, through the following argument: in general

$$\frac{\bar{n}_0}{N} = \frac{\bar{n}_0}{\Omega}\frac{\Omega}{N} = \left(\frac{1}{\omega} - \frac{1}{\lambda^3} g_3(z) \right) \omega = 1 - \frac{\omega}{\lambda^3} g_3(z)$$

and for $z = 1$ the same relation must hold, therefore

$$\frac{\bar{n}_0}{N} = 1 - \frac{\omega_c}{\lambda^3} g_3(1) = 1 - \left(\frac{T}{T_c} \right)^{3/2} = 1 - \frac{\omega}{\omega_c} \tag{D.76}$$

For $T > T_c$ all states $\mathbf{k}$ are occupied but $\bar{n}_\mathbf{k}/N$ is infinitesimal; for $T < T_c$ the same is true for all $\mathbf{k} \neq 0$, but $\bar{n}_0/N$ is finite and for $T = 0$ the entire system has condensed to the $\mathbf{k} = 0$ state. Similar statements apply for the occupation of states for $\omega > \omega_c$ or $\omega < \omega_c$. Below the transition point (either in T or ω) the value of z is fixed at unity, or the occupation of the $\mathbf{k} = 0$ state would not be macroscopic.

The BEC transition in the ideal Bose gas has the characteristics of a first order phase transition. To show this, we first examine the equation of state for the Bose gas above and below the transition. Above the transition, $z < 1$ and the term $\ln(1 - z)$ in the equation of state, Eq. (D.71), is finite, so when it is divided by the macroscopic variable Ω it vanishes. Therefore the equation of state above the transition reads

$$\frac{P}{k_B T} = \frac{1}{\lambda^3} g_5(z), \quad \text{for } \omega \geq \omega_c$$

Below the transition we have shown that the average occupation of the $\mathbf{k} = 0$ state divided by Ω must be a finite quantity, which implies that

$$\frac{\bar{n}_0}{\Omega} = \frac{z}{(1-z)\Omega} : \text{finite} \implies \frac{z}{1-z} \sim \Omega$$

But we have also found that below the transition $z = 1$, so taking logarithms of both sides in the last equation we find

$$\ln(1-z) \sim -\ln(\Omega) \implies \frac{\ln(1-z)}{\Omega} \sim -\frac{\ln(\Omega)}{\Omega} \to 0 \text{ for } \Omega \to \infty$$

and therefore the equation of state below the transition reads

$$\frac{P}{k_B T} = \frac{1}{\lambda^3} g_5(1), \quad \text{for } \omega < \omega_c$$

If we consider the transition as taking place at constant temperature by decreasing the value of ω, then below ω_c the transition pressure is given by

$$P_t(T) = \frac{k_B T}{\lambda^3} g_5(1) = \left(\frac{m k_B}{2\pi\hbar^2}\right)^{3/2} k_B T^{5/2} g_5(1)$$

which is only a function of the temperature T. From this expression we obtain

$$\frac{dP_t(T)}{dT} = \frac{5}{2}\frac{k_B g_5(1)}{\lambda^3} = \frac{1}{T\omega_c}\left(k_B T \frac{5 g_5(1)}{2 g_3(1)}\right)$$

where we have used Eq. (D.75) to rewrite the last expression in terms of ω_c and the ratio of the functions g_5 and g_3 both evaluated at $z = 1$. We have also included a factor of T in the denominator and a compensating factor in the numerator, in order to produce an expression that has the form of the Clausius–Clapeyron equation, Eq. (C.28) with the latent heat and the change in volume given by

$$L = \frac{5 k_B T g_5(1)}{2 g_3(1)}, \qquad \Delta\omega = \omega_c$$

The presence of a latent heat means that the transition is first order (see also Problem 3).

D.3.3 Spins in an external magnetic field

The simplest case of quantum spins in an external magnetic field is that of free fermions. The response of this system to a magnetic field is called Pauli paramagnetism. Another standard model for physical systems which can respond to an external magnetic field is that of spins on a lattice. The value of the spin depends on the constituents of the physical system. The simplest models deal with spins that can point either in the direction along the external field (they are aligned with or parallel to the field) or in the opposite direction (anti-aligned or antiparallel). In terms of a microscopic description of the system, this corresponds to spin-1/2 particles with their z component taken along the direction of the field, and having as possible values $s_z = +1/2$ (spins aligned with the field) or $s_z = -1/2$

(spins pointing in the direction opposite to the field). The interaction between the spins and the field can be dominant, with spin–spin interactions vanishingly small, or the interaction among the spins can be dominant, with the spin–external field interaction of minor importance. We discuss these two extreme cases, and give examples of their physical realization, following the discussion of Pauli paramagnetism.

Free fermions in a magnetic field: Pauli paramagnetism

We take the magnetic moment of the fermions to be m_0, hence their energy in an external magnetic field H will be given by

$$\epsilon_{\mathbf{k}\uparrow} = \frac{\hbar^2\mathbf{k}^2}{2m} - m_0 H, \quad \epsilon_{\mathbf{k}\downarrow} = \frac{\hbar^2\mathbf{k}^2}{2m} + m_0 H$$

where we have assigned the lowest energy to the spin pointing along the direction of the field (spin $\uparrow$) and the highest energy to the spin pointing against the direction of the field (spin $\downarrow$). The occupation of state with wave-vector $\mathbf{k}$ and spin $\uparrow$ will be denoted by $n_{\mathbf{k}\uparrow}$, and the total number of particles with spin up will be denoted by $N_\uparrow$; $n_{\mathbf{k}\downarrow}$ and $N_\downarrow$ are the corresponding quantities for spin-down particles. The total energy of the system in a configuration with $N_\uparrow$ spins pointing up and $N_\downarrow$ spins pointing down is given by

$$E(N_\uparrow, N_\downarrow) = \sum_{\mathbf{k}} (n_{\mathbf{k}\uparrow} + n_{\mathbf{k}\downarrow}) \frac{\hbar^2\mathbf{k}^2}{2m} - m_0 H(N_\uparrow - N_\downarrow)$$

$$\sum_{\mathbf{k}} n_{\mathbf{k}\uparrow} = N_\uparrow, \quad \sum_{\mathbf{k}} n_{\mathbf{k}\downarrow} = N_\downarrow, \quad \sum_{\mathbf{k}} (n_{\mathbf{k}\uparrow} + n_{\mathbf{k}\downarrow}) = N$$

The canonical partition function will be given as the sum over all configurations consistent with the above expressions, which is equivalent to summing over all values of $N_\uparrow$ from zero to N, with $N_\downarrow = N - N_\uparrow$:

$$Z_N(H, T) = \sum_{N_\uparrow=0}^{N} \mathrm{e}^{\beta m_0 H(2N_\uparrow - N)} \left[\sum_{\{n_{\mathbf{k}\uparrow}\}} \mathrm{e}^{-\beta \sum_{\mathbf{k}} n_{\mathbf{k}\uparrow} \hbar^2\mathbf{k}^2/2m} \sum_{\{n_{\mathbf{k}\downarrow}\}} \mathrm{e}^{-\beta \sum_{\mathbf{k}} n_{\mathbf{k}\downarrow} \hbar^2\mathbf{k}^2/2m} \right]$$

We can take advantage of the definition of the partition function of spinless fermions,

$$Z_N^{(0)} = \sum_{\{n_{\mathbf{k}}\}} \mathrm{e}^{-\beta \sum_{\mathbf{k}} n_{\mathbf{k}} \hbar^2\mathbf{k}^2/2m}, \quad \sum_{\mathbf{k}} n_{\mathbf{k}} = N$$

to rewrite the partition function $Z_N(H, T)$ as

$$Z_N(H, T) = \mathrm{e}^{-\beta m_0 H N} \sum_{N_\uparrow=0}^{N} \mathrm{e}^{2\beta m_0 H N_\uparrow} Z_{N_\uparrow}^{(0)} Z_{N-N_\uparrow}^{(0)}$$

and from this we obtain for the expression $(\ln Z_N)/N$ which appears in thermodynamic averages,

$$\frac{1}{N} Z_N(H, T) = -\beta m_0 H + \frac{1}{N} \ln \left[\sum_{N_\uparrow=0}^{N} \mathrm{e}^{2\beta m_0 H N_\uparrow + \ln\left(Z_{N_\uparrow}^{(0)}\right) + \ln\left(Z_{N-N_\uparrow}^{(0)}\right)} \right]$$

We will approximate the above summation over $N_\uparrow$ by the largest term in it, with the usual assumption that in the thermodynamic limit this term dominates; to find the value of $N_\uparrow$

that corresponds to this term, we take the derivative of the exponent with respect to $N_\uparrow$ and set it equal to zero, obtaining

$$2m_0H + \frac{1}{\beta}\frac{\partial}{\partial N_\uparrow}\ln\left(Z^{(0)}_{N_\uparrow}\right) - \frac{1}{\beta}\frac{\partial}{\partial(N - N_\uparrow)}\ln\left(Z^{(0)}_{N-N_\uparrow}\right) = 0$$

where in the second partial derivative we have changed variables from $N_\uparrow$ to $N - N_\uparrow$. From Eqs. (D.39), we can identify the two partial derivatives in the above expression with the negative chemical potentials, thus obtaining the relation

$$\mu^{(0)}(\bar{N}_\uparrow) - \mu^{(0)}(N - \bar{N}_\uparrow) = 2m_0H$$

where we have denoted by $\bar{N}_\uparrow$ the value corresponding to the largest term in the partition function. We define the average magnetization per particle m as

$$m(T) = \frac{M}{N} = \frac{m_0(\bar{N}_\uparrow - \bar{N}_\downarrow)}{N}$$

in terms of which we can express $N_\uparrow$ and $N_\downarrow$ as

$$\bar{N}_\uparrow = \frac{1 + m/m_0}{2}N, \quad \bar{N}_\downarrow = \frac{1 - m/m_0}{2}N$$

With these expressions, the relation we derived above for the chemical potentials becomes

$$\mu^{(0)}\left(\frac{1 + m/m_0}{2}N\right) - \mu^{(0)}\left(\frac{1 - m/m_0}{2}N\right) = 2m_0H$$

For small external field, we expect $N_\uparrow$ and $N_\downarrow$ to be close to $N/2$ and the average magnetization m to be very small, therefore we can use Taylor expansions around $N/2$ to obtain

$$\left[\frac{\partial\mu^{(0)}(qN)}{\partial q}\right]_{q=1/2}\left(\frac{m}{2m_0}\right) - \left[\frac{\partial\mu^{(0)}(qN)}{\partial q}\right]_{q=1/2}\left(-\frac{m}{2m_0}\right) = 2m_0H$$

$$\Longrightarrow m(T) = 2m_0^2H\left[\frac{\partial\mu^{(0)}(qN)}{\partial q}\right]^{-1}_{q=1/2}$$

With this expression, the magnetic susceptibility per particle χ takes the form

$$\chi(T) \equiv \frac{\partial m}{\partial H} = 2m_0^2\left[\frac{\partial\mu^{(0)}(qN)}{\partial q}\right]^{-1}_{q=1/2}$$

while the magnetic susceptibility per volume is simply $\chi N/\Omega = n\chi$.

In the low-temperature limit we use Eq. (D.69) and the first term in the corresponding expansion of $h_3(z)$,

$$h_3(z) = \frac{4}{3\sqrt{\pi}}\left[(\ln z)^{3/2} + \frac{\pi^2}{8}(\ln z)^{-1/2} + \cdots\right]$$

as well as the relation between the fugacity z and the chemical potential $\mu^{(0)}$, that is, $z = \exp(\beta\mu^{(0)})$ to obtain

$$\mu^{(0)}(N) = k_BT\left(\frac{3\sqrt{\pi}}{4}\frac{N}{\Omega}\lambda^3\right)^{2/3} \Longrightarrow \left[\frac{\partial\mu^{(0)}(qN)}{\partial q}\right]_{q=1/2} = \frac{4}{3}\frac{h^2}{2m}\left(\frac{3N}{8\pi\Omega}\right)^{2/3}$$

with λ the thermal wavelength. For spin-1/2 fermions, the Fermi energy is given by

$$\epsilon_{\mathrm{F}} = \frac{h^2}{2m}\left(\frac{3N}{8\pi\Omega}\right)^{2/3}$$

which, together with the previous result, gives for the magnetic susceptibility per particle at zero temperature

$$\chi(T=0) = \frac{3m_0^2}{2\epsilon_{\mathrm{F}}}$$

Keeping the first two terms in the low-temperature expansion of $h_3(z)$ we obtain the next term in the susceptibility per particle:

$$\chi(T) = \frac{3m_0^2}{2\epsilon_{\mathrm{F}}}\left[1 - \frac{\pi^2}{12}\left(\frac{k_{\mathrm{B}}T}{\epsilon_{\mathrm{F}}}\right)^2\right], \quad \text{low-}T\text{ limit}$$

In the high-temperature limit we use the corresponding expansion of $h_3(z)$, which is

$$h_3(z) = z - \frac{z^2}{2^{3/2}} + \cdots$$

With the lowest term only, we obtain

$$\mu^{(0)}(N) = k_{\mathrm{B}}T \ln\left(\frac{N}{\Omega}\lambda^3\right) \Longrightarrow \left[\frac{\partial\mu^{(0)}(qN)}{\partial q}\right]_{q=1/2} = 2k_{\mathrm{B}}T$$

which gives for the susceptibility per particle

$$\chi(T\to\infty) = \frac{m_0^2}{k_{\mathrm{B}}T}$$

Keeping the first two terms in the high-temperature expansion of $h_3(z)$, we obtain the next term in the susceptibility per particle:

$$\chi(T) = \frac{m_0^2}{k_{\mathrm{B}}T}\left[1 - \left(\frac{N\lambda^3}{\Omega 2^{5/2}}\right)\right], \quad \text{high-}T\text{ limit}$$

Non-interacting spins on a lattice: negative temperature

We consider the following model: a number N of magnetic dipoles are subjected to an external magnetic field H. The dipoles have magnetic moment m_0 and can be oriented either along or against the magnetic field. Moreover, we assume that the dipoles do not interact among themselves and are distinguishable, by virtue of being situated at fixed positions in space, for example at the sites of a crystalline lattice. Since each dipole can be oriented either parallel or antiparallel to the external field, its energy can have only two values. We assign to the lower energy of the two states (which by convention corresponds to the dipole pointing along the external field), the value $-\epsilon = -m_0H$, and to the higher energy of the two states we assign the value $\epsilon = m_0H$.

We assume that in a given configuration of the system there are n dipoles oriented along the direction of the field, and therefore $N - n$ dipoles are pointing against the direction of the field. The total internal energy of the system $\mathcal{E}$ and its magnetization $\mathcal{M}$ are given by

$$\begin{aligned} \mathcal{E} &= -n\epsilon + (N-n)\epsilon = (N-2n)m_0H \\ \mathcal{M} &= nm_0 - (N-n)m_0 = -(N-2n)m_0 \end{aligned} \tag{D.77}$$

with the direction of the magnetic field H taken to define the positive direction of magnetization (for instance, the $+\hat{\mathbf{z}}$ axis). The partition function for this system in the canonical ensemble is given by

$$Z_N = \sum_{n,N-n} \frac{N!}{n!(N-n)!} \mathrm{e}^{-\beta(-n\epsilon+(N-n)\epsilon)}$$
$$= \sum_{n=0}^{N} \frac{(N-n+1)(N-n+2)\cdots N}{n!} \left(\mathrm{e}^{-\beta\epsilon}\right)^n \left(\mathrm{e}^{\beta\epsilon}\right)^{N-n} \quad \text{(D.78)}$$

where the first sum implies summation over all the possible values of n and $N-n$, which is equivalent to summing over all values of n from zero to N. In the last expression we recognize the familiar binomial expansion (see Appendix G, Eq. (G.39)), with the help of which we obtain

$$Z_N = (\mathrm{e}^{\beta\epsilon} + \mathrm{e}^{-\beta\epsilon})^N = (2\cosh(\beta\epsilon))^N \quad \text{(D.79)}$$

Having calculated the partition function, we can use it to obtain the thermodynamics of the system with the standard expressions we have derived in the context of ensemble theory. We first calculate the free energy F from the usual definition:

$$F = -k_{\mathrm{B}}T \ln Z_N = -Nk_{\mathrm{B}}T \ln(2\cosh(\beta\epsilon)) \quad \text{(D.80)}$$

and from that the entropy through

$$S = -\left(\frac{\partial F}{\partial T}\right)_H = Nk_{\mathrm{B}}\left[\ln(2\cosh(\beta\epsilon)) - \beta\epsilon\tanh(\beta\epsilon)\right] \quad \text{(D.81)}$$

We can use the calculated free energy and entropy to obtain the average internal energy E through

$$E = F + TS = -N\epsilon\tanh(\beta\epsilon) \quad \text{(D.82)}$$

The same result is obtained by calculating the average of $\mathcal{E}$ with the use of the partition function:

$$E = \overline{\mathcal{E}} = \frac{1}{Z_N}\sum_{n,N-n} \frac{N!}{n!(N-n)!}(-n\epsilon + (N-n)\epsilon)\mathrm{e}^{-\beta(-n\epsilon+(N-n)\epsilon)}$$
$$= -\left(\frac{\partial}{\partial\beta}\right)_H \ln Z_N = -N\epsilon\tanh(\beta\epsilon) \quad \text{(D.83)}$$

By the same method we can calculate the average magnetization M as the average of $\mathcal{M}$:

$$M = \overline{\mathcal{M}} = \frac{1}{Z_N}\sum_{n,N-n} \frac{N!}{n!(N-n)!}(nm_0 - (N-n)m_0)\mathrm{e}^{-\beta(-nm_0H+(N-n)m_0H)}$$
$$= \frac{1}{\beta}\left(\frac{\partial}{\partial H}\right)_T \ln Z_N = Nm_0\tanh(\beta\epsilon) \quad \text{(D.84)}$$

We notice that the last equality can also be written in terms of a partial derivative of the free energy, defined in Eq. (D.80), as

$$M = \frac{1}{\beta}\left(\frac{\partial \ln Z_N}{\partial H}\right)_T = k_{\mathrm{B}}T\left(\frac{\partial \ln Z_N}{\partial H}\right)_T = -\left(\frac{\partial F}{\partial H}\right)_T \quad \text{(D.85)}$$

We can now determine the relation between average magnetization and internal energy, using $\epsilon = m_0 H$:

$$E = -N\epsilon \tanh(\beta\epsilon) = -N m_0 H \tanh(\beta\epsilon) = -HM \tag{D.86}$$

which is what we would expect because of the relation between the internal energy and magnetization $\mathcal{E} = -H\mathcal{M}$, as is evident from Eq. (D.77).

If we wished to complete the description of the system in terms of thermodynamic potentials and their derivatives, then the above results imply the following relations:

$$E = -MH \Longrightarrow \mathrm{d}E = -M\mathrm{d}H - H\mathrm{d}M$$

$$F = E - TS \Longrightarrow \mathrm{d}F = \mathrm{d}E - T\mathrm{d}S - S\mathrm{d}T = -M\mathrm{d}H - H\mathrm{d}M - T\mathrm{d}S - S\mathrm{d}T$$

We have also shown that

$$M = -\left(\frac{\partial F}{\partial H}\right)_T, \quad S = -\left(\frac{\partial F}{\partial T}\right)_H$$

which together require that

$$\mathrm{d}F = -M\mathrm{d}H - S\mathrm{d}T \Longrightarrow -H\mathrm{d}M - T\mathrm{d}S = 0 \Longrightarrow T\mathrm{d}S = -H\mathrm{d}M$$

This last result tells us that the usual expression for the change in internal energy as the difference between the heat absorbed and the work done by the system must be expressed as

$$\mathrm{d}E = T\mathrm{d}S - M\mathrm{d}H$$

which compels us to identify H as the equivalent of the volume of the system and M as the equivalent of the pressure, so that $M\mathrm{d}H$ becomes the equivalent of the mechanical work $P\mathrm{d}\Omega$ done by the system. This equivalence is counterintuitive, because the magnetic field is the intensive variable and the magnetization is the extensive variable, so the reverse equivalence (H with P and M with Ω) would seem more natural. What is peculiar about this system is that its internal energy is directly related to the work done on the system $E = -MH$, while in the standard example of the ideal gas we have seen that the internal energy is a function of the temperature alone, $E = C_\Omega T$, and the work done on the system is an independent term $P\Omega$. Although it is counterintuitive, upon some reflection the correspondence between the magnetic field and volume and between the magnetization and pressure appears more reasonable: the magnetic field is the actual constraint that we impose on the system, just like confining a gas of atoms within the volume of a container; the magnetization is the response of the system to the external constraint, just like the exertion of pressure on the walls of the container is the response of the gas to its confinement within the volume of the container. An additional clue that H should be identified with the volume in this case is the fact that the partition function depends only on N and H, the two variables that we consider fixed in the canonical ensemble; in general, the canonical ensemble consists of fixing the number of particles and volume of the system. Taking these analogies one step further, we might wish to define the Gibbs free energy for this system as

$$G = F + HM = E - TS + HM = -TS$$

which in turn gives

$$\mathrm{d}G = -S\mathrm{d}T - T\mathrm{d}S = -S\mathrm{d}T + H\mathrm{d}M$$

and from this we can obtain the entropy and the magnetic field as partial derivatives of G with respect to T and M

$$S = \left(\frac{\partial G}{\partial T}\right)_M, \quad H = \left(\frac{\partial G}{\partial M}\right)_T$$

which completes the description of the system in terms of thermodynamic potentials, fields and variables. As a final step, we can calculate the specific heat at constant magnetic field which, as we argued above, should correspond to the specific heat at constant volume in the general case:

$$C_H = \left(\frac{\partial E}{\partial T}\right)_H = \frac{\partial \beta}{\partial T}\left(\frac{\partial E}{\partial T}\right)_H = N k_{\mathrm{B}} \beta^2 \epsilon^2 \frac{1}{\cosh^2(\beta\epsilon)}$$

In order to discuss the behavior of this system in more detail, it is convenient to define the variable

$$u = \frac{E}{N\epsilon}$$

and express the entropy S and the inverse temperature multiplied by the energy per dipole, that is, $\beta\epsilon$, as functions of u. We find from Eqs. (D.81) and (D.82)

$$\beta\epsilon = \tanh^{-1}(-u) = \frac{1}{2}\ln\left(\frac{1-u}{1+u}\right)$$

$$\frac{S}{N k_{\mathrm{B}}} = \beta\epsilon u + \ln 2 - \frac{1}{2}\ln\left(1-u^2\right)$$

$$= \ln 2 - \frac{1}{2}\left[(1-u)\ln(1-u) + (1+u)\ln(1+u)\right]$$

It is obvious from its definition, and from the fact that the lowest and highest possible values for the internal energy are $-N\epsilon$ and $N\epsilon$, that u lies in the interval $u \in [-1, 1]$. The plots of $\beta\epsilon$ and S/Nk_{B} as functions of u are given in Fig. D.3. These plots suggest the intriguing possibility that in this system we might observe *negative* temperatures, while the entropy has the unusual behavior that it rises and then falls with increasing u, taking

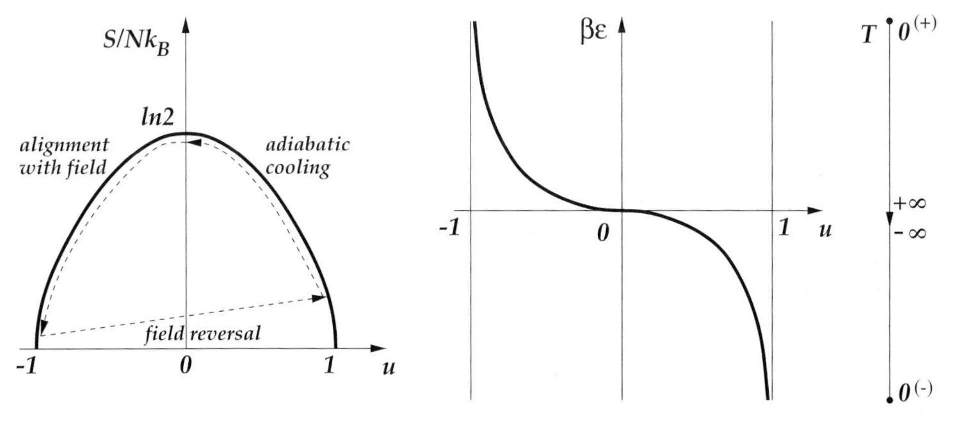

Figure D.3. The entropy per dipole S/Nk_{B} and the inverse temperature $\beta\epsilon$ as functions of $u = E/N\epsilon$ for the model of dipoles in an external magnetic field H.

its maximum value for $u = 0$. Let us elaborate on this, at first sight perplexing, behavior. We consider the behavior of the entropy first. Near the extreme values of u we have

$$u \to -1^{(+)} \Longrightarrow S = 0^{(+)}, \quad u \to +1^{(-)} \Longrightarrow S = 0^{(+)}$$

where the $(-)$ and $(+)$ superscripts denote approach from below or from above, respectively. This is reassuring, because it is at least consistent with the third law of thermodynamics: the entropy goes to zero from above, when the energy of the system reaches its lowest or its highest value. Moreover, in the limit of very high temperature, or very low external field, u must be zero: in either limit, the natural state of the system is to have half of the dipoles oriented along the direction of the field and the other half against it, giving a total energy equal to zero. This is also the state of maximum entropy, as we would expect. Indeed, in this state $n = N/2 \Longrightarrow N - n = N/2$, and the number of configurations associated with it is

$$\frac{N!}{n!(N-n)!} = \frac{N!}{(\frac{N}{2})!(\frac{N}{2})!}$$

which is larger than for any other value of n and $N - n$ different than $N/2$. Using Stirling's formula we find that, in the microcanonical ensemble at fixed energy $E = 0$, the entropy of this state is

$$S = k_B \ln\left[\frac{N!}{(\frac{N}{2})!(\frac{N}{2})!}\right] \approx k_B\left[N \ln N - N - 2\left(\frac{N}{2}\ln\left(\frac{N}{2}\right) - \frac{N}{2}\right)\right] = Nk_B \ln 2$$

as expected. From this state, which also has zero magnetization, the system can be brought to a state of higher magnetization or lower energy either by increasing the magnetic field or by lowering the temperature. In either case it will be moving from the maximum of the entropy value at $u = 0$ toward lower values of the entropy in the range of $u \in [-1, 0]$, since the value of u is decreasing as the energy of the system decreases. This is consistent with the second law of thermodynamics, with the entropy decreasing because work is being done on the system in order to align the dipoles to the external field.

As far as the temperature is concerned, we have

$$\text{for } u \to -1^{(+)} : \beta\epsilon = \frac{1}{2}\ln\left(\frac{2^{(-)}}{0^{(+)}}\right) \to +\infty \Longrightarrow T \to 0^{(+)}, \quad T > 0$$

$$\text{for } u \to +1^{(-)} : \beta\epsilon = \frac{1}{2}\ln\left(\frac{0^{(+)}}{2^{(-)}}\right) \to -\infty \Longrightarrow T \to 0^{(-)}, \quad T < 0$$

that is, for $-1 \le u \le 0$ the temperature is *positive*, ranging from $0^{(+)}$ to $+\infty$, while for $0 \le u \le +1$ the temperature is *negative*, ranging from $-\infty$ to $0^{(-)}$. The question now is, what is the meaning of a negative temperature? In the microcanonical ensemble picture, the relation between temperature and entropy is given by

$$\frac{1}{T} = \frac{\partial S}{\partial E}$$

If the system is in the state of maximum entropy discussed above, for which $E = 0$, increasing the energy by a small amount will decrease the entropy and will produce a large negative temperature. Indeed, if an infinitesimal excess of dipoles point in the direction against the external magnetic field, which corresponds to an increase in the energy ($\delta E > 0, \quad u > 0$), the number of states available to the system is smaller than in the case of zero energy, and the accompanying infinitesimal change in the entropy ($\delta S < 0$) will yield a temperature $T \to -\infty$. Since this situation involves an increase of the energy, it must be interpreted as corresponding to a temperature *higher* than $T \to +\infty$, which represented the case of maximum entropy and zero energy

$$T \to +\infty: \quad S = Nk_{\mathrm{B}} \ln 2, \quad E = 0$$

As the energy of the system increases further, $u \in [0, 1]$, the temperature will continue to *increase*, from $-\infty$ to $0^{(-)}$. Thus, the proper way to interpret the temperature scale is by considering $T \to 0^{(+)}$ to lie at the beginning of the scale (lowest values), $T \to +\infty$ to lie in the middle of the scale and be immediately followed by $T \to -\infty$, and $T \to 0^{(-)}$ to be at the end of the scale (highest values), as illustrated in Fig. D.3. The reason for this seemingly peculiar scale is that the natural variable is actually the inverse temperature $\beta = 1/k_{\mathrm{B}}T$, which on the same scale is monotonically decreasing: it starts at $+\infty$ and goes to zero from positive values, then continues to decrease by assuming negative values just below zero and going all the way to $-\infty$.

Finally, we note that even though the situation described above may seem strange, it can actually be realized in physical systems. This was demonstrated first by Purcell and Pound for nuclear spins (represented by the dipoles in our model) in an insulator [235], and more recently for nuclear spins in metals [236]. The way experiment works is by first inducing a large magnetization to the system by aligning the nuclear spins with the external field and then suddenly reversing the external field direction, as indicated in Fig. D.3. When the direction of the field is reversed, the system is in the range of negative temperatures and in a state of very high energy and low entropy. The success of the experiment relies on weak interactions between the dipoles so that the sudden reversal of the external field does not change their state; in the simple model discussed above, interactions between the dipoles are neglected altogether. Once the external field has been reversed, the system can be allowed to cool adiabatically, along the negative temperature scale from $0^{(-)}$ to $-\infty$. When the temperature has cooled to $-\infty$ the system is in the zero energy, zero magnetization, maximum entropy state, from which it can be again magnetized by applying an external magnetic field. In all stages of the process, the second law of thermodynamics is obeyed, with entropy increasing when no work is done on the system.

Interacting spins on a lattice: the Ising model

The Ising model is an important paradigm because it captures the physics of several physical systems with short-range interactions. The standard model is a system of spins on a lattice with nearest neighbor interactions. The Ising model is also relevant to the lattice gas model and the binary alloy model. The lattice gas is a system of particles that can be situated at discrete positions only, but exhibit otherwise the behavior of a gas in the sense that they move freely between positions which are not already occupied by other particles; each particle experiences the presence of other particles only if they are nearest neighbors at the discrete sites which they can occupy, which can be taken as points on a lattice. The binary alloy model assumes that there exists a regular lattice whose sites are occupied by one of two types of atoms, and the interactions between atoms at nearest neighbor sites depend on the type of atoms. Both models can be formally mapped onto the Ising model.

The Ising model in its simplest form is described by the hamiltonian

$$\mathcal{H} = -J \sum_{\langle ij \rangle} s_i s_j - H \sum_i s_i \tag{D.87}$$

where J and H are constants, and the s_i's are spins situated at regular lattice sites. The notation $\langle ij \rangle$ implies summation over nearest neighbor pairs only. The constant J, called the coupling constant, represents the interaction between nearest neighbors. If J is positive it tends to make nearest neighbor spins parallel, that is, $s_i s_j = |s_i||s_j|$ because this gives the lowest energy contribution; if J is negative, the lowest energy contribution comes from configurations in which nearest neighbors are antiparallel, that is, $s_i s_j = -|s_i||s_j|$. H represents an external field which applies to all spins on the lattice; the second term in the hamiltonian tends to align spins parallel to the field since this lowers its contribution to the energy. The values of s_i's depend on the spin of the particles: For spin-1/2 particles, $s_i = \pm 1/2$, but a factor of 1/4 can be incorporated in the value of the coupling constant and a factor of 1/2 in the value of the external field, leaving as the relevant values of $s_i = \pm 1$. Similarly, for spin-1 particles, $s_i = \pm 1, 0$. The simple version of the model described above can be extended to considerably more complicated forms. For instance, the coupling constant may depend on the sites it couples, in which case J_{ij} assumes a different value for each pair of indices ij. The range of interactions can be extended beyond nearest neighbors. The external field can be non-homogeneous, that is, it can have different values at different sites, H_i. Finally, the lattice on which the spins are assumed to reside plays an important role: the dimensionality of the lattice and its connectivity (number of nearest neighbors of a site) are essential features of the model.

In the present discussion we concentrate on the simplest version of the model, with positive coupling constant J between nearest neighbors only, and homogeneous external field H. We also consider spins that take values $s_i = \pm 1$, that is, we deal with spin-1/2 particles. Our objective is to determine the stable phases of the system in the thermodynamic limit, that is, when the number of spins $N \to \infty$ and the external field $H \to 0$. In this sense, the physics of the model is dominated by the interactions between nearest neighbors rather than by the presense of the external field; the latter only breaks the symmetry between spins pointing up (along the field) or down (opposite to the field). The phases of the system are characterized by the average value M of the total magnetization $\mathcal{M}$, at a given temperature:

$$M(T) = \overline{\mathcal{M}}, \quad \mathcal{M} = \sum_i s_i$$

From the expression for the free energy of the system at temperature T,

$$F = E - TS = \overline{\mathcal{H}} - TS$$

we can argue that there are two possibilities. First, that the stable phase of the system has average magnetization zero, which we call the disordered phase; this phase is favored by the entropy term. Second, that the stable phase of the system has non-zero average magnetization, which we call the ordered phase; this phase is favored by the internal energy term. The behavior of the system as a function of temperature reflects a competition between these two terms. At low enough temperature we expect the internal energy to win, and the system to be ordered; at zero temperature the obvious stable phase is one with all the spins parallel to each other. At high enough temperature we expect the entropy term to win, and the system to be disordered. There are then two important questions:

(i) is there a transition between the two phases at finite temperature; and
(ii) if there is a transition, what is its character?

If there is a transition at finite temperature, we would expect the magnetization to behave in the manner illustrated in Fig. D.4: at the critical temperature T_c order starts to develop

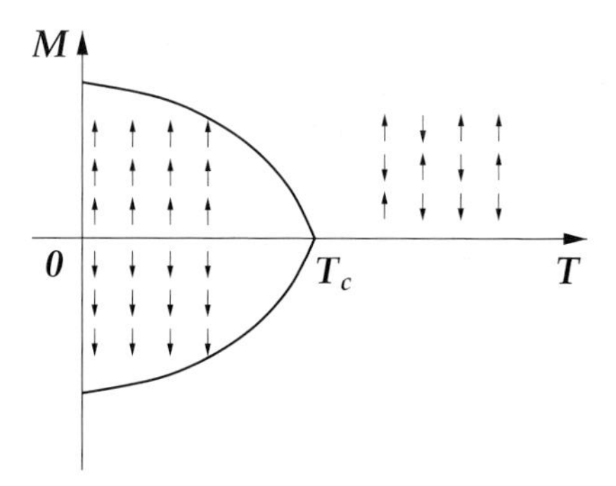

Figure D.4. The average magnetization M as a function of temperature for the Ising model. The sets of up–down arrows illustrate the state of the system in the different regions of temperature.

reflected in the magnetization acquiring a non-zero value, which increases steadily as the temperature is lowered and assumes its highest value at $T = 0$, when all spins have been aligned. The role of the weak external magnetic field, $H \to 0$, is to choose one of the two possible orientations for alignment of the spins, up (positive magnetization) or down (negative magnetization), but otherwise does not affect the behavior of the system.

In one dimension, the nearest neighbor Ising model does not exhibit a phase transition at finite temperature. This can be established by a simple argument: consider the lowest energy configuration in which all spins point in the same direction. We choose some arbitrary point between a pair of spins in the chain of N spins (with $N \to \infty$) and flip all the spins to the right; we call this point a domain wall since it splits the chain into two domains with all spins parallel within each domain. The change in the free energy introduced by the domain wall is given by

$$\Delta F = \Delta E - T\Delta S = 2J - Tk_{\rm B}\ln(N-1)$$

where the first term comes from the change in internal energy and the second comes from the change in the entropy: the entropy of the initial configuration is zero, because there is only one state with all spins pointing up; the entropy of a configuration with all the spins to the right flipped is $k_{\rm B}\ln(N-1)$, because there are $N-1$ choices of where the domain wall can be introduced. It is evident that for any finite temperature, in the thermodynamic limit the entropy term wins. Since this argument can be repeated for each domain, the stable phase of the system cannot be ordered because the introduction of domain walls always lowers the free energy due to the entropy term. For the 1D system it is also possible to find an analytical solution to the problem, which again shows the absence of a phase transition for finite temperature (see Problem 4).

In two or higher dimensions the Ising model does exhibit a phase transition at finite temperature. The existence of a phase transition at finite temperature was first established by Peierls. The problem has been solved analytically by Onsager in two dimensions, but remains unsolved in more than two dimensions. In the latter case one resorts to approximate or numerical methods to obtain a solution.

An instructive approximate method for finding the critical temperature is the so called mean field (also known as the Bragg–Williams) approximation. Since this is a general approach that can be applied to many situations, we discuss it in some detail. The starting point is to consider a system of N spins on a lattice with q nearest neighbors per site, and assume that it has average magnetization M, the result of $N_\uparrow$ spins pointing up and $N_\downarrow$

spins pointing down. Our goal is to analyze this system with the usual tools of statistical mechanics, that is, by calculating the partition function in the canonical ensemble since we are dealing with a fixed number of particles. We introduce the average magnetization per spin $m = M/N$ and express the various quantities that appear in the partition function in terms of m and N:

$$m = \frac{M}{N} = \frac{N_\uparrow - N_\downarrow}{N} \quad \text{and} \quad N_\uparrow + N_\downarrow = N \Longrightarrow N_\uparrow = \frac{N}{2}(1+m), \quad N_\downarrow = \frac{N}{2}(1-m)$$

The energy of the configuration with average magnetization per spin m is given by

$$\mathcal{E}(m) = -\left(\frac{q}{2}Jm^2 + Hm\right)N$$

where the first term is the contribution, on average, of the $N/2$ pairs of spins. The degeneracy of this configuration is

$$\mathcal{Q}(m) = \frac{N!}{N_\uparrow! N_\downarrow!} = \frac{N!}{\left(\frac{N}{2}(1+m)\right)!\left(\frac{N}{2}(1-m)\right)!}$$

Using these results we can express the canonical partition function as

$$Z_N = \sum_m \mathcal{Q}(m)\mathrm{e}^{-\beta\mathcal{E}(m)} \tag{D.88}$$

The sum over all the values of m ranging from zero to unity in increments of $2/N$ is difficult to evaluate, so we replace it by the largest term and assume that its contribution is dominant.[2] We can then use the Stirling formula, Eq. (D.2), to obtain

$$\frac{1}{N}\ln Z_N = \beta\left(\frac{q}{2}J(m^*)^2 + Hm^*\right) - \frac{1+m^*}{2}\ln\left(\frac{1+m^*}{2}\right) - \frac{1-m^*}{2}\ln\left(\frac{1-m^*}{2}\right)$$

where we have used m^* to denote the value of the average magnetization per spin that gives the largest term in the sum of the canonical partition function. In order to determine the value of m^* which maximizes the summand in Eq. (D.88), we take its derivative with respect to m and evaluate it at $m = m^*$. It is actually more convenient to deal with the logarithm of this term, since the logarithm is a monotonic function; therefore we apply this procedure to the logarithm of the summand and obtain

$$m^* = \tanh\left[\beta(qJm^* + H)\right] \tag{D.89}$$

This equation can be solved graphically by finding the intersection of the function $y_1(m) = \tanh(\beta qJm)$ with the straight line $y_0(m) = m$, as shown in Fig. D.5. For high enough T (small enough β), the equation has only one root, $m^* = 0$, which corresponds to the disordered phase. For low enough T (large enough β), the equation has three roots, $m^* = 0, \pm m_0$. The value of the temperature that separates the two cases, called the critical temperature, can be determined by the requiring that the slopes of the two functions $y_1(m)$ and $y_0(m)$ are equal at the origin:

$$\frac{\mathrm{d}}{\mathrm{d}m}\tanh(\beta qJm) = 1 \Longrightarrow k_\mathrm{B}T_c = Jq \tag{D.90}$$

This is of course an approximate result, based on the mean field approach; for instance, for the 2D square lattice this result gives $k_\mathrm{B}T_c = 4J$, whereas the exact solution due to

[2] This is an argument often invoked in dealing with partition functions; it is based on the fact that in the thermodynamic limit $N \to \infty$ the most favorable configuration of the system has a large weight, which falls off extremely rapidly for other configurations.

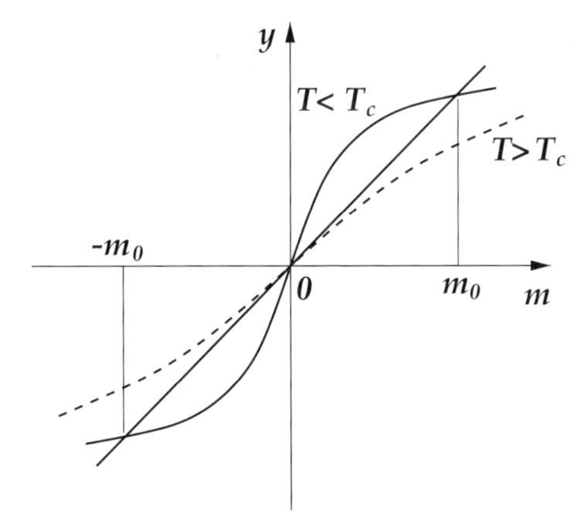

Figure D.5. Graphical solution of Eq. (D.89), which determines the value of the magnetization m^* for $T > T_c$ (dashed line) and $T < T_c$ (solid line).

Onsager gives $k_B T_c = 2.27J$. Of the three roots of Eq. (D.89) for $T < T_c$, we must choose the one that minimizes the free energy per particle F_N/N, obtained in the usual way from the partition function,

$$\frac{F_N}{N} = -\frac{1}{\beta N} \ln Z_N = \frac{q}{2} J(m^*)^2 + \frac{1}{2\beta} \ln\left[1 - (m^*)^2\right] - \frac{1}{\beta} \ln 2$$

where we have used the fact that for the solutions of Eq. (D.89)

$$\beta q J m^* + H = \frac{1}{2} \ln\left(\frac{1+m^*}{1-m^*}\right)$$

which can be easily proved from the definition of $\tanh(x) = (\mathrm{e}^x - \mathrm{e}^{-x})/(\mathrm{e}^x + \mathrm{e}^{-x})$. From the expression of the free energy we then find that it is minimized for $m^* = \pm m_0$ when Eq. (D.89) admits three roots, with $T < T_c$. This can be shown by the following considerations: with the help of Eqs. (D.89) and (D.90) we can express the free energy per particle, in units of $k_B T_c$, as

$$\frac{F_N}{N k_B T_c} = \frac{m^2}{2} - \left(\frac{T}{T_c}\right) \ln\left[\cosh\left(\frac{T_c}{T} m\right)\right] - \left(\frac{T}{T_c}\right) \ln 2 \qquad \text{(D.91)}$$

where we have used m for the average magnetization per particle for simplicity. A plot of $F_N/(N k_B T_c)$ as a function of m for various values of T/T_c is shown in Fig. D.6. For $T > T_c$ the free energy has only one minimum at $m^* = 0$. For $T < T_c$, the free energy has two minima at $m^* = \pm m_0$, as expected. With decreasing T, the value of m_0 approaches unity, and it becomes exactly one for $T = 0$. At $T = T_c$, the curve is very flat near $m = 0$, signifying the onset of the phase transition. The magnetization m plays the role of the *order parameter*, which determines the nature of the phases above and below the critical temperature. This type of behavior of the free energy as a function of the order parameter is common in second order phase transitions, as illustrated in Fig. D.6. The point at which the free energy changes behavior from having a single minimum to having two minima can be used to determine the critical temperature. An analysis of the phase transition in terms of the behavior of the free energy as a function of the order parameter is referred to as the Ginzburg–Landau theory. The presence of the external magnetic field H would

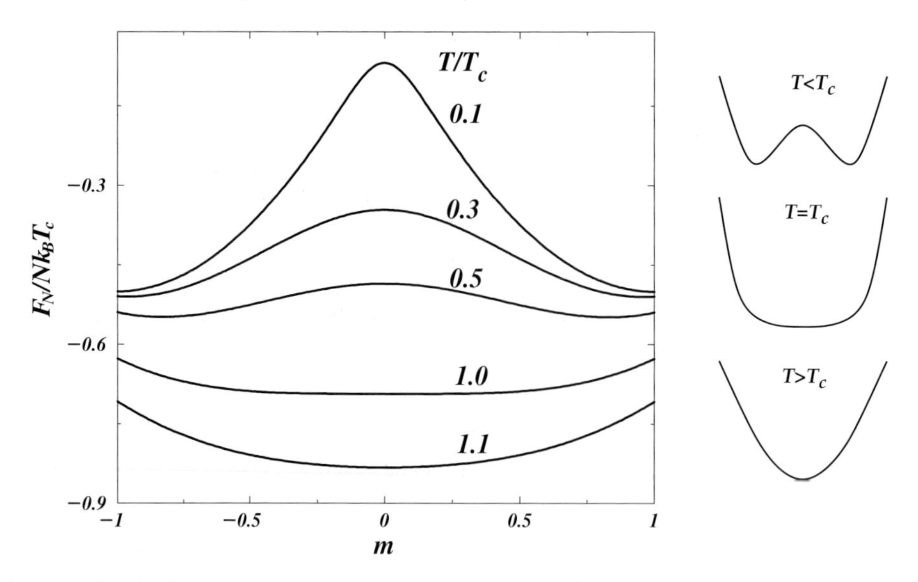

Figure D.6. The free energy per particle F_N/N in units of $k_B T_c$, as a function of the average magnetization per particle m, for various values of T/T_c, from Eq. (D.91). For $T > T_c$ there is only one minimum at $m = 0$, while for $T < T_c$ there are two symmetric minima at $\pm m_0$. The value of m_0 approaches unity as T approaches zero. Notice the very flat behavior of the free energy for $T = T_c$ near $m = 0$, signifying the onset of the phase transition. The diagram on the right illustrates the general behavior of the free energy as a function of the order parameter for $T <, =, > T_c$.

break the symmetry of the free energy with respect to $m \to -m$, and would make one of the two minima at $\pm m_0$ lower than the other. Thus, the weak external magnetic field $H \to 0$ serves to pick one of the two equivalent ground states of the system.

It remains to elucidate the character of the transition. This is usually done through a set of *critical exponents* which describe the behavior of various physical quantities close to the critical temperature. These quantities are expressed as powers of $|T - T_c|$ for $T \to T_c$. Some examples of critical exponents are:

$$\text{order parameter} : m \sim |T_c - T|^{\alpha}$$
$$\text{susceptibility} : \chi = \frac{\partial m}{\partial H} \sim |T_c - T|^{-\gamma}$$

We will calculate the values of α and γ for the Ising model in the mean field approximation. From Eqs. (D.89) and (D.90) we obtain the following expression for the value of the average magnetization at temperature T below T_c, in the limit $H \to 0$:

$$m_0 = \frac{\exp\left[\left(\frac{T_c}{T}\right) m_0\right] - \exp\left[-\left(\frac{T_c}{T}\right) m_0\right]}{\exp\left[\left(\frac{T_c}{T}\right) m_0\right] + \exp\left[-\left(\frac{T_c}{T}\right) m_0\right]}$$

Near $T = T_c$ the value of m_0 is small, so we can use the lowest terms in the Taylor expansions of the exponentials to obtain

$$m_0 = \frac{\left(\frac{T_c}{T}\right) m_0 + \frac{1}{6}\left(\frac{T_c}{T}\right)^3 m_0^3}{1 + \frac{1}{2}\left(\frac{T_c}{T}\right)^2 m_0^2}$$

which, with the definition $t = (T - T_c)/T$, gives

$$m_0^2 = 2t\left(-\frac{t}{3} - \frac{2}{3}\right)^{-1}(t+1)^2$$

We are only interested in the behavior of m_0 very close to the critical point, so we keep only the lowest order term of t in the above expression to obtain

$$m_0^2 \sim -3t = 3\frac{T_c - T}{T} \Longrightarrow \alpha = \frac{1}{2}$$

For the calculation of γ we keep the external magnetic field, which is also a small quantity, in Eq. (D.89) to find

$$m_0 = \frac{\left(\frac{T_c}{T}m_0 + \frac{H}{k_B T}\right) + \frac{1}{6}\left(\frac{T_c}{T}m_0 + \frac{H}{k_B T}\right)^3}{1 + \frac{1}{2}\left(\frac{T_c}{T}m_0 + \frac{H}{k_B T}\right)^2}$$

Differentiating this expression with respect to H and taking the limit $H \to 0$ we find

$$\chi = \frac{\mathrm{d}m_0}{\mathrm{d}H} = \left(\frac{T_c}{T}\chi + \frac{1}{k_B T}\right)\left[1 - \left(\frac{T_c}{T}m_0\right) + \frac{1}{2}\left(\frac{T_c}{T}m_0\right)^2\right]$$

where we have also kept only lowest order terms in m_0. We can use $m_0 \sim t^{1/2}$ from our previous result to substitute into the above equation, which is then solved for χ to produce to lowest order in $T - T_c$

$$\chi \sim \frac{1}{k_B(T - T_c)} \Longrightarrow \gamma = 1$$

The study of critical exponents for various models is an interesting subfield of statistical mechanics. Models that are at first sight different but have the same set of critical exponents are referred to as belonging to the same *universality class*, which implies that their physics is essentially the same.

Further reading

1. *Statistical Mechanics*, K. Huang (2nd edn, J. Wiley, New York, 1987).
   This is a standard treatment of the subject at an advanced level, on which most of the discussion in this Appendix is based.

Problems

1. Using the expression for the grand partition function of the spinless Fermi gas, Eq. (D.59), the definition of the corresponding free energy, Eq. (D.38), and the general relation between the free energy and entropy, Eq. (D.39), show that the entropy of a spinless Fermi gas is given by

$$S = k_B \sum_{\mathbf{k}} [\bar{n}_{\mathbf{k}} \ln \bar{n}_{\mathbf{k}} + (1 - \bar{n}_{\mathbf{k}}) \ln(1 - \bar{n}_{\mathbf{k}})]$$

   where $\bar{n}_{\mathbf{k}}$ is the average occupation number of state $\mathbf{k}$, as given in Eq. (D.61).

2. We want to investigate the behavior of the Fermi gas at low and high temperatures. This can be done by comparing the thermal wavelength λ with the average inter-particle distance $(\Omega/N)^{1/3} = \omega^{1/3}$ (here we use N for the average number of particles for simplicity). The two limits correspond to

$$\frac{\lambda^3}{\omega} << 1 : \quad \text{low-temperature, classical limit}$$
$$\frac{\lambda^3}{\omega} >> 1 : \quad \text{high-temperature, quantum limit}$$

Classical limit From Eq. (D.69) and the series expansion of $h_3(z)$, Eq. (D.70), it is evident that the high-temperature limit also corresponds to z small. Using Eq. (D.68) and the series expansion for $h_3(z)$ show that in the high-temperature limit

$$\bar{n}_{\mathbf{k}}^{(F)} \approx z e^{-\beta\epsilon_{\mathbf{k}}} = \frac{\lambda^3}{\omega} e^{-\beta\epsilon_{\mathbf{k}}}$$

which is the Boltzmann distribution, as expected. Next prove the following relation:

$$\frac{P\Omega}{Nk_BT} = \frac{P\omega}{k_BT} = \frac{h_5(z)}{h_3(z)}$$

and from this show that the equation of state for the free fermion gas at high temperature is given by

$$\frac{P\Omega}{Nk_BT} = 1 + \frac{\lambda^3}{\sqrt{32}}\frac{N}{\Omega} + \cdots$$

The first term on the right-hand side gives the expected classical result, while the second term is a correction that arises from the effective interaction due to the quantum mechanical nature of the particles, as discussed in the text (this is a repulsive interaction).

Quantum limit For large z the behavior of the function $h_3(z)$ is approximated by the expansion

$$h_3(z) \approx \frac{4}{3\sqrt{\pi}}\left[(\ln z)^{3/2} + \frac{\pi^2}{8}(\ln z)^{-1/2} + \cdots\right]$$

Show that by keeping the lowest order term in this expansion on the right-hand side of Eq. (D.69) gives for the Fermi chemical potential μ_F (also referred to as the Fermi energy ϵ_F)

$$\mu_F = \epsilon_F = \frac{\hbar^2}{2m}\left(\frac{6\pi^2}{\omega}\right)^{2/3}$$

for spinless fermions. The total internal energy of the system in this limit can be obtained from

$$E = \sum_{\mathbf{k}} \epsilon_{\mathbf{k}}\bar{n}_{\mathbf{k}} \to \Omega \int \frac{d\mathbf{k}}{(2\pi)^3}\frac{\hbar^2\mathbf{k}^2}{2m}\frac{1}{\exp\left[-\beta\left(\frac{\hbar^2\mathbf{k}^2}{2m} - \epsilon_F\right)\right] + 1}$$

Using integration by parts, show that

$$E = \frac{\Omega\hbar^2}{20\pi^2 m}\int_0^\infty k^5\left(-\frac{\partial}{\partial k}\bar{n}_{\mathbf{k}}\right)\mathrm{d}k$$

$$= \frac{\beta\Omega\hbar^4}{20\pi^2 m^2}\int_0^\infty k^6 \frac{\exp\left[\beta\left(\frac{\hbar^2k^2}{2m}-\epsilon_{\mathrm{F}}\right)\right]}{\left(\exp\left[\beta\left(\frac{\hbar^2k^2}{2m}-\epsilon_{\mathrm{F}}\right)\right]+1\right)^2}\mathrm{d}k$$

In order to evaluate this integral we will use the following trick: the average occupation number $\bar{n}_{\mathbf{k}}$ drops off very sharply from unity to zero near ϵ_{F} as a function of the energy, or equivalently near k_{F} as a function of the magnitude of the wave-vector $k = |\mathbf{k}|$, where k_{F} is the Fermi momentum that appeared in Eqs. (D.9), (D.10) and (D.11). We can therefore expand k^6 in a Taylor series around k_{F} to obtain

$$k^6 \approx k_{\mathrm{F}}^6 + 6k_{\mathrm{F}}^5(k - k_{\mathrm{F}}) + \cdots$$

Show that when this is substituted into the integrand of the above expression for the internal energy it gives

$$E = \frac{3}{5}N\epsilon_{\mathrm{F}}\left[1 + \frac{5\pi^2}{12}\left(\frac{k_{\mathrm{B}}T}{\epsilon_{\mathrm{F}}}\right)^2 + \cdots\right]$$

The first term is the result for zero temperature derived in the text, Eq. (D.12); the second term is the lowest order correction at low temperature. From this expression show that the low-temperature specific heat at constant volume for this system is given by

$$\frac{C_\Omega}{Nk_{\mathrm{B}}} = \frac{\pi^2 k_{\mathrm{B}}T}{2\epsilon_{\mathrm{F}}}$$

Also show that the equation of state takes the form

$$P\Omega = \frac{2}{5}N\epsilon_{\mathrm{F}}\left[1 + \frac{5\pi^2}{12}\left(\frac{k_{\mathrm{B}}T}{\epsilon_{\mathrm{F}}}\right)^2 + \cdots\right]$$

3. (a) For the ideal Bose gas, prove that the internal energy is given by

$$E = \frac{3}{2}\frac{k_{\mathrm{B}}T\Omega}{\lambda^3}g_5(z) = \frac{3}{2}P\Omega$$

where λ is the thermal wavelength and z the fugacity, for both $z = 1$ and $z < 1$.

(b) Using the expression for the internal energy, calculate the specific heat at constant volume C_Ω; for this, it will be useful to show first that the following relations hold:

$$g_{n-1}(z) = z\frac{\partial g_n(z)}{\partial z}, \quad \frac{\partial g_3(z)}{\partial T} = -\frac{3}{2T}g_3(z), \quad \frac{\partial z}{\partial T} = \frac{\partial g_3(z)}{\partial T}\left[\frac{\partial g_3(z)}{\partial z}\right]^{-1}$$

which yield for the specific heat at constant volume

$$\frac{C_\Omega}{Nk_{\mathrm{B}}} = \frac{15\omega}{4\lambda^3}g_5(1), \qquad \text{for } z = 1$$

$$= \frac{15\omega}{4\lambda^3}g_5(z) - \frac{9g_3(z)}{4g_1(z)}, \qquad \text{for } z < 1$$

with $\omega = \Omega/N$.

(c) Show that C_Ω is continuous as a function of T but that its first derivative is discontinuous, and that the discontinuity occurs at the critical temperature T_c for the BEC. This is further evidence that the BEC is a first order transition in the ideal Bose gas. Calculate the discontinuity in the first derivative of C_Ω as a function of T and show that it can be expressed in terms of the critical temperature T_c as

$$\Delta\left(\frac{\partial C_\Omega}{\partial T}\right)_{T_c} = \frac{27Nk_B}{16\pi T_c}[g_3(1)]^2$$

(d) Show that for $z = 1$ the specific heat at constant volume can be expressed as

$$\frac{C_\Omega}{Nk_B} = \frac{15}{4}\frac{g_5(1)}{g_3(1)}\left(1 - \frac{\bar{n}_0}{N}\right)$$

where $\bar{n}_0$ is the average occupation of the state $\mathbf{k} = 0$. From this we may deduce that below the critical temperature only the normal component of the gas, that is, the part that has not condensed to the $\mathbf{k} = 0$ state, contributes to the specific heat. Finally, show that for $z = 1$ the specific heat at constant volume can be expressed as

$$\frac{C_\Omega}{Nk_B} = \frac{15}{4}\frac{g_5(1)}{g_3(1)}\left(\frac{T}{T_c}\right)^{3/2}$$

from which we can extract the low-temperature behavior of the specific heat to be $\sim T^{3/2}$.

4. In 1D we can solve the nearest neighbor spin-1/2 Ising model analytically by the transfer matrix method; the hamiltonian of the system is defined in Eq. (D.87). Consider a chain of N spins with periodic boundary conditions, that is, the right neighbor of the last spin is the first spin, and take for simplicity the spins to have values $s_i = \pm 1$.
(a) For $N = 2$, show that the matrix with entries

$$\mathcal{T}_{ij}(s_1, s_2) = \exp\left(-\beta\mathcal{H}(s_1, s_2)\right)$$

with $\mathcal{H}(s_1, s_2)$ the hamiltonian of the two-spin system, is given by

$$\mathcal{T} = \begin{pmatrix} \exp(\beta(J+H)) & \exp(-\beta J) \\ \exp(-\beta J) & \exp(\beta(J-H)) \end{pmatrix}$$

where the columns correspond to $s_1 = \pm 1$ and the rows correspond to $s_2 = \pm 1$. From this, show that the partition function of the two-spin system is

$$Z_2 = \sum_{s_1, s_2} \exp\left(-\beta\mathcal{H}(s_1, s_2)\right) = \mathrm{Tr}[\mathcal{T}^2]$$

(b) Show that the hamiltonian of the N spin system with periodic boundary conditions can be written as

$$\mathcal{H}(s_1, \ldots, s_N) = -\left(\frac{H}{2}s_1 + Js_1s_2 + \frac{H}{2}s_2\right) - \left(\frac{H}{2}s_2 + Js_2s_3 + \frac{H}{2}s_3\right) - \cdots$$
$$- \left(\frac{H}{2}s_{N-1} + Js_{N-1}s_N + \frac{H}{2}s_N\right) - \left(\frac{H}{2}s_N + Js_Ns_1 + \frac{H}{2}s_1\right)$$

and therefore that the partition function of the system can be written as

$$Z_N = \sum_{s_1,\ldots,s_N} \exp\left(-\beta\mathcal{H}(s_1,\ldots,s_N)\right) = \mathrm{Tr}[\mathcal{T}^N] \tag{D.92}$$

(c) The trace of a matrix $\mathcal{T}$ is unchanged if $\mathcal{T}$ is multiplied by another matrix $\mathcal{S}$ from the right and its inverse $\mathcal{S}^{-1}$ from the left:

$$\mathrm{Tr}[\mathcal{T}] = \mathrm{Tr}[\mathcal{S}^{-1}\mathcal{T}\mathcal{S}]$$

Show that this general statement allows us to multiply each power of $\mathcal{T}$ in Eq. (D.92) by $\mathcal{S}$ on the right and $\mathcal{S}^{-1}$ on the left without changing the value of the trace, to obtain

$$Z_N = \mathrm{Tr}\left[\left(\mathcal{S}^{-1}\mathcal{T}\mathcal{S}\right)^N\right]$$

We can then choose $\mathcal{S}$ to be the matrix of eigenvectors of $\mathcal{T}$, so that the product $\mathcal{S}^{-1}\mathcal{T}\mathcal{S}$ becomes a diagonal matrix with entries equal to the two eigenvalues of $\mathcal{T}$, which we denote as τ_1, τ_2. Thus, show that the partition function takes the form

$$Z_N = \mathrm{Tr}\left[\begin{pmatrix} \tau_1 & 0 \\ 0 & \tau_2 \end{pmatrix}^N\right] = \tau_1^N + \tau_2^N$$

(d) Using the partition function, show that the free energy of the system is equal to

$$F_N = -k_{\mathrm{B}}TN\ln\tau_1 - k_{\mathrm{B}}T\ln\left[1 + \left(\frac{\tau_2}{\tau_1}\right)^N\right]$$

Since $\mathcal{T}$ is a real symmetric matrix, it has two real eigenvalues, which in general are not equal. Assuming that $|\tau_2| < |\tau_1|$, show that in the thermodynamic limit $N \to \infty$ the free energy per particle takes the form

$$\frac{F_N}{N} = -k_{\mathrm{B}}T\ln\tau_1 = -J - k_{\mathrm{B}}T\ln\left[\cosh(\beta H) + \sqrt{\sinh^2(\beta H) + \mathrm{e}^{-4\beta J}}\right]$$

(e) Show that the average magnetization per particle is given by

$$m(H,T) = -\beta\frac{\partial}{\partial H}\left(\frac{F_N}{N}\right)_T = \frac{\sinh(\beta H)}{\sqrt{\sinh^2(\beta H) + \mathrm{e}^{-4\beta J}}}$$

From this expression argue that the average magnetization per particle for $H \to 0$ goes to zero as long as the temperature is finite, and therefore that there is no phase transition at any finite temperature.

Appendix E

Elements of elasticity theory

Elasticity theory considers solids from a macroscopic point of view, and deals with them as an elastic continuum. The basic assumption is that a solid can be divided into small elements, each of which is considered to be of macroscopic size, that is, very large on the atomic scale. Moreover, it is also assumed that we are dealing with small changes in the state of the solid with respect to a reference configuration, so that the response of the solid is well within the elastic regime; in other words, the amount of deformation is proportional to the applied force, just as in a spring. Although these assumptions may seem very restrictive, limiting the applicability of elasticity theory to very large scales, in most cases this theory is essentially correct all the way down to scales of order a few atomic distances. This is due to the fact that it takes a lot of energy to distort solids far from their equilibrium reference state, and for small deviations from that state solids behave in a similar way to elastic springs.

E.1 The strain tensor

We begin with the definitions of the strain and stress tensors in a solid. The reference configuration is usually taken to be the equilibrium structure of the solid, on which there are no external forces. We define strain as the amount by which a small element of the solid is distorted with respect to the reference configuration. The arbitrary point in the solid at (x, y, z) cartesian coordinates moves to $(x + u, y + v, z + w)$ when the solid is strained, as illustrated in Fig. E.1. Each of the displacement fields u, v, w is a function of the position in the solid: $u(x, y, z)$, $v(x, y, z)$, $w(x, y, z)$.

The normal components of the strain are defined to be

$$\epsilon_{xx} = \frac{\partial u}{\partial x}, \quad \epsilon_{yy} = \frac{\partial v}{\partial y}, \quad \epsilon_{zz} = \frac{\partial w}{\partial z} \tag{E.1}$$

These quantities represent the simple stretch of the solid in one direction in response to a force acting in this direction, if there were no deformation in the other two directions. A solid, however, can be deformed in various ways, beyond the simple stretch along the chosen axes x, y, z. In particular, it can be sheared as well as stretched. Stretching and shearing are usually coupled, that is, when a solid is pulled (or pushed) not only does it elongate (or contract) by stretching, but it also changes size in the direction perpendicular to the pulling (or pushing). The shear itself will induce strain in directions other than the

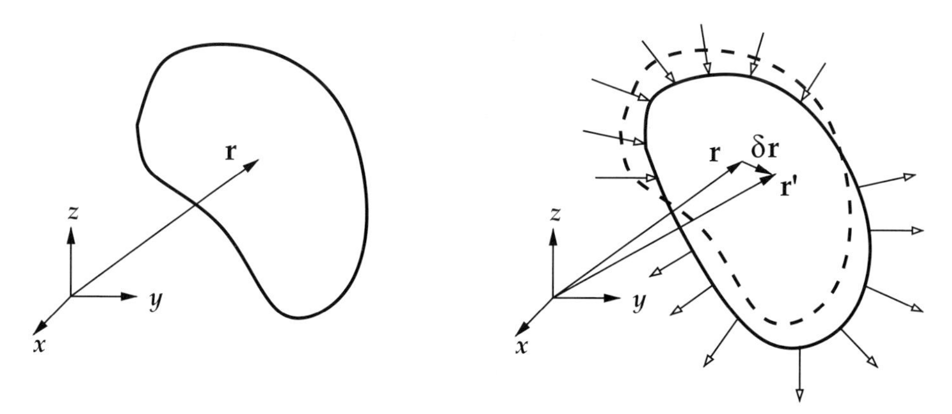

Figure E.1. Definition of the strain tensor for the cartesian (x, y, z) coordinate system: the arbitrary point in the solid at $\mathbf{r} = x\hat{\mathbf{x}} + y\hat{\mathbf{y}} + z\hat{\mathbf{z}}$ in the reference configuration moves to $\mathbf{r}' = \mathbf{r} + \delta\mathbf{r} = (x + u)\hat{\mathbf{x}} + (y + v)\hat{\mathbf{y}} + (z + w)\hat{\mathbf{z}}$ under the action of forces on the solid. (In this example the top part is under compression, the bottom part is under tension.)

direction in which the shearing force acts. To describe this type of deformation we introduce the shear strain components given by

$$\epsilon_{xy} = \frac{1}{2}\left(\frac{\partial u}{\partial y} + \frac{\partial v}{\partial x}\right), \quad \epsilon_{yz} = \frac{1}{2}\left(\frac{\partial v}{\partial z} + \frac{\partial w}{\partial y}\right), \quad \epsilon_{zx} = \frac{1}{2}\left(\frac{\partial w}{\partial x} + \frac{\partial u}{\partial z}\right) \tag{E.2}$$

which are symmetric, $\epsilon_{yx} = \epsilon_{xy}, \epsilon_{zy} = \epsilon_{yz}, \epsilon_{xz} = \epsilon_{zx}$. Thus, we need a second rank tensor to describe the state of strain, which is a symmetric tensor with diagonal elements $\epsilon_{ii}(i = x, y, z)$, and off-diagonal elements $\epsilon_{ij}, (i, j = x, y, z)$. We also introduce, for future use, the rotation tensor

$$\omega_{xy} = \frac{1}{2}\left(\frac{\partial u}{\partial y} - \frac{\partial v}{\partial x}\right), \quad \omega_{yz} = \frac{1}{2}\left(\frac{\partial v}{\partial z} - \frac{\partial w}{\partial y}\right), \quad \omega_{zx} = \frac{1}{2}\left(\frac{\partial w}{\partial x} - \frac{\partial u}{\partial z}\right) \tag{E.3}$$

which describes the infinitesimal rotation of the volume element at (x, y, z). This is an antisymmetric tensor, $\omega_{ji} = -\omega_{ij}, (i, j = x, y, z)$.

There is a different notation which makes the symmetric or antisymmetric nature of the strain and rotation tensors more evident: we define the three cartesian axes as x_1, x_2, x_3 (instead of x, y, z) and the three displacement fields as u_1, u_2, u_3 (instead of u, v, w). Then the strain tensor takes the form

$$\epsilon_{ij} = \frac{1}{2}\left(\frac{\partial u_i}{\partial x_j} + \frac{\partial u_j}{\partial x_i}\right) \tag{E.4}$$

which holds for both the diagonal and off-diagonal components. Similarly, the rotation tensor with this alternative notation takes the form

$$\omega_{ij} = \frac{1}{2}\left(\frac{\partial u_i}{\partial x_j} - \frac{\partial u_j}{\partial x_i}\right) \tag{E.5}$$

which is obviously an antisymmetric tensor. This notation becomes particularly useful in the calculation of physical quantities when other multiple-index tensors are also involved, as we will see below. For the moment, we will use the original notation in terms of x, y, z

and u, v, w, which makes it easier to associate the symbols in three-dimensional space with the corresponding physical quantities.

It is a straightforward geometrical exercise to show that, if the coordinate axes x and y are rotated around the axis z by an angle θ, then the strain components in the new coordinate frame $(x'y'z')$ are given by

$$\begin{aligned}
\epsilon_{x'x'} &= \frac{1}{2}(\epsilon_{xx} + \epsilon_{yy}) + \frac{1}{2}(\epsilon_{xx} - \epsilon_{yy}) \cos 2\theta + \epsilon_{xy} \sin 2\theta \\
\epsilon_{y'y'} &= \frac{1}{2}(\epsilon_{xx} + \epsilon_{yy}) - \frac{1}{2}(\epsilon_{xx} - \epsilon_{yy}) \cos 2\theta - \epsilon_{xy} \sin 2\theta \\
\epsilon_{x'y'} &= \frac{1}{2}(\epsilon_{yy} - \epsilon_{xx}) \sin 2\theta + \epsilon_{xy} \cos 2\theta
\end{aligned} \tag{E.6}$$

The last expression is useful, because it can be turned into the following equation:

$$\tan 2\theta = \frac{\epsilon_{xy}}{\epsilon_{xx} - \epsilon_{yy}} \tag{E.7}$$

through which we can identify the rotation around the z axis which will make the shear components of the strain ϵ_{xy} vanish. For this rotation, only normal strains will survive; the rotated axes are called the principal strain axes.

E.2 The stress tensor

Next we define the stress tensor. The definition of stress is a generalization of the definition of pressure, i.e. force per unit area. We take the small element of the solid on which the external forces act to be an orthogonal parallelepiped with the directions normal to its surfaces defining the axes directions $\hat{\mathbf{x}}$, $\hat{\mathbf{y}}$, $\hat{\mathbf{z}}$. The forces that act on these surfaces can have arbitrary directions: for example, the force on the surface identified by $+\hat{\mathbf{x}}$ can have F_x, F_y, F_z components. Consequently, we need a second rank tensor to describe stress, just as for strain. We define the diagonal elements of the stress tensor as $\sigma_{ii} = F_i/A_i (i = x, y, z)$ where F_i is the component of the force that acts on the surface in the $\hat{\mathbf{i}}$ direction and A_i is the area of this surface of the volume element (see Fig. E.2). Similarly, the off-diagonal components of the stress are denoted by σ_{ij} and are given by $\sigma_{ij} = F_j/A_i$. Care must be taken in defining the sign of the stress components: the normal directions to the surfaces of the volume element are always taken to point outward from the volume enclosed by the parallelepiped. Thus, if the force applied to a surface points outward, the stress is positive (called tensile stress), while if it is pointing inward, the stress is negative (called compressive stress). The stress tensor is also symmetric, $\sigma_{ji} = \sigma_{ij}, (i, j = x, y, z)$, which is a consequence of torque equilibrium. Similar definitions and relations hold for the cylindrical polar coordinate system (r, θ, z), which is related to the cartesian coordinate system (x, y, z) through $r = \sqrt{x^2 + y^2}$ and $\theta = \tan^{-1}(y/x)$.

Just as in the analysis of strains, a rotation of the x, y coordinate axes by an angle θ around the z axis gives the following relation between stresses in the old $(\sigma_{xx}, \sigma_{yy}, \sigma_{xy})$

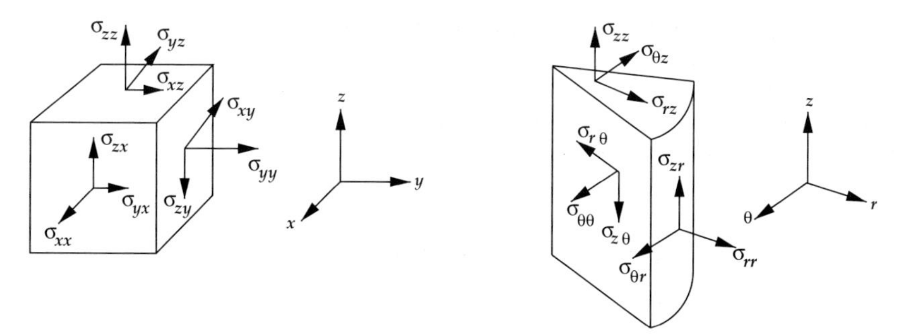

Figure E.2. Definition of the stress tensor for the cartesian (x, y, z) coordinate system, and the polar (r, θ, z) coordinate system.

and new $(\sigma_{x'x'}, \sigma_{y'y'}, \sigma_{x'y'})$ coordinate systems:

$$\sigma_{x'x'} = \frac{1}{2}(\sigma_{xx} + \sigma_{yy}) + \frac{1}{2}(\sigma_{xx} - \sigma_{yy})\cos 2\theta + \sigma_{xy}\sin 2\theta$$
$$\sigma_{y'y'} = \frac{1}{2}(\sigma_{xx} + \sigma_{yy}) - \frac{1}{2}(\sigma_{xx} - \sigma_{yy})\cos 2\theta - \sigma_{xy}\sin 2\theta$$
$$\sigma_{x'y'} = \frac{1}{2}(\sigma_{yy} - \sigma_{xx})\sin 2\theta + \sigma_{xy}\cos 2\theta \quad \text{(E.8)}$$

From the last equation we can determine the angle θ that will give vanishing shear stress $\sigma_{x'y'}$ in the new coordinate system, through the relation

$$\tan 2\theta = \frac{2\sigma_{xy}}{\sigma_{xx} - \sigma_{yy}} \quad \text{(E.9)}$$

which determines the principal stress axes. On the planes that are perpendicular to the principal stress axes, the shear stress is zero, and only the normal component of the stress survives. There are two possible values of the angle θ that can satisfy the equation for the principal stress, which differ by $\pi/2$. The two values of the normal stress from Eq. (E.8) are given by

$$\sigma_{x'x'} = \frac{1}{2}(\sigma_{xx} + \sigma_{yy}) + \left(\frac{1}{4}(\sigma_{xx} - \sigma_{yy})^2 + \sigma_{xy}^2\right)^{1/2}$$
$$\sigma_{y'y'} = \frac{1}{2}(\sigma_{xx} + \sigma_{yy}) - \left(\frac{1}{4}(\sigma_{xx} - \sigma_{yy})^2 + \sigma_{xy}^2\right)^{1/2} \quad \text{(E.10)}$$

which turn out to be the maximum and minimum values of the stress.

The condition of equilibrium, for an element on which an arbitrary force with components (f_x, f_y, f_z) is exerted, is

$$\frac{\partial \sigma_{xx}}{\partial x} + \frac{\partial \sigma_{xy}}{\partial y} + \frac{\partial \sigma_{xz}}{\partial z} = -f_x$$
$$\frac{\partial \sigma_{yx}}{\partial x} + \frac{\partial \sigma_{yy}}{\partial y} + \frac{\partial \sigma_{yz}}{\partial z} = -f_y$$
$$\frac{\partial \sigma_{zx}}{\partial x} + \frac{\partial \sigma_{zy}}{\partial y} + \frac{\partial \sigma_{zz}}{\partial z} = -f_z \quad \text{(E.11)}$$

where the vector **f** has dimensions of force/volume. The equilibrium in terms of torque is assured by the fact that the stress tensor is symmetric, $\sigma_{ij} = \sigma_{ji}(i, j = x, y, z)$.

E.3 Stress–strain relations

The stress and strain tensors are useful in describing the response of a solid to an external load. In the simplest expression, we assume that only a normal stress is acting in the x direction, in response to which the solid is deformed by

$$\epsilon_{xx} = \frac{1}{Y}\sigma_{xx} \tag{E.12}$$

that is, if the stress is compressive (negative) the solid is compressed in the x direction; if the stress is tensile (positive) the solid is elongated. The amount of elongation or contraction is determined by the positive coefficient $1/Y$, where the constant Y is called Young's modulus.[1] The above equation is the familiar Hooke's law for elastic strain. For an arbitrary stress applied to the solid, we need to invoke all of the strain components to describe its state of deformation; the corresponding general form of Hooke's law is referred to as a constitutive relation. The traditional way of doing this is through the elastic constants of the solid. These are represented by a tensor of rank 4, C_{ijkl}, with each index taking three values corresponding to x, y, z. A contracted version of this tensor is a matrix with only two indices C_{ij}, which take six values each (this is referred to as Voigt notation). The connection between the two sets of elastic constants is made through the following identification of indices:

$$1 \to xx, \quad 2 \to yy, \quad 3 \to zz, \quad 4 \to yz, \quad 5 \to zx, \quad 6 \to xy \tag{E.13}$$

In terms of the contracted elastic constants, the constitutive law relates stress to strain as follows:

$$\begin{aligned}
\sigma_{xx} &= C_{11}\epsilon_{xx} + C_{12}\epsilon_{yy} + C_{13}\epsilon_{zz} + C_{14}\epsilon_{xy} + C_{15}\epsilon_{yz} + C_{16}\epsilon_{zx} \\
\sigma_{yy} &= C_{21}\epsilon_{xx} + C_{22}\epsilon_{yy} + C_{23}\epsilon_{zz} + C_{24}\epsilon_{xy} + C_{25}\epsilon_{yz} + C_{26}\epsilon_{zx} \\
\sigma_{zz} &= C_{31}\epsilon_{xx} + C_{32}\epsilon_{yy} + C_{33}\epsilon_{zz} + C_{34}\epsilon_{xy} + C_{35}\epsilon_{yz} + C_{36}\epsilon_{zx} \\
\sigma_{yz} &= C_{41}\epsilon_{xx} + C_{42}\epsilon_{yy} + C_{43}\epsilon_{zz} + C_{44}\epsilon_{xy} + C_{45}\epsilon_{yz} + C_{46}\epsilon_{zx} \\
\sigma_{zx} &= C_{51}\epsilon_{xx} + C_{52}\epsilon_{yy} + C_{53}\epsilon_{zz} + C_{54}\epsilon_{xy} + C_{55}\epsilon_{yz} + C_{56}\epsilon_{zx} \\
\sigma_{xy} &= C_{61}\epsilon_{xx} + C_{62}\epsilon_{yy} + C_{63}\epsilon_{zz} + C_{64}\epsilon_{xy} + C_{65}\epsilon_{yz} + C_{66}\epsilon_{zx}
\end{aligned} \tag{E.14}$$

Of the 36 components of the matrix C_{ij} only 21 are independent in the most general case, the six diagonal ones (corresponding to $i = j$) and the 15 off-diagonal on the upper or lower triangle (corresponding either to $j < i$ or to $j > i$). These equations imply that the solid is linear elastic, that is, its deformation is linear in the applied stress. This is valid for many physical situations if the deformation is sufficiently small.

The use of the more general tensor of rank 4, C_{ijkl}, simplifies the constitutive equations when the Einstein repeated-index summation convention is employed; this convention consists of summing over all the allowed values of indices which appear more than once on one side of an equation. For example, the above mentioned linear elastic constitutive

[1] In the literature, Young's modulus is usually denoted by the symbol E, but here we reserve this symbol for the energy as in previous chapters, and adopt instead the symbol Y.

relations are written as

$$\sigma_{ij} = C_{ijkl}\epsilon_{kl} \tag{E.15}$$

where summation over the values of the indices k and l on the right-hand side is implied because they appear twice (they are repeated) on the same side of the equation. Of course, not all of the values of C_{ijkl} are independent, because of symmetry; specifically, the symmetries we have already mentioned, $\epsilon_{ij} = \epsilon_{ji}$ and $\sigma_{ij} = \sigma_{ji}$, imply that $C_{ijkl} = C_{jikl} = C_{ijlk} = C_{jilk}$, which makes many of the components of the four-index tensor C_{ijkl} identical. In the most general case there are 21 independent components, just as for the contracted notation with two indices. Additional symmetries of the solid further reduce the number of independent values of the elastic constants. For example, in an isotropic solid many of the components vanish, and of the non-vanishing ones only two are independent. Similarly, in a solid of cubic symmetry, there are only three independent elastic constants. We will discuss these important examples below.

E.4 Strain energy density

Next, we make the connection between energy, stress and strain. We will calculate the energy per unit volume in terms of the applied stress and the corresponding strain, which is called the strain energy density (the repeated-index summation is implied in the following derivation). We assume that the energy is at its lowest value, taken to be zero, at the initial state of the system before stress is applied. The rate of change of the energy will be given by

$$\frac{\mathrm{d}E}{\mathrm{d}t} = \int_{\Omega} f_i \frac{\partial u_i}{\partial t}\mathrm{d}\Omega + \int_{S} \sigma_{ij} n_j \frac{\partial u_i}{\partial t}\mathrm{d}S \tag{E.16}$$

where we have used the alternative notation $u_1 = u$, $u_2 = v$, $u_3 = w$ and $x_1 = x$, $x_2 = y$, $x_3 = z$, employed in Eqs. (E.4) and (E.5). The first term in Eq. (E.16) is due to volume distortions, with f_i the i component of the force on the volume element $\mathrm{d}\Omega$. The second term in Eq. (E.16) is due to surface distortions: the surface contribution involves the vector $\mathbf{n}$ with components n_i $(i = 1, 2, 3)$, which is the unit vector normal to a surface element $\mathrm{d}S$; the product $\sigma_{ij} n_j$ represents the i component of the force on the surface element. By turning the surface integral into a volume one through the divergence theorem, we obtain

$$\frac{\mathrm{d}E}{\mathrm{d}t} = \int_{\Omega} \left[f_i \frac{\partial u_i}{\partial t} + \frac{\partial}{\partial x_j}\left(\sigma_{ij}\frac{\partial u_i}{\partial t}\right)\right]\mathrm{d}\Omega \tag{E.17}$$

Newton's law of motion $\mathbf{F} = m\mathrm{d}^2\mathbf{r}/\mathrm{d}t^2$ applied to the volume element takes the form

$$\rho\frac{\partial^2 u_i}{\partial t^2} = f_i + \frac{\partial \sigma_{ij}}{\partial x_j} \tag{E.18}$$

where ρ is the density. In this equation, f_i is the component of the external force exerted on the volume element and $\partial\sigma_{ij}/\partial x_j$ is the contribution of the stress, which also amounts to a force on the volume element (see also Eq. (E.11), where the corresponding relation for equilibrium conditions is given). From the definitions of the strain and rotation tensors, Eqs. (E.4) and (E.5), we have

$$\frac{\partial}{\partial t}\left(\frac{\partial u_i}{\partial x_j}\right) = \frac{\partial}{\partial t}\left(\epsilon_{ij} + \omega_{ij}\right) \tag{E.19}$$

Using Eqs. (E.18) and (E.19) to simplify the expression for the change in the energy, Eq. (E.17), we obtain

$$\frac{\mathrm{d}E}{\mathrm{d}t} = \frac{\mathrm{d}}{\mathrm{d}t}(K+U) = \int_{\Omega}\left[\rho\frac{\partial^2 u_i}{\partial t^2}\frac{\partial u_i}{\partial t}\right]\mathrm{d}\Omega + \int_{\Omega}\left[\sigma_{ij}\frac{\partial \epsilon_{ij}}{\partial t}\right]\mathrm{d}\Omega \tag{E.20}$$

where we have taken advantage of the symmetric and antisymmetric nature of the stress and the rotation tensors to set $\sigma_{ij}\omega_{ij} = 0$. In Eq. (E.20), the first term on the right-hand side represents the kinetic energy K and the second represents the potential energy U. To obtain the kinetic energy K, we integrate over time the first term on the right-hand side of the last equation:

$$K = \int_{\Omega}\rho\left[\frac{\partial u_i}{\partial t}\right]^2\mathrm{d}\Omega \tag{E.21}$$

Depending on the conditions, we can identify U as the internal energy if the loading process is adiabatic, that is, without exchange of heat, or as the Helmholtz free energy if the loading process is isothermal, that is, quasistatic reversible under constant temperature. For the strain energy density W, defined as the potential energy per unit volume, we obtain

$$U = \int_{\Omega} W\mathrm{d}\Omega \Rightarrow \frac{\partial W}{\partial t} = \sigma_{ij}\frac{\partial \epsilon_{ij}}{\partial t} \Rightarrow \sigma_{ij} = \frac{\partial W}{\partial \epsilon_{ij}} \tag{E.22}$$

Now we can use the Taylor expansion for the strain energy density in terms of small strains,

$$W = W_0 + \frac{1}{2}\left[\frac{\partial^2 W}{\partial \epsilon_{ij}\partial \epsilon_{kl}}\right]_0 \epsilon_{ij}\epsilon_{kl} \tag{E.23}$$

where the first order term is omitted on the assumption that we are referring to changes from an equilibrium state. Taking a derivative of this expression with respect to ϵ_{mn} we obtain

$$\begin{aligned}\sigma_{mn} &= \frac{\partial W}{\partial \epsilon_{mn}} = \frac{1}{2}\left[\frac{\partial^2 W}{\partial \epsilon_{ij}\partial \epsilon_{kl}}\right]_0 (\delta_{im}\delta_{jn}\epsilon_{kl} + \delta_{km}\delta_{ln}\epsilon_{ij})\\ &= \frac{1}{2}\left[\frac{\partial^2 W}{\partial \epsilon_{mn}\partial \epsilon_{kl}}\right]_0 \epsilon_{kl} + \frac{1}{2}\left[\frac{\partial^2 W}{\partial \epsilon_{mn}\partial \epsilon_{ij}}\right]_0 \epsilon_{ij} = \left[\frac{\partial^2 W}{\partial \epsilon_{mn}\partial \epsilon_{ij}}\right]_0 \epsilon_{ij}\end{aligned} \tag{E.24}$$

which, by comparison to Eq. (E.15), yields

$$C_{ijkl} = \left[\frac{\partial^2 W}{\partial \epsilon_{ij}\partial \epsilon_{kl}}\right]_0 \tag{E.25}$$

This last equation demonstrates that the elastic constant tensor can be obtained by calculating the variations to second order in the potential energy density with respect to small strains. Using this expression, we can write the strain energy density as

$$W = W_0 + \frac{1}{2}C_{ijkl}\epsilon_{ij}\epsilon_{kl} = W_0 + \frac{1}{2}\sigma_{ij}\epsilon_{ij} \tag{E.26}$$

which shows that the strain energy density due to elastic deformation is obtained by multiplying the stress with the strain tensors. Another interesting result that this expression reveals is that the strain energy density is quadratic in the strain tensor ϵ_{ij}. We collect in Table E.1 the relations we have just described within continuum elasticity and

Table E.1. *Continuum elasticity equations and classical mechanics.*

The correspondence between the continuum elasticity equations developed for the stress and strain tensors and the strain energy density, and the general relations between the force **F** and time derivative of the position **r**, or the spatial derivative of the potential energy as given by classical mechanics. The last two equations correspond to the special case of a harmonic potential energy, which is implicit in the linear relation between stress and strain.

General expression	Continuum elasticity	
$m\frac{\partial^2 \mathbf{r}}{\partial t^2} = \mathbf{F}$	$\rho\frac{\partial^2 u_i}{\partial t^2} = f_i + \frac{\partial \sigma_{ij}}{\partial x_j}$	(E.18)
$\mathbf{F} = -\nabla_{\mathbf{r}} U(\mathbf{r})$	$\sigma_{ij} = \frac{\partial W}{\partial \epsilon_{ij}}$	(E.22)
$\mathbf{F} = -\kappa(\mathbf{r} - \mathbf{r}_0)$	$\sigma_{ij} = C_{ijkl}\epsilon_{kl}$	(E.25)
$U(\mathbf{r}) = U_0 + \frac{1}{2}\kappa(\mathbf{r} - \mathbf{r}_0)^2$	$W = W_0 + \frac{1}{2}C_{ijkl}\epsilon_{ij}\epsilon_{kl}$	(E.26)

the corresponding general relations between forces **F** and the time derivatives of the position **r** or spatial derivatives of the potential energy $U(\mathbf{r})$.

E.5 Applications of elasticity theory

E.5.1 Isotropic elastic solid

We examine next the example of an isotropic, elastic solid. First, when a normal strain σ_{xx} is applied in the x direction, the solid not only deforms according to Eq. (E.12), but it also deforms in the y, z directions according to

$$\epsilon_{yy} = \epsilon_{zz} = -\nu\epsilon_{xx} = -\nu\frac{\sigma_{xx}}{Y} \tag{E.27}$$

where ν is called Poisson's ratio. Poisson's ratio is a positive quantity: when a material is pulled in one direction, its cross-section perpendicular to the pulling direction becomes smaller (a rod gets thinner when pulled along its length).[2] From this we conclude that if three normal stresses σ_{xx}, σ_{yy}, σ_{zz} are applied to an element which is an orthogonal parallelepiped with faces normal to the x, y, z axes, its state of strain will be given by

$$\begin{aligned}
\epsilon_{xx} &= \frac{1}{Y}[\sigma_{xx} - \nu(\sigma_{yy} + \sigma_{zz})] \\
\epsilon_{yy} &= \frac{1}{Y}[\sigma_{yy} - \nu(\sigma_{xx} + \sigma_{zz})] \\
\epsilon_{zz} &= \frac{1}{Y}[\sigma_{zz} - \nu(\sigma_{xx} + \sigma_{yy})]
\end{aligned} \tag{E.28}$$

[2] It is often mentioned that in certain exceptional solids, such as cork, Poisson's ratio can be negative. Cork has actually a very small (near zero) Poisson's ratio, but recently some materials have been synthesized that do have a negative Poisson's ratio.

An analogous expression holds for shear stresses. Namely, if a shear stress σ_{xy} is applied to a solid, the corresponding shear strain is given by

$$\epsilon_{xy} = \frac{1}{\mu}\sigma_{xy} \tag{E.29}$$

where μ is the shear modulus. Two similar equations hold for the other components of shear strain and stress. Of the three elastic constants introduced so far, Young's modulus Y, Poisson's ratio ν and shear modulus μ, only two are independent, because they obey the relation

$$\mu = \frac{Y}{2(1+\nu)} \tag{E.30}$$

The relations for the strain can be written compactly with the use of the Kronecker δ and the Einstein summation convention, as

$$\epsilon_{ij} = \frac{1+\nu}{Y}\sigma_{ij} - \delta_{ij}\frac{\nu}{Y}\sigma_{kk}, \quad i, j, k = x, y, z \tag{E.31}$$

where σ_{kk} stands for $\sigma_{xx} + \sigma_{yy} + \sigma_{zz}$. These equations can be inverted to give the stresses in terms of the strains:

$$\sigma_{ij} = 2\mu\epsilon_{ij} + \delta_{ij}\lambda\epsilon_{kk} \tag{E.32}$$

where the quantity λ is defined as

$$\lambda = \frac{Y\nu}{(1+\nu)(1-2\nu)} \tag{E.33}$$

and is known as Lamé's constant. For example, in cartesian coordinates, the above equation represents the following set of relations:

$$\begin{aligned}
\sigma_{xx} &= 2\mu\epsilon_{xx} + \lambda(\epsilon_{xx} + \epsilon_{yy} + \epsilon_{zz}) \\
\sigma_{yy} &= 2\mu\epsilon_{yy} + \lambda(\epsilon_{xx} + \epsilon_{yy} + \epsilon_{zz}) \\
\sigma_{zz} &= 2\mu\epsilon_{zz} + \lambda(\epsilon_{xx} + \epsilon_{yy} + \epsilon_{zz}) \\
\sigma_{xy} &= 2\mu\epsilon_{xy}, \quad \sigma_{yz} = 2\mu\epsilon_{yz}, \quad \sigma_{zx} = 2\mu\epsilon_{yz}
\end{aligned}$$

Values of the shear modulus, Lamé's constant and Poisson's ratio for representative solids are given in Table E.2.

We use the expressions derived so far to analyze first the familiar case of hydrostatic pressure applied to an isotropic solid. We consider an orthogonal parallelepiped element in the solid which experiences stresses

$$\sigma_{xx} = \sigma_{yy} = \sigma_{zz} = -p, \quad \sigma_{xy} = \sigma_{yz} = \sigma_{zx} = 0 \tag{E.34}$$

with the hydrostatic pressure $p > 0$. The ensuing strains are then given by

$$\epsilon_{xx} = \epsilon_{yy} = \epsilon_{zz} = -\frac{1-2\nu}{Y}p, \quad \epsilon_{xy} = \epsilon_{yz} = \epsilon_{zx} = 0 \tag{E.35}$$

The change in volume of the element will be given by

$$\mathrm{d}x(1+\epsilon_{xx}) \times \mathrm{d}y(1+\epsilon_{yy}) \times \mathrm{d}z(1+\epsilon_{zz}) = (1 + \epsilon_{xx} + \epsilon_{yy} + \epsilon_{zz})\mathrm{d}x\mathrm{d}y\mathrm{d}z \tag{E.36}$$

Table E.2. *Elastic constants of isotropic solids.*

Values are given for the shear modulus μ, Lamé's constant λ and Poisson's ratio ν for representative covalent and ionic solids and metals. μ and λ are given in Mbar (10^{11} Pa); ν is dimensionless. The crystals in each category are ordered by their shear modulus value in decreasing order.

Covalent and ionic solids				Metals			
Crystal	μ	λ	ν	Crystal	μ	λ	ν
				W	1.6	2.01	0.278
C	5.36	0.85	0.068	Mo	1.23	1.89	0.305
Si	0.681	0.524	0.218	Cr	1.21	0.778	0.130
Ge	0.564	0.376	0.200	Ni	0.947	1.17	0.276
				Fe	0.860	1.21	0.291
MgO	1.29	0.68	0.173	Cu	0.546	1.006	0.324
LiF	0.515	0.307	0.187	Ag	0.338	0.811	0.354
PbS	0.343	0.393	0.267	Au	0.310	1.46	0.412
NaCl	0.148	0.146	0.248	Al	0.265	0.593	0.347
KCl	0.105	0.104	0.250	Pb	0.101	0.348	0.387
AgBr	0.087	0.345	0.401	Na	0.038	0.025	0.201
				K	0.017	0.029	0.312

Source: Hirth and Lothe, mentioned in the Further reading section, chapter 10.

where we neglect terms higher than first order in the strains. We define the total change in volume $\epsilon_\Omega = \epsilon_{xx} + \epsilon_{yy} + \epsilon_{zz}$, and find from the above equation that

$$\epsilon_\Omega = -\frac{3}{Y}(1 - 2\nu)p = -\frac{1}{B}p \tag{E.37}$$

where B is by definition the bulk modulus of the solid. This relationship reveals that the bulk modulus is expressed in terms of Young's modulus and Poisson's ratio as

$$B = \frac{Y}{3(1 - 2\nu)} \tag{E.38}$$

By adding the three equations for the normal strains in an isotropic solid, Eq. (E.28), we obtain

$$\epsilon_\Omega = \frac{1 - 2\nu}{Y} 3\sigma_\Omega = \frac{1}{B}\sigma_\Omega \tag{E.39}$$

where we have defined the volume stress $\sigma_\Omega = (\sigma_{xx} + \sigma_{yy} + \sigma_{zz})/3$, which is called the hydrostatic component of stress. Apparently, the result of Eq. (E.39) applies to any state of stress, and the quantities ϵ_Ω, σ_Ω are both invariant with respect to orthogonal transformations of the coordinate axes. The reason for this simple relation is that the shear strains, which in the present example are zero, do not contribute to any changes in the volume.

We can also use the stress–strain relations, Eq. (E.32), and the general expression Eq. (E.26), to arrive at the following expression for the strain energy density of the isotropic solid (we take the constant $W_0 = 0$ for simplicity):

$$W = \frac{1}{2}B(\epsilon_{kk})^2 + \mu\left[\epsilon_{ij}\epsilon_{ij} - \frac{1}{3}(\epsilon_{kk})^2\right] \tag{E.40}$$

where we have also used Eqs. (E.30), (E.33) and (E.38), to introduce the bulk modulus B in the expression for W. Finally, using the same relations we can rewrite the general equilibrium condition of Eq. (E.18), for the case of an isotropic solid with no external forces, as

$$\left(B + \frac{1}{3}\mu\right)\frac{\partial^2 u_j}{\partial x_i \partial x_j} + \mu\frac{\partial^2 u_i}{\partial x_j \partial x_j} = \rho\frac{\partial^2 u_i}{\partial t^2}$$

or equivalently, in vector notation,

$$\left(B + \frac{1}{3}\mu\right)\nabla_{\mathbf{r}}(\nabla_{\mathbf{r}} \cdot \mathbf{u}) + \mu\nabla_{\mathbf{r}}^2\mathbf{u} = \rho\frac{\partial^2 \mathbf{u}}{\partial t^2} \tag{E.41}$$

This last expression allows us to find the speed with which waves can propagate in an isotropic solid. We assume that the displacement field $\mathbf{u}$ has a waveform

$$\mathbf{u}(\mathbf{r}, t) = \mathbf{p}f(\mathbf{n}\cdot\mathbf{r} - ct) \tag{E.42}$$

where $\mathbf{p}$ is the polarization of the wave, $\mathbf{n}$ is the direction of propagation and c is its speed. For dilatational waves, $\mathbf{p} \parallel \mathbf{n}$, we find that the speed of the wave is given by

$$c_d = \sqrt{\left(B + \frac{4}{3}\mu\right)\frac{1}{\rho}} \tag{E.43}$$

while for shear waves, $\mathbf{p} \perp \mathbf{n}$, the speed is given by

$$c_s = \sqrt{\frac{\mu}{\rho}} \tag{E.44}$$

We note that on the surface of the solid there can only exist shear waves; the speed of propagation of these waves, called the Rayleigh speed c_R, is typically $c_R \approx 0.9c_s$.

E.5.2 Plane strain

Of particular interest are situations where the strain tensor is essentially two-dimensional; such situations are often encountered in cross-sections of solids which can be considered infinite in one dimension, by convention assigned the z axis or the x_3 axis, and due to the symmetry of the problem only the strain on a plane perpendicular to this direction matters. This is referred to as "plane strain". We discuss the form that the stress–strain relations take in the case of plane strain for an isotropic solid, using the notation x_3 for the special axis and x_1, x_2 for the plane perpendicular to it.

First, by assumption we will have $\epsilon_{33} = 0$, that is, there is no strain in the special direction. Moreover, all physical quantities have no variation with respect to x_3. Thus, all derivatives with respect to x_3, as well as the stress components σ_{i3}, $i = 1, 2$, are set to zero. With these conditions, the general equilibrium equations, Eq. (E.18), with no external forces, reduce to

$$\frac{\partial \sigma_{11}}{\partial x_1} + \frac{\partial \sigma_{21}}{\partial x_2} = 0, \quad \frac{\partial \sigma_{12}}{\partial x_1} + \frac{\partial \sigma_{22}}{\partial x_2} = 0 \tag{E.45}$$

while from the definition of the strain tensor, Eq. (E.4), we find by differentiation of the various components with respect to x_1 and x_2:

$$\frac{\partial^2 \epsilon_{12}}{\partial x_1 \partial x_2} = \frac{1}{2}\left(\frac{\partial^2 \epsilon_{11}}{\partial x_2^2} + \frac{\partial^2 \epsilon_{22}}{\partial x_1^2}\right) \tag{E.46}$$

From the general stress–strain relations of an isotropic solid, Eq. (E.31), and the fact that in plane strain $\epsilon_{33} = 0$, we obtain

$$\sigma_{33} = \nu(\sigma_{11} + \sigma_{22}) \tag{E.47}$$

Using this last result in the expressions for ϵ_{11} and ϵ_{22} of Eq. (E.31), taking the second derivatives of these expressions with x_2 and x_1, respectively, and the cross-derivative of ϵ_{12} with x_1 and x_2, and combining with Eqs. (E.45) and (E.46), we arrive at

$$\nabla_{12}^2(\sigma_{11} + \sigma_{22}) = 0 \tag{E.48}$$

where we have written the laplacian with the index 12 to emphasize that only the x_1 and x_2 second derivatives are involved. In order to solve this equation, we can write the stress as

$$\sigma_{ij} = -\frac{\partial^2 A}{\partial x_i \partial x_j} + \delta_{ij}\nabla_{12}^2 A \tag{E.49}$$

where A is a function of the variables x_1, x_2, known as the Airy function. It is easy to verify that this choice for the stress automatically satisfies the equilibrium conditions Eq. (E.45).

In order to solve the problem of plane strain, all that remains to do is find the form of the Airy function. Inserting the expression of Eq. (E.49) into Eq. (E.48) we obtain

$$\nabla_{12}^2(\nabla_{12}^2 A) = 0 \tag{E.50}$$

It is often useful to employ the polar coordinates (r, θ, z), with (r, θ) the in-plane coordinates and z the special axis of symmetry. With this notation, the various components of the stress take the form

$$\sigma_{rr} = \nabla_{12}^2 A - \frac{\partial^2 A}{\partial r^2}, \quad \sigma_{\theta\theta} = \frac{\partial^2 A}{\partial r^2}, \quad \sigma_{r\theta} = -\frac{\partial}{\partial r}\left[\frac{1}{r}\frac{\partial A}{\partial \theta}\right] \tag{E.51}$$

with A a function of r and θ, and the laplacian in the plane expressed as

$$\nabla_{12}^2 = \frac{\partial^2}{\partial r^2} + \frac{1}{r}\frac{\partial}{\partial r} + \frac{1}{r^2}\frac{\partial^2}{\partial \theta^2}$$

(see also Appendix G). A useful observation is that the general solution of Eq. (E.50) is

$$A(r, \theta) = r^2 f(r, \theta) + g(r, \theta) \tag{E.52}$$

where the functions $f(r, \theta)$, $g(r, \theta)$ each satisfy Laplace's equation in the plane:

$$\nabla_{12}^2 f(r, \theta) = 0, \quad \nabla_{12}^2 g(r, \theta) = 0.$$

The form of f and g depends on the boundary conditions relevant to the problem under consideration.

E.5.3 Solid with cubic symmetry

As a final example, we discuss the case of a solid which has cubic symmetry. The minimal symmetry in such a case consists of axes of three-fold rotation, which are the major axes of the cube, as illustrated in Fig. 3.12(a). The rotations by $2\pi/3$ around these symmetry axes lead to the following permutations of the x, y, z axes:

$$\begin{aligned} 1: &\quad x \to y \to z \to x \\ 2: &\quad x \to z \to -y \to x \\ 3: &\quad x \to -z \to y \to x \\ 4: &\quad x \to -y \to -z \to x \end{aligned} \tag{E.53}$$

as described in Table 3.4. Now we can use these general symmetries of the solid to deduce relations between the elastic constants.

The simplest way to derive these relations is to consider the strain energy density expression of Eq. (E.26), which must be invariant under any of the symmetry operations. This expression for the energy is quadratic in the strains, so that the terms that appear in it can have one of the following forms:

$$\epsilon_{ii}^2, \quad \epsilon_{ii}\epsilon_{jj}, \quad \epsilon_{ij}^2, \quad \epsilon_{ii}\epsilon_{ij}, \quad \epsilon_{ii}\epsilon_{jk}, \quad \epsilon_{ij}\epsilon_{jk} \tag{E.54}$$

with $(i, j, k) = (x, y, z)$. Each type of the first three terms in Eq. (E.54) gives a contribution to the energy that must be the same for all the possible values of the indices, since the three axes x, y, z are equivalent due to the cubic symmetry. Specifically, the contribution of ϵ_{xx}^2 must be the same as that of ϵ_{yy}^2 and that of ϵ_{zz}^2, and similarly for the terms ϵ_{ij}^2 and $\epsilon_{ii}\epsilon_{jj}$. These considerations imply that

$$\begin{aligned} &C_{11} = C_{22} = C_{33} \\ &C_{12} = C_{13} = C_{21} = C_{23} = C_{31} = C_{32} \\ &C_{44} = C_{55} = C_{66} \end{aligned}$$

so that there are at least three independent elastic constants, as claimed earlier. Moreover, the first three terms in Eq. (E.54), when summed over the values of the indices, are unchanged under the symmetry operations of Eq. (E.53), while the last three terms change sign. For example, the term

$$\epsilon_{xx}^2 + \epsilon_{yy}^2 + \epsilon_{zz}^2$$

(which corresponds to the first term in Eq. (E.54), since $C_{11} = C_{22} = C_{33}$) is unchanged under any of the operations in Eq. (E.53). In contrast to this, the term $\epsilon_{xx}\epsilon_{xy}$ under the third operation in Eq. (E.53) is transformed into

$$\epsilon_{xx}\epsilon_{xy} \to \epsilon_{-z-z}\epsilon_{-zx} = -\epsilon_{zz}\epsilon_{zx}$$

which is another term of the same character but opposite sign. The coefficients of terms that change sign must vanish, since otherwise, for an arbitrary state of strain, the energy would not be invariant under the symmetry operations. This implies the vanishing of the following coefficients:

$$\begin{aligned} &C_{14} = C_{15} = C_{16} = C_{41} = C_{51} = C_{61} = 0 \\ &C_{24} = C_{25} = C_{26} = C_{42} = C_{52} = C_{62} = 0 \\ &C_{34} = C_{35} = C_{36} = C_{43} = C_{53} = C_{63} = 0 \\ &C_{45} = C_{46} = C_{54} = C_{64} = C_{56} = C_{65} = 0 \end{aligned}$$

Table E.3. *Elastic constants for representative cubic crystals.*

Values are given at room temperature, in units of Mbar (10^{11} Pa). The names of crystal structures are DIA = diamond, ZBL = zincblende, CUB = cubic, FCC = face-centered cubic, BCC = body-centered cubic (for details see chapter 1).

Crystal		C_{11}	C_{12}	C_{44}	Crystal		C_{11}	C_{12}	C_{44}
C	(DIA)	10.76	1.250	5.760	Ir	(FCC)	5.800	2.420	2.560
Si	(DIA)	1.658	0.639	0.796	W	(BCC)	5.224	2.044	1.608
Ge	(DIA)	1.284	0.482	0.667	Mo	(BCC)	4.637	1.578	1.092
					Pt	(FCC)	3.467	2.507	0.765
AlSb	(ZBL)	0.894	0.443	0.416	Cr	(BCC)	3.398	0.586	0.990
GaP	(ZBL)	1.412	0.625	0.705	Ta	(BCC)	2.602	1.545	0.826
GaAs	(ZBL)	1.188	0.537	0.594	Ni	(FCC)	2.481	1.549	1.242
GaSb	(ZBL)	0.884	0.403	0.432	Nb	(BCC)	2.465	1.345	0.287
InP	(ZBL)	1.022	0.576	0.460	Fe	(BCC)	2.260	1.400	1.160
InAs	(ZBL)	0.833	0.453	0.396	V	(BCC)	2.287	1.190	0.432
InSb	(ZBL)	0.672	0.367	0.302	Pd	(FCC)	2.271	1.760	0.717
					Cu	(FCC)	1.683	1.221	0.757
ZnS	(ZBL)	1.046	0.653	0.461	Ag	(FCC)	1.240	0.937	0.461
ZnSe	(ZBL)	0.807	0.488	0.441	Au	(FCC)	1.924	1.630	0.420
ZnTe	(ZBL)	0.713	0.408	0.312	Al	(FCC)	1.067	0.604	0.283
					Pb	(FCC)	0.497	0.423	0.150
LiCl	(CUB)	0.493	0.231	0.250	Li	(BCC)	0.135	0.114	0.088
NaCl	(CUB)	0.495	0.129	0.129	Na	(BCC)	0.074	0.062	0.042
KCl	(CUB)	0.407	0.071	0.063	K	(BCC)	0.037	0.031	0.019
RbCl	(CUB)	0.361	0.062	0.047	Rb	(BCC)	0.030	0.025	0.017
CsCl	(CUB)	0.364	0.088	0.080	Cs	(BCC)	0.025	0.021	0.015

Source: Ref. [74].

leaving as the only independent coefficients C_{11}, C_{12}, C_{44}. This result leads to the following expression for the strain energy density of a solid with cubic symmetry:

$$W_{cubic} = \frac{1}{2}C_{11}\left(\epsilon_{xx}^2 + \epsilon_{yy}^2 + \epsilon_{zz}^2\right) + \frac{1}{2}C_{44}\left(\epsilon_{xy}^2 + \epsilon_{yz}^2 + \epsilon_{zx}^2\right) + C_{12}\left(\epsilon_{xx}\epsilon_{yy} + \epsilon_{yy}\epsilon_{zz} + \epsilon_{zz}\epsilon_{xx}\right) \quad \text{(E.55)}$$

where we have taken the constant term W_0 to be zero for simplicity.

This expression is also useful for calculating the elastic constants C_{11}, C_{12}, C_{44}, using total-energy calculations of the type discussed in detail in chapter 6. Specifically, for C_{11} it suffices to distort the solid so that of all the strain components only ϵ_{xx} is non-zero, and then calculate the total energy as a function of the magnitude of ϵ_{xx} and fit it with a second order term in this variable: the coefficient of this term is $\frac{1}{2}C_{11}$. In a similar fashion, one can obtain values for C_{12} and C_{44}. As we discussed earlier for the isotropic solid, the diagonal components of the strain correspond to the change in the volume, so that ϵ_{xx} corresponds to a volume change due to elongation or contraction of the solid in one direction only. In this sense, the elastic constant C_{11} is related to the response of the solid to elongation or contraction in one direction only. The strains ϵ_{ij} describe pure shear distortions of the solid without changes in the volume, so that the elastic constant C_{44} is related to the response of the solid to shear forces. The elastic constant C_{12} involves two different diagonal components of the strain, so that it is related to simultaneous distortions (elongation or contraction but not shear) along two different orthogonal axes. In Table E.3 we give the values of the elastic constants of some representative cubic solids.

Further reading

1. *The Mathematical Theory of Elasticity*, A.E.H. Love (Cambridge University Press, Cambridge, 1927).
   This is a classic book on the theory of elasticity, with advanced treatments of many important topics.
2. *Elasticity: Tensor, Dyadic and Engineering Approaches*, P.C. Cou and N.J. Pagano (Van Nostrand Co., Princeton, 1967).
   This is an accessible and concise treatment of the theory of elasticity.

Problems

1. Using the definitions of the normal and shear strain, Eqs. (E.1) and (E.2), show that the following relations hold:

$$\frac{\partial^2 \epsilon_{xx}}{\partial y^2} + \frac{\partial^2 \epsilon_{yy}}{\partial x^2} = 2\frac{\partial^2 \epsilon_{xy}}{\partial x \partial y} \tag{E.56}$$

plus two more similar relations obtained by cyclical permutations of the indices; also:

$$\frac{\partial^2 \epsilon_{xx}}{\partial y \partial z} = \frac{\partial}{\partial x}\left(-\frac{\partial \epsilon_{yz}}{\partial x} + \frac{\partial \epsilon_{xz}}{\partial y} + \frac{\partial \epsilon_{xy}}{\partial z}\right) \tag{E.57}$$

plus two more similar relations obtained by cyclical permutations of the indices. These are called Saint–Venant strain compatibility relations.

2. Prove the strain rotation relations, Eq. (E.6). To this end, you need to draw a simple diagram with strains identified for a rectangular element on the xy plane, and to calculate the corresponding strains in the $x'y'$ coordinate system, which is rotated by θ with respect to the original coordinate system. You also need to assume that strains are small, so that second order terms can be neglected.
3. Show that the normal strains along the principal strain axes are given by

$$\epsilon_{x'x'} = \frac{\epsilon_{xx} + \epsilon_{yy}}{2} + \left(\left(\frac{\epsilon_{xx} - \epsilon_{yy}}{2}\right)^2 + \epsilon_{xy}^2\right)^{1/2}$$

$$\epsilon_{y'y'} = \frac{\epsilon_{xx} + \epsilon_{yy}}{2} - \left(\left(\frac{\epsilon_{xx} - \epsilon_{yy}}{2}\right)^2 + \epsilon_{xy}^2\right)^{1/2} \tag{E.58}$$

where $\epsilon_{xx}, \epsilon_{yy}, \epsilon_{xy}$ are the strain components for an arbitrary set of coordinate axes. For this, you will need the expression that determines the angle of rotation which identifies the principal axes, Eq. (E.7).

4. Prove that the normal stress values given by Eq. (E.10) are indeed the maximum and minimum values. To do this, start with Eq. (E.8), and calculate its second derivative at the values of the angle given by Eq. (E.9).
5. Prove the relationship between the three elastic constants of an isotropic solid, Eq. (E.30). To do this, consider a two dimensional section of an isotropic solid on the xy plane, and assume

that $\sigma_{xz} = \sigma_{yz} = \sigma_{zz} = 0$; express the strains $\epsilon_{xx}, \epsilon_{yy}, \epsilon_{xy}$ in terms of the stresses for this case. Next, change coordinate system to $x'y'$, which is related to the original one by a rotation through θ around the z axis, and re-express the new strains in terms of the new stresses in the primed coordinate system. By manipulating the resulting equations, Eq. (E.30) can be easily derived.

6. Derive the expression for the strain energy density of an isotropic solid, Eq. (E.40), starting with the general expression Eq. (E.26) and using the relevant stress–strain relations, Eq. (E.32).
7. Derive the equations of motion for an isotropic solid without any external forces, Eq. (E.41), starting with the general relations Eq. (E.18) and using the relevant stress–strain relations, Eq. (E.32).
8. Show that the waveform assumed for the displacement field $\mathbf{u}(\mathbf{r}, t)$ in Eq. (E.42) leads to the following relation, when employed in the stress equation of motion, Eq. (E.41):

$$\left(B + \frac{1}{3}\mu\right) f''(\mathbf{n} \cdot \mathbf{r} - ct)(\mathbf{p} \cdot \mathbf{n})\mathbf{n} + \mu \mathbf{p} f''(\mathbf{n} \cdot \mathbf{r} - ct)|\mathbf{n}|^2 = c^2 \rho \mathbf{p} f''(\mathbf{n} \cdot \mathbf{r} - ct),$$

where $f''(x)$ is the second derivative of $f(x)$ with respect to its argument. From this, show that the speeds of dilatational and shear waves in an isotropic elastic medium are given by the expressions of Eq. (E.43) and Eq. (E.44), respectively.

Appendix F

The Madelung energy

We provide here a detailed calculation of the ion–ion interaction contributions to the total energy of a solid, the so called Madelung energy:

$$U^{ion-ion} = \frac{1}{2}\sum_{I \neq J} \frac{Q_I Q_J}{|\mathbf{R}_I - \mathbf{R}_J|} \tag{F.1}$$

where Q_I, Q_J are point charges located at $\mathbf{R}_I$, $\mathbf{R}_J$, which are not necessarily lattice vectors. We saw in chapter 1 that this sum converges slowly and cannot be easily calculated by truncating the summation. The periodic nature of crystals allows us to address this problem in an efficient manner, through a clever manipulation of the sum. This consists essentially of adding and subtracting artificial charges to the real ionic charges, which produces two series that converge faster than the original one (see for example Frank's method [237]).

The method we will discuss here for obtaining accurate values of the Madelung energy is due to Ewald; it relies on exploiting Poisson's equation

$$\nabla_{\mathbf{r}}^2 \Phi(\mathbf{r}) = -4\pi\rho(\mathbf{r})$$

with $\Phi(\mathbf{r})$ the electrostatic potential due to the charge distribution $\rho(\mathbf{r})$, and takes advantage of the reciprocal-space vectors $\mathbf{G}$ to express the electrostatic potential of a set of point charges. The Madelung energy can be expressed as

$$U^{ion-ion} = \frac{1}{2}\sum_{J} Q_J \Phi_J(\mathbf{R}_J) \tag{F.2}$$

where $\Phi_J(\mathbf{r})$ is the electrostatic potential due to the ions

$$\Phi_J(\mathbf{r}) = \sum_{I \neq J} \frac{Q_I}{|\mathbf{R}_I - \mathbf{r}|} \tag{F.3}$$

with the contribution of ion J excluded because it will lead to an infinity when evaluated at $\mathbf{r} = \mathbf{R}_J$. For crystals bonded through metallic or ionic bonds, the Madelung energy is positive (repulsive) and the point charges represent the ions stripped of their valence electrons, which have been distributed to bonds. In the opposite extreme of a purely ionic crystal this energy is negative (attractive), and is the reason for the stability of the solid. In either case, the crystal is considered to be overall neutral when the charges of the ions and the valence electrons are taken into account, so that the average charge and the average value of the electrostatic potential are both zero. In the Fourier transform of the potential, the term that corresponds to $\mathbf{G} = 0$ represents the average potential, so this term can be taken to be zero.

F.1 Potential of a gaussian function

We seek first to calculate the electrostatic potential of a charge distribution described by a normalized gaussian function, that is, a gaussian function which integrates to unity. This result is a central piece of the calculation of the Madelung energy through the Ewald method. We consider the charge distribution of a normalized gaussian centered at the origin of the coordinate axes:

$$\rho_g(\mathbf{r}) = \left(\frac{\alpha}{\sqrt{\pi}}\right)^3 \mathrm{e}^{-\alpha^2|\mathbf{r}|^2} \tag{F.4}$$

as determined in Eq. (1.73). The potential generated by this charge distribution is

$$\Phi_g(\mathbf{r}) = \left(\frac{\alpha}{\sqrt{\pi}}\right)^3 \int \frac{\mathrm{e}^{-\alpha^2|\mathbf{r}'|^2}}{|\mathbf{r}' - \mathbf{r}|} \mathrm{d}\mathbf{r}' \tag{F.5}$$

Changing variables to $\mathbf{t} = \mathbf{r}' - \mathbf{r} \Rightarrow \mathbf{r}' = \mathbf{t} + \mathbf{r}$ in the above equation, we obtain

$$\Phi_g(\mathbf{r}) = \left(\frac{\alpha}{\sqrt{\pi}}\right)^3 \int \frac{\mathrm{e}^{-\alpha^2(t^2+r^2+2\mathbf{t}\cdot\mathbf{r})}}{t} \mathrm{d}\mathbf{t} \tag{F.6}$$

We use spherical coordinates to perform the integration $\mathrm{d}\mathbf{t} = t^2 \mathrm{d}t \sin\theta \mathrm{d}\theta \mathrm{d}\phi$, with the convenient choice of $\mathbf{r}$ defining the z axis of the integration variable $\mathbf{t}$ which gives $\mathbf{t} \cdot \mathbf{r} = tr\cos\theta$, to obtain

$$\begin{aligned}
\Phi_g(\mathbf{r}) &= \left(\frac{\alpha}{\sqrt{\pi}}\right)^3 \int_0^{2\pi} \int_0^{\pi} \int_0^{\infty} \frac{\mathrm{e}^{-\alpha^2(t^2+r^2+2tr\cos\theta)}}{t} t^2 \mathrm{d}t \sin\theta \mathrm{d}\theta \mathrm{d}\phi \\
&= \left(\frac{\alpha}{\sqrt{\pi}}\right)^3 2\pi \mathrm{e}^{-\alpha^2 r^2} \int_0^{\infty} \mathrm{e}^{-\alpha^2 t^2} \left[\int_{-1}^{1} \mathrm{e}^{-2\alpha^2 trw} \mathrm{d}w\right] t\mathrm{d}t \\
&= -\frac{\alpha}{r\sqrt{\pi}} \left[\int_0^{\infty} \mathrm{e}^{-\alpha^2(t+r)^2} \mathrm{d}t - \int_0^{\infty} \mathrm{e}^{-\alpha^2(t-r)^2} \mathrm{d}t\right] \\
&= -\frac{\alpha}{r\sqrt{\pi}} \left[\int_r^{\infty} \mathrm{e}^{-\alpha^2 t^2} \mathrm{d}t - \int_{-r}^{\infty} \mathrm{e}^{-\alpha^2 t^2} \mathrm{d}t\right] = \frac{\alpha}{r\sqrt{\pi}} \int_{-r}^{r} \mathrm{e}^{-\alpha^2 t^2} \mathrm{d}t \\
&= \frac{2\alpha}{r\sqrt{\pi}} \int_0^{r} \mathrm{e}^{-\alpha^2 t^2} \mathrm{d}t
\end{aligned}$$

The last expression is related to the error function, the definition of which is

$$\mathrm{erf}(r) \equiv \frac{2}{\sqrt{\pi}} \int_0^r \mathrm{e}^{-t^2} \mathrm{d}t \tag{F.7}$$

Notice that for large values of r the error function becomes

$$\mathrm{erf}(r \to \infty) \to \frac{2}{\sqrt{\pi}} \int_0^{\infty} \mathrm{e}^{-t^2} \mathrm{d}t = 1$$

Thus, we can express the potential of the gaussian using the error function as

$$\Phi_g(\mathbf{r}) = \frac{\mathrm{erf}(\alpha r)}{r} \tag{F.8}$$

F.2 The Ewald method

We begin with the Fourier Transform of the charge distribution ρ which, for a crystal with reciprocal-space vectors $\mathbf{G}$, is given by

$$\rho(\mathbf{r}) = \sum_{\mathbf{G}} \rho(\mathbf{G}) \mathrm{e}^{\mathrm{i}\mathbf{G}\cdot\mathbf{r}} \rightarrow \rho(\mathbf{G}) = \frac{1}{\Omega} \int \mathrm{d}\mathbf{r} \rho(\mathbf{r}) \mathrm{e}^{-\mathrm{i}\mathbf{G}\cdot\mathbf{r}} \tag{F.9}$$

with Ω the volume of the crystal. When this is substituted into the Poisson equation it leads to

$$\Phi(\mathbf{r}) = 4\pi \sum_{\mathbf{G}} \rho(\mathbf{G}) \mathrm{e}^{\mathrm{i}\mathbf{G}\cdot\mathbf{r}} \frac{1}{|\mathbf{G}|^2} \tag{F.10}$$

For a set of point charges Q_I located at $\mathbf{R}_I$, these expressions give for the charge density

$$\rho(\mathbf{r}) = \sum_I Q_I \delta(\mathbf{r} - \mathbf{R}_I) \rightarrow \rho(\mathbf{G}) = \frac{1}{\Omega} \sum_I Q_I \mathrm{e}^{-\mathrm{i}\mathbf{G}\cdot\mathbf{R}_I} \tag{F.11}$$

while the total potential takes the form

$$\Phi(\mathbf{r}) = \frac{4\pi}{\Omega} \sum_I \sum_{\mathbf{G}\neq 0} Q_I \frac{\mathrm{e}^{\mathrm{i}\mathbf{G}\cdot(\mathbf{r}-\mathbf{R}_I)}}{|\mathbf{G}|^2} \tag{F.12}$$

where we have excluded the $\mathbf{G} = 0$ term from the summation, consistent with our earlier analysis of the average charge and potential, which both vanish. If we wanted to calculate the electrostatic energy using this expression for the potential, we would obtain

$$U^{ion--ion} = \frac{1}{2} \sum_J Q_J \Phi(\mathbf{R}_J) = \frac{2\pi}{\Omega} \sum_{I,J} \sum_{\mathbf{G}\neq 0} Q_I Q_J \frac{\mathrm{e}^{\mathrm{i}\mathbf{G}\cdot(\mathbf{R}_J - \mathbf{R}_I)}}{|\mathbf{G}|^2} \tag{F.13}$$

where now, having expressed the potential by its Fourier transform, we need not explicitly exclude the $I = J$ term since it does not lead to infinities. This expression does not converge any faster than the one in Eq. (F.1), because the summation over $\mathbf{G}$ can be thought of as equivalent to an integral,

$$\sum_{\mathbf{G}} \rightarrow \int \mathrm{d}\mathbf{G} = \int |\mathbf{G}|^2 \mathrm{d}|\mathbf{G}| \mathrm{d}\hat{\mathbf{G}}$$

so that the $1/|\mathbf{G}|^2$ term in Eq. (F.13) is canceled out and the remaining summation over I, J converges again slowly (the factor $\exp[\mathrm{i}\mathbf{G}\cdot(\mathbf{R}_J - \mathbf{R}_I)]$ has magnitude unity).

Ewald's trick was to replace the ionic point charges with gaussians normalized to unity:

$$Q_I \delta(\mathbf{r} - \mathbf{R}_I) \rightarrow Q_I \left(\frac{\alpha}{\sqrt{\pi}}\right)^3 \mathrm{e}^{-\alpha^2|\mathbf{r}-\mathbf{R}_I|^2} \tag{F.14}$$

where α is a parameter; with $\alpha \rightarrow \infty$ the normalized gaussian approaches a δ-function (see Appendix G). With this change, the Fourier transform of the charge density and the corresponding potential become

$$\rho_\alpha(\mathbf{G}) = \frac{1}{\Omega} \sum_I Q_I \mathrm{e}^{-|\mathbf{G}|^2/4\alpha^2} \mathrm{e}^{-\mathrm{i}\mathbf{G}\cdot\mathbf{R}}$$

$$\Phi_\alpha(\mathbf{r}) = \frac{4\pi}{\Omega} \sum_I \sum_{\mathbf{G}\neq 0} Q_I \frac{\mathrm{e}^{-|\mathbf{G}|^2/4\alpha^2}}{|\mathbf{G}|^2} \mathrm{e}^{\mathrm{i}\mathbf{G}\cdot(\mathbf{r}-\mathbf{R}_I)} \tag{F.15}$$

which both depend on the parameter α. In the limit $\alpha \to \infty$, these expressions reduce exactly to the ones derived above, Eqs. (F.11) and (F.12). If α is finite, we need to add the proper terms to restore the physical situation of point charges. This is accomplished by adding to the potential, for each ion Q_I at $\mathbf{R}_I$, a term equal to the contribution of a point charge $Q_I/|\mathbf{r} - \mathbf{R}_I|$ plus the contribution of a *negative* gaussian, that is, the opposite of the term in Eq. (F.14), to cancel the contribution of that artificial term. The potential generated by a gaussian charge distribution is described by the error function, as shown in section F.1, Eq. (F.8). Consequently, the potential of the point charge Q_I and the accompanying negative gaussian, both centered at $\mathbf{R}_I$, takes the form

$$Q_I \frac{1 - \mathrm{erf}(\alpha|\mathbf{r} - \mathbf{R}_I|)}{|\mathbf{r} - \mathbf{R}_I|} \tag{F.16}$$

This gives for the total potential

$$\Phi(\mathbf{r}) = \frac{4\pi}{\Omega} \sum_I \sum_{\mathbf{G}\neq 0} Q_I \frac{\mathrm{e}^{-|\mathbf{G}|^2/4\alpha^2} \mathrm{e}^{\mathrm{i}\mathbf{G}\cdot(\mathbf{r}-\mathbf{R}_I)}}{|\mathbf{G}|^2} + \sum_I Q_I \frac{1 - \mathrm{erf}(\alpha|\mathbf{r} - \mathbf{R}_I|)}{|\mathbf{r} - \mathbf{R}_I|} \tag{F.17}$$

which is a generally useful expression because it is valid for every point $\mathbf{r}$ in the crystal, except at the position of the ions because of the infinities introduced by the term $1/|\mathbf{r} - \mathbf{R}_I|$ in the second sum. It is also a convenient form for actual calculations of the Madelung energy. The reason is that the potential in Eq. (F.17) involves two sums, one over $\mathbf{G}$ and I and one over only the ionic positions $\mathbf{R}_I$, and both summands in these sums converge very quickly, because the terms

$$\frac{\mathrm{e}^{-|\mathbf{G}|^2/4\alpha^2}}{|\mathbf{G}|^2}, \quad \frac{1 - \mathrm{erf}(\alpha r)}{r}$$

fall off very fast with increasing $|\mathbf{G}|$ and r, for finite α. The advantage of this approach is that the Madelung energy can now be obtained accurately by including a relatively small number of terms in the summations over $\mathbf{G}$ and I. The number of terms to be included is determined by the shells in reciprocal space (sets of reciprocal-space vectors of equal magnitude) and the shells in real space (sets of Bravais lattice vectors of equal magnitude) required to achieve the desired level of convergence. Typical values of α for such calculations are between 1 and 10.

In the final expression for the Madelung energy, we must take care to exclude the contribution to the energy of the true Coulomb potential due to ion I evaluated at $\mathbf{R}_I$, to avoid infinities. This term arises from the $1/|\mathbf{r} - \mathbf{R}_I|$ part of the second sum in Eq. (F.17), which must be dropped. However, for convergence purposes, and in order to be consistent with the extension of the first sum in Eq. (F.17) over all values of I, we must include separately the term which has $\mathbf{r} = \mathbf{R}_I = \mathbf{R}_J$ in the argument of the error function in the second sum. In essence, we do not need to worry about infinities once we have smoothed the charge distribution to gaussians, so that we can extend the sums over all values of the indices I and J, but we must do this consistently in the two sums of Eq. (F.17). From the definition of the error function, we have

$$\mathrm{erf}(r) = \frac{2}{\sqrt{\pi}} \int_0^r \mathrm{e}^{-t^2} \mathrm{d}t = \frac{2}{\sqrt{\pi}} \int_0^r (1 - t^2 + \cdots) \mathrm{d}t = \frac{2}{\sqrt{\pi}} \left[r - \frac{1}{3} r^3 + \cdots \right]$$

which is valid in the limit of $r \to 0$; this shows that the term in question becomes

$$\lim_{\mathbf{r} \to \mathbf{R}_J} \left[\frac{\mathrm{erf}(\alpha |\mathbf{r} - \mathbf{R}_J|)}{|\mathbf{r} - \mathbf{R}_J|} \right] \to \frac{2}{\sqrt{\pi}} \alpha$$

Putting this result together with the previous expressions, we obtain for the Madelung energy

$$U^{ion-ion} = \frac{2\pi}{\Omega} \sum_{I,J} \sum_{\mathbf{G} \neq 0} Q_I Q_J \frac{\mathrm{e}^{-|\mathbf{G}|^2/4\alpha^2} \mathrm{e}^{\mathrm{i}\mathbf{G} \cdot (\mathbf{R}_J - \mathbf{R}_I)}}{|\mathbf{G}|^2} - \sum_J Q_J^2 \frac{\alpha}{\sqrt{\pi}}$$
$$+ \frac{1}{2} \sum_{I \neq J} Q_I Q_J \frac{1 - \mathrm{erf}(\alpha |\mathbf{R}_J - \mathbf{R}_I|)}{|\mathbf{R}_J - \mathbf{R}_I|} \qquad \text{(F.18)}$$

As a final step, we replace the summations over the positions of ions in the entire crystal by summations over Bravais lattice vectors $\mathbf{R}$, $\mathbf{R}'$ and positions of ions $\mathbf{t}_I$, $\mathbf{t}_J$ *within* the primitive unit cell (PUC):

$$\mathbf{R}_I \to \mathbf{R} + \mathbf{t}_I, \quad \mathbf{R}_J \to \mathbf{R}' + \mathbf{t}_J, \quad \mathbf{t}_I, \mathbf{t}_J \in \mathrm{PUC}.$$

Since for all Bravais lattice vectors $\mathbf{R}$ and all reciprocal lattice vectors $\mathbf{G}$ we have $\exp(\pm\mathrm{i}\mathbf{G} \cdot \mathbf{R}) = 1$, the expression for the Madelung energy takes the form

$$U^{ion-ion} = \frac{2\pi}{\Omega} \sum_{\mathbf{R},\mathbf{R}'} \sum_{I,J \in PUC} \sum_{\mathbf{G} \neq 0} Q_I Q_J \frac{\mathrm{e}^{-|\mathbf{G}|^2/4\alpha^2} \mathrm{e}^{\mathrm{i}\mathbf{G} \cdot (\mathbf{t}_J - \mathbf{t}_I)}}{|\mathbf{G}|^2} - \sum_J Q_J^2 \frac{\alpha}{\sqrt{\pi}}$$
$$+ \frac{1}{2} \sum_{\mathbf{R},\mathbf{R}'} \sum_{I,J \in PUC}{}' Q_I Q_J \frac{1 - \mathrm{erf}(\alpha |(\mathbf{R}' + \mathbf{t}_J) - (\mathbf{R} + \mathbf{t}_I)|)}{|(\mathbf{R}' + \mathbf{t}_J) - (\mathbf{R} + \mathbf{t}_I)|}$$

where the symbol $\sum'$ implies summation over values of I, J within the PUC, such that $I \neq J$ when $\mathbf{R} = \mathbf{R}'$ (the pair $I = J$ is allowed for $\mathbf{R} \neq \mathbf{R}'$). Since the vector $\mathbf{R} - \mathbf{R}'$ is simply another Bravais lattice, we can express the Madelung energy per PUC as

$$\frac{U^{ion-ion}}{N} = \frac{2\pi}{\Omega_{PUC}} \sum_{I,J \in PUC} \sum_{\mathbf{G} \neq 0} Q_I Q_J \frac{\mathrm{e}^{-|\mathbf{G}|^2/4\alpha^2} \mathrm{e}^{\mathrm{i}\mathbf{G} \cdot (\mathbf{t}_J - \mathbf{t}_I)}}{|\mathbf{G}|^2} - \sum_{J \in PUC} Q_J^2 \frac{\alpha}{\sqrt{\pi}}$$
$$+ \frac{1}{2} \sum_{\mathbf{R}} \sum_{I,J \in PUC}{}' Q_I Q_J \frac{1 - \mathrm{erf}(\alpha |(\mathbf{t}_J - \mathbf{t}_I) + \mathbf{R}|)}{|(\mathbf{t}_J - \mathbf{t}_I) + \mathbf{R}|}$$

with $\sum'$ implying summation over values of $I \neq J$ for $\mathbf{R} = 0$; in the above expressions N is the total number of PUCs in the crystal and $\Omega_{PUC} = \Omega/N$ the volume of the primitive unit cell.

Problems

1. Show that the potential calculated with the Ewald method for a set of ions Q_I at positions $\mathbf{R}_I$, Eq. (F.17), is indeed independent of the parameter α, when the summations over reciprocal

lattice vectors and ion shells are truncated. (Hint: calculate $d\Phi/d\alpha$ for the expression in Eq. (F.17) and show that it vanishes.)

2. Show that an equivalent expression to Eq. (F.18) for the calculation of the Madelung energy is

$$U^{ion-ion} = \frac{2\pi}{\Omega} \sum_{I \neq J} \sum_{\mathbf{G} \neq 0} Q_I Q_J \frac{e^{-|\mathbf{G}|^2/4\alpha^2} e^{i\mathbf{G}\cdot(\mathbf{R}_J - \mathbf{R}_I)}}{|\mathbf{G}|^2} + \frac{1}{2} \sum_{I \neq J} Q_I Q_J \frac{1 - \text{erf}(\alpha |\mathbf{R}_J - \mathbf{R}_I|)}{|\mathbf{R}_J - \mathbf{R}_I|}$$

The reason Eq. (F.18) is often preferred is that it is based on the total potential $\Phi(\mathbf{r})$ obtained from Eq. (F.17), which is valid everywhere in the crystal except at the positions of the ions.

3. Use the Ewald summation method to calculate the Madelung energy per unit cell of the NaCl crystal. How many shells do you need to include for convergence of the Madelung energy to one part in 10^3? Verify, by using a few different values of α, that the final answer does not depend on this parameter (but the number of terms required to reach the desired convergence may depend on it).

Appendix G

Mathematical tools

In this Appendix we collect a number of useful mathematical results that are employed frequently throughout the text. The emphasis is not on careful derivations but on the motivation behind the results. Our purpose is to provide a reminder and a handy reference for the reader, whom we assume to have been exposed to the underlying mathematical concepts, but whom we do not expect to recall the relevant formulae each instant the author needs to employ them.

G.1 Differential operators

The differential operators encountered often in the description of the physical properties of solids are the gradient of a scalar field $\nabla_{\mathbf{r}}\Phi(\mathbf{r})$, the divergence of a vector field $\nabla_{\mathbf{r}} \cdot \mathbf{F}(\mathbf{r})$, the curl of a vector field $\nabla_{\mathbf{r}} \times \mathbf{F}(\mathbf{r})$, and the laplacian of a scalar field $\nabla_{\mathbf{r}}^2\Phi(\mathbf{r})$ (the laplacian of a vector field is simply the vector addition of the laplacian of its components, $\nabla_{\mathbf{r}}^2\mathbf{F} = \nabla_{\mathbf{r}}^2 F_x\hat{\mathbf{x}} + \nabla_{\mathbf{r}}^2 F_y\hat{\mathbf{y}} + \nabla_{\mathbf{r}}^2 F_z\hat{\mathbf{z}}$). These operators in three dimensions are expressed in the different sets of coordinates as follows.

In cartesian coordinates (x, y, z):

$$\mathbf{r} = x\hat{\mathbf{x}} + y\hat{\mathbf{y}} + z\hat{\mathbf{z}} \tag{G.1}$$

$$\mathbf{F}(\mathbf{r}) = F_x(\mathbf{r})\hat{\mathbf{x}} + F_y(\mathbf{r})\hat{\mathbf{y}} + F_z(\mathbf{r})\hat{\mathbf{z}} \tag{G.2}$$

$$\nabla_{\mathbf{r}}\Phi(\mathbf{r}) = \frac{\partial\Phi}{\partial x}\hat{\mathbf{x}} + \frac{\partial\Phi}{\partial y}\hat{\mathbf{y}} + \frac{\partial\Phi}{\partial z}\hat{\mathbf{z}} \tag{G.3}$$

$$\nabla_{\mathbf{r}} \cdot \mathbf{F}(\mathbf{r}) = \frac{\partial F_x}{\partial x} + \frac{\partial F_y}{\partial y} + \frac{\partial F_z}{\partial z} \tag{G.4}$$

$$\nabla_{\mathbf{r}} \times \mathbf{F}(\mathbf{r}) = \left(\frac{\partial F_z}{\partial y} - \frac{\partial F_y}{\partial z}\right)\hat{\mathbf{x}} + \left(\frac{\partial F_x}{\partial z} - \frac{\partial F_z}{\partial x}\right)\hat{\mathbf{y}} + \left(\frac{\partial F_y}{\partial x} - \frac{\partial F_x}{\partial y}\right)\hat{\mathbf{z}} \tag{G.5}$$

$$\nabla_{\mathbf{r}}^2\Phi(\mathbf{r}) = \frac{\partial^2\Phi}{\partial x^2} + \frac{\partial^2\Phi}{\partial y^2} + \frac{\partial^2\Phi}{\partial z^2} \tag{G.6}$$

In polar coordinates (r, θ, z):

$$r = \sqrt{x^2 + y^2},\ \theta = \tan^{-1}(y/x), z \longleftrightarrow x = r\cos\theta,\ y = r\sin\theta, z \tag{G.7}$$

$$\mathbf{F}(\mathbf{r}) = F_r(\mathbf{r})\hat{\mathbf{r}} + F_\theta(\mathbf{r})\hat{\theta} + F_z(\mathbf{r})\hat{\mathbf{z}} \tag{G.8}$$

$$\nabla_{\mathbf{r}}\Phi(\mathbf{r}) = \frac{\partial\Phi}{\partial r}\hat{\mathbf{r}} + \frac{1}{r}\frac{\partial\Phi}{\partial\theta}\hat{\theta} + \frac{\partial\Phi}{\partial z}\hat{\mathbf{z}} \tag{G.9}$$

$$\nabla_{\mathbf{r}} \cdot \mathbf{F}(\mathbf{r}) = \frac{1}{r}\frac{\partial}{\partial r}(rF_r) + \frac{1}{r}\frac{\partial F_\theta}{\partial\theta} + \frac{\partial F_z}{\partial z} \tag{G.10}$$

$$\nabla_{\mathbf{r}} \times \mathbf{F}(\mathbf{r}) = \left(\frac{1}{r}\frac{\partial F_z}{\partial\theta} - \frac{\partial F_\theta}{\partial z}\right)\hat{\mathbf{r}} + \left(\frac{\partial F_r}{\partial z} - \frac{\partial F_z}{\partial r}\right)\hat{\theta} + \left(\frac{\partial F_\theta}{\partial r} - \frac{1}{r}\frac{\partial F_r}{\partial\theta} + \frac{F_\theta}{r}\right)\hat{\mathbf{z}} \tag{G.11}$$

$$\nabla^2_{\mathbf{r}}\Phi(\mathbf{r}) = \frac{1}{r}\frac{\partial}{\partial r}\left(r\frac{\partial\Phi}{\partial r}\right) + \frac{1}{r^2}\frac{\partial^2\Phi}{\partial\theta^2} + \frac{\partial^2\Phi}{\partial z^2} \tag{G.12}$$

In spherical coordinates (r, θ, ϕ):

$$r = \sqrt{x^2+y^2+z^2},\ \theta = \cos^{-1}(z/r),\ \phi = \cos^{-1}(x/\sqrt{x^2+y^2}) \longleftrightarrow x = r\sin\theta\cos\phi,\ y = r\sin\theta\sin\phi,\ z = r\cos\theta \tag{G.13}$$

$$\mathbf{F}(\mathbf{r}) = F_r(\mathbf{r})\hat{\mathbf{r}} + F_\theta(\mathbf{r})\hat{\theta} + F_\phi(\mathbf{r})\hat{\phi} \tag{G.14}$$

$$\nabla_{\mathbf{r}}\Phi(\mathbf{r}) = \frac{\partial\Phi}{\partial r}\hat{\mathbf{r}} + \frac{1}{r}\frac{\partial\Phi}{\partial\theta}\hat{\theta} + \frac{1}{r\sin\theta}\frac{\partial\Phi}{\partial\phi}\hat{\phi} \tag{G.15}$$

$$\nabla_{\mathbf{r}} \cdot \mathbf{F}(\mathbf{r}) = \frac{1}{r^2}\frac{\partial}{\partial r}(r^2F_r) + \frac{1}{r\sin\theta}\frac{\partial}{\partial\theta}(\sin\theta F_\theta) + \frac{1}{r\sin\theta}\frac{\partial F_\phi}{\partial\phi} \tag{G.16}$$

$$\nabla_{\mathbf{r}} \times \mathbf{F}(\mathbf{r}) = \frac{1}{r\sin\theta}\left[\frac{\partial}{\partial\theta}(\sin\theta F_\phi) - \frac{\partial F_\theta}{\partial\phi}\right]\hat{\mathbf{r}} + \left[\frac{1}{r\sin\theta}\frac{\partial F_r}{\partial\phi} - \frac{1}{r}\frac{\partial(rF_\phi)}{\partial r}\right]\hat{\theta} + \frac{1}{r}\left[\frac{\partial(rF_\theta)}{\partial r} - \frac{\partial F_r}{\partial\theta}\right]\hat{\phi} \tag{G.17}$$

$$\nabla^2_{\mathbf{r}}\Phi(\mathbf{r}) = \frac{1}{r^2}\frac{\partial}{\partial r}\left(r^2\frac{\partial\Phi}{\partial r}\right) + \frac{1}{r^2\sin\theta}\frac{\partial}{\partial\theta}\left(\sin\theta\frac{\partial\Phi}{\partial\theta}\right) + \frac{1}{r^2\sin^2\theta}\frac{\partial^2\Phi}{\partial\phi^2} \tag{G.18}$$

A useful relation that can be easily proved in any set of coordinates is that the curl of the gradient of scalar field is identically zero:

$$\nabla_{\mathbf{r}} \times \nabla_{\mathbf{r}}\Phi(\mathbf{r}) = 0 \tag{G.19}$$

For example, if we define $\mathbf{F}(\mathbf{r}) = \nabla_{\mathbf{r}}\Phi(\mathbf{r})$, in cartesian coordinates we will have

$$F_x = \frac{\partial\Phi}{\partial x}, \quad F_y = \frac{\partial\Phi}{\partial y}, \quad F_z = \frac{\partial\Phi}{\partial z}$$

in terms of which the x component of the curl will be

$$\left(\frac{\partial F_z}{\partial y} - \frac{\partial F_y}{\partial z}\right) = \left(\frac{\partial^2\Phi}{\partial z\partial y} - \frac{\partial^2\Phi}{\partial y\partial z}\right) = 0$$

and similarly for the y and z components. Another relation of the same type is that the divergence of the curl of a vector field vanishes identically:

$$\nabla_{\mathbf{r}} \cdot \nabla_{\mathbf{r}} \times \mathbf{F} = \frac{\partial}{\partial x}\left(\frac{\partial F_z}{\partial y} - \frac{\partial F_y}{\partial z}\right) + \frac{\partial}{\partial y}\left(\frac{\partial F_x}{\partial z} - \frac{\partial F_z}{\partial x}\right) + \frac{\partial}{\partial z}\left(\frac{\partial F_y}{\partial x} - \frac{\partial F_x}{\partial y}\right)$$

$$\Rightarrow \nabla_{\mathbf{r}} \cdot (\nabla_{\mathbf{r}} \times \mathbf{F}(\mathbf{r})) = 0 \tag{G.20}$$

A third useful relation allows us to express the curl of the curl of a vector field in terms of its divergence and its laplacian:

$$\nabla_{\mathbf{r}} \times (\nabla_{\mathbf{r}} \times \mathbf{F}(\mathbf{r})) = \nabla_{\mathbf{r}}(\nabla_{\mathbf{r}} \cdot \mathbf{F}(\mathbf{r})) - \nabla_{\mathbf{r}}^2 \mathbf{F}(\mathbf{r}) \tag{G.21}$$

the proof of which is left as an exercise.

Lastly, we mention the two fundamental theorems that are often invoked in relation to integrals of the divergence or the curl of a vector field. The first, known as the divergence or **Gauss's theorem**, relates the integral of the divergence of a vector field over a volume Ω to the integral of the field over the surface S which encloses the volume of integration:

$$\int_{\Omega} \nabla_{\mathbf{r}} \cdot \mathbf{F}(\mathbf{r}) \mathrm{d}\mathbf{r} = \oint_S \mathbf{F}(\mathbf{r}) \cdot \hat{\mathbf{n}}_S \, \mathrm{d}S \tag{G.22}$$

where $\hat{\mathbf{n}}_S$ is the surface-normal unit vector on element $\mathrm{d}S$. The second, known as the curl or **Stokes's theorem**, relates the integral of the curl of a vector field over a surface S to the line integral of the field over the contour C which encloses the surface of integration:

$$\int_S (\nabla_{\mathbf{r}} \times \mathbf{F}(\mathbf{r})) \cdot \hat{\mathbf{n}}_S \, \mathrm{d}S = \oint_C \mathbf{F}(\mathbf{r}) \cdot \mathrm{d}\mathbf{l} \tag{G.23}$$

where $\hat{\mathbf{n}}_S$ is the surface-normal unit vector on element $\mathrm{d}S$ and $\mathrm{d}\mathbf{l}$ is the differential vector along the contour C.

G.2 Power series expansions

The Taylor series expansion of a continuous and infinitely differentiable function of x around x_0 in powers of $(x - x_0)$ is

$$f(x) = f(x_0) + \frac{1}{1!}\left[\frac{\partial f}{\partial x}\right]_{x_0}(x - x_0) + \frac{1}{2!}\left[\frac{\partial^2 f}{\partial x^2}\right]_{x_0}(x - x_0)^2$$

$$+ \frac{1}{3!}\left[\frac{\partial^3 f}{\partial x^3}\right]_{x_0}(x - x_0)^3 + \cdots \tag{G.24}$$

with the first, second, third, ... derivatives evaluated at $x = x_0$. We have written the derivatives as partials to allow for the possibility that f depends on other variables as well; the above expression is easily generalized to multivariable functions. For x close to x_0, the expansion can be truncated to the first few terms, giving a very good approximation for $f(x)$ in terms of the function and its lowest few derivatives evaluated at $x = x_0$.

Using the general expression for the Taylor series, we obtain for the common exponential, logarithmic, trigonometric and hyperbolic functions:

$$\mathrm{e}^x = 1 + x + \frac{1}{2}x^2 + \frac{1}{6}x^3 + \cdots, \quad x_0 = 0 \tag{G.25}$$

$$\log(1+x) = x - \frac{1}{2}x^2 + \frac{1}{3}x^3 + \cdots, \quad x_0 = 1 \tag{G.26}$$

$$\cos x = 1 - \frac{1}{2}x^2 + \frac{1}{24}x^4 + \cdots, \quad x_0 = 0 \tag{G.27}$$

$$\cos(\pi + x) = -1 + \frac{1}{2}x^2 - \frac{1}{24}x^4 + \cdots, \quad x_0 = \pi \tag{G.28}$$

$$\sin x = x - \frac{1}{6}x^3 + \frac{1}{120}x^5 + \cdots, \quad x_0 = 0 \tag{G.29}$$

$$\sin(\pi + x) = -x + \frac{1}{6}x^3 - \frac{1}{120}x^5 + \cdots, \quad x_0 = \pi \tag{G.30}$$

$$\tan x \equiv \frac{\sin x}{\cos x} = x + \frac{1}{3}x^3 + \frac{2}{15}x^5 + \cdots, \quad x_0 = 0 \tag{G.31}$$

$$\cot x \equiv \frac{\cos x}{\sin x} = \frac{1}{x} - \frac{1}{3}x - \frac{1}{45}x^3 + \cdots, \quad x_0 = 0 \tag{G.32}$$

$$\cosh x \equiv \frac{e^x + e^{-x}}{2} = 1 + \frac{1}{2}x^2 + \frac{1}{24}x^4 + \cdots, \quad x_0 = 0 \tag{G.33}$$

$$\sinh x \equiv \frac{e^x - e^{-x}}{2} = x + \frac{1}{6}x^3 + \frac{1}{120}x^5 + \cdots, \quad x_0 = 0 \tag{G.34}$$

$$\tanh x \equiv \frac{\sinh x}{\cosh x} = x - \frac{1}{3}x^3 + \frac{2}{15}x^5 + \cdots, \quad x_0 = 0 \tag{G.35}$$

$$\coth x \equiv \frac{\cosh x}{\sinh x} = \frac{1}{x} + \frac{1}{3}x - \frac{1}{45}x^3 + \cdots, \quad x_0 = 0 \tag{G.36}$$

$$\frac{1}{1 \pm x} = 1 \mp x + x^2 \mp x^3 + \cdots, \quad |x| < 1 \tag{G.37}$$

$$\frac{1}{\sqrt{1 \pm x}} = 1 \mp \frac{1}{2}x + \frac{3}{8}x^2 \mp \frac{5}{16}x^3 + \cdots, \quad |x| < 1 \tag{G.38}$$

The last expansion is useful for the so called multi-pole expansion of $|\mathbf{r} - \mathbf{r}'|^{-1}$ for $r' = |\mathbf{r}'| \ll r = |\mathbf{r}|$:

$$\frac{1}{|\mathbf{r} - \mathbf{r}'|} = \frac{1}{r\sqrt{1 + r'^2/r^2 - 2\mathbf{r}\cdot\mathbf{r}'/r^2}}$$
$$= \frac{1}{r} + \frac{\mathbf{r}\cdot\mathbf{r}'}{r^3} + \frac{3(\mathbf{r}\cdot\mathbf{r}')^2 - r'^2 r^2}{2r^5} + \frac{5(\mathbf{r}\cdot\mathbf{r}')^3 - 3(\mathbf{r}\cdot\mathbf{r}')r'^2 r^2}{2r^7} + \cdots$$

with the first term referred to as the monopole term, the second the dipole term, the third the quadrupole term, the fourth the octupole term, etc.

Another very useful power series expansion is the binomial expansion:

$$(a+b)^p = a^p + \frac{p}{1!}a^{p-1}b + \frac{p(p-1)}{2!}a^{p-2}b^2 + \frac{p(p-1)(p-2)}{3!}a^{p-3}b^3 + \cdots$$
$$+ \frac{p(p-1)(p-2)\cdots(p-(n-1))}{n!}a^{p-n}b^n + \cdots \tag{G.39}$$

with $a > 0$, b, p real. This can be easily obtained as the Taylor series expansion of the function $f(x) = a^p(1+x)^p$, with $x = (b/a)$, expanded around $x_0 = 0$. It is evident from this expression that for $p = n$: an integer, the binomial expansion is finite and has a total number of $(n + 1)$ terms; if p is not an integer it has infinite terms.

G.3 Functional derivatives

The derivative of a definite integral over a variable x with respect to a variable y that appears either in the limits of integration or in the integrand is given by

$$\frac{\partial}{\partial y}\int_{a(y)}^{b(y)} f(x, y)\mathrm{d}x = \int_{a(y)}^{b(y)} \frac{\partial f}{\partial y}(x, y)\mathrm{d}x + \frac{\mathrm{d}b(y)}{\mathrm{d}y} f(x = b, y) - \frac{\mathrm{d}a(y)}{\mathrm{d}y} f(x = a, y) \quad \text{(G.40)}$$

where $\partial f/\partial y$ is a function of both x and y. This expression can be used as the basis for performing functional derivatives: suppose that we have a functional $F[y]$ of a function $y(x)$ which involves the integration of another function of x and y over the variable x:

$$F[y] = \int_a^b f(x, y)\mathrm{d}x \quad \text{(G.41)}$$

where the limits of integration are constant. We want to find the condition that will give an extremum (maximum or minimum) of $F[y]$ with respect to changes in $y(x)$ in the interval $x \in [a, b]$. Suppose we change y to y' by adding the function $\delta y(x) = \epsilon\eta(x)$:

$$y'(x) = y(x) + \delta y(x) = y(x) + \epsilon\eta(x),$$

where ϵ is an infinitesimal quantity ($\epsilon \to 0$) and $\eta(x)$ is an arbitrary function that satisfies the conditions $\eta(a) = \eta(b) = 0$ so that we do not introduce changes in the limits of integration. The corresponding F will be given by

$$F[y'] = \int_a^b f(x, y')\mathrm{d}x = \int_a^b f(x, y + \epsilon\eta)\mathrm{d}x$$

Taking the derivative of this expression with respect to ϵ and setting it equal to zero to obtain an extremum, we find

$$\frac{\mathrm{d}F[y']}{\mathrm{d}\epsilon} = \int_a^b \frac{\partial f}{\partial y'}\frac{\mathrm{d}y'}{\mathrm{d}\epsilon}\mathrm{d}x = \int_a^b \frac{\partial f}{\partial y'}\eta(x)\mathrm{d}x = 0$$

and since we want this to be true for $\epsilon \to 0$ and arbitrary $\eta(x)$, we conclude that we must have

$$\frac{\delta F}{\delta y} \equiv \lim_{\epsilon\to 0}\frac{\partial f}{\partial y'}(x, y) = \frac{\partial f}{\partial y}(x, y) = 0$$

Thus, the necessary condition for finding the extremum of $F[y]$ with respect to variations in $y(x)$, when $F[y]$ has the form of Eq. (G.41), which we define as the variational functional derivative $\delta F/\delta y$, is the vanishing of the partial derivative with respect to y of the *integrand* appearing in F.

G.4 Fourier and inverse Fourier transforms

We derive the Fourier transform relations for one-dimensional functions of a single variable x and the corresponding Fourier space (also referred to as "reciprocal" space) functions with variable k. We start with the definition of the Fourier expansion of a function $f(x)$ with period $2L$ in terms of complex exponentials:

$$f(x) = \sum_{n=-\infty}^{\infty} f_n \mathrm{e}^{\mathrm{i}\frac{n\pi}{L}x} \tag{G.42}$$

with n taking all integer values, which obviously satisfies the periodic relation $f(x + 2L) = f(x)$. Using the orthogonality of the complex exponentials

$$\int_{-L}^{L} \mathrm{e}^{-\mathrm{i}\frac{m\pi}{L}x} \mathrm{e}^{\mathrm{i}\frac{n\pi}{L}x} \mathrm{d}x = (2L)\delta_{nm} \tag{G.43}$$

with δ_{nm} the Kronecker delta,

$$\delta_{nm} = \begin{cases} 1 & \text{for } n = m \\ 0 & \text{for } n \neq m \end{cases} \tag{G.44}$$

we can multiply both sides of Eq. (G.43) by $\exp(-\mathrm{i}m\pi x/L)$ and integrate over x to obtain

$$f_n = \frac{1}{2L} \int_{-L}^{L} f(x) \mathrm{e}^{-\mathrm{i}\frac{n\pi}{L}x} \tag{G.45}$$

We next define the reciprocal-space variable k through

$$k = \frac{n\pi}{L} \Rightarrow \mathrm{d}k = \frac{\pi}{L} \tag{G.46}$$

and take the limit $L \to \infty$, in which case k becomes a continuous variable, to obtain the Fourier transform $\tilde{f}(k)$:

$$\tilde{f}(k) \equiv \lim_{L\to\infty} [f_n 2L] = \int_{-\infty}^{\infty} f(x) \mathrm{e}^{-\mathrm{i}kx} \mathrm{d}x \tag{G.47}$$

Using Eq. (G.46) in Eq. (G.42), in the limit $L \to \infty$ we obtain the inverse Fourier transform:

$$f(x) = \lim_{L\to\infty} \sum_{n=-\infty}^{\infty} (f_n 2L) \frac{1}{2L} \mathrm{e}^{\mathrm{i}\frac{n\pi}{L}x} = \frac{1}{2\pi} \int_{-\infty}^{\infty} \tilde{f}(k) \mathrm{e}^{\mathrm{i}kx} \mathrm{d}k \tag{G.48}$$

The expressions for the Fourier and inverse Fourier transforms in 3D are straightforward generalizations of Eqs. (G.47) and (G.48) with x replaced by a vector $\mathbf{r}$ and k replaced by a vector $\mathbf{k}$:

$$\tilde{f}(\mathbf{k}) = \int f(\mathbf{r}) \mathrm{e}^{-\mathrm{i}\mathbf{k}\cdot\mathbf{r}} \mathrm{d}\mathbf{r} \Rightarrow f(\mathbf{r}) = \frac{1}{(2\pi)^3} \int \tilde{f}(\mathbf{k}) \mathrm{e}^{\mathrm{i}\mathbf{k}\cdot\mathbf{r}} \mathrm{d}\mathbf{k} \tag{G.49}$$

Example As an illustration, we calculate the Fourier and inverse Fourier transforms of the function $f(\mathbf{r}) = |\mathbf{r} - \mathbf{r}'|^{-1}$:

$$\int \mathrm{e}^{-\mathrm{i}\mathbf{k}\cdot\mathbf{r}} \frac{1}{|\mathbf{r} - \mathbf{r}'|} \mathrm{d}\mathbf{r} = \frac{4\pi}{k^2} \mathrm{e}^{-\mathrm{i}\mathbf{k}\cdot\mathbf{r}'} \Leftrightarrow \int \frac{\mathrm{d}\mathbf{k}}{(2\pi)^3} \frac{4\pi}{k^2} \mathrm{e}^{\mathrm{i}\mathbf{k}\cdot(\mathbf{r}-\mathbf{r}')} = \frac{1}{|\mathbf{r} - \mathbf{r}'|} \tag{G.50}$$

We notice first that we can shift the origin by $\mathbf{r}'$, so that, from the definition of the Fourier transform in 3D space Eq. (G.49), we need to calculate the following integral:

$$\int \mathrm{e}^{-\mathrm{i}\mathbf{k}\cdot\mathbf{r}} \frac{1}{|\mathbf{r}-\mathbf{r}'|} \mathrm{d}\mathbf{r} = \mathrm{e}^{-\mathrm{i}\mathbf{k}\cdot\mathbf{r}'} \int \mathrm{e}^{-\mathrm{i}\mathbf{k}\cdot(\mathbf{r}-\mathbf{r}')} \frac{1}{|\mathbf{r}-\mathbf{r}'|} \mathrm{d}\mathbf{r} = \mathrm{e}^{-\mathrm{i}\mathbf{k}\cdot\mathbf{r}'} \int \mathrm{e}^{-\mathrm{i}\mathbf{k}\cdot\mathbf{r}} \frac{1}{r} \mathrm{d}\mathbf{r}$$

The Fourier transform can now be calculated with a useful general trick: we multiply the function $1/r$ by $\exp(-\eta r)$, where $\eta > 0$, and in the end we will take the limit $\eta \to 0$. Using spherical coordinates $\mathrm{d}\mathbf{r} = r^2 \mathrm{d}r \sin\theta \mathrm{d}\theta \mathrm{d}\phi = \mathrm{d}x \mathrm{d}y \mathrm{d}z$, with the convenient choice of the z axis along the $\mathbf{k}$ vector (so that $\mathbf{k}\cdot\mathbf{r} = kr\cos(\theta)$), we obtain

$$\begin{aligned}
\int \frac{\mathrm{e}^{-\eta r}}{r} \mathrm{e}^{-\mathrm{i}\mathbf{k}\cdot\mathbf{r}} \mathrm{d}\mathbf{r} &= \int_0^{2\pi}\int_0^{\pi}\int_0^{\infty} \left[r^2 \sin\theta \frac{\mathrm{e}^{-\eta r}}{r} \mathrm{e}^{-\mathrm{i}kr\cos\theta} \right] \mathrm{d}r \mathrm{d}\theta \mathrm{d}\phi \\
&= 2\pi \int_0^{\infty}\int_{-1}^{1} \left[r\mathrm{e}^{-(\mathrm{i}kw+\eta)r} \right] \mathrm{d}w \mathrm{d}r \\
&= \frac{2\pi}{-\mathrm{i}k} \left(\left[\frac{1}{-\mathrm{i}k-\eta} \mathrm{e}^{(-\mathrm{i}k-\eta)r} \right]_0^{\infty} - \left[\frac{1}{\mathrm{i}k-\eta} \mathrm{e}^{(\mathrm{i}k-\eta)r} \right]_0^{\infty} \right) \\
&= \frac{4\pi}{k^2+\eta^2}
\end{aligned}$$

which, in the limit $\eta \to 0$, gives the desired result.

G.5 The δ-function and its Fourier transform

G.5.1 The δ-function and the θ-function

The standard definition of the δ-function, also known as the Dirac function, is

$$\delta(0) \to \infty, \quad \delta(x \neq 0) = 0, \quad \int_{-\infty}^{\infty} \delta(x - x')\mathrm{d}x' = 1 \tag{G.51}$$

that is, it is a function with an infinite peak at the zero of its argument, it is zero everywhere else, and it integrates to unity. From this definition, it follows that the product of the δ-function with an arbitrary function $f(x)$ integrated over all values of x must satisfy

$$\int_{-\infty}^{\infty} f(x')\delta(x - x')\mathrm{d}x' = f(x) \tag{G.52}$$

The δ-function is not a function in the usual sense, it is a function represented by a limit of usual functions. For example, a simple generalization of the Kronecker δ is

$$w(a; x) = \begin{cases} \frac{1}{2a} & \text{for } |x| \leq a \\ 0 & \text{for } |x| > a \end{cases} \tag{G.53}$$

We will refer to this as the "window" function, since its product with any other function picks out the values of that function in the range $-a \leq x \leq a$, that is, in a window of width $2a$ in x. Taking the limit of $w(a; x)$ for $a \to 0$ produces a function with the desired behavior to represent the δ-function. Typically, we are interested in smooth, continuous functions whose appropriate limit can represent a δ-function. We give below a few examples of such functions.

A useful representation of the δ-function is a gaussian

$$\delta(x - x') = \lim_{\beta \to 0} \left(\frac{1}{\beta\sqrt{\pi}} \mathrm{e}^{-(x-x')^2/\beta^2} \right), \tag{G.54}$$

To prove that this function has the proper behavior, we note that around $x = x'$ it has width β which vanishes for $\beta \to 0$, its value at $x = x'$ is $1/\beta\sqrt{\pi}$ which is infinite for $\beta \to 0$, and its integral over all values of x' is unity for any value of β (see discussion on normalized gaussians in section G.6).

Another useful representation of the δ-function is a lorentzian,

$$\delta(x - x') = \lim_{\beta \to 0} \left(\frac{1}{\pi} \frac{\beta}{(x - x')^2 + \beta^2} \right) \tag{G.55}$$

To prove that this function has the proper behavior, we note that around $x = x'$ it has width β, which vanishes for $\beta \to 0$, its value at $x = x'$ is $1/\pi\beta$ which is infinite for $\beta \to 0$, and its integral over all values of x' is unity for any value of β, which is easily shown by contour integration (the function has simple poles at $x = x' \pm \mathrm{i}\beta$).

Yet another useful representation of the δ-function is the following:

$$\delta(x - x') = \lim_{\beta \to 0} \left[\frac{1}{\pi\beta} \sin\left(\frac{x - x'}{\beta} \right) \frac{\beta}{x - x'} \right] \tag{G.56}$$

To prove that this function has the proper behavior, we note that the function $\sin(y)/y \to 1$ for $y \to 0$ and its width around $y = 0$ is determined by half the interval between y values at which it first becomes zero, which occurs at $y = \pm\pi$, that is, its width in the variable y is π. With $y = (x - x')/\beta$, we conclude that around $x = x'$ the function on the right-hand side of Eq. (G.56) has width $\beta\pi$ in the variable x, which vanishes for $\beta \to 0$, while its value at $x = x'$ is $1/\pi\beta$ which is infinite for $\beta \to 0$. Finally, its integral over all values of x' is unity for any value of β, which is easily shown by expressing

$$\sin\left(\frac{x - x'}{\beta} \right) \frac{1}{x - x'} = \frac{\mathrm{e}^{\mathrm{i}(x-x')/\beta}}{2\mathrm{i}(x - x')} - \frac{\mathrm{e}^{-\mathrm{i}(x-x')/\beta}}{2\mathrm{i}(x - x')}$$

and using contour integration (each term in the integrand has a simple pole at $x = x'$).

In order to construct δ-functions in more than one dimension we simply multiply δ-functions in each of the independent dimensions. For example, in cartesian coordinates we would have

$$\delta(\mathbf{r} - \mathbf{r}') = \delta(x - x')\delta(y - y')\delta(z - z')$$

where $\mathbf{r} = x\hat{\mathbf{x}} + y\hat{\mathbf{y}} + z\hat{\mathbf{z}}$ and $\mathbf{r}' = x'\hat{\mathbf{x}} + y'\hat{\mathbf{y}} + z'\hat{\mathbf{z}}$ are two 3D vectors. A useful representation of the δ-function in 3D is given by

$$\delta(\mathbf{r} - \mathbf{r}') = \frac{1}{4\pi} \nabla_{\mathbf{r}} \cdot \left(\frac{\mathbf{r} - \mathbf{r}'}{|\mathbf{r} - \mathbf{r}'|^3} \right) \tag{G.57}$$

This can be proved as follows: the expression on the right-hand side of this equation is identically zero everywhere except at $x = x'$, $y = y'$, $z = z'$, at which point it becomes infinite (both claims can be easily shown by direct application of the divergence expression, Eq. (G.4)); these are the two essential features of the δ-function. Moreover, using the divergence theorem, we have:

$$\int \nabla_{\mathbf{r}} \cdot \left(\frac{\mathbf{r} - \mathbf{r}'}{|\mathbf{r} - \mathbf{r}'|^3} \right) \mathrm{d}\mathbf{r}' = \oint_{S'} \left(\frac{\mathbf{r} - \mathbf{r}'}{|\mathbf{r} - \mathbf{r}'|^3} \right) \cdot \hat{\mathbf{n}}_{S'} \mathrm{d}S'$$

where S' is a surface enclosing the volume of integration and $\hat{\mathbf{n}}_{S'}$ is the surface-normal unit vector. Choosing the volume of integration to be a sphere centered at $\mathbf{r}$, in which case the surface-normal unit vector takes the form

$$\hat{\mathbf{n}}_{S'} = \frac{\mathbf{r} - \mathbf{r}'}{|\mathbf{r} - \mathbf{r}'|}$$

we obtain for the surface integral

$$\oint_{S'} \left(\frac{\mathbf{r} - \mathbf{r}'}{|\mathbf{r} - \mathbf{r}'|^3} \right) \cdot \hat{\mathbf{n}}_{S'} \mathrm{d}S' = \int_0^{2\pi} \int_0^{\pi} \frac{1}{|\mathbf{r} - \mathbf{r}'|^2} |\mathbf{r} - \mathbf{r}'|^2 \sin\theta \mathrm{d}\theta \mathrm{d}\phi = 4\pi$$

This completes the proof that the right-hand side of Eq. (G.57) is a properly normalized δ-function in 3D.

A function closely related to the δ-function is the so called θ-function or step function, also known as the Heavyside function:

$$\theta(x - x') = 0, \ \text{for } x < x', \quad \theta(x - x') = 1, \ \text{for } x > x' \tag{G.58}$$

A useful representation of the θ-function is

$$\theta(x - x') = \lim_{\beta \to 0} \left(\frac{1}{\mathrm{e}^{-(x-x')/\beta} + 1} \right) \tag{G.59}$$

The δ-function and the θ-function are functions of their argument only in the limiting sense given, for example, in Eqs. (G.54) and (G.59). The behavior of these functions for several values of β is shown in Fig. G.1. From the definition of the θ-function, we can see that its derivative must be a δ-function:

$$\frac{\mathrm{d}}{\mathrm{d}x}\theta(x - x') = \delta(x - x')$$

It is easy to show that the derivative of the representation of the θ-function given in Eq. (G.59)

$$\lim_{\beta \to 0} \left(\frac{\mathrm{e}^{-(x-x')/\beta}}{\beta(\mathrm{e}^{-(x-x')/\beta} + 1)^2} \right)$$

is indeed another representation of the δ-function, as expected.

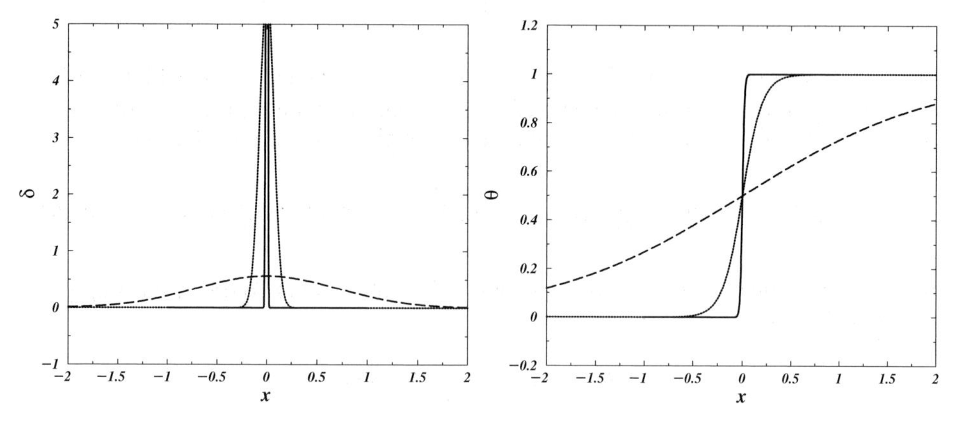

Figure G.1. **Left:** the δ-function as represented by a normalized gaussian Eq. (G.54), for $\beta = 1$ (dashed line), 0.1 (dotted line) and 0.01 (solid line). **Right:** the θ-function as represented by Eq. (G.59), for $\beta = 1$ (dashed line), 0.1 (dotted line) and 0.01 (solid line).

An expression often arises in which the δ-function has as its argument another function, $f(x)$, and the integral of its product with yet another function, $g(x)$, must be evaluated:

$$I = \int_{-\infty}^{\infty} g(x)\delta(f(x))\mathrm{d}x$$

We consider the result of this evaluation, assuming that $f(x)$ has only simple roots, that is, $f(x_0) = 0$, $f'(x_0) \neq 0$, with $f'(x)$ the derivative of $f(x)$. Since the δ-function is zero when its argument is not zero, we work first only in the neighborhood of one simple root x_0 of $f(x)$, that is, $x \in [x_1, x_2]$, $x_1 < x_0 < x_2$. We define the variable $u = f(x)$, which at $x = x_0, x_1, x_2$ takes the values $0, u_1, u_2$, respectively. From this definition we have

$$\mathrm{d}u = f'(x)\mathrm{d}x \Longrightarrow \mathrm{d}x = \left(f'(x)\right)^{-1}\mathrm{d}u, \quad x = f^{-1}(u)$$

These expressions, when inserted into the integral under consideration with x in the neighborhood of x_0, give

$$I_0 = \int_{u_1}^{u_2} g(f^{-1}(u))\delta(u)\left(f'(f^{-1}(u))\right)^{-1}\mathrm{d}u$$

In order to evaluate this integral, we only need to consider the order of the limits: with x_1, x_2 close to x_0, we will have

$$f'(x_0) > 0 \Rightarrow f(x_1) = u_1 < f(x_0) = 0 < f(x_2) = u_2$$
$$f'(x_0) < 0 \Rightarrow f(x_1) = u_1 > f(x_0) = 0 > f(x_2) = u_2$$

which gives in turn

$$f'(x_0) > 0 \Rightarrow I_0 = \int_{u_1}^{u_2} g(f^{-1}(u))\delta(u)\left(f'(f^{-1}(u))\right)^{-1}\mathrm{d}u = g(x_0)\left|f'(x_0)\right|^{-1}$$

$$f'(x_0) < 0 \Rightarrow I_0 = -\int_{u_2}^{u_1} g(f^{-1}(u))\delta(u)\left[-\left|f'(f^{-1}(u))\right|^{-1}\right]\mathrm{d}u = g(x_0)\left|f'(x_0)\right|^{-1}$$

Thus, in both cases we find that the integral is equal to the product of $g(x)$ with $\left|f'(x)\right|^{-1}$ evaluated at $x = x_0$. Since the argument of $\delta(f(x))$ vanishes at all the roots $x_0^{(i)}$ of $f(x)$, we will obtain similar contributions from each root, which gives as a final result

$$\int_{-\infty}^{\infty} g(x)\delta(f(x))\mathrm{d}x = \sum_i g(x_0^{(i)})\left|f'(x_0^{(i)})\right|^{-1}, \quad \text{where } f(x_0^{(i)}) = 0 \tag{G.60}$$

with the summation on i running over all the simple roots of $f(x)$. These results imply directly the following relations:

$$\delta(a(x - x_0)) = \frac{1}{|a|}\delta(x - x_0)$$

$$\delta(f(x)) = \sum_i \left|f'(x_0^{(i)})\right|^{-1}\delta(x - x_0), \quad \text{where } f(x_0^{(i)}) = 0$$

G.5.2 Fourier transform of the δ-function

To derive the Fourier transform of the δ-function we start with its Fourier expansion representation as in Eq. (G.42):

$$\delta(x) = \sum_{n=-\infty}^{\infty} \mathrm{d}_n \mathrm{e}^{\mathrm{i}\frac{n\pi}{L}x} \Rightarrow \mathrm{d}_n = \frac{1}{2L}\int_{-L}^{L} \delta(x)\mathrm{e}^{-\mathrm{i}\frac{n\pi}{L}x}\mathrm{d}x = \frac{1}{2L} \tag{G.61}$$

which in the limit $L \to \infty$ produces

$$\tilde{\delta}(k) = \lim_{L\to\infty}(2Ld_n) = 1 \tag{G.62}$$

Therefore, the inverse Fourier transform of the δ-function gives

$$\delta(x - x') = \frac{1}{2\pi}\int_{-\infty}^{\infty} \tilde{\delta}(k)\mathrm{e}^{\mathrm{i}k(x-x')}\mathrm{d}k = \frac{1}{2\pi}\int_{-\infty}^{\infty} \mathrm{e}^{\mathrm{i}k(x-x')}\mathrm{d}k \tag{G.63}$$

These relations for a function of the one-dimensional variable x can be straightforwardly generalized to functions of three-dimensional variables $\mathbf{r} = x\hat{\mathbf{x}} + y\hat{\mathbf{y}} + z\hat{\mathbf{z}}$ and $\mathbf{k} = k_x\hat{\mathbf{x}} + k_y\hat{\mathbf{y}} + k_z\hat{\mathbf{z}}$:

$$\delta(\mathbf{r} - \mathbf{r}') = \frac{1}{(2\pi)^3}\int \mathrm{e}^{\mathrm{i}(\mathbf{r}-\mathbf{r}')\cdot\mathbf{k}}\mathrm{d}\mathbf{k} \tag{G.64}$$

or equivalently, by inverting the roles of $\mathbf{k}$ and $\mathbf{r}$, we obtain

$$\delta(\mathbf{k} - \mathbf{k}') = \frac{1}{(2\pi)^3}\int \mathrm{e}^{-\mathrm{i}(\mathbf{k}-\mathbf{k}')\cdot\mathbf{r}}\mathrm{d}\mathbf{r} \tag{G.65}$$

Notice that both δ-functions are symmetric:

$$\delta(\mathbf{r} - \mathbf{r}') = \delta(\mathbf{r}' - \mathbf{r}), \quad \delta(\mathbf{k} - \mathbf{k}') = \delta(\mathbf{k}' - \mathbf{k})$$

G.5.3 The δ-function sums for crystals

Consider the sum over all points in the first Brillouin Zone (BZ): $\sum_{\mathbf{k}\in BZ}\exp(\mathrm{i}\mathbf{k}\cdot\mathbf{R})$. Let us assume that all $\mathbf{k}$ vectors lie within the first BZ. If we shift $\mathbf{k}$ by an arbitrary vector $\mathbf{k}_0 \in BZ$, this sum does not change, because for a vector $\mathbf{k} + \mathbf{k}_0$ which lies outside the first BZ there is a vector within the first BZ that differs from it by $\mathbf{G}$ and $\exp(\mathrm{i}\mathbf{G}\cdot\mathbf{R}) = 1$. Moreover, all points translated outside the first BZ by the addition of $\mathbf{k}_0$ have exactly one image by translation inside the first BZ through $\mathbf{G}$, since all BZs have the same volume. Therefore we have

$$\sum_{\mathbf{k}\in BZ}\mathrm{e}^{\mathrm{i}\mathbf{k}\cdot\mathbf{R}} = \sum_{\mathbf{k}\in BZ}\mathrm{e}^{\mathrm{i}(\mathbf{k}+\mathbf{k}_0)\cdot\mathbf{R}} = \mathrm{e}^{\mathrm{i}\mathbf{k}_0\cdot\mathbf{R}}\left[\sum_{\mathbf{k}\in BZ}\mathrm{e}^{\mathrm{i}\mathbf{k}\cdot\mathbf{R}}\right] \Rightarrow \mathrm{e}^{\mathrm{i}\mathbf{k}_0\cdot\mathbf{R}} = 1 \Rightarrow \mathbf{R} = 0 \tag{G.66}$$

where the last equality must hold because the relation $\exp(\mathrm{i}\mathbf{k}_0\cdot\mathbf{R}) = 1$ must hold for any $\mathbf{k}_0 \in BZ$, assuming that the sum itself is not zero. Thus we have proved that the sum $\sum_{\mathbf{k}\in BZ}\exp(\mathrm{i}\mathbf{k}\cdot\mathbf{R})$, if it is not equal to zero, necessitates that $\mathbf{R} = 0$, in which case all terms in the sum are unity and the sum itself becomes equal to N, the total number of PUCs in the crystal. If $\mathbf{R} \neq 0$ then $\exp(\mathrm{i}\mathbf{k}_0\cdot\mathbf{R}) \neq 1$ for an arbitrary value of $\mathbf{k}_0$, and therefore the sum itself must be zero in order for the original equality in Eq. (G.66) to hold. These results put together imply

$$\frac{1}{N}\sum_{\mathbf{k}\in BZ} \mathrm{e}^{\mathrm{i}\mathbf{k}\cdot\mathbf{R}} = \delta(\mathbf{R}) \tag{G.67}$$

where $\delta(\mathbf{R})$ represents a Kronecker δ, defined in Eq. (G.44): $\delta(\mathbf{R}) = 1$ for $\mathbf{R} = 0$ and $\delta(\mathbf{R}) = 0$ for $\mathbf{R} \neq 0$. This is consistent with the fact that, strictly speaking, the summation over $\mathbf{k}$ vectors is a discrete sum over the N values that $\mathbf{k}$ takes in the BZ. In the limit of an infinite crystal, $N \to \infty$, it is customary to think of $\mathbf{k}$ as a continuous variable and to treat $\delta(\mathbf{k})$ as a δ-function rather than a Kronecker δ. The definition of $\delta(\mathbf{R})$ implies that for an arbitrary function $f(\mathbf{R})$ we will have

$$\frac{1}{N}\sum_{\mathbf{R}}\sum_{\mathbf{k}\in BZ} f(\mathbf{R})\mathrm{e}^{\mathrm{i}\mathbf{k}\cdot\mathbf{R}} = \sum_{\mathbf{R}} f(\mathbf{R})\delta(\mathbf{R}) = f(0) \tag{G.68}$$

By a similar line of argument we can show that

$$\frac{1}{N}\sum_{\mathbf{R}} \mathrm{e}^{\mathrm{i}\mathbf{k}\cdot\mathbf{R}} = \delta(\mathbf{k} - \mathbf{G}) \tag{G.69}$$

where we treat the summation over $\mathbf{R}$ vectors as a discrete sum, in which case the right-hand side is a Kronecker δ. To prove this, we change every vector $\mathbf{R}$ in the argument of the exponential of the sum by $\mathbf{R}_0$; but each new vector $\mathbf{R} + \mathbf{R}_0$ is also a translational vector, so that the sum does not change because the summation extends already over all possible translations, that is:

$$\sum_{\mathbf{R}} \mathrm{e}^{\mathrm{i}\mathbf{k}\cdot\mathbf{R}} = \sum_{\mathbf{R}} \mathrm{e}^{\mathrm{i}\mathbf{k}\cdot(\mathbf{R}+\mathbf{R}_0)} = \mathrm{e}^{\mathrm{i}\mathbf{k}\cdot\mathbf{R}_0}\left[\sum_{\mathbf{R}} \mathrm{e}^{\mathrm{i}\mathbf{k}\cdot\mathbf{R}}\right] \Rightarrow \mathrm{e}^{\mathrm{i}\mathbf{k}\cdot\mathbf{R}_0} = 1 \Rightarrow \mathbf{k} = \mathbf{G} \tag{G.70}$$

where the last equation is a consequence of the fact that the relation $\exp(\mathrm{i}\mathbf{k}\cdot\mathbf{R}_0) = 1$ must hold for an arbitrary translational vector $\mathbf{R}_0$, assuming that the sum itself is not zero. If $\mathbf{k} = \mathbf{G}$, a reciprocal-space vector, then all terms in the above sum are unity, and the sum itself is equal to N (there are N terms in the sum, N being the total number of PUCs in the crystal). If $\mathbf{k} \neq \mathbf{G}$, then the sum itself must be zero, so that the original equality in Eq. (G.70) can still hold with $\mathbf{R}_0$ an arbitrary translational vector. The two results together imply the relation Eq. (G.69), where $\delta(\mathbf{k})$ represents a Kronecker δ, defined in Eq. (G.44): $\delta(\mathbf{k} - \mathbf{G}) = 1$ for $\mathbf{k} = \mathbf{G}$ and $\delta(\mathbf{k} - \mathbf{G}) = 0$ for $\mathbf{k} \neq \mathbf{G}$. This is consistent with the fact that the summation over $\mathbf{R}$ vectors is a discrete sum over the N values that $\mathbf{R}$ takes for the entire crystal. The definition of $\delta(\mathbf{k} - \mathbf{G})$ implies that for an arbitrary function $g(\mathbf{k})$, if $\mathbf{k}$ is restricted to the first Brillouin Zone we will have:

$$\frac{1}{N}\sum_{\mathbf{k}\in BZ}\sum_{\mathbf{R}} g(\mathbf{k})\mathrm{e}^{\mathrm{i}\mathbf{k}\cdot\mathbf{R}} = \sum_{\mathbf{k}\in BZ} g(\mathbf{k})\delta(\mathbf{k}) = g(0) \tag{G.71}$$

because the only reciprocal lattice vector contained in the first BZ is $\mathbf{G} = 0$.

G.6 Normalized gaussians

In order to find the constant factor which will ensure normalization of the gaussian function we must calculate its integral over all values of the variable. This integral is most easily performed by using Gauss's trick of taking the square root of the square of the integral and turning the integration over cartesian coordinates $dxdy$ into one over polar coordinates $rdrd\theta$:

$$
\begin{aligned}
\int_{-\infty}^{\infty} e^{-\alpha^2 x^2} dx &= \left[\int_{-\infty}^{\infty} e^{-\alpha^2 x^2} dx \int_{-\infty}^{\infty} e^{-\alpha^2 y^2} dy \right]^{\frac{1}{2}} \\
&= \left[\int_{-\infty}^{\infty} \int_{-\infty}^{\infty} e^{-\alpha^2 (x^2+y^2)} dxdy \right]^{\frac{1}{2}} = \left[\int_{0}^{2\pi} \int_{0}^{\infty} e^{-\alpha^2 r^2} rdrd\theta \right]^{\frac{1}{2}} \\
&= \left[2\pi \int_{0}^{\infty} e^{-\alpha^2 r^2} \frac{1}{2} dr^2 \right]^{\frac{1}{2}} = \frac{\sqrt{\pi}}{\alpha} \qquad \text{(G.72)}
\end{aligned}
$$

so that in three dimensions the normalized gaussian centered ar $\mathbf{r}_0$ is

$$
\left(\frac{\alpha}{\sqrt{\pi}} \right)^3 e^{-\alpha^2 |\mathbf{r}-\mathbf{r}_0|^2} \Longrightarrow \left(\frac{\alpha}{\sqrt{\pi}} \right)^3 \int e^{-\alpha^2 |\mathbf{r}-\mathbf{r}_0|^2} d\mathbf{r} = 1 \qquad \text{(G.73)}
$$

In the limit of $\alpha \to \infty$ this function tends to an infinitely sharp spike centered at $\mathbf{r} = \mathbf{r}_0$ which integrates to unity, that is, a properly defined δ-function (compare with Eq. (G.54), with $\beta = 1/\alpha$).

The above derivation provides an easy way of calculating the even moments of the normalized gaussian, by simply taking derivatives with respect to α^2 of both sides of Eq. (G.72):

$$
\begin{aligned}
\int_{-\infty}^{\infty} x^2 e^{-\alpha^2 x^2} dx &= -\frac{\partial}{\partial(\alpha^2)} \int_{-\infty}^{\infty} e^{-\alpha^2 x^2} dx = \frac{1}{2} \frac{\sqrt{\pi}}{\alpha^3} \\
\int_{-\infty}^{\infty} x^4 e^{-\alpha^2 x^2} dx &= -\frac{\partial}{\partial(\alpha^2)} \int_{-\infty}^{\infty} x^2 e^{-\alpha^2 x^2} dx = \frac{3}{4} \frac{\sqrt{\pi}}{\alpha^5} \\
\int_{-\infty}^{\infty} x^{2n} e^{-\alpha^2 x^2} dx &= -\frac{\partial^n}{\partial(\alpha^2)^n} \int_{-\infty}^{\infty} e^{-\alpha^2 x^2} dx = \frac{(2n-1)!!}{2^n} \frac{\sqrt{\pi}}{\alpha^{2n+1}}
\end{aligned}
$$

where the symbol $(2n - 1)!!$ denotes the product of all odd integers up to $(2n - 1)$.

Appendix H

Nobel prize citations

- 1954 Nobel prize for Chemistry: Linus Carl Pauling, for his research into the nature of the chemical bond and its application to the elucidation of the structure of complex substances.
- 1956 Nobel prize for Physics: William Shockley, John Bardeen and Walter Houser Brattain, for their research in semiconductors and their discovery of the transistor effect.
- 1962 Nobel prize for Medicine: Francis Harry Compton Crick, James Dewey Watson and Maurice Hugh Frederick Wilkins, for their discoveries concerning the molecular structure of nucleic acids and its significance for information transfer in living material.
- 1968 Nobel prize for Medicine: Robert W. Holley, Hav Gobind Khorana and Marshall W. Nirenberg, for their interpretation of the genetic code and its function in protein synthesis.
- 1972 Nobel prize for Physics: John Bardeen, Leon N. Cooper and J. Robert Schrieffer, for their jointly developed theory of superconductivity, usually called the BCS theory.
- 1977 Nobel prize for Physics: Philip W. Anderson, Sir Nevill F. Mott and John H. van Vleck, for their fundamental theoretical investigations of the electronic structure of magnetic and disordered systems.
- 1985 Nobel prize for Chemistry: Herbert A. Hauptman and Jerome Karle, for their outstanding achievements in the development of direct methods for the determination of crystal structures.
- 1985 Nobel prize for Physics: Klaus von Klitzing, for the discovery of the quantized Hall effect.
- 1986 Nobel prize for Physics: Gerd Binning and Heinrich Rohrer, for their design of the scanning tunneling microscope; Ernst Ruska, for his fundamental work in electron optics, and for the design of the first electron microscope.
- 1987 Nobel prize for Physics: J. Georg Bednorz and K. Alexander Müller, for their important breakthrough in the discovery of superconductivity in ceramic materials.
- 1991 Nobel prize for Physics: Pierre-Gilles de Gennes, for discovering that methods developed for studying order phenomena in simple systems can be generalized to more complex forms of matter, in particular liquid crystals and polymers.
- 1996 Nobel prize for Chemistry: Robert F. Curl, Harold W. Kroto and Richard E. Smalley, for their discovery of fullerenes.
- 1998 Nobel prize for Physics: R.B. Laughlin, D.C. Tsui and H.L. Störmer, for their discovery of a new form of quantum fluid with fractionally charged excitations.

- 1998 Nobel prize for Chemistry: Walter Khon for his development of density functional theory, and John A. Pople for his development of computational methods in quantum chemistry.
- 2000 Nobel prize for Physics: Zhores I. Alferov and Herbert Kroemer for developing semiconductor heterostructures used in high-speed and opto-electronics and Jack St. Clair Kilby, for his part in the invention of the integrated circuit.
- 2000 Nobel prize for Chemistry: Alan J. Heeger, Alan G. MacDiarmid and Hideki Shirakawa, for the development of conducting polymers.

Appendix I

Units and symbols

Table I.1. *Units of physical quantities relevant to the physics of solids.*

Quantity	Unit	Symbol	Relations
Charge	electron charge	e	$e = 1.60219 \times 10^{-19}$ C
Mass	electron mass	m_e	$m_e = 9.1094 \times 10^{-28}$ g
	proton mass	m_p	$m_p = 1836.15\ m_e$
Length	Bohr radius	a_0	$1\ a_0 = 0.529177 \times 10^{-8}$ cm
	angstrom	Å	1 Å $= 10^{-8}$ cm
Energy	Hartree $= e^2/a_0$	H	$1\ \mathrm{H} = 4.3597 \times 10^{-11}$ erg
	Rydberg $= \hbar^2/2m_e a_0^2$	Ry	1 H = 2 Ry
	electron volt	eV	1 Ry = 13.6058 eV
	kcal/mol	kcal/mol	1 eV = 23.06 kcal/mol
	k_{B} kelvin	K	$1\ \mathrm{eV}/k_{\mathrm{B}} = 1.1604 \times 10^4$ K
	$\hbar$ hertz $= (2\pi\hbar)\mathrm{s}^{-1}$	Hz	$1\ \mathrm{eV}/\hbar = 2.4179 \times 10^{14}$ Hz
Pressure	H/a_0^3		294.21×10^{12} dyn/cm^2
	megabar	Mbar	$1\ \mathrm{H}/a_0^3 = 294.21$ Mbar
	pascal	Pa	1 Mbar = 100 GPa
Resistance	$\hbar a_0/e^2$		21.73985 μΩ· cm

Table I.2. *Names of the power-ten multiples and fractions of a unit.*

Name (symbol)	femto (f)	pico (p)	nano (n)	micro (μ)	milli (m)	kilo (k)	mega (M)	giga (G)	tera (T)	peta (P)
	10^{-15}	10^{-12}	10^{-9}	10^{-6}	10^{-3}	10^{3}	10^{6}	10^{9}	10^{12}	10^{15}

References

[1] J. Friedel, J. de Phys. (Paris) **35**, L-59 (1974).
[2] F.C. Frank, Proc. Roy. Soc. A **215**, 43 (1952).
[3] F.C. Frank and J.S. Kasper, Acta Cryst. **11**, 184 (1958).
[4] F.C. Frank and J.S. Kasper, Acta Cryst. **12**, 483 (1959).
[5] D. Schechtman, I. Blech, D. Gratias and J.W. Cahn, Phys. Rev. Lett. **53**, 1951 (1984).
[6] N.H. Fletcher, *The Chemical Physics of Ice* (Cambridge University Press, Cambridge, 1970).
[7] J.D. Bernal and R.H. Fowler, J. Chem. Phys. **1**, 515–548 (1933).
[8] L. Pauling, J. Am. Chem. Soc. **57**, 2680 (1935).
[9] E.D. Isaacs, A. Shukla, P.M. Platzman, D.R. Hamann, B. Barbiellini and C.A. Tulk, Phys. Rev. Lett. **82**, 600 (1999).
[10] E. Wigner and H.B. Huntington, J. Chem. Phys. **3**, 764 (1935).
[11] I.F. Silvera, Rev. Mod. Phys. **52**, 393 (1980).
[12] I.F. Silvera, Proceedings of the Finnish Physical Society Annual Meeting – Arkhimedes **2**, 108 (1992).
[13] H.K. Mao and R.J. Hemley, American Scientist **80**, 234 (1992).
[14] N.W. Ashcroft and N.D. Mermin, *Solid State Physics* (Saunders College Publishing, Philadelphia, 1976).
[15] H.A. Bethe and R.W. Jackiw, *Intermediate Quantum Mechanics* (Benjamin/Cummings, Reading, Massachusetts, 1968).
[16] J.C. Slater, Phys. Rev. **34**, 1293 (1929).
[17] A.L. Fetter and J.D. Walecka, *Quantum Theory of Many-Particle Systems* (McGraw-Hill, New York, 1971).
[18] J.C. Slater, Phys. Rev. **87**, 385 (1951).
[19] J. Hubbard, Proc. Roy. Soc. A **276**, 238 (1963).
[20] J. Hubbard, Proc. Roy. Soc. A **277**, 237 (1964).
[21] J. Hubbard, Proc. Roy. Soc. A **281**, 401 (1964).
[22] P. Hohenberg and W. Kohn, Phys. Rev. **136**, B864 (1964).
[23] W. Kohn and L. Sham, Phys. Rev. **140**, A1133 (1965).
[24] L. Hedin and B.I. Lundqvist, J. Phys. C **4**, 2064 (1971).
[25] J.P. Perdew and A. Zunger, Phys. Rev. B **23**, 5048 (1981).
[26] D. M. Ceperley and B. J. Alder, Phys. Rev. Lett. **45**, 566 (1980).
[27] J.P. Perdew and Y. Wang, Phys. Rev. B **33**, 8800 (1986).
[28] J.C. Phillips and L. Kleinman, Phys. Rev. **116**, 287 (1959).
[29] D.R. Hamann, M. Schlüter and C. Chang, Phys. Rev. Lett. **43**, 1494 (1979).

[30] G. Bachelet, D.R. Hamann and M. Schlüter, Phys. Rev. B **26**, 4199 (1982).
[31] A.M. Rappe, K.M. Rabe, E. Kaxiras and J.D. Joannopoulos, Phys. Rev. B **41**, 1227 (1990).
[32] D. Vanderbilt, Phys. Rev. B **41**, 7892 (1990).
[33] N. Troullier and J.L. Martins, Phys. Rev. B **43**, 8861 (1991).
[34] A. Baldereschi, Phys. Rev. B **7**, 5212 (1973).
[35] D.J. Chadi and M.L. Cohen, Phys. Rev. B **7**, 692 (1973).
[36] H.J. Monkhorst and J.D. Pack, Phys. Rev. B **13**, 5188 (1976).
[37] W.A. Harrison, *Electronic Structure and the Properties of Solids* (W.H. Freeman, San Francisco, 1980).
[38] P. Blaudeck, Th. Frauenheim, D. Porezag, G. Seifert and E. Fromm, J. Phys.: Cond. Mat. C **4**, 6389 (1992).
[39] R.E. Cohen, M.J. Mehl, and D.A. Papaconstantopoulos, Phys. Rev. B **50**, 14694 (1994).
[40] M.J. Mehl and D.A. Papaconstantopoulos, Phys. Rev. B **54**, 4519 (1996).
[41] D.A. Papaconstantopoulos, *Handbook of the Band Structure of Elemental Solids* (Plenum Press, New York, 1989).
[42] D.J. Chadi, Phys. Rev. B **29**, 785 (1984).
[43] M. Elstner, D. Porezag, M. Haugk, J. Elsner, Th. Frauenheim, G. Seifert and S. Suhai, Phys. Rev. B **58**, 7260 (1998).
[44] E. Wigner and F. Seitz, Phys. Rev. **43**, 804 (1933).
[45] J.C. Slater, Phys. Rev. **51**, 846 (1937).
[46] D. Singh, *Planewaves, Pseudopotentials and the LAPW Method* (Kluwer Academic, Boston, 1994).
[47] C. Herring, Phys. Rev. **57**, 1169 (1940).
[48] M. Hybertsen and S. Louie, Phys. Rev. Lett **55** 1418 (1985); Phys. Rev. B **34** 5390 (1986).
[49] R. Godby, M. Schlüter, and L. Sham, Phys. Rev. Lett. **56**, 2415, (1986); Phys. Rev. B **37**, 10159, (1988).
[50] L. Fritsche, Phys. Rev. B **33**, 3976 (1986); Physica B **172**, 7 (1991).
[51] I. N. Remediakis and E. Kaxiras, Phys. Rev. B **59**, 5536 (1999).
[52] Z.W. Lu, A. Zunger and M. Deusch, Phys. Rev. B **47**, 9385 (1993).
[53] L.F. Mattheis, Phys. Rev. Lett. **58**, 1028 (1987).
[54] O. Jepsen and O.K. Andersen, Solid State Comm. **9**, 1763 (1971).
[55] G. Lehmann and M. Taut, Phys. Stat. Solidi **54**, 469 (1972).
[56] H. Ehrenreich and M.H. Cohen, Phys. Rev. **115**, 786 (1959).
[57] R. Kubo, J. Phys. Soc. Japan **12** 570 (1957); D. A. Greenwood, Proc. Phys. Soc. (London) **A71**, 585 (1958).
[58] H. Ehrenreich and H.R. Philipp, Phys. Rev. **128**, 1622 (1962).
[59] H.R. Philipp and H. Ehrenreich, Phys. Rev. **129**, 1550 (1963).
[60] H.R. Philipp and H. Ehrenreich, in *Semiconductors and Semimetals*, edited by R.K. Willardson and A.C. Beer, vol. 3, p. 93 (Academic Press, 1967).
[61] H. Ehrenreich, IEEE Spectrum **2**, 162 (1965).
[62] J. Frenkel, Phys. Rev. **37**, 1276 (1931).
[63] N. Mott, Trans. Faraday Soc. **34**, 500 (1938); Wannier, Phys. Rev. **52**, 191 (1937).
[64] C. Kittel, *Introduction to Solid State Physics* (Seventh Edition, J. Wiley, New York, 1996).
[65] J.R. Chelikowsky and M.L. Cohen, "Ab initio pseudopotentials and the structural properties of semiconductors", in *Handbook on Semiconductors*, edited by P.T. Landsberg, vol. 1, p. 59 (North Holland, 1992).

[66] J. Ihm, A. Zunger and M.L. Cohen, J. Phys. C: Sol. St. Phys. **12**, 4409 (1979).
[67] M. T. Yin and M. L. Cohen, Phys. Rev. B **26**, 5668 (1982).
[68] L.L. Boyer, E. Kaxiras, J.L. Feldman, J.Q. Broughton and M.J. Mehl, Phys. Rev. Lett. **67**, 715 (1991); E. Kaxiras and L.L. Boyer, Modelling and Simulation in Materials Science and Engineering **1**, 91 (1992).
[69] M.L. Cohen, Phys. Rev. B **32**, 7988 (1985).
[70] J.C. Phillips, *Bonds and Bands in Semiconductors* (Academic Press, New York, 1973).
[71] J.H. Rose, J.R. Smith, F. Guinea and J. Ferrante, Phys. Rev. B **29**, 2963 (1984).
[72] A.Y. Liu and M.L. Cohen, Phys. Rev. B **41**, 10727 (1990).
[73] R. Car and M. Parrinello, Phys. Rev. Lett. **55**, 2471 (1985).
[74] D.R. Lide, *CRC Handbook of Chemistry and Physics* (CRC Press, Boca Raton, 1999–2000).
[75] V.L. Moruzzi, J.F. Janak and A.R. Williams, *Calculated Electronic Properties of Metals* (Pergamon Press, New York, 1978).
[76] *Semiconductors*, Landolt-Börnstein: Numerical Data and Functional Relationships in Science and Technology, New Series, O. Madelung, M. Scholz and H. Weiss, **17**, 135 (Springer-Verlag, Berlin, 1982).
[77] M. T. Yin and M. L. Cohen, Phys. Rev. B **26**, 3259 (1982).
[78] R. Stedman, L. Almqvist, and G. Nilsson, Phys. Rev. **162**, 549 (1967).
[79] J. de Launay, *Solid State Physics* **2**, edited by F. Seitz and D. Turnbul (Academic Press, Boston, 1956).
[80] R.E. Peierls, *Quantum Theory of Solids* (Oxford Univeristy Press, Oxford, 1955).
[81] E.C. Stoner, Proc. Roy. Soc. A **165**, 372 (1938); Proc. Roy. Soc. A **169**, 339 (1939).
[82] W. Jones and N.H. March, *Theoretical Solid State Physics* vol. 1 and 2 (J. Wiley, London, 1973).
[83] H. Vosko and J.J. Perdew, Can. J. Phys. **53**, 1385 (1975).
[84] O. Gunnarsson, J. Phys. F **6**, 587 (1976).
[85] J.F. Janak, Phys. Rev. B **16**, 255 (1977).
[86] N.D. Mermin and H. Wagner, Phys. Rev. Lett. **17**, 1133 (1966).
[87] K. Huang and E. Manousakis, Phys. Rev. B **36**, 8302 (1987).
[88] A.B. Harris and R.V. Lange, Phys. Rev. **157**, 279 (1967).
[89] W.F. Brinkman and T.M. Rice, Phys. Rev. B **2**, 1324 (1970).
[90] V. J. Emery, Phys. Rev. B **14**, 2989 (1976).
[91] P.W. Anderson, Science **235**, 1196 (1987).
[92] P.W. Anderson, Phys. Rev. **86**, 694 (1952).
[93] R. Kubo, Phys. Rev. **87**, 568 (1952).
[94] L. Onsager, Phil. Mag. **43**, 1006 (1952).
[95] K. von Klitzing, G. Dorda and M. Pepper, Phys. Rev. Lett. **45**, 494 (1980).
[96] R.B. Laughlin, Phys. Rev. B **23**, 5632 (1981).
[97] B.I. Halperin, Phys. Rev. B **25**, 2185 (1982).
[98] D.C. Tsui, H.L. Störmer and A.C. Gossard, Phys. Rev. Lett. **48**, 1559 (1982); Phys. Rev. B **25**, 1405 (1982).
[99] J.K. Jain, Phys. Rev. Lett. **63**, 199 (1989); Phys. Rev. B **40**, 8079 (1989).
[100] R.B. Laughlin, Phys. Rev. Lett. **50**, 1395 (1983).
[101] B.I. Halperin, P.A. Lee and N. Read, Phys. Rev. B **47**, 7312 (1993).
[102] K. Onnes, Leiden Communications **120** (1911).

[103] W. Meissner and R. Oschenfeld, Naturwissenschaften **21**, 787 (1933).
[104] P.W. Anderson, Phys. Rev. Lett. **9**, 309 (1962).
[105] H. Frolich, Phys. Rev. **79**, 845 (1950).
[106] F. London and H. London, Proc. Roy. Soc. A **149**, 71 (1935).
[107] A.B. Pippard, Proc. Roy. Soc. A **216**, 547 (1953).
[108] V.L. Ginzburg and L.D. Landau, Zh. Eksp. Teor. Fiz. **20**, 1064 (1950).
[109] A.A. Abrikosov, Zh. Eksp. Teor. Fiz. **32**, 1442 (1957).
[110] J. Bardeen, L.N. Cooper, J.R. Schrieffer, Phys. Rev. **108**, 1175 (1957).
[111] H. Frolich, Proc. Roy. Soc. A **215**, 291 (1952).
[112] J. Bardeen and D. Pines, Phys. Rev. **99**, 1140 (1955).
[113] L.N. Cooper, Phys. Rev. **104**, 1189 (1956).
[114] W.G. McMillan, Phys. Rev. **167**, 331 (1968).
[115] K.J. Chang and M.L. Cohen, Phys. Rev. B **31**, 7819 (1985).
[116] G.H. Vineyard, J. Phys. Chem. Solids **3**, 121 (1957).
[117] K.C. Pandey and E. Kaxiras, Phys. Rev. Lett. **66**, 915 (1991); E. Kaxiras and K.C. Pandey, Phys. Rev. B **47**, 1659 (1993).
[118] M.J. Aziz, J. Appl. Phys. **70**, 2810 (1997).
[119] M. Lannoo, "Deep and shallow impurities in semiconductors", in *Handbook on Semiconductors*, edited by P.T. Landsberg, vol. 1, p. 113 (North Holland, 1992).
[120] W. Schottky, Naturwissenschaften **26**, 843 (1938).
[121] S.G. Louie, J. Chelikowsky and M.L. Cohen, Phys. Rev. B **15**, 2154 (1977).
[122] V. Volterra, Ann. Ecole Norm. Super. **24**, 400 (1907).
[123] E. Orowan, Z. Phys. **89**, 605, 634 (1934).
[124] M. Polanyi, Z. Phys. **89**, 660 (1934).
[125] G.I. Taylor, Proc. Roy. Soc. A **145**, 362 (1934).
[126] J.M. Burgers, Proc. Kon. Ned. Akad. Wetenschap. **42**, 293, 378 (1939).
[127] J.C.H. Spence, *Experimental High Resolution Electron Microscopy* (Oxford University Press, New York, 1988).
[128] J.M. Gibson, Materials Research Society Bulletin, March 1991, pp. 27-35.
[129] R.E. Peierls, Proc. Roy. Soc. **52**, 34 (1940).
[130] F.R.N. Nabarro, Proc. Roy. Soc. **59**, 256 (1947).
[131] J.D. Eshelby, Phil. Mag. **40**, 903 (1949).
[132] V. Vitek, Phys. Stat. Soli. **18**, 683 (1966); Phys. Stat. Sol. **22**, 453 (1967); Phil. Mag. **18**, 773 (1968).
[133] V. Vitek and M. Yamaguchi, J. Phys. F **3**, 537 (1973); M. Yamaguchi and V. Vitek, J. Phys. F **5**, 11 (1975).
[134] M.S. Duesbery and V. Vitek, Acta Mater. **46**, 1481 (1998).
[135] Y. Juan and E. Kaxiras, Phil. Mag. A **74**, 1367 (1996).
[136] J. Frenkel, Z. Phys. **37**, 572 (1926).
[137] J.R. Rice, J. Mech. Phys. Solids **40**, 239 (1992).
[138] R.W. Armstrong, Mater. Sci. Eng. **1**, 251 (1966).
[139] A. Kelly, W.R. Tyson, A.H. Cotrell, Phil. Mag. **15**, 567 (1967).
[140] A.A. Griffith, Phil. Trans. Roy. Soc. (London), **A221**, 163 (1920).
[141] J.R. Rice and R.M. Thomson, Phil. Mag. **29**, 73 (1974).
[142] J.R. Rice, G.E. Beltz and Y. Sun, in *Topics in Fracture and Fatigue*, edited by A.S. Argon (Springer-Verlag, Berlin, 1992).
[143] P.B. Hirsch and S.G. Roberts, Phil. Mag. A **64**, 55 (1991).
[144] M. Khantha, D.P. Pope, and V. Vitek, Phys. Rev. Lett. **73**, 684 (1994).
[145] A. Hartmaier and P. Gumbsch, Phys. Status Solidi B **202**, R1 (1997).
[146] A.E. Carlsson and R.M. Thomson, Phil. Mag. A **75**, 749 (1997).

[147] G. Xu, A.S. Argon and M. Ortiz, Phil. Mag. A **75**, 341 (1997).
[148] Y. Juan, Y. Sun and E. Kaxiras, Phil. Mag. Lett. **73**, 233 (1996).
[149] U.V. Waghmare, E. Kaxiras, V. Bulatov and M.S. Duesbery, Modelling and Simulation in Materials Science and Engineering, **6**, 483 (1998).
[150] V. Bulatov, F.F. Abraham, L. Kubin, B. Devincre and S. Yip, Nature **391** 669 (1998).
[151] V.V. Bulatov and L.P. Kubin, in *Modeling and Simulation of Solids*, edited by E. Kaxiras and S. Yip, Current Opinion in Solid State and Materials Science, **3**, 558 (1998).
[152] E.O. Hall, Proc. Phys. Soc. Lond. B **64**, 747 (1951); N.J. Petch, J. Iron Steel Inst. **174**, 25 (1953).
[153] S. Yip, Nature **391**, 532 (1998); J. Schiotz, F.D. Di Tolla, K.W. Jacobsen, Nature **391**, 561 (1998).
[154] F.F. Abraham, J.Q. Broughton, N. Bernstein and E. Kaxiras, Computers in Physics **12**, 538 (1998).
[155] R. Phillips, in *Modeling and Simulation of Solids*, edited by E. Kaxiras and S. Yip, Current Opinion in Solid State and Materials Science, **3**, 526 (1998).
[156] J.M. Kosterlitz and D.J. Thouless, J. Phys. C **6**, 1181 (1973).
[157] D.R. Nelson and B.I. Halperin, Phys. Rev. **19**, 2457 (1979).
[158] J. Tersoff and D.R. Hamann, Phys. Rev. Lett. **50**, 1998 (1983); Phys. Rev. B **31**, 805 (1985); J. Tersoff, Phys. Rev. B **41**, 1235 (1990).
[159] C.J. Chen, Phys. Rev. B **42**, 8841 (1990); Mod. Phys. Lett. B **5**, 107 (1991).
[160] J. Bardeen, Phys. Rev. Lett. **6**, 57 (1961).
[161] O. Leifeld, D. Grützmacher, B. Müller, K. Kern, E. Kaxiras and P. Kelires, Phys. Rev. Lett. **82**, 972 (1999); W. Barvosa-Carter, A.S. Bracker, J.C. Culbetrson, B.Z. Nosko, B.V. Shanabrook, L.J. Whitman, H. Kim, N.A. Modine and E. Kaxiras, Phys. Rev. Lett. **84**, 4649 (2000).
[162] D.M. Eigler and E.K. Schweizer, Nature **344**, 524 (1990).
[163] L.J. Whitman, J.A. Stroscio, R.A. Dragoset and R.J. Celotta, Science **251**, 1206 (1991).
[164] I.-W. Lyo and Ph. Avouris, Science **253**, 173 (1991).
[165] J. J. Boland, Science **262**, 1703 (1993).
[166] M.F. Cromie, C.P. Lutz and D.M. Eigler, Science **262**, 218 (1993).
[167] Ph. Avouris and I.-W. Lyo, Science **264**, 942 (1994).
[168] N.D. Lang and W. Kohn, Phys. Rev. B **1**, 4555 (1970).
[169] R.M. Feenstra, J.A. Stroscio, J. Tersoff and A.P. Fein, Phys. Rev. Lett. **58**, 1192 (1987).
[170] E. Kaxiras, Phys. Rev. B **43**, 6824 (1991).
[171] D. Vanderbilt, Phys. Rev. B **36**, 6209 (1987); R. D. Meade and D. Vanderbilt, Phys. Rev. B **40**, 3905 (1989).
[172] G. Binnig, H. Rohrer, Ch. Gerber, and E. Weibel, Phys. Rev. Lett. **50**, 120 (1983).
[173] K. Takayanagi, Y. Tanishiro, M. Takahashi, and S. Takahashi, J. Vac. Sci. Technol. A **3**, 1502 (1985).
[174] R.J. Hamers, R.M. Tromp and J.E. Demuth, Phys. Rev. Lett. **56**, 1972 (1986).
[175] Y.-W. Mo, B.S. Swartzentruber, R. Kariotis, M.B. Webb and M.G. Lagally, Phys. Rev. Lett. **63**, 2393 (1989).
[176] Y.W. Mo, J. Kleiner, M.B. Webb and M.G. Lagally, Phys. Rev. Lett. **67**, 1998 (1991).
[177] Z. Zhang, F. Wu, H.J.W. Zandvliet, B. Poelsema, H. Metiu, and M. Lagally, Phys. Rev. Lett. **74**, 3644 (1995).

[178] Z. Zhang, Y. T. Lu and H. Metiu, Surf. Sci. **80**, 1248 (1991); Phys. Rev. B **46**, 1917 (1992).
[179] S.F. Edwards and D.R. Wilkinson, Proc. Roy. Soc. (London) **A381**, 17 (1982).
[180] C. Herring, J. Appl. Phys. **21**, 437 (1950).
[181] D.E. Wolf and J. Villain, Europhys. Lett. **13**, 389 (1990).
[182] M. Kardar, G. Parisi and Y.-C. Zhang, Phys. Rev. Lett. **56**, 889 (1986).
[183] V. Heine, Phys. Rev. **138A**, 1689 (1965).
[184] S.G. Louie and M.L. Cohen, Phys. Rev. B **13**, 2461 (1976).
[185] D.E. Polk, J. Noncrystalline Solids **5**, 365 (1971); D.E. Polk and D.S. Boudreaux, Phys. Rev. Lett. **31**, 92 (1973).
[186] J. F. Justo, M. Z. Bazant, E. Kaxiras, V. V. Bulatov, and S. Yip, Phys. Rev. B, **58**, 2539 (1998).
[187] R. Car and M. Parrinello, Phys. Rev. Lett. **60**, 204 (1988); I. Stich, R. Car and M. Parrinello, Phys. Rev. B **44**, 11092 (1991).
[188] F. Wooten, K. Winer and D. Wearie, Phys. Rev. Lett. **54**, 1392 (1985).
[189] S.T. Pantelides, Phys. Rev. Lett. **57**, 2979 (1986).
[190] P.W. Anderson, Phys. Rev. **109**, 1492 (1958); Rev. Mod. Phys. **50**, 191 (1978).
[191] W.P. Su, J.R. Schrieffer and A.J. Heeger, Phys. Rev. Lett. **42**, 1698 (1979).
[192] M.J. Rice and E.J. Mele, Phys. Rev. B **25**, 1339 (1982).
[193] A. Uhlherr and D.N. Theodorou, in *Modeling and Simulation of Solids*, edited by E. Kaxiras and S. Yip, Current Opinion in Solid State and Materials Science, **3**, 544 (1998).
[194] H.-W. Fink and C. Schönenberger, Nature, **398**, 407 (1999).
[195] D. Porath, A. Bezryadin, S. de Vries and C. Dekker, Nature, **403**, 635 (2000).
[196] W.D. Knight, K. Clemenger, W.A. de Heer, W. Saunders, M.Y. Chou and M.L. Cohen, Phys. Rev. Lett. **52**, 2141 (1984).
[197] W.A. de Heer, W.D. Knight, M.Y. Chou and M.L. Cohen, in *Solid State Physics*, edited by H. Ehrenreich, F. Seitz and D. Turnbull, vol. 40, p. 93 (Academic Press, New York, 1987).
[198] S. Bjornholm, J. Borggreen, O. Echt, K. Hansen, J. Pedersen and H.D. Rasmussen, Phys. Rev. Lett. **65**, 1627 (1990).
[199] H. Nishioka, K.. Hansen, B.R. Mottelson, Phys. Rev. B **42**, 9377 (1990).
[200] J.A. Alonso, M.D. Glossman and M.P. Iniguez, Inter. J. of Modern Phys. B **6**, 3613 (1992).
[201] J.-Y. Yi, D.J. Oh, J. Bernholc and R. Car, Chem. Phys. Lett. **174**, 461 (1990).
[202] H.-P. Cheng, R.S. Berry and R.L. Whetten, Phys. Rev. B **43**, 10647 (1991).
[203] L.L. Boyer, M.R. Pederson, K.A. Jackson, and J.Q. Broughton, in *Clusters and Cluster-Assembled Materials*, edited by R.S. Averback, J. Bernholc and D.L. Nelson, MRS Symposia Proceedings, **206**, 253 (1991).
[204] Q.-M. Zhang, J.-Y. Yi, C.J. Brabec, E.B. Anderson, B.N. Davidson, S.A. Kajihara, and J. Bernholc, Inter. J. of Modern Phys. B **6**, 3667 (1992).
[205] A.P. Seitsonen, M.J. Puska, M. Alatalo, R.M. Nieminen, V. Milman, and M.C. Payne, Phys. Rev. B **48**, 1981 (1993).
[206] S.N. Khana and P. Jena, Chem. Phys. Lett. **218**, 383 (1994).
[207] H.W. Kroto, J.R. Heath, S.C. O'Brien, R.F. Curl and R.E. Smalley, Nature **318**, 162 (1985).
[208] W. Krätschmer, L.D. Lamb, K. Fostiropoulos, and D.R. Huffman, Nature **347**, 359 (1990).
[209] F. Chung and S. Sternberg, American Scientist **81**, 56 (1993).
[210] R.C. Haddon, Accounts of Chemical Research **25**, 127 (1992).

[211] W.E. Pickett, in *Solid State Physics*, edited by H. Ehrenreich and F. Spaepen, vol. 48, p. 225 (Academic Press, New York, 1994).
[212] C.M. Lieber and C.C. Chen in *Solid State Physics*, edited by H. Ehrenreich and F. Spaepen, vol. 48, p. 109 (Academic Press, New York, 1994).
[213] D. Bylander and K. Kleinman, Phys. Rev. B **47**, 10967 (1993).
[214] E. Kaxiras *et al.*, Phys. Rev. B **49**, 8446 (1994).
[215] R.F. Curl and R.E. Smalley, Science **242**, 1017 (1988).
[216] Y. Chai *et al.*, J. Chem. Phys. **95**, 7564 (1991).
[217] J.H. Weaver *et al.*, Chem. Phys. Lett. **190**, 460 (1991).
[218] S.C. O'Brien, J.R. Heath, R.F. Curl and R.E. Smalley, J. Chem. Phys. **88**, 220 (1988).
[219] F.D. Weiss, S.C. O'Brien, J.L. Eklund, R.F. Curl and R.E. Smalley, J. Am. Chem. Soc. **110**, 4464 (1988).
[220] U. Zimmermann, N. Malinowski, U. Näher, S. Frank and T.P. Martin, Phys. Rev. Lett. **72**, 3542 (1994).
[221] F. Tast, N. Malinowski, S. Frank, M. Heinebrodt, I.M.L. Billas and T.P. Martin, Phys. Rev. Lett. **77**, 3529 (1996).
[222] M.R. Pederson, C.V. Porezag, D.C. Patton and E. Kaxiras, Chem. Phys. Lett. **303**, 373 (1999).
[223] S. Iijima, Nature **354**, 56 (1991).
[224] D. Ugarte, Nature **359**, 707 (1992).
[225] J.L. Elkind, J.M. Alford, F.D. Weiss, R.T. Laaksonen and R.E. Smalley, J. Chem. Phys. **87**, 2397 (1987).
[226] M.F. Jarrold, J.E. Bower and K.M. Creegan, J. Chem. Phys. **90**, 3615 (1989).
[227] E. Kaxiras, Chem. Phys. Lett. **163**, 323 (1989); Phys. Rev. Lett. **64**, 551 (1990); Phys. Rev. B **56**, 13455 (1997).
[228] N. Binggeli, J.L. Martins and J.R. Chelikowsky, Phys. Rev. Lett. **68**, 2956 (1992).
[229] U. Röthlisberger, W. Andreoni and M. Parinello, Phys. Rev. Lett. **72**, 665 (1994).
[230] B.C. Guo, K.P. Kerns and A.W. Castleman, Jr., Science **255**, 1411 (1992); *ibid* **256**, 515 (1992); *ibid* **256**, 818 (1992).
[231] "The Double Helix :a Personal Account of the Discovery of the Structure of DNA" J.D. Watson (Simon & Schuster, New York, 1998).
[232] P.J. de Pablo, F. Moreno-Herrero, J. Colchero, *et al.*, Phys. Rev. Lett. **85**, 4992 (2000).
[233] D.A. Doyle, J. Morais Cabral, R.A. Pfuetzner, A. Kuo, J.M. Gulbis, S.L. Cohen, B.T. Chait and R. MacKinnon, Science, **280**, 69 (1998).
[234] P. O. Lowdin, Phys. Rev. **97**, 1490 (1955).
[235] E.M. Purcell and R.V. Pound, Phys. Rev. **81**, 279 (1951).
[236] P. Hakonen and O.V. Lounasmaa, Science **265**, 1821 (1994).
[237] F.C. Frank, Phil. Mag. **41** (1950).

Index